Engineering Mathematics

Wind-induced collapse of oil storage tanks at Haydock, Lancashire, England, in 1967

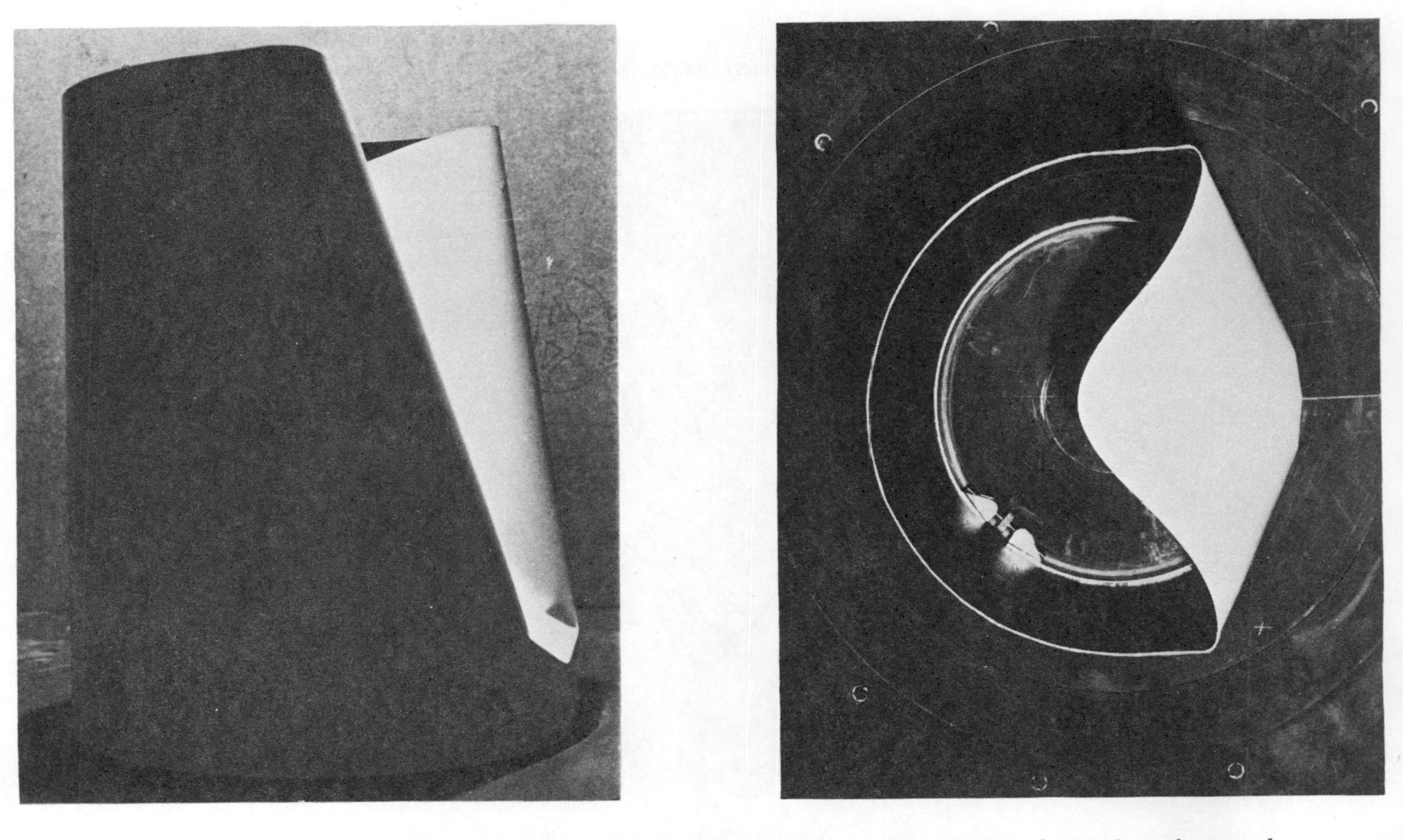

Photographs showing the wind tunnel simulation of the full-scale collapse of the shells in the previous photograph

Engineering Mathematics

Second Edition

A. C. Bajpai
L. R. Mustoe
D. Walker
Loughborough University of Technology

In collaboration with
W. T. Martin
Massachusetts Institute of Technology

JOHN WILEY & SONS
Chichester . New York . Brisbane . Toronto . Singapore

Reprinted August 1990

British Library Cataloguing in Publication Data:

Bajpai, A. C. (Avinash Chandra), *1925–*
Engineering mathematics. – 2nd ed.
1. Mathematics, – For engineering
I. Title II. Mustoe, L. R. (Leslie R)
III. Walker, D. (Dennis), *1932–*
IV. Martin, W. T.
510

ISBN 0 471 92283 8

Printed and bound in Great Britain by Courier International Ltd, Tiptree, Essex

PREFACE

It is well accepted that a good mathematical grounding is essential for all engineers and scientists. This book is designed to provide this grounding, starting from a fairly elementary level and is aimed at first year undergraduate science and engineering students in universities, polytechnics and colleges in all parts of the world. It would also be useful for students preparing for the Engineering Council examinations in mathematics at Part 1 standard.

The basic concept of the book is that it should provide a motivation for the student. Thus, wherever possible, a topic is introduced by considering a real example and formulating the mathematical model for the problem; its solution is considered by both analytical and numerical techniques. In this way, it is hoped to integrate the two approaches, whereas many texts have regarded the analytical and numerical methods as separate entities. As a consequence, students have failed to realise the possibilities of the different methods or that on occasions a combination of both analytical and numerical techniques is needed. Indeed, in most practical cases met by the engineer and scientist the desired answer is a set of numbers; even if the solution can be obtained completely by analytical methods, the final process is often to obtain discrete values from the analytical expression.

The authors believe that some proofs are necessary where basic principles are involved. However, in other cases where it is thought that the proof is too difficult for students at this stage, it has been omitted or only outlined. For the numerical techniques, the approach has been to form a heuristically derived algorithm to illustrate how it is used and a formal justification is given only in the simpler cases.

Where a computer approach is helpful, a flow diagram is provided; in some instances a listing of the appropriate Basic computer program is provided. Many books are devoted solely to the teaching of computer programming; we therefore do not attempt to cover this topic in detail. It is hoped that students will either have prior knowledge of programming or will be studying this in a parallel course.

Throughout the text there is a generous supply of worked examples which illustrate both the theory and its application. Supplementary problems are provided at the end of each chapter.

Finally, we sincerely hope that both students and lecturers read the Open Letters which set the ethos and philosophy for the book.

Since the first edition appeared the authors have received much helpful comment and criticism. The overwhelming reaction, fortunately, was favourable. However, the advent of the pocket calculator and the microcomputer and the shifting in emphasis of parts of the syllabus meant that a new edition of this book was needed. Accordingly, taking the views of our readers, both staff and students, into account, the text material was completely revised. The opportunity was taken to remove some mathematical 'dead wood' and to omit some material which is now covered more thoroughly at secondary level; this made room for some new items.

Changes in the revised edition

The present edition differs from its predecessor in the following respects.

1 *Chapters* The text material has been re-organised into twenty-five chapters instead of the thirteen of the first edition. This makes for a more modular approach to teaching the material.

2 *Problem sets* These now appear at the end of each chapter instead of at the end of each section. The problems have been modified and have included more up to date examples from Engineering Council examinations.

3 *Numerical methods* More emphasis is placed on pocket calculator methods and microcomputer solutions. Sample programs are included in the text at the relevant places.

4 *Discrete mathematics* A new chapter has been written to capture the growing importance of discrete mathematics in the engineering curriculum.

5 *Transform methods* The material on Laplace transforms has been augmented to include step functions and periodic functions.

The authors would like to thank all staff and students who have made helpful suggestions for the improvement of the text. In addition they acknowledge with pleasure a debt of gratitude to the following:

Staff and students of Loughborough University of Technology and other institutions who have participated in the development of the original text and the revisions incorporated in this edition.

Mr J Mountfield of Warrington, Lancashire, for permission to use his photograph of the collapsed oil storage tanks.

Dr D J Johns, Founding Director of the City Polytechnic of Hong Kong and Vice-Chancellor (Elect) of the University of Bradford for providing the photographs of the wind tunnel tests on models of the oil tanks.

John Wiley & Sons Ltd for their help and co-operation.

The University of London and The Engineering Council for permission to use questions from their past examination papers. (These are denoted by LU and EC respectively.)

Miss H A Wyatt for typing this book.

CONTENTS

0

OPEN LETTERS

Open Letter to Students

This is a book with a difference. It is different in that we do not seek simply to give you a grounding in those mathematical techniques that you will need in your studies. Rather we hope that you will be encouraged to think mathematically. By this, we mean that after following through the text you will be able to look at some practical problem, think about forming the problem in mathematical terms, consider the possible ways of obtaining an answer to this mathematical problem and choose the most suitable way. Then you should be able to find an answer and furthermore interpret this answer in the context of the original problem. This is not to say that we do not place a great emphasis on building up your knowledge and skills in mathematical techniques. It is essential for you to be able to handle and manipulate mathematical formulae and equations: indeed, much of our book is devoted to the development of such mathematical ability. However, we regard this as only one part of the story and at the end of your study we hope that you will not have to ask the usual questions, *What use is all this mathematics? Why do we have to study such abstract ideas?* and *How does this tie in with my other subjects?*

The whole scene is set in Chapter 1 and its sole purpose is to establish a way of thinking. For this reason there will appear to be little conventional mathematics in this chapter but we strongly plead with you to follow it closely. It will not be a waste of time we can assure you, and a careful study of the ideas involved will be repaid in that you will get a feeling for the relevance of the mathematical work and in that you will already be thinking along the right lines. Constant reference to this chapter will be made in the rest of the book and the whole development of the text hinges on a clear understanding of the principles expounded here.

We hope that you will enjoy using this book and wish you good luck in your studies.

Open Letter to Teachers

Perhaps you will have read the preface and the open letter to students. We should like to address a few remarks to you especially. This book is not for the lecturer who is content to approach engineering mathematics either as a collection of "cookbook techniques" or as a watered-down version of honours mathematics. It tries to present an integrated study in two ways. Often the complaint has been made that mathematics is isolated from the engineering subjects and seems to bear little relevance; abstract ideas are studied with little attempt to link them to the engineering world. This book seeks, where possible, to introduce the techniques via practical examples. The second feature we want to emphasise is the welding together of numerical and analytical methods. We feel that the separation is artificial and we have tried to present a problem-oriented view in which the techniques that seem most suitable in a particular situation are employed; in any event, numerical methods, now well established, often give a solution where the analytical techniques have failed and it is unrealistic to treat them as second-best. Indeed, some topics, for example the behaviour of sequences, are better approached from a numerical view-point.

You will have noticed that we have asked the students to read carefully Chapter 1. We cannot emphasise the importance of this too strongly as without so doing they will not really appreciate the flavour of the book. We are relying on your help in this matter. We hope that you will enjoy teaching your course along the lines we advocate; certainly we have found it a challenging and stimulating experience in the years we have taught via this approach.

1

WHY MATHEMATICS?

In your previous mathematical work, the emphasis will have been placed most probably on the development of manipulative skills. The subject will have been broken down into fairly water-tight compartments and often it is not clear how one compartment impinges on the next. Some of the techniques studied may have appeared to be 'tricks' and this impression is encouraged by the fact that the numbers chosen are usually those that make the answers come out easily and exactly. This is not necessarily a criticism of the way you have learned mathematics up to this stage: in order to proceed it was necessary that you should have mastered many basic techniques and acquired much background information; this is the case in any other discipline, be it science, history, geography or languages.

However, the role of mathematics in the study of scientific or engineering problems goes deeper and wider than this. In such problems it may be that one specific answer is required or, more generally, the nature of the relationship between two or more of the variable quantities involved is sought in order that certain deductions may be drawn.

1.1 Mathematical Models

To clarify ideas, let us discuss four simple experiments.

(i) A uniform beam is simply supported near its ends and a load placed near its centre; the beam will be deflected. We may seek the maximum load that the beam can support before breaking or we may be interested in the profile that the beam adopts. More likely we shall be interested in predicting these features for beams at the design stage.

(ii) A liquid that has been heated and removed from the source of heat will cool; we shall be interested in the rate at which it cools, with a possible view to predicting the time that elapses before a specified temperature is reached.

(iii) If a loaded spring is set in motion, the position of the weight varies about its equilibrium position, sometimes below and sometimes above. We might wish to know the greatest depth the load reaches or how long it will be before the amplitude of the oscillations decreases below a certain amount.

(iv) A simple electrical circuit is set up in which the potential difference across the ends of a conductor is varied and the current passing through it is measured. The object might be to study the relationship between the potential difference and the current or to predict the current which flows through the conductor for a non-observed potential difference.

We can examine this last example more closely. Figure 1.1 below shows a typical set of results from the experiment.

We call the potential difference V and the current i. The results suggest a straight line relationship of the form $V = V_0 + Ri$ where V_0 and R are constants.

The values of V_0 and R can be obtained graphically or by using a method employing directly the numerical values of the observations. The danger in using the graphical approach to fitting the straight line is that it is a subjective process and different people might obtain different results. However, the graphical approach does have one advantage sometimes in that a wrong observation can be eliminated from consideration.

This relationship is the **mathematical model** for the physical situation and is an **empirical** formula, since it is based on experimental data and not on a background theory. However, this model can be used to predict values of i for non-observed values of V and vice-versa, and, provided the model gives reasonable predictions, it will suffice.

These predictions can be made graphically or by using a numerical formula, i.e. a formula which involves the specific numerical values of the observations.

There is little virtue in deducing a relationship $V = V_0 + Ri + 10^{-6}Ri^2$ if the extra term plays no part in the predictions, at least to the accuracy to which we are working. However, we must bear in mind that we can only safely predict within the range of observations – a process known as **interpolation**; the dangers attendant on predicting outside the range (**extrapolation**) are shown in Figure 1.2.

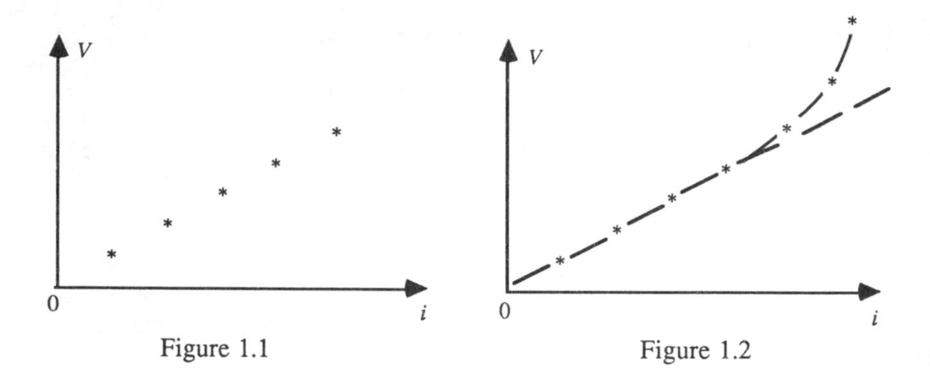

Figure 1.1 Figure 1.2

If we had an underlying theory based on certain assumptions, for example, concerning the forces on electrons, then we might know the limitations of the predictions from this theoretical model. One further point concerning observations is worth making and we illustrate this with the third experiment.

The nature of the relationship between the position of the load on the spring, z, relative to its equilibrium position is shown. Had the observations been only those circled in Figure 1.3, a different impression could have emerged if there was no reference to the original problem.

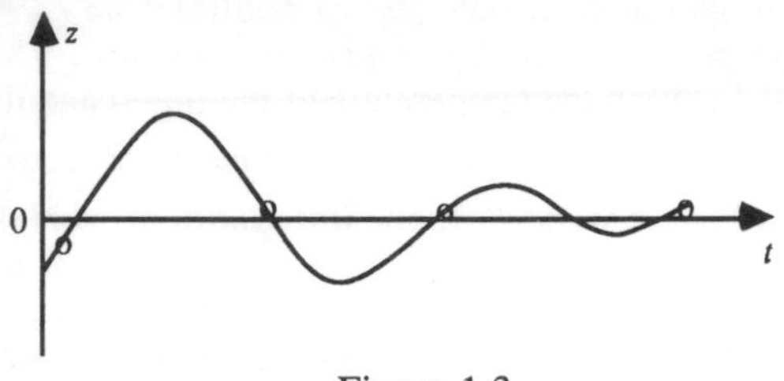

Figure 1.3

Models and Assumptions

We now turn our attention to the cooling liquid. A typical set of observations is shown in Figure 1.4, where θ is the temperature of the liquid at time t.

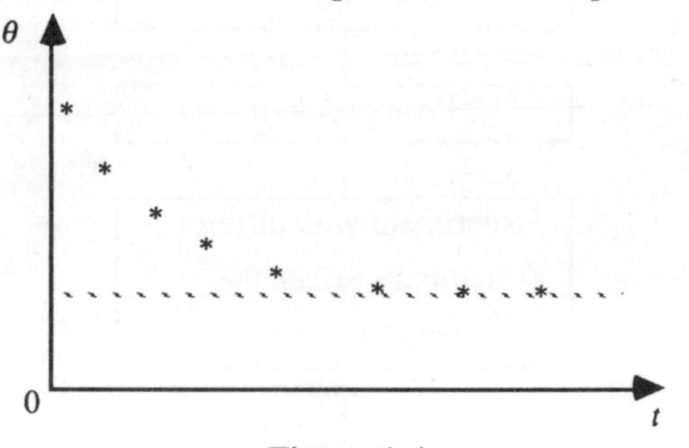

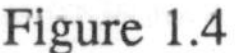
Figure 1.4

It is clear that a linear relationship between θ and t is not suggested. What empirical formula can we try to fit to the data? We could try

$$\theta = \frac{A}{t} + C \quad \text{or} \quad \theta = \frac{B}{t^2} + C \qquad \text{where } A, B \text{ and } C \text{ are constants.}$$

The graphs of these relationships have shapes similar to that suggested by the observations, but this is not a very satisfactory approach. We must build a **model** based on physical **assumptions** and then interpret this mathematically. From observations we conclude that the rate of cooling is rapid at first and gradually becomes less rapid until it becomes very small. What factors do we think affect the cooling process? It cannot be the temperature of the liquid alone, since if a beaker of liquid helium were placed in a room at average temperature, the liquid would boil away. You will no doubt have remarked that this is hardly an example of a hot liquid; this example was used to emphasise that it is important to specify conditions correctly: by *hot* we mean *hot compared to the surrounding atmosphere.*

What would seem to be relevant is the excess temperature of the liquid over that of its surroundings. But is this all? What about air currents or the shape of the container or the material from which it is constructed? From experiments Newton concluded that, provided air currents were not abnormally large, the dominating factor was the excess temperature of the liquid over that of its surroundings and proposed Newton's law of cooling: *the rate of loss of heat from a liquid is proportional to the excess temperature.* This law gives an implicit relationship between the temperature of the liquid and time. We should like to obtain a formula giving the temperature of the liquid, θ, explicitly in terms of time, t.† In such a case we say t is the **independent variable** and θ is the **dependent variable**. The use of such a formula is that it usually allows a qualitative prediction of the relationship (drawing a picture) as well as a quantitative one. Sometimes a formula is so complicated that it does not lend itself to easy interpretation either qualitatively or quantitatively and in such cases it is probably best to estimate particular values of the dependent variable at specified values of the independent variable by a method which sets out to do just this.

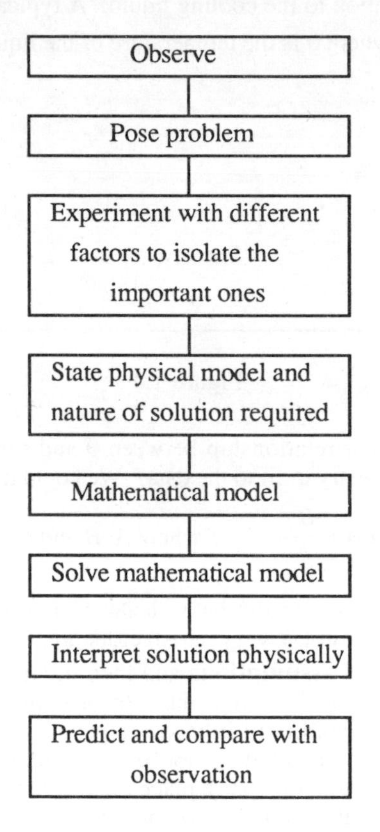

Figure 1.5

†In the formula $x^2 + y^3 = 2y$ the relationship between x and y is **implicit**, whereas in the formula $y=x^2 + 2$, the variable y is given **explicitly** in terms of x.

The mathematical statement of Newton's Law is

$$\frac{d\theta}{dt} = -k(\theta - \theta_s)$$

where θ_s is the temperature of the surroundings and k is a constant for the liquid which has to be experimentally determined. If we add the condition that at $t = 0$, $\theta = \theta_0$, then we have a mathematical model.

From this mathematical model we can obtain an expression for θ in terms of t, although we have not yet acquired the necessary mathematical knowledge. We shall study the techniques required later, and return to a discussion of this problem in Chapter 21.

It has been found that this model does compare reasonably well with observation, and gives a good representation of the temperature variation. As we have taken a very simple model, it is worth asking whether the addition of other factors might give a better representation. The difficulty is how to allow for these other factors in the formulation of a new model and we should bear in mind that the resulting model might be too complicated to yield a useful formula.

The **block diagram** above, Figure 1.5, shows the simplified process we have followed so far.

Let us now examine the problem of the loaded beam (Figure 1.6) with a view to studying more closely the assumptions made in building a model.

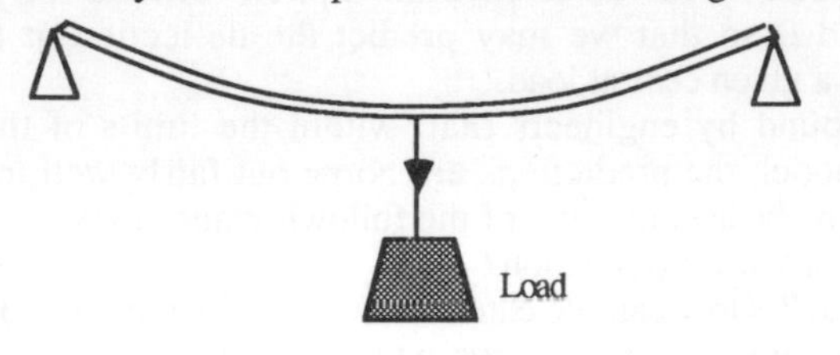

Figure 1.6

Suppose that we restrict our problem to that of predicting how far a given beam will deflect when a known load is suspended from its mid-point. We then carry out experiments with different shapes of beam, different beam materials and different loads. From these experiments graphs could be plotted showing the variation of the deflection with each factor separately. It is impossible in practice to draw graphs for every type of beam and load; therefore we try to find a mathematical formula covering all possible cases. This will involve going back to the theory of elasticity and forming a physical model which states an implicit relationship between the deflection at any point of the beam and certain physical properties of the beam. From this physical model represented in Figure 1.7, the following equation can be produced:

$$EI\frac{d^4y}{dx^4} = \frac{W}{l}$$

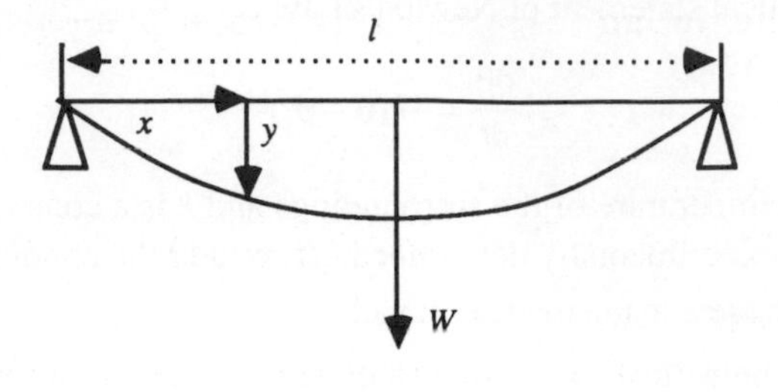

Figure 1.7

where W is the weight of the beam of length l, E is the Young's modulus of elasticity for the material of the beam and I is the relevant moment of inertia of its cross-section. We must point out that several important simplifying assumptions, given below, have been made in order to produce this model:

(i) the deflection is small
(ii) the beam is thin
(iii) the beam is of uniform cross-section
(iv) the beam is homogeneous
(v) the supports are point supports (knife edges) and are exactly at the ends of the beam.

The equation can be solved to give an explicit formula for y in terms of x (and E, I, W and l) so that we may predict the deflection at any point of a given beam with a given central load.

It has been found by engineers that, within the limits of the assumptions governing the model, the predictions are borne out fairly well in practice. But you should by now be asking some of the following questions:

(i) How small is a small deflection?
(ii) What is 'thin'? How can we cater for beams which are not 'thin'?
(iii) Can we take into account beams which are of non-uniform cross-section and which are non-homogeneous?
(iv) Does it matter if the supports are not knife edges and are not exactly at the ends of the beam?

There are two kinds of question here. The first kind involves a careful definition of the limitations of the model and the second is concerned with generalising the model to make it more widely applicable (e.g. non-uniform cross-section). One way of testing for the first is to check with experiment or subsequent observations: this approach could have embarrassing results. A second way is not to make the simplifying assumption of, for example, thinness, which is really in the realms of the second kind of question, i.e. generalising the model. Then one can look at the generalised solution and see how the thickness of the beam enters into it and under what circumstances its effects can be ignored.

All this assumes that a neat solution to the mathematical model exists. This is by no means always the case. It is time to consider an example which illustrates the need for a revision of the block diagram.

The Pitot Tube in an Air Speed Indicator: Sophistication of a Mathematical Model

We choose this example to show how a model can progressively be made more sophisticated to obtain better results, but where the question may be asked: *Is this extra sophistication a worthwhile exercise?*

You may not have met the *Pitot Tube* before – it is a commonly used device for measuring the speed of an aircraft or a body moving in a fluid. When the speed of the aircraft is constant, the problem of measuring its speed forward into still air is equivalent to measuring the speed of air past a stationary aircraft. That is, we imagine the air to be flowing past the body at rest rather than the body moving and the air at rest.

The pitot tube, as shown in Figure 1.8, consists of a slender tubular body aligned with the stream of air whose speed is to be measured. It is of such a shape that it does not interfere unduly with the airflow and therefore at some distance downstream from the nose of the tube the speed of the air will be almost the same as the speed upstream and the pressure will also be practically that of the free stream. At the nose, the stream meets the tube and is brought to rest so that there is a build-up of pressure and we get a pressure p_1 inside the tube at A, called the **stagnation pressure**. The pressure p just outside the tube at A will be equal to the pressure at B, which is measured at the static opening and is called the **static pressure**. The difference in pressures between A and B is $(p_1 - p)$, which can be measured using a manometer. This difference in pressure is used to calculate the speed of the air stream which is usually read off automatically on an air speed indicator (differential pressure gauge).

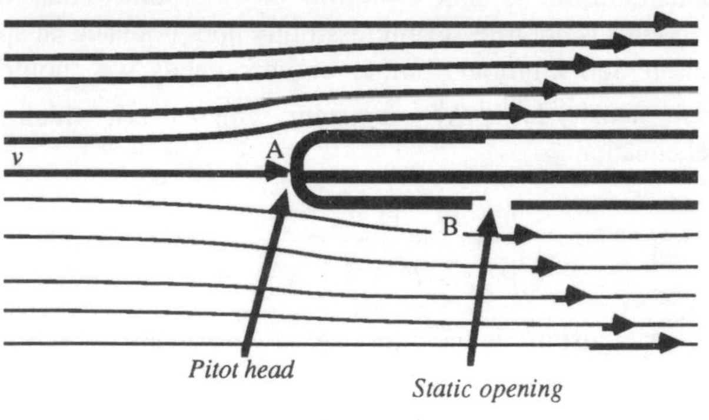

Figure 1.8

The most simple model of airflow is one in which the air is assumed to be incompressible. This model gives rise to the energy equation

$$p + \frac{1}{2}\rho v^2 = p_1$$

where ρ is the density of the air and v its speed. Solving, we obtain

$$v = \sqrt{\frac{2(p_1 - p)}{\rho}} \tag{1.1}$$

Air speed indicators for some low-speed commercial air planes are calibrated according to this equation using a value for ρ of 1.2256 kg/m^3. (The density of air is taken to be that for a standard atmosphere at sea-level.)

Is (1.1) a good formula and how accurate is it? First of all, since the density used is that of the density of air at sea-level, we see that this formula will only be correct under sea-level conditions. It is well-known that the air gets *thinner* the higher we go above the earth, which means that ρ decreases as the altitude increases. Therefore the true air speed at altitude from this formula is

$$v = \sqrt{\frac{2(p_1 - p)}{\rho_{\text{altitude}}}} = \sqrt{\frac{2(p_1 - p)}{\rho_{\text{sea-level}}}} \cdot \sqrt{\frac{\rho_{\text{sea-level}}}{\rho_{\text{altitude}}}} \tag{1.2}$$

At 6000m the density is roughly 0.6597 kg/m^3.
Hence the error factor at 6000m is

$$\sqrt{\frac{\rho_{\text{sea-level}}}{\rho_{\text{altitude}}}} = \sqrt{\frac{1.2256}{0.6597}} = 1.36 \text{ (3 significant figures)}$$

An air speed indicator which is calibrated using (1.1) will always give an air speed which is too low.

A second consideration is that equation (1.1) assumes that the air is incompressible and this is not true. Compressibility does not have an appreciable effect below about 320 km/hour, but at higher speeds we should use the *adiabatic* pressure-density law p/ρ^γ = constant, with $\gamma = 1.4$, which gives rise to the new model equation

$$\frac{p}{\rho} + \frac{(\gamma - 1)}{\gamma}\frac{v^2}{2} = \frac{p_1}{\rho_1}$$

where ρ_1 is the density just inside the pitot tube at A. Solving for v, we get

$$v^2 = \frac{2\gamma}{\gamma - 1}\left[\frac{p_1}{\rho_1} - \frac{p}{\rho}\right]$$

and since

$$\frac{p_1}{\rho_1^\gamma} = \frac{p}{\rho^\gamma}$$

then $$\rho_1 = \rho\left[\frac{p_1}{p}\right]^{1/\gamma}$$

therefore $$v^2 = \frac{2\gamma}{\gamma-1}\left[\frac{p_1}{\rho}\left[\frac{p}{p_1}\right]^{1/\gamma} - \frac{p}{\rho}\right]$$

$$= \frac{2\gamma}{\gamma-1}\,\frac{p^{(1/\gamma)}}{\rho}\left[p_1^{(\gamma-1)/\gamma} - p^{(\gamma-1)/\gamma}\right]$$

thus $$v = \sqrt{\frac{2\gamma}{\gamma-1}\,\frac{p^{(1/\gamma)}}{\rho}\left[p_1^{(\gamma-1)/\gamma} - p^{(\gamma-1)/\gamma}\right]}$$

and with $\gamma = 1.4$ $$v = \sqrt{7\frac{p^{5/7}}{\rho}\left[p_1^{2/7} - p^{2/7}\right]} \qquad (1.3)$$

Compare equations (1.1) and (1.3) and see the difficulty introduced by making our model more sophisticated. We can easily measure $(p_1 - p)$ with a manometer, but to measure $(p_1{}^{2/7} - p^{2/7})$ is very much more difficult, and is the error introduced by using the simpler equation (1.1) appreciable at sea-level?

It can be shown that there is a 5% error at 770 km/hour which increases to 10% at about 1100 km/hour.

Thus, the model has become very sophisticated in order to gain an accuracy of about 5% at 800 km/hour and it is a matter of judgement (and usually expense) as to whether the extra sophistication is worthwhile.

We have not told the whole story, because in modern high speed aircraft the Mach number†, M, is measured rather than the air speed and it can be shown that

$$M^2 = \frac{2}{\gamma-1}\left[\left[\frac{p_1-p}{p}+1\right]^{(\gamma-1)/\gamma} + 1\right]$$

which is used for calibration of the Machmeter with the quantity $(p_1 - p)$ being measured directly by the instrument. Again, an error is introduced in this model by taking p in the denominator of the term $(p_1 - p)/p$ to have its value at sea-level.

We can expand these ideas in a block diagram, Figure 1.9a, which illustrates the way in which the modelling process works in general.

†The Mach number is the ratio of the speed of the aircraft to the local speed of sound.

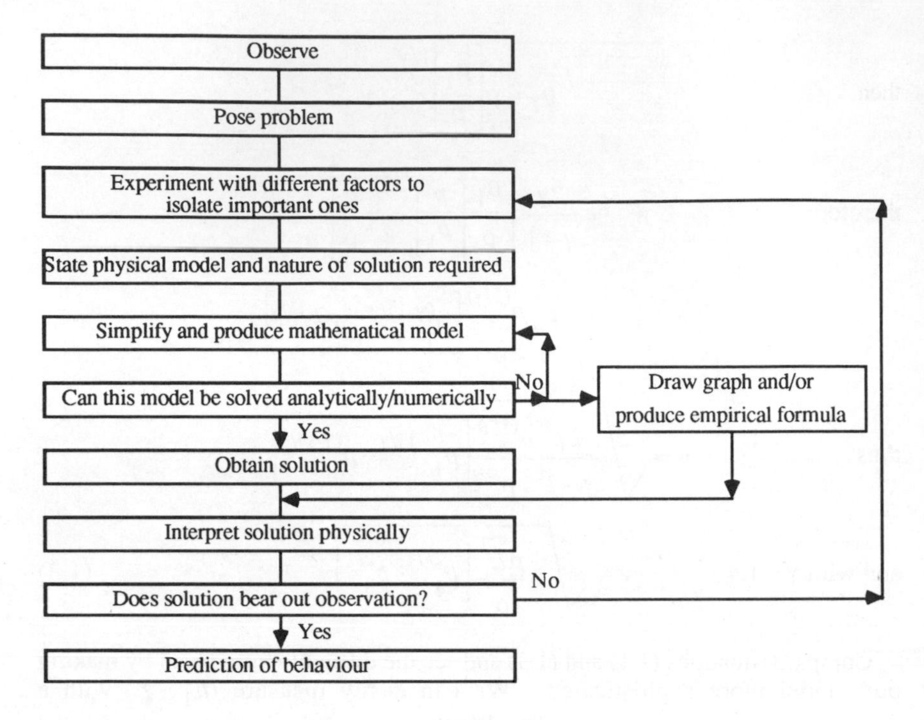

Figure 1.9a

A revised block diagram which has been developed by a colleague, Peter Armstrong, since Figure 1.9a was first published is shown in Figure 1.9b illustrating an algorithm of the mathematical modelling process.

1.2 Solutions to Mathematical Models

Sometimes a mathematical model can be solved exactly in the sense that it yields either a set of values for the dependent variables involved or an accurate formula: by accurate, we mean as far as the model is concerned. Sometimes, however, the model does not yield this information because the necessary techniques of solution have not been developed. If simplification of the model yields results which are far from those found in practice, then we must look again at our model to see whether we can squeeze information out of it by different means.

In the first category, where we seek one answer or set of answers, sometimes an algebraic formula for solution exists as is the case with a quadratic equation: but such a formula is not always possible, e.g. $x^5 - 7x^2 + 3 = 0$; one method of solving this equation would be to draw a graph of $x^5 - 7x^2 + 3$ and find where this crosses the x-axis; alternatively, we can obtain successively

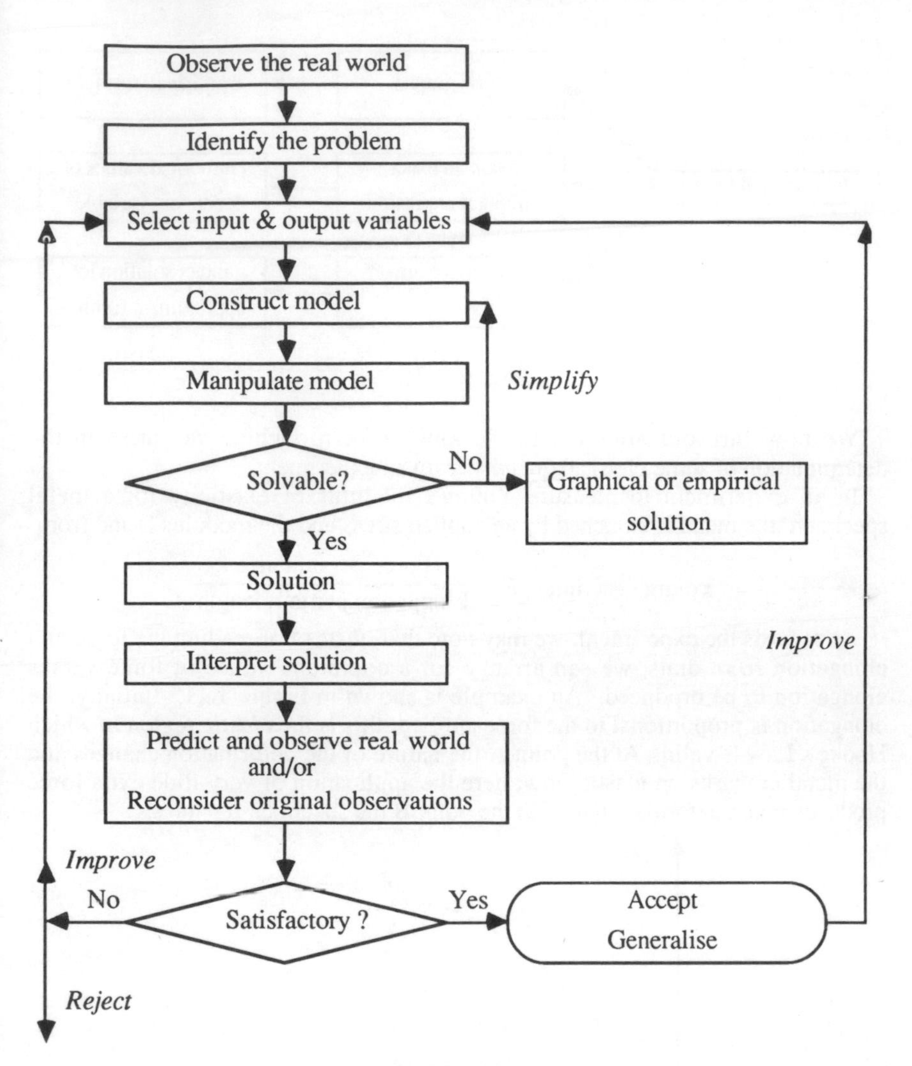

Figure 1.9b

better approximations to the correct answers by a so-called iterative technique which we shall discuss in Chapter 3.

In the second category, where the ideal would be a formula expressing the dependent variable explicitly in terms of the independent variables, the possibilities open are as shown in Figure 1.10.

Any one of these approaches provides information about the dependent variable and is regarded by us as a solution to the model, albeit an *approximate solution.* In any event, an exact formula may still have to yield numerical values for purposes of prediction and it is not always possible to get an exact number from the formula.

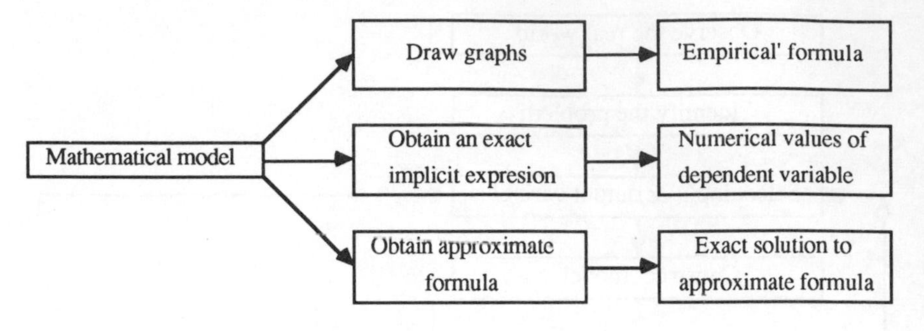

Figure 1.10

We now turn our attention to the kinds of error which can arise in the determination of some physical quantity from an experiment.

In an experiment to measure Young's modulus of elasticity for a metal specimen, the metal is stretched by an applied stress and the modulus found from

$$\text{Young's modulus, } E = \frac{\text{Force per unit area}}{\text{Elongation per unit length}}$$

As regards the experiment, we may note that, instead of reading the force and elongation from dials, we can arrange for a continuous trace of force versus elongation to be produced. An example is shown in Figure 1.11. Initially, the elongation is proportional to the force applied; this is the *elastic* region in which Hooke's Law is valid. At the point A the nature of the deformation changes and the metal embarks on plastic flow; here the application of very little extra force produces much deformation until at the point B the specimen fractures.

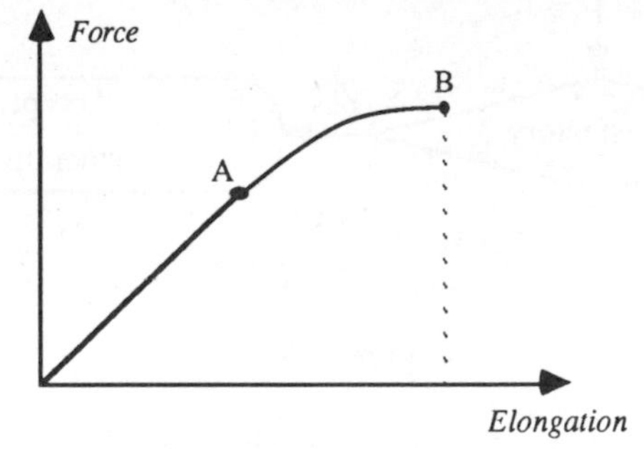

Figure 1.11

In the elastic region we aim to find Young's modulus and this can be partly achieved by finding the slope of the graph in the region. The slope can be found either by measurement from a trace produced by a sophisticated machine or by plotting observations manually on a graph and then making direct measurements from the graph by using the observations themselves or by using a numerical formula.

The cross-sectional area of the specimen which undergoes deformation has to be measured: in the case of a rectangular specimen this involves measuring average breadth and thickness using a micrometer screw gauge. As we shall see, one measurement may not suffice, even if the specimen is of constant area (to the accuracy of measurement possible). The quantities can be inserted into the formula

$$E = \frac{\text{STRESS}}{\text{STRAIN}} = \frac{\text{LOAD APPLIED} \div \text{AREA}}{\text{EXTENSION} \div \text{ORIGINAL LENGTH}}$$

$$= 7.031 \times 10^9 \text{ kN/m}^2, \text{ say,}$$

and a result obtained. But just how accurate is the result, and how can we estimate the likely error? As quoted, we should expect to rely on the last figure given, i.e. we should have 4 significant figures accuracy.

Where might errors have occurred? There are **three** main areas: experimental error, observational error and calculational error.

1 *Errors in Experimenting*

Suppose we determine the resistance of a wire by passing different measured currents through it and measuring the potential difference between its ends. The usual method would be to start with a small current and gradually increase it. But the wire heats proportionally to the square of the current and as it heats, its resistance changes with temperature. We may attempt to minimise this effect by decreasing the current again to average the readings but this may not be enough.

This source of error is, however, not within our scope and we shall not discuss it further.

2 *Errors in Observation*

In taking any measurement an error is often made in the last figure read. One can aim for greater **precision**† by various devices, e.g. the vernier.

The principle of the vernier is employed in the micrometer screw gauge, which is used to measure the thickness and breadth of the metal specimen. Even if these quantities are obtained by taking several readings of each, the discrepancies in the last figure could be attributable to actual random variations in the specimen or to error in observation. In some cases, the least step in the micrometer is only a factor of 10 smaller than the diameter to be measured.

3 *Errors in Calculation*

In the formula quoted for Young's modulus, assuming for the moment that the component values are accurate, could we obtain an accurate result? Clearly the answer is that we cannot, since even the simplest division, e.g. 2/3 produces an infinite decimal which we can either truncate or round-off.

Truncation means that to 4 stored places, 2/3 = 0.6666, i.e. we chop off all figures after the ones we store.

Round-off would produce 0.6667 (i.e. we quote the answer in the fourth decimal place to the nearest whole number, which is 7).

In either case we have not got the result exactly, and, if in the course of a

†Note: For π, 3.1 is accurate to 1 d.p., but not very precise
3.14158 is more precise but not accurate to 5 d.p. (3.14159)

lengthy computation there are many such arithmetic operations, the cumulative effects could be serious.

Examples

We consider some further examples:

3.323 is 3.32 rounded to 2 decimal places or truncated to 2 decimal places.

It is conventional to round up a 5, e.g.
2.105 is 2.11 and
–2.105 is –2.11 both rounded to 2 decimal places.

It is always essential when a rounded answer is given to quote the decimal places of accuracy. If this is not done the answer could be taken as exact.

We could get an answer more exactly using a pocket calculator which can store 8 figures, but even this has its limitations, since

$$\frac{1}{3} \times 3 = 0.9999999 \neq 1$$

Even a digital computer has problems of how to store numbers which are both large and small. Let us consider a very simplified example as depicted below. Here we can store numbers to 2 places of decimals and two pre-decimal point digits, four spaces in all. The effect of trying to store four numbers is shown. The second is stored to within 1/2000 of its value, the third to within 1/350 of its value, but the fourth is lost altogether and the first has completely lost any meaning. Yet the computer wants to store very large and very small numbers to the same precision and the clue here is significant figures rather than decimal places.

1234	56.00		No significant figures kept
	12.34	56	3 s.f. accuracy
	0.12	3456	2 s.f. accuracy
	0.00	123456	None

The mass of an object is determined by different methods according to its magnitude. A small amount of a chemical may be weighed correct to 0.00001 gram on a good balance and a quoted mass of 0.86723 gram may be correct to 5 significant figures. On the other hand, the mass of a bag of potatoes may be accurate only to 0.02kg and it may be quoted as 2.54kg, where we have only two significant figures accuracy. Were we to quote the result as 2.543kg this would be implying greater accuracy than the problem demands. Again, if we multiply a mass correct to 0.1 gram by g (= 9.81) to obtain its weight, this is reasonable; there could be little justification in using a value for g correct to 6 d.p.

Floating-point Numbers

We alleviate the problem by the following artifice. In the following list of numbers, we have rewritten the four numbers by **floating** the decimal point and then by coding **times ten to the power of** as E; in this way we have achieved

6 significant figure accuracy by using only 10 spaces of storage, and we can handle numbers as small as 10^{-9} and as large as 10^9. Most reasonably-sized digital computers usually handle numbers in the range 10^{-76} to 10^{76}, can cope with 11 significant figures and can allow for negative numbers; they, in fact, work with binary arithmetic. Greater precision (almost double) can be obtained under special circumstances.

1.23456×10^5	1.23456 E + 5
1.23456×10^1	1.23456 E + 1
1.23456×10^{-1}	1.23456 E – 1
1.23456×10^{-3}	1.23456 E – 3

Suppose we wish to add together two numbers in floating-point form:

$$1.23 \times 10^{-1} + 4.862 \times 10^{-2}$$

The numbers are 0.123 and 0.04862, and their sum is 0.17162 or 1.7162×10^{-1}; if we could not justify a fourth significant figure in the first number, we would have to write the sum as 1.72×10^{-1}. If we wish to multiply two such numbers, we proceed as follows:

$$(1.2 \times 10^{-1}) \times (1.3 \times 10^4) = (1.2 \times 1.3) \times 10^{-1+4} = 1.56 \times 10^3$$

1.3 Algorithms and Flow Charts

An algorithm may be defined as a set of rules which unambiguously lays down the steps by which a problem can be solved or reports that no solution is possible. The formulation of an algorithm is a key stage in the process of solving a problem via digital computation. It is independent of the subsequent computer code and can therefore be used on different machines with different computer languages.

Two qualities looked for in a successful algorithm are efficiency and robustness. The more efficient an algorithm is, the less unnecessary computer storage and computer time it consumes. The more robust the algorithm, the more it is able to cope with abnormal situations which arise in the class of problems it is designed to handle.

Example 1

An example where an algorithm is used without most people realising that it underlies the set of rules they follow, is in the addition of two sums of money. We might express such rules as:

(i) Add the second digits of the pence together

(ii) If this sum is less than 10, record it, and proceed to step (iii) or step (iv) as appropriate; if not, record the second digit and make a note to 'carry 1'

(iii) Add the first digits of the pence together and add on the carry if required. Repeat step (ii)

(iv) Add the last digits of the pounds together and add on the carry if any; repeat step (ii)

(v) Repeat step (iv) with the next digits of the pounds until no more numbers require to be added.

If we follow these rules through with two particular sums of money we shall see how they are used. Let the sums of money be £12.45 and £4.38

(i) $5 + 8 = 13$ (ii) Record 3 and carry 1 £ . 3
(iii) $4 + 3 + 1 = 8$ (ii) Record 8 £ .83
(iv) $2 + 4 = 6$ (ii) Record 6 £ 6.83
(v) $1 = 1$ (ii) Record 1 £16.83

STOP

We see that, even for simple problems, the written description of an algorithm can be complicated. A **flow chart** presents a semi-pictorial representation which acts as a more efficient means of communication. It plays a similar role to a blue-print. In constructing a flow chart we use several shapes of boxes connected by lines.

The *beginning* and *end* of the sequence of statements under consideration are indicated by an oval box:

Statements of action, i.e. executable statements are indicated by a rectangular box:

Decisions are indicated by a lozenge- or diamond-shaped box:

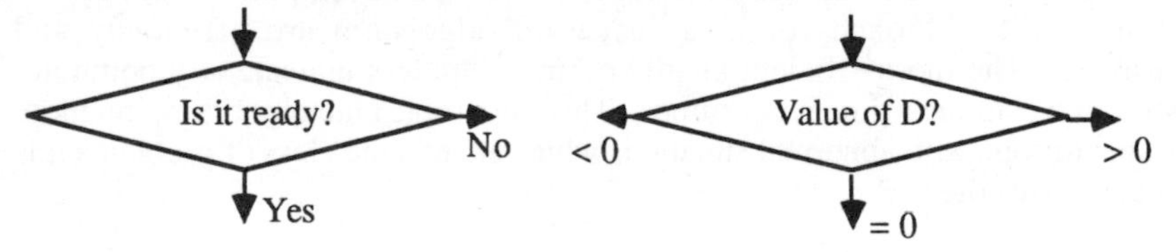

For their use in explaining computer programs, other symbols are employed: a commonly-used box for *input* and *output* is:

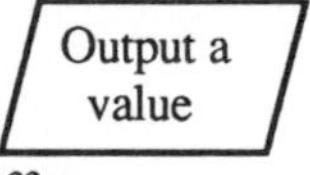

Conventions for these symbols differ.

Example 2

The quadratic equation $ax^2 + bx + c = 0$ can be solved by using the formula

$$x = \frac{-b \pm \sqrt{b^2 - 4ac}}{2a}$$

We wish to write out a logical flow chart of operations to give the solution of this quadratic equation. The crux of the use of the formula is whether the discriminant $D = (b^2 - 4ac)$ is positive, zero or negative and we shall have three branches as in Figure 1.12.

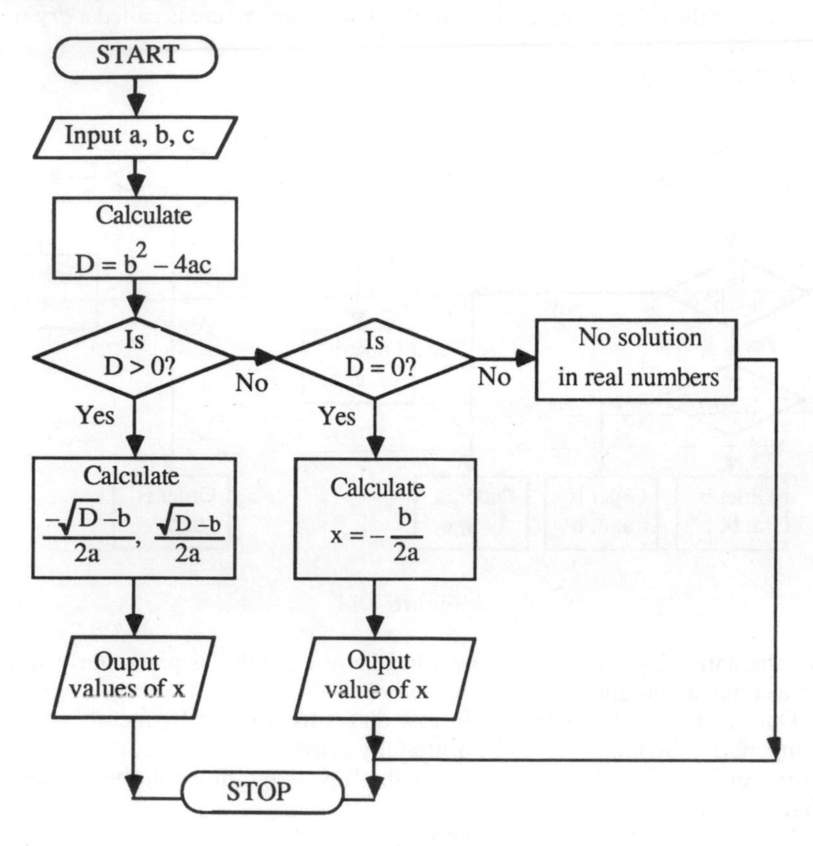

Figure 1.12

We could deal with the branching in one go and we reproduce in Figure 1.13 the relevant portion of the flow chart.

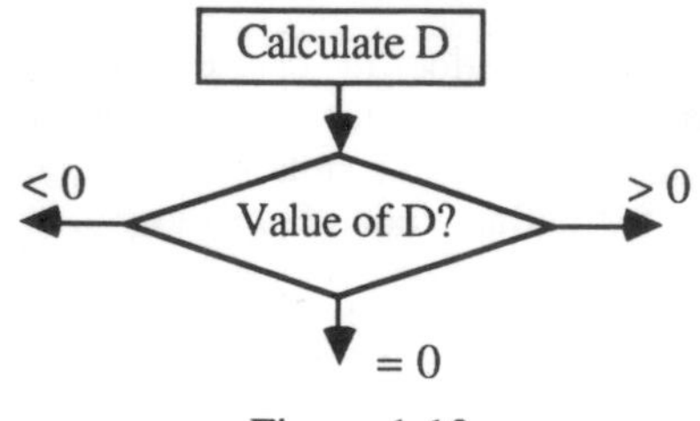

Figure 1.13

Example 3
A problem which appears simple, but which needs careful thought before programming on a computer is that of placing a set of three numbers in ascending order.

Follow through the chart, Figure 1.14, with several sets of three numbers, performing the calculations as you go; this kind of procedure is called a *dry run*.

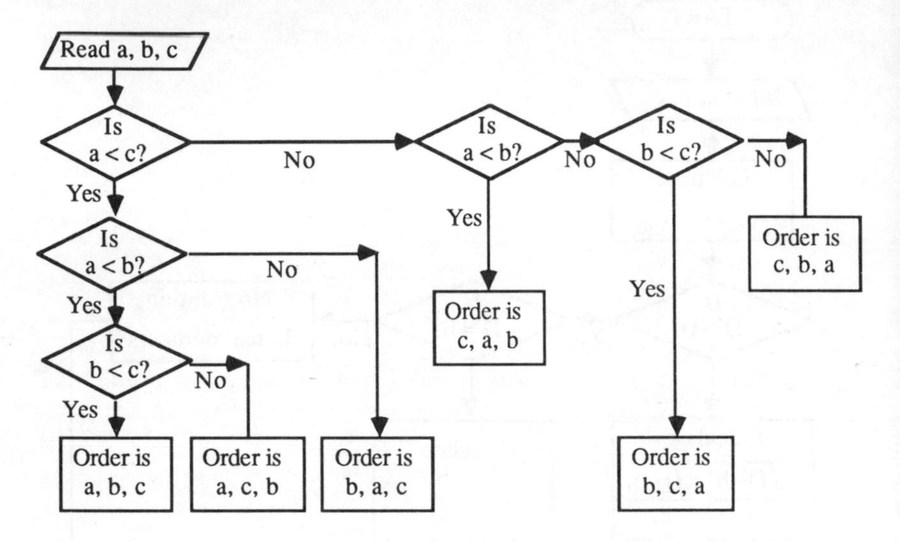

Figure 1.14

An alternative lay-out is now shown in Figure 1.15: this is particularly suitable for a computer program.

One fundamental problem with flow diagrams is that a frequently occurring component – the loop – has to be artificially constructed. Hence, if we wished to carry out a set of instructions N times the flow diagram would be as shown in Figure 1.16.

Unfortunately, the most efficiently constructed code in a high-level language would not correlate too well with the diagram.

```
    FORTRAN 77                     BBC BASIC
DO  10    I = 1,N             30   FOR I = 1 to N
    Set of instructions       40   Set of instructions
10  CONTINUE                  50   NEXT I
```

We see that in neither language does the check for I = N take place explicitly.

A second approach is to use a natural programming language. A major advantage of phrasing an algorithm in a suitably developed natural programming language is its closeness to the final code. For example, to evaluate

$$c_{ij} = \sum_{k=1}^{m} a_{ik} b_{kj} \qquad \text{for } i = 1, \ldots, n \text{ and } j = 1, \ldots, r$$

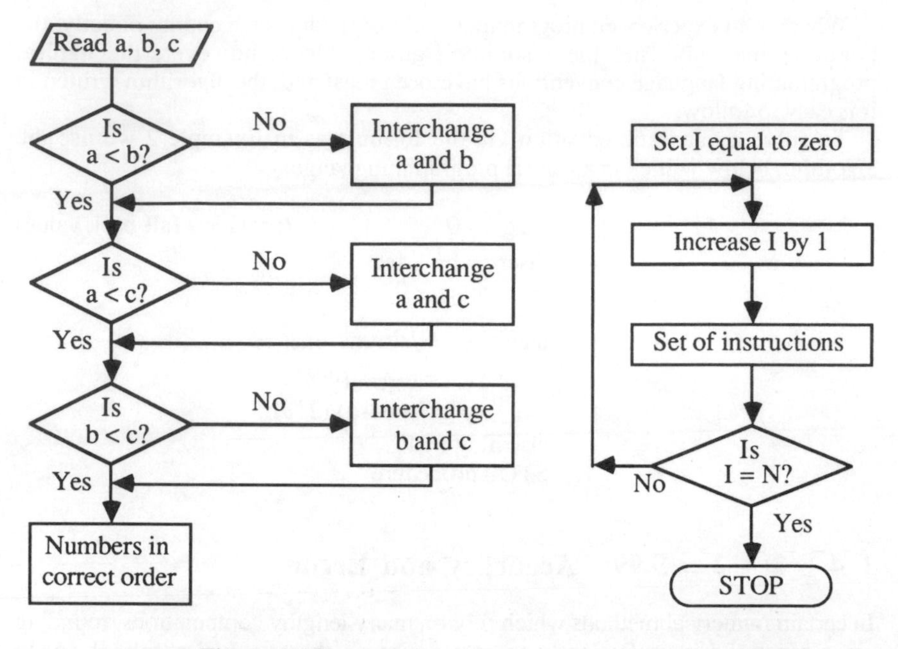

Figure 1.15 Figure 1.16

it would be sufficient to write the natural programming language statements

```
for i := 1 to n step 1
    for j := 1 to r step 1
```

$$c_{ij} = \sum_{k=1}^{m} a_{ik} * b_{kj}$$

```
    end of j – loop
end of i – loop
```

This assumes that the summation can be coded relatively simply.

Coding up this example gives

FORTRAN 77

```
   DO 30  I = 1,N
      DO 20 J = 1,R
         C(I,J) = 0.0
         DO 10 K = 1,M
         C(I,J) = C(I,J)+A(I,K)*B(K,J)
10       CONTINUE
20    CONTINUE
30 CONTINUE
```

BBC BASIC

```
10FOR I = 1 to N
20 FOR J = 1 to R
30  C(I,J) = 0.0
40  FOR K = 1 to M
50  C(I,J)=C(I,J)+A(I,K)*B(K,J)
60  NEXT K
70 NEXT J
80NEXT I
```

Whereas an experienced programmer will understand such coding directly, the beginner may not find them so intelligible. Once, however, the natural programming language conventions have been mastered, the algorithm written in it is easy to follow.

To solve a quadratic equation via the formula as in Example 2 we use the algorithm below written in a natural programming language

```
iflag := 0                        (provide a fall-back value)
discr := b^2 – 4ac
if    discr ≥ 0

then sq  := √discr
     x1  := (–b+sq)/(2*a)
     x2  := (–b–sq)/(2*a)
else iflag := 1
STOP procedure
```

1.4 2 + 2 = 3.99: Accuracy and Error

In certain numerical methods which rely on many lengthy computations, rounding error can build up sufficiently to cast doubt on the eventual result, if not to destroy totally its validity. We must have some means of estimating these errors, and, whilst the error analysis of some numerical methods is a very involved (and unsatisfactory) process, we can at least attempt a brief study. Usually we adopt the most pessimistic approach, e.g. the possible error of the difference between two numbers is the sum of the possible errors in each number.

Arithmetic of Errors

Let N be the exact value of some quantity and n an approximate value of the quantity such that

$$n = N + e$$

Then

e	is the **Error**
$\lvert e\rvert$	is the **Absolute error** (i.e. $= -e$ if $e < 0$)
$\lvert e\rvert/\lvert N\rvert$	is the **Relative error**
$(\lvert e\rvert/\lvert N\rvert) \times 100$	is the **Percentage error**

Addition

Let n_1, n_2 be approximations to N_1, N_2 such that

$$n_1 = N_1 + e_1, \quad n_2 = N_2 + e_2$$

Hence, $n_1 + n_2 = (N_1 + N_2) + (e_1 + e_2)$

The absolute error is $|e_1 + e_2|$ which can be shown to be $\leq |e_1| + |e_2|$

For example, we let $N_1 = 2.6$, $N_2 = 3.1$, $n_1 = 2.55$ and $n_2 = 3.16$;

therefore $e_1 = -0.05$, $e_2 = +0.06$. Now $n_1 + n_2 = 5.71$, $N_1 + N_2 = 5.7$;

then $|e_1 + e_2| = 0.01$, and $|e_1| + |e_2| = 0.05 + 0.06 = 0.11$

Subtraction
Using the above notation,

$$n_1 - n_2 = (N_1 - N_2) + (e_1 - e_2)$$

and the absolute error is $|e_1 - e_2| \le |e_1| + |e_2|$ (not proved here)

With the example, $n_1 - n_2 = -0.61$; $N_1 - N_2 = -0.5$, $|e_1 - e_2| = 0.11$

We may summarise by saying that, in addition and subtraction, the absolute error in the result is, at most, the sum of the absolute errors.

Multiplication

$$n_1 n_2 \doteqdot N_1 N_2 + (N_2 e_1 + N_1 e_2)$$

We neglect the term $e_1 e_2$ since it is of the second order of smallness, assuming the individual errors are small.

$$\text{The relative error} \doteqdot \frac{|N_2 e_1 + N_1 e_2|}{|N_1 N_2|} = \left|\frac{e_1}{N_1} + \frac{e_2}{N_2}\right| \le \left|\frac{e_1}{N_1}\right| + \left|\frac{e_2}{N_2}\right|$$

Thus the relative error in the result is, at most, the sum of the individual relative errors.

In the current example, $n_1 n_2 = 8.058$

$$N_1 N_2 = 8.06, \text{ relative error} = \frac{0.002}{8.06} = 0.00025 \quad (5 \text{ d.p.});$$

$$\frac{|N_2 e_1 + N_1 e_2|}{|N_1 N_2|} = \frac{|3.1 \times -0.05 + 2.6 \times 0.06|}{2.6 \times 3.1} = \frac{0.001}{8.06} = 0.00012 \quad (5 \text{ d.p.})$$

$$\left|\frac{e_1}{N_1}\right| + \left|\frac{e_2}{N_2}\right| = \left|\frac{-0.05}{2.6}\right| + \left|\frac{0.06}{3.1}\right| = 0.0386 \quad (4 \text{ d.p.})$$

Division

A similar result applies here, i.e. the relative error is, at most, the sum of the individual relative errors.

The proof of this may be omitted at a first reading. We have

$$\frac{n_1}{n_2} = \frac{N_1 + e_1}{N_2 + e_2} = [N_1 + e_1]\frac{1}{N_2}\left[1 + \frac{e_2}{N_2}\right]^{-1}$$

$$\cong [N_1 + e_1] \frac{1}{N_2} \left[1 - \frac{e_2}{N_2} \right] \quad \text{by the Binomial expansion}$$

$$= \left[\frac{N_1}{N_2} + \frac{e_1}{N_2} \right] \left[1 - \frac{e_2}{N_2} \right]$$

$$\cong \frac{N_1}{N_2} + \frac{e_1}{N_2} - \frac{N_1}{N_2} \frac{e_2}{N_2} \quad \text{ignoring the term } e_1 e_2$$

$$\text{Therefore, relative error} = \left| \frac{e_1}{N_2} - \frac{N_1}{N_2} \frac{e_2}{N_2} \right| \Big/ |N_1/N_2|$$

$$= \left| \frac{e_1}{N_1} - \frac{e_2}{N_2} \right|$$

$$\leq \left| \frac{e_1}{N_1} \right| + \left| \frac{e_2}{N_2} \right| \quad \text{as suggested.}$$

Note that all these results apply approximately and only if e_1, e_2 are small.

With the current example,

$$\frac{n_1}{n_2} = 0.8070 \quad (4 \text{ d.p.}), \qquad \frac{N_1}{N_2} = 0.8387 \quad (4 \text{ d.p.});$$

$$\text{and relative error} = \frac{0.0317}{0.8070} = 0.0393 \quad (4 \text{ d.p.})$$

Example

Let us consider now the error involved in finding the hypotenuse, a, of a right-angled triangle using Pythagoras' Theorem:

$$a^2 = b^2 + c^2$$

Suppose errors δb, δc are made in b, c respectively, giving an error δa^2 in a^2 or an error Δa in a;

Then $\quad a^2 + \delta a^2 = (b + \delta b)^2 + (c + \delta c)^2 \doteqdot b^2 + c^2 + 2b\,\delta b + 2c\delta c$

i.e. $\quad \delta a^2 = 2b\,\delta b + 2c\delta c$

Now $\quad a + \Delta a \cong (b^2 + c^2 + 2b\,\delta b + 2c\delta c)^{1/2}$

$$= (a^2 + 2b\,\delta b + 2c\delta c)^{1/2}$$

$$= a\left[1 + \frac{2b}{a^2}\delta b + \frac{2c}{a^2}\delta c\right]^{1/2}$$

$$\cong a\left[1 + \frac{b}{a^2}\delta b + \frac{c}{a^2}\delta c\right] \quad \text{by the Binomial expansion}$$

$$= a + \frac{b}{a}\delta b + \frac{c}{a}\delta c$$

Thus $\qquad \Delta a \cong \dfrac{b}{a}\delta b + \dfrac{c}{a}\delta c$ and

$$\left[\frac{\Delta a}{a} \times 100\right]\% \cong \left[\frac{b}{a^2}\delta b + \frac{c}{a^2}\delta c\right] \times 100\ \% \qquad (1.4)$$

If $b = 3$ and $c = 4$, then $a = 5$; but if b were measured as 3.01 and c as 3.98, a would be found as 4.990 (3 d.p.). This gives a percentage error of 0.2%.

From formula (1.4), the percentage error is

$$\left[\frac{3}{5^2} \times 0.01 + \frac{4}{5^2} \times (-0.02)\right] \times 100\ \% = \left[\frac{-0.05}{25} \times 100\right]\% = 0.2\%$$

Avoidance or Minimisation of Errors

We glanced at the problem of overcoming systematic experimental errors: we now briefly examine the other categories.

To minimise random error, a larger number of readings may be taken. Since the probable error $\propto 1/\sqrt{n}$, where n is the number of readings, one needs 100 times as many readings to give ten times the accuracy.

Often rounding errors can be minimised by low cunning; the troubles are at their worst when two approximately equal numbers are subtracted.

Consider

$$Z = \sqrt{X+1} - \sqrt{X} \quad \text{where } X = 80$$

If we work to 6 s.f., we have $Z = 9 - 8.94427 = 0.05573$

If we now are only able to work to 3 s.f. we find $Z = 9 - 8.94 = 0.06$.

Using the identity $(\sqrt{X+1} - \sqrt{X})(\sqrt{X+1} + \sqrt{X}) \equiv 1$, we see that

$Z = \dfrac{1}{\sqrt{X+1} + \sqrt{X}} = \dfrac{1}{17.9} = 0.0559$, which is more accurate, even though we are only working with 3 s.f.

Similarly we can overcome the problem of finding accurately the roots of a quadratic equation when one of those roots is very small.

Example
We work to 2 d.p. in the following example.

$$x^2 - 500x + 1 = 0$$

Solving by the formula method, we obtain

$$x = \frac{500 \pm \sqrt{250\,000 - 4.0}}{2} \doteqdot \frac{500 \pm \sqrt{250\,000}}{2}$$

i.e. $x_1 = 500$, $x_2 = 0$; and these do not satisfy the equation, but x_1 is probably more accurate than x_2. (If x_1, x_2 are the roots of $ax^2 + bx + c = 0$, then $x_1x_2 = c/a$). In this example, $x_1x_2 = 1$

$$\therefore \quad x_2 \cong \frac{1}{500} = 2.0 \times 10^{-3}$$

In fact, the values are $x_1 = 499.996$ (6 s.f.) and $x_2 = 2.000 \times 10^{-3}$ (4 s.f.) so we have clearly overcome our troubles.

A consideration of the formula used in experimental data may give a guide to the accuracy required in certain measurements. In the case of the simple pendulum the period of oscillation,

$$T = 2\pi\sqrt{\frac{l}{g}}$$

where l is the length of the pendulum; we assume that this is an exact model for the moment.

If this is used to estimate g, we note that $g = 4\pi^2\dfrac{l}{T^2}$

Now an error of $x\%$ in the reading of T will have the same effect on the estimate of g as an error of $2x\%$ in the reading of l. Usually, if we cannot measure l to less than 0.5% accuracy, there is little virtue in concentrating on achieving a reading of T to 0.005% accuracy.

(As an aside note one might mention as a general rule that it is wise to work to one more decimal place than the accuracy quoted.)

It is perhaps necessary to re-emphasise at this juncture that even an *exact* analytical formula resulting from solving an equation of a model is only as exact as the precision that can be obtained upon substituting numerical values into it. For example, if $y = \sin(\ln x^2)$ then at each of the three operations: *square*, *take logs* and *find the sine*, accuracy may be lost.

Ill-Conditioning
Sometimes there are errors inherent in a system and no simple way of overcoming such trouble exists. If we examine the two pairs of equations

$$\left.\begin{aligned} x + y \quad &= 1 \\ x + 1.001y &= 0 \end{aligned}\right\} \quad (1.5) \qquad \text{and} \qquad \left.\begin{aligned} x + y \quad &= 1 \\ x + 0.999y &= 0 \end{aligned}\right\} \quad (1.6)$$

we see that the solutions are respectively $x = +1001$, $y = -1000$ and $x = -999$, $y = +1000$. A change in one of the coefficients of about 0.2% has caused a radical change in the nature of the solution.

If we look at Figure 1.17, we see that the two lines in each set are nearly parallel; we have exaggerated the situations to show that the two lines in each case do meet. The fact that their slopes are so nearly the same means that only a slight change in the slope of one causes it to overbalance in much the same way as the addition of a small mass to one pan of a chemical balance which is comparing almost equal masses could cause a reversal of the relative positions of the pans.

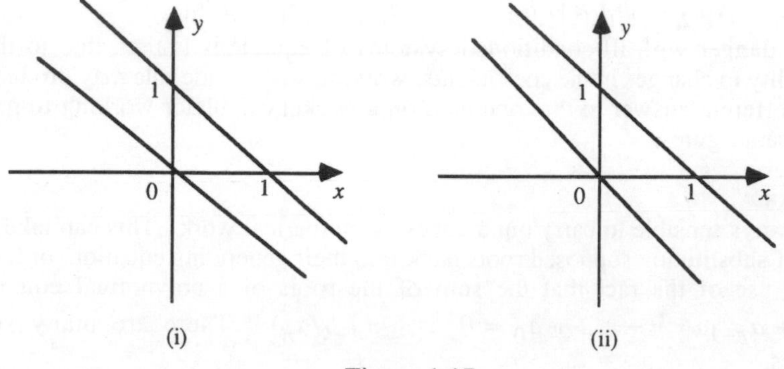

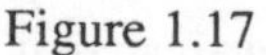
Figure 1.17

In our example the slopes are (to 3 d.p.) –1 and –0.999 for the first pair of equations (1.5) and –1 and –1.001 for the second pair (1.6). The difference in slopes for each pair is small. For the pair of equations

$$\left.\begin{aligned} a_1x + b_1y &= c_1 \\ a_2x + b_2y &= c_2 \end{aligned}\right\} \qquad (1.7)$$

where $a_1, b_1, a_2, b_2, c_1, c_2$ are constants, the solution may be written

$$x = \frac{c_1b_2 - c_2b_1}{a_1b_2 - a_2b_1} \qquad\qquad y = \frac{c_2a_1 - c_1a_2}{a_1b_2 - a_2b_1}$$

and it is clear that if $a_1b_2 = a_2b_1$ then we have no solution. Ignoring special cases where b_2 or b_1 are zero, this condition becomes

$$a_1/b_1 = a_2/b_2$$

which is equivalent to saying that the lines are parallel.

If $a_1b_2 \cong a_2b_1$ then we might suppose that we are dealing with the *ill-conditioned* cases above.

However, consider the same two sets of equations (1.5) and (1.6), rewritten as follows:

$$\left.\begin{aligned} 1000x + 1000y &= 1000 \\ 1000x + 1001y &= 0 \end{aligned}\right\} \quad (1.8) \qquad \left.\begin{aligned} 1000x + 1000y &= 1000 \\ 1000x + 999y &= 0 \end{aligned}\right\} \quad (1.9)$$

The quantity $(a_1b_2 - a_2b_1)$ is 1000 for the first pair (1.8) and –1000 for the second (1.9) and can hardly be considered small, yet these sets of equations are precisely the same as the sets (1.5) and (1.6) considered earlier. The clue here is that the value of $(a_1b_2 - a_2b_1)$ is small compared with either a_1b_2 or a_2b_1; note that we are not interested in the sign of these quantities, merely their size.

We use the symbol $\ll$, meaning **very much less than.**

We may interpret our condition as:

$$|a_1b_2 - a_2b_1| \ll |a_1b_2| \quad \text{or} \quad |a_1b_2 - a_2b_1| \ll |a_2b_1|$$

The danger with ill-conditioned systems of equations is that, due to their sensibility to changes in the coefficients, working with a slide rule may produce a quite different answer to that obtained on a pocket calculator working to more significant figures.

Checks

It is always sensible to carry out a check on numerical work. This can take the form of substituting supposed roots back into their generating equation, or it can be the use of the fact that the sum of the roots of a polynomial equation $a_nx^n + a_{n-1}x^{n-1} + \ldots.. + a_0 = 0$ is $(-a_{n-1}/a_n)$. There are many such artifices.

Often when a fraction is being evaluated, it is wise to estimate the order of the magnitude of the fraction. For example,

$$\frac{\pi^2\, 19.3}{4.8 \times 5.1} \cong \frac{10 \times 20}{5 \times 5} = 8$$

and this compares with the value 7.78 (2 d.p.) obtained with a calculator.

1.5 Build-Up of Knowledge

From an engineering point of view the aims of our studies of mathematics are two-fold.

First of all we want to be able to solve physical problems. As pointed out earlier, a solution to a problem can take many forms and can be obtained by various means – either analytical or numerical with appropriate use of the computer. However, at the present time we do not have the necessary techniques and skills to solve many problems nor do we have sufficient knowledge of the functions and numbers involved in solutions. Therefore, before we can try to solve most problems we have got to build up our knowledge of mathematics with the associated skills and techniques available. In the following chapters we shall be studying various functions, number systems, analytical and numerical techniques and only when we have acquired a thorough working knowledge of these will it be possible to tackle the more demanding engineering problems. At all times we shall try, wherever possible, to link the mathematics to practical problems, but at intermediate stages we shall have to study topics which at first sight will appear to be abstract and of very little practical use. It is just like learning a language: we have got to build up our vocabulary and syntax. A question always asked by engineering students is: *Where shall I use this, and do I need to know such abstract mathematical ideas?* Well, we hope to show you, as the pages go by, just where you can use your mathematics and how useful it can be in your work.

Build-up of knowledge is also important in another way, in that the more we know the more sophisticated we can make our mathematical models for use in design work and the more we shall appreciate the limitations of our models. It was lack of knowledge which caused the well-known Tacoma Narrows Bridge failure where a wind speed of 42 mph caused the bridge to collapse, although it had been designed to resist a steady wind of at least 100 mph. It was certainly lack of knowledge which caused the wind-induced collapse of full-scale oil storage tanks in the dramatic first picture of our frontispiece. The expense of such a failure can be realised by comparing the size of the tanks with the size of a workmen's hut in the foreground. The collapse was subsequently modelled successfully in a series of wind-tunnel studies; see the second page of the frontispiece. The aim of the tests was to understand and measure the form of the collapse mechanism and to produce an empirical criterion against which a detailed analytical treatment could be compared. Had these tests been performed and a mathematical model been available, the design of the tanks would have been improved.

Although you, as an engineer, will perhaps not be working on the mathematical solutions and mathematics involved in design, you will probably be working in a team and it will be necessary for you to be able to converse with those who are formulating mathematical models and providing solutions. You will also be able to understand the assumptions made in forming these models and thence the limitations of their use. We think, therefore, that you should build up your knowledge of mathematics and become, if not an expert in the language, certainly competent enough to be able to read on to more advanced work.

Summary

After reading through this chapter you should be aware of the role played by mathematics in your engineering subjects. The concepts of a mathematical model and a solution to such a model should be understood; in particular you must realise that a mathematical model is built on a number of assumptions and, therefore, any solution is valid only as long as these assumptions hold.

The value of flowcharting a problem should be recognised; it is a means of analysing the nature of the problem. Then you ought to appreciate the kinds of error that arise in numerical calculations and the ways in which we can estimate the error in a result; you should, further have an idea of the methods for cutting down errors, where possible.

Unlike other chapters in this book, this one has emphasised the idea of mathematical modelling rather than dwelling on techniques.

Problems

Section 1.2

1 Divide 4.2 by 2.3 assuming errors in each number of ± 0.05. Consider the largest and the smallest values the fraction can take and check against the relative error as expressed in section 1.2.

2 Likewise for +, × and – with these two numbers.

3 If, in the determination of the hypotenuse, a, of a right-angled triangle, b and c are read as 3.02 and 3.98 respectively, where each reading could be in error by ±0.03, calculate an approximate value for a and quote the actual and percentage errors.

Section 1.3

4 Write an algorithm for multiplying together two decimal numbers.

5 Write an algorithm for sorting cards of one suit into ascending order.

6 In a digital computer, numbers are stored in specific locations; if one number is read into a location already occupied, the first number there is obliterated and the new one takes its place. Devise a flow chart to interchange two numbers, a and b.

7 Construct a flow chart to solve the simultaneous equations

$$a_1x + b_1y = c_1, \quad a_2x + b_2y = c_2$$

taking special care with zero coefficients.

8 The polynomial $y = ax^3 + bx^2 + cx + d$ can be evaluated as

$$y = [(ax + b)x + c]x + d$$

Why is this a preferable method of evaluation on a computer? Produce a flow chart to read in the values of the variables and carry out the evaluation.

9 Develop a flow chart to read in two sets of numbers, each set in ascending order, and merge them into one set of numbers also in ascending order. Carry out a dry run on suitable sets of numbers.

10 Construct a flow chart to read in a set of numbers and find their average; since the number of members of the set may be a variable quantity this must be allowed for.

11 What could the flow chart in Figure 1.18 map out?

Hint: If (x_1,y_1) and (x_2,y_2) are the coordinates of two points in a plane, $(y_2 - y_1)/(x_2 - x_1)$ is the slope of the straight line joining them.

12 Write algorithms in a natural programming language for Problems 7,8,9,10 and 11.

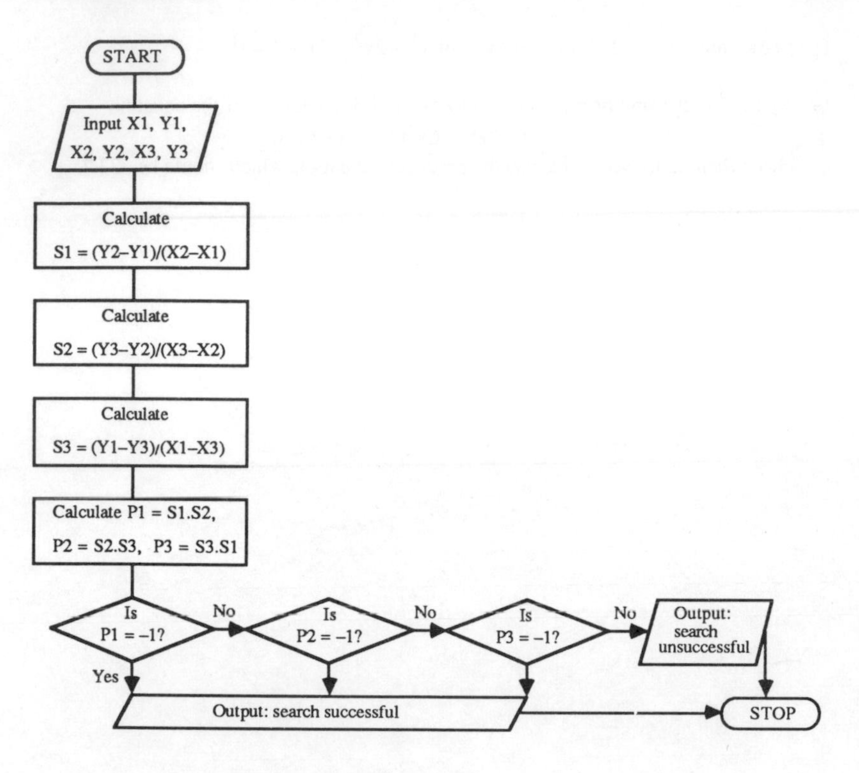

Figure 1.18

Section 1.4

13 Consider the special cases where b_2 or b_1 is zero.

14 Show that the pair of equations

$$\begin{aligned} x - y &= 2 \\ 30x - 30.01y &= 60.02 \end{aligned}$$

is ill-conditioned. Solve working to 1 decimal place, 2 d.p. and 4 d.p. Examine separately the results of changing the –30.01 to –30 and the 60.02 to 60.

15 Repeat Problem 14 for the equations

$$\begin{aligned} x + 2y &= 5 \\ 100x + 199.8y &= 299.6 \end{aligned}$$

Section 1.5

16 Estimate the values of $\dfrac{4.1 \times 16.9}{2.3 \times 0.95}$ and $\dfrac{3.61}{1.52\pi}$.

Compare with values obtained by a calculator.

17 Repeat Problem 16 for $\dfrac{17.2 \times 28.9}{-4.4 \times 81.8}$, $\dfrac{\pi^3 \times 41.3}{11.2 \times 7.63}$, $\dfrac{-1.62}{1.85 \times 74.2}$, ln 18

18 Show that $x = 1, -2, 3$ are the roots of $x^3 - 2x^2 - 5x + 6 = 0$.

19 By finding the sum of the supposed roots $x = 1.31, 4.8, -5.2$ of the equation

$$x^3 - 0.97x^2 + 25.43x + 33.7 = 0$$

check their validity. Check also the product of the roots which should be -33.7.

2

FUNCTIONS AND SETS

2.1 Number Systems and Inequalities

The set of real numbers has as its elements the rational numbers and the irrational numbers. On a Venn diagram they could be represented as follows:

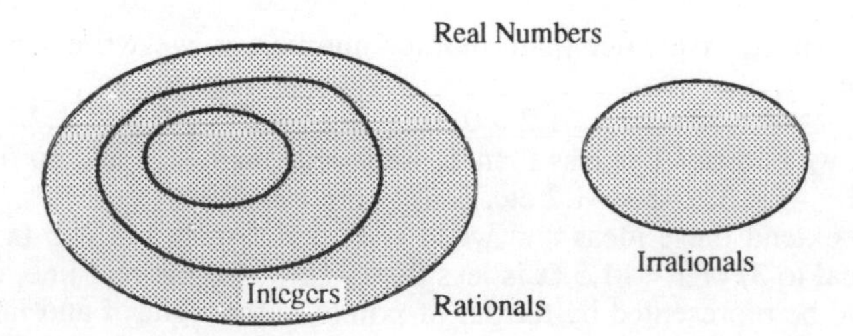

Notice that Real numbers R = Rationals ∪ Irrationals, and
Cardinals ⊂ Integers ⊂ Rationals

The set of rational numbers contains integers and non-integers; the integers contain the natural numbers 1, 2, 3, 4, ...

Contained in the set of integers is the subset of **prime numbers** which are defined as being only divisible by 1 and the number itself. The set of positive prime numbers is {1, 2, 3, 5, 7, 11, 13, 17, 19, ...}.

We can represent all the real numbers by points on a line: the **Real Line**.

Considering first the subset of the integers, we can plot these on the real line as follows:

We say that there is a **one-to-one correspondence** between the integers and these points on the real line. By this we mean that

(i) to every integer there corresponds one, and only one, point on the real line

(ii) every one of the plotted points represents one, and only one, of the integers.

We can now turn to the rational numbers and mark off points on the real line corresponding to all the elements of this set besides the integers. This is much more difficult because between any two integers there is an infinite number of rationals. Even though there is an infinite number of rationals there will still be gaps on the line where no rational exists. It is shown in higher mathematical texts that **all** these remaining points are taken up by the irrational numbers. That is, there is a one-to-one correspondence between the set of real numbers and the set of points on the real line. This means that

(i) to **every** real number, there corresponds one, and only one, point on the real line

(ii) **every** point on the real line represents one, and only one, real number.

Inequalities

If we choose any two numbers on the real line then you will see that the right-hand one is the greater.

- 4 - 3 - 2 - 1 0 1 2 3 4

If one number, x, is greater than another number, y, we write $x > y$, so that for example

$$2 > 1, \quad 3 > -1, \quad 1.2 > 0.9, \quad -1.2 > -2, \quad -1.1 > -1.2$$

Of course, we can say 1 is less than 2, –2 is less than –1.2 and so on, and we write $1 < 2$, $-1 < 3$, $-2 < -1.2$ etc.

We can extend these ideas and write statements such as $x \geq 3$ (x is greater than or equal to 3) or $x < -1.5$ (x is less than –1.5). On the real line, the first of these would be represented by the set of points to the right of and including 3, and the second would be represented by the set of points to the left of –1.5 and not including 1.5. In each case the sets are said to represent semi-infinite intervals of the real line since they each extend to infinity in one direction only.

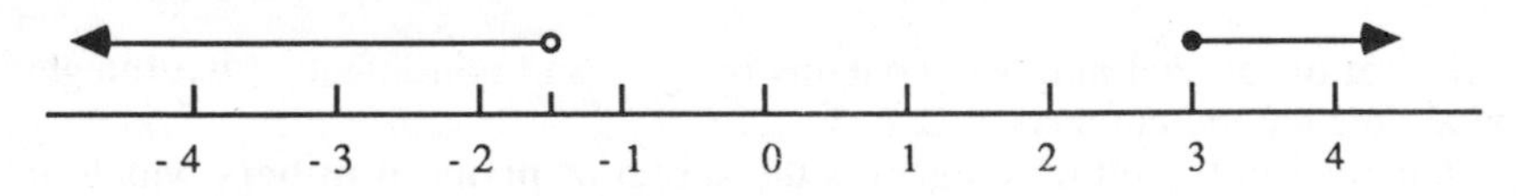

Again we can write statements such as $-3 \leq x \leq -2$ or $-1 < x < 1$ or perhaps $3 \leq x < 5.5$. These sets of points would be represented on the real line by **finite intervals**, the • indicating that the point is included, ∘ indicating that the point is not included in the set.

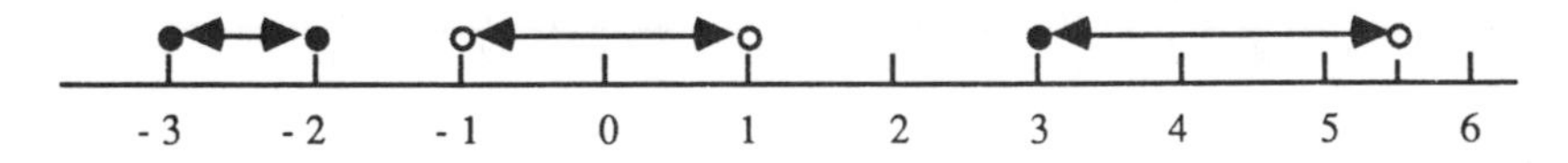

Notice that the interval for $-1 < x < 1$ is symmetrical about the origin and represents points which are at a distance less than one from the origin. We write this as $|x| < 1$ and say the modulus of x (or mod x) is less than 1. Alternatively it can be read as the absolute value of x is less than 1 – this means more because it says the magnitude of x, whatever its sign, is less than 1. We can generalise to statements like $|x - 3| \leq 5$, which means that the distance of x from the number 3 is less than or equal to 5, and on the real line this is represented by the set of points in the interval $-2 \leq x \leq 8$.

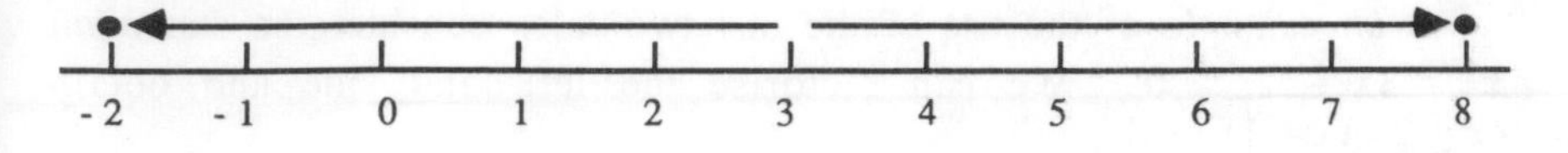

We introduce now a notation which is often used to denote intervals. We write $x \in [a, b]$ for the statement $a \leq x \leq b$. The interval $[a, b]$ is called a **closed** interval.

In a similar way we write $x \in (a, b)$ for the statement $a < x < b$ and the interval (a, b) which does not contain a or b is called an **open** interval. It should be perfectly clear what we mean by the intervals $(a, b]$ and $[a, b)$. The first of these does *not* contain a while the second does *not* contain b.

Sometimes it is not easy, in the first instance, to see what an inequality statement means and we have to unravel the statement to understand its implication; this necessitates rules for handling inequalities. We summarise these rules below:

Inequality Rules

(i) $a > b$ if and only if $a - b$ is some positive number.
For example, $-1 > -2$ since $-1 - (-2) = +1$, a positive number

(ii) If $a > b$, then $a + c > b + c$. This says we can add a real number c (positive or negative) to each side of the inequality and not change the sense of the inequality.
For example, if we are given $2x - 7 > x + 2$
then $2x - 7 + 7 > x + 2 + 7$
or $2x > x + 9$
so that $2x - x > x + 9 - x$
or $x > 9$

(iii) If $a > b$ and $\lambda > 0$, then $a\lambda > b\lambda$. That is, we can multiply by a positive number without changing the sense of the inequality.
For example, $3x - 4 < x - 8 \Rightarrow 2x < -4 \Rightarrow x < -2$

(iv) If $a > b$ and $\lambda < 0$, then $a\lambda < b\lambda$. That is, multiplying by a negative number *changes* the sense of the inequality.
For example, $x - 5 < 5x - 2 \Rightarrow -4x < 3 \Rightarrow x > -3/4$

(v) If $a > b$ and $c > d$, then $a + c > b + d$.

(vi) If $ab > 0$, then *either* $a > 0$ and $b > 0$
or $a < 0$ and $b < 0$

(vii) If $ab < 0$, then *either* $a > 0$ and $b < 0$
or $a < 0$ and $b > 0$

All these rules are still valid if $>$ is replaced by $\geq$ and $<$ by $\leq$.

As an example of the use of the last two rules consider the statement $x^2 - 3x + 2 > 0$. We can factorise the left-hand side and obtain $(x - 2)(x - 1) > 0$.

Hence *either* (a) $x - 2 > 0$ and $x - 1 > 0$ *or* (b) $x - 2 < 0$ and $x - 1 < 0$

On the real line we have for (a):

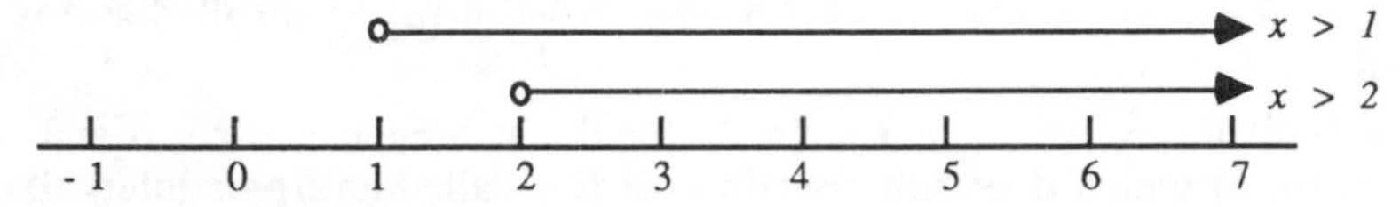

To satisfy both statements that $x > 2$ *and* $x > 1$ we must have $x > 2$.

On the real line we have for (b):

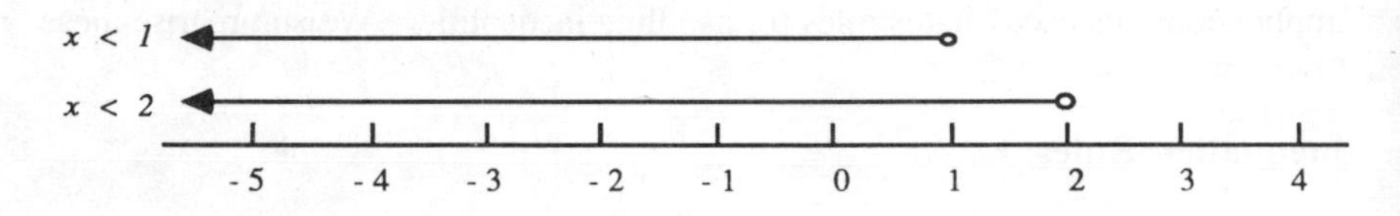

To satisfy both these statements we must have $x < 1$.

Hence we have $x > 2$ or $x < 1$ if $x^2 - 3x + 2 > 0$. (Try $x = 1.5$ in the l.h.s.)

Suppose we have been given that $x^2 - 3x + 2 < 0$.

Then $(x - 1)(x - 2) < 0 \Rightarrow$ either (i) $(x - 1) > 0$ and $(x - 2) < 0$
or (ii) $(x - 1) < 0$ and $(x - 2) > 0$

(i)

We must have $1 < x < 2$ to satisfy both these.

(ii)

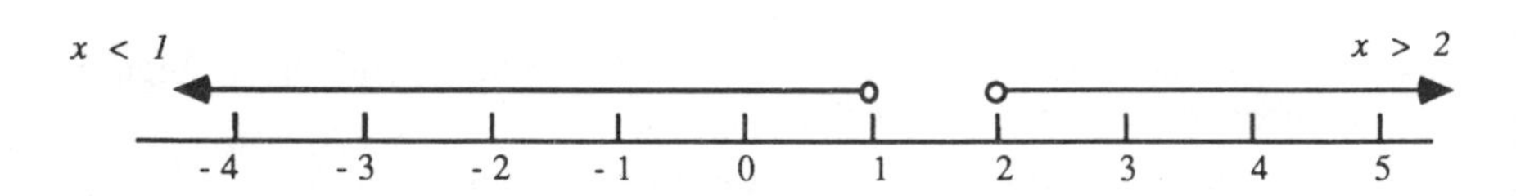

The solution set here is empty – there are no points such that $x < 1$ and $x > 2$.

Thus for $x^2 - 3x + 2 < 0$, $1 < x < 2$. Alternatively we can write $x \in (1, 2)$.

2.2 Relations and Functions

The idea of a function is paramount in our formulation of the interaction between natural phenomena. For example, we talk loosely of the profit of a company as being a function of raw materials, plant, labour employed, taxation, and so on. In mechanics we speak of the position of a body, its velocity and acceleration as being all functions of time. *What do we mean by saying that some quantity is a function of another?*

We start by considering **Relations**. We make a *definition:*

A relation is a set of ordered pairs.

Example 1

The speed of a car (υ m.p.h.) was measured at various times (t minutes) after starting from rest and gave

t (minutes)	0	1	2	3	4	5	6
υ (m.p.h.)	0	20	36	50	60	60	60

We can write this relation as a set of ordered pairs

$$R_1 = \{(0, 0)\ (1, 20)\ (2, 36)\ (3, 50)\ (4, 60)\ (5, 60)\ (6, 60)\}$$

Example 2

Let Z^+ be the set of positive integers and consider the set of ordered pairs $\{(m, n)\}$ such that $m \in Z^+$, $n \in Z^+$ and $m + n \leq 5$.

The relation is a set of ordered pairs

$$R_2 = \{(1, 1)\ (1, 2)\ (1, 3)\ (1, 4)\ (2, 1)\ (2, 2)\ (2, 3)\ (3, 1)\ (3, 2)\ (4,1)\}$$

Example 3

A car factory produces 4 different models and the numbers of each model produced on a certain day is given by

Model 1	Model 2	Model 3	Model 4
200	500	150	400

This gives the relation

$$R_3 = \{(1, 200)\ (2, 500)\ (3, 150)\ (4, 400)\}$$

We can represent relations graphically by using coordinate axes and plotting the ordered pair (x, y) as a point.

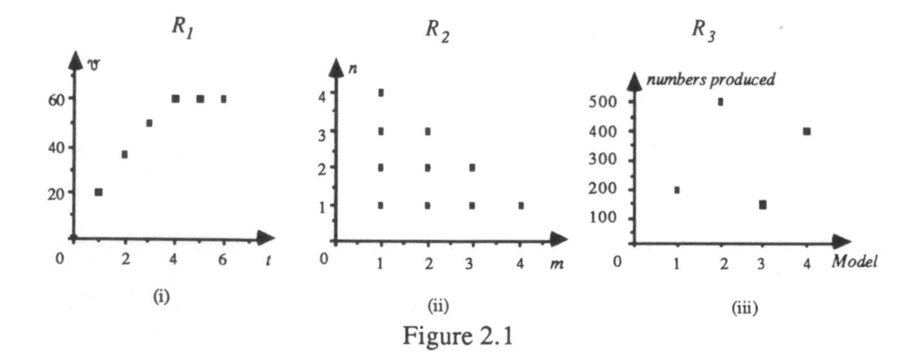

Figure 2.1

The relations R_1, R_2, R_3, given in the examples are represented by the preceding diagrams, Figure 2.1 (i) – (iii).

These are all examples of **Graphs**. The x values (on the horizontal line) constitute a set X called the **Domain** and the set of y values constitute a set Y called the **Range** or **Co-domain.**

If each element of the domain is related to a single element of the range, the relation is said to be **one-to-one**. If one element of the domain is related to more than one element of the range the relation is **one-many**. Finally, if two or more elements of the domain are related to the same element of the range, the relation is **many-one**. Other relations are called **many-many**. Looking back at our examples we can see that R_1 is many-one since members 4, 5, 6 of the domain are all related to the member 60 of the range. R_2 is many-many, R_3 is one-to-one.

Functions

Several different definitions will serve to illustrate the ideas of functions although the definitions are equivalent.

Definition 1

A function is a relation in which no two of its ordered pairs have the same first member but different second members. Looking back again at our examples, R_1 is a function because all the first members of the ordered pairs are different. R_2 is not a function; R_3 is a function and, in words, we would say that the daily number of vehicles is a function of the models produced.

Definition 2

A function is a mapping that exists between two sets X and Y such that to each element $x \in X$ there is associated one and only one element $y \in Y$.

To illustrate the mapping idea we can use diagrams and in such a representation (Figure 2.2), relation R_1, which is a function, would be

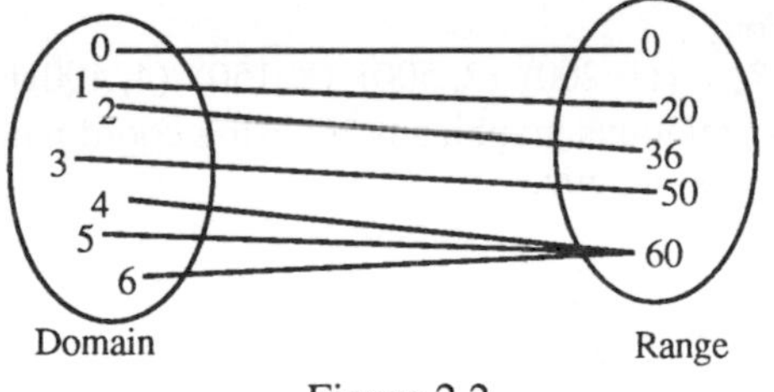

Figure 2.2

Notice elements 4, 5, 6 all related to 60, but this does not violate the definition.

We can draw similar diagrams (Figure 2.3) for R_2 and R_3.

Definition 3

A function is a relation which is many-one or one-one. A relation which is many-many or one-many is *not* a function.

Compare this with the previous definitions and satisfy yourself that it is equivalent to these.

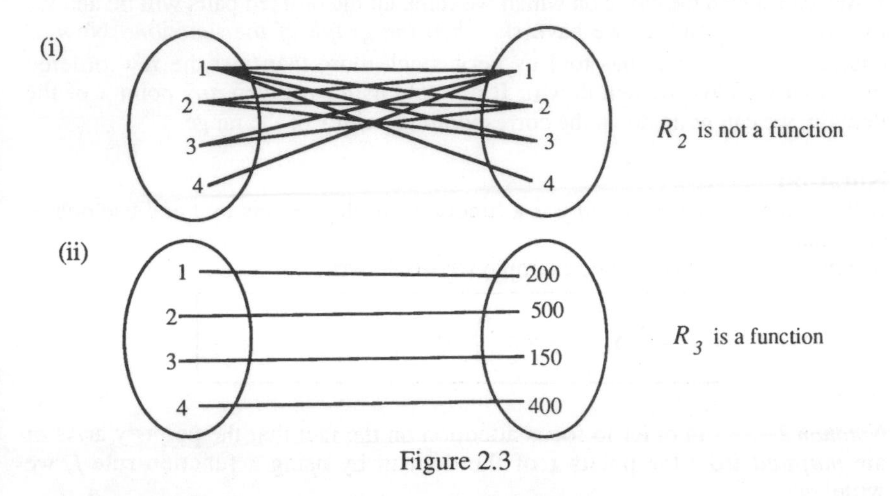

Figure 2.3

We shall primarily be concerned in this book with functions where the domain and range are subsets of the set of real numbers. Such functions are called **Real Valued Functions of a Real Variable.**

Example 4

Very often we are given some rule by which we can obtain the ordered pairs of the function. For example, we might be given that the function is the set of ordered pairs $\{(x, y)\}$ such that $x \in R^+, y \in R^+$ (+ve reals) and $y = x^2$. We could never write down all the ordered pairs that are elements of this function since the set of positive real numbers is infinite. We could write down enough to draw a **graph** of the function (Figure 2.4) – (1, 1), (2, 4), (3, 9), $(\frac{1}{4}, \frac{1}{16})$, $(\frac{1}{2}, \frac{1}{4})$, $(1\frac{1}{2}, 2\frac{1}{4})$, (4, 16), $(2\frac{1}{2}, 6\frac{1}{4})$, $(3\frac{1}{2}, 12\frac{1}{4})$.

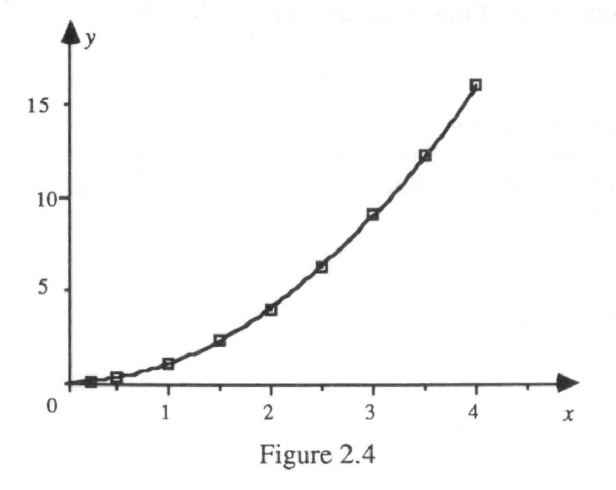

Figure 2.4

We can sketch the curve on which we think all the ordered pairs will lie and we use the expression that we have *sketched the graph of the function.* Now of course the rule $y = x^2$ has told us very much more than just the few ordered pairs that we have written down. It has told us how, given *any* point x of the domain, we can write down the corresponding point y of the range.

Notation

If the relation between x and y is a function, we denote this by f and use one of two notations

Notation 1: In the above example we would write

$$f = \left\{(x, y) : y = x^2\right\} \text{ or for short } f : y = x^2$$

Notation 2: In order to focus attention on the fact that the points y arise or are ***mapped*** from the points x of the domain by using a function rule f, we write

$$y = f(x) = x^2$$

[read "y equals f of x"]

The first notation will have been met by students who have followed a "modern maths" syllabus. However, from now on we shall use the second notation, unless further clarity is desired at some point.

As was mentioned in Chapter 1, when considering certain relations (these were in fact functions) the symbol denoting an arbitrary point in the domain is called the **independent variable** of f while a symbol standing for a point in the range is called the **dependent variable**. Here x is the independent variable, y the dependent variable.

Further Examples of Functions and their Graphs

Example 5

$y = f(x) = 2x + 1, \quad x, y \in R$

We can work out a few of the ordered pairs giving, for example, (0, 1) (1, 3) (2, 5) (3, 7) (4, 9) and we sketch the graph of the function as a straight line, as in Figure 2.5(a).

Example 6

$y = f(x) = |x|$. See Figure 2.5(b)

Example 7

$y = f(x) = \dfrac{|x|}{x}, \; x \neq 0$. See Figure 2.5(c)

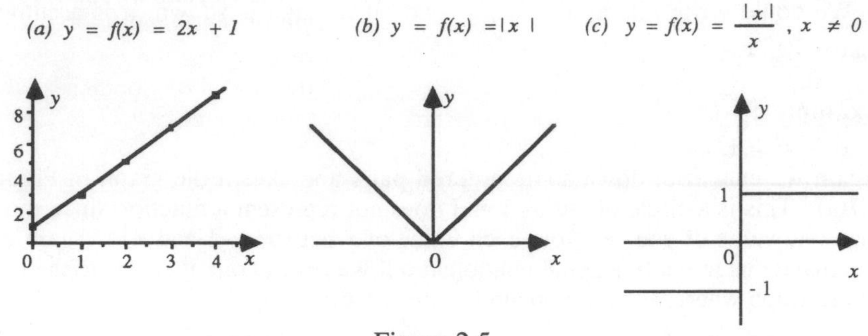

Figure 2.5

Examples of Relations that are not Functions

We may be given a formula defining a relation for which we can draw a graph but which is not a function.

Example 8

$y = \pm\sqrt{x}$

Working out a few ordered pairs we shall get (0, 0) (1, 1) (1, –1) (4, 2) (4, –2) and we can see straight away that this is not a function since the relation is one-many. The square root $\sqrt{x}$ gives + or – the absolute value of $\sqrt{x}$. The graph is as shown in Figure 2.6(a).

If we restrict the range to say $y \in R^+$ then this is a function and we only get the positive square root so that the relation is now one-to-one. We could then write $y = f(x) = +\sqrt{x}$; refer to Figure 2.6(b).

In the same way we could restrict the range by saying $y \in R^-$ so that we only consider the negative square root which then has the graph in Figure 2.6(c).

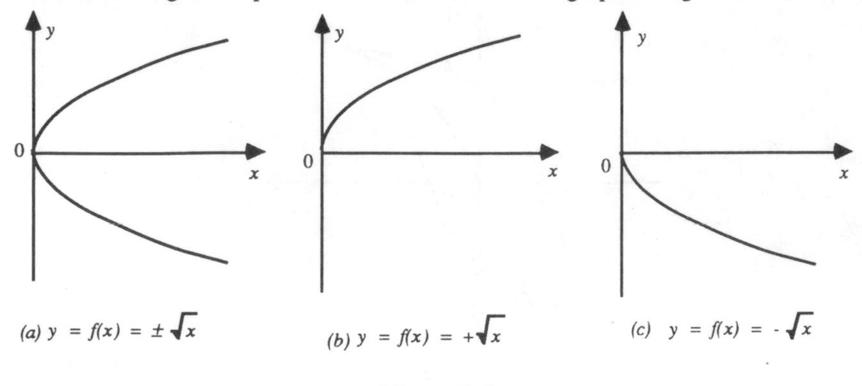

Figure 2.6

We could write this as $y = f(x) = -\sqrt{x}$. If no sign is shown, it is assumed that $y = +\sqrt{x}$.

Example 9

$x^2 + y^2 = 1$

Again we can write down some ordered pairs and sketch the graph in Figure 2.7(a). This is a circle of radius 1 and does not represent a function since more than one value of y arises from each value of x between -1 and $+1$. Again we can make this into a functional relationship if we restrict our attention to the part of the graph where $y \geq 0$; we obtain Figure 2.7(b)

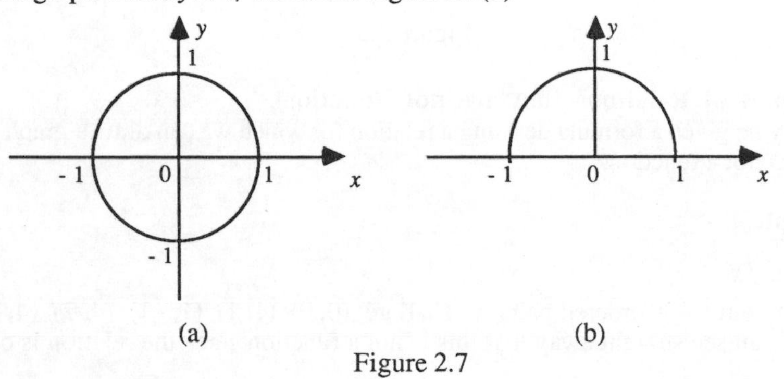

Figure 2.7

Example 10

Hysteresis Loop (refer to Figure 2.8)

A ferromagnetic specimen is originally unmagnetised, i.e. $\mathbf{B} = \mathbf{0}$, at point a; an increasing magnetic field $\mathbf{H}$ is applied until point b is reached, when the field is then decreased to zero, but there is a *remanance* of magnetisation at c; a further *coercive force* is necessary to demagnetise the specimen. Continuation to point e, and a further reversal of $\mathbf{H}$, allows us to trace out a closed loop: the *hysteresis loop*. The relation between $\mathbf{B}$ and $\mathbf{H}$ is many-many.

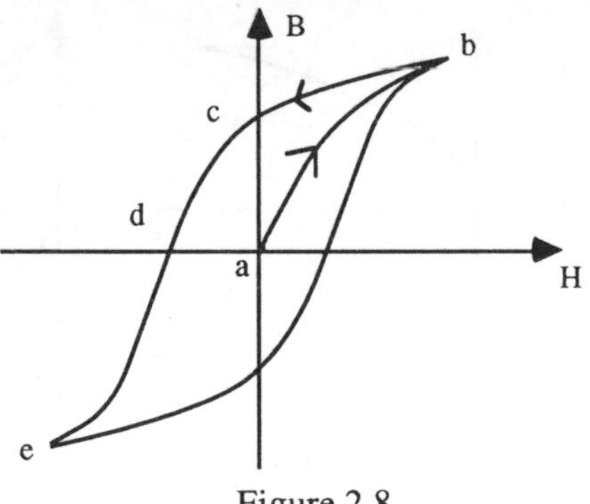

Figure 2.8

It is crucial to get used to the notation $f(x)$ and to understand it fully. Consider the function $f(x)$ shown in Figure 2.9.

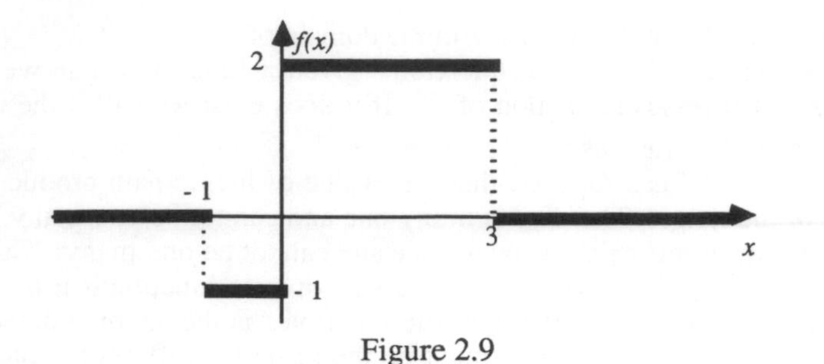

Figure 2.9

It can be described by the formulae

$$f(x) = \begin{cases} 2, & 0 \leq x < 3 \\ -1, & -1 \leq x < 0 \\ 0, & \text{otherwise} \end{cases}$$

In Figures 2.10(i) and (ii) we show respectively the graphs of $f(x - 4)$ and $f(2x)$.

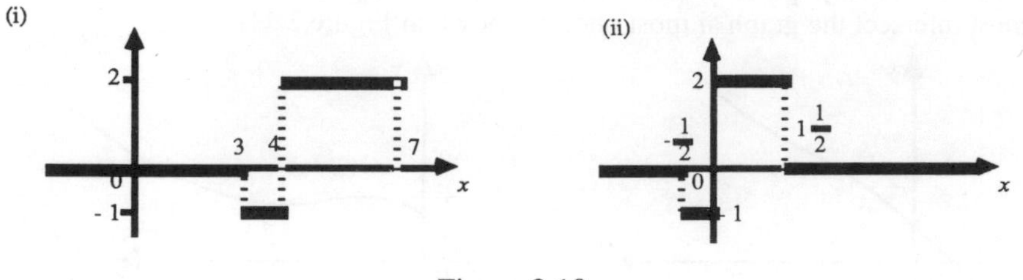

Figure 2.10

Inverse Functions

In Physics, Boyle's Law is an idealised law which says "the pressure and volume of a gas held at constant temperature obey the law PV = constant". In other words, if we were doing an experiment in which we measured the volume of a gas for different pressures we would find that V is a function of P, namely

$$V = f(P) = \frac{\text{constant}}{P}$$

Written in this way, P is the independent variable and V is the dependent variable. However, we could always have done the experiment in reverse: that is, we could measure the different pressures arising from different volumes and we would then obtain

$$P = g(V) = \frac{\text{constant}}{V}$$

(g used here to distinguish between the two functions)

This time V is the independent variable and P is the dependent variable. Of course the nature of the function g is dependent on the function f.

The function g is called the **Inverse Function** of f and must satisfy the identity $f[g(V)] \equiv V$ for all volumes V in the domain of g.

We now consider the general question: given a function f, can we find a function g that reverses the action of f? If it does exist we call it the inverse function and often write it as f^{-1}.

We know that if f is a function then one value of the domain produces only one value of the range. That is, f is many-one and cannot be one-many. If the function g exists, it must also be many-one and cannot be one-many. Therefore it follows that a function f will only possess an inverse function if it is **one-to-one**. Put another way, we know that the function f is the set of ordered pairs (x, y) and none of the values of x must be the same for different values of y. The inverse function must interchange the numbers in the ordered pair since it interchanges the dependent and independent variables giving the set of ordered pairs (y, x). Now none of the values of y must be the same for different values of x. Taken together we now have

$$\begin{cases} \text{one value of } x \text{ produces only one value of } y \\ \text{one value of } y \text{ produces only one value of } x \end{cases}$$

Therefore the relationship between y and x is one-to-one. We can tell straight away from the graph of a function if it is one-to-one because each horizontal line must intersect the graph at most once as shown in Figure 2.11.

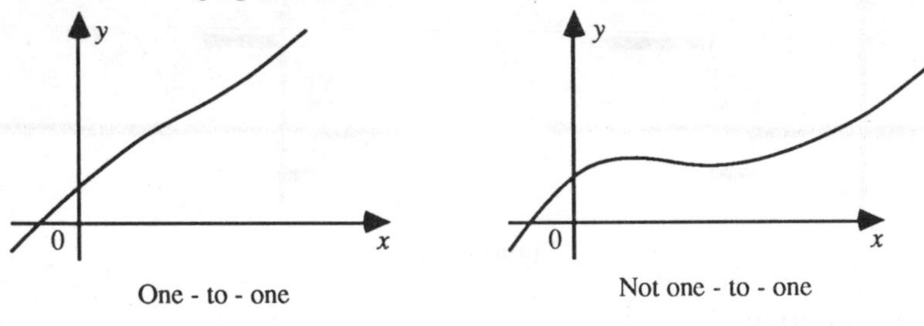

Figure 2.11

We have already met in Example 8 a function which did not have an inverse which is also a function unless we restrict the range in some way. In Example 8 we considered $y = \sqrt{x}$ which is the inverse of the function $y = x^2$. The graphs of these are shown in Figure 2.12

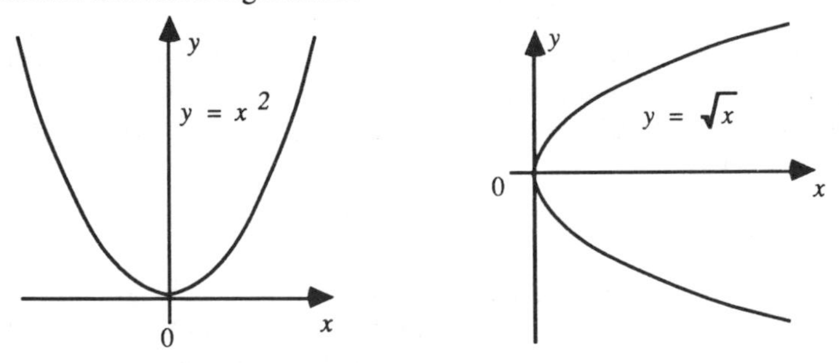

Figure 2.12

The relationships are certainly not one-to-one and we can only call $y = \sqrt{x}$ the inverse *function* if we restrict the range to $y \geq 0$ and write $y = +\sqrt{x}$, meaning the positive square root.

Consider $y = \sin^{-1} x$ which is the inverse of the function $y = \sin x$. The graphs of these functions are shown in Figure 2.13.

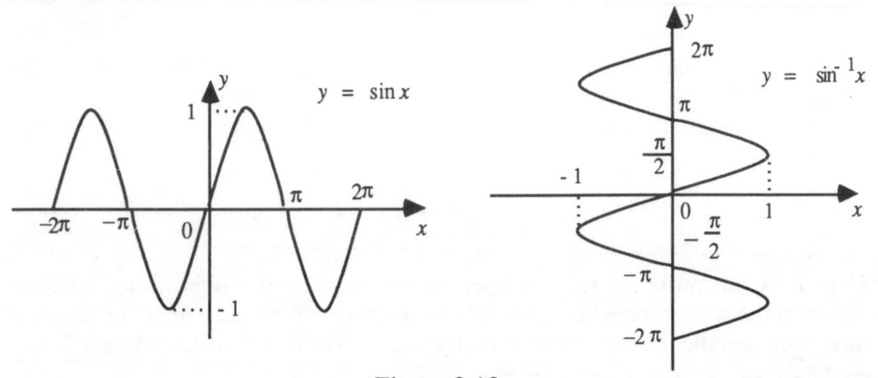

Figure 2.13

Again the inverse is not a *function* unless we restrict the range of $y = \sin^{-1} x$ and we usually take $-\pi/2 \leq y \leq \pi/2$.

Geometrically the graph of the inverse function f^{-1} is drawn by reflecting the graph of $y = f(x)$ in the line $y = x$, as illustrated in Figure 2.14.

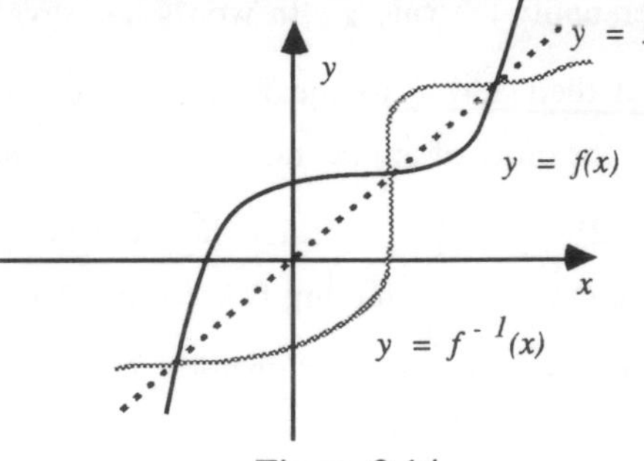

Figure 2.14

You should check that this is so for the graphs of the examples above. We have to be careful that the inverse is a *function*, since we cannot proceed to find derivatives and integrals unless we are considering functions.

Combination of Functions

Suppose we consider the two functions $f(x) = x^2$ and $g(x) = 3x - 2$. Then it is natural to define the **sum** of the two functions as $h(x) = x^2 + 3x - 2$. In function symbols we could say that $h = f + g$. However, we must bear in mind that such a definition is valid only if the two functions have the same domain.

Likewise we could define the **difference** as $k(x) = x^2 - 3x + 2$, and the **product** as $p(x) = x^2(3x - 2)$; note that we are really defining the arithmetic operations of +, − and . when applied to the functions rules f and g as $f + g$, $f - g$ and $f.g$. We have deliberately left out the definition of **quotient** because there is an additional difficulty. We should like to say

$$q(x) \equiv \frac{f(x)}{g(x)} = \frac{x^2}{(3x-2)}$$

but the denominator vanishes at $x = 2/3$ and so we must define $q(x)$ on the domain which is the set of real numbers with the element 2/3 removed.

These ideas clearly generalise to more complicated function combinations. For example,

$$f(x) = \frac{2x^3 + \sqrt{x+2}}{x+1}$$

Here it is not quite so easy to specify its domain (the importance of such specification will become apparent in later chapters). We could break the function down into smaller components in many ways. We could write f as $(g + h) \div k$ where $g(x) = 2x^3$, $h(x) = \sqrt{x+2}$ and $k(x) = x + 1$. Notice that for h to be defined, x must be ≥ -2 and in order that the division can occur, x must not equal −1. The domain is then $\{x : x \neq -1 \text{ and } x \geq -2\}$.

Another way of combining functions is to apply them successively. Consider the two functions $f(x) = \frac{1}{2}x$(halve) and $g(x) = x + 2$ (add two). Suppose we first form $f(x)$ and then apply the rule g. In words this says that we halve the given value of x and then add 2 to the result, i.e. $g[f(x)] = \frac{1}{2}x + 2$ †. However, if we reverse the order of application, i.e. add two then halve the result we get $f[g(x)] = \frac{1}{2}(x + 2) = \frac{1}{2}x + 1$ †. Clearly the order of application matters. And there is a further complication: having first formed $f(x)$ we can only form $g[f(x)]$ if $f(x)$ lies in the domain of g. Consider $f(x) = \sqrt{x}$ (positive square root) and $g(x) = 4x$; f has as domain the set of positive real numbers. Now the co-domain of f is the set of positive real numbers and g can certainly be applied to this set to give $g[f(x)] = 4\sqrt{x}$, where x belongs to the set of positive real numbers. However, g has all real numbers in the domain, and it is possible for $g(x)$ to be negative. Now $f[g(-4)]$ cannot be defined and so care is needed; even if we restrict x to the positive real numbers, the domain of $g[f(x)]$ is not necessarily the same as that of $f\,[g(x)]$.

† Sometimes $g[f(x)]$ is written $(fog)(x)$ and $f[g(x)]$ is written $(gof)(x)$

Consider now $f(x) = \sqrt{x}$, $g(x) = x - 2$. Here $f[g(x)] = \sqrt{x-2}$ while $g[f(x)] = \sqrt{x} - 2$. Considering $f[g(x)]$, we see that x must be real and ≥ 2, while for $g[f(x)]$ we must have x as real and ≥ 0. Thus in this case the domains of $f[g(x)]$ and $g[f(x)]$ are not the same.

We have already met the idea of *undoing* a function to see whether the inverse rule is itself a function. With a composite function, we must undo the functions in reverse order, as one would take off first the outermost of several wrappings comprising a parcel.

If $f(x) = 2x$, $g(x) = x + 1$ then $f[g(x)] = 2(x + 1)$ and $g[f(x)] = 2x + 1$. To undo $f[g(x)]$* we note that it was formed by *adding one* then *doubling*; the inverse operations are *subtracting one* and *halving* and if these are applied in the reverse order, x becomes first $\frac{1}{2}x$ and then $\frac{1}{2}x - 1$. Thus $f[g(x)]^{-1} = \frac{1}{2}x - 1$. This is written $g^{-1}[f^{-1}(x)] = \frac{1}{2}x - 1$. In this case the inverse mapping is a function. Similarly the inverse of $g[f(x)]$ is written $f^{-1}[g^{-1}(x)]$; again, the inverse mapping is a function.

Troubles with attempted inversion arise in a similar way to the inversion of simple functions.

2.3 Standard Functions

We now consider the functions based on sine, cosine and tangent.

(i) We define the **Sine Function** as $y = f(x) = \sin x$ where the domain of x is $\{x : x \text{ is real}\}$. Further, x is always measured in radians for calculus. The graph of $y = f(x) = \sin x$ is shown in Figure 2.15

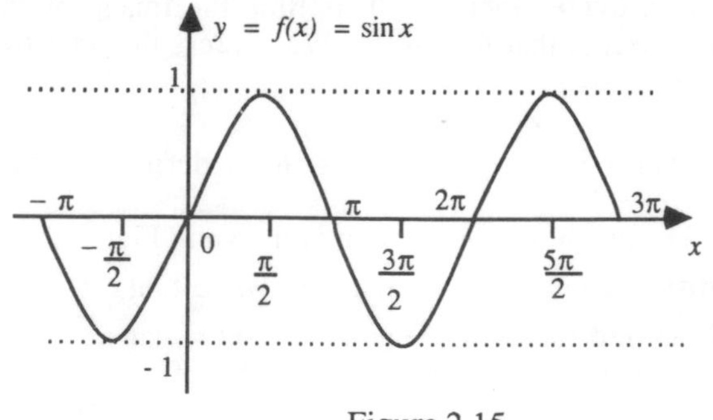

Figure 2.15

This is a **Periodic Function**: the basic shape is repeated over and over again. Its **Period** is 2π – that is, it repeats itself every 2π. It is an **odd**

* The inverse mapping is more easily seen as $(gof)^{-1}$ and hence as $f^{-1} og^{-1}$

function. By this we mean that if the y-axis is regarded as a mirror, then the image in the mirror of the graph for $x > 0$ is exactly the same shape but is multiplied by (-1). For example the following is an odd function

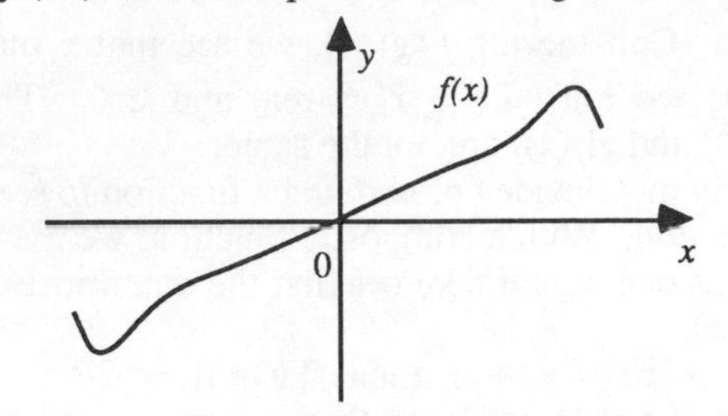

Note also that the range of the sine function is $-1 \leq f(x) \leq 1$.

(ii) $y = f(x) = \cos x$ for $x \in R$ (Real Numbers) is defined as the **Cosine Function**. The graph of this function (Figure 2.16) is

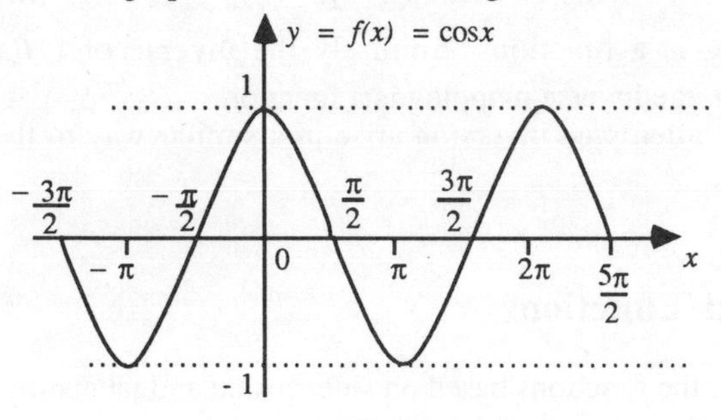

Figure 2.16

It is a *periodic function* of period 2π and $\cos(x + 2n\pi) = \cos x$ for $n \in Z$. It is an **even** function. This time the image of the graph for $x \geq 0$ in the y-axis is that for $x < 0$. We express this fact by saying that $\cos(-x) = \cos x$.

(iii) $y = f(x) = \tan x = \left\{\dfrac{\sin x}{\cos x}\right\}$ for $x \in R$ is defined as the **Tangent Function**. The graph of this function is shown in Figure 2.17.

It is **periodic** of period π so that we write $\tan(x + n\pi) = \tan x$ for $n \in Z$. It is an **odd** function. That is, $\tan(-x) = -\tan x$.

As an exercise, you should draw the graphs of $y = f(x) = 1/(\sin x) = \operatorname{cosec} x$, $y = f(x) = 1/(\cos x) = \sec x$ and $y = f(x) = 1/(\tan x) = \cot x$. You will find the graphs of these functions in the Appendix at the end of the book.

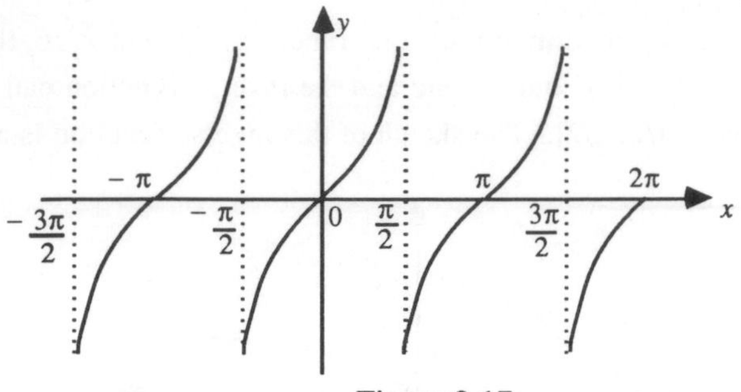

Figure 2.17

The inverse trigonometric functions

(i) $y = f(x) = \sin^{-1} x$ (sometimes expressed as arc sin x)
Notice that the domain of the function is the set of numbers x such that $-1 \leq x \leq 1$. For $|x| > 1$, the function is not defined. It is **not** *periodic*.
It is an *odd* function, so that $\sin^{-1}(-x) = -\sin^{-1} x$.
For work in the calculus we should quote the answer in radian form. The reason for this will become clear when you reach the sections on differentiation and integration.
Examples: $\sin^{-1}(0.4) = 0.4117$, $\sin^{-1}(-0.2) = -0.1993$

(ii) $y = f(x) = \cos^{-1} x$(sometimes expressed as y = arc cos x)
This is read as 'y is the angle (in radians) whose cosine is x'. We choose an interval of the domain of $y = \cos x$ where $\cos x$ is one-to-one. Usually this is taken as $[0, \pi]$. Then the inverse function has a graph as in Figure 2.18 with range $[0, \pi]$ (the dotted parts only being shown to show the relationship to $y = \cos x$).

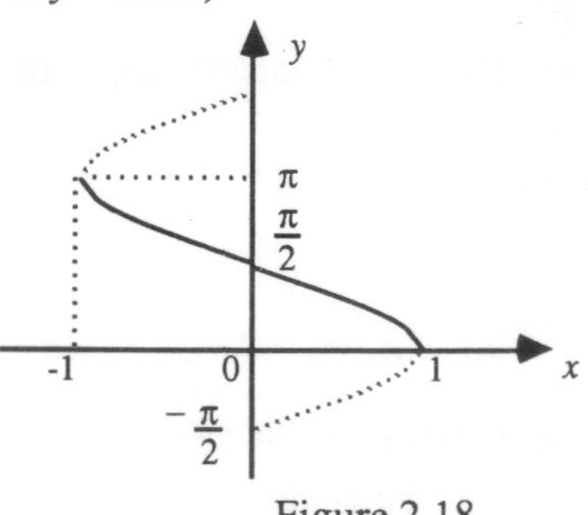

Figure 2.18

Notice that $y = \cos^{-1} x$ is defined only for x such that $-1 \leq x \leq 1$. It is **not** *periodic*.
It is neither *odd nor even*.
Examples: $\cos^{-1}(0.5) = 1.0472$, $\cos^{-1}(-0.3) = 2.8394$.

(iii) $y = f(x) = \tan^{-1} x$ (or $y = \text{arc} \tan x$)
If we restrict the domain of the function $y = \tan x$ to the interval $[-\pi/2, \pi/2]$ then it is one-to-one and the inverse function can be defined with range $[-\pi/2, \pi/2]$. The sketch of this inverse function is as in Figure 2.19.

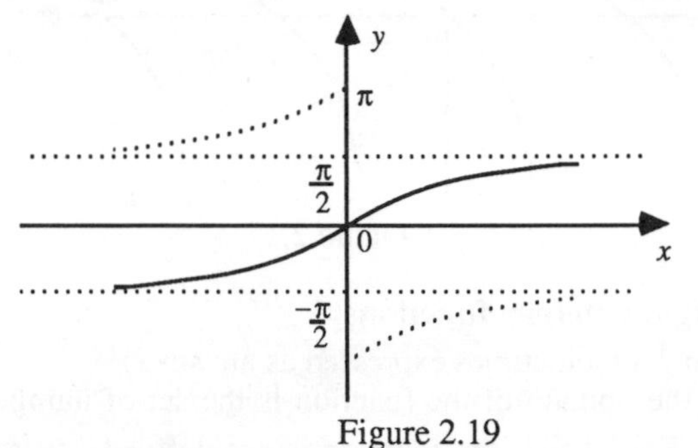

Figure 2.19

Notice that x may take any value from $-\infty$ to $+\infty$. It is **not** *periodic*. It is an *odd* function so that $\tan^{-1}(-x) = -\tan^{-1} x$.
Numerical examples are $\tan^{-1}(-191) = -1.5656$, $\tan^{-1}(2) = 1.1071$. Note again the use of radians!
You should now consider the inverse functions $y = f(x) = \text{cosec}^{-1} x$, $y = f(x) = \sec^{-1}(x)$, $y = f(x) = \cot^{-1}(x)$. Define a suitable interval and state whether they are periodic functions and whether they are even or odd or neither. With your definition then find $\text{cosec}^{-1}(0.3)$, $\sec^{-1}(\pm 0.4)$ and $\cot^{-1}(\pm 0.7)$.
You will find the graphs of these inverse functions in the Appendix.

The logarithmic function

You may have already met the use of logarithms to the base α. They have properties:

$$\log_\alpha (a \times b) = \log_\alpha a + \log_\alpha b,$$

$$\log_\alpha(a^n) = n \log_\alpha a,$$

$$\log_\alpha(1) = 0,$$

$$\log_\alpha(\alpha) = 1$$

The whole theory of logarithms hinges on the fact that

$$\text{if } x = \log_\alpha a \text{ then } a = \alpha^x$$

A special case is where the base is e, which in the section on sequences is defined by

$$e = \lim_{n \to \infty} \left[1 + \frac{1}{n}\right]^n$$

The value of e is 2.718.... Logarithms to the base e arise in many physical situations and are called **Natural Logarithms**.

Thus we define the natural logarithmic function as

$$y = \log_e x \quad \text{with} \quad e = 2.718...$$

In science and engineering the notation $y = \ln x$ is often used. The rules become

$$\ln x_1 + \ln x_2 = \ln (x_1 x_2)$$

$$\ln x^n = n \ln x$$

$$\ln(1) = 0, \quad \ln(e) = 1$$

$$\text{and} \qquad \text{if } y = \ln x, \text{ then } x = e^y.$$

Its graph is sketched in Figure 2.20. Notice that $y = \ln x$ is only defined for $x \geq 0$ and that as x approaches 0 through positive values of x (usually written $x \downarrow 0$) then ln tends to $-\infty$. Notice that it is a one-to-one relation and as such must possess an inverse which will be the inverse function. This inverse function is called the **Exponential Function** and details are given in the next sub-section.

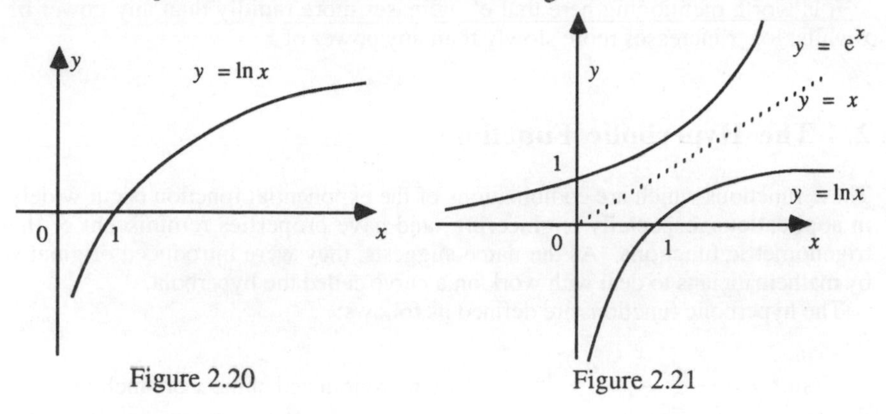

Figure 2.20 Figure 2.21

The exponential function

Remembering the result that if $y = \ln x$ then $x = e^y$ we conclude that the function which will *undo* the function $y = \ln x$ is $y = e^x$. This is the **exponential function**. (For convenience in printing and for large expressions we often write $y = \exp(x)$ instead of $y = e^x$.)

A sketch of its graph can be obtained by reflecting the graph of $y = \ln x$ in the line $y = x$. It is shown in Figure 2.21

Notice that $y = e^x$ is defined for all values of x and that as x tends to a large negative value, e^x tends to zero. Besides the usual indicial law that $e^{x_1}.e^{x_2} = e^{x_1+x_2}$ we have the very important result that

$$e^{\ln x} = e^{\log_e x} = x$$

We mention also the associated function $y = e^{-x} = 1/e^x = (1/e)^x$, very often called the *curve of exponential decay*, which has a graph as shown in Figure 2.22

Exponential decays occur widely in physical situations – indeed we have already met this shape in Chapter 1, when looking at the cooling of a liquid.

One other associated curve is that of the function y = a^x for some number

$a > 1$. Its graph, Figure 2.23, is of the same general shape as $y = e^x$.

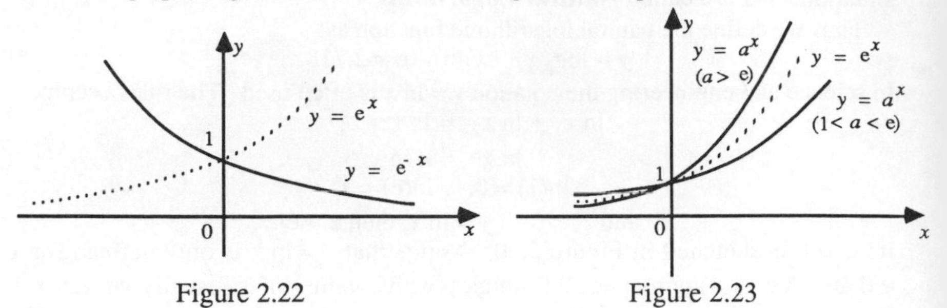

Figure 2.22 Figure 2.23

We note that $y = a^x = e^{x\log_e a} = e^{x\ln a}$, so that the functions are related.

It is worth mentioning here that e^x increases more rapidly than any power of x whilst $\log_e x$ increases more slowly than any power of x.

2.4 The Hyperbolic Functions

These functions which are combinations of the exponential function occur widely in applications, especially engineering, and have properties reminiscent of the trigonometric functions. As the name suggests, they were introduced originally by mathematicians to deal with work on a curve called the hyperbola.

The hyperbolic functions are defined as follows:

$$\sinh x = \frac{e^x - e^{-x}}{2}$$ (often pronounced shine x or sinch x)

$$\cosh x = \frac{e^x + e^{-x}}{2}$$ (pronounced cosh x)

$$\tanh x = \frac{\sinh x}{\cosh x} = \frac{e^x - e^{-x}}{e^x + e^{-x}}$$ (pronounced than x or tanch x)

$$\operatorname{cosech} x = \frac{1}{\sinh x} \qquad \operatorname{sech} x = \frac{1}{\cosh x} \qquad \coth x = \frac{1}{\tanh x}$$

We shall draw the graphs of $y = \sinh x$, $y = \cosh x$ and $y = \tanh x$ only, leaving the others as exercises for the reader.

(i) $y = \sinh x = \frac{1}{2}(e^x - e^{-x})$

We can draw the graph as shown in Figure 2.24 by drawing the curve half-way between the graphs of e^x and $-e^{-x}$ since $\sinh x = \frac{1}{2}[(e^x + (-e^{-x})]$.

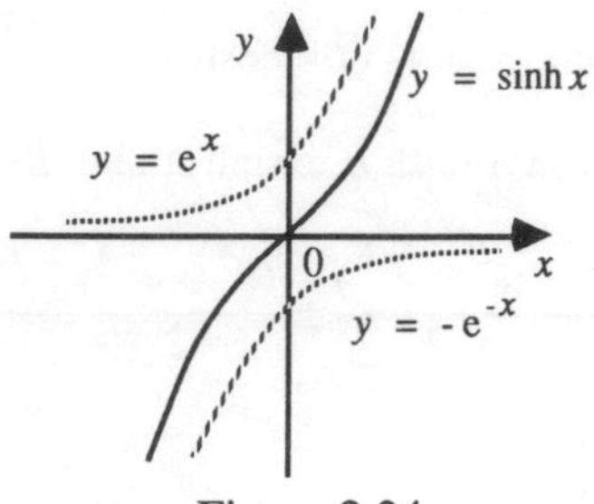

Figure 2.24

Notice that
(a) $y = \sinh x$ is an odd function, $\sinh(-x) = -\sinh x$
(b) $\sinh(0) = 0$
(c) $\sinh(A \pm B) = \sinh A \cosh B \pm \cosh A \sinh B$

$$\left[\sinh(A+B) = \frac{e^{A+B} - e^{-A-B}}{2} = \frac{e^{A}.e^{B} - e^{-A}.e^{-B}}{2} \right.$$

$$= \frac{(e^{A} - e^{-A})}{2}\frac{(e^{B} + e^{-B})}{2} + \frac{(e^{A} + e^{-A})}{2}\frac{(e^{B} - e^{-B})}{2}$$

$$\left. = \sinh A \cosh B + \cosh A \sinh B \right]$$

There are other similarities to the sine function which we shall meet later. However, two major differences are that the function is **not periodic** and also that it is a one-to-one relation.

(ii) $y = \cosh x = \dfrac{e^{x} + e^{-x}}{2}$

The sketch of the graph is thus obtained, as in Figure 2.25, by taking the average of e^{x} and e^{-x}.

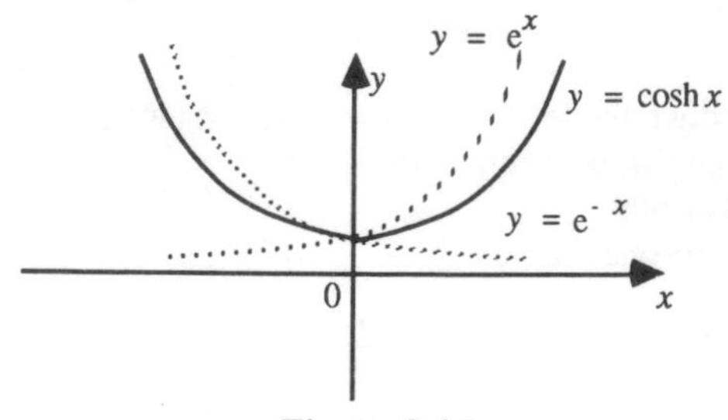

Figure 2.25

Notice that

(a) $y = \cosh x$ is **even**, $\cosh(-x) = \cosh x$

(b) $\cosh(0) = 1$

(c) $\cosh(A \pm B) = \cosh A \cosh B \pm \sinh A \sinh B$

$$\left[\text{R.H.S} = \frac{(e^{A}+e^{-A})}{2}\frac{(e^{B}+e^{-B})}{2} \pm \frac{(e^{A}-e^{-A})}{2}\frac{(e^{B}-e^{-B})}{2} = \frac{e^{A\pm B}+e^{-A\pm B}}{2}\right.$$

$$\left. = \text{L.H.S.}\right]$$

Compare this with $\cos(A + B)$ but note the difference in sign. The major difference again is that the function $y = \cosh x$ is **not periodic**. We shall have to be careful when defining its inverse since the relation is not one-to-one.

(iii) $y = \tanh x = \dfrac{e^{x}-e^{-x}}{e^{x}+e^{-x}}$

By writing this in the form $y = (e^{2x} - 1)/(e^{2x} + 1)$ or in the alternative form $y = (1 - e^{-2x})/(1 + e^{-2x})$, you should show that

(a) when $x = 0$, $y = 0$

(b) when x tends to a large negative value, y tends towards -1

(c) when x tends to a large positive value, y tends to $+1$

and that the graph of $y = \tanh x$ is as shown in Figure 2.26

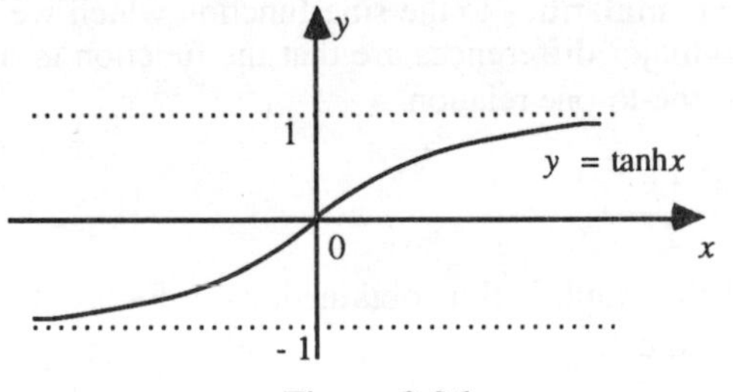

Figure 2.26

We can also deduce the following properties of $\tanh x$

(a) $y = \tanh x$ is an **odd** function

(b) it is **not** periodic

(c) it is a one-to-one relation

(d) $\tanh(A \pm B) = \dfrac{\tanh A \pm \tanh B}{1 \pm \tanh A \tanh B}$

Perhaps the most important results are the connections between $\sinh x$, $\cosh x$ and $\tanh x$ which are stated without proof.

(i) $\cosh^2 A - \sinh^2 A = 1 \Leftrightarrow \cosh^2 A = 1 + \sinh^2 A$

$\Leftrightarrow \sinh^2 A = \cosh^2 A - 1$ ($a \Leftrightarrow b$ means $a \Rightarrow b$ **and** $b \Rightarrow a$)

(ii) $\sinh 2A = 2 \sinh A \cosh A$

(iii) $\cosh 2A = 2 \cosh^2 A - 1 = 1 + 2 \sinh^2 A$

(iv) $\cosh A + \sinh A = e^A$

(v) $\cosh A - \sinh A = e^{-A}$

(vi) $(\cosh A \pm \sinh A)^n = \cosh nA \pm \sinh nA$

(vii) $\text{sech}^2 A = 1 - \tanh^2 A$

(viii) $\text{cosech}^2 A = \coth^2 A - 1$

Again, note the similarity with the trigonometric results apart from an occassional difference in sign. In fact we can write down any hyperbolic identity from the corresponding trigonometric identity by using the following rule (we cannot prove this yet) "Cosine is replaced by cosh, sine is replaced by sinh and tangent is replaced by tanh, but if there is product sine × sine, this must be replaced by –sinh × sinh".

Let us take an example

$$\cos 3A = 4 \cos^3 A - 3 \cos A$$
$$\sin 3A = 3 \sin A - 4 \sin^3 A$$

The corresponding hyperbolic identities are

$$\cosh 3A = 4 \cosh^3 A - 3 \cosh A$$
$$\sinh 3A = 3 \sinh A + 4 \sinh^3 A$$

Note $\sin^3 A = (\sin A \times \sin A) \times \sin A$ which has an implied sine × sine therefore gives $(-\sinh A \times \sinh A) \times \sinh A$.

Now write down the hyperbolic equivalents of the following

(a) $\tan 2A = 2 \tan A/(1 - \tan^2 A)$

(b) $\sin 4A = 4 \sin A \cos A - 8 \sin^3 A \cos A$

The inverse hyperbolic functions

(i) $y = \sinh^{-1} x$

y is the number whose sinh is x and an equivalent expression is $x = \sinh y$. Looking back at the graph of $y = \sinh x$ we see that $y = \sinh x$ is a one-to-one relationship. Thus the inverse function exists without any restriction on the range. A sketch of its graph is obtained by a mirror image of $y = \sinh x$ in the line $y = x$ and is shown in Figure 2.27.

Logarithmic equivalent

If $\qquad y = \sinh^{-1} x$

then $\qquad \sinh y = x$

that is, $\qquad \dfrac{e^y - e^{-y}}{2} = x$

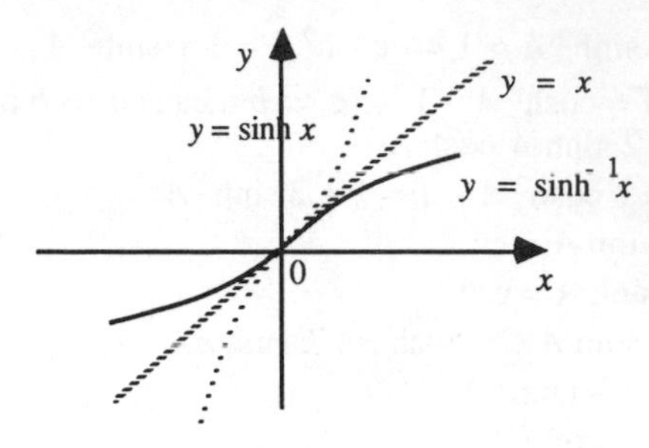

Figure 2.27

Multiplying by $2e^y$ and rearranging we get

$$(e^y)^2 - 2x\,e^y - 1 = 0$$

Solving this quadratic in e^y yields

$$e^y = x \pm \sqrt{x^2 + 1}$$

Now $\sqrt{x^2 + 1} > x$ and the negative sign would give a negative value for e^y which is impossible. Thus only the positive sign is admissible and

$$e^y = x + \sqrt{x^2 + 1}$$

Thus $$y = \ln(x + \sqrt{x^2 + 1})$$

i.e. $$\sinh^{-1} x = \ln(x + \sqrt{x^2 + 1})$$

Example: $$\sinh^{-1}\left(\tfrac{1}{2}\right) = \ln\left(\tfrac{1}{2} + \sqrt{5/4}\right) = \ln\frac{1 + \sqrt{5}}{2} = 0.4810 \quad (4 \text{ d.p.})$$

Hence the logarithmic equivalent is a much more convenient method of evaluating the inverse function.

(ii) $y = \cosh^{-1} x$

The relationship $y = \cosh x$ is not one-to-one and we shall have to restrict the domain to define the inverse curve. We choose only the positive x-axis as in Figure 2.28(i). The inverse function $y = \cosh^{-1} x$ is then only defined for $y \geq 0$ and has a graph as shown in Figure 2.28(ii).

Logarithmic equivalent

If $$y = \cosh^{-1} x \quad (x \geq 1)$$

then $$\cosh y = x$$

or
$$\frac{e^y + e^{-y}}{2} = x$$

Multiplying by $2e^y$ and rearranging
$$(e^y)^2 - 2x\, e^y + 1 = 0$$

Solving we get $\quad e^y = x \pm \sqrt{x^2 - 1}$

The positive sign must be taken to obtain a point on the inverse function curve. Therefore $\quad e^y = x + \sqrt{x^2 - 1}$

or $\quad y = \ln(x + \sqrt{x^2 - 1})$

Hence $\quad \cosh^{-1} x = \ln(x + \sqrt{x^2 - 1})$

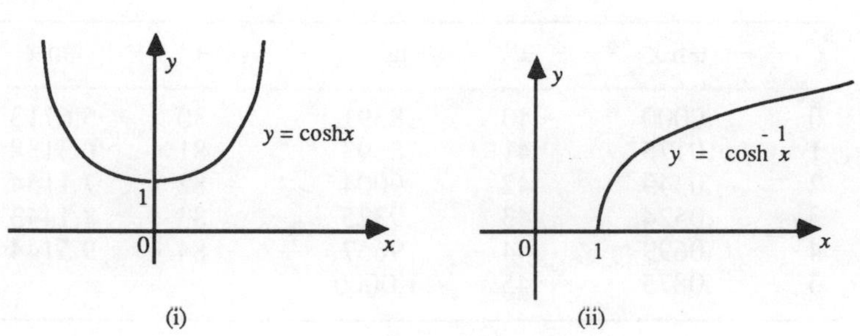

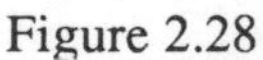
Figure 2.28

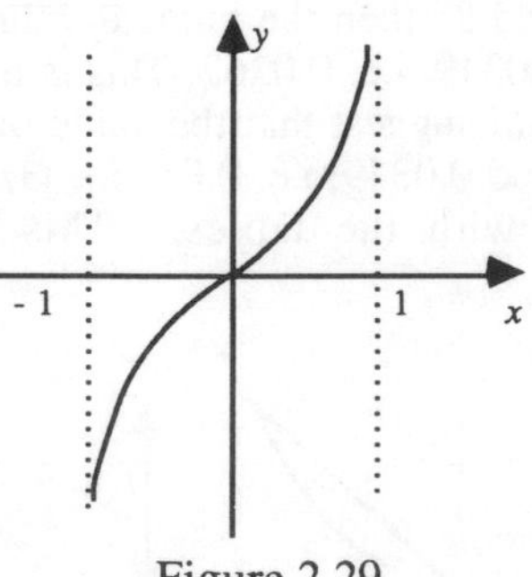

Figure 2.29

(iii) $y = \tanh^{-1} x$
  (a) the inverse function exists without modification of the range
  (b) its graph is of the form shown in Figure 2.29
  (c) the function only exists for $|x| < 1$
  (d) $\tanh^{-1} x = \frac{1}{2}\log[(1 + x)/(1 - x)]$

2.5 Tabulation of Functions

Engineers work frequently with tables. These can be of a mathematical nature (e.g. tables of logarithms or sines) or can be tables obtained by experiment for design purposes. The extract below, Table 2.1, is taken from a table of values for the pressure of saturated water in bars, P, as a function of temperature in degrees centigrade, T.

Table 2.1

T	10	11	12	13	14	15
P	0.01227	0.01312	0.01401	0.01497	0.01597	0.01704

Often we wish to estimate a value that has not been tabulated.
If the function does not change rapidly over the interval in which we want to **interpolate** we may assume that the function behaves like a straight line.

Let us focus our attention on three extracts from a table of tangents (see Table 2.2)

Table 2.2

x^o	tan x	x^o	tan x	x^o	tan x
0	.0000	40	.8391	80	5.6713
1	.0175	41	.8693	81	6.3138
2	.0349	42	.9004	82	7.1154
3	.0524	43	.9325	83	8.1443
4	.0699	44	.9657	84	9.5144
5	.0875	45	1.0000		

Suppose we wish to estimate tan 1° 30'. We may say that since 1° 30' lies halfway between 1° and 2° then the tangent value lies approximately halfway between 0.0175 and 0.0349, i.e. 0.0262. This *is* the value of tan 1° 30' to 4 d.p. Similar reasoning would suggest that the value of tan 1° 20' is one third of the way between 0.0175 and 0.0349, i.e. $0.0175 + (1/3)(0.0349 - 0.0175) = 0.0233$ which again agrees with the tables. This process is called **Linear Interpolation.**

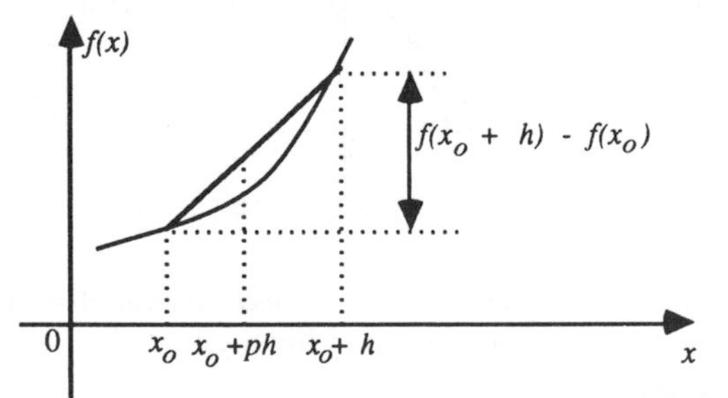

Figure 2.30

For linear interpolation, if $x = x_0 + ph$, where h is the (constant) table spacing in x, and p the fraction of the interval under consideration (see Figure 2.30) then

$$f(x_0 + ph) \equiv f(x_0) + p[f(x_0 + h) - f(x_0)]$$

Will this process always produce such good results? Let us now estimate tan 40° 30'. We obtain

$$0.8391 + \tfrac{1}{2}(0.8693 - 0.8391) = 0.8542$$

against the tabulated value of 0.8451. Estimating tan 81° 30' produces 6.7146 against the tabulated value of 6.6912. The assumption of linear behaviour is not justified in these cases and a more accurate method must be found.

Since we are dogged by round-off in most practical tables, the main features tend to be somewhat obscured. For the moment we turn our attention to tables for simple mathematical expressions where the tabulated values are exact.

Consider the table of values for $f(x) = 2x - 3$ in Table 2.3.

Table 2.3

x	$f(x)$	Differences
4	5	
		2
5	7	
		2
6	9	
		2
7	11	

Here the differences are **constant**. The values in the column of differences as obtained in these two examples are called **First Differences**. We can extend such tables to obtain differences of differences (second differences). For $f(x) = x^2$, we obtain Table 2.4.

Table 2.4

x	x^2	First Differences	Second Differences
11	121		
		23	
12	144		2
		25	
13	169		2
		27	
14	196		2
		29	
15	225		

This time the **second** differences are constant.

Consider $f(x) = 2x - x^2$

Table 2.5

x^o	$f(x)$	1st	2nd
0	0		
		1	
1	1		–2
		–1	
2	0		–2
		–3	
3	–3		–2
		–5	
4	–8		–2
		–7	
5	–15		

The second differences are again constant. Is this a general result? We put the process on a formal basis; if the difference between $f(x_0)$ and $f(x_0 + h)$ is denoted $\Delta f(x_0)$, that is, $\Delta f(x_0) = f(x_0 + h) - f(x_0)$, then suppose $f(x) = a + bx$, where a and b are constants.

$$\text{Now } f(x_0) = a + bx_0, \quad f(x_0 + h) = a + b(x_0 + h) = a + bx_0 + bh$$

therefore $f(x_0 + h) - f(x_0) \equiv \Delta f(x_0) = bh$ which is constant if the spacing of values, h, is constant.

Similarly, let us take $f(x) = a + bx + cx^2$, where a, b, c are constants.

Then $\quad f(x_0) = a + bx_0 + cx_0^2$

and $\quad f(x_0 + h) = a + b(x_0 + h) + c(x_0 + h)^2$

so that $\quad \Delta f(x_0) \equiv f(x_0 + h) - f(x_0) \equiv 2cx_0h + ch^2 + bh$

Likewise $\quad f(x_0 + 2h) = a + b(x_0 + 2h) + c(x_0 + 2h)^2$

and $\quad \Delta f(x_0 + h) \equiv f(x_0 + 2h) - f(x_0 + h) = 2cx_0h + 3ch^2 + bh$

The second difference, $\Delta f(x_0 + h) - \Delta f(x_0)$, denoted by $\Delta^2 f(x_0)$, is given by $\Delta^2 f(x_0) \equiv 2ch^2$ and is constant for constant spacing.

Thus the second differences of a second degree polynomial are constant. It would be tempting to conclude that the third differences of a cubic are constant. (Prove this algebraically and form a suitable table of values; you should consider only $y = dx^3$ for d constant – why can we ignore the terms in x^2, x and the constant term?)

Consider now the tables overleaf for $f(x) = 2x - x^2$ for non-integral values of x (Table 2.6 and Table 2.7).

In Table 2.7 the differences are recorded as integers for ease; when used in calculations they must be reinstated to their true values.

Table 2.6

x	$f(x)$	First differences	Second differences
1.0	1.00		
		−0.01	
1.1	0.99		−0.02
		−0.03	
1.2	0.96		−0.02
		−0.05	
1.3	0.91		−0.02
		−0.07	
1.4	0.84		−0.02
		−0.09	
1.5	0.75		

Table 2.7

x	$f(x)$	First differences	Second differences
1.0	1.00		
		−1	
1.1	0.99		−2
		−3	
1.2	0.96		−2
		−5	
1.3	0.91		−2
		−7	
1.4	0.84		−2
		−9	
1.5	0.75		

Example 1

Let us now examine the dangers of round-off. We have produced below two tables for $f(x) = x^3 - 2x^2 + 1$, the first keeping exact values and the second using values of $f(x)$ correct to 2 d.p.

Table 2.8

x	$f(x)$	1st diff	2nd diff	*3rd* diff
0	1.000			
		−19		
0.1	0.981		−34	
		−53		6
0.2	0.928		−28	
		−81		6
0.3	0.847		−22	
		−103		6
0.4	0.744		−16	
		−119		6
0.5	0.625		−10	
		−129		6
0.6	0.496		−4	
		−133		6
0.7	0.363		2	
		−131		6
0.8	0.232		8	
		−123		6
0.9	0.109		14	
		−109		
1.0	0.000			

x	$f(x)$	1st	2nd	3rd	4th	5th	6th
0	1.00						
		−2					
0.1	0.98		−3				
		−5		0			
0.2	0.93		−3		0		
		−8		0		3	
0.3	0.85		−3		3		−11
		−11		3		−8	
0.4	0.74		0		−5		16
		−11		−2		8	
0.5	0.63		−2		3		−10
		−13		1		−2	
0.6	0.50		−1		1		−1
		−14		2		−3	
0.7	0.36		1		−2		5
		−13		0		2	
0.8	0.23		1		0		
		−12		0			
0.9	0.11		1				
		−11					
1.0	0.00						

Note:

(i) In the second table the third differences are not constant, and hence the fourth and higher differences are not zero.

(ii) The values of the fourth and higher differences are due to round-off errors.

(iii) The magnitudes of third differences are smaller than those to the left and to the right.

(iv) The magnitude of differences higher than the fourth tend to increase and signs follow no discernible pattern.

The general rule is that differences which are larger in magnitude than the predecessors should be ignored as should all higher differences.

We can analyse the situation as follows: rounding errors have their greatest effect when entries in a table have errors which are alternatively $\frac{1}{2}$ and $-\frac{1}{2}$ in the last decimal place shown.

Errors	1st diff	2nd diff	3rd diff	4th diff	5th diff
$\frac{1}{2}$					
	−1				
$-\frac{1}{2}$		2			
	1		−4		
$\frac{1}{2}$		−2		8	
	−1		4		−16
$-\frac{1}{2}$		2		−8	
	1		−4		
$\frac{1}{2}$		−2			
	−1				
$-\frac{1}{2}$					

In general, the magnitude of the maximum error in the nth difference is 2^{n-1}. Thus, in our table of rounded-off values, the third differences have an absolute bound due to errors of 2^2, i.e. 4, in the last place shown. This means that if they have values less than 4 they could be due entirely to round-off and so might not be worth including in calculations. We are really looking for differences which could not have arisen by errors alone; note that if the second differences were below 2 this could also be due to round-off. However, the actual bounds are seldom reached in practice and we are being over-generous in our criteria.

Example 2

Let us pursue a second example. In Table 2.9, values of tan x have been taken from 4 figure tables and rounded to 3 d.p.

We conclude here that second and higher differences should be neglected.

Table 2.9

x°	tan x	1st	2nd	3rd	4th	5th
0	0.000					
		18				
1	0.018		–1			
		17		1		
2	0.035		0		0	
		17		1		–2
3	0.052		1		–2	
		18		–1		
4	0.070		0			
		18				
5	0.088					
		(1)	(2)	(4)	(8)	(16)

Absolute error bound is shown in parentheses, thus: ( )

Extension of a Table using Differences
If we have a table where a constant set of differences has been reached then we may extend the table in either direction. Table 2.10 is exact. We wish to estimate $f(1.0)$.

Table 2.10

x	$f(x)$			
0	0			
		24		
0.2	0.024		144	
		168		144
0.4	0.192		288	
		456		144
0.6	0.648		432	
		888		**144**
0.8	1.536		**576**	
		1464		
1.0	**3.000**			

The numbers below the bold lines are added from right to left.

Now try and show that $f(-0.2) = -0.024$.
If the table is inexact due to round-off, the answers will be approximate. If the table has not got almost constant differences, extrapolation will be poor.

Errors in Tabulated Values
There is another source of error in a difference table: due to a mistake in a value for $f(x)$. We shall not dwell too deeply on this at this stage.

Problems

Section 2.1

1 Represent on the real line the sets

(a) $\{-3, 2, 6, -2\}$ (b) $\{x : x < 2\}$ (c) $\{x : -1 < x \leq 4\}$

2 Solve and represent the solutions to

(a) $6 - 3a > 2a + 1$ (b) $\frac{1}{3}(b + 3) \geq \frac{3}{4}$ (c) $5(3 - 2p) < -6$

(d) $(2x - 1)(4 - x) < 0$ (e) $(y + 2)(y - 5) < 0$ (f) $|x| \leq 0$

3 Rewrite without the absolute value signs and in closed or open interval forms

(a) $|x| < 4$ (b) $|x - 3| < 2$ (c) $|3x + 2| < 9$

4 Rewrite with absolute value signs

(a) $-3 < x < 7$ (b) $3 < x < 9$

5 Find a positive integer M such that $\frac{1}{m + 2} < 0.001$ for all integers m greater than M.

6 Show that for all values of x

(a) $x^2 + 2x + 8 > 0$ (b) $2x^2 - 3x + 4 > 0$ (c) $2x^2 - x > -2$

(d) $(x - 1)^2 > x - 4$

7 Determine the set of numbers x for which the following conditions are satisfied

(a) $|2x^2 + 3x - 2| < 1$ (b) $|6x - 3x^2| \leq 4$ (c) $|x^3 + x| \geq 2$

(d) $\frac{1}{x} < \frac{1}{4}$ (e) $100 < \frac{1}{x} < 200$

Section 2.2

8 Draw graphs of the following relations (Z is the set of integers)

(a) $\{(1, 1)\ (1, 2)\ (3, 5)\ (4, 6)\ (5, 20)\}$

(b) $\{(1, -1)\ (1, 8)\ (2, 6)\ (3, 6)\ (3, -2)\}$

(c) $\{(1.5, 2)\ (1.6, 3)\ (1.7, 5)\ (1.8, 7)\ (1.9, 9)\}$

(d) {The ordered pairs (x, y) such that $x < y$ $(x, y \in Z)$}

(e) {The ordered pairs (x, y) such that $|x - y| < 6$ $(x, y \in Z)$}

9 What are the domains and ranges of the above relations?

10 Classify the above relations as being one-one, many-one, one-many or many-many.

11 If $A = \{x \in R : 0 \leq x \leq 10\}$, $B = \{x \in R : 5 \leq x \leq 15\}$ where R is the set of real numbers, which of the following sets define a function with domain in A and range in B?

(a) $\{(0, 6)\ (2, 6)\ (3, 8)\ (5, 14)\}$ (b) $\{(2, 5)\ (3, 9)\ (10, 6)\ (2, 9)\}$

(c) $\{(0, 6)\ (4, 3)\ (7, 11)\ (8, 4)\}$ (d) $\{(2, 10)\ (5, 11)\ (6, 13)\ (2, 12)\ (7, 20)\}$

12 Write down the largest domain for which each of the following functions is defined

(a) $f(x) = x^3 + 3$ (b) $f(x) = x^2 + \sqrt{x - 1}$ (c) $f(x) = x^2 + \sqrt{1 - x^2}$

(d) $f(x) = \frac{1}{x} + x$ (e) $f(x) = \sqrt{x^2 - 1}$

13 State whether the mappings defined by the functions in Problem 12 are one-one or many-one.

14 Sketch the graphs of the following functions in the stated domains of definition
(a) $f(x) = x^3$ for $x \in [-3, 3]$ (b) $f(x) = x^2 - 3x + 2$ for $x \in [1, 2]$
(c) $f(x) = |x| - x$ for $x \in [-3, 3]$ (d) $f(x) = x + \frac{1}{x}$ for $x \in [\frac{1}{2}, 5]$

15 Which of the following in Figure 2.31 are the graphs of functions?

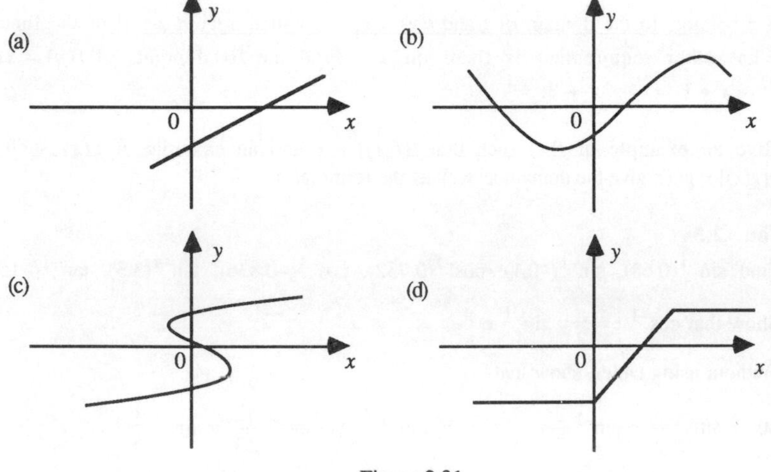

Figure 2.31

16 Complete the entries in this table

f	Domain	Is mapping one-one?	f^{-1} when it exists
x	$[-5, 2]$		
$\frac{1}{2 + x}$	$[1, 4]$		
$\sin x$	$[-\pi/4, \pi/2]$		
$x^2 - 60x + 50$	$[0, 60]$		

17 Which of the following functions has an inverse function?
(a) $f(x) = 6x^2 + 2$ (b) $f(x) = x^3 - 3$ (c) $f(x) = x, \ x \in [-1, +1]$

(d) $\begin{cases} f(x) = -1, & x < -1 \\ f(x) = +1, & x \geq +1 \end{cases}$ (e) $f(x) = \sqrt{1 - x^2}$

18 Find, if it exists, the inverse function of each of the functions defined in Problem 17.

19 Give formulae for the functions $f + g, f - g, f.g, f \div g$ and $g \div f$ with the appropriate domains when $f(x) = x - 3$, $g(x) = x^2 - 1$.

20 Repeat Problem 19 for $f(x) = x^2 - 2x + 1$, $g(x) = x - 1$.

21 If $f(x)$, $g(x)$ are as in Problems 19 and 20, give formulae for $f[g(x)]$ and $g[f(x)]$ stating the domains in each case.

22 If x belongs to the domain of f and $f(x) = x$, it is called a *fixed point* of the function. What other requirement is there on x? Find the fixed points of $f(x) = x, x^2, x^2 - x + 1, x^2 - 3x + 3$.

23 Give an example of $f(x)$ such that $f[f(x)] = x$ and an example of $g(x)$ such that $g[g(x)] = g(x)$; give the domain as well as the formula.

Section 2.3

24 Find $\sin^{-1}(0.65)$, $\sin^{-1}(-0.1)$, $\cos^{-1}(0.732)$, $\cos^{-1}(-0.836)$, $\tan^{-1}(3.3)$, $\tan^{-1}(-15)$

25 Show that $\cos^{-1} x = \frac{\pi}{2} - \sin^{-1} x$

26 Without using tables, show that

(a) $2 \sin^{-1} \frac{3}{5} = \sin^{-1} \frac{24}{25}$ (b) $\tan^{-1} \frac{63}{16} - \cos^{-1} \frac{12}{13} = \sin^{-1} \frac{4}{5}$

27 Show that $\log_9 27 = \frac{3}{2}$

Section 2.4

28 Prove, using basic definitions, that

(a) $\cosh 2A = 2 \cosh^2 A - 1 = 1 + 2 \sinh^2 A$

(b) $(\cosh A + \sinh A)^n = \cosh nA + \sinh nA$

(c) $\text{sech}^2 A = 1 - \tanh^2 A$

(d) $2 \sinh A \sinh B = \cosh(A + B) - \cosh(A - B)$

29 Evaluate $\cosh(2)$, $\sinh(0.5)$, $\tanh(-0.3)$, $\sinh^{-1}(1.5)$, $\cosh^{-1}(3)$, $\tanh^{-1}(1/4)$, $\tanh^{-1}(-0.25)$, $\cosh^{-1}(1.4)$, $\sinh^{-1}(-2)$
(Hint: for inverse functions – use log equivalents)

30 If $\text{cosech}\, x = \frac{-9}{40}$, find $\cosh x$ and $\tanh x$

31 Prove that

(a) $\text{sech}^{-1}\, x = \ln \dfrac{1 + \sqrt{1 - x^2}}{x}$ (b) $\text{cosech}^{-1}\, x = \ln \dfrac{1 + \sqrt{1 + x^2}}{x}$

32 Prove that $\tanh^{-1} \dfrac{x^2 - 1}{x^2 + 1} = \ln x$

Section 2.5

33 Tabulate the values of cos x for $x = \pi/2, 3\pi/2, 5\pi/2$ radians and calculate $\cos 11\pi/12$ by linear interpolation. Comment.
Now tabulate the values of cos x for $x = \pi/2, 2\pi/3, 5\pi/6, \pi$, and repeat likewise for $x =$ 160°, 170°. Conclusions?

34 From a four figure table of reciprocals, obtain 1/1.817 by linear interpolation and compare with the value calculated using mean differences. Why the discrepancy?

35 Form a difference table from the table of pressure of saturated water P and temperature T, as far as the differences seem useful (Table 2.1).

36 To evaluate the sum of the squares of the first n natural numbers (we can take $n = 0, 1, \ldots, 6$) form a difference table and deduce the degree of the polynomial it represents. Fit suitable data to show that it is $[1/6][n(n + 1)(2n + 1)]$.

37 Tabulate $f(x) = 3x^3 - 2x^2 + x$ for $x = 0(2)10$† until constant differences are reached.

38 Tabulate $4.2x^3 - 0.3x^2 + 2.7x - 0.8$ exactly for $x = 0(0.1)1.0$; extend the table to $x = -0.2$ and $x = -0.1$. Now write the table correct to 2 d.p. and form a difference table. What are the highest differences that can be used?

39 The results from a tensile test are shown: estimate the stress at a strain of 2500×10^{-6} mm/mm using linear interpolation. Why can we use linear interpolation here?

Stress MN/m^2	Strain mm/mm
0	0
30	1003×10^{-6}
60	2007×10^{-6}
90	3009×10^{-6}
120	4013×10^{-6}
150	5017×10^{-6}

40 Produce a difference table from data for air density ρ and estimate density at altitude 5500 and 20 000 m.

† $x = 0$ to 10 in steps of 2

Altitude m(10^3)	ρ kg/mm^3
0	1.2256
1	1.1121
2	1.0069
3	0.9095
4	0.8194
5	0.7364
6	0.6599
7	0.5897
8	0.5253
9	0.4665
10	0.4128

3

ELEMENTARY IDEAS ON LIMITS

Introduction

In the study of a slender column of length l built in at its lower end A, pinned and laterally supported at its upper end B, as shown in Figure 3.1, it is necessary to solve the equation

$$\tan u = u \qquad (3.1)$$

where $u^2 = \frac{Pl^2}{EI}$, P is the *buckling load*, EI is the *flexural rigidity*.

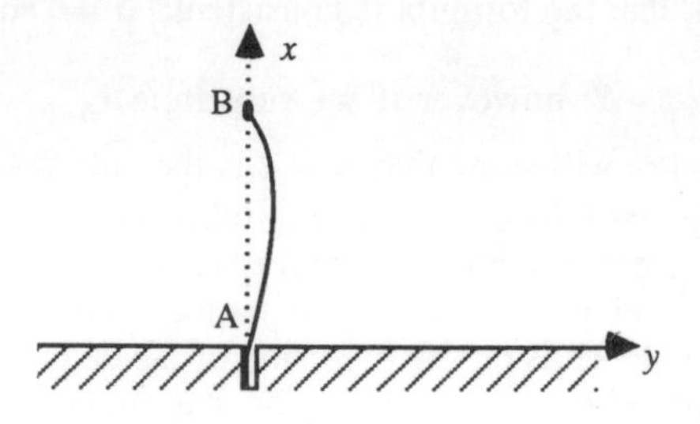

Figure 3.1

How can we obtain the root of equation (3.1), if no formula exists? It may seem naïve to suggest that we make a guess or even a succession of guesses; however, as long as we do not guess wildly but use the information from previous guesses in an organised way we can tie down the value of the root quite accurately. In order to see clearly the way that the mechanism for guesswork behaves, we shall take some simple abstract problems to which the answers are already known; this has two advantages: firstly, it prevents the physical aspects of a problem obscuring the basic mathematical processes, secondly, it allows us to see how the guesses compare with the known answer. This latter advantage has a wider importance for, if we can test a method against a known result, we can evaluate its strengths and weaknesses and hence apply it confidently in situations where the result is not known by other means.

We find $\sqrt{4}$ by a method of successive guesses. It is helpful to consider the geometrical equivalent: the determination of the length of the side of a square of area 4 units. (In algebraic terms, we seek the positive root of $x^2 = 4$.)

Suppose we guess that $x = 1$. Instead of a square we have a rectangle of area 4 with sides of length 1 and 4. Clearly the sides of length 4 are too long, those of length 1 too short. The length we requirc is somewhere in between and we choose to guess at the arithmetic mean, 2.5. To test the accuracy of this guess, we divide the area 4 by this length and obtain 1.6. Clearly the number we seek lies between 2.5 and 1.6 and we choose a next estimate as $\frac{1}{2}(2.5 + 1.6) = 2.05$. We seem to be approaching the true answer of 2 but we cannot be sure.

To investigate further, we must put the method on an algebraic footing. Let x_0 be the initial guess of 1. Then the next guess, $x_1 = \frac{1}{2}(x_0 + 4/x_0)$ and the guess of 2.05 is $x_2 = \frac{1}{2}(x_1 + 4/x_1)$. We can thus define the method as $x_0 = 1$, $x_{n+1} = \frac{1}{2}(x_n + 4/x_n)$; we have given an initial guess and a rule for generating successive guesses. In Table 3.1 we show the results to 9 d.p. in the left-hand column (using a calculator) and rounded to 2 d.p. in the right-hand column. Notice that rounding helps in getting the right answer.

After 4 applications of the rule we have got very close to the exact value of the root, but would we be able to get it exactly if we applied the rule more times? First, we must check that the formula is consistent: if we substitute $x_n = 2$ then you can see that $x_{n+1} = 2$; however if we substitute $x_{n+1} = 2$ then $2 = \frac{1}{2}(x_n + 4/x_n)$ and a little algebra will show that $x_n = 2$ is the only result. What does this tell us? Only if we guess the exact answer initially will the sequence of guesses provide the exact answer; otherwise we shall never obtain it. Of course, this is a theoretical statement and in practice we can get the value correct to a number of decimal places which depends on the precision of the calculating aid we employ and the build-up of round-off error. (It may be that round-off prevents us getting the value we seek accurate to the full number of decimal places the aid can provide.)

Table 3.1

n	x_n	x_n
0	1	1
1	2.5	2.5
2	2.05	2.05
3	2.000609756	2.00
4	2.000000092	2.00

NOTE: $(2.000000092)^2 = 4.000000368$

Table 3.2

n	x_n (to 4 d.p.)	x_n
0	10	64
1	5.2	32.0625
2	2.9846	16.0936
3	2.1624	8.1710
4	2.0060	4.3302
5	2.0000	2.6269
6	2.0000	2.0748

Why the almost linear decrease early on?

It may be argued that there is a danger in taking such a simple problem on which to demonstrate the technique: for example, it may be the case with a practical problem that we cannot make an initial guess which is reasonably close to the value sought. Let us investigate the effect of other initial guesses. Obviously, $x_0 = 0$ is not acceptable (why?) but in Table 3.2 the results using initial guesses of 10 and 64 are shown; you try using –3.

It would seem that the initial guess merely delays the achievement of a particular accuracy. This can be shown theoretically, but we must always bear in mind that in practice round-off may spoil matters.

Let us now investigate the application of this approach to finding $\sqrt{2}$. You may know that $\sqrt{2}$ is irrational and hence cannot be exactly represented by a finite decimal; since the approximation method will not give it exactly anyway, this does not matter. However, we were able to stop the sequence of approximations to $\sqrt{4}$ when the last approximation was as close to the known result of 2 as was required; we can here assume that we have only a vague idea of the value of $\sqrt{2}$ and consequently cannot be sure when we have the value correct to, say, 4 d.p. If we examine Tables 3.1 and 3.2 again we see that the approximations, as well as approaching 2, are getting closer together. This will be our criterion for stopping the sequence of guesses: if we require an accuracy of 4 d.p. we shall stop when two consecutive approximations differ by less than .0001. We start with $x_0 = 1$ and use the **recurrence formula** $x_{n+1} = \frac{1}{2}(x_n + 2/x_n)$.

Table 3.3

n	x_n
0	1
1	1.5
2	1.416666666
3	1.414215700
4	1.414213562
5	1.414213562

Table 3.3 shows the sequence of approximations as obtained on a calculator. We see that the last two approximations agree to 9 d.p. and we could quote this value as the result required.

Suppose we generalise this method to one which finds the positive square root of any non-negative number, A; the *iterative formula* becomes

$$x_{n+1} = \frac{1}{2}\left[x_n + \frac{A}{x_n}\right] \tag{3.2}$$

It is valuable at this stage to flow chart the process (Figure 3.2). We let XOLD and XNEW take the role of x_n and x_{n+1}, TOL stand for the accuracy (tolerance) required. You should modify the flow chart to incorporate a check that the number of times the formula is used does not exceed a pre-specified number.

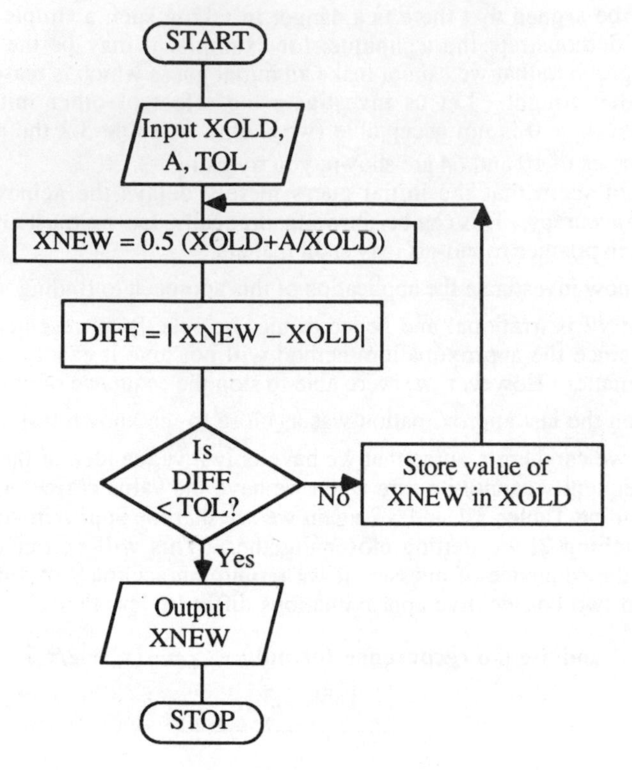

Figure 3.2

3.1 Sequences and Limits

We have obviously touched on a wealth of problems that need answering. First we make a definition: a set of objects (these may not necessarily be all distinct) which are arranged in a definite order is called a **sequence**. Examples are

(i) the days in a week
(ii) the letters and digits forming a postal code
(iii) the tasks performed to start a car
(iv) the integers 1 to 5 arranged in ascending order
(v) the numbers 1, 4, 7, 10, 13, ...

We confine our attention to sequences of numbers. An *infinite sequence* has an infinite number of elements and can be specified by an incomplete list as in Example (v) above, or by a rule which gives the nth term in the sequence [eg, $u_n = 3n - 2$ for (v)] or by a *recurrence relation* which tells us how to generate one term from others [e.g., $u_{n+1} = u_n + 3$ with $u_1 = 1$ for (v)].

Consider some more examples:

(vi) the sequences 3, 8, 15, 24,... can be specified by $u_n = n(n + 2)$; from this formula we can say that $u_{n+1} = (n + 1)(n + 3)$, replacing n by $(n + 1)$

(vii) the sequence specified by $u_{n+1} = 2u_n$, $u_1 = 2$ can also be written as

$$u_{n+1} = 2^2u_{n-1} = 2^3u_{n-2} \ldots\ldots = 2^nu_1 = 2^{n+1} \quad \text{or} \quad u_n = 2^n;$$

further, it can be specified as 2, 4, 8, 16,... This sequence is an example of a *geometric progression* $a, ar, ar^2,\ldots$ where one term is generated from its predecessor by multiplication by a common ratio r.

We saw earlier that each sequence of approximations to $\sqrt{4}$ approached the value of 2 and we may refer to the number 2 as the **LIMIT** of that (infinite) sequence: we say that the sequence **converges** to the limit 2. We shall need some general definitions which will allow us to describe the behaviour of infinite sequences. We choose some simple, abstract examples:

(viii) the sequence $u_n = (1/2)^n$, i.e. 1/2, 1/4, 1/8, 1/16, ...

(ix) the sequence $u_n = 1 - (1/2)^n = 1/2, 3/4, 7/8, 15/16, 31/32, 63/64, \ldots$

(x) the sequence 1, 0, 1, 0, 1, ...

(xi) the sequence 1, –2, 3, –4, 5, ...

(xii) the sequence 1/2, 1/2, 3/4, 3/4, 3/4, 7/8, 7/8, 7/8, 7/8, 15/16, ...

(xiii) the sequence –1, –2, –4, –8, ...

The first concept we must define is that of an **increasing** sequence. If $u_{n+1} \geq u_n$ all through the sequence, it is said to be *monotonically* increasing. If $u_{n+1} > u_n$ all through the sequence, it is said to be *strictly* increasing.

Hence sequences (ix) and (xii) are both monotonically increasing and, in addition, sequence (ix) is strictly increasing. Similarly, we say that sequence (viii) is strictly decreasing. Sequences (x) and (xi) are said to *oscillate*.

The second concept is that of **boundedness**. A sequence is said to have an *upper bound M* if no term $u_n > M$. For sequence (viii), the number 47 is clearly an upper bound and so are 7, 2, 2/3, etc. We shall define a *least upper bound* (l.u.b.) for a sequence to be that number M^* such that

(a) $u_n \leq M^*$ for all terms u_n of the sequence, and

(b) if we take any number $M^+ \leq M^*$ we can find at least one term of the sequence which exceeds M^+ (in other words, nothing smaller will do).

For sequence (viii) the l.u.b. is 1/2 and equality is achieved with the initial term; for sequence (ix) the l.u.b. is 1 (convince yourself by taking a number just less than 1 and find a value for n such that $u_n, u_{n+1}, u_{n+2}, \ldots$ are all greater than your number).

The l.u.b.'s for sequences (x) and (xii) are both 1, that for (xiii) is –1.

In a similar fashion we can define a *lower bound m* and a *greatest lower bound* (g.l.b.), m^*, and you should verify that the g.l.b.'s for sequences (viii), (ix), (x) and (xii) are respectively 0, 1/2, 0, 1/2. Sequences (vi) and (vii) both have g.l.b.'s but neither has an upper bound. Such sequences are said to **diverge**. Sequence (xi) has neither an upper nor a lower bound and is said to oscillate infinitely in contrast to (x) which oscillates finitely. Sequence (xiii) also diverges (to what?).

In Figure 3.3 are shown the schematic behaviour of some of these sequences, identified by the sequence number shown on the graph. The terms are shown by bold dots and are joined by broken lines for convenience in discerning any trends. The diagrams are **not** to scale.

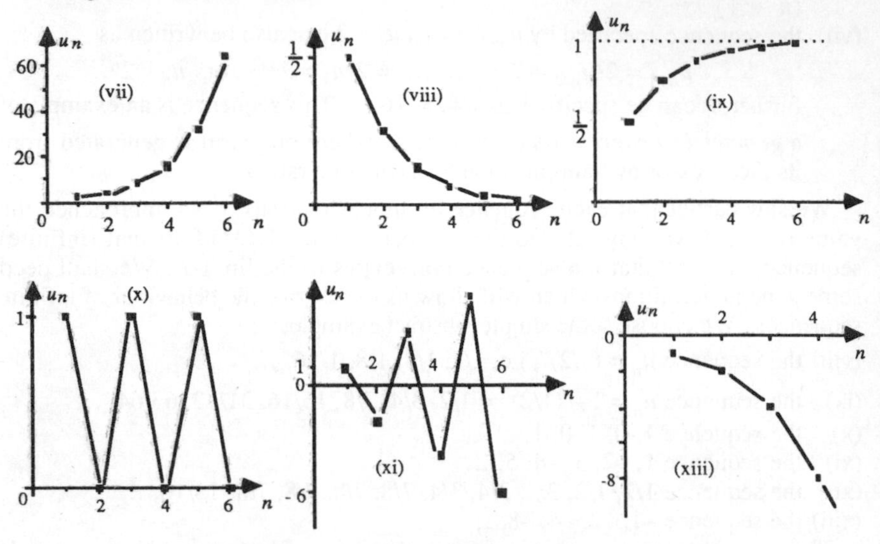

Figure 3.3

It is now time to tie down the concept of the **limit of a sequence**. The sequence whose typical term is u_n (often written $\{u_n\}$) is said to have a limit L (or to converge to the limit L) if for every positive number ε there is a term u_N in the sequence after which every term is within the range of values $(L - \varepsilon, L + \varepsilon)$. Notice three points: first, that the criterion must hold for every positive number ε no matter how small; second, that the critical term of the sequence, u_N, will not necessarily be the same for all possible ε; third, we are only interested in the long-term behaviour of the sequence since, with an infinite number of terms, the earlier ones are but a minute fraction of the total. We illustrate these three points by examples.

The **probability** of an event may be regarded as a pre-assigned number, based on theoretical grounds, or as long-range relative frequency. The results of tossing a presumed unbiased coin are tabulated below in Table 3.4. You can see that after a 'jerky' start, things settle down to a frequency close to 0.5. Whether the figure of 0.5 is justified is discussed in Chapter 4.

Table 3.4

Number of *tosses* to date	1	2	3	4	5	10	100	1000	10000
Number of *heads*	0	1	2	3	3	6	52	511	5054
Relative frequency	0	0.5	0.67	0.75	0.6	0.6	0.52	0.511	0.5054

It is important to consider the *tail* of a sequence since the first few terms may be atypical and, as in this example, more subject to influence by single events.

Look again at sequence (x), i.e. 1, 0, 1, 0, 1, ... Certainly if $\varepsilon = 2$ we might try and claim a limit of 0.5, since all terms of the sequence lie between $0.5 - 2$ and $0.5 + 2$, i.e. in the range (–1.5 to 2.5). Indeed we can take $\varepsilon = 1$ to obtain the range (–0.5 to 1.5); but consider what happens if $\varepsilon = 1/4$: all terms lie outside the range (0.25, 0.75) and the criterion must hold for **all** $\varepsilon > 0$. Hence the sequence does not converge.

If we turn again to sequence (viii), $u_n = (1/2)^n$, we have said that the limit is 0. We attempt to justify this. Suppose we choose $\varepsilon = 1/64$; we look for a critical term such that thereafter, terms of the sequence lie in the interval [–1/64, 1/64]. But the sixth term is 1/64 and clearly, all succeeding terms lie in the required interval. Similarly, if we take $\varepsilon = 1/256$ we note that all terms after the eighth are in the range of values [–1/256, 1/256].

To generalise this argument, first suggest how you would choose N if ε was of the form $1/2^r$ where r is a positive integer. Then decide what you would do for a value of ε in the interval $(1/2^{r+1}, 1/2^r)$; for example, if $\varepsilon = 1/230$ then $1/256 < \varepsilon < 1/128$ and $r = 7$.

Sequence (ix) needs a little more care. We claim that the limit is 1. Now take $\varepsilon = 1/16$; we seek that term after which all terms lie in the range of values (15/16, 17/16) and we see that the critical term, u_N, is the fourth. If we take $\varepsilon = 1/64$ we must advance past the sixth term. You should try and show that if $\varepsilon = 1/2^r$ then $N = r$, i.e. $N = \log \varepsilon / \log \frac{1}{2}$. This is a complicated formula for N but the fact that as ε decreases N increases is clear as it was for sequence (viii).

Alternative ways of writing the statement that the limit of $\{u_n\}$ is L are

$$\lim_{n\to\infty} u_n = L$$

and

given any $\varepsilon > 0$, *there exists an integer* N *such that* $|u_n - L| < \varepsilon$ *when* $n > N$.

Obviously, from a practical viewpoint, the rate at which convergence takes place is important and will be studied later.

It would be tedious if the limit of every sequence had to be verified by resort to the definition. To aid us in this task we quote some rules for sequences in general:

Let $\{u_n\}$ and $\{v_n\}$ be sequences such that $\lim_{n\to\infty} u_n = U$, $\lim_{n\to\infty} v_n = V$.

1 The sequence $\{u_n + v_n\}$ which has terms $u_1 + v_1$, $u_2 + v_2$ etc, has limit $U + V$.

Examples

(i) The sequence $2\frac{1}{2}$, $2\frac{1}{4}$, $2\frac{1}{8}$, $2\frac{1}{16}$,... can be regarded as the sum of the sequences 2, 2, 2, 2, ... and $\frac{1}{2}$, $\frac{1}{4}$, $\frac{1}{8}$, $\frac{1}{16}$, ... Since these have limits 2 and 0 respectively, the sum sequence converges and has limit 2

(ii) The sequences $2\frac{1}{2}$, $2\frac{1}{4}$, $2\frac{1}{8}$, $2\frac{1}{16}$, ... and $\frac{1}{2}$, $\frac{3}{4}$, $\frac{7}{8}$, $\frac{15}{16}$,... have limits 2 and 1 respectively. The difference sequence 2, $1\frac{1}{2}$, $1\frac{1}{4}$, $1\frac{1}{8}$, ... has limit $2 - 1 = 1$

(iii) The sequence $2\frac{1}{2}$, $3\frac{1}{3}$, $4\frac{1}{4}$, $5\frac{1}{5}$, ... is the sum of the sequences 2, 3, 4, 5, ... and $\frac{1}{2}$, $\frac{1}{3}$, $\frac{1}{4}$, $\frac{1}{5}$, ... Since the first of these has no limit, the sum sequence has no limit.

2 The sequence $\{\alpha u_n\}$ which has terms αu_1, αu_2, etc, has limit αU.
Example
With $\alpha = 3$, we see that the sequence $\frac{3}{2}$, $\frac{3}{4}$, $\frac{3}{8}$, $\frac{3}{16}$, ... converges and has limit

$$3 \lim_{n\to\infty} \{\tfrac{1}{2}, \tfrac{1}{4}, \tfrac{1}{8}, \tfrac{1}{16}, \ldots\} = 3 \times 0 = 0$$

3 The sequence $\{u_n v_n\}$ has limit UV.
Example
The sequences $1\frac{1}{2}$, $1\frac{3}{4}$, $1\frac{7}{8}$, $1\frac{15}{16}$, and $2\frac{1}{2}$, $2\frac{1}{4}$, $2\frac{1}{8}$, $2\frac{1}{16}$, ... both have limit 2. The product sequence 15/4, 63/16, 255/64, 1023/256, ... has limit 4.

[Note that (2) follows from (3) by taking $\{v_n\}$ to be the constant sequence α, α, α,...]

4 The sequence $\{u_n/v_n\}$ has limit U/V PROVIDED $V \neq 0$
Example
With $\{u_n\} = 1, 1, 1, 1, \ldots$ and $\{v_n\} = 1/2, 3/4, 7/8, 15/16, \ldots$. We find $\{u_n/v_n\} = 2/1, 4/3, 8/7, 16/15$ has limit 1.

Note that we can say nothing about the limit of the sequence 2/1, 4/1, 8/1, 16/1 from these rules, where $\{v_n\} = 1, 1, 1, \ldots$ though common sense suggests it diverges. (We use the symbolism $u_n \to \infty$ to indicate that the terms in the sequence increase without limit.)

Examples

(i) $\lim_{n\to\infty} \dfrac{2n - 1}{3n + 4}$

Here it is useless to write the nth term as the quotient of $2n - 1$ and $3n + 4$ since neither $\{2n - 1\}$ nor $\{3n + 4\}$ converges. However, if we write $(2n - 1)/(3n + 4)$ as $(2 - 1/n)/(3 + 4/n)$ we may use Rule 1 to say that the limit of the numerator $2 - 1/n$ is 2 and that of the denominator is 3, then by Rule 4, the limit we seek is 2/3.

(ii) $$\lim_{n\to\infty} \frac{n^2 + 3n + 1}{2n^2 - 4n - 2}$$

Division by n^2 in both numerator and denominator gives the nth term as $(1 + 3/n + 1/n^2)/(2 - 4/n - 2/n^2)$ and, by extension of the above argument, the limit of the sequence is 1/2.

(iii) $$\lim_{n\to\infty} \frac{2n^{-1} + 3n^{-4}}{5n^{-1} - 2n^{-2}}$$

Multiplication of top and bottom by n gives the nth term as $(2 + 3n^{-3})/(5 - 2n^{-1})$ and the limit as 2/5.

(iv) $$\lim_{n\to\infty} \frac{1 + 2 + 3 + \ldots + n}{n} = \lim_{n\to\infty} \frac{\frac{1}{2}n(n+1)}{n} = \lim_{n\to\infty} \frac{1}{2}(n+1)$$

Whilst the limit does not exist (as a finite quantity), since $\frac{1}{2}(n + 1) \to \infty$ as $n \to \infty$, it is customary to write $\lim\limits_{n\to\infty} \frac{1}{2}(n + 1) = \infty$.

Notice that we use the symbols $\to \infty$ since infinity is a state to be approached rather than achieved. Note that

(a) an increasing sequence bounded above tends to a limit

(b) a decreasing sequence bounded below tends to a limit.

Examples

(a) The sequence $1, 1\frac{1}{2}, 1\frac{3}{4}, 1\frac{7}{8}, 1\frac{15}{16}, \ldots$ is increasing and bounded above by 2; its limit is 2.

(b) The sequence $1, \frac{1}{2}, \frac{1}{4}, \frac{1}{8}, \frac{1}{16}, \ldots$ is decreasing and bounded below by 0; its limit is 0.

Finally, we return to the formula for approximations to $\sqrt{2}$:

$$x_{n+1} = \frac{1}{2}\left[x_n + \frac{2}{x_n}\right]$$

Suppose that the limit of the sequence $\{x_n\}$ is L: then, clearly, the limit of the sequence $\{x_{n+1}\}$ is L. By applying the rules for combinations of sequences we see that

$$\lim_{n\to\infty} \frac{1}{2}\left[x_n + \frac{2}{x_n}\right] = \frac{1}{2}\lim_{n\to\infty}\left[x_n + \frac{2}{x_n}\right]$$

$$= \frac{1}{2}\lim_{n\to\infty} x_n + \frac{1}{2}\lim_{n\to\infty}\frac{2}{x_n} = \frac{1}{2}L + \frac{1}{2}\cdot\frac{2}{L}$$

But this limit must be also $\lim_{n\to\infty} x_{n+1}$ which is L. Then $L = \frac{1}{2}L + 1/L$, i.e. $\frac{1}{2}L = 1/L$ or $L^2 = 2$, giving $L = \pm\sqrt{2}$. If x_0 is positive we get a sequence of positive numbers, and $L = +\sqrt{2}$, whereas, if x_0 is negative, $L = -\sqrt{2}$.

You should follow this argument replacing the iteration formula by the general one for finding $\sqrt{A}$.

The number e

The following limit, denoted e, is of importance

$$\lim_{n\to\infty}\left[\left(1 + \frac{1}{n}\right)^n\right] \tag{3.3}$$

Write a computer program to evaluate the first 20 terms of this sequence; the limit value is 2.7183 to 4 d.p. You will see that convergence is slow. Convergence to a limit may be shown by demonstrating that the sequence is increasing and bounded above by 3.

3.2 Functions of a Discrete Variable – Induction

Sometimes the variables being studied are **continuous**, e.g. time, temperature, angle of twist; but sometimes they are **discrete**, e.g. if the total voltage in a circuit is provided by a number of batteries each of a given voltage V then the total voltage can only be an integral multiple of V: (V, $2V$, $3V$, ...). Another important class of examples arises in the consideration of probabilities of discrete events, e.g. the probabilities of the various possible scores with two dice; the domain is the set {2, 3, 4, ..., 12}. A special case is when the domain of the function is the set of natural numbers: {1, 2, 3, 4, ...}. We can regard a sequence as a function of this type so that $\{u_n\}$ where $u_n = 2n - 1$ may be written $f(n) = 2n - 1$.

We consider a class of situations where the problem is expressed in terms of integers. It often happens that we need to verify that a conjecture holds for all values of a set; for example, it can be claimed that the formula $S_n \equiv 1 + 2 + 3 + \dots + n = \frac{1}{2}n(n + 1)$ is valid for all positive integral values of n. It would be impossible to prove the conjecture true for each value of n in turn and it would be dangerous to prove it to be true for a few values of n and hope it is true for other values.

The proof by mathematical induction is in two parts: first we show the conjecture to be true for the smallest value of n possible (usually $n = 1$), then we show that if the conjecture holds for a particular, but unspecified, value of n, say p, then it holds for the next value of n, i.e. $p + 1$. Two examples should help fix ideas.

Example 1

We start with the conjecture that

$$S_n \equiv 1 + 2 + 3 + \dots + n = \tfrac{1}{2}n(n + 1)$$

When $n = 1$ the sum consists of the term 1 only and the formula yields $\frac{1}{2}\cdot 1(1 + 1) = 1$, hence the conjecture is true in this case. Now we assume it to be true for $n = p$, i.e. $1 + 2 + \dots + p = \frac{1}{2}p(p + 1)$. To investigate whether it holds true for $n = p + 1$, add the number $(p + 1)$ to both sides of this equation: $1 + 2 + \dots + p + (p + 1) = \frac{1}{2}p(p + 1) + (p + 1)$. Now the right-hand side may be simplified to $[\frac{1}{2}(p + 1)(p + 2)]$ and this is just the result we would have got by substituting $(p + 1)$ for n in the formula for S_n. Hence, if the conjecture is true when $n = p$, it is true when $n = p + 1$. But we have shown directly that the conjecture is true when $n = 1$ and this second part has shown that the conjecture is true when $n = 2$ (put unspecified p equal to 1). Since this result is true when $n = 2$, a further application of the second part shows the conjecture to be true when $n = 3$ (put $p = 2$). In this way we prove the conjecture true for $n = 4, 5, 6, \dots$. We have thus moved from the particular to the general – as opposed to the usual mathematical procedure.

Example 2

As a second example, consider the possibility that $n! > 2^n$. It is certainly not true when $n = 1$ or when $n = 2$ or 3. But when $n = 4$, $4! = 24 > 16 = 2^4$. This is the smallest value of n for which the inequality holds. Now let us suppose that $p! > 2^p$ for some unspecified $p \geq 4$.

Multiply both sides of the inequality by $(p + 1)$; then $(p + 1)p! > (p + 1)2^p$, i.e. $(p + 1)! > (p + 1)2^p$ and since $p + 1 > 2$,

$(p + 1)2^p > 2^{p+1}$; hence $(p + 1)! > (p + 1)2^p > 2^{p+1}$ and so $(p + 1)! > 2^{p+1}$, proving the conjecture true when $n = p + 1$.

We conclude that $n! > 2^n$ for $n = 4, 5, 6, \ldots$

Notice that

(i) such conjectures may be proved by other means

(ii) we must have a conjecture in terms of n before the method of mathematical induction is applicable. A problem such as the determination of $1^2 + 2^2 + 3^2 + \ldots + n^2$ may best be tackled by a different approach.

3.3 Limits of Functions and Continuity

We have said that a sequence could be regarded as a function defined on the natural numbers and we have discussed the limits of sequences. We might ask whether any useful information could be obtained from studying the application of the limit process to other functions: this is the analytical approach to the study of functions.

A keystone in this study are the fundamental ideas of **continuity** and **discontinuity**. These concepts occur in physical situations: across a shock front in a gas the velocity and pressure of the gas change abruptly; the depth of water in a channel changes abruptly at a sluice gate, see Figure 3.4. In such cases discontinuities are said to occur.

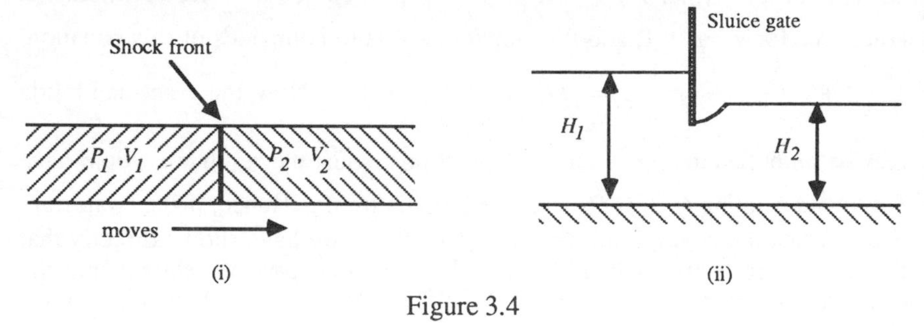

Figure 3.4

We can see that $\lim_{n\to\infty}(1 - \frac{1}{n}) = 1$

and it would seem natural to suppose that $\lim_{x\to\infty}(1 - 1/x) = 1$;

we note that whereas n approached infinity via a discrete set of values, x moves through a continuous range of values. Notice that $1 - (1/x)$ is not defined at $x = 0$ (see Figure 3.5). We may liken the special case of finding the limit of the sequence $\{1 - (1/n)\}$ to walking on stepping stones placed on those values of x which are integral, instead of continuously proceeding to the limit as we do when considering the function $f(x) = 1 - 1/x$, defined for all $x \neq 0$. But other possibilities open up for the function $1 - 1/x$; what happens to $1 - 1/x$ as x

becomes more and more negative? Since $-1/x$ is positive and becomes smaller, $1 - 1/x$ approaches the value 1 from above. In this example, 1 is said to be an **asymptote** to the function or the function is said to be **asymptotic** to $f(x) = 1$ for large positive and large negative values of x. We shall study asymptotes further in the next chapter. We are often interested in values of the function for less extreme values of x and we may ask what is $\lim_{x \to 3} f(x)$;

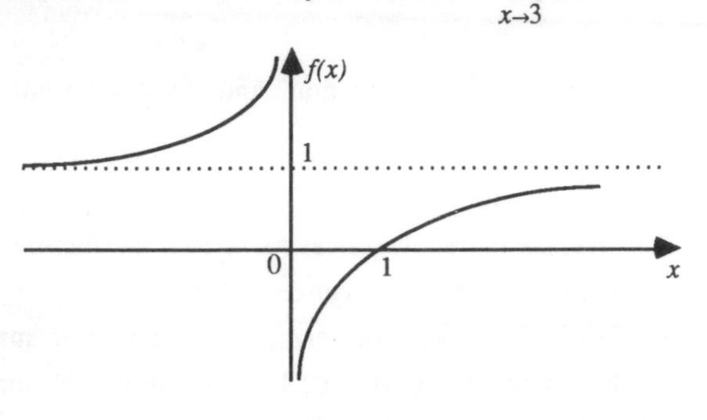

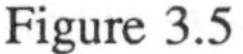
Figure 3.5

in this case the answer is 2/3, which is $f(3)$. Again $\lim_{x \to 1} f(x) = 0 = f(1)$, but this pattern breaks down for $\lim_{x \to 0} f(x)$ since $f(0)$ is not defined.

Some further examples should help clarify ideas.

1 If $f(x) = x^2 - 2$, then we would expect $\lim_{x \to 3} f(x) = 7 = f(3)$

2 If $f(x) = \dfrac{x^2 - 1}{x - 1}$ then the function is not defined at $x = 1$.

In order to evaluate $\lim_{x \to 1} f(x)$ we must first simplify $f(x)$ to $(x + 1)$, which we can do as long as $x \neq 1$, and then take the limit to obtain the result 2. However, we could expect $\lim_{x \to 3} f(x)$ to be $(3^2 - 1)/(3 - 1)$, i.e. 4, which is equal to $f(3)$.

What we have been doing is applying the rules for sequences in a slightly different form; we may state these modified rules as follows. Assuming that all limits quoted exist, then if

$$\lim_{x \to a} f(x) = A, \quad \lim_{x \to a} g(x) = B,$$

(i) $\lim_{x \to a} [f(x) + g(x)] = A + B$

(ii) $\lim_{x \to a} [f(x).g(x)] = A.B$

[Note the special case where $f(x)$ is a constant C, then $\lim_{x \to a} f(x) = C$]

(iii) $\lim_{x \to a} [f(x)/g(x)] = A/B$, *provided* $B \neq 0$

We have to some extent jumped the gun, and what we really need is a definition of what is meant by $\lim_{x \to a} f(x) = A$.

We effectively want to say that the closer x approaches the value a then the closer $f(x)$ is to A. We measure closeness of x to a by means of the gap between them: $|x - a|$; similarly, a measure of the closeness of $f(x)$ to A is $|f(x) - A|$. We want to say that given any error $\varepsilon > 0$, however small, we can confine $f(x)$ within the range $(A - \varepsilon, A + \varepsilon)$ by confining x to some interval $(a - \delta, a + \delta)$; the less freedom we allow $f(x)$, the narrower this last interval must be. The important point is that we can always find a value for δ. As with sequences, we must guess A first.

Consider $f(x) = 1 + 4x^2$. We shall assert that

$$\lim_{x \to 0} f(x) = 1,$$

and try to prove this. From the definition we must take $A = 1$ and consider $|f(x) - 1| = |4x^2|$. Suppose we allow an error of 0.01, then we must confine x to some range about 0 so that $|4x^2| < 0.01$, i.e. $4x^2 < 0.01$, i.e. $x^2 < 0.0025$ and hence x lies in the interval $(-0.05, 0.05)$. Again, were we to allow an error of 0.0001 we should be forced to restrict x to the interval $(-0.005, 0.005)$ and if we take a general error ε we must restrict x to the interval $(-\frac{1}{2}\sqrt{\varepsilon}, \frac{1}{2}\sqrt{\varepsilon})$. It must be emphasised that, in general, the application of the formal definition is a process involving much algebraic manipulation which we shall not pursue here.

Returning to the graph of $f(x) = 1 - 1/x$ we see that the definition of $\lim_{x \to 0} f(x)$ has said nothing about the direction from which we approach 0. Someone journeying to $x = 0$ along the left-hand portion of the graph would obtain a very different impression of the behaviour of $f(x)$ near $x = 1$ to a traveller approaching that value of x from the right. Clearly the graph is discontinuous and we say that $f(x)$ has an infinite discontinuity at $x = 0$. A less dramatic example is the graph (Figure 3.6) of the current in a simple circuit before and after its direction has been instantaneously reversed.

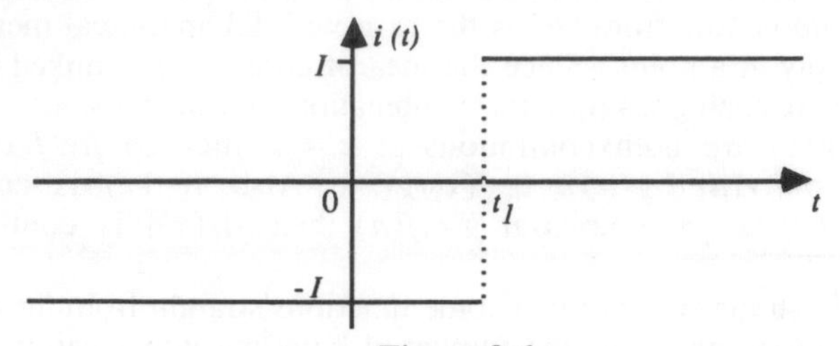

Figure 3.6

In order to complete the definition of the function, let us suppose that

$$i(t)\begin{cases} = -I & \text{for } t \le t_1 \\ = I & \text{for } t > t_1 \end{cases}$$

This function has a finite discontinuity at $t = t_1$.

We can see that the limit of $i(t)$ as t approaches t_1 from below is $-I$; this is written

$$\lim_{t \uparrow t_1} i(t) = -I$$

and the limit of $i(t)$ as t approaches t_1 from above is I, written as

$$\lim_{t \downarrow t_1} i(t) = I$$

and only the former limit is actually equal to $i(t_1)$. Now we give some graphs in Figure 3.7 to 'invented' functions to illustrate the possibilities of discontinuity.

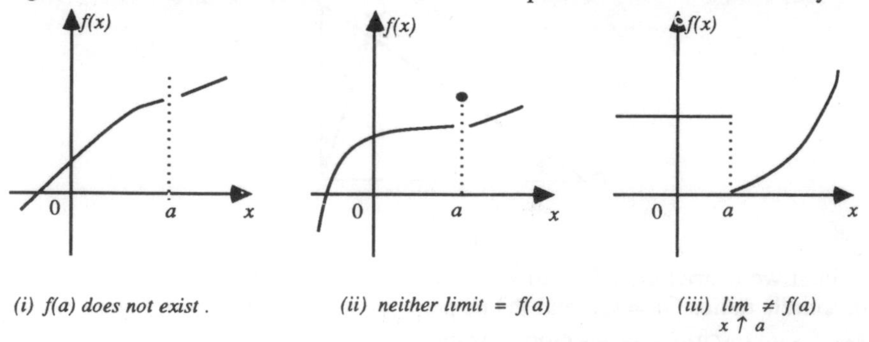

(i) $f(a)$ does not exist. *(ii) neither limit $= f(a)$* *(iii) $\lim_{x \uparrow a} \ne f(a)$*

Figure 3.7

We say that $f(x)$ is **continuous** at $x = a$ if $f(x)$ is defined for all values of x in some interval containing a and if $\lim_{x \to a} f(x) = f(a)$. This means, as for sequences, that we are only concerned with the tail of function values, i.e. we need concern ourselves only with the immediate vicinity of a. Hence the function $i(t)$ is continuous at all points save $t = t_1$.

A function $f(x)$ which is continuous at all points on which it is defined is called a **continuous function**. As is the case with all analytical methods, the basic results apply at a point. Since the idea of continuity is linked to that of sequences there are analogous rules for combinations of functions.

If $f(x)$ and $g(x)$ are both continuous at $x = a$, then so are $f(x) + g(x)$, $f(x).g(x)$ and, *provided* $g(a) \neq 0$, $f(x)/g(x)$. Also, if $f(x)$ is continuous at $x = a$ and $g(y)$ is continuous at $y = f(a)$ then $g[f(x)]$ is continuous at $x = a$.

We can establish the continuity of some functions straight from the definition and by use of these rules extend the number of functions which can be shown to be continuous. For example, we can show directly that $f(x) = C$, a constant, and that $g(x) = x$ are continuous and by repeated application of the rules any polynomial $a_n x^n + \ldots + a_1 x + a_0$, where a_i are constants, is continuous. We state some useful results.

1 $\sin x$, $\cos x$ are continuous everywhere
2 $\tan x$ and $\sec x$ are continuous everywhere except at odd multiples of $\pi/2$
3 $\cot x$ and $\operatorname{cosec} x$ are continuous except at multiples of π
4 $1/(x - a)$ is continuous everywhere except at $x = a$.

You should remember that the idea of continuity is a simple one but that the mathematician has to formalise its definition to be rigorous. For our purposes, the simple-minded approach will suffice.

We conclude by examining an important limit:

$$\lim_{x \to 0} \frac{\sin x}{x}$$

where x is measured in radians.

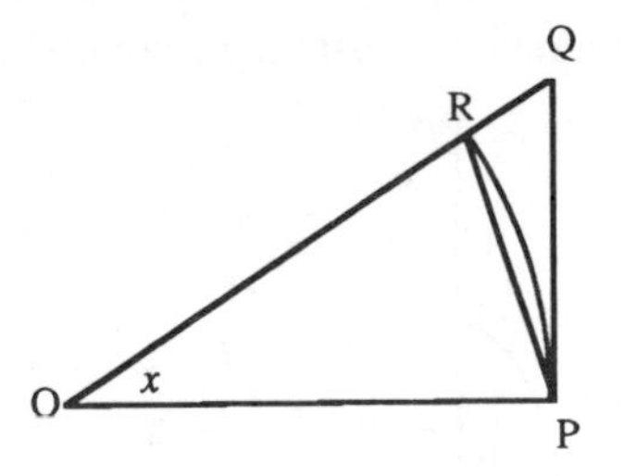

Figure 3.8

First we restrict ourselves to values of x in $0 < x < \pi/2$. Let O be the centre of a circle radius OP = OR and PQ be perpendicular to OP. From Figure 3.8 we can see that ΔOPR < sector OPR < ΔOPQ,

i.e. $\frac{1}{2}\text{OP.OR} \sin x < \frac{1}{2}(\text{OP})^2 x < \frac{1}{2}\text{OP.PQ}$.

But PQ = OP $\tan x$, and on dividing by $\frac{1}{2}(\text{OP})^2$,

$$\sin x < x < \frac{\sin x}{\cos x}$$

so that if we restrict ourselves to acute angles

$$1 < \frac{x}{\sin x} < \frac{1}{\cos x} \tag{3.4}$$

Now $\cos x$ is a continuous function and as $x \downarrow 0$, $\cos x \to \cos 0 = 1$ and so $1/\cos x \to 1$. If in equation (3.4) we let $x \downarrow 0$, then $x/\sin x$ is squeezed between 1 and something which is approaching 1; therefore its limiting value must also be 1. Hence $\lim_{x\downarrow 0} x/\sin x = 1$ and by the rule for quotients $\lim_{x\downarrow 0}(\sin x)/x = 1$. Since $[\sin(-x)]/-x = (\sin x)/x$ we have $\lim_{x\uparrow 0}(\sin x)/x = 1$.

It follows that $\lim_{x\to 0} (\sin x)/x = 1$.

It would be interesting for you to support these ideas by writing a computer program to print out values of $(\sin x)/x$ for $x = 1, 0.99, 0.98, \ldots 0.01$.

3.4 Rates of Change and Differentiation

Consider the experimental data in Figure 3.9(i) in which the results of a tensile test are shown. The graph is effectively a straight line and if we seek the average rate of change of strain with stress we note that this is almost constant over the range of observations. We may read the slope of the graph anywhere, but we note that the strain rises from 0 to 5017×10^{-6} while the stress increases from 0 to 150; the average rate of change is then $(5017 \times 10^{-6})/150$. Such straight line relationships are rare and the relationship between the current passing through a varistor and the potential difference across its ends is an example of a non-linear equation. Experimental data is shown in Table 3.5 and graphed in Figure 3.9(ii).

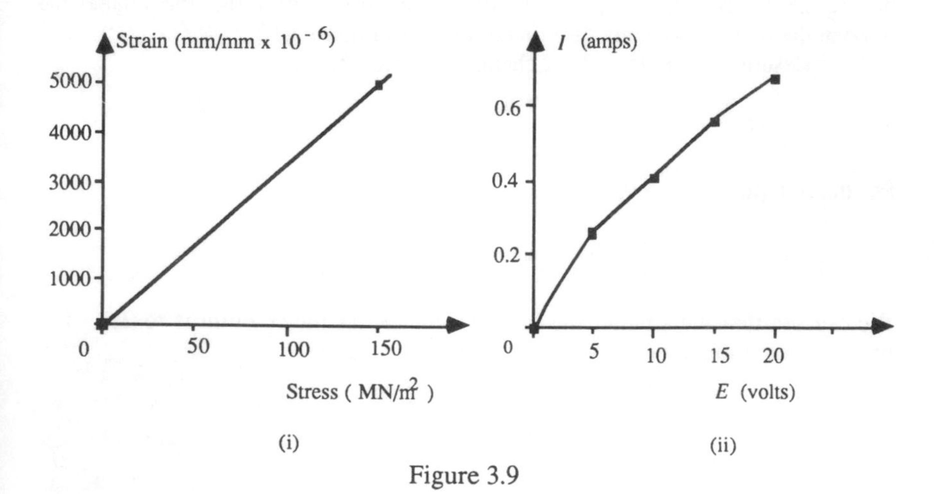

Figure 3.9

Table 3.5

E (volts)	0	5	10	15	20
I (amps)	0	0.26	0.41	0.56	0.68

The average rates of change of I with E are 0.052 in the range 0 to 5 volts, 0.03 in the range 5 to 10 volts, 0.03 in the range 10 to 15 volts and 0.024 in the range 15 to 20 volts.

If we calculate the average rates of change in the intervals 0 to 10 volts, and 10 to 20 volts, the results are 0.041 and 0.027 respectively. The average rate of change over the whole interval is 0.034. Therefore in the region of 8 volts, we get different estimates depending on the interval chosen. It is clear that the smaller the interval, the more accurate the estimate, but the ideal is the (unattainable) local rate of change.

We now consider a second example. The speed of water in a channel was measured at various depths and tabulated as in Table 3.6.

Table 3.6

Depth, cm	0	25	75	150	200
Speed, cm/sec	30	33	37	29	11

The shear stress in the water requires the local rate of change of speed with depth. To estimate this at 75cm we could take the average rate of change over the interval 0 to 150cm, i.e. –0.007 (3 d.p.). But the average from the interval 0 to 75 is 0.093 (3 d.p.) and the average from the interval 75 to 150 is –0.107 (3 d.p.). Which of these figures, if any, is a realistic estimate of the rate of change we seek? We might suppose that the average must be taken over an interval which has 75 near its centre and that the smaller the interval the better. However, if we have to estimate the rate of change at 200cm we are forced to take an interval having 200 as its right-hand end, but we would still suppose that the smaller the interval the better. To generalise matters refer to Figure 3.10.

We measure the average rate of change over the interval

$$[x_0, x_0 + h] \text{ by } \frac{f(x_0 + h) - f(x_0)}{x_0 + h - x_0} \quad \text{i.e.} \quad \frac{f(x_0 + h) - f(x_0)}{h}$$

For the function $f(x) = x^2$ we obtain

$$\frac{(x_0 + h)^2 - x_0^2}{h} = \frac{2x_0h + h^2}{h} = 2x_0 + h$$

We can see that the smaller the interval, the closer the fraction is to $2x_0$. The limitation on accuracy is imposed by the tabular spacing, h.

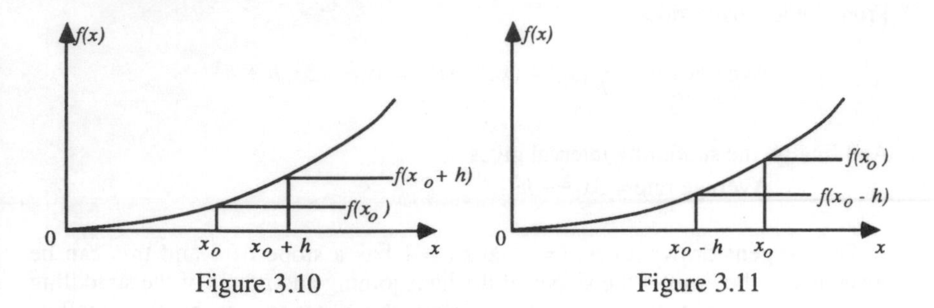

Figure 3.10 Figure 3.11

Refer now to Figure 3.11. Repeating the exercise for the average rate over the interval $[x_0 - h, x_0]$ we obtain the fraction

$$\frac{f(x_0) - f(x_0 - h)}{h} = \frac{2x_0h - h^2}{h} = 2x_0 - h$$

With $x_0 = 1$, we obtain $2 - h$ and, this time, as h gets smaller, the average rate approaches 2 from below.

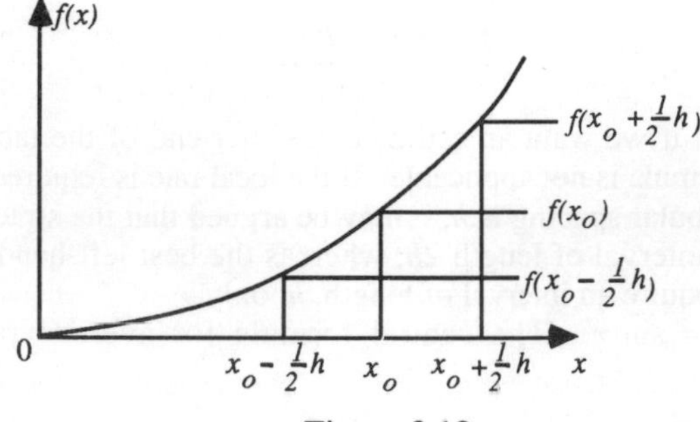

Figure 3.12

Finally we consider Figure 3.12 which represents the case of an interval with x_0 as its mid-point. The average rate is

$$\frac{f(x_0 + \frac{1}{2}h) - f(x_0 - \frac{1}{2}h)}{h} = \frac{2x_0h}{h}$$
$$= 2x_0$$

Hence, at $x_0 = 1$, this rate is 2, irrespective of the length of the interval. This seems a speciality of $f(x) = x^2$ so let us consider $f(x) = x^3$ and repeat the procedure.

From the right-hand side

$$\text{Average rate} = \frac{(x_0 + h)^3 - x_0^3}{h} = 3x_0^2 + 3x_0h + h^2$$

From the left-hand side

$$\text{Average rate} = \frac{1}{h}[x_0^3 - (x_0 - h)^3] = 3x_0^2 - 3x_0h + h^2$$

And finally, the straddling interval gives

$$\text{Average rate} = 3x_0^2 + h^2$$

The tangent drawn to $f(x) = x^3$ at $x_0 = 1$ has a slope of 3 and this can be imagined as the limit of the slopes of the lines joining end points of the straddling interval. Ideally then we should estimate the average rate by the smallest straddling value that the table of values will permit. Note further that if the tabular spacing is h the function values to hand will be $f(x_0 - h)$, $f(x_0)$ and $f(x_0 + h)$ and if we require an estimate of the local rate at x_0 we must use

$$\frac{f(x_0 + h) - f(x_0 - h)}{2h} \qquad (3.5)$$

We shall refer to this as the *central formula*. If we require the estimate half-way between two tabulated points $f(x_0)$ and $f(x_0+h)$, the formula becomes

$$\frac{f(x_0 + h) - f(x_0)}{h}$$

Further notice that if we want an estimate at either end of the tabulated values, then the central formula is not applicable. If the local rate is required at a tabulated point where the tabular spacing is h, it may be argued that the straddling formula would require an interval of length $2h$, whereas the best left-hand or right-hand estimates would require an interval of length, h, only.

Consider $f(x) = \sin x$. The central formula for average rate of change becomes

$$\frac{\sin(x_0 + h) - \sin(x_0 - h)}{2h}$$

$$= \frac{\sin x_0 \cos h + \cos x_0 \sin h - (\sin x_0 \cos h - \cos x_0 \sin h)}{2h}$$

$$= \frac{2 \cos x_0 \sin h}{2h} = \cos x_0 . \frac{\sin h}{h}$$

Then, as $h \to 0$, $\dfrac{\sin h}{h} \to 1$ and so the rate of change becomes $\cos x_0$. (This result applied when h was measured in radians, consequently the same restriction holds for the whole derivation.)

We note that we have proceeded to the limit at a point x_0. As with continuity of a function we define the property at a point and hope it generalises to all points. We assume that a limit exists from either side and that the values are the same. We have a definition: the function $f(x)$ is **differentiable** at $x = x_0$ if

$$\lim_{x \to x_0} \frac{f(x) - f(x_0)}{x - x_0}$$

exists; the value of the limit is called the **derivative** of $f(x)$ at $x = x_0$, and the process of finding it is called **differentiation**. Observe that we have used a notation for limit which allows it to be approached from either side and which assumes the same value for the two approaches.

Further, if the function is differentiable at every point in its domain it is called a **differentiable function**. In such a case, to each x_0 in the domain we may associate a number: the value of the derivative at x_0. This will produce a new function which is called the **derived function**. For $f(x) = x^2$ we have seen that its derivative at x_0 is $2x_0$ and so the derived function, written $f'(x)$, is $2x$. For $f(x) = x^3, f'(x) = 3x^2$ and for $f(x) = \sin x, f'(x) = \cos x$. It would be tedious if the derived function had to be developed from the definition directly but as the underlying concept is that of a limit the usual rules apply, e.g. *the sum of two differentiable functions is differentiable and the value of the derivative of the sum at any point is the sum of the separate derivatives*, that is $\{[f(x) + g(x)]' = f'(x) + g'(x)\}$. We take up these rules further in Chapter 12.

If the function is not differentiable at just one point it ceases to be called differentiable (c.f. continuity); the breakdown can occur if the limit does not exist at all, or if different values are obtained from the two directions.

It is important to note that the derivative $f'(x_0)$ at each point x_0 at which it exists gives the value of the slope of the graph of $f(x)$ at that point.

Snags

We first examine

$$f(x) = \begin{cases} 1, & x \geq 0 \\ -1, & x < 0 \end{cases}$$

which is, by our criterion, a function. Its graph is shown in Figure 3.13.

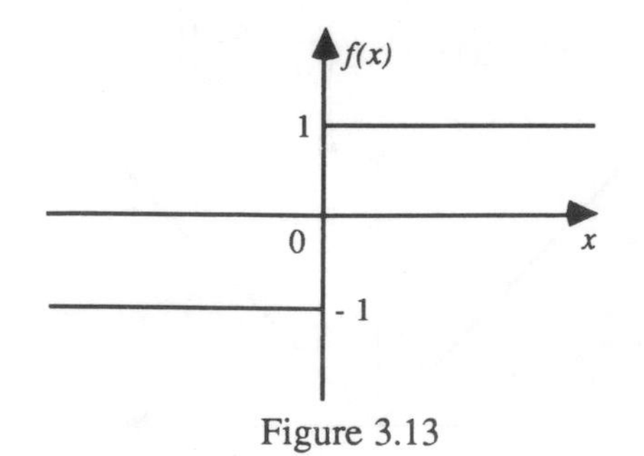

Figure 3.13

The local rate of change at all points $x_0 > 0$ is 0 as it is for $x_0 < 0$. The trouble occurs at $x_0 = 0$ because a *finite* change (2 units) takes place at a *point*, i.e. over an interval of zero length.

The right-hand approximation, $\dfrac{f(x_0)-f(0)}{x_0-0}=\dfrac{1-1}{x_0}=0$

and, as with $x_0 \downarrow 0$, the fraction stays zero. The left-hand approximation,

$$\frac{f(0)-f(x_0)}{0-x_0}=\frac{1-(-1)}{x_0}=\frac{2}{x_0}$$

has no limit as $x_0 \uparrow 0$.

This is an illustration of a general result that *if a function is not continuous at a point it is not differentiable there.* Danger will also exist in practice if a relationship is almost discontinuous.

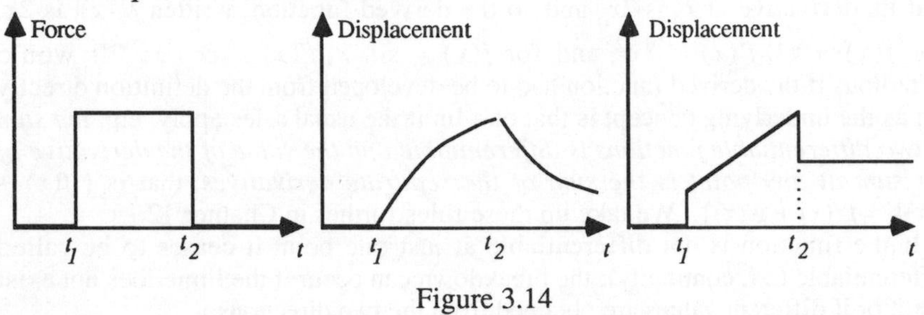

Figure 3.14

Figure 3.14(ii) shows the relationship between displacements of a specimen of asphalt subjected to a compressive force behaving as in Figure 3.14(i). The rapid changes at t_1 and t_2 cause very high rates of change. The force is suddenly applied at $t = t_1$ and removed at $t = t_2$.

An idealised model of the situation would behave as in Figure 3.14(iii). This idealised relationship is discontinuous at $t = t_1$ and $t = t_2$. However, we must not rush into the trap that all functions which are continuous are differentiable. Reflect on $f(x) = |x|$ which is shown in Figure 3.15. This can also be written

$$f(x)=\begin{cases} x, & x \geq 0 \\ -x, & x \leq 0 \end{cases}$$

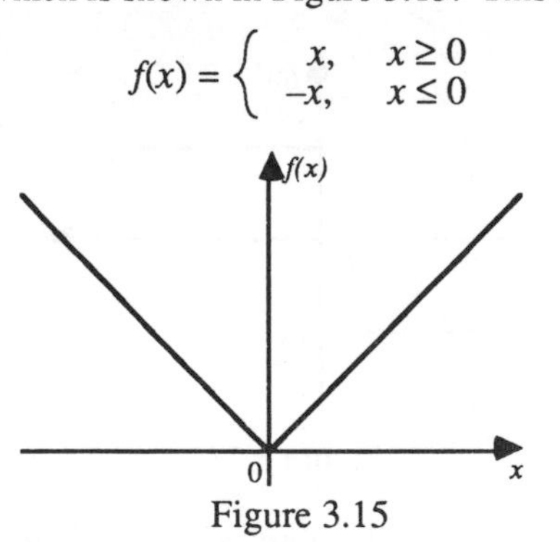

Figure 3.15

It is clear that the derivative of $f(x)$ is 1 for $x > 0$ and -1 for $x < 0$. Trouble is present at $x = 0$; notice that if $x_0 > 0$

$$\frac{f(x_0)-f(0)}{x_0-0}=\frac{x_0}{x_0}=1$$

whilst if $x_0 < 0$,

$$\frac{f(0)-f(x_0)}{0-x_0}=\frac{-(-x_0)}{-x_0}=-1$$

Conflict has occurred and so we conclude that $f(x)$ is not differentiable at $x = 0$ despite being continuous there. (It is true, in general, that a differentiable function is continuous.)

It is worth mentioning that the derived function (excluding $x = 0$ from consideration) of $f(x) = |x|$ is the function

$$f(x)=\begin{cases} 1, & x>0 \\ -1, & x<0 \end{cases}$$

The value of $f(0)$ does not exist.

It is a general principle that the process of differentiation tends to produce a more badly behaved function than the original and numerically is a process to be avoided where possible.

3.5 Elementary Integration

In this section we consider two classes of problem:

(i) that of estimating areas, volumes and other geometrically-based quantities

(ii) that of *undoing* differentiation. For example, given an equation for the velocity of the piston in Section 12.1, determining its displacement at any time, or given the bending moment equation for a loaded beam (Section 12.3), finding its deflected profile.

We shall show later in this section that the underlying process involved in classes (i) and (ii) is fundamentally the same.

Estimation of Areas

It is desired to level a piece of uneven ground by taking earth from the higher portions and placing it in the lower regions: the method of *cut and fill*. This clearly requires the estimation of the volume of soil in an irregularly-shaped piece of land. As a simpler problem let us consider a cross-section through such a piece of land and try to estimate the cross-sectional area. We superimpose suitable axes on the profile, taking the lowest point as being at zero height. The typical measurements which might be taken (each subject to error ±0.005m) are tabulated in Table 3.7 and graphed in Figure 3.16(i).

Table 3.7

x (metres)	0	2	4	6	8	10
y (metres)	0	1.62	5.24	9.31	10.29	10.83

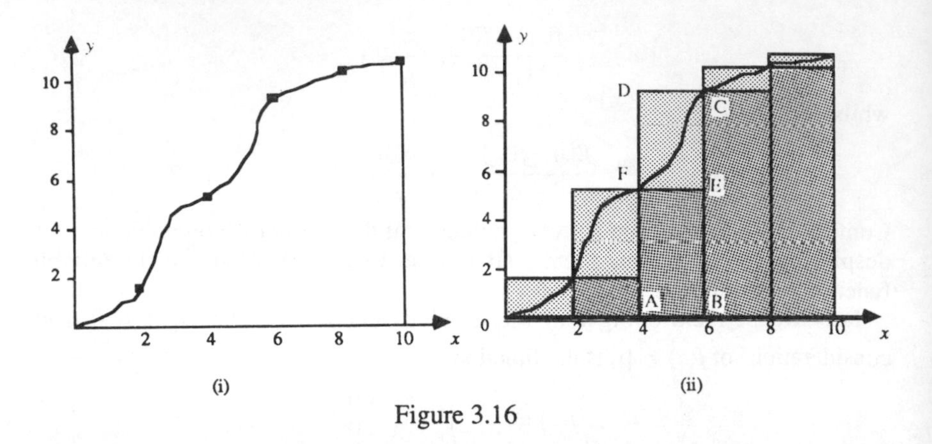

Figure 3.16

It would be of little use to say that the area of the cross-section is between 0 and $108.3m^2$. A sensible thing to do would be to use the measurements and divide the area into strips as shown in Figure 3.16(ii). We assume that the function representing the profile is monotonically increasing.

Since our basic idea of area is in terms of squares or rectangles we can estimate the area of each strip by the rectangles shown. If we choose the larger rectangles (ABCD is a typical one), their sum is an **over-estimate** for the required cross-section area and the smaller rectangles (ABEF is typical) provide an **under-estimate**. The value of the over-estimate is

$$(2 \times 1.62) + (2 \times 5.24) + (2 \times 9.31) + (2 \times 10.29) + (2 \times 10.83) = 74.58m^2$$

and the under-estimate is

$$(2 \times 0) + (2 \times 1.62) + (2 \times 5.24) + (2 \times 9.31) + (2 \times 10.29) = 52.92m^2$$

Notice that the difference is $(2 \times 10.83) - (2 \times 0) = 21.66m^2$, which is the sum of the differences of the rectangles for each strip (and in this case is the area of the largest rectangle: why?) We could quote the results as

$\frac{1}{2}(74.58 + 52.92) \pm \frac{1}{2}21.66 = (63.75 \pm 10.83)m^2$. We have not yet taken into account the inexact measurements which will cause errors.

The worst that could happen is that all errors could reinforce; the maximum error in the product is

$$(0.005)(1.62 + 5.24 + 9.31 + 10.29 + 10.83) + 0.005(2 + 2 + 2 + 2 + 2)$$
$$= 0.235 \quad (3 \text{ d.p.})$$

Therefore we must modify our estimate to $(63.75 \pm 11.07)m^2$.

From now on, we ignore the aspect of inaccurate measurements and assume that the strips we take are narrow enough to ensure that the function is monotonic in each strip. If the function is not monotonic overall, then the error in the result will be much less than the estimate. (We can always split the area into monotonic regions.)

Round-off apart, it would seem that if n is increased indefinitely then the estimate of the area becomes progressively better. It is tempting to think that in the limit as $n \to \infty$ the error will tend to zero and that the estimate will become exact for **continuous** functions. Consider any continuous function, $f(x)$, in some interval $[a, b]$. We seek the shaded area shown in Figure 3.17.

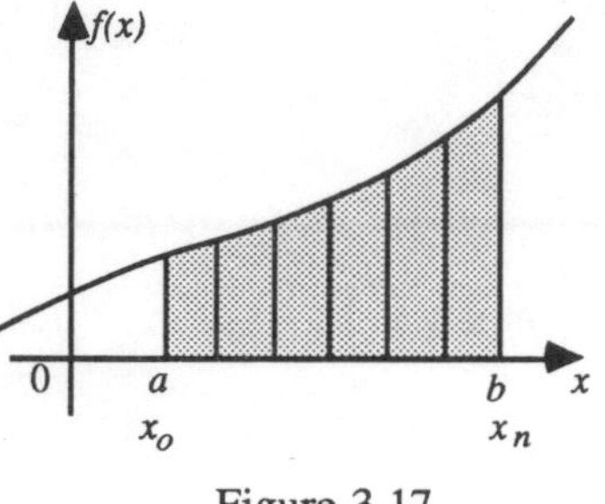

Figure 3.17

We divide the interval into n equal strips of width $h = (b - a)/n$.

We denote by $\overline{S}_n$ the total area of the larger rectangles and $\underline{S}_n$ the total area of the smaller rectangles. Formally,

$$\overline{S}_n = \sum_{i=1}^{n} \bar{f}(x_i)\delta x$$

where $\qquad \bar{f}(x_i)$ is $\max\limits_{x \in [x_{i-1}, x_i]} f(x)$

and $\qquad \delta x = x_i - x_{i-1}$

with $\underline{S}_n$ defined similarly.

Then the true area A is sandwiched between the two values: $\underline{S}_n < A < \overline{S}_n$.

$\underline{S}_n$ is called the *lower sum* and $\overline{S}_n$ the *upper sum*. By what we have seen earlier, if we increase the number of strips the estimates improve, i.e.

$$\underline{S}_n < \underline{S}_{n+1} < \ldots < A < \ldots < \overline{S}_{n+1} < \overline{S}_n$$

The true area A is sandwiched between two sequences $\{\underline{S}_n\}$ and $\{\overline{S}_n\}$.

If each of these sequences tends to a limiting value and the value of the limit is the same in each case, the function is said to be **integrable** *over* the interval $[a, b]$ and the common limit is the **definite integral** of $f(x)$ over $[a, b]$, written

$$\int_a^b f(x)\mathrm{d}x \tag{3.6}$$

and $f(x)$ is called the **integrand**.

We shall assume the result that all continuous functions are integrable over the region of continuity. In the same way that differentiability is a stronger condition than continuity, this latter is stronger than integrability.

Properties of the definite integral
Consequent upon the rules for limits we have the following results:

(i) $$\int_a^b \lambda f(x)\mathrm{d}x = \lambda \int_a^b f(x)\mathrm{d}x \tag{3.7}$$

(ii) $$\int_a^b \{f(x) + g(x)\}\mathrm{d}x = \int_a^b f(x)\mathrm{d}x + \int_a^b g(x)\mathrm{d}x \tag{3.8}$$

(iii) $$\int_a^c f(x)\mathrm{d}x + \int_c^b f(x)\mathrm{d}x = \int_a^b f(x)\mathrm{d}x \tag{3.9}$$

It is assumed that $f(x)$ and $g(x)$ are continuous in the interval $[a, b]$, that λ is any constant (note the case $\lambda = -1$) and that $a < c < b$. If at any part of $[a, b]$ $f(x) < 0$ then the integral of that region is taken as negative.

We have assumed that $a < b$ so far but in the course of manipulating integrals it may happen that $a > b$; we then use the result

(iv) $$\int_b^a f(x)\mathrm{d}x = -\int_a^b f(x)\mathrm{d}x \tag{3.10}$$

If for all x in $[a, b]$, $g(x) \le f(x) \le h(x)$ where the three functions are integrable over $[a, b]$ then

$$\int_a^b g(x)\mathrm{d}x \le \int_a^b f(x)\mathrm{d}x \le \int_a^b h(x)\mathrm{d}x \tag{3.11}$$

In particular, if the maximum and minimum values of $f(x)$ on $[a, b]$ are M and m respectively then

$$m(b-a) \le \int_a^b f(x)\mathrm{d}x \le M(b-a) \tag{3.12}$$

Now $$m \le \frac{1}{b-a}\int_a^b f(x)\mathrm{d}x \le M$$

But if $f(x)$ is continuous on $[a, b]$ then, by the intermediate value theorem, there is a point ξ where $a \le \xi \le b$ such that

$$\frac{1}{b-a}\int_a^b f(x)\mathrm{d}x = f(\xi) \tag{3.13}$$

This is the **First Mean Value Theorem** for integrals and it implies that there is a point of average height (in the sense indicated) in $[a, b]$.

$F(x)$ is called an **indefinite integral** or a **primitive function** of $f(x)$ if

$$\int_a^b f(x)\,\mathrm{d}x = F(b) - F(a) \tag{3.14}$$

We say *an* indefinite integral rather than *the* indefinite integral since there is more than one such integral for a given function $f(x)$.

It would be fairly obvious that an indefinite integral of x^2 is $x^3/3$, but so is $(x^3/3) + 6$, $(x^3/3) - 8.2$ or $(x^3/3) + C$, where C is any constant, since the constant cancels out on subtraction of $F(a)$ from $F(b)$. We have then a whole family of indefinite integrals differing from each other only by a constant, i.e. the graph of one can be shifted into that of another by moving it parallel to the y-axis (Figure 3.18).

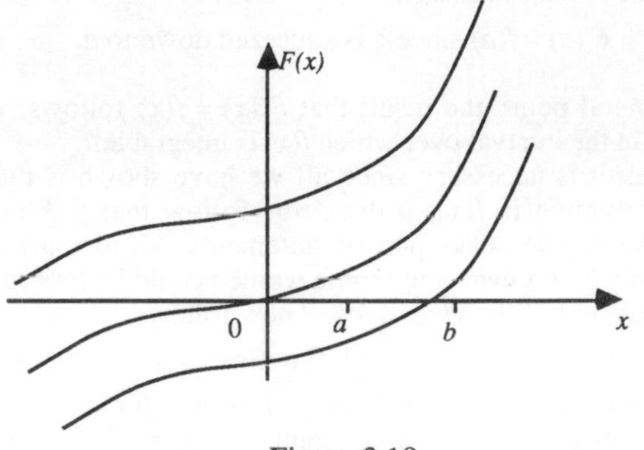

Figure 3.18

Every point whose x-coordinate is between a and b lies on one (and only one) of these curves. It should be fairly clear that there is certainly one curve which crosses the x-axis at $x = a$.

We denote the primitive function which takes a zero value at $x = a$ by $G(x)$; then

$$\int_a^b f(x)\mathrm{d}x = G(b) - G(a) = G(b)$$

If we hold the lower limit fixed and vary b, then

$$\int_a^x f(x)\mathrm{d}x = G(x)$$

We can therefore represent the indefinite integral as a function.

We now show two important results

I. If $F(x)$ is a primitive function of $f(x)$, then $F'(x) = f(x)$. (3.15)

II. Any two primitives of $f(x)$ differ only by a constant. (3.16)

The proof of the first is as follows:

Now
$$F(a+h) - F(a) = \int_a^{a+h} f(x)\mathrm{d}x$$

and so
$$\frac{F(a+h) - F(a)}{h} = \frac{1}{h}\int_a^{a+h} f(x)\mathrm{d}x$$

$$= f(\xi) \text{ where } a \le \xi \le a+h \quad \text{(by 3.13)}$$

If we take the limit of both sides as $h \to 0$ we have

$$F'(a) = f(a) \text{ since } \xi \text{ is squeezed down to } a.$$

Since 'a' is a general point, the result that $F'(x) = f(x)$ follows, if we assume x to be any point in the interval over which $f(x)$ is integrable.

The second result is necessary since all we have shown is that if $F(x)$ is a primitive, its derivative is $f(x)$; it does **not** follow that if $F'(x) = f(x)$ then $F(x)$ is a primitive of $f(x)$. (The pair of statements 'All men are fools', 'X is a fool' does not imply X is a man since some women could be fools also.)

A sketch of the proof of the second result now follows.

Let $F(x)$, $H(x)$ both have derived functions $f(x)$. Then define $Q(x) \equiv F(x) - H(x)$. We have $Q'(x) \equiv F'(x) - H'(x) \equiv f(x) - f(x) = 0$. Since $Q'(x) = 0$ it follows that $Q(x)$ is a constant function. [We are in danger of going round in circles: how do we know there is not a non-constant function whose derivative is identically zero? You must take this result on trust since we are not proving the result rigorously, merely giving a plausible argument. Rigorous proofs can be found in the books listed in the bibliography.]

To sum up, we know that a primitive function exists and any function which differs from it by a constant is also a primitive; hence any function whose derivative is $f(x)$ is also a primitive.

Theorems (3.15) and (3.16) constitute the **Fundamental Theorem of the Calculus**, the link between differentiation and integration. We see that all we need for integration is to find a primitive of $f(x)$ to *undo* the differentiation that led to $f(x)$. Since $x^3/3$ differentiates to x^2, then $x^3/3$ is a primitive of x^2 sometimes called **antiderivative** or more loosely an **indefinite integral**. The process of finding a primitive is called **(indefinite) integration**.

We usually write this simply as

$$\int f(x)\mathrm{d}x.$$

Examples of Indefinite Integration

Here we have $\mathrm{d}y/\mathrm{d}x = f(x)$. This is an example of an ordinary differential equation (discussed in Chapter 21), and our task is to find y as a function of x; e.g. $\mathrm{d}y/\mathrm{d}x = 4x^2$. We might be tempted to write $y = (4/3)x^3$ which does indeed satisfy the given equations, but so also does $y = (4/3)x^3 + 1$; as does $y = (4/3)x^3 + 476.2$ and, in general, $y = (4/3)x^3 + C$, where C is any constant, often called an **arbitrary constant**.

To decide on the particular value of C we need a **boundary condition** such as $y = 3$ when $x = 2$; then

$$3 = \frac{4}{3}(2)^3 + C$$

therefore
$$C = 3 - \frac{32}{3} = \frac{-23}{3}$$

and the particular solution is
$$y = \frac{4}{3}x^3 - \frac{23}{3}$$

[Note that if y is a function of time, $y = y(t)$, and the condition is given for $t = 0$, it is known as an **initial condition**.]

In fact, all ordinary differential equations require the use of integration for their solution. We can make a start at these solutions by writing down a list of standard results from differentiation and working back to form a list of standard integrals.

In the following C is the arbitrary constant.

$$\frac{\mathrm{d}}{\mathrm{d}x}(x^n) = nx^{n-1} \qquad \int x^n\mathrm{d}x = \frac{x^{n+1}}{n+1} + C \quad (n \neq -1)$$

$$\frac{\mathrm{d}}{\mathrm{d}x}(\ln x) = \frac{1}{x} \qquad \int \frac{1}{x}\mathrm{d}x = \ln |x| + C$$

$$\frac{\mathrm{d}}{\mathrm{d}x}[\ln f(x)] = \frac{f'(x)}{f(x)} \qquad \int \frac{f'(x)\mathrm{d}x}{f(x)} = \ln |f(x)| + C$$

$$\frac{\mathrm{d}}{\mathrm{d}x}(\mathrm{e}^{kx}) = k\mathrm{e}^{kx}\ (k\text{ constant}) \qquad \int \mathrm{e}^{kx}\,\mathrm{d}x = \frac{\mathrm{e}^{kx}}{k} + C$$

$$\frac{\mathrm{d}}{\mathrm{d}x}(\sin x) = \cos x \qquad \int \cos x\,\mathrm{d}x = \sin x + C$$

$$\frac{d}{dx}(\cos x) = -\sin x \qquad \int \sin x \, dx = -\cos x + C$$

$$\frac{d}{dx}(\tan x) = \sec^2 x \qquad \int \sec^2 x \, dx = \tan x + C$$

$$\frac{d}{dx}(\cosh x) = \sinh x \qquad \int \sinh x \, dx = \cosh x + C$$

$$\frac{d}{dx}(\sinh x) = \cosh x \qquad \int \cosh x \, dx = \sinh x + C$$

$$\frac{d}{dx}(\sin^{-1} x) = \frac{1}{\sqrt{1-x^2}} \qquad \int \frac{1}{\sqrt{1-x^2}} dx = \sin^{-1} x + C$$

We use some of these entries with the rules of Section 3.5.

Examples

1 $\int 2 \cos x \, dx = 2 \int \cos x \, dx$ by (3.7)

$= 2 \sin x + 2C$

But if C is any constant, $2C$ can be rewritten as C without loss of generality.

2 $\int (x^4 + e^{2x}) dx = \int x^4 \, dx + \int e^{2x} \, dx$ by (3.8)

$$= \frac{x^5}{5} + \frac{e^{2x}}{2} + C$$

Here we have *absorbed* the two arbitrary constants into one.

3 $\int -e^{4x} \, dx = -\int e^{4x} \, dx$ by (3.9)

$$= -\frac{e^{4x}}{4} + C$$

Problems

Introduction

1 The *geometric mean* of two positive numbers, a, b, is defined to be $\sqrt{ab}$. Show that the iterative formula (3.2) for square roots approximates at each step the geometric mean by the arithmetic mean. Show that, in general, the geometric mean of two numbers is less than or equal to their arithmetic mean; under what circumstances is equality achieved? Comment on the inequality by reference to the successive approximations in Tables 3.1 and 3.3.

2 Verify that the formula $x_{n+1} = \frac{1}{2}[x_n + (2/x_n)]$ yields $\sqrt{2}$ as an approximation only if $\sqrt{2}$ is fed in as x_n. In the ideal state, $x_{n+1} = x_n$; show that by writing $x_{n+1} = x_n = x$ the resulting equation has $\sqrt{2}$ as its positive root. Where does the other root come from?

3 Starting from an equivalent geometrical problem, develop two possible *iterative formulae* for finding $\sqrt[3]{A}$ where A is a positive number. Test both on the case $A = 9$ and then use the preferred one to calculate $\sqrt[3]{10}$ to 2 d.p. Write and run a computer program to calculate $\sqrt[3]{10}$ to 5 d.p. and output the result. Arrange for the result to be cubed as a check on its accuracy. Consider the iteration formula $x_{n+1} = 8/x_n^2$; start with guesses 1.8 and 2.2 and apply the formula three times in each case. What conclusion can you draw?

Section 3.1

4 Write, in each case, as a list, the sequence specified by
(i) $u_n = 2n(2n - 1)$ (ii) $u_{n+1} = 3u_n - u_{n-1}$ given $u_1 = 0, u_2 = 2$

5 List the first 4 elements of the sequence $u_n = (-1)^n + 3^n$

6 Graph the first 6 terms of the sequences

(i) $u_n = 3$ (ii) $u_n = \dfrac{2n-1}{n}$ (iii) $u_n = n(-1)^n$ (iv) $u_n = 2 + \dfrac{(-1)^n}{n}$

State whether the sequences are bounded above or below, whether they are increasing, decreasing or oscillatory. If any is convergent, state its limit.

7 Repeat Problem 6 with the sequences:

(i) $u_n = 2n$ (ii) $u_n = \dfrac{2}{n}$ (iii) $u_n = 1 - \dfrac{1}{n}$ (iv) $u_n = 10^{-4} + \dfrac{1}{n}$

(v) $u_n = \dfrac{2n+2}{n-1}$ (vi) $u_n = \left[n + \dfrac{1}{n}\right]^2$ (vii) $u_n = (-n)^n$

8 Find an interval which contains all but the first 10 terms of the sequence $\{1 + (-2)^n/n\}$. Find a similar interval which excludes the first 1000 terms.

9 Let $\{u_n\} = 1\frac{1}{2}, 1\frac{1}{3}, 1\frac{1}{4}, 1\frac{1}{5}, \ldots$ and $\{v_n\} = 1, 1, 1, 1, \ldots$ What is $\{u_n + v_n\}$ and what are the limits of the three sequences? Find the limits of $\{u_n . v_n\}$ and $\{u_n/v_n\}$.

Repeat for $\{v_n\} = 1, 3, 5, 7, \ldots$ and for $\{v_n\} = 3, 2\frac{1}{3}, 2\frac{1}{5}, 2\frac{1}{7}, \ldots$

Section 3.2

10 Consider $S_n = a + (a + d) + \ldots + a + (n - 1)d$: the sum of the first n terms of an arithmetic progression. Write down the terms of the right-hand side in reverse order and obtain an expression for $2S_n$ which can be simplified by taking a common factor from each term. Hence deduce $S_n = \frac{1}{2}n[2a + (n - 1)d]$. Verify this formula by induction.

11 Consider $S_n = a + ar + ar^2 + \ldots + ar^{n-1}$. Write down expressions for rS_n and $(1 - r)S_n$ and hence show $S_n = a(1 - r^n)/(1 - r)$. Verify the formula by induction.

12 Show by induction

(i) $1^2 + 2^2 + 3^2 + \ldots + n^2 = \frac{1}{6}n(n+ 1)(2n + 1)$

(ii) $1^3 + 2^3 + \ldots + n^3 = \frac{1}{4}n^2(n + 1)^2$

(iii) $n^{n-1} \geq n!$

(iv) $$(a + b)^n = \sum_{r=0}^{n} \binom{n}{r} a^{n-r} b^r \equiv a^n + na^{n-1}b + \frac{n(n-1)}{2!} a^{n-2}b^2 + \ldots + \frac{n(n-1)\ldots(n-r+1)}{r!} a^{n-r} b^r + \ldots + nab^{n-1} + b^n$$

Section 3.3

13 Determine $\lim\limits_{x\to 2}(x^2 + 2)$; $\lim\limits_{x\to -1} \dfrac{x^2 + 4x + 3}{x + 1}$; $\lim\limits_{x\to -1} \dfrac{x^2 + 4x + 4}{x + 1}$

14 Examine for discontinuities $\dfrac{1}{x}$, $\dfrac{2x - 4}{x^2 - 3x + 2}$

$f(x) = |x|$ $\quad g(x) = \begin{cases} x, & x \geq 0 \\ 0, & x < 0 \end{cases}$ $\quad h(x) = \begin{cases} x + 1, & x > 1 \\ 2 - x, & x \leq 1 \end{cases}$

15 The function sign(x) takes the value 1 if $x > 0$, -1 if $x < 0$ and sign(0) = 0. Draw its graph.

16 Outline the steps by which you might show that $x^2 \sin x + 2 \cos x$ is a continuous function.

17 Show that $\lim_{x \to 0} \frac{\sin \lambda x}{x} = \lambda$, and by noting that $\cos \lambda x = 1 - 2 \sin^2 \frac{1}{2}\lambda x$ deduce that

$$\lim_{x \to 0} \frac{1 - \cos \lambda x}{x^2} = \frac{\lambda^2}{2}$$

18 $f(x)$ is **of the order of** $g(x)$, written $f(x)$ is $O[g(x)]$, in the region of $x = a$ if $f(x)/g(x)$ remains finite as $x \to a$. Thus since

$$\frac{x^3 + x - 1}{x^3 + 2} = \frac{1 + (1/x^2) - (1/x^3)}{1 + (2/x^3)} \to 1 \text{ as } x \to \infty$$

we say $x^3 + x - 1$ is $O(x^3 + 2)$ as $x \to \infty$.
More usually we compare with powers of x and we could say $x^3 + x - 1$ is $O(x^3)$ as $x \to \infty$. (Note that we choose the smallest power of x for which the statement is true.) Let $O(1)$ denote a constant.
Show $3 \cos x$ is $O(1)$ as $x \to 0$, $x^2 + 0.001x^4$ is $0(x^4)$ as $x \to \infty$, that

$(x - 1)(2x + 3)$ is $O(x^2)$ and hence $\frac{2x^2 + x - 3}{x^2 - 4} \to 2$ as $x \to \infty$.

19 Show $2 + (1/x^2)$ is $O(1)$ for large x and $(6x^3 + 2)/(x^2 - 1)$ is $O(x)$ for large x: investigate the behaviour for large x of

$$\frac{x}{2\sqrt{x^3 + 2}}, \quad \frac{2x^2 \cos x}{x^2 + 1}$$

20 $f(x)$ is of **smaller order** than $g(x)$, written $f(x)$ *is* $o[g(x)]$ as $x \to a$, if in the region of $x = a$, $f(x)/g(x) \to 0$ as $x \to a$.
Usually we take a to be zero. For example, $x^4 + 4$ is $o(x^3)$ as $x \to 0$. Investigate the behaviour of the functions in Problem 19 as $x \to 0$.

Section 3.4

21 As accurately as possible, estimate the rate of change of strain with stress at stresses of 0, 50, 150 MN/m^2 from the values in Problem 39 on page 65.

22 In the study of consolidation of soil the following readings were obtained for the *time factor* T_v against the mean degree of consolidation $\overline{U}$. Establish the rates of change of T_v with $\overline{U}$ for $\overline{U} = 0, 0.3, 0.8$

$\overline{U}$	0	0.1	0.2	0.3	0.4	0.5	0.6	0.7	0.8	0.9	1.0
T_v	0	0.009	0.032	0.072	0.130	0.201	0.294	0.413	0.581	0.865	∞

23 Estimate as accurately as possible the rate of change of density of air with altitude from the table in Problem 40 on page 65. Make estimations at sea-level, 5000 metres and 10 000 metres.

24 The angle turned through by a shaft was measured and the results tabulated as follows:

t (seconds)	0.0	0.2	0.4	0.6	0.8	1.0
θ (radians)	−0.003	0.057	0.148	0.282	0.470	0.706

Estimate the angular velocity $\dot{\theta}$ when $t = 0, 0.4, 0.8, 1.0$

25 Consider the following which is an extract from a table of Naperian logarithms.

x	3.6	3.8	3.9	4.0	4.1	4.2	4.4
$\ln x$	1.2809	1.3350	1.3610	1.3863	1.4110	1.4351	1.4816

From the interval [3.6, 4.4] the average rate of change is 0.251 (3 d.p.)
From the interval [3.8, 4.2] the average rate of change is 0.250 (3 d.p.)
From the interval [3.9, 4.1] the average rate of change is 0.250 (3 d.p.)
Repeat for similar entries around 2.0, 3.0, 5.0. What do you conjecture about the derived function of $\ln x$?

26 Repeat Problem 25 using a table of reciprocals and suggest a suitable derived function for $f(x) = 1/x$.

27 Tabulate $f(x) = x^3$ for $x = -\frac{1}{4}, -\frac{1}{2}, -\frac{3}{4}, -1, -1\frac{1}{4}, -1\frac{1}{2}, -1\frac{3}{4}, -2$. Estimate $f'(0)$ by the average rates of change over the intervals $[-\frac{1}{4}, \frac{1}{4}]$, $[-\frac{1}{2}, \frac{1}{2}]$, ..., $[-2, 2]$. Compare with the values from intervals $[0, \frac{1}{4}]$, $[0, \frac{1}{2}]$, ... $[0, 2]$. Comment.

28 Tabulate $f(x) = \sinh x$ for $x = 1.1(0.1)2.0$. Produce successive approximations to $f'(1.5)$ and compare with $\cosh 1.5$. Repeat for $f(x) = \cosh x$ and compare with $\sinh 1.5$.

29 A number a is part of the data for an experiment. In the course of some calculations, it is necessary to form a^2. If the original number was subject to a small data error ε the value of a^2 is in error by $2a\varepsilon$ (to first order). The function of squaring has thus multiplied the error by a **scale factor** of $2a$. Suppose we wish to solve $x^3 = 9$. We know that the true value is nearer to 2 than 3. Calculate s, the *scale factor* for the function $f(x) = x^3$ at $x = 2$, hoping that this is a good approximation to the average scale factor over the interval $[2, \sqrt[3]{9}]$. {Note that $f(x) = x^3$ maps $[2, \sqrt[3]{9}]$ into [8, 9].}

If the true value of $\sqrt[3]{9}$ is $2 + \varepsilon$, then the 'error' ε is scaled into an error of $s\varepsilon$;

but $s\varepsilon = 9 - 8 = 1$, hence we can find ε approximately and obtain a better estimate of $\sqrt[3]{9}$.

Starting with this new value, repeat the process to obtain a second revised estimate of $\sqrt[3]{9}$.

Repeat, if necessary, until $\sqrt[3]{9}$ is obtained correct to 2 d.p. (compare with tables).

30 Repeat Problem 29 to find $\sqrt{17}$ to 2 d.p.

31 Obtain from the definition the derived function of $f(x) = ax^3 + bx^2 + cx + d$, where a, b, c, d are constants.

32 Obtain the derived function of each of the following: $\cos x$, $\sin 2x$, $\cos 3x$

33 We seek the derived function of $1/x$. Obtain the expression $-(1/a)(a + h)^{-1}$ as an approximation to the derivative at $x = a$. Write $(a + h)^{-1}$ as

$$\frac{1}{a}\left[1 + \frac{h}{a}\right]^{-1}$$

and expand by the binomial theorem. Hence deduce the derived function as $-1/x^2$. By a similar procedure deduce that the derived function of $1/x^2$ is $-2/x^3$.

34 Draw a flow chart and write a computer program to calculate the approximation $(-1/a)(a + h)^{-1}$ for $h = 1.0(0.05)0.5$, and $a = 2.0$. Compare with the known derivative.

35 Create a table of sines and compute the best estimate of the derivative at the points 0.96(0.02)1.04. These should be close to values of the cosines of these angles. If we start with these values and repeat the process we shall obtain 3 values which are approximations to $-\sin 0.98$, $-\sin 1.0$ and $-\sin 1.02$ (the derivative of $\cos x$ is $-\sin x$). Repeating the process twice more should leave us with one value, approximately $\sin 1.0$ (the derivative of $-\sin x$ is $-\cos x$ and the derivative of $-\cos x$ is $\sin x$).
What do you observe about the accuracy of the two estimates of cos 1.0 and the two estimates of sin 1.0?

36 It is clear that the methods for estimating $f'(a)$ replace the tangent by straight line chords. It is claimed that the central formula gives the same estimate as would a parabola through the points D, E, F (see Figure 3.19). To investigate this claim, let the parabola be of the form $g(x) = \alpha(x - x_2)(x - x_1) + \beta(x - x_1) + \gamma$.
Find $g'(x)$ and hence $g'(x_2)$. Since $g(x) = f(x)$ at x_1, x_2, x_3, determine α, β and γ, and hence show that $g'(x_2)$ is the central formula. Why should that make this formula more accurate than the others?

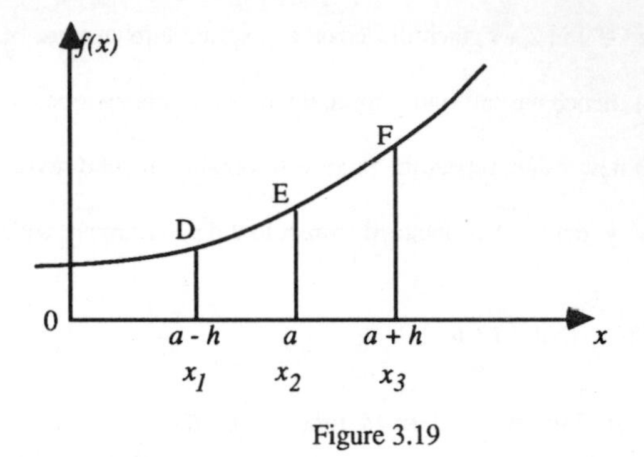

Figure 3.19

Section 3.5

37 Develop over- and under-estimates for the area under the curve for $y = 80 - x^3$ in the range $x = 0$ to $x = 4$ with 4 strips, and 8 strips. If possible, write and run a computer program to cope with 16, 32 and 64 strips. What conclusions can you draw?

38 Repeat for $y = 400 - x^4$ over the range $x = 0$ to $x = 4$. How many strips would you need for accuracy of 1 d.p., 2 d.p.? Comment.

39 Assume that $\displaystyle\int_a^b \frac{1}{(x-2)^2}\,dx = \frac{-1}{b-2} - \frac{-1}{a-2}$

If $a = 1$, $b = 3$, use the estimation method with 3, 5, 7, 9 strips and compare results. What is your conclusion?

40 From properties given in (3.7) to (3.10), prove

(i) $\displaystyle\int_a^b f(x)\,dx = \int_a^b f(a+b-x)\,dx$ (Hint: put $a + b - x = t$)

(ii) $\displaystyle\int_{-a}^{a} f(x)\,dx = \begin{cases} 0 & \text{if } f \text{ is odd} \\ 2\displaystyle\int_0^a f(x)\,dx & \text{if } f \text{ is even} \end{cases}$

(iii) $\displaystyle\int_0^{2a} f(x)\,dx = \int_0^a [f(x) + f(2a-x)]\,dx$

41 Find bounds on the following:

$$\int_0^{\pi/2} \cos x\,dx, \quad \int_0^{1/2} (1-x^2)\,dx, \quad \int_0^3 \frac{1}{1+x^2}\,dx$$

42 Find the point ξ in property (3.13) for

$$\int_0^1 x\,dx, \qquad \int_0^1 x^2\,dx, \qquad \int_0^4 x^3\,dx$$

43 Verify the following indefinite integrals by means of differentiation, C being any constant.

(i) $$\int \frac{dx}{a^2 - x^2} = \frac{1}{2a}\ln\left\{\frac{a+x}{a-x}\right\} + C$$

(ii) $$\int \frac{dx}{\sqrt{a^2 - x^2}} = \sin^{-1}\frac{x}{a} + C$$

(iii) $$\int \sin x \cos x\,dx = -\frac{1}{4}\cos 2x + C$$

(iv) $$\int \frac{3x\,dx}{x^2 + 2} = \frac{3}{2}\ln(x^2 + 2) + C$$

(v) $$\int (e^{-3x} + e^{2x})dx = -\frac{1}{3}e^{-3x} + \frac{1}{2}e^{2x} + C$$

(vi) $$\int \cos x\sqrt{\sin x}\,dx = \frac{2}{3}(\sin x)^{3/2} + C$$

(vii) $$\int \frac{dx}{x\sqrt{x^2 - 4}} = \frac{1}{2}\cos^{-1}\frac{2}{x} + C$$

(viii) $$\int x^2(x^3 + 1)^3 dx = \frac{1}{12}(x^3 + 1)^4 + C$$

(ix) $$\int \frac{x^2\,dx}{\sqrt{1 - x^2}} = -\frac{x}{2}\sqrt{1 - x^2} + \frac{1}{2}\sin^{-1}x + C$$

(x) $$\int x \sin x\,dx = \sin x - x\cos x + C$$

(xi) $$\int x\,e^{3x}dx = \frac{e^{3x}}{3}\left[x - \frac{1}{3}\right] + C$$

(xii) $$\int x^2 \sinh x\,dx = (x^2 + 2)\cosh x - 2x \sinh x + C$$

4

INTRODUCTION TO STATISTICAL METHODS

4.1 The Role of Statistics

We have seen already in Section 1.4, that the repeated taking of a measurement produced a set of values which could be grouped and presented as a histogram. The presentation of the data thus collected and the drawing of conclusions as to the actual value of the quantity being measured fall in the realm of statistics, and in this section we look briefly at the application of the statistical method which will be pursued in later chapters of this book.

The first stage is the **collection of data**. Data is collected on a vast scale in every sphere of human activity: the national census, local and national surveys, university and college records - everywhere information is collected and analysed. The information can be of two kinds: *qualitative* (e.g. colour of people's eyes) or *quantitative* (e.g. heights of people) with a 'grey area' of information which is semi-quantitative (e.g. television programme ratings). It is necessary to take care in collecting the information. If a market survey is carried out, the organisers must ensure that the people asked are representative of the population as a whole in respect of the questions asked, and the choice and wording of the questions itself needs careful scrutiny. In this context, the statistician uses the term **population** to mean the *set of all possible measurements*. This may mean all the people who might be concerned in some issue, for example, the electorate or the house-owners in an urban area, or it might mean the diameters of all bolts manufactured by a given process or the lengths of life of all light bulbs made by a particular factory. If we wish to make some estimate of the behaviour of a population it may be necessary or desirable (on grounds of expense, or difficulty otherwise) to take measurements from a **sample**, that is, a *subset of the population*. Opinion polls take samples of the electorate since it would be pointless, and costly, to interview the whole population; inspectors in factories cannot always test non-destructively their product and so they test a sample of the output. The choice of size of a sample is discussed later in this section and the choice of the members of the sample is taken up in Section 4.6.

Having collected the data, the next stage is to present the information in such a way that patterns and trends are discernible. This constitutes **descriptive statistics**. The information can be summarised pictorially (see the next section) and this is sometimes sufficient by itself; more usually, graphs and charts serve as adjuncts to numerical summaries. From quantitative data, certain numbers can be calculated which help describe various aspects of the data; such numbers are known as **sample statistics**. An average may be calculated to give an idea of the *middle* value of a set of readings (see Section 4.4) and a measure of the spread of values about this average is often useful (see Section 4.5); for example the controller of a process designed to produce 1 kg bags of some commodity will hope that the average weight in each bag will be over 1 kg but will not want the weights to vary too much since this will produce some seriously underweight bags (bad for sales) and some seriously overweight ones (bad for profits).

The final, and most important, aspect of statistical work is the drawing of inferences from the data: **inferential statistics**. It must be borne in mind that *one only uses statistics in this aspect if there is insufficient data to draw firm conclusions*. Consequently, with *every* conclusion that is stated there must be a *risk* quoted: the risk that the conclusion is wrong. Sometimes the conclusion may call for further investigation; sometimes the evidence of the data is so strong that action can be taken on that evidence alone; but one must emphasise that *no amount of evidence is absolutely conclusive* unless the sample taken consists of the whole population. Let us consider an example of this sampling procedure to see how it works.

Items are produced by a manufacturing process at the rate of 200 per hour. In order to monitor the product, a sample is to be taken at specified times and some statistic measured from the sample which will give a clue as to the behaviour of the population (of items manufactured). How often should the sample be taken? How big should the sample be? The question of timing revolves around how continuous a monitoring is desired. That of sample size is somewhat easier to handle. Clearly the larger the sample, the more representative it is likely to be; a sample of 1 could be quite atypical, but the collective information from a sample of 20 is likely to be near to the population behaviour.

In order to make any statement about the quality of the product we need to compare the sample statistic with an expected value of the corresponding **population parameter** which is calculated from a *model* (in the case of observational errors, this is the **Gaussian**, or **normal** model). There will, in general, be a discrepancy between the observed and expected values of this parameter and it is the role of statistics to aid us in deciding how much of this discrepancy can be attributed to the fluctuations between samples (i.e. we may possibly have got a sample containing the highest values - this *can* happen however good our sample technique) and how much can be attributed to the underlying difference in the values. The basis of the method is to assume there is **no** discrepancy and calculate the chance of the observed discrepancy occurring. It is then up to the control inspector to decide whether the chance is acceptable in the light of his experience. Should he decide that the chance is too small to be acceptable (and in advance he sets a *critical level* below which he decides the chance is unacceptably low) then he could be rejecting when he should be accepting, if the sample was atypical of the population; in this case, he can improve the situation by choosing a low critical level. If he does this, he increases the danger of accepting that which should be rejected. We see here that the statistical method is not a clear-cut *yes* or *no* approach. We can merely state risks of conclusions being wrong, and leave the decision to experienced workers in the appropriate field. On the other hand, we can be told in advance the acceptable risk, and work within that.

In this chapter, we look at descriptive statistics and the idea of a random sample, in Chapter 19 we examine models of chance and in Chapter 20 we study sampling and its applications.

4.2 Graphical Representation of Data

We now look at some methods of displaying data which can be classified in categories rather than grouping according to ranges of values. The first method is the **pie chart** which shows the relationship of each constituent to the total: if a category takes a third of the whole then the area of the corresponding sector is one third of that of the circle or 'pie'. In Figure 4.1 is shown the distribution of seats in the House of Commons, according to party, after the General Election of 1987.

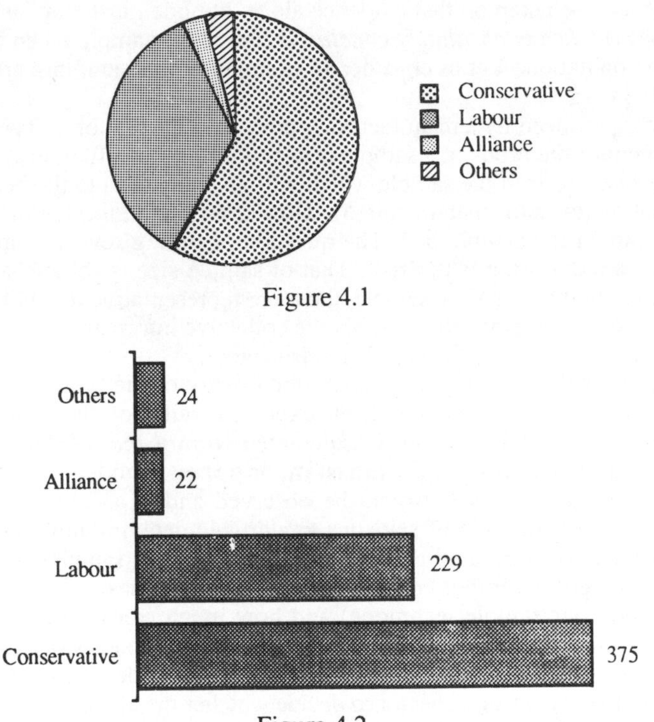

Figure 4.1

Figure 4.2

A second method is by means of the **bar chart** where the bars may be vertical or horizontal (if there are few categories and a wide range of values to be displayed). In Figure 4.2 the same data as for the pie chart is displayed, the widths of each bar are equal and the length is proportional to frequency.

Sometimes the bar chart may be modified to accommodate subdivisions. If the MPs in each party are split according to sex we obtain Figure 4.3.

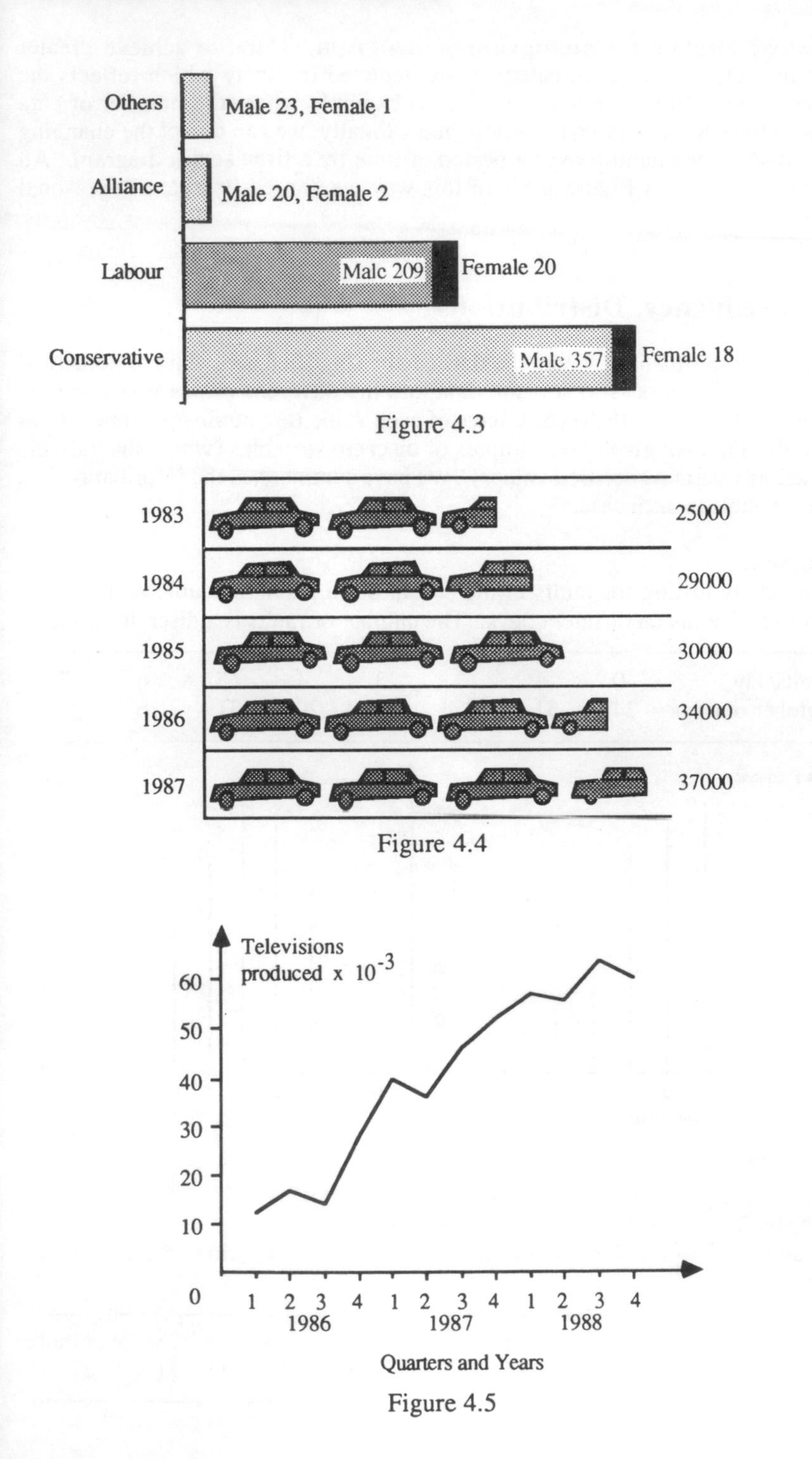

Figure 4.3

Figure 4.4

Figure 4.5

Next we mention the **pictogram** or **ideogram**. Here, to achieve greater visual impact, units in each category are depicted in a way which reflects the subject matter. In Figure 4.4 we show an example where the number of cars produced by a certain company is depicted. Finally, we can depict the changing situation of some quantity over a period of time by a **time series** diagram. An example is shown in Figure 4.5. In this way we can easily see any seasonal trends.

4.3 Frequency Distributions

We now turn to situations where the data can be classified according to numerical values. The first task is to sort the data into the different values which can be assumed and calculate the **frequencies** of each value (the number of times it has occurred). First we give two examples of **discrete** variables (where the variable assumes only certain specified values). We have summarised the information in a frequency table in each case.

Example 1
A year's daily testing for faults in the sheath of an insulated cable revealed the number of of faults on different days. The number of faults is a discrete variable.

Faults/day	0	1	2	3	4	5	6	7
Number of days	24	51	87	82	70	31	16	4

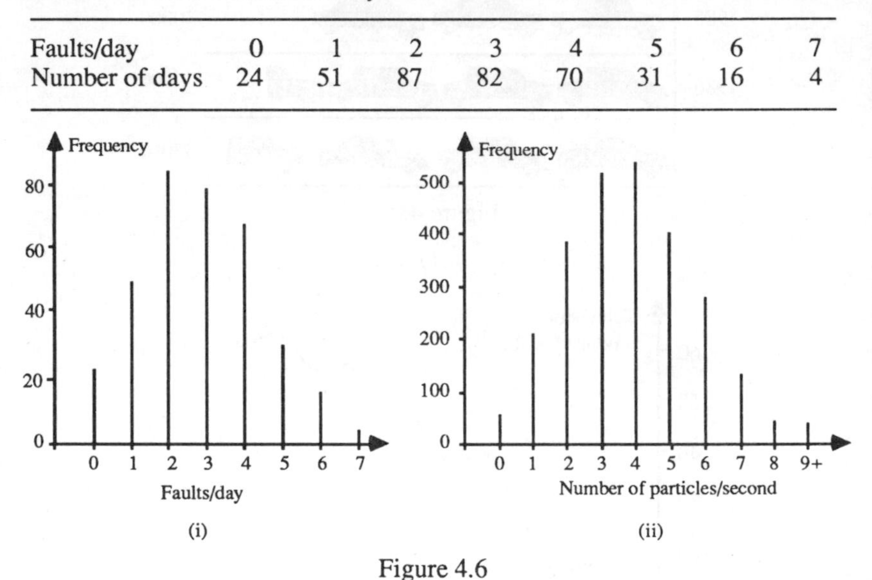

Figure 4.6

Example 2
The number of α-particles emitted in each second over a period of one hour was counted.

Number in a second	0	1	2	3	4	5	6	7	8	9 or more
Frequency	53	207	385	523	538	402	279	133	41	37

We can display each set of data by a line diagram as illustrated in Figures 4.6 (i) and (ii).

In certain cases the variable is **continuous**; for example, the time taken to perform a given task; since the variable can only be measured to a given accuracy, however, this gives it the appearance of being discrete. Since it would be senseless to record every observation even to the accuracy possible, it is customary to **group** the data and count the frequency of observations falling in each group; these are sometimes loosely called the frequencies of each group.

For both discrete and continuous variables the definitions of each group and their associated frequencies form the **frequency distribution**.

Example 3
A sample of 80 ball-bearings was taken from a machine's output and the diameters measured to 0.001 mm. The results are shown in Table 4.1.

Table 4.1

4.350	4.366	4.374	4.381	4.375	4.382	4.358	4.366	4.396	4.351
4.366	4.374	4.375	4.360	4.367	4.376	4.381	4.396	4.351	4.366
4.383	4.361	4.369	4.377	4.374	4.382	4.354	4.366	4.383	4.377
4.369	4.361	4.374	4.382	4.354	4.366	4.369	4.361	4.378	4.383
4.384	4.378	4.370	4.362	4.362	4.365	4.373	4.388	4.365	4.373
4.380	4.390	4.363	4.372	4.373	4.387	4.384	4.362	4.372	4.378
4.379	4.372	4.379	4.364	4.384	4.389	4.373	4.362	4.385	4.364
4.379	4.373	4.372	4.379	4.384	4.385	4.385	4.365	4.372	4.379

In Table 4.2 we first group the data and extend the end points of the groups to make them match at the end points even though the extensions contain no observations.

Table 4.2

GROUP			RELATIVE FREQUENCY
end points included	*extended end points*		*2 d.p.*
4.350 – 4.354	4.3495 – 4.3545	5	.06
4.355 – 4.359	4.3545 – 4.3595	1	.01
4.360 – 4.364	4.3595 – 4.3645	11	.14
4.365 – 4.369	4.3645 – 4.3695	13	.16
4.370 – 4.374	4.3695 – 4.3745	15	.19
4.375 – 4.379	4.3745 – 4.3795	13	.16
4.380 – 4.384	4.3795 – 4.3845	13	.16
4.385 – 4.389	4.3845 – 4.3895	6	.07(5)
4.390 – 4.394	4.3895 – 4.3945	1	.01
4.395 – 4.399	4.3945 – 4.3995	2	.02(5)
		80	

We can represent the data graphically by a **histogram** in Figure 4.7 – where the centre of each bar is the mid-point of each group; notice that the variable concerned is continuous, even though the measurements quoted give an impression of its being discrete.

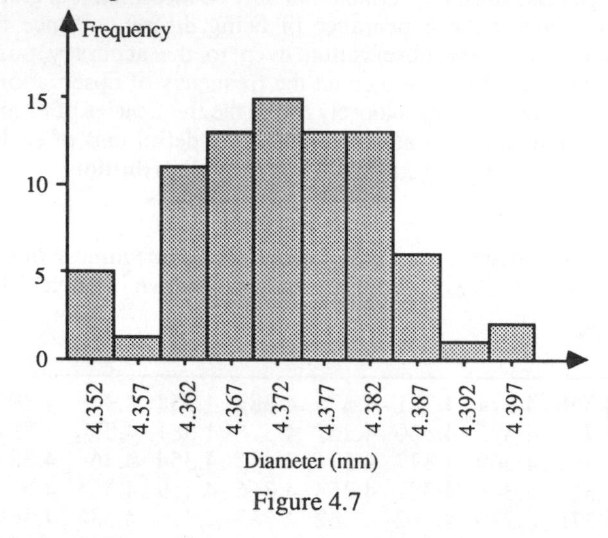

Figure 4.7

You should notice that in grouping the data, some detail has been sacrificed in order to achieve some clear picture of the behaviour of the data.

Two other terms can usefully be introduced here. First, the concept of **relative frequency**: this is simply the frequency of a certain variable or of a group divided by the overall total (see Table 4.2). Notice that the total relative frequency (which should be 1) is 0.99, due to rounding error. It is customary to denote the value taken by a discrete variable by x_i (this may refer to the group as we shall see in the next section) and its associated frequency by f_i. If there are N observations in all, the relative frequencies are f_i/N. Since

$$\sum_{i=1}^{n} f_i \equiv f_1 + f_2 + f_3 + \ldots.. f_n = N, \qquad \sum_{i=1}^{n} \frac{f_i}{N} = \frac{N}{N} = 1$$

The second concept is that of **cumulative frequency distribution**. For this we accumulate frequencies so that we find the frequency of all values up to and including the one under consideration. In Example 3 the cumulative frequencies are 5, 6, 17, 30, 45, 58, 71, 77, 78, 80. These can be plotted on a **cumulative frequency diagram** or **ogive** and this is shown in Figure 4.8.

The choice of the number of groups and the group widths is partly a subjective matter, but can be helped by a guide-line: it is usually best to take between 6 and 15 groups, depending on the amount of data.

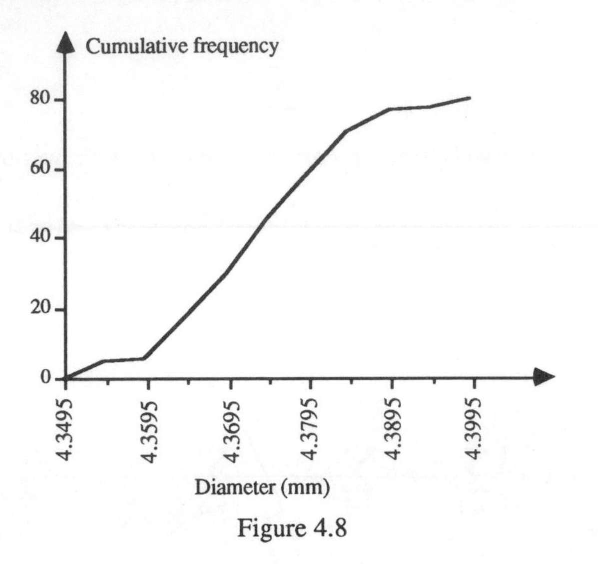

Figure 4.8

4.4 Measures of Central Tendency

Given a frequency table we wish to provide an indication of the 'middle' or 'average' value. In this section we examine three possible measures.

Arithmetic Mean

This is the 'layman's' average. The sum of all the values is divided by the total number of values. Suppose we had five observations x_1, x_2, x_3, x_4 and x_5; the **arithmetic mean** (often shortened to *mean*) is

$$\tfrac{1}{5}\,(x_1 + x_2 + x_3 + x_4 + x_5) \quad \text{or} \quad \frac{1}{5}\sum_{i=1}^{5} x_i$$

In general, the mean of n observations $x_1, x_2, \ldots\ldots, x_n$ is

$$\frac{1}{n}\sum_{i=1}^{n} x_i$$

and is written $\bar{x}$ if the observations are regarded as a sample or μ if regarded as a population. The sample mean $\bar{x}$ can be used as an estimate of the population mean μ. For this chapter we shall use $\bar{x}$. We have so far said nothing about the observations being distinct. Suppose the five observations were x_1, x_1, x_1, x_2 and x_2 ($x_1 \neq x_2$); then the above principles would apply, but we can use the fact that if the frequency of x_1 is f_1 (= 3 in this case) and that of x_2 is f_2 (= 2), then

$$x_1 + x_1 + x_1 + x_2 + x_2 = f_1 x_1 + f_2 x_2$$

and we have

$$\bar{x} = \frac{1}{5}\sum_{r=1}^{2} f_r x_r$$

We say $\bar{x}$ is the **weighted mean** of x_1 and x_2 with weights $f_1/5$ and $f_2/5$.

In general, if n observations fall into distinct values $x_1, x_2, \ldots\ldots, x_s$ with frequencies $f_1, f_2, \ldots\ldots, f_s$ then, since

$$\sum_{r=1}^{s} f_r = n$$

we have

$$\bar{x} = \frac{1}{n}\sum_{r=1}^{s} f_r x_r = \sum_{r=1}^{s} f_r x_r \Big/ \sum_{r=1}^{s} f_r \tag{4.1}$$

Note that

$$\sum_{i=1}^{n} (x_i - \bar{x}) = \sum_{i=1}^{n} x_i - \sum_{i=1}^{n} \bar{x} = n\,\bar{x} - \bar{x}\,n = 0$$

and this property, that the sum of discrepancies from the mean is zero, may be used as a check in calculations of the mean. (c.f. the physical analogue of centre of mass.) The mean is best used to describe, for example, the typical tensile strength of a material, the typical diameter of washers produced by a machine or the typical life of a component.

Mode

The **mode** is that value which occurs most frequently. If the data is grouped we speak of the **modal group**.

In the three examples of Section 4.3 the modes of the first two are respectively 2 faults/day and 4 α-particles a second, and the modal group of the third is 4.370 – 4.374.

If a distribution has one mode it is called **unimodal**; if two, it is **bimodal**. In considering the donations to a retirement present, or the wages of members in a firm it may be more relevant to ask which value occurs most frequently rather than a mean value. In the latter case the use of the mean would take directors' salaries into account and one such salary might be equivalent to the wages of twenty factory hands. The mode is also useful for non-quantitative variables.

Median

First, the values of the variable are arranged in increasing or decreasing order; if the number of values is odd, the median is the middle value, whereas if the number of values is even it is the mean of the two middle values. For example, the median of [3,3,4,12,21] is 4 and of [3,3,4,12,21,25] is 8 and this value is [4 + 12]/2.

The median divides the values in the population into two equal, or almost equal, parts. Half the values are less than or equal to the median. Note that the value 25 above could have been altered to, say, 135 without changing the median, whereas the mean would have altered. In the example on ball-bearings there are 80 observations, both the 40th and 41st occurring in the group 4.370 – 4.374 and this is the **median group**.

Choice of the appropriate measure

Where the choice of appropriate measure of central tendency is not obvious the following comments may help. The mean is most readily understood and lends itself more readily to further statistical analysis; further, it takes all the values into account. It has the disadvantage of being affected by extreme values and where these occur the median may be a more useful measure. If these measures cannot readily be applied the mode can come into play.

Relationships between the measures

If the distribution is symmetrical then the mean, median and mode coincide; if the distribution is not symmetrical but not too heavily *skewed*, then we have the approximate relationship (mean – mode) ≅ 3(mean – median) – see Figure 4.9.

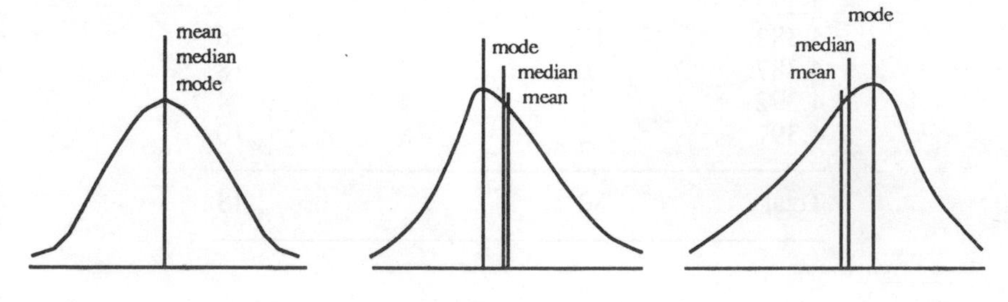

Figure 4.9

Calculation of the mean

We return to the ball-bearing example and calculate the mean of the values which are grouped.

We could tabulate f_i and $f_i x_i$, compute their totals and then calculate $\bar{x}$ as

$$\sum_{i=1}^{n} f_i x_i \Big/ \sum_{i=1}^{n} f_i$$

As a check we could compute

$$\sum_{i=1}^{n} f_i(x_i - \bar{x})$$

and this should be zero; in practice it may not be exactly zero, because of the grouping of the values.

However, with grouped data, it is always advisable to **code** or transform the variable x_i into a new variable u_i. This transformation has the effect of simplifying the arithmetic and comprises two stages. First we guess or assume a mean somewhere in the middle of the values of x_i; in the following calculations resulting in Table 4.3 we guess the mean of the ball-bearing diameters to be 4.372, one of the group central values. In the second stage we divide (x_i – 4.372) by the group width 0.005 to obtain the new variable

$$u_i = \frac{(x_i - 4.372)}{0.005}$$

Note that some of the values of u_i are negative.

Table 4.3

x_i	f_i	u_i	$f_i u_i$
4.352	5	–4	–20
4.357	1	–3	–3
4.362	11	–2	–22
4.367	13	–1	–13
4.372	15	0	0
4.377	13	1	13
4.382	13	2	26
4.387	6	3	18
4.392	1	4	4
4.397	2	5	10
Total	80		13

The values of u_i are also integral, for example if $x_i = 4.352$ then

$$u_i = \frac{4.352 - 4.372}{0.005} = -\frac{0.020}{0.005} = -4$$

We have chosen a new origin for the measurements and scaled their values. The calculations now proceed as in Table 4.3.

Now

$$\bar{u} = \frac{1}{n}\sum f_i u_i = 13/80 = 0.1625$$

But

$$\bar{u} = \frac{1}{n}\left[\sum f_i (x_i - 4.372)/0.005\right]$$

$$= \frac{1}{0.005}\frac{1}{n}\left[\sum f_i x_i - n\ 4.372\right]$$

$$= \frac{1}{0.005}[\ \bar{x} - 4.372]$$

therefore

$$\bar{x} = 0.005\ \bar{u}\ + 4.372 = 4.3728 \quad (4\text{ d.p.})$$

The same result for $\bar{x}$ has been obtained here with much simpler arithmetic and this second approach would often give a more accurate result. In general if the **coded variable** $u = (x - a)/b$ then

$$\bar{u} = (\ \bar{x} - a)/b \quad \text{or} \quad \bar{x} = b\ \bar{u} + a \tag{4.2}$$

4.5 Measures of Variability

A manufacturer of bolts would hope for two things with regard to their diameter: a mean close to the desired value and not too wide a spread of values about that mean. In this section we look at measures of spread about the mean.

Range
This is simply the quantity (largest value – smallest value). For Example 1 of Section 4.3 the range is (7 – 0) = 7. For Example 3 it is (4.396 – 4.350) = 0.046.

Inter-Quartile Range
We can find three values which divide the ordered values of the variable into four groups each having the same frequency. The central value is then the median and the remaining two are the **first** and **third quartiles**. The **inter-quartile range** is (3rd quartile – 1st quartile). It has the advantage over the range in that it measures a more typical spread since it does not rely on extreme values only. It should be noted that the inter-quartile range contains *half* the total frequency.

Mean Deviation
To take all the values into account we measure

$$\frac{1}{n}\sum_{i=1}^{n} |\, x_i - \bar{x}\,| \quad \text{or} \quad \frac{1}{n}\sum_{r=1}^{s} f_r\, |\, x_r - \bar{x}\,|$$

The modulus sign causes difficulty in manipulation, however.

Variance
To overcome this difficulty we use the **variance**, which is the mean squared deviation about the mean.

Variance is written var(x) or $V(x)$

For a population of size n

$$V(x) = \frac{1}{n}\sum_{r=1}^{s} f_r\,(x_r - \mu)^2$$

This has the disadvantage that extremes are emphasised. For example, if we take the numbers 4, +1, 0, –1, the mean is 1, and

$$\begin{aligned} V(x) &= \tfrac{1}{4}[(4-1)^2 + (1-1)^2 + (0-1)^2 + (-1-1)^2] \\ &= \tfrac{1}{4}[9 + 0 + 1 + 4] \\ &= 3.5 \end{aligned}$$

Now if the reading of 4 had been in error it should have been ignored, yet it completely dominates the variance. We must emphasise that the variance is a particularly useful measure of variability.

Warning. There is a distinction to be drawn here between **population variance** σ^2 and **sample variance** s^2.

When we are dealing with n observations regarded as a sample we calculate

$$s^2 = \frac{1}{n-1}\sum_{i=1}^{n} f_i(x_i - \bar{x})^2$$

The reason for division by $(n - 1)$ is connected with the estimation of the population variance σ^2 from the sample variance s^2.

For large samples, division by $(n - 1)$ may be approximated by division by n. You should, however, make it quite clear whether you are treating the numbers in an example as a population or as a sample and find the appropriate variance.

Standard Deviation

In order to achieve a parity between the units of spread and the units of the values we take standard deviation = s.d. (or σ or s) as $\sqrt{\text{variance}}$. If the numbers of the previous example were lengths in cm we would have

$$\sigma = \sqrt{3.5} = 1.87 \text{ cm} \quad (3 \text{ s.f.})$$

It should be clear that the bigger σ, the larger the spread of values. Its physical analogy is radius of gyration.

We deduce a useful equivalent form for $V(x)$ which saves many calculations and sometimes yields a more accurate result.

We deal first with a population.

Now

$$\begin{aligned} nV(x) &= \sum_{i=1}^{n} (x_i - \mu)^2 \\ &= \sum_{i=1}^{n} (x_i^2 - 2\mu x_i + \mu^2) \end{aligned}$$

$$= \sum_{i=1}^{n} x_i^2 - \sum_{i=1}^{n} 2\mu x_i + \sum_{i=1}^{n} \mu^2$$

$$= \sum_{i=1}^{n} x_i^2 - 2\mu \sum_{i=1}^{n} x_i + n\mu^2$$

$$= \sum_{i=1}^{n} x_i^2 - 2\mu . n\mu + n\mu^2$$

$$= \sum_{i=1}^{n} x_i^2 - n\mu^2$$

therefore

$$V(x) = \left[\frac{1}{n} \sum_{i=1}^{n} x_i^2 \right] - \mu^2 \tag{4.3}$$

(Mean of the square – square of the mean)

The result also holds for grouped data.
For a sample, the result is

$$V(x) = \frac{1}{n-1} \sum_{i=1}^{n} x_i^2 - \frac{n}{n-1} \bar{x}^2 \tag{4.4}$$

Example
We return to Example 3 of Section 4.3. We shall treat the observations as a sample and use the formula for variance given at the bottom of page 115. Then we have

$$V(x) = \frac{1}{79} \cdot 8.225 \times 10^{-3} = 1.041 \times 10^{-4}$$

and so

$$\sigma = 1.020 \times 10^{-2} = 0.010 \quad \text{(3 d.p.)}$$

(Notice we quote σ to as many d.p. as the original observations.)
If we use (4.4) we have

$$V(x) = \frac{1}{79}(1529.727) - \left[\frac{349.825}{80} \right]^2 \cdot \frac{80}{79}$$

Working with a pocket calculator with 8 digit capacity gives

$$V(x) = 19.363632 - 19.363532 = 0.000100$$

The rearrangement

$$\frac{1529.727 - (349.825)^2/80}{79}$$

gives

$$V(x) = 0.0001041 \quad \text{(more precision)}$$

which yields

$$\sigma = 0.0102 \quad \text{(4 d.p.)}$$

It is always advisable to use coded variables. It can be shown that for

$$u = \left[\frac{x-a}{b}\right]$$

the population variance

$$V(u) = \frac{1}{n}\sum_{i=1}^{n}(u_i - \bar{u})^2 = \frac{1}{n}\sum_{i=1}^{n}\left[\frac{x_i - a}{b} - \frac{\bar{x}-a}{b}\right]^2$$

$$= \frac{1}{n}\sum_{i=1}^{n}\left[\frac{x_i - \bar{x}}{b}\right]^2$$

$$= \frac{1}{b^2}\frac{1}{n}\sum_{i=1}^{n}(x_i - \bar{x})^2$$

$$= \frac{1}{b^2}V(x)$$

The same result holds for sample variance [despite division by $(n-1)$ instead of n] and we summarise to say that if $u = (x-a)/b$

$$V(u) = \frac{1}{b^2}V(x) \tag{4.5}$$

We shall find $V(u)$ as

$$\frac{1}{n-1}\sum_{i=1}^{n} f_i u_i^2 - \bar{u}^2\left[\frac{n}{n-1}\right]$$

We tabulate the necessary information in Table 4.4; the coding is as before ($a = 4.372$, $b = 0.005$).

Table 4.4

Group central value x_i	f_i	u_i	f_iu_i	$f_iu_i^2$
4.352	5	−4	−20	80
4.357	1	−3	−3	9
4.362	11	−2	−22	44
4.367	13	−1	−13	13
4.372	15	0	0	0
4.377	13	1	13	13
4.382	13	2	26	52
4.387	6	3	18	54
4.392	1	4	4	16
4.397	2	5	10	50
Totals	80		13	331

Then

$$V(u) = \frac{1}{79} \times 331 - \left[\frac{13}{80}\right]^2 \cdot \frac{80}{79}$$

$$= \frac{331 - 169/80}{79}$$

$$= 4.1631 \quad \text{(4 d.p.)}$$

Hence $V(x) = b^2V(u) = (0.005)^2 \times 4.1631$ and so

$$s = 0.005 \times \sqrt{4.1631} = 0.0102 \quad \text{(4 d.p.)}$$

We can see that we have retained accuracy by a modified approach; relationship (4.3) can be used if a calculating aid with good precision is available, but in general coding the variables is preferable.

4.6 Random Samples

We have already remarked that a sample is a subset of a population and that to make some statement about the population it is necessary to infer from the behaviour of a sample. We hope that the estimate from a sample is not too far from the truth and that the average of the estimate from all possible samples is the value we seek. It is little use choosing a sample of the electorate in a particular constituency by choosing names from a telephone directory since this sample would reflect the better-paid members of the electorate.

We seek a technique of sampling which will give each member of the population an equal chance of selection. Put another way, if we fix on a sample size N from a given population there are many such possible samples; we aim to ensure that each possible sample has an equal chance of being chosen. We call a sample so chosen a **simple random sample**. If we wished to select 16 people from a population of 160, we could write each of the 160 names on a separate slip of paper, mix the slips in a container and draw 16 of them. This is a somewhat tedious process.

Alternatively we can make use of a table of random numbers. Random numbers are the digits 0, 1, 2, ..., 9 arranged, with repetitions, in no particular order. These digits can be produced by a computer in such a way that the sequence repeats only after several millions. A sample of such digits is shown in Table 4.5 (see page 126).

We now give an illustration of how to use the table.

Example

Explain how to choose a random sample of 4 from populations of 10, 7, 25, 155 respectively.

The first task is always to label the members of the population serially.

(i) For a population of size 10 we use single digits 0, 1, ..., 9 (0 corresponds to 10) and take a sequence of digits from the table starting anywhere and continue until we have met four different ones. Suppose we take the block 17512, we ignore the second 1 and obtain 1, 7, 5, 2; the items thus labelled will be the ones chosen.

(ii) For a population of size 7, the easiest way is to ignore the digits 0, 8 and 9 and repeat the above. If the sequence is 73570, we have only obtained 7, 3 and 5 and we take the next block 86860 and pick up 6 to complete the sample.

(iii) For a population of size 25 we could take digits in pairs and ignore all combinations 00 and above 26 (there are 100 combinations: 00 to 99); this would waste a possible three-quarters of the digits and, if we wanted a larger sample, could prove time-consuming. A useful alternative is to divide 25 into 100 and note we have an exact result of 4. Then, in taking pairs of digits, divide by 25 and take the remainder: the result is the member of the population chosen. This second approach can be illustrated by the block 91 30 76 42 20; remainders after division by 25 are: 16, 5, 1, 17, 20 and we take the members 16, 5, 1 and 17.

(iv) For a population of size 155 we need triples of digits. The second method above now becomes: divide 1000 by 155 to get 6+; now $6 \times 155 = 930$ and so the combinations 000 and above 930 are ignored. The triples of random numbers are divided by 155 and the remainders taken. Thus the blocks 235 954 329 410 479 give 80, ignore, 19, 100 and 14 as the sample.

Finally we mention that it may be desirable to use a different form of sampling. If it is wished to divide a population according to strata, depending on some feature, e.g. geographical area or income group, we take simple random samples of appropriate size from each stratum to give an overall **stratified random sample**. On the other hand, practical sampling may be too difficult, in which case, possibly with the aid of a digital computer, samples may be theoretically chosen by a technique known as **simulated sampling**.

Problems

Section 4.2

1 Draw pie charts and bar charts to illustrate the following data:

(a)

Continent	*Area (millions of square miles)*
Africa	11.7
America	16.3
Antarctica	5.3
Asia	17.0
Australia	3.0
Europe	3.9

(b)

Religion	*Followers (millions)*
Buddhism	269
Confucianism	280
Eastern Orthodox	220
Hinduism	300
Muslim	400
Protestant	210
Roman Catholic	420
Others	160

2 Draw a pictogram to illustrate the following data for the weight of fish caught by a certain fishing boat. (Take one drawing of a fish to represent 5000 tonnes.)

Year	*Weight of fish caught (tonnes)*
1965	20 000
1966	25 700
1967	25 500
1968	30 000
1969	35 000
1970	33 000
1971	30 000
1972	40 000

3 Draw a time series graph for the data of Problem 2.

4 Draw a time series graph for the following data obtained from a weather station.

Year		1965	1966	1967	1968	1969	1970
	1	100.0	90.0	105.0	95.0	107.5	117.5
Rainfall	2	50.0	65.0	52.5	47.5	35.0	32.5
for quarters	3	37.5	35.0	57.5	37.5	42.5	30.0
(in cm)	4	75.0	100.0	155.0	77.5	92.5	82.5

Section 4.3

5 Separate units of cotton yarn were tested for the count, which is a measure of the fineness of the yarn (length per unit weight), giving the following results:

35.0	39.6	38.1	37.4	37.4	38.4	36.5	38.4	38.4	37.2
36.6	35.1	39.6	38.2	38.2	37.8	37.3	36.2	38.9	37.9
37.4	36.6	35.1	35.4	35.4	37.0	38.0	37.2	37.3	38.4
38.1	37.4	36.6	36.6	36.6	36.2	39.0	37.8	36.2	38.5
37.5	37.5	38.3	38.3	36.9	36.2	36.3	37.9	38.5	38.6
38.2	36.0	36.1	37.7	36.1	36.5	37.2	37.2	36.4	36.5
35.8	36.7	36.9	36.9	37.8	37.3	37.3	37.9	37.9	37.2
36.6	37.6	37.7	36.1	38.3	38.8	38.7	36.4	37.3	37.9

Draw up a frequency table taking groups 35 – 35.4, 35.5 – 35.9, ..., 39.5 – 39.9.
Draw a histogram, using the true end points 34.95 – 35.45, 35.45 – 35.95, ..., 39.45 – 39.95.
Repeat for the grouping 35.0 to 35.9, 36.0 to 36.9, etc. Has this grouping lost any information or altered the histogram? Comment on your results.

6 For each of the tables of data given below
(i) determine the greatest and least observation
(ii) decide on a suitable number of groups
(iii) form a frequency table
(iv) draw a histogram and a cumulative frequency diagram.

(a) Lengths of screws (to nearest 1/1000 cm)

3.239	2.671	3.114	3.078	2.830	2.700	3.314	3.055
3.060	3.155	3.211	2.965	2.956	2.611	2.933	2.942
3.130	2.792	3.353	3.149	2.996	2.964	2.765	2.800
2.713	3.125	2.654	2.773	3.142	3.070	2.931	3.202
2.921	3.198	3.054	3.197	3.153	3.201	3.474	3.394
3.111	3.084	2.755	2.516	3.194	2.907	2.905	3.061
3.186	3.228	3.026	3.427	2.821	2.865	2.852	3.018
2.850	3.166	3.439	2.819	2.716	2.861	2.754	3.231
2.751	3.221	2.631	3.020	3.258	3.333	3.111	3.371
2.800	2.836	2.723	3.051	3.242	3.245	3.063	3.211

(b) Masses of bags of sugar (kilogrammes)

1.48	1.54	1.48	1.51	1.44	1.77	1.68	1.53	1.33	1.51
1.43	1.59	1.55	1.51	1.63	1.36	1.52	1.21	1.75	1.68
1.52	1.66	1.25	1.49	1.55	1.40	1.58	1.66	1.32	1.35
1.50	1.34	1.61	1.42	1.64	1.67	1.50	1.67	1.52	1.30
1.31	1.45	1.62	1.35	1.61	1.63	1.43	1.45	1.51	1.62

Section 4.4

7 (i) The number of members and means of 3 distributions are as follows:

Number of members	280	350	170
Mean	45	54	50

Find the mean of the combined distribution.

(ii) A candidate's examination marks in three subjects were 86%, 47% and 63%; if these subjects were weighted in the ratio 1 : 3 : 2 find his weighted mean percentage mark.

8 The numbers of rejects from articles produced by a machine in 36 consecutive periods were:

3, 1, 2, 4, 0, 1, 2, 2, 3, 2, 1, 2, 5, 0, 1, 2, 1, 3,
2, 3, 4, 0, 1, 1, 2, 2, 3, 0, 3, 1, 2, 0, 1, 4, 3, 2.

Draw up a frequency table showing the numbers of periods with 0, 1, 2, 3, 4 and 5 rejects.
Draw up a cumulative frequency table.
Find (i) the median
(ii) the mode
and (iii) the arithmetic mean.

9 For each of the problems on Section 4.3 find the mean, mode and median.

10 The following table gives the distribution for the bursting strength of 225 samples of vinyl coated nylon

Bursting strength (lb)	*Number of samples which burst in stated range*
61 – 70	2
71 – 80	8
81 – 90	17
91 – 100	28
101 – 110	36
111 – 120	40
121 – 130	34
131 – 140	29
141 – 150	19
151 – 160	10
161 – 170	2

Draw a histogram to represent the distribution.
Taking 115.5 as a new origin and the group width as a unit, calculate the arithmetic mean.
Construct a cumulative frequency table.

11 The mean of 13 numbers is 11 and the mean of 23 other numbers is 7. Find the mean of the 36 numbers taken together.

Section 4.5

12 The following table shows 80 measurements of the iron solution index of tin plate samples, used to estimate the corrosion resistance of tin plated steel.

Interval	0-.20	.21-.40	.41-.60	.61-.80	.81-1.00	1.01-1.30	1.31-1.60	1.61-2.00
Frequency	2	2	12	26	15	14	7	2

(a) Construct a suitable frequency diagram to illustrate the data.
(b) Draw a cumulative frequency diagram and estimate the percentage of the indices which exceed 1.25.
(c) Estimate the median and the inter-quartile range for the index.

13 During a piece of work study the normalised time for a certain operation was recorded on 46 occasions and these values were subgrouped as follows:

t (secs)	9.0	9.5	10.0	10.5	11.0	11.5	12.0	12.5	13.0
f	2	5	9	11	8	6	3	1	1

By introducing a suitable coded variable for the time, calculate the mean value, the mean deviation and the standard deviation.

14 The breaking load in N of 200 specimens of a certain material was measured and the results arranged in the following grouped frequency distribution.

Breaking load in N (mid-interval values)	*Number of Specimens*
982.5	3
987.5	10
992.5	27
997.5	62
1002.5	56
1007.5	24
1012.5	12
1017.5	5
1022.5	1

Determine the mean and standard deviation of this distribution. State the modal group and the median group. Find also the mean deviation and inter-quartile range.

15 The breaking strengths in N/m^2 of 100 metal specimens are given in the following table, with a class interval of 0.2 N/m^2.

Centre of interval (N/m^2)	32.0	32.2	32.4	32.6	32.8	33.0	33.2	33.4	33.6	33.8
No. of specimens	1	4	11	13	19	26	12	9	3	2

Using the method of taking an assumed mean, calculate the mean and standard deviation. Define the median of a frequency distribution and using a graphical method (or otherwise) calculate it for the above distribution. (EC)

16 (a) You are given a set of fixed values $x_1, x_2, x_3, \ldots, x_n$ and a separate value A which can vary. The function S is defined as $S = \sum_{i=1}^{n} (x_i - A)^2$.

Find dS/dA and show that the least value of S occurs when

$$A = \frac{x_1 + x_2 + x_3 + \ldots + x_n}{n}$$

(b) The mean value of a given set $\{x_i\}$ of 30 numerical values is 6.213 and the standard deviation is 0.264. Find the values of

$$\sum_{i=1}^{i=30} x_i \quad \text{and} \quad \sum_{i=1}^{i=30} x_i^2$$

Two additional values become available: 6.425, 6.346. Find the mean value and the standard deviation of the whole set of 32 values. (EC)

17 If $\bar{x}$ is the arithmetic mean of n values of x, show that

$$\sum (x - \bar{x})^2 = \sum x^2 - \frac{(\Sigma x)^2}{n}$$

In a traffic count, the number of vehicles passing a check point in a two-minute interval was recorded for 63 such intervals giving the following results:

No. of vehicles passing in 2 min. interval	0	1	2	3	4	5	6	7	8
No. of intervals	1	11	12	11	10	9	4	2	3

Find the mean and standard deviation of this distribution.

18 100 readings recorded in a certain experiment had a mean value of 0.913 and standard deviation of 0.156. If the following further 10 readings are then obtained, what would be the mean and standard deviation of the complete set of results?
0.875 0.898 0.912 0.913 0.917 0.920 0.923 0.924 0.928 0.937

19 The following frequency table records the diameter of 200 grains of moulding sand (in appropriate units)

Diameter	8	10	12	14	16	18	20
Frequency	25	40	67	35	23	7	3

State the modal group and median group and draw a histogram for the distribution. Choose a suitable assumed mean and hence calculate the mean and standard deviation, σ, of the distribution. The standard deviation calculated from the original measurements (before grouping) was 2.88. Why is there a discrepancy between this and your value for σ? (EC)

20 The times taken by 35 people to perform a particular calculation were as follows (given in minutes and seconds)

2.48	2.00	2.34	1.44	1.58	2.27	2.25
2.24	1.57	1.41	2.32	3.13	3.39	1.41
1.11	2.00	2.50	3.27	1.04	1.25	1.55
4.24	1.26	2.32	2.43	3.11	2.14	3.52
1.36	1.47	3.21	3.04	2.30	1.43	1.05

Group the data into a frequency table, draw a histogram and calculate the sample mean, sample standard deviation, median group, modal group, mean deviation and inter-quartile range. Justify your choice of grouping and of coding the variable.

21 A quantity sometimes calculated is the **coefficient of variation** which is defined to be the ratio of the standard deviation to the mean. Find the coefficient of variation for Problems 13, 14, 15 and 17.

22 Grouping data leads to inaccuracies in the calculation of mean and standard deviation. **Sheppard's correction** partially compensates for the error in the standard deviation. It may be stated:

$$\text{corrected variance} = \text{calculated variance} - \frac{1}{12}(\text{group width})^2$$

Apply this to the data of Problem 6(b), and check its effect by calculating the variance of the original 50 observations and the variance of the data when grouped.

23 Choose a random sample of size 8 from the data (a) of Problem 6. Repeat for sample sizes 6, 14.

24 Repeat Problem 23 for the data (b) of Problem 6.

25 Arrange the numbers 1 to 8 in random order; then arrange the numbers 1 to 16 in random order.

Table 4.5
A thousand random numbers

17512	73570	86860	91307	64220	23595	43294	10479
76841	09058	01305	60495	13421	71688	04120	80918
11052	32848	14058	88001	94641	70167	40104	35255
34311	42935	36458	04201	71573	37722	58698	46115
54641	26072	04705	27077	34834	14491	53407	22248
45749	23937	57052	53045	02583	30298	59306	50144
63243	59906	74883	31145	20350	47412	35309	02287
21051	96604	33444	52746	11929	77340	95053	84498
95766	17077	96760	96507	57473	22620	30675	76773
38800	29448	56232	61173	91526	86160	97255	79578
68736	08852	78657	91294	84045	76828	49909	80634
13607	36975	76285	95314	19047	07958	77110	95166
07671	34747	67528	90777	61004	04959	83438	57088
40736	06846	73412	59487	02897	09274	46440	13225
51282	73638	10025	54990	29162	38279	13792	09391
26501	43588	81906	69802	68634	16651	11125	77249
01557	13374	29465	27171	15987	59264	37949	03338
32264	32023	37468	04735	13468	21383	36507	77813
46169	29950	93968	48856	37585	86315	05745	76432
69995	63424	18966	12377	17669	07622	74232	23604
62212	31191	77839	72307	55308	54250	39561	25338
16413	40686	34371	52671	28770	36396	06696	28522
81999	34374	23814	38043	22069	95938	70678	64351
76141	69928	81966	03285	65663	10845	68747	29214
59349	34499	61480	12054	61602	09961	09195	28880

5

SETS AND PROBABILITY

At the outset, the theory of sets may seem to have very little practical significance. We are going to study this theory for two reasons. Firstly it is basic to many branches of mathematics and is particularly useful in the initial ideas of functions. Secondly the ideas of set algebra are fundamental in probability theory which we shall meet in Section 5.2. Set theory has practical applications in the theory of switching circuits and Boolean Algebra.

5.1 Sets and Venn Diagrams

Definition of a Set

A set comprises a collection of objects or elements, each of these objects possessing some property which is chosen to define membership of the set. The definition of membership can either be by a list of members or by a rule which unambiguously determines membership. Here are some examples:

(i) the numbers 1, 2, 6, 9
(ii) the consonants of the English alphabet
(iii) all integers
(iv) all living people
(v) all Shakespeare plays
(vi) all people attending classes at a Technical College in the year 1988
(vii) 1, 3, 5, 7, 9, ...
(viii) the solutions of the equation $x^2 - 3x + 2 = 0$
(ix) 6
(x) all right-angled triangles.

Several points arise from these examples. We see that (i), (ix) and, to a certain extent, (vii) define membership by a list of members whereas the others prescribe rules for membership. (vii) could have been defined alternatively by the rule *all odd positive integers*.

Examples (i), (ii), (iv), (v), (vi), (viii) and (ix) are all **finite sets** having a finite number of members, whereas (iii), (vii) and (x) are **infinite sets** containing an infinite number of members.

Notation
Members of a set are denoted by lower case letters $a, b, c, \ldots x, \ldots$ and the sets themselves by capital letters $A, B, C, \ldots$.

We write $A = \{1, 3, 5, \ldots\}$ which says *the set A consists of the odd positive integers*. Any odd positive integer belongs to the set A which we write as $x \in A$. If an object does not belong to a set A we write $x \notin A$.

Example 1

If we take I as the set of all integers, then $3 \in I$ but $\pi \notin I$.

It is often useful to have a notation indicating the rule by which the set is defined. Suppose we wish to say: If I is the set of positive integers then the set A consists of those numbers x belonging to I such that $x^2 + 1$ is less than or equal to 17. We would write:

$$\text{If } I = \{1, 2, 3, \ldots\}, \quad A = \{x \in I : x^2 + 1 \leq 17\}$$

(The *colon* : stands for *such that.*)

Of course in this case we can list the members of the set and write $A = \{1, 2, 3, 4\}$. Notice $5 \notin A$ since $5^2 + 1 > 17$, and similarly for any larger integer.

Example 2

(viii) could be written as

$$R = \{\text{Real numbers}\}, \quad A = \{x \in R : x^2 - 3x + 2 = 0\}$$

We don't need always to have the first bracket, and indeed x does not have to be a number. For example (v) could be written

$$A = \{x : x \text{ is a Shakespeare play}\}$$

Try writing the other examples in one of these forms.

The empty or null set

It is possible for a set to contain no members. It is then called the **empty** or **null set** and is denoted by $\emptyset$. For example, if $R_+ = \{\text{Positive real numbers}\}$, $A = \{x \in R_+ : x^2 + 3x + 2 = 0\}$, then $A = \emptyset$ since the roots of $x^2 + 3x + 2 = 0$ are -1 and -2 and no x satisfies both requirements.

Equality of two sets

Two sets A and B are said to be equal if A and B contain precisely the same elements. We then write $A = B$. The following conditions must both be satisfied:

(i) if $x \in A$, then $x \in B$ $(x \in A \Rightarrow x \in B)$ ($\Rightarrow$ is read *implies that*)

(ii) if $x \in B$, then $x \in A$ $(x \in B \Rightarrow x \in A)$

Example 3

$$I = \{1, 2, 3, \ldots\}, \quad A = \{x \in I : x^2 + 1 \leq 17\}$$
$$B = \{x \in I : 1 \leq x \leq 4\}$$

Both A and B have only the members $\{1, 2, 3, 4\}$. Therefore $A = B$.

Subsets
If we have two sets A and B which are such that all the elements of A are also members of B (but not necessarily vice-versa) then we say A is a subset of B and write $A \subset B$.

For example: If A = {cardinals} and B = {integers}, then $A \subset B$.
If C = {rationals} and R = {real numbers}, then $C \subset R$.

Note that if $A \subset B$ and $B \subset C$ then $A \subset C$. You should convince yourself of this by considering a few examples. If equality of two such sets is possible we write $B \subseteq C$. Examine examples (i) – (x) and write a list showing which set is a subset of one of the others.

Disjoint sets
If two sets have no elements in common we say that they are **disjoint**; examples (ii) and (iii) are disjoint as are (v) and (vi). This leads us to the idea of the common membership of two sets called the intersection of A and B.

Intersection
The intersection of two sets A and B is defined as the set of elements which belong to both A and B. This is written $A \cap B$ (read as A *intersection* B). We could write $A \cap B = \{x: x \in A \text{ and } x \in B\}$.

Thus, for example, if $A = \{1, 2, 3, 4, 5\}$, $B = \{2, 3, 4, 6, 8, 9\}$ then $A \cap B = \{2, 3, 4\}$. For this example show, by considering each side separately, that

$$A \cap (A \cap B) = A \cap B \text{ and } B \cap (A \cap B) = A \cap B$$

Union of two sets
The union of two sets, A and B, is defined as the set consisting of members belonging to either A or B or to both A and B. This is written $A \cup B$ (read as A *union* B). Again, if $A = \{1, 2, 3, 4, 5\}$, $B = \{2, 3, 4, 6, 8, 9\}$ then $A \cup B = \{1, 2, 3, 4, 5, 6, 8, 9,\}$.

Show for these sets that $A \cap (A \cup B) = A$ $\quad A \cup (A \cap B) = A$
$B \cap (A \cup B) = B$ $\quad B \cup (A \cap B) = B$

Venn diagrams

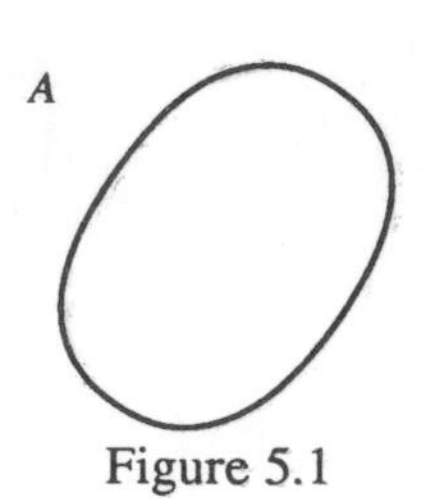

Figure 5.1

The ideas of unions, intersection and other combinations of sets may be illustrated by means of **Venn diagrams**. The members of a set are represented by the points within a closed curve, the shape of the curve being immaterial, but usually drawn as an oval, as shown in Figure 5.1.

The relationship between the curves then gives the relationship between the sets. In Figure 5.2 we represent, using Venn diagrams

(i) disjoint sets
(ii) the intersection of two sets
(iii) union
(iv) subset

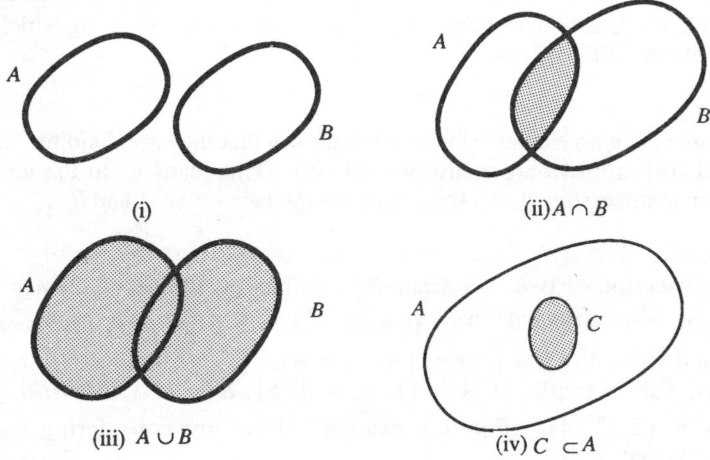

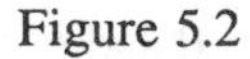
Figure 5.2

Notice that the shaded portions represent the appropriate set. Let us try to construct $A \cap (B \cup C)$ in two stages (see Figure 5.3). First of all $B \cup C$ is as shown in Figure 5.3(i). Now $A \cap (B \cup C)$ means the set of elements which belong to both $B \cup C$ and A and this is shown in Figure 5.3(ii).

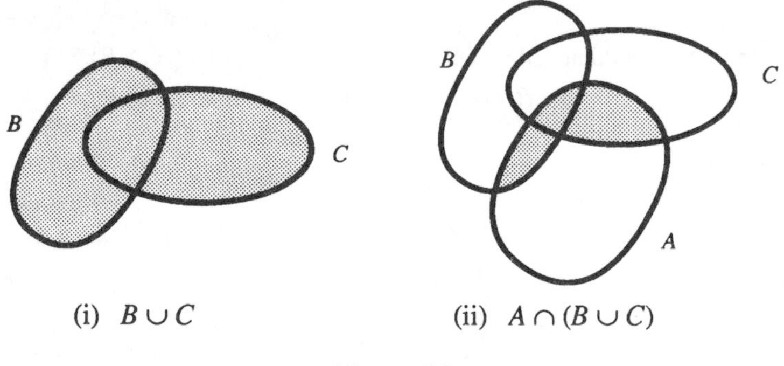

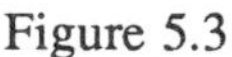
Figure 5.3

Using Venn diagrams we can also show the equality of two expressions. For example, $(A \cup B) \cap (A \cup C) = A \cup (B \cap C)$. The left-hand side is made up of the intersection of $A \cup B$ and $A \cup C$ which is built up as follows in Figure 5.4.

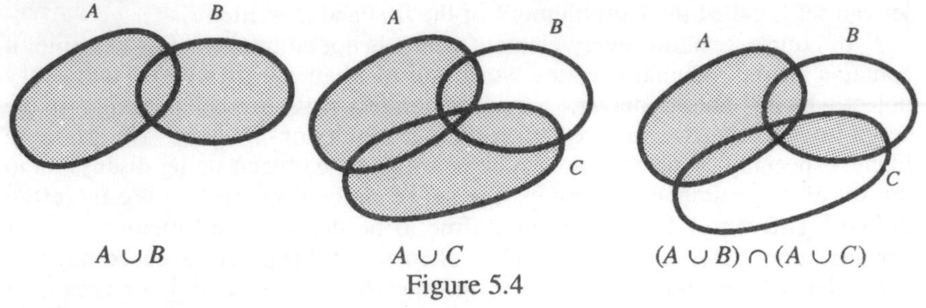

Figure 5.4

The right-hand side is the union of A and $B \cap C$ which is built up as follows in Figure 5.5.

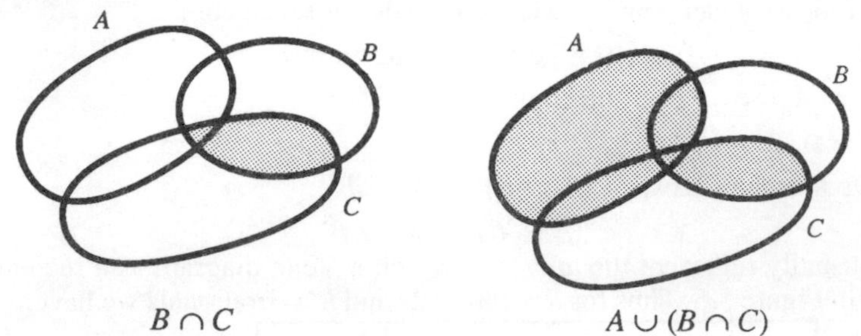

Figure 5.5

It is seen that the two expressions represent the same region and are, therefore, equal.

By considering Venn diagrams it is easy to show the following **Basic Set Operations** and you should try to prove each one.

1 $A \cup A = A \cap A = A$

2 $A \cap B = B \cap A$ (order does not matter)

3 $A \cup B = B \cup A$ (order does not matter)

4 $(A \cup B) \cup C = A \cup (B \cup C)$ (brackets can be moved and there will be no confusion if we write $A \cup B \cup C$)

5 $(A \cap B) \cap C = A \cap (B \cap C)$ (we can write this as $A \cap B \cap C$)

6 $A \cup (B \cap C) = (A \cup B) \cap (A \cup C)$ (notice brackets cannot be moved here)

7 $A \cap (B \cup C) = (A \cap B) \cup (A \cap C)$ (notice brackets cannot be moved here)

8 $A \cup (A \cap B) = A$

9 $A \cap (A \cup B) = A$

Universal set and Complementation

Defining a set A not only determines those elements which belong to A, but also determines those which do not belong to A. For example, if I is the set of integers, then I also defines another set of elements which are not integers. The second set is called the **Complement** of the first and is written I'.

I', of course, contains every element which is not an integer. For example, it contains all the mountains of the world, all of Shakespeare's plays, and every object which is not an integer. Now it is unlikely if we are considering I as the set of integers that we require I' to contain all such non-integers. We are more likely, especially in mathematics, to be restricting the objects under discussion to the set of real numbers or perhaps the set of rational numbers. We therefore define a **Universal Set** which we define to be the set of all elements under consideration. Having determined what elements of this universal set belong to a particular set, the elements remaining make up the complement. For example, if R is the set of rational numbers and the universal set is the set of real numbers, then R' is the set of real numbers which are not rational. In other words, R' is the set of irrational numbers.

The following can now be added to the laws of set algebra.

(i) $A \cup A' = I$ (I is here the universal set)

$A \cap A' = \emptyset$ (the null set)

(ii) $I' = \emptyset, \emptyset' = I$

(iii) De Morgan's Laws $(A \cup B)' = A' \cap B'$

$(A \cap B)' = A' \cup B'$

We usually represent the universal set on a Venn diagram as a rectangle, as shown in Figure 5.6 Thus for R = rationals and R' = irrationals we have:

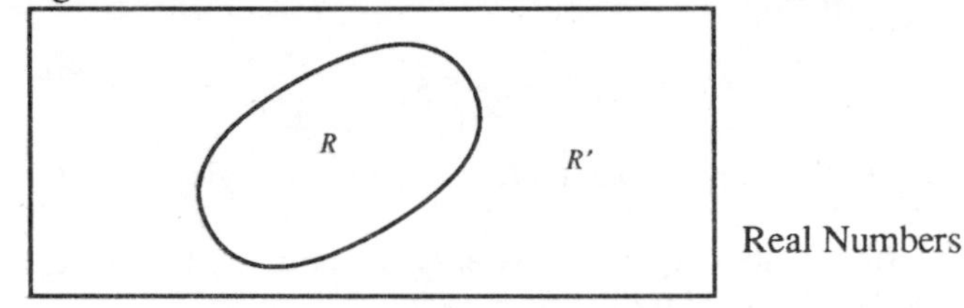

Figure 5.6

You should show, using Venn diagrams, the validity of De Morgan's laws.

The **cardinality** of a finite set is the number of its members. The cardinality of the empty set is clearly zero.

In general two sets S and T have the same cardinality if there is a one-one correspondence between their members. Hence the cardinality of the even positive integers and that of the natural numbers are the same.

A set is *countable* if its members can be put in one-one correspondence with a subset of the natural numbers. A set which is infinite and non-countable is called *uncountable*. As an example the even integers can be associated with the natural numbers by the rule $x \to x/2$. It can be shown that all the real numbers between 0 and 1 form an uncountable set.

Cartesian Product

If A and B are both subsets of some set S then we define the Cartesian product of A and B, written $A \times B$, as the set of ordered pairs (x, y) where $x \in A$ and $y \in B$. Often, we write $A \times A$ as A^2 with obvious extensions for A^3, A^4, etc.

Example

If $A = \{1, 2, 3\}$ and $B = \{4, 5\}$ then

$A \times B = \{(1, 4), (1, 5), (2, 4), (2, 5), (3, 4), (3, 5)\}$ and

$B^3 = \{(4, 4, 4), (4, 4, 5), (4, 5, 5), (5, 5, 5), (4, 5, 4), (5, 5, 4), (5, 4, 5), (5, 4, 4)\}$

As a less abstract example we could consider the construction of a database on cars comprising $A = \{\text{model}\}$, $B = \{\text{body type}\}$, $C = \{\text{engine size}\}$ and then $A \times B \times C$ would give the maximum possible set of different car types. Again, if $A = \{"0", "1", "2", ..., "9"\}$ then A^4 could represent the set of the last four digits of a telephone number.

5.2 Probability and Chance

We first conduct a **statistical experiment** which will have one or more possible **outcomes** which can be specified in advance.

Some examples are:

(i) tossing a coin with outcomes — head, tail
(ii) throwing a die with outcomes — 1, 2, 3, 4, 5, 6
(iii) aiming at a target with outcomes — success, failure
(iv) testing an electrical component — defective, non-defective

In each case, before the experiment is carried out, we are uncertain as to the outcome.

The set S of all possible outcomes of an experiment is called a **sample space** (or, sometimes, **outcome space**). The members of the set are called **sample points**. In the throwing of a die, the sample space is $S = \{1, 2, 3, 4, 5, 6\}$ and typical sample points are 2 and 5. Likewise in tossing a coin the sample space is $S = \{\text{head, tail}\}$. We take two much-used examples to consolidate ideas.

Example 1

A pair of dice are thrown and the total score observed. There are 36 possible outcomes as shown in Table 5.1.

Typical sample points are (4, 3), (2, 5), (6, 6).

Example 2

Three coins are tossed once each. The sample space of the experiment is $S = \{HHH, HHT, HTH, HTT, THH, THT, TTH, TTT\}$ where H and T represent Heads and Tails respectively. If we are interested in those outcomes where there are more heads than tails we see that there are four possible outcomes: $\{HHH, HHT, HTH, THH\}$. These comprise a subset of the sample space. Any subset of a sample space is called an **event**. The subset $\{TTT\}$ is the event that exactly three tails occur and the subset $\{HHT, HTH, THH\}$ is the event that exactly one tail occurs.

Table 5.1

Score on 1st die	Score on 2nd die					
	1	2	3	4	5	6
1	(1, 1)	(1, 2)	(1, 3)	(1, 4)	(1, 5)	(1, 6)
2	(2, 1)	(2, 2)	(2, 3)	(2, 4)	(2, 5)	(2, 6)
3	(3, 1)	(3, 2)	(3, 3)	(3, 4)	(3, 5)	(3, 6)
4	(4, 1)	(4, 2)	(4, 3)	(4, 4)	(4, 5)	(4, 6)
5	(5, 1)	(5, 2)	(5, 3)	(5, 4)	(5, 5)	(5, 6)
6	(6, 1)	(6, 2)	(6, 3)	(6, 4)	(6, 5)	(6, 6)

Combination of events

We may now revert to Example 1 and note that the event that the total score is 8 is the subset of the outcomes $A = \{(2, 6), (3, 5), (4, 4), (5, 3), (6, 2)\}$; the event that the second die scores a six is the subset $B = \{(1, 6), (2, 6), (3, 6), (4, 6), (5, 6), (6, 6)\}$. The event that the total score is 8 **or** the second die scores 6 is the subset $A \cup B = \{(2, 6), (3, 5), (4, 4), (5, 3), (6, 2), (1, 6), (3, 6), (4, 6), (5, 6), (6, 6)\}$; the event that the total score is 8 **and** the second die scores 6 is the subset $A \cap B = \{(2, 6)\}$.

Relevant to the sample space S of Example 1, the **complement** of the event that there are more heads than tails is found by removing the outcomes of the first event from the sample space. This yields the complementary event $\{HTT, THT, TTH, TTT\}$.

Further ideas

Suppose we toss a coin many times, then this experiment consists of several simple experiments. If we decide to toss the coin 10 times and count the number of heads then we should find that if this experiment were carried out on different occasions, the number of heads would vary. The number of heads is a **discrete random variable**; it is discrete because it can only take *specific* values and it is random because we cannot predict its value in advance. What we can hope to do is to estimate the chances of a particular value occurring and to do this we shall need to form a mathematical model. We loosely define probability of an event as a measure of the likelihood of its occurrence and define it more rigorously in the next section. The model is called a **probability model**. The sequence $\{H, T, H, H, H, T, T, \ldots\}$ is called a **random sequence**.

An example of a **continuous random variable** would be the angle made with a fixed direction by a spun pointer mounted on a support. The angle could take any value in the range [0, 360°] and cannot be predicted in advance; again, we must point out that the inability to measure with infinite precision does not destroy the theoretical continuous aspect.

In a more realistic situation, suppose we are trying to build a model of how an epidemic would behave if it broke out in a town. We might be interested, for example, in how long it would last or whether it would affect more than half the population. To do this we must make assumptions as to what are the chances of a person being affected in all circumstances. Perhaps we could run several computer simulations and test the model. In any case we need an understanding of probability and this we develop in the next section.

There are two ways of looking at probability: as a long-range relative frequency or as a pre-assigned entity. In the example of the tossing of a coin we may assume that the probability of a head occurring on any one throw is $\frac{1}{2}$ and then say, for example, in 10 trials we expect 5 heads, in 1000 we expect 500, etc, attributing departures from these numbers to chance fluctuations. Alternatively, we may observe the relative frequency of heads as we toss the coin 1 000 000 times and note that, after early fluctuations, this relative frequency settles down and if after 1 000 000 trials we observe 501 211 heads then we could say that the probability of a head on each throw is 0.501211 and accept this is a better guide than the value $\frac{1}{2}$. We shall adopt the former approach for ease of model building.

Each sample point is assigned a weight or measure of the chance of it occurring – the **probability**. This is a number p where $0 \leq p \leq 1$ such that the sum of the probabilities of all points in the sample space is 1. An outcome which has assigned to it a probability of 1 is a **certainty** and one which has assigned to it a probability of 0 is an **impossibility**. Hence the smaller the probability of an outcome, the less likely it is to occur.

The probability of an event A, written $p(A)$, is the sum of the probabilities of all sample points in A.

We shall restrict ourselves to discrete finite sample spaces for our illustrations in this section. In many situations we assume that each outcome is equally likely and assign to them equal probabilities. Then $p(A)$ = (number of outcomes in A)/(total number of outcomes). We shall obtain several results in this section which do not rely on this assumption unless so stated.

Example 1

If four 'fair coins' are tossed once each, what is the probability of obtaining exactly two tails?

By a 'fair' coin we mean that on each throw $p(\text{head}) = \frac{1}{2} = p(\text{tail})$. The sample space consists of 16 equally likely outcomes (check this) of which the six favourable ones are $HHTT, HTHT, HTTH, THHT, THTH, TTHH$ and so the probability we seek is

$$\frac{6}{16} = \frac{3}{8} \quad (\text{or is } \frac{1}{16} + \frac{1}{16} + \frac{1}{16} + \frac{1}{16} + \frac{1}{16} + \frac{1}{16})$$

Example 2

If two fair dice are rolled, what is the probability of a total score of 4?

The sample space has 36 equally likely outcomes and the probability of each is

1/36. The subset A = total score of 4 = {(1, 3), (2, 2), (3, 1)} and so $p(A)$ = 3/36 = 1/12.

We now quote some important rules. Here, $\emptyset$ is the null set and S is the sample space.

(i) $p(S) = 1$

(ii) $p(\emptyset) = 0$

(iii) $p(A') = 1 - p(A)$

(If there are 4 defective fuses in a box of 100 then the probability of a fuse selected at random being defective is 4/100 = 1/25 and hence the probability that it is not defective is 24/25.)

5.3 Compound Events

We now look at the probability of events compounded from simpler ones.

General rule of addition

We illustrate by example. A pack of 52 playing cards is shuffled and a card drawn out; what is the probability that this card is either an ace or a red card?

Now the sample space comprises 52 equally likely outcomes and the event A = *either an ace or a red card* is the union of two other events B = *card is an ace* and C = *card is red.* There are 4 sample points in B and 26 in C, but only 28 in $B \cup C$ since we do not count the Ace of Hearts and Ace of Diamonds twice, i.e. number of sample points in A = number of aces + number of red cards – number of red aces

$$= 4 + 26 - 2$$

Then $$p(A) = \frac{4}{52} + \frac{26}{52} - \frac{2}{52} = \frac{28}{52}$$

The event "card is a red ace" is $B \cap C$.

In general, if B and C are events in S,

$$p(B \cup C) = p(B) + p(C) - p(B \cap C) \tag{5.1}$$

Example

The probability that a man watches TV in any one evening is 0.6; the probability that he listens to the radio is 0.3 and the probability that he does both is 0.15. What is the probability that he does neither?

Let V be the event *he watches TV*, R be the event *he listens to radio*, then

$$\begin{aligned} p(V \cup R) &= p(V) + p(R) - p(V \cap R) \\ &= 0.6 + 0.3 - 0.15 \\ &= 0.75 \end{aligned}$$

therefore $p(\text{he does neither}) = 1 - p(V \cup R) = 0.25$

We show the results on the Venn diagram (Figure 5.7). Note that the probability that the man listens to radio but doesn't watch TV is 0.15 and the probability that he watches TV but but doesn't listen to the radio is 0.45.

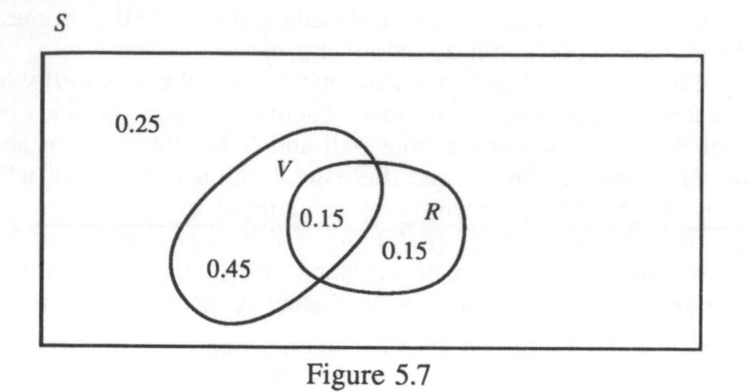

Figure 5.7

Addition rule for exclusive events

If the two events B and C are **mutually exclusive**, i.e. $B \cap C = \emptyset$, then $p(B \cap C) = 0$ and so, if B and C cannot occur simultaneously,

$$p(B \cup C) = p(B) + p(C) \qquad (5.2)$$

Example

If a fair die is rolled, what is the probability of obtaining a 1 or a 6? The events B = *obtaining 1* and C = *obtaining 6* are mutually exclusive and each has probability 1/6; by applying (5.2) we have

$$p(B \cup C) = \frac{1}{6} + \frac{1}{6} = \frac{1}{3}$$

Both addition rules can be extended to more than two events. For example, for three events

$$p(A \cup B \cup C) = p(A) + p(B) + p(C) - p(A \cap B) - p(B \cap C)$$
$$- p(C \cap A) + p(A \cap B \cap C)$$

You can check this for yourselves by drawing a suitable Venn diagram.

5.4 Conditional Probability

If two events follow each other then the outcome of the second may depend on the outcome of the first. For example, if we draw a card from a pack and then draw a second card, the probability of the second being a heart is clearly affected by whether the first card was a heart or not. We make a plausible definition by a non-practical example, chosen for its simplicity.

	numbered	blank	Totals
Red	3	5	8
Blue	4	1	5
Totals	7	6	13

Suppose we have a box containing red balls and blue balls. Some balls are numbered and the composition is as shown above.

We draw a ball from the bag and we are interested in the probability of the ball being blue knowing that it bears a number. There are 13 equally likely outcomes. Let B be the event of selecting a blue ball and N be the event of selecting a numbered ball. Now we can see that there are 7 numbered balls of which 4 are blue and we say that the probability of selecting a blue ball given that it is numbered is 4/7. We write this $p(B/N)$ and read it as *probability of B given N*. Now the probability $p(B)$ given no advance information is 5/13 and it does make a difference in this case if we know whether event N has occurred.

Further $p(N) = 7/13$ and $p(B \cap N) = 4/13$, since only four balls are both blue and numbered. In this example we can see that

$$\frac{p(B \cap N)}{p(N)} = \frac{4}{7}$$

and we might ask whether it is coincidence that this value is also $p(B/N)$. When we are told N has occurred we can cut down our sample space from the original 13 points each of which has probability 1/13 to a sample space of 7 members which are still equally likely and must have probability 1/7; we effectively have to reweigh them by scaling the weights by a factor 13/7 and this is done by dividing by $p(N)$. The members we require are in $B \cap N$ and this leads us to identify $p(B \cap N)/ p(N)$ with $p(B/N)$.

In general, we have the result that if two events A and B are such that $p(B) \neq 0$, then the **conditional probability** of A given B is

$$p(A/B) = \frac{p(A \cap B)}{p(B)} \tag{5.3}$$

Try to argue this equation from the Venn diagram.
We consolidate these ideas by three examples.

Example 1

The probability $p(B)$ that it rains on 15 July in Loughborough is 0.6; the probability $p(A)$ that it rains there on 15 and 16 July is 0.35. If we know that it has rained there on 15 July, the probability of rain on the following day is

$$p(A/B) = \frac{0.35}{0.6} = \frac{7}{12}$$

Example 2

Two dice are rolled. What is the probability of the total score exceeding 8 given that the first die shows:

(a) 6 (b) 4 (c) 2?

(a) We have a sample space of 36 equally likely outcomes and we let A be the event *total score exceeds 8* and B the event *first die shows 6*.
Then
$B = \{(6,1), (6,2), (6,3), (6,4), (6,5), (6,6)\}$ and $p(B) = 6/36$.

$A = \{(3,6), (4,5), (4,6), (5,4), (5,5), (5,6), (6,3), (6,4), (6,5), (6,6)\}$ and $p(A) = 10/36$.

Now $A \cap B = \{(6, 3), (6, 4), (6, 5), (6, 6)\}$ and $p(A \cap B) = 4/36$.

Then $p(A/B) = p(A \cap B)/p(B) = 4/36 \div 6/36 = 4/6 = 2/3$.

We can see directly that of the 6 outcomes in B, 4 give a total greater than 8. From (5.3) we can deduce that $p(A \cap B) = p(A/B).p(B)$. Equally well, we can say that $p(A \cap B) = p(B/A).p(A)$.

In words, the first of these relationships can be stated that *the probability of both A and B occurring is the probability that B occurs multiplied by the probability that A occurs given that B has occurred.*

(b) In this case B is the event *first die shows 4*. Then
$B = \{(4, 1), (4, 2), (4, 3), (4, 4), (4, 5), (4, 6)\}$ and $p(B) = 6/36$.
$A \cap B = \{(4, 5), (4, 6)\}$ and therefore $p(A \cap B) = 2/36$
Then $p(A/B) = 2/36 \div 6/36 = 1/3$

(c) Here, $A \cap B = \emptyset$ and, therefore $p(A/B) = 0$

Example 3

Two cards are drawn without replacement from a pack. What is the probability that

(i) both cards are kings (ii) one is a king and one is an ace?

(i) Let B be the event the *first card is a king*, A = *second card is a king*.

Then $$p(A \cap B) = p(B).p(A/B) = \frac{4}{52}.\frac{3}{51}$$

since if a king is drawn first we have 51 cards left, containing 3 kings.

Hence $$p(A \cap B) = \frac{4}{52\,.\,17} \cong 0.0045 \quad (4 \text{ d.p.})$$

(ii) Here we have two exclusive events: (ace, king) and (king, ace).

$$p(\text{king on 1st and ace on 2nd}) = \frac{4}{52}\cdot\frac{4}{51}$$
$$= p(\text{ace on 1st and king on 2nd})$$

Hence $$p(\text{king and ace}) = \frac{16}{52\,.\,51} + \frac{16}{52\,.\,51} = \frac{32}{52\,.\,51} \cong 0.0124 \quad (4 \text{ d.p.})$$

Independent events

Suppose we consider two experiments: tossing a coin, and drawing a card from a pack; it would be difficult to imagine any interaction between them. Consequently, if we define events H = *tossing a head* and D = *drawing an ace* we would not expect $p(D)$ to be affected by whether H had occurred or not. In

other words, in this situation $p(D/H) = p(D) = 4/52$.
We say that the two events are **statistically independent.**

If two events A and B are such that $p(A/B) = p(A)$ and $p(B/A) = p(B)$ then

$$p(A \cap B) = p(A).p(B) \tag{5.4}$$

Notice that *physical independence implies statistical independence* and, conversely, *statistical dependence implies physical dependence.*

Example
A die is rolled twice. What is the probability of two successive sixes?

$$p(6 \cap 6) = p(6).p(6) = \frac{1}{6} \cdot \frac{1}{6} = \frac{1}{36}$$

General Examples
All the rules can be extended to more than two events. We now consider some examples bringing in the rules but in an informal way. They serve to illustrate the basic ideas.

Example 1
A lot of 16 articles consists of 10 good ones, 4 with only minor defects and 2 with major defects.

(i) An article is drawn at random. Find the probability that
  (a) it has no defects
  (b) it has no major defects
  (c) it is either good or has major defects.

(ii) Two articles are selected at random. Find the probability that
  (a) both are good
  (b) both have major defects
  (c) at least one is good
  (d) at most one is good
  (e) exactly one is good
  (f) neither has major defects
  (g) neither is good.

(iii) Suppose a lot of 16 articles is to be accepted if three articles selected at random have no major defect. What is the probability that the lot described above is rejected?

(i) Let G = *article selected is good*, J = *article has major defect*, M = *article has minor defect.*

(a) We require $$p(G) = \frac{10}{16} = \frac{5}{8}$$

(b) This is $$p(G \cup M) = p(G) + p(M) = \frac{10}{16} + \frac{4}{16} = \frac{14}{16} = \frac{7}{8}$$

Alternatively $p(J') = 1 - p(J) = 1 - \dfrac{2}{16} = \dfrac{14}{16}$ as before

(c) This is $p(G \cup J) = p(G) + p(J) = \frac{10}{16} + \frac{2}{16} = \frac{3}{4}$

(ii) Let G_1 = *first article selected is good*, J_2 = *second article has a major defect*, and so on.

(a) $p(G_1 \cap G_2) = p(G_1).p(G_2/G_1) = \frac{10}{16} \cdot \frac{9}{15} = \frac{3}{8}$

(b) $p(J_1 \cap J_2) = p(J_1).p(J_2/J_1) = \frac{2}{16} \cdot \frac{1}{15} = \frac{1}{120}$

(c) If at least 1 is good, then we have p(1 is good) + p(2 are good).

$$\begin{aligned} p(\text{1 is good}) &= p(G_1 \cap G_2' + G_1' \cap G_2) \\ &= p(G_1 \cap G_2') + p(G_1' \cap G_2) \\ &= \frac{10}{16} \cdot \frac{6}{15} + \frac{6}{16} \cdot \frac{10}{15} = \frac{1}{4} + \frac{1}{4} = \frac{1}{2} \end{aligned}$$

therefore $\quad p(\text{at least 1 is good}) = \frac{1}{2} + \frac{3}{8} = \frac{7}{8}$

Alternatively, $p(\text{at least 1 is good}) = 1 - p(\text{neither is good})$

$$= 1 - \frac{6}{16} \cdot \frac{5}{15} = 1 - \frac{2}{16} = \frac{7}{8}$$

(d) $$\begin{aligned} p(\text{at most 1 is good}) &= p(\text{1 is good or neither is good}) \\ &= 1 - p(\text{both are good}) \\ &= 1 - \frac{3}{8} = \frac{5}{8} \end{aligned}$$

(e) $p(\text{1 is good}) = \frac{1}{2}$ [From (c)]

(f) $p(\text{neither has major defects}) = p(J_1' \cap J_2') = \frac{14}{16} \cdot \frac{13}{15} = \frac{91}{120}$

(g) $p(\text{neither is good}) = \frac{1}{8}$ [From (c)]

(iii) $p(J_1' \cap J_2' \cap J_3') = \frac{14}{16} \cdot \frac{13}{15} \cdot \frac{12}{14} = \frac{13}{20}$

therefore

$$p(\text{lot is rejected}) = 1 - \frac{13}{20} = \frac{7}{20}$$

Example 2

A circuit has three components in parallel; the probability of one of them failing is 0.05. If the circuit will work providing at least one of the components does not fail, what is the probability of the circuit working?

$$p(\text{all three components failing}) = (0.05)^3$$

therefore $p(\text{circuit working}) = 1 - (0.05)^3 = 1 - 0.000125 = 0.999875$

Example 3

In a bag there are 40 balls, 3 of which are red. 5 balls are drawn without replacement. What is

(a) p(no red ball is drawn)
(b) p(the first four are not red)
(c) p(red ball appears on the fifth drawing only)?

With the notation R_i = red ball in draw i,

(a) $p(R_1' \cap R_2' \cap R_3' \cap R_4' \cap R_5') = \dfrac{37.36.35.34.33}{40.39.38.37.36} = 0.662$ (3 d.p.)

(b) $p(R_1' \cap R_2' \cap R_3' \cap R_4') = \dfrac{37}{40} \cdot \dfrac{36}{39} \cdot \dfrac{35}{38} \cdot \dfrac{34}{37} = 0.723$ (3 d.p.)

(c) $p(R_1' \cap R_2' \cap R_3' \cap R_4' \cap R_5') = \dfrac{37}{40} \cdot \dfrac{36}{39} \cdot \dfrac{35}{38} \cdot \dfrac{34}{37} \cdot \dfrac{3}{36} = 0.060$ (3 d.p.)

Example 4

A circuit consists of three independent components A, B, C in series. Let A = *component A is defective*, etc, then

$$p(A) = 0.03, \quad p(B) = 0.15, \quad p(C) = 0.01.$$

The circuit fails if any component is defective.

What is (a) p(circuit fails)
(b) p(B alone fails)?

(a) $p(\text{circuit fails})$
$$\begin{aligned} &= 1 - p(\text{all components work}) \\ &= 1 - p(A' \cap B' \cap C') \\ &= 1 - p(A') \,.\, p(B') \,.\, p(C') \\ &= 1 - (0.97) \,.\, (0.85) \,.\, (0.99) \\ &= 1 - 0.816 = 0.184 \quad (3 \text{ d.p.}) \end{aligned}$$

(b) $p(A' \cap B \cap C') = p(A') \cdot p(B) \cdot p(C')$
$= (0.97) \cdot (0.15) \cdot (0.99) = 0.144 \quad (3 \text{ d.p.})$

Tree diagrams
A useful way of enumerating probabilities when there are several outcomes arising from a compounding of events is by means of a so-called **tree diagram**. We illustrate by an example.

Three dice are thrown and a success occurs with each die if the score exceeds 2. Then the probability of success is $p = 2/3$ and that of failure is $q = 1 - p = 1/3$ (see Figure 5.8).

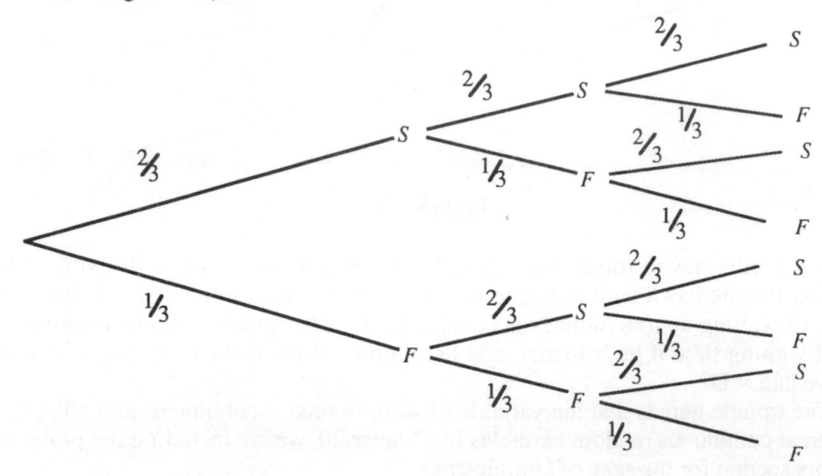

Figure 5.8

To evaluate the probability of an outcome we seek, such as *SFS*, we proceed along the relevant branches of the tree and then multiply the probabilities encountered (assuming the three dice give independent results) to obtain

$$\left[\frac{2}{3} \cdot \frac{1}{3} \cdot \frac{2}{3}\right] = \frac{4}{27}$$

If we do this for all outcomes we obtain Figure 5.9.

We check that these probabilities sum to 1. If we require the probability of at least two successes, we can see that the relevant mutually exclusive outcomes are *SSS, SSF, SFS, FSS* and their probabilities sum to $\left[\frac{8}{27} + \frac{4}{27} + \frac{4}{27} + \frac{4}{27}\right] = \frac{20}{27}$

Diagram Solution
Finally, we mention one other technique for solving probability problems which we include for its simplicity despite its limited application. Again an example suffices to explain the method.

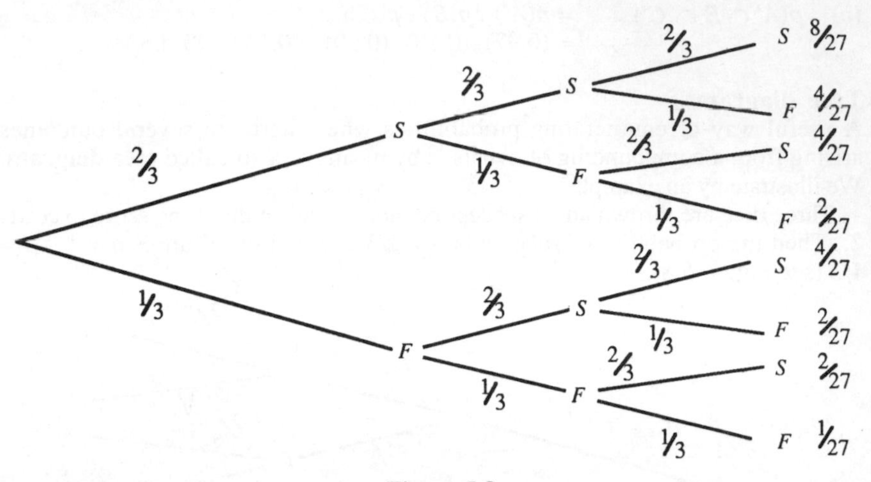

Figure 5.9

A traveller has to travel from A to D changing buses at stops B and C. He knows that he has a wait at B and one at C of at most 8 minutes each, but any time of waiting up to 8 minutes is equally likely. He can afford up to 13 minutes total waiting time if he is to arrive at D on time. What is the probability he will arrive late at D?

The trouble here is that the variable of waiting time is continuous and although we treat continuous random variables in Chapter 20, we are including the problem in this section for the sake of completeness.

Let the waiting time at B be x and that at C be y; we seek $p(x + y > 13)$. We draw a box as shown in Figure 5.10 and the coordinates (x, y) of any point in the box are the two waiting times. The line $x + y = 13$ is depicted and any point in the shaded area is equally likely.

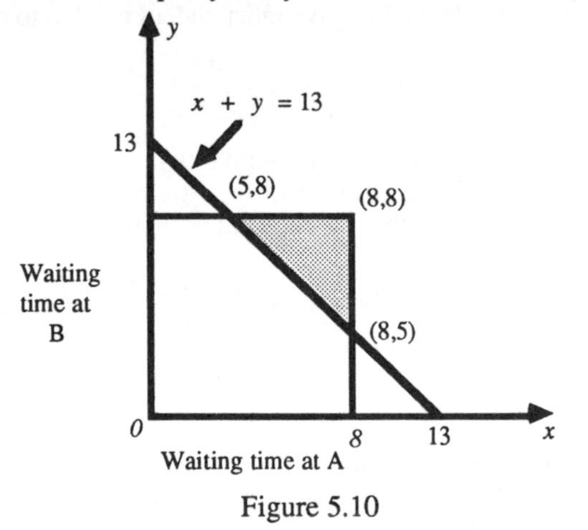

Figure 5.10

We can say probability of a region of outcomes is proportional to its area and since the totality of outcomes has area 64 units we note that the shaded area is $\frac{1}{2}(3^2) = 9/2$ units, which gives $p(\text{late at D}) = \frac{9}{2} \div 64 = \frac{9}{128}$

Problems

Section 5.1

1 Enumerate the elements in the following sets, where I is the set of all integers both positive and negative.

(a) $S = \{x \in I: 5 < x^2 < 46\}$ (b) $S = \{x \in I: 4 < x^3 + 1 \leq 102\}$

(c) $S = \{x \in I \text{ and } y \in I: 12 < x^2 + y^2 \leq 25\}$

(d) $S = \{x \in I: x^2 - 4x + 3 = 0\}$

2 Form the union and intersection of the sets A and B if

(a) $A = \{1, 3, 5, 7, 9, 11\}$, $B = \{1, 2, 3, 4, 5, 6, 7, 8\}$

(b) $A = \{x \in I: 3 < x < 10\}$, $B = \{x \in I: 0 < x^2 + x + 1 < 42\}$

(c) $A = \{0, \sqrt{2}, 6, \pi, 5, 8, 9, 10\}$, $B = \{x \in I: 0 < x + 5 < 20\}$

3 Verify the basic set operations (8) and (9) for the sets A and B in Problem 2.

4 Given that A, B, C are arbitrary sets, which of the following are true?

(i) If $A \subseteq B$ and $B \subseteq A$ then $A = B$ (ii) If $A \subset B$ and $B \subseteq C$ then $A \subset C$

(iii) If $A \neq B$ and $B \neq C$ then $A \neq C$

5 The *difference* of two sets A and B, written $A - B$ is defined as $A - B = A \cap B'$. Illustrate $A - B$ in a Venn diagram. If $A = \{a, b, c, d\}$, $B = \{c, e, g\}$, $C = \{a, d, e, f\}$ are subsets of $S = \{a, b, c, d, e, f, g\}$, find

(i) $A \cap C$ (ii) $B \cup C$ (iii) B'

(iv) $A \cap B \cap C$ (v) $A - C$ (vi) $C - A$

(vii) $B - C$ (viii) $A - B$ (ix) $(A \cup C)'$

(x) $A \times B$ (xi) $(A \cup B) \cap C'$

6 For each statement following, state under what conditions it is true.

(i) $A \cup B = A$ (ii) $A \cap B = A$ (iii) $A \cup \emptyset = \emptyset$

(iv) $A \cap \emptyset = \emptyset$ (v) $A - B = \emptyset$ (vi) $A \cup B \subseteq A \cap B$

7 The *symmetric difference* of two sets is defined by $A \oplus B = (A - B) \cup (B - A)$.

(i) Draw a Venn diagram to illustrate the definition.

(ii) Show that $A \oplus B = (A \cup B) - (A \cap B)$.

(iii) What is $A \oplus A$?

(iv) For $A = \{1, 3, 5, 7\}$ and $B = \{1, 2, 3, 4, 5\}$ what are $A \oplus B$, $B \oplus A$?

8 If $A = \{1, 2\}, B = \{c, d\}, C = \{4, 5\}$ find
(i) $A \times (A \cup C)$ (ii) $(A \times B) \cup (A \times C)$ (iii) $A \times (B \cap C)$
(iv) $(A \times B) \cap (A \times C)$

9 If $A = \{1, 2, 4, 5\}$, $B = \{2, 3, 5, 6\}$, $C = \{4, 5, 6, 7\}$, show that
(a) $(A \cup B) \cup C = A \cup (B \cup C)$ (b) $A \cup (A \cap B) = A$
(c) $A \cup (B \cap C) = (A \cup B) \cap (A \cup C)$ (d) $A \cap (B \cup C) = (A \cap B) \cup (A \cap C)$

Section 5.2

10 Represent by points on a graph the 36 possible results of throwing two dice. Identify the sets of points for which the sum of scores of the two dice is
(a) equal to 9 (b) greater than or equal to 7

11 A card is drawn from a pack. The sample space will be the set of all possible outcomes. Identify on a diagram the subsets that give the events
(a) the drawn card is either a club or a queen, or both
(b) the drawn card is a card lower than the 4 (ace is high)
(c) the drawn card is a card higher than the 3.
Identify the complements of the subsets in (a), (b) and (c).

Section 5.4

12 Four articles are distributed at random among six containers. What is the probability that
(i) all objects are in the same container (ii) no two objects are in the same container?

13 A lot consists of 12 good articles, 6 articles with only minor defects and 2 with major defects.
(i) One article is drawn at random. Find the probability that
(a) it has no defects (b) it has no major defects
(c) it is either good or has major defects.

(ii) Two articles are selected at random. Find the probability that
(a) both are good (b) both have major defects
(c) at least one is good (d) at most one is good
(e) exactly one is good (f) neither has major defects
(g) neither is good.

(iii) Suppose a lot of twenty articles is to be accepted if three articles selected at random have no major defect. What is the probability that the lot described in this question is
(a) accepted (b) rejected?

14 In a bolt factory machines A, B, C manufacture 25, 35 and 40 per cent respectively of the total output. Of their outputs, 5, 4, and 2 per cent respectively are defective bolts. A bolt is drawn at random and found to be defective. What are the probabilities that it was manufactured by machines A, B or C respectively?

15 Two cards are drawn from a pack of 52 cards. What is the probability of drawing two aces if
(i) the first card is returned to the pack before the second is drawn
(ii) the first card is not returned to the pack?

16 A tester smokes cigarettes of each of three different brands, A, B and C. He then assigns the name A to one cigarette, B to another and C to the remaining one. If he were unable to discriminate between the brands (i.e. if the names were assigned at random) what would be the probability of just one brand being correctly identified?

17 A box contains 4 bad tubes and 5 good tubes.
(i) Two are drawn together and one of them is tested and found to be good. What is the probability that the other is also good?
(ii) If the tubes are checked by drawing a tube at random, testing it and repeating the process (not replacing the tube tested) until all four bad tubes are located, what is the probability that the fourth bad tube will be found
(a) at the fifth test (b) at the tenth test?

18 A product is made up of three parts, A, B and C. The manufacturing process is such that the probability of A being defective is 0.2, of B being defective is 0.1, and of C being defective is 0.3. What is the probability that the assembled product will contain at least two defective parts?

19 An operator drilling for oil has options with six companies and must decide at the end of the year whether to take them up. The probability that he will strike oil by that time is 0.0125. If he does strike oil he will take up five of the options at random, and if he does not he will take up one at random. What is the probability that a particular company A will have its option taken up?

20 In a rifle shooting contest the target consists of a bull and outer, the contestants scoring 2 points for a bull, 1 point for an outer and zero otherwise. If the chance of a bull is 50% and an outer 30%, what are the probabilities that with five shots a contestant will score
(a) 10 points (b) 8 points (c) less than 8 points?

21 A product is made up of three parts, A, B and C. The manufacturing processes are such that the probability of A being defective is 0.07, of B being defective is 0.05 and of C being defective is 0.08. What is the probability that the assembled product will contain
(a) no defectives (b) at least 2 defectives?

22 A box contains 3 white, 4 blue and 6 black balls. Two balls are drawn consecutively from the box without replacement. Find the probability that both are the same colour.

23 The odds for a particular event A are 5 to 1 against; for an independent event B the odds are 3 to 1 against, and for an independent event C the odds are 2 to 1 on. Find the probability that
(a) all 3 events occur (b) only event A occurs (c) only one event occurs.

24 The probability that parents with a trace of a certain blue pigment in otherwise brown eyes will have a child with blue eyes is 1 in 4. If there are six children in a family what is the probability that at least half of them will have blue eyes? (EC)

25 (i) A die is thrown repeatedly until a 6 appears. What is the probability that a 6 does not appear until
(a) the 4th throw (b) the 10th throw (c) the rth throw?
(ii) Four players in turn draw (without replacement) a card from a pack to decide who will deal; the highest wins, aces count high. The first player draws the nine of hearts. What is the probability that he will deal?

26 (i) A, B and C take turns to throw a coin. Show that the probabilities of each of them throwing the first head are 4/7, 2/7 and 1/7 respectively.
(ii) Two dice are thrown together repeatedly. Can you find the probability that a total score of 9 appears before a total score of 8?

27 Three bolts are selected at random from a box of six bolts and three nuts are selected at random from a box of six nuts. If each box contains 2 faulty items find
(i) the probability that at least one of the six selected items is faulty
(ii) the mean number of satisfactory nut and bolt pairs that can be obtained from the six selected items. (LU)

28 (i) Given that the probability that any one of 4 telephone lines is engaged at an instant is 1/3, calculate the probability
(a) that two of the four lines are engaged and the other two are free
(b) that at least one of the four lines is engaged. (LU)

(ii) A machine is powered by three similar storage batteries; it will function satisfactorily only if at least two of these batteries are serviceable. The probability of any one battery becoming unserviceable in less than 50 hours is 0.2, and of becoming unserviceable in less than 100 hours is 0.6. Find the probability that the machine will function satisfactorily for
(a) at least 50 hours (b) between 50 and 100 hours. (LU)

29 A pond contains a large unknown number N of fish. A random sample of r fish is taken, marked and put back in the pond. A week later, a random sample of s fish is taken. Find the probability of the second sample containing n marked fish and estimate the value of N which maximises this probability. Hence estimate the likely value of N in the case where $r = 20$, $s = 25$ and $n = 2$.

30 The probability that a particular component of a machine needs replacing in a five-year period is 0.25 and that of another component is 0.20. Find the probability that in a five-year period
(i) both components need replacing (ii) neither component needs replacing
(iii) just one needs replacing. (EC)

31 (a) It is known that 10% of the output of a certain manufacturing process is unsatisfactory and considered defective. Find the probability that out of a large batch
(i) a sample of 20 contains exactly 2 defective items
(ii) a sample of 10 contains not more than 1 defective item.

(b) It has been found that the probability of a particular component of a machine needing replacement in a 5 year period is $\frac{1}{8}$, and that for another component is $\frac{1}{6}$. Assuming independence, find the probability that in a particular 5 year period

(i) both components need replacing (ii) neither need replacing

(iii) either one or the other or both need replacing. (EC)

32 A machine is powered by three similar batteries and will function satisfactorily if at least two of the batteries are operational. It is known that a battery becomes non-operational in less than 40 hours with a probability of 0.25. Similarly a battery becomes non-operational in less than 80 hours with a probability of 0.7.

Find the probability that the machine functions satisfactorily for

(i) at least 40 hours (ii) between 40 and 80 hours. (EC)

6

DISCRETE MATHEMATICS

6.1 Propositions and Propositional Logic

In mathematics we use *statements* that can be shown to be *true* or *false* and we call these **propositions**. An example of such a proposition is: "If x is an even number and y is an even number, then $x + y$ is an even number". This is a statement that can be shown to be true. Another example is: "There are two solutions to the equation $x^2 + 2x + n^2 = 0$ where n is an integer and they are both integers". This is a proposition but is obviously false.

Examples of statements which are not propositions are:

(a) "She is very talented" (b) "$x + y = 6$"

since we need further information in each case to decide whether the statement is true or false. Note that when claiming that a statement is a proposition we do not need to know whether it is true or false; we only need to know that it is one or the other.

Notation

We shall use lower case letters to represent propositions and write

p: If x is an even integer, x^2 is an even integer

which is read "The proposition p is that: If x is an even integer, x^2 is an even integer". Another example (which is, of course, false) is:

q: If $i > 4$, then $i^2 + 4 < 20$.

This is the proposition q that if i is greater than 4, $i^2 + 4$ will be less than 20.

We often use p, q and so on as *propositional variables*. That is they could represent any proposition rather than standing for any particular proposition.

Connectives and Truth Values

Many propositions are the combination of other simple propositions with *connecting* words. The two most common connecting words (**connectives**) are 'AND' and 'OR'. We make statements such as:

"Harold Wilson was once Prime Minister AND so was Winston Churchill"

"$x > 4$ AND $y \leq 3$"

The symbol $\wedge$ is used to denote the connective AND. Thus we can write $p \wedge q$.

Truth Tables

Truth Tables are often used to determine the truth or falsehood of compound propositions and to provide definitions of the connectives.

The following truth table defines the connective AND.

p	q	$p \wedge q$
T	T	T
T	F	F
F	T	F
F	F	F

The columns give the possible combinations of true and false which occur in the propositions p and q with the end result $p \wedge q$.

Hence if p is true and q is true, then $p \wedge q$ would be true; and

if p is false and q is true, then $p \wedge q$ would be false.

The two other rows can be interpreted similarly.

Notice that a proposition of the form p AND q is false unless **both** p and q are true.

The OR *connective* is denoted by the symbol $\vee$ and we write $p \vee q$. The intent of the OR connective is to produce a statement which is true when at least one of its constituent parts is true. Hence the truth table for OR is

p	q	$p \vee q$
T	T	T
T	F	T
F	T	T
F	F	F

The statement $p \wedge q$ is called the **conjunction** of p and q while the statement $p \vee q$ is called the **disjunction** of statements p and q.

The Negation

One other basic operation, NOT, used in propositional logic is the **negation** of a proposition and is the proposition which states that the original proposition is false. In words an example is: "It is not true that 6 is a prime number".

In symbolic form p: 6 is a prime number (which is false).

Its negation, symbolized by $\sim p$ is that 6 is not a prime number (which is true).

In general, the truth table for NOT is given by

p	$\sim p$
T	F
F	T

The **digital computer** uses the **binary system** since most electronic equipment is bimodal (switches are off or on, voltages are high or low, etc). Thus a computer store has two states, one used to represent 0 and the other to represent 1.

Each 0 and 1 represented is called a **bit** and a cluster of 8 bits is called a **byte**. **Words** are made up of one or more clusters of bytes.

Machine language in a computer works on information 'bit by bit' with 1 being regarded as true and 0 as false. Operations work on information clustered into bytes and words. If a byte contained the data 10101010 and another byte the data 01110101, the result of applying the AND operator would be 00100000 while the OR operator would give 11111111.

Complex Propositions

Using only the connectives AND, OR and the negation we can construct complex propositions such as

$$[(p \vee q) \wedge r] \wedge [\sim (p \wedge q)]$$

and we immediately ask the question "Can we simplify such a proposition?" That is, can we find an equivalent but simpler proposition which has, under the same circumstances, the same outcome of false or true as the original statement?

We shall learn later how to do this, but for the moment all we can do is to draw up a truth table showing the outcomes of the proposition as follows:

p	q	r	$p \vee q$	$(p \vee q) \wedge r$	$p \wedge q$	$\sim(p \wedge q)$	$[(p \vee q) \wedge r] \wedge [\sim(p \wedge q)]$
T	T	T	T	T	T	F	F
T	T	F	T	F	T	F	F
T	F	T	T	T	F	T	T
T	F	F	T	F	F	T	F
F	T	T	T	T	F	T	T
F	T	F	T	F	F	T	F
F	F	T	F	F	F	T	F
F	F	F	F	F	F	T	F

combination of possible truth values

Suppose $p: x = 4$, $q: y > 6$, $r: x \leq 2$. Then given $x = 4$, $y = 8$, p is true, q is true but r is false. Hence working our way across the second line of the truth table we see that the complex proposition [($x = 4$ OR $y > 6$) AND ($x \leq 2$)] AND [NOT ($x = 4$ AND $y > 6$)] is false.

But if we are given $x = 1$, $y = 10$ then p is false, q is true and r is true, so that using the fifth line of the truth table, the complex proposition is true. Work through the form in words and decide whether the proposition is false.

Now $x = 4$ OR $y > 6$ is **true** when $x = 1$, $y = 10$; also $x \leq 2$.
Hence [($x = 4$ OR $y > 6$) AND ($x \leq 2$)] is **true**.
But $x = 4$ AND $y > 6$ is false, hence [NOT ($x = 4$ AND $y > 6$)] is **true**.
Finally, [($x = 4$ OR $y > 6$) AND ($x \leq 2$)] AND [NOT ($x = 4$ AND $y > 6$)] is **true**.

Examples of such complex propositions occur in computer programs. We might have for example

```
IF ((x < 5.0) AND (2x < 10.7)) OR (SQRT(5x) > 5.1)
THEN WRITE LN(A)
```

If successive input values of x are 1.0, 5.1, 2.4, 7.2, 5.3 we find the output values.

Let $p: x < 5.0$, $q: 2x < 10.7$, r: SQRT($5x$) > 5.1

Writing a truth table for ((x < 5.0) AND ($2x$ < 10.7)) OR (SQRT($5x$) > 5.1) is the same as truth table for $(p \wedge q) \vee r$, namely

p	q	r	$p \wedge q$	$(p \wedge q) \vee r$
T	T	T	T	T
T	T	F	T	T
T	F	T	F	T
T	F	F	F	F
F	T	T	F	T
F	T	F	F	F
F	F	T	F	T
F	F	F	F	F

If x = 1.0, p is true, q is true, r is false:- proposition is true; output is LN(1.0), which is zero.
If x = 5.1, p is false, q is true, r is false:- proposition is false; no output will be printed.
If x = 2.4, p is true, q is true, r is false:- proposition is true; output is LN(2.4) = 0.87547
If x = 7.2, p is false, q is false, r is true:- proposition is true; output is LN(7.2) = 1.97408
If x = 5.3, p is false, q is true, r is true:- proposition is true; output is LN(5.3) = 1.66771

Another example is

IF (((X < Y) AND (Y = 2) OR (Z = 10)) AND (NOT ((X < Y) AND (Z = 10)))
THEN WRITE 'DATA OUT OF RANGE'

What will happen if X is 3, Y = 2, Z = 15? The message will be printed only if the compound proposition is false.

Writing $p: X < Y$, $q: Y = 2$, $r: Z = 10$.

Then we need to examine the compound proposition $[((p \wedge q) \vee r] \wedge [\sim(p \wedge r)]$ by forming the truth table.

p	q	r	$p \wedge q$	$(p \wedge q) \vee r$	$p \wedge r$	$\sim(p \wedge r)$	$[(p \wedge q) \vee r] \wedge [\sim(p \wedge r)]$
T	T	T	T	T	T	F	F
T	T	F	T	T	F	T	T
T	F	T	F	T	T	F	F
T	F	F	F	F	F	T	F
F	T	T	F	T	F	T	T
F	T	F	F	F	F	T	F
F	F	T	F	T	F	T	T
F	F	F	F	F	F	T	F

Now when $X = 3, Y = 2, Z = 15,$ p is false, q is true, and r is false. Hence from the sixth line of the truth table the proposition is false so that the message would not be printed. Verify for yourself that in the cases
(i) $X = 6, Y = 7, Z = 10$ (ii) $X = 1, Y = 2, Z = 8$
(iii) $X = 4, Y = 2, Z = 10$
the message is printed out in the last two only.

Algorithms

We can summarise the process of setting up a truth table as the following of a set of instructions which can be carried out mechanically in order to perform a particular task. This is an example of an **algorithm** which can be given to a computer to perform.

Tautology

A tautology is a complex proposition which has the property that, regardless of the truth values of the basic propositions it contains, it is always true.

For example, the proposition $p\vee\sim(p\wedge q)$ is a tautology. Constructing the truth table:

p	q	$p\wedge q$	$\sim(p\wedge q)$	$p\vee\sim(p\wedge q)$
T	T	T	F	T
T	F	F	T	T
F	T	F	T	T
F	F	F	T	T

The table shows that for all values of p and q, $p\vee\sim(p\wedge q)$ is true.

Contradiction

A proposition which is always false is a **contradiction**. This, of course, is necessarily a negation of a tautology.

For example $(p\wedge q)\wedge\sim(p\vee q)$ is a contradiction, since the truth table is

p	q	$p\wedge q$	$p\vee q$	$\sim(p\vee q)$	$(p\wedge q)\wedge\sim(p\vee q)$
T	T	T	T	F	F
T	F	F	T	F	F
F	T	F	T	F	F
F	F	F	F	T	F

so that the proposition is false for all values of p and q. It follows that the proposition $\sim[(p\wedge q)\wedge\sim(p\vee q)]$ must be a tautology.

Equivalent Propositions

Two propositions are logically equivalent if they have exactly the same truth values under all circumstances; $\equiv$ is the shorthand used for 'is equivalent to'. As an example, we show that $[(p\wedge q)\vee r]\wedge[\sim(p\wedge r)] \equiv [(\sim p\wedge r)]\vee[(p\wedge q)\wedge\sim r]$

Constructing the truth table of the left-hand side we obtain

p	q	r	$p \wedge q$	$(p \wedge q) \vee r$	$\sim(p \wedge r)$	$[(p \wedge q) \vee r] \wedge [\sim(p \wedge r)]$
T	T	T	T	T	F	F
T	T	F	T	T	T	T
T	F	T	F	T	F	F
T	F	F	F	F	T	F
F	T	T	F	T	T	T
F	T	F	F	F	T	F
F	F	T	F	T	T	T
F	F	F	F	F	T	F

Similarly, constructing the truth table for the right-hand side gives

p	q	r	$\sim p \wedge r$	$p \wedge q$	$(p \wedge q) \wedge \sim r$	$[\sim p \wedge r] \vee [(p \wedge q) \wedge \sim r]$
T	T	T	F	T	F	F
T	T	F	F	T	T	T
T	F	T	F	F	F	F
T	F	F	F	F	F	F
F	T	T	T	F	F	T
F	T	F	F	F	F	F
F	F	T	T	F	F	T
F	F	F	F	F	F	F

The propositions have the same truth tables and are therefore logically equivalent.

Algebra of Propositions

There are a number of fundamental logically equivalent expressions which are especially useful in proofs. These are as follows where t can take only the value T and f can take only the value F.

1 Idempotent Laws:

(a) $p \vee p \equiv p$ (b) $p \wedge p \equiv p$

2 Associative Laws:

(a) $(p \vee q) \vee r \equiv p \vee (q \vee r)$ (b) $(p \wedge q) \wedge r \equiv p \wedge (q \wedge r)$

3 Commutative Laws:

(a) $p \vee q \equiv q \vee p$ (b) $p \wedge q \equiv q \wedge p$

4 Distributive Laws:

(a) $p \vee (q \wedge r) \equiv (p \vee q) \wedge (p \vee r)$ (b) $p \wedge (q \vee r) \equiv (p \wedge q) \vee (p \wedge r)$

5 De Morgan's Laws:

(a) $\sim(p \vee q) \equiv \sim p \wedge \sim q$ (b) $\sim(p \wedge q) \equiv \sim p \vee \sim q$

6 Identity Laws:

(a) $p \vee f \equiv p$ (b) $p \wedge t \equiv p$

7 Identity Laws:

(a) $p \vee t \equiv t$ (b) $p \wedge f \equiv f$

8 Complement Laws:

(a) $p \vee {\sim}p \equiv t$ (b) $p \wedge {\sim}p \equiv f$

9 Complement Laws:

(a) ${\sim}{\sim}p \equiv p$ (b) ${\sim}t \equiv f$, ${\sim}f \equiv t$

We shall prove 4(a) and 5(b) but leave the rest for you to prove. We construct the truth table for the left-hand side of 4(a):

p	q	r	$q \wedge r$	$p \vee (q \wedge r)$
T	T	T	T	T
T	T	F	F	T
T	F	T	F	T
T	F	F	F	T
F	T	T	T	T
F	T	F	F	F
F	F	T	F	F
F	F	F	F	F

Construction of the truth table for the right-hand side gives

p	q	r	$p \vee q$	$p \vee r$	$(p \vee q) \wedge (p \vee r)$
T	T	T	T	T	T
T	T	F	T	T	T
T	F	T	T	T	T
T	F	F	T	T	T
F	T	T	T	T	T
F	T	F	T	F	F
F	F	T	F	T	F
F	F	F	F	F	F

Hence the left-hand side and right-hand side are equivalent.

Taking 5(b), we construct the truth tables for both sides to give

p	q	${\sim}(p \wedge q)$	${\sim}p$	${\sim}q$	${\sim}p \vee {\sim}q$
T	T	F	F	F	F
T	F	T	F	T	T
F	T	T	T	F	T
F	F	T	T	T	T
	LHS			RHS	

Once again, the left-hand and right-hand sides are equivalent.

Simplification of a Proposition

We can use the algebra of propositions to simplify a given proposition.

For example, $(p \vee q) \wedge \sim p \equiv \sim p \wedge (p \vee q)$ (using 3(b))

$\equiv (\sim p \wedge p) \vee (\sim p \wedge q)$ (using 4(b))

$\equiv f \vee (\sim p \wedge q)$ (using 8(b))

$\equiv \sim p \wedge q$ (using 6(a))

Conditional Statements

(a) **Conditional** $p \rightarrow q$

Many statements, particularly in mathematics, are of the form 'If p then q'. We define the truth table for this structure as

p	q	$p \rightarrow q$
T	T	T
T	F	F
F	T	T
F	F	T

The third line is to be interpreted that if p is false, whatever happens to q must be true. It is like saying "If you can get a first class degree, I will give you £100". What we don't say is what will happen if you don't get a first class degree – I still might give you £100; on the other hand I may not.

Having noted this we can see that the statement $p \rightarrow q$ is true does not mean that p *causes* q to be true. But it is noted that $p \rightarrow q$ is often read as one of the following:

p implies q, p only if q, q if p, p is sufficient for q, q is necessary for p.

The following propositions are equivalent to $p \rightarrow q$

(i) $\sim p \vee q$ (ii) $\sim q \rightarrow \sim p$ (iii) $\sim(p \wedge \sim q)$

(b) **Biconditional** $p \leftrightarrow q$

Another common statement in mathematics is of the form 'p if and only if q' or simply 'p iff q' and is written $p \leftrightarrow q$. The truth table for this is

p	q	$p \leftrightarrow q$
T	T	T
T	F	F
F	T	F
F	F	T

As the notation suggests, $p \leftrightarrow q \equiv (p \rightarrow q) \wedge (q \rightarrow p)$ as can readily be seen from the truth table that can be constructed for the right-hand side.
The truth table is shown on the following page.

p	q	$p \to q$	$q \to p$	$(p \to q) \wedge (q \to p)$
T	T	T	T	T
T	F	F	T	F
F	T	T	F	F
F	F	T	T	T

We can also write $p \leftrightarrow q$ in terms of the fundamental connectives $\wedge$, $\vee$ and $\sim$ and it is easy to show that

$$(p \leftrightarrow q) \equiv (\sim p \vee q) \wedge (\sim q \vee p)$$

Finally, we have equivalent statements which give the negation of the conditional and biconditional. They are respectively

(i) $\sim(p \to q) \equiv p \wedge \sim q$

(ii) $\sim(p \leftrightarrow q) \equiv p \leftrightarrow \sim q \equiv \sim p \leftrightarrow q$

Converse, Inverse and Contrapositive

If we are given the conditional proposition $p \to q$ then we define the **converse** proposition as $q \to p$, the **inverse** proposition as $\sim p \to \sim q$ and the **contrapositive** proposition as $\sim q \to \sim p$. Their truth tables are

		Conditional	Converse	Inverse	Contrapositive
p	q	$p \to q$	$q \to p$	$\sim p \to \sim q$	$\sim q \to \sim p$
T	T	T	T	T	T
T	F	F	T	T	F
F	T	T	F	F	T
F	F	T	T	T	T

It is seen that the conditional $p \to q$ is logically equivalent to its contrapositive $\sim q \to \sim p$; and that the converse and the inverse are logically equivalent to each other.

Examples of the equivalence of $p \to q$ and $\sim q \to \sim p$ are

p: the triangle A is equilateral,q: the triangle A is isosceles

$p \to q$: an equilateral triangle is isosceles

$\sim q \to \sim p$: a triangle which is not isosceles is not equilateral.

Write down the converse and inverse statements and show that they are false.

6.2 Arguments and Proof, Logical Implication

An **argument** is an assertion that a given set of propositions $p_1, p_2, \ldots, p_n$ (called **premises**) yields (has as a consequence) another proposition q (called the **conclusion**).

We use the notation $\quad p_1, p_2, \ldots, p_n \vdash q.$

As an example, consider the following argument: "If a number is divisible by 6, then the number is divisible by 3". The proposition given is p_1: x is a number divisible by 6.

Now we know from our mathematical experience that the following propositions are true:

p_2: $x = k.6$ where k is integer	(definition of divisibility)
p_3: $6 = 3.2$	(number fact)
p_4: $x = k.3.2$	(substitution)
p_5: $x = (2k).3$	(property of multiplication)
p_6: $2k$ is an integer	(known fact)

We therefore get the conclusion q: x is divisible by 3.

We say the **argument** $p_1, p_2, \ldots, p_n \vdash q$ is true, if q is true whenever **all** the premises $p_1, p_2, \ldots, p_n$ are true; otherwise the argument is false.

Thus an argument is a statement which has a truth value. It is a **valid** argument if the argument is true. It is a **fallacy** if the argument is false.

As an example, a well-known argument (called *modus ponens* or the Law of Detachment) is $p, p \to q \vdash q$.

This argument is valid as can be seen from the following truth table

p	q	$p \to q$
T	T	T
T	F	F
F	T	T
F	F	T

Notice when p and $p \to q$ are both true, then q is true (line 1). Hence the argument is valid. Line 1, in fact, is the only one where both p and $p \to q$ are true.

Another example is the argument (called *syllogism*)

$$p \to q, q \to r \vdash p \to r$$

The truth table below shows that when $p \to q$ and $q \to r$ are both true (lines 1, 5, 7, 8) then so is $p \to r$. Hence the argument is valid.

p	q	r	$p \to q$	$q \to r$	$p \to r$
T	T	T	T	T	T
T	T	F	T	F	F
T	F	T	F	T	T
T	F	F	F	T	F
F	T	T	T	T	T
F	T	F	T	F	T
F	F	T	T	T	T
F	F	F	T	T	T

An example of a fallacy is $q, p \to q \vdash p$

The truth table below shows that when **both** q and $p \rightarrow q$ are true in lines 1 and 3, then p is only true in line 1 and false in line 3. Hence the argument is a fallacy; for it to be valid, p must be true for all the cases when both q and $p \rightarrow q$ are true.

p	q	$p \rightarrow q$	p
T	T	T	T
T	F	F	T
F	T	T	F
F	F	T	F

An example of such a fallacy in words is:

"Computers must cost less than £300 to be in the average home".
"Walkers are selling computers at £290 in Loughborough."
"Hence computers are in the average home in Loughborough."

Calling p: computers are in the average home in Loughborough
q: the price of a computer is less than £300

we obtain the example above as the fallacy $q, p \rightarrow q \vdash p$.

Now the propositions $p_1, p_2, \ldots, p_n$ are true simultaneously if and only if $p_1 \wedge p_2 \wedge \ldots, \wedge p_n$ is true. Thus the argument $p_1, p_2, \ldots, p_n \vdash q$ is valid if and only if q is true whenever $p_1 \wedge p_2 \wedge \ldots, \wedge p_n$ is true. Another way of saying this is that $(p_1 \wedge p_2 \wedge \ldots, \wedge p_n) \rightarrow q$ is a tautology. We therefore have the following:

Theorem

The argument $p_1, p_2, \ldots, p_n \vdash q$ *is valid if and only if the proposition* $(p_1 \wedge p_2 \wedge \ldots, \wedge p_n) \rightarrow q$ *is a tautology.*

As an example, consider the argument (often called *modus tollens*)

$p \rightarrow q, \sim q \vdash \sim p$

The truth table for $[(p \rightarrow q) \wedge \sim q] \rightarrow \sim p$ is

p	q	$p \rightarrow q$	$\sim q$	$(p \rightarrow q) \wedge \sim q$	$\sim p$	$[(p \rightarrow q) \wedge \sim q] \rightarrow \sim p$
T	T	T	F	F	F	T
T	F	F	T	F	F	T
F	T	T	F	F	T	T
F	F	T	T	T	T	T

This is a tautology and the argument is valid.

The argument $p \rightarrow q, \sim p \vdash \sim q$ is a fallacy, as can be seen from the truth table on the next page.

We can demonstrate the validity of an argument by an alternative method. An example follows after the truth table.

p	q	$p \rightarrow q$	$(p \rightarrow q) \wedge \sim p$	$[p \rightarrow q \wedge \sim p] \rightarrow \sim q$
T	T	T	F	T
T	F	F	T	T
F	T	T	T	F
F	F	T	T	T

Example

Prove that $p \rightarrow q, \sim r \rightarrow \sim q, \sim r \vdash \sim p$.
We state respectively

(1)	$p \rightarrow q$	given
(2)	$\sim q \rightarrow \sim p$	contrapositive of (1)
(3)	$\sim r \rightarrow \sim q$	given
(4)	$\sim r \rightarrow \sim p$	syllogism; (3) and (2)
(5)	$\sim r$	given
(6)	$\sim p$	modus ponens; (4) and (5)

We quote a list of the commonly used rules:

$p, p \rightarrow q \vdash q$	modus ponens, or law of detachment
$p \rightarrow q, q \rightarrow r \vdash p \rightarrow r$	syllogism
$p \rightarrow q, \sim q \vdash \sim p$	modus tollens
$p \vdash p \vee q$	addition
$p \wedge q \vdash p$	specialisation
$p, q \vdash p \wedge q$	conjunction
$p \rightarrow q \vdash \sim q \rightarrow \sim p$	contrapositive

Logical Implication

Let p and q be propositions involving propositions $p_1, p_2, \ldots, p_n$. The proposition $p(p_1, p_2, \ldots)$ is said to *logically imply* a proposition $q(p_1, p_2, \ldots)$ if $q(p_1, p_2, \ldots)$ is true whenever $p(p_1, p_2, \ldots)$ is true.

Now if $q(p_1, p_2, \ldots)$ is true whenever $p(p_1, p_2, \ldots)$ is true then the argument $p(p_1, p_2, \ldots) \vdash q(p_1, p_2, \ldots)$ is valid and furthermore $p \vdash q$ is valid if and only if the conditional statement $p \rightarrow q$ is a **tautology**.

Example 1

Show that $p \wedge q$ logically implies $p \leftrightarrow q$.
We have to show that either

(i) $p \leftrightarrow q$ is true whenever $p \rightarrow q$ is true, *or*

(ii) the argument $p \wedge q \vdash p \leftrightarrow q$ is valid, *or*

(iii) $p \wedge q \rightarrow p \leftrightarrow q$ is a tautology.

(i) Setting up the truth table we obtain the first four columns below and we can see that whenever $p \wedge q$ is true (first line), so is $p \leftrightarrow q$.

p	q	$p \wedge q$	$p \leftrightarrow q$	$p \wedge q \rightarrow p \leftrightarrow q$
T	T	T	T	T
T	F	F	F	T
F	T	F	F	T
F	F	F	T	T

(ii) From the table we see that the argument $p \wedge q \mid\!\!-\, p \leftrightarrow q$ is valid.

(iii) Examining the fifth column of the truth table confirms that $p \wedge q \rightarrow p \leftrightarrow q$ is a tautology.

Example 2

Prove that $(p \vee q) \wedge (\sim q \vee r)$ logically implies $p \vee r$

The truth table is

p	q	r	$p \vee q$	$\sim q \vee r$	$(p \vee q) \wedge (\sim q \vee r)$	$p \vee r$	$(\alpha) \rightarrow (\beta)$
T	T	T	T	T	T	T	T
T	T	F	T	F	F	T	T
T	F	T	T	T	T	T	T
T	F	F	T	T	T	T	T
F	T	T	T	T	T	T	T
F	T	F	T	F	F	F	T
F	F	T	F	T	F	T	T
F	F	F	F	T	F	F	T
					(α)	(β)	

and we see that

(i) $p \vee r$ is true whenever $(p \vee q) \wedge (\sim q \vee r)$ is true

so that (ii) the argument $(p \vee q) \wedge (\sim q \vee r) \mid\!\!-\, p \vee r$ is valid

and (iii) $(p \vee q) \wedge (\sim q \vee r) \rightarrow p \vee r$ is a tautology.

6.3 Predicates and Predicate Logic

Many statements are not simple propositions and we define a **predicate** as an expression containing variable symbols such that when the variables are replaced by numerical values, the expression becomes a proposition.

The following are examples of predicates:

(a) If $x < y$, then $x^2 < y^2$ (b) $x + y = 5$ (c) $A \geq 60$

(d) age > 60 (e) If $x < 3$, then $y = 5$, else $y = 10$

A predicate which involves one variable is called a **unary** (or one-place) predicate whilst one which involves two variables is called a **binary** (or two-place) predicate and, in general, a predicate which involves n variables is **n-ary**.

We denote predicates by capital letters P, Q, R, ... and list the variables in parentheses. For example, we could have

$$\mathrm{P}(x), \quad \mathrm{Q}(x, y) \quad \mathrm{R}(x, y, z)$$

A predicate can be formed from other predicates using logical connectives just as in the case of propositions. For example

$$A(x) \rightarrow B(x), \quad B(x, y) \leftrightarrow C(x, y)$$

The terminology used is that a **predicate** is **satisfiable**, provided that there is an **n-tuple** of values which satisfies it. If all n-tuples satisfy it, it is said to be **valid.**

Two predicates are **equivalent** if they have the same truth value for all possible values of their variables; note that $P(x)$ and $Q(x)$ are equivalent if $P(x) \leftrightarrow Q(x)$ is valid.

Quantifiers

Quantifiers are phrases that tell us in some sense what values may be taken by the variables specified in a predicate.

The **universal** quantifier $\forall$ is interpreted as "for all", "for each" or "for every". Thus $\forall x(x > 0)$ is read "for every x, x is greater than zero". Similarly $\forall x\, \forall y\, P(x, y)$ is read "for all x and y, $P(x, y)$", where $P(x, y)$ is some unspecified property of x and y. An example is $\forall x\, P(x)$ where x is a book in the University Library and $P(x)$ is the property that x has a green cover. Thus $\forall x\, P(x)$ says that every book in the University Library has a green cover.

The **existential** quantifier $\exists$ is read "there exists one". Thus $\exists x(x > 0)$ is read "there exists an x such that x is greater than zero". Similarly, $\exists x\, \exists y\, (x + y < 10)$ is read "there exist x and y such that $x + y$ is less than 10". If x is a book in the University Library and $P(x)$ is the property that x has a green cover then $\exists x\, P(x)$ means that there exists one book in the University Library which has a green cover. Note that we really mean 'at least one'.

We can combine $\forall$ and $\exists$: for example, $\forall x\, \exists y\, P(x, y)$ which is read "for all x, there exists a y such that $P(x, y)$". You should consider whether this is the same as $\exists y\, \forall x\, P(x, y)$ which says "there exists one y so that for all x, $P(x, y)$ is true".

Examples

Consider the following predicates

(a) $P(x)$: x is an intelligent student (b) $Q(x)$: x is a student who likes music

(c) $R(x, y)$: x is a student older than student y.

Then we have

$\forall x\, P(x)$: "for all students $P(x)$ is true", i.e. "All students are intelligent"

$\exists x\, P(x) \wedge Q(x)$: "there is a student who is intelligent and likes music"

$\forall x\, \exists y\, R(x, y) \rightarrow Q(x)$: "Considering all students, there is at least one student y such that if x is older than y then x likes music".

Negations of Quantified Predicates

Consider $\sim\forall x\, P(x)$ which says that "for not all x is $P(x)$ true" which is the same as "There exists some x such that P(x) is not true" which is $\exists x \sim P(x)$;

hence $\sim\forall x\, P(x)$ is equivalent to $\exists x \sim P(x)$

Similarly $\sim\exists x\, P(x)$ is equivalent to $\forall x \sim P(x)$

or in words "there does not exist an x such that P(x) is true" with the equivalent "P(x) is not true for all x".

Predicates and Computer Science

Modern computing now uses predicates at both the programming and applications level. The language PROLOG allows the use of predicates and the logical manipulation of predicates. At the applications level, relational databases permit the storage of information about various aspects of an item. For example, a car database can store data such as model, body type, sun roof, power steering, central locking, etc. The user can pose questions to the database which are in the form of predicates such as "Can I have a hatch-back with 2000 cc engine, including power steering and central locking?" The database can then print out all data items for which the given predicate is true.

Predicate Logic

Propositional logic discussed earlier is just the special case of predicate logic in which all the predicates have no variables. The logic of predicates works in just the same way as propositional logic except that we need to add the rules of inference which relate to quantifiers. These are as follows:

1 $P(a) \vdash \exists x\, P(x)$ (a is a constant here)

2 If a is arbitrary, $P(a) \vdash \forall x\, P(x)$

3 $\exists x\, P(x) \vdash P(a)$ (a is a special element of the universe of x)

4 $\forall x\, P(x) \vdash P(a)$ (a is *any* element of the universe of x)

We also need the relationships between quantifiers and logical connectives. These are

5 $\forall x[P(x) \wedge Q(x)]$ is equivalent to $\forall x\, P(x) \wedge \forall x\, Q(x)$

6 $\exists x[P(x) \vee Q(x)]$ is equivalent to $\exists x\, P(x) \vee \exists x\, Q(x)$

7 $\forall x\, P(x) \vee \forall x\, Q(x) \vdash \forall x\, [P(x) \vee Q(x)]$ (these are not equivalent)

8 $\exists x[P(x) \wedge Q(x)] \vdash \exists x\, P(x) \wedge \exists x\, Q(x)$ (these are not equivalent)

To show the validity of an argument we often proceed as follows:
Take (5) above as an example. Then

Given $\forall x[P(x) \wedge Q(x)]$, for an arbitrarily chosen x we have $P(x) \wedge Q(x)$ and hence $P(x)$.

But x is arbitrarily chosen, so that $\forall x\, P(x)$ and likewise $\forall x\, Q(x)$. Hence $\forall x\, P(x) \wedge \forall x\, Q(x)$.

6.4 Boolean Algebra

In the first section of this chapter we studied propositional logic and found the laws that follow:

1 Commutative laws:

$p \vee q \leftrightarrow q \vee p$ $\quad$ $p \wedge q \leftrightarrow q \wedge p$

2 Associative Laws:

$(p \vee q) \vee r \leftrightarrow p \vee (q \vee r)$ $\quad$ $(p \wedge q) \wedge r \leftrightarrow p \wedge (q \wedge r)$

3 Distributive Laws:

$p \wedge (q \vee r) \leftrightarrow (p \wedge q) \vee (p \wedge r)$ $\quad$ $p \vee (q \wedge r) \leftrightarrow (p \vee q) \wedge (p \vee r)$

4 Identity Laws:

$p \vee f \leftrightarrow p$ $\quad$ $p \wedge t \leftrightarrow p$

5 Negation Laws:

$p \vee \sim p \leftrightarrow t$ $\quad$ $p \wedge \sim p \leftrightarrow f$

Added to these we were able to prove De Morgan's laws:

$\sim(p \wedge q) \leftrightarrow \sim p \vee \sim q$

$\sim(p \vee q) \leftrightarrow \sim p \wedge \sim q$

Similarly we found in Chapter 5 for *Set Theory* the laws

1 Commutative laws:

$A \cup B = B \cup A$ $\quad$ $A \cap B = B \cap A$

2 Associative laws:

$(A \cup B) \cup C = A \cup (B \cup C)$ $\quad$ $(A \cap B) \cap C = A \cap (B \cap C)$

3 Distributive laws:

$A \cap (B \cup C) = (A \cap B) \cup (A \cap C$

$A \cup (B \cap C) = (A \cup B) \cap (A \cup C)$

4 Identity laws:

$A \cup \emptyset = A$ ($\emptyset$ is the empty set) $\quad$ $A \cap U = A$ (U is the universal set)

5 Complement laws:

$A \cup A' = U$ $\quad$ $A \cap A' = \emptyset$

Again, we found De Morgan's laws:

$(A \cup B)' = A' \cap B'$ $\quad$ $(A \cap B)' = A' \cup B'$

The laws for both propositional logic and set theory are seen to be exactly parallel, the only difference being in the notation. They are indeed special instances of a more general theory known as **Boolean Algebra**.

Definition: A Boolean Algebra is an ordered 6-tuple $(S, 0, 1, +, *, ')$ in which S is a set of elements, containing the two distinct elements 0 and 1, + and

$*$ are binary operations on S and $'$ is a unary operation on S such that the following properties hold:

1 Commutative laws:
$x + y = y + x$ $\qquad$ $x * y = y * x$

2 Associative laws:
$(x + y) + z = x + (y + z)$ $\qquad$ $(x * y) * z = x * (y * z)$

3 Distributive laws:
$x * (y + z) = (x * y) + (x * z)$ $\qquad$ $x + (y * z) = (x + y) * (x + z)$

4 Identity laws:
$x + 0 = x$ $\qquad$ $x * 1 = x$

5 Complement laws:
$x + x' = 1$ $\qquad$ $x * x' = 0$

Clearly these laws cover those of both propositional logic and set algebra.
The following properties are easily proved:

Idempotent laws:
$x + x = x$ $\qquad$ $x * x = x$

Null laws:
$x + 1 = 1$ $\qquad$ $x * 0 = 0$

Absorption laws:
$x + (x * y) = x$ $\qquad$ $x * (x + y) = x$

Involution law:
$(x')' = x$

Complement:
$0' = 1$ $\qquad$ $1' = 0$

De Morgan's laws:
$(x + y)' = x' * y'$ $\qquad$ $(x * y)' = x' + y'$

Boolean Algebra and Computer Logic

There is a relationship between Boolean Algebra and the wiring diagrams for the electronic circuits in computers, calculators, control devices and so on. Indeed we shall see that truth functions, the operations of Boolean Algebra and the wiring diagrams are all related. As a result we find that we can simplify wiring diagrams by using the properties of Boolean Algebra.

Electronic Gates

Electronic switches or gates are designed to operate in the same way as logical connectives. We have an OR gate which works as the logical connective $\vee$ – it receives input from two wires carrying voltages and gives an output which depends on the input at a given instant. The convention adopted is that a low voltage is treated as 0 while a high voltage is treated as 1 (0 and 1 behaving like F and T we used earlier). The OR gate or device behaves as described in the truth table below, x and y being the inputs and $x + y$ the output. The actual physical make-up of the gate does not concern us here – only how it behaves.

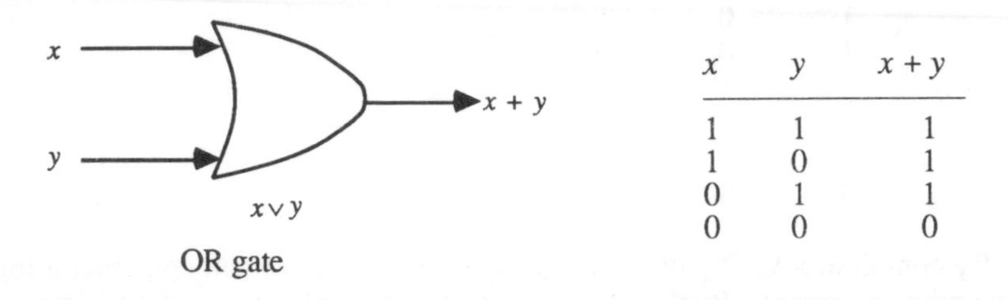

x	y	$x + y$
1	1	1
1	0	1
0	1	1
0	0	0

OR gate

In the same way it is possible to construct a gate which behaves like $\wedge$ (called the AND gate) and one behaving like ~ (called an **inverter**); these are operators of logic with truth tables as given.

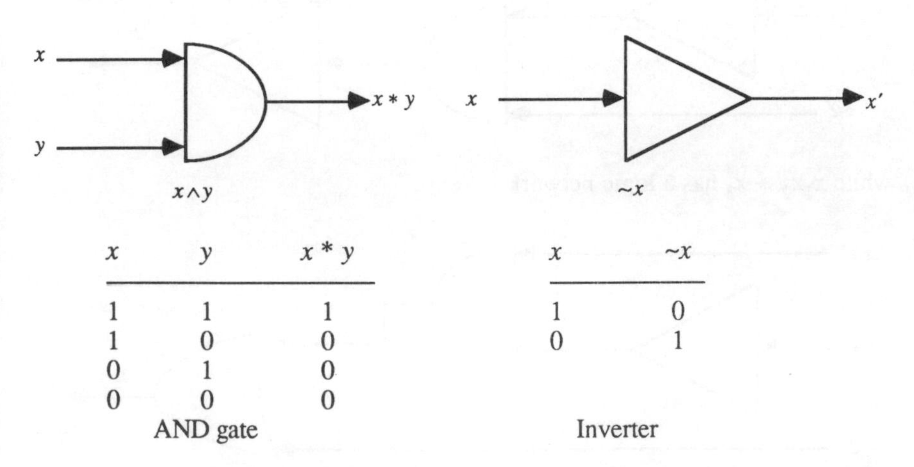

x	y	$x * y$
1	1	1
1	0	0
0	1	0
0	0	0

AND gate

x	$\sim x$
1	0
0	1

Inverter

Some engineers use multiplicative notation instead of * and so the AND gate produces output $x.y$

Each of the devices above is an example of a **Boolean** expression which is defined as any finite string of symbols $x_1, x_2, \ldots x_n$ formed by the operators +, . , and '. To avoid excessive use of parentheses, the convention is adopted that . takes precedence over +, so that $x_1 + x_2.x_3$ stands for $x_1 + (x_2.x_3)$. Furthermore the symbol . is often omitted and so we write $x_1 + x_2x_3$.

Further examples of Boolean expressions are

$$x_1x_2 + x_1'x_2', (x_1'x_2)'(x_1 + x_3) \text{ and } x_1x_2x_3 + x_1x_2x_3' + x_1x_2'x_3'$$

The truth table of a Boolean expression contains 0's and 1's and defines a **truth function** f whose domain is a set of n-tuples of 0's and 1's and whose range is {0, 1}.

Any Boolean expression defines a unique truth function. For example, $x_1x_2' + x_3$ defines the truth function given by the table below.

x_1	x_2	x_3	x_1x_2'	$x_1x_2' + x_3$
1	1	1	0	1
1	1	0	0	0
1	0	1	1	1
1	0	0	1	1
0	1	1	0	1
0	1	0	0	0
0	0	1	0	1
0	0	0	0	0

By combining AND gates, OR gates and inverters we can construct a logic network to represent a Boolean expression. For example, the Boolean expression $(x_1'x_2)'$ has a logic network

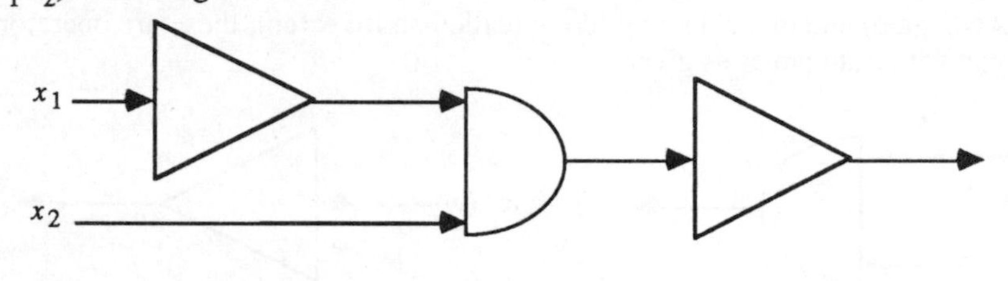

while $x_1x_2' + x_3$ has a logic network

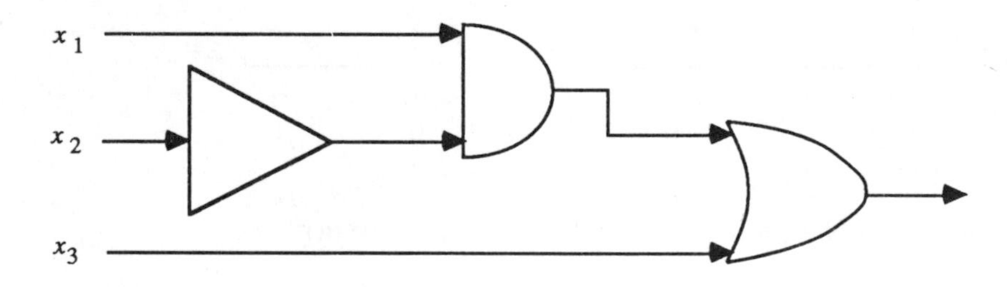

Conversely we can write a Boolean expression with the same truth table as a given logic network. For example, the logic network shown in Figure 6.1 has a Boolean expression $x'y + z'w + y'w$.

So far then we can carry out the procedures

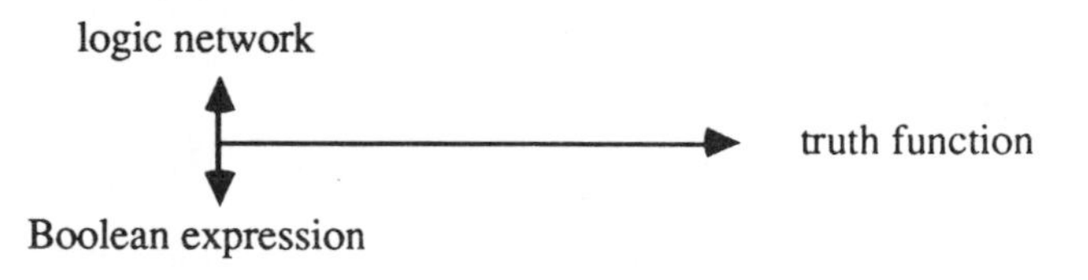

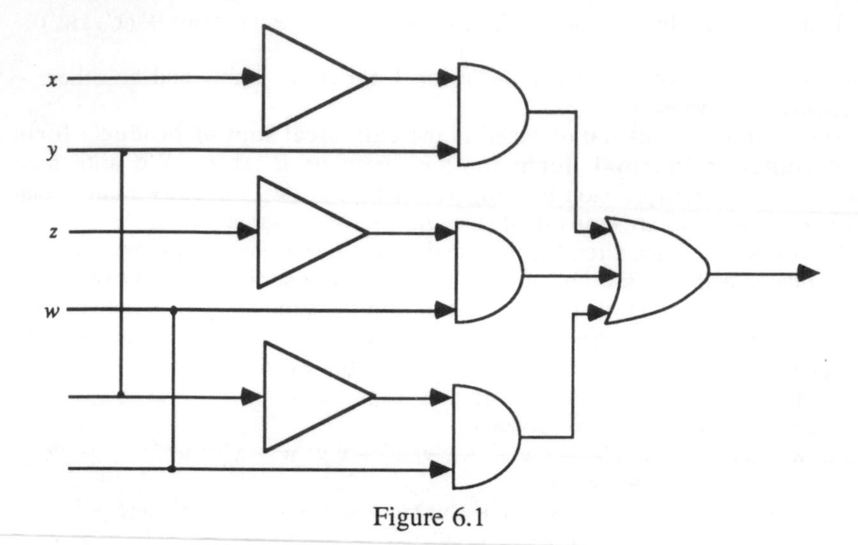

Figure 6.1

The truth function for $x'y + z'w + y'w$ is as follows:

x	y	z	w	$x'y$	$z'w$	$y'w$	$x'y+z'w+y'w$
1	1	1	1	0	0	0	0
1	1	1	0	0	0	0	0
1	1	0	1	0	1	0	1
1	1	0	0	0	0	0	0
1	0	1	1	0	0	1	1
1	0	1	0	0	0	0	0
1	0	0	1	0	1	1	1
1	0	0	0	0	0	0	0
0	1	1	1	1	0	0	1
0	1	1	0	1	0	0	1
0	1	0	1	1	1	0	1
0	1	0	0	1	0	0	1
0	0	1	1	0	0	1	1
0	0	1	0	0	0	0	0
0	0	0	1	0	1	1	1
0	0	0	0	0	0	0	0

What we cannot do at the moment is to take a truth function and obtain an equivalent Boolean expression (and hence a logic network) having the same truth table. The procedure for doing this is simple. We find that we need only to look at the rows of the truth table which have value equal to 1. In the current example this occurs in rows 3, 5, 7, 9, 10, 11, 12, 13 and 15. We first write down the expression of the form

$$(\quad)+(\quad)+(\quad)+(\quad)+(\quad)+(\quad)+(\quad)+(\quad)+(\quad)$$

such that the first term has value 1 for the input values of row 3, the second term has value 1 for the input values of row 5 and so on.

Thus we obtain $(xyz'w) + (xy'zw) + (xy'z'w) + (x'yzw) + (x'yzw') + (x'yz'w) + (x'yz'w') + (x'y'zw) + (x'y'z'w)$
Thus the entire expression has the value 1 for these inputs and no others – precisely what we want.

The form of expression obtained is the **canonical sum of products form** or **disjunctive normal form** for the given truth table. We note how complicated this expression is compared with $x'y+z'w+y'w$ but we know that the two expressions are **equivalent** since they have the same truth table.

Of course, the logic circuit represented by the former expression will be very complicated and it is desirable to find how to **minimise** the Boolean expression for a Boolean function; that is we want to find the simplest logical network for that function. We could try to use all the laws of Boolean algebra together with the properties given earlier to try to simplify the expression but we have no guarantee of getting to the required simplest expression.

For example the expression above is

$$\begin{aligned}
&xyz'w + xy'zw + xy'z'w + x'yzw + x'yzw' + x'yz'w + x'yz'w' \\
&\qquad + x'y'zw + x'y'z'w \\
&= xz'w(y + y') + x'yz(w + w') + x'yz'(w + w') + xy'zw + x'y'w(z + z') \\
&= xz'w + x'yz + x'yz' + x'y'w + xy'zw \\
&= xz'w + x'y(z + z') + y'w(x' + xz) \\
&= xz'w + x'y + y'w(x' + z) \\
&= xz'w + x'y + y'wx' + y'wz
\end{aligned}$$

This is a simpler expression but not the simplest.

6.5 Minimisation of Boolean Expressions

There are two well-known approaches to minimisation – the Karnaugh Map and the Quine-McCluskey algorithm. We shall consider only the former. First we look at two other logic elements which are used in integrated circuits rather than the AND and OR gates and inverters. They are as follows.

I **The NAND gate** which is denoted

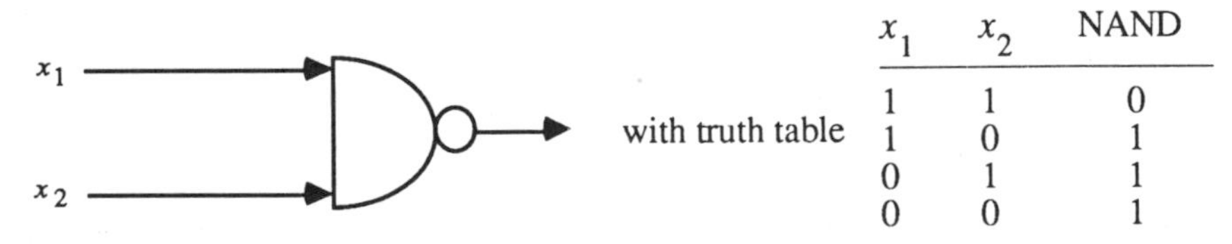

with truth table

x_1	x_2	NAND
1	1	0
1	0	1
0	1	1
0	0	1

It is easy to show that the NAND gate is represented by the Boolean expression $(x_1\,x_2)' = x_1' + x_2'$.

The NAND gate alone is sufficient to provide the same function as inverters, AND and OR gates, as shown below:

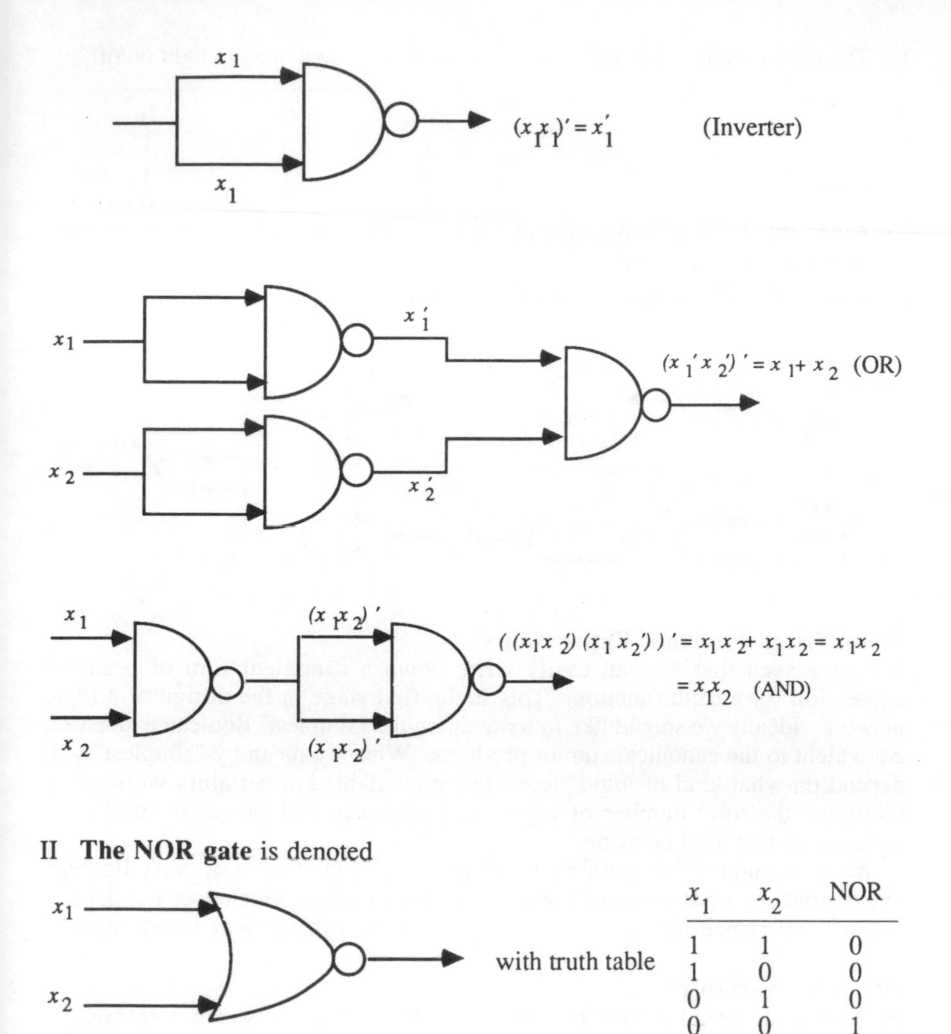

II **The NOR gate** is denoted

x_1	x_2	NOR
1	1	0
1	0	0
0	1	0
0	0	1

The NOR gate has Boolean expression $(x_1 + x_2)'$.

Just as in the case of the NAND gate, networks can be constructed using NOR elements only that can replace an inverter, an OR gate and an AND gate. You should try to do this.

Example

A stair light is controlled by two light switches, one upstairs and one downstairs. Find (a) a truth function (b) a Boolean expression and (c) a logic network using NAND only that allows the light to be switched on or off by either switch.

(a) The truth function is as follows

switch x_1	switch x_2	light on/off
0	0	0
1	0	1
0	1	1
1	1	0

(b) A possible Boolean expression (not necessarily the best) is the canonical sum of products $x_1 x_2' + x_1' x_2$

(c) Using only NAND gates we get the network:

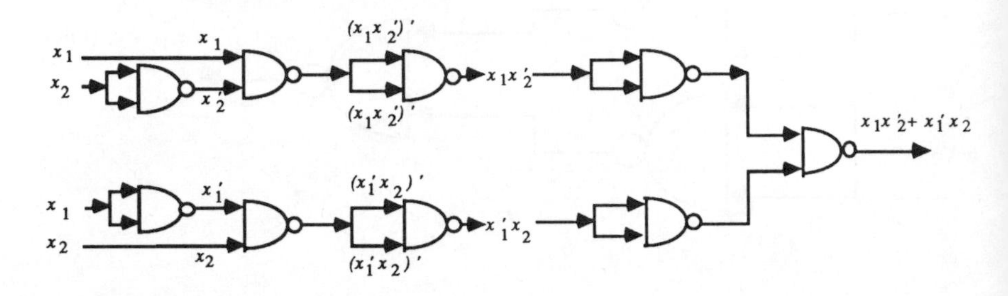

Simplifying Boolean Expressions

We have seen that we can easily write down a canonical sum of products expression for a truth function. This is the first stage in the design of a logic network. Ideally we should like to write down the "simplest" Boolean expression equivalent to the canonical sum of products. What we mean by "simplest" will depend on what kind of logic elements are available but certainly we want to minimize the total number of logic elements used and the total number of connections that must be made.

We have said that it is possible with experience and guile to simplify Boolean expressions but what we want is a mechanical procedure which does not depend on such experience and guile. One such procedure is that of **Karnaugh maps.**

(a) Two variables

In the canonical sum of products form for a truth function, we are interested in values of the input variables that produce outputs of 1. The idea behind a Karnaugh Map is to record the 1's of a function in an array that forces products of inputs differing by only one factor to be adjacent. The array for two variables is as follows:

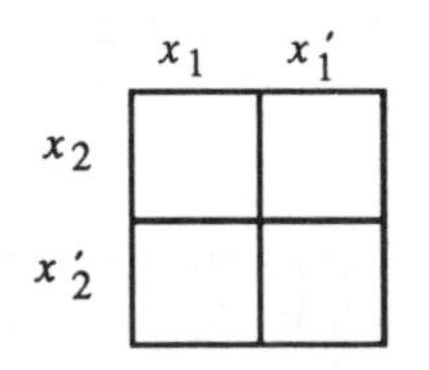

This has adjacent elements x_2x_1 and x_2x_1', x_1x_2 and x_1x_2', x_2x_1' and $x_2'x_1'$, all differing by one factor only.

The truth function

x_1	x_2	$f(x_1, x_2)$
1	1	1
1	0	1
0	1	0
0	0	0

has canonical sum of products form $x_1 x_2 + x_1 x_2'$ and a Karnaugh map

	x_1	x_1'
x_2	1	0
x_2'	1	0

We see that there are two adjacent 1's and that these occur in the first column – the column under x_1. In this column it is x_2 that changes and can be eliminated. Thus the Boolean expression is reduced to x_1.
We know this from Boolean algebra because

$$x_1 x_2 + x_1 x_2' = x_1(x_2 + x_2') = x_1.1 = x_1$$

Similarly, for the truth function

x_1	x_2	$f(x_1, x_2)$
1	1	1
1	0	0
0	1	1
0	0	0

the canonical sum of products is $x_1 x_2 + x_1' x_2$

with Karnaugh map

	x_1	x_1'
x_2	1	1
x_2'	0	0

This time the 1's occur in the row alongside x_2. In this row it is x_1 that changes and can be eliminated. Thus the Boolean expression is reduced to x_2.

Note that $\quad x_1 x_2 + x_1' x_2 = (x_1 + x_1')x_2 = 1.x_2 = x_2$

(b) Three variables

The array form for a function of 3 variables is as follows:

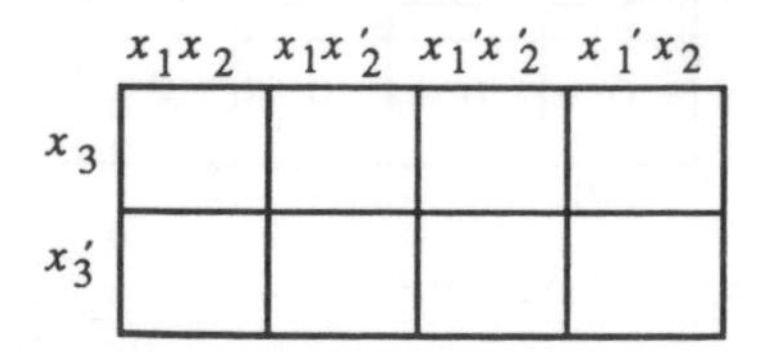

Adjacent cells along the top row are $x_1x_2x_3$, $x_1x_2'x_3$, $x_1'x_2'x_3$, $x_1'x_2x_3$; all differing by only one factor, the first and the last also differing by one change only. (You can imagine the map wrapped round a circular drum with the left-hand and right-hand edges glued together.) Similarly for the bottom row and so in the array, adjacent squares differ by only one factor.

If two adjacent squares are marked with 1's, one variable can be eliminated, just as before. If four adjacent squares are marked with 1's, either in a single row or a square, two variables may be eliminated.

As an example, we are given the truth function

x_1	x_2	x_3	$f(x_1, x_2, x_3)$
1	1	1	1
1	1	0	1
1	0	1	1
1	0	0	1
0	1	1	0
0	1	0	0
0	0	1	0
0	0	0	0

The Karnaugh map is

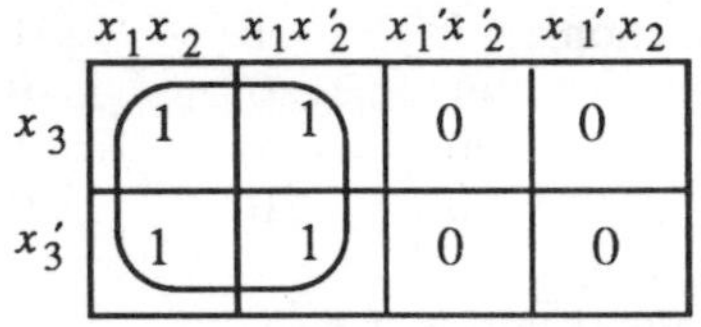

	x_1x_2	x_1x_2'	$x_1'x_2'$	$x_1'x_2$
x_3	1	1	0	0
x_3'	1	1	0	0

with four adjacent squares.

The canonical sum of products is

$$\begin{aligned} x_1x_2x_3 + x_1x_2'x_3 + x_1x_2'x_3' + x_1x_2x_3' &= x_1x_3(x_2 + x_2') + x_1x_3'(x_2 + x_2') \\ &= x_1x_3 + x_1x_3' \\ &= x_1(x_3 + x_3') \\ &= x_1 \end{aligned}$$

The four adjacent squares have a common x_1 which does not change.

Similarly

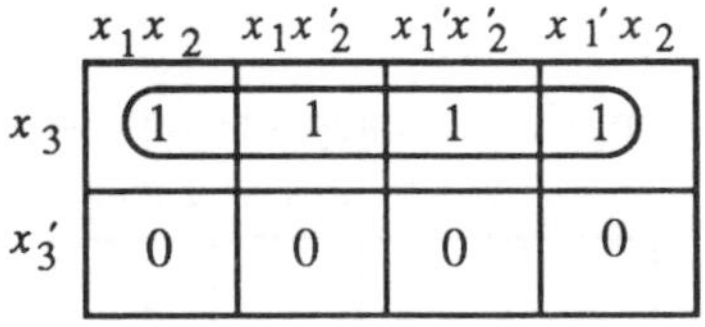

	x_1x_2	x_1x_2'	$x_1'x_2'$	$x_1'x_2$
x_3	1	1	1	1
x_3'	0	0	0	0

would have Boolean expression x_3, while

	x_1x_2	x_1x_2'	$x_1'x_2'$	$x_1'x_2$
x_3	1	0	0	1
x_3'	1	0	0	1

would have Boolean expression x_2.

Two adjacent squares can occur in many ways; examples are

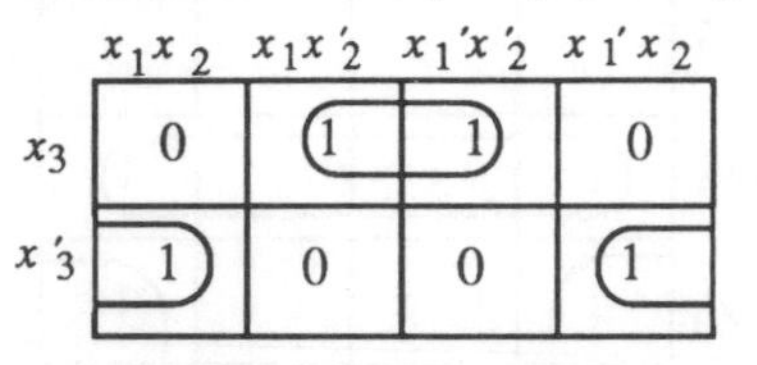

which would have Boolean expression $x_3x_2' + x_3'x_2$ (look for the variables that stay constant).

Of course, groups of adjacent squares can overlap. Let us draw the Karnaugh map of $x_1x_2x_3 + x_1x_2'x_3 + x_1x_2x_3' + x_1x_2'x_3' + x_1'x_2'x_3$:

	x_1x_2	x_1x_2'	$x_1'x_2'$	$x_1'x_2$
x_3	1	1	1	0
x_3'	1	1	0	0

We can therefore write the expression as $x_1 + x_2'x_3$

(c) Four variables

The array for a function of four variables is

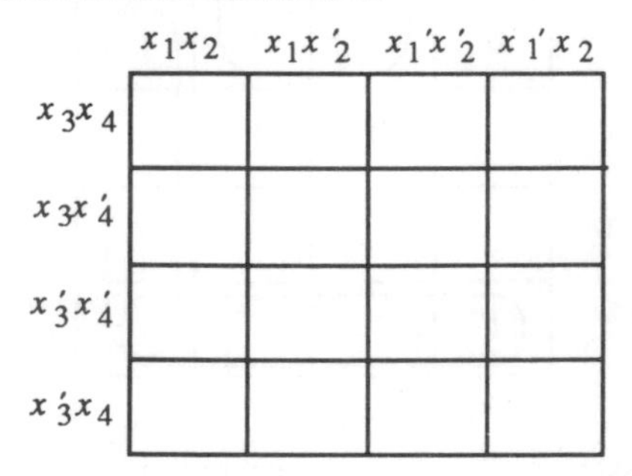

Two adjacent squares marked with a 1 allow one variable to be eliminated, four adjacent squares marked with a 1 allow two variables to be eliminated, while eight adjacent squares marked with a 1 allow 3 variables to be eliminated.

Taking the example

$$x_1x_2x_3x_4 + x_1x_2x_3'x_4 + x_1'x_2x_3x_4 + x_1'x_2x_3'x_4 + x_1'x_2'x_3'x_4'$$
$$+ x_1x_2'x_3'x_4' + x_1'x_2x_3'x_4' + x_1x_2x_3'x_4' + x_1'x_2'x_3x_4$$

we draw the Karnaugh map as follows:

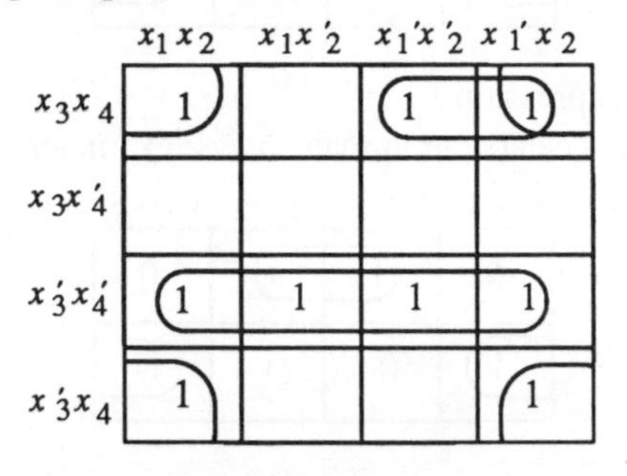

The expression equivalent to the above is $x_3'x_4' + x_2x_4 + x_1'x_3x_4$
Notice the pattern with four adjacent squares.

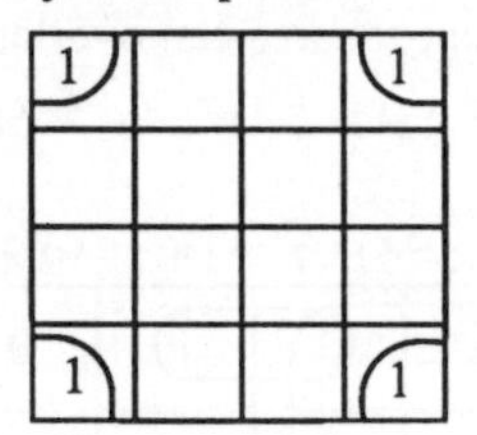

Returning to the example on page 170, where we have the expression

$$xyz'w + xy'zw + x'yzw' + xy'z'w + x'yz'w' + x'yz'w + x'y'zw$$
$$+ x'y'z'w + x'yzw$$

The Karnaugh map is

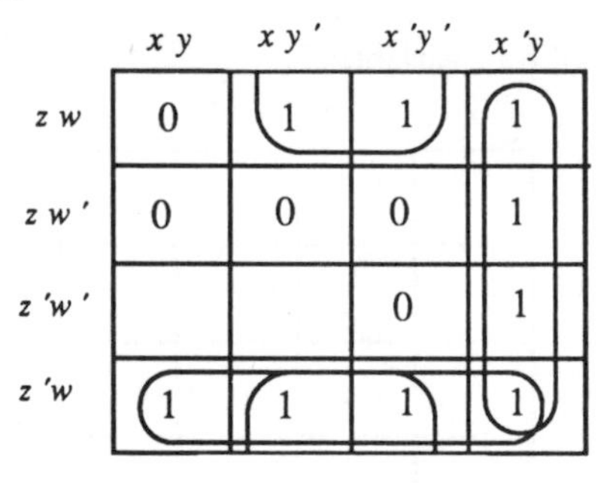

The equivalent expression is $x'y + z'w + y'w$ which is the minimal expression.

In the minimisation process we must use every square marked with a 1 in such a way as to obtain the largest combination of marked squares possible but attempt to find a "covering" which uses as few blocks as possible.

Let us take a further example. The expression given is

$$x_1x_2x_3x_4 + x_1x_2x_3x_4' + x_1x_2'x_3x_4' + x_1'x_2'x_3x_4' + x_1x_2x_3'x_4'$$
$$+ x_1x_2'x_3'x_4' + x_1'x_2'x_3'x_4' + x_1x_2x_3'x_4 + x_1'x_2x_3'x_4$$

with Karnaugh map:

	x_1x_2	x_1x_2'	$x_1'x_2'$	$x_1'x_2$
x_3x_4	1			
x_3x_4'	1	1	1	
$x_3'x_4'$	1	1	1	
$x_3'x_4$	1			1

We first proceed to mark blocks of size 8 – there are none. Then we mark all blocks of size 4 – there are two; the first column and centre block.
Next we mark all blocks of size 2 that are not already covered by a block of larger size and find the 1 in the lower left is adjacent to the block in the lower right. Finally we mark any single boxes not yet marked – there are none. Thus we have:

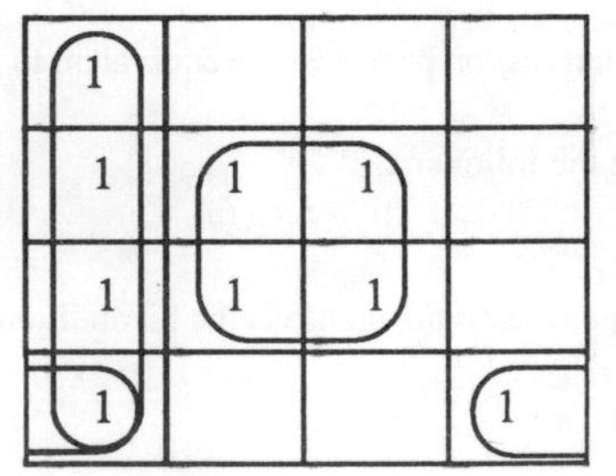

giving the minimal expression $x_1x_2 + x_2'x_4' + x_2x_3'x_4$

It is possible to use Karnaugh maps for functions of five, six and more variables but these need three-dimensional drawings and are usually very complicated. In such cases, the Quine-McCluskey procedure allied to a systematic computerized algorithm is more appropriate.

Problems

Section 6.1

1 Which of the following is a proposition?
 (i) The population of the UK is 60 million. (ii) $x > y$
 (iii) If n is an even integer, $(n + 1)$ is odd.

2 Let p be "It is hot" and q be "I am tired". Convert to English statements
 (i) $\sim p$ (ii) $p \wedge q$ (iii) $p \vee q$ (iv) $q \vee \sim p$ (v) $\sim p \wedge \sim q$

3 Let p be "He is old" and q be "He is clever". Write in symbolic form:
 (i) He is old and clever. (ii) He is old but not clever.
 (iii) He is neither old nor clever. (iv) It is not true that he is young or not clever.

4 Find truth tables for the following
 (i) $\sim p \wedge q$ (ii) $\sim(p \wedge q)$ (iii) $\sim(p \vee \sim q)$ (iv) $\sim p \vee \sim q$ (v) $p \vee (q \wedge r)$
 (vi) $(p \vee q) \wedge r$ (vii) $(p \vee q) \wedge (p \vee r)$ (viii) $\sim (p \vee (q \wedge r))$ (ix) $\sim p \wedge (\sim q \vee \sim r)$

5 Using truth tables find whether the following are tautologies or contradictions or neither.
 (i) $\sim(p \wedge q) \vee (\sim p \vee \sim q)$ (ii) $(\sim p \wedge q) \vee (\sim(\sim p \wedge q))$
 (iii) $(\sim p \vee q) \wedge (p \vee \sim q)$ (iv) $(p \wedge \sim q) \wedge (\sim p \vee q)$

6 Prove the propositions on pages 155 and 156, excepting 4(a) and 5(b).

7 Show that $(p \vee q) \wedge r$ is not equivalent to $p \vee (q \wedge r)$ but that $p \vee \sim(q \wedge r)$ is equivalent to $(p \vee \sim q) \vee \sim r$.

8 Show that the following pairs of statements are equivalent.
 (i) (($n = 6$) OR ($a > 4$)) AND ($x = 1$); (($n = 6$) AND ($x = 1$)) OR (($a > 4$) AND ($x = 1$))
 (ii) NOT (($n = 6$) AND ($a \leq 4$)); (($n <> 6$) OR ($a > 4$))

9 Show that forms (i), (ii) and (iii) on page 157 are equivalent to $p \rightarrow q$

10 Construct truth tables for the following:
 (i) $((p \rightarrow q) \wedge p) \rightarrow q$ (ii) $p \leftrightarrow (q \rightarrow r)$ (iii) $(p \leftrightarrow q) \leftrightarrow r$

11 Show that the following pairs are equivalent via the biconditional.
 (i) $(p \vee q) \wedge r$; $(p \wedge r) \vee (q \wedge r)$ (ii) $(p \rightarrow q) \rightarrow r$; $\sim(p \wedge \sim q) \rightarrow r$
 (iii) $(p \wedge q) \vee (\sim p \wedge \sim q)$; $p \leftrightarrow q$

12 Write down the converse, inverse and contrapositive to the following statements.
 (i) Every geometric figure with four right angles is a square.
 (ii) All engineers have practical skills or are good at mathematics.

Section 6.2

13 Show via truth tables that the following arguments are valid.
 (i) $p \vdash p \vee q$ (addition) (ii) $p \wedge q \mid\!\rightarrow p$ (specialization)
 (iii) $p, q \vdash p \wedge q$ (conjunction) (iv) $p \leftrightarrow q, q \vdash p$
 (v) $p \rightarrow q, \sim q \vdash \sim p$

14 Show that the argument $p \rightarrow \sim q, r \rightarrow q, r \vdash \sim p$ is valid
 (i) via a truth table (ii) by an argument similar to that on page 161.

15 Test the validity of the following arguments.
 (i) $\sim p \rightarrow q, p \vdash \sim q$ (ii) $\sim p \rightarrow \sim q, q \vdash p$ (iii) $p \rightarrow q, r \vee \sim q, \sim r \vdash \sim p$
 (iv) If I work hard I will pass mathematics. I did not work hard. I failed mathematics.

(v) If 6 is even, then 2 does not divide 7. Either 5 is not prime or 2 divides 7. However, 5 is prime. Therefore 6 is not even.

16 Show that $p \leftrightarrow q$ logically implies $p \rightarrow q$, but that $p \leftrightarrow \sim q$ does not logically imply $p \rightarrow q$

17 Show that $p \wedge (q \vee r)$ logically implies $(p \wedge q) \vee r$

Section 6.3

18 What is the truth value of each of the following where the domain consists of the integers?

(i) $(\forall x)(\exists y)(x + y = y)$ (ii) $(\exists y)(\forall x)(x + y = y)$
(iii) $(\forall x)(\exists y)(x + y = 0)$ (iv) $(\exists y)(\forall x)(x + y = 0)$
(v) $(\exists x)(\exists y)(y = x^2)$ (vi) $(\forall x)(x^4 > 0)$

19 Prove (6), (7), (8) on page 164.

20 If $R(x)$ is "x is a Rolls-Royce", $M(x)$ is "x is a Metro", $J(x)$ is "x is a Jaguar", $F(x, y)$ is "x is faster than y", then write each of the following as a symbolic statement.

(i) Only Rolls-Royces are faster than Metros.
(ii) All Jaguars are faster than some Rolls-Royces.
(iii) If there is a Rolls-Royce that is faster than a Jaguar then all Rolls-Royces are faster than all Jaguars.

21 Prove that each of the following is a theorem of predicate logic.

(i) $(\forall x)P(x) \rightarrow (\forall x)(P(x) \vee Q(x))$
(ii) $(\exists x)(\exists y)P(x, y) \rightarrow (\exists y)(\exists x)P(x, y)$
(iii) $(\forall x)(P(x) \rightarrow Q(x)) \rightarrow ((\exists x)(P(x) \rightarrow (\exists x)Q(x))$

22 Use predicate logic to verify the following arguments.

(i) Every computer has an input device. The Prime is a computer. Hence the Prime has an input device.
(ii) 2 is an integer. The square root of 2 is an irrational number. Hence some integers have irrational numbers as their square root.
(iii) All members of a panel come from either industry or the Civil Service. Everyone in the Civil Service with a history degree is in favour of a particular motion. Nigel is not from industry but does have a history degree. Therefore, if Nigel is a member of the panel he is in favour of the motion.

Section 6.4

23 Prove the properties on page 166.

24 Construct logic networks for the following Boolean expressions using AND gates, OR gates and inverters.

(i) $(x_1 + x_2')x_3$ (ii) $(x_1 + x_2)' + x_2'x_3$
(iii) $x_1x_2' + (x_1x_2)'$ (iv) $(x_1 + x_3)'x_2 + x_2'$

25 A logic network for a computer printer control is shown below

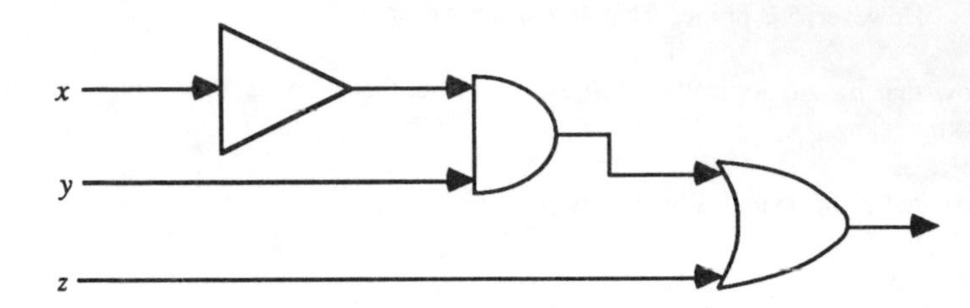

where x represents the "paper is out", y: "data is present", z: "override switch is on". Explain how the network operates and represent it in symbolic form.

26 Find the truth functions for the Boolean expressions

$$(x_1 + x_2)(x_1' + x_2)(x_2 + x_3) \quad \text{and} \quad (x_1 x_2) + (x_1' x_3)$$

Can you draw any conclusions?

Section 6.5

27 Construct a logical network corresponding to $x_1 x_3 + x_2'$ using NAND elements.

28 Via truth tables show that (i) $\sim p$ is equivalent to p NAND q (ii) $p \vee q$ is equivalent to (p NAND p) NAND (q NAND q).

29 Produce Karnaugh maps for

(i) $(x_1 x_2 + x_1 x_2')'$

(ii) $x_1 x_2' x_3' + x_1 x_2 + x_3$

(iii) $x_1 x_2 x_3 + x_1 x_2' x_3 + x_1 x_2 x_3'$

(iv) $x_1 + x_1' x_2' + x_1 x_2 x_3' + x_4$

(v) $x_1 x_2 x_3 x_4 + x_1 x_2 x_3' x_4 + x_1' x_2' x_3' x_4'$

30 Via Karnaugh maps simplify the following Boolean expressions

(i) $x_1 x_2' x_3' + x_1 x_2' x_3$

(ii) $x_1 x_2 x_3 + x_1 x_2 x_3' + x_1 x_2' x_3$

(iii) $x_1 x_2 x_3 + x_1 x_2 x_3' + x_1' x_2 x_3' + x_1' x_2' x_3'$

(iv) $x_1' x_2 x_3' + x_1' x_2' x_3' + x_1 x_2' x_3' + x_1 x_2 x_3 + x_1 x_2' x_3$

(v) $x_1 x_2 x_3 x_4 + x_1' x_2 x_3 x_4' + x_1 x_2' x_3 x_4 + x_1 x_2' x_3 x_4 + x_3' x_4'$

7

GEOMETRY AND CURVES

7.1 Coordinate Geometry of the Plane

It is assumed that the reader has already met some of the standard results in elementary coordinate geometry. We collect a few of them here for revision purposes.

Distance between two points
By Pythagoras' Theorem, the distance between the two points (x_1, y_1) and (x_2, y_2) is

$$d = \sqrt{(x_1 - x_2)^2 + (y_1 - y_2)^2} \tag{7.1}$$

Gradient of a line
The **gradient** or slope, m, of the line joining (x_1, y_1) and (x_2, y_2) is defined by

$$m = (y_2 - y_1)/(x_2 - x_1) \tag{7.2}$$

Note that if $x_1 = x_2$ the line is parallel to the y-axis (and vice versa).

The straight line
The equation of a straight line can take one of many forms, depending on the information given.

(i) *Line through two points, (x_1, y_1) and (x_2, y_2)*

$$y - y_1 = \left[\frac{y_2 - y_1}{x_2 - x_1}\right](x - x_1) \tag{7.3}$$

(ii) *Line through a given point (x_1, y_1) with a given slope m*

$$y - y_1 = m(x - x_1) \tag{7.4}$$

(iii) *Line making known intercepts on the axes*
Suppose the intercepts are $(a, 0)$ and $(0, b)$. The equation is

$$\frac{x}{a}+\frac{y}{b}=1 \qquad (7.5)$$

(iv) *Line with slope m and intercept c on the y-axis*

$$y = mx + c \qquad (7.6)$$

(v) *General form of the equation*

Other forms exist, but they are all rearrangements of the general equation

$$ax + by + c = 0 \qquad (7.7)$$

Writing this as $y = -ax/b - c/b$ and comparing with (7.6) we see that this has slope $-a/b$ and intercept on the y-axis $-c/b$.

Angle between two lines

For the lines $y = m_1x + c_1$ and $y = m_2x + c_2$; let θ be the angle between the lines, then

$$\tan\theta = \frac{m_2 - m_1}{1 + m_2 m_1} \qquad (7.8)$$

If $m_1 = m_2$ then $\tan\theta = 0$ and the lines are parallel.

If $m_1m_2 \to -1$ $\tan\theta \to \infty$ and $\theta \to \pi/2$; i.e. the condition for two lines to be perpendicular is that the product of their gradients is equal to -1.

For the lines $a_1x + b_1y + c_1 = 0$ and $a_2x + b_2y + c_2 = 0$ the angle is given by

$$\cos\theta = \frac{a_1 a_2 + b_1 b_2}{\sqrt{(a_1^2 + b_1^2)(a_2^2 + b_2^2)}} \qquad (7.9)$$

These lines are parallel if

$$a_1b_2 = a_2b_1 \qquad (7.10a)$$

and perpendicular if

$$a_1a_2 + b_1b_2 = 0 \qquad (7.10b)$$

Notice that these results do not let us down in special cases, e.g. $a_1 = 0$.

Shortest distance of a point from a line

If the point is (x_1, y_1) and the line is $ax + by + c = 0$ then the required distance is d where

$$d^2 = (ax_1 + by_1 + c)/(a^2 + b^2) \qquad (7.11)$$

Equation of a circle

The equation of a circle centred at (x_0, y_0) and of radius r is

$$(x - x_0)^2 + (y - y_0)^2 = r^2 \qquad (7.12)$$

Equation of an ellipse
The ellipse shown in Figure 7.1 has equation

$$\frac{x^2}{a^2}+\frac{y^2}{b^2}-1=0 \tag{7.13}$$

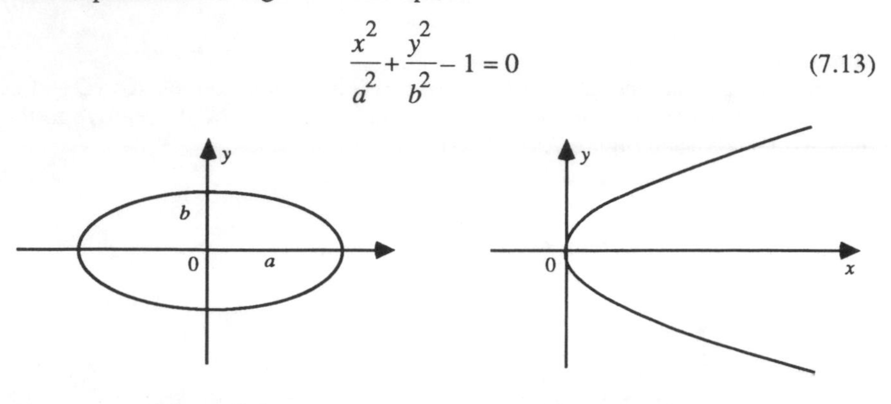

Figure 7.1 Figure 7.2

Equation of a parabola
The parabola shown in Figure 7.2 has equation

$$y^2-4ax=0 \tag{7.14}$$

Notice that the curve $y = Kx^2$ where K is constant, is also a parabola.

Equation of a hyperbola
The hyperbola with equation

$$\frac{x^2}{a^2}-\frac{y^2}{b^2}-1=0 \tag{7.15}$$

is shown in Figure 7.3(a).

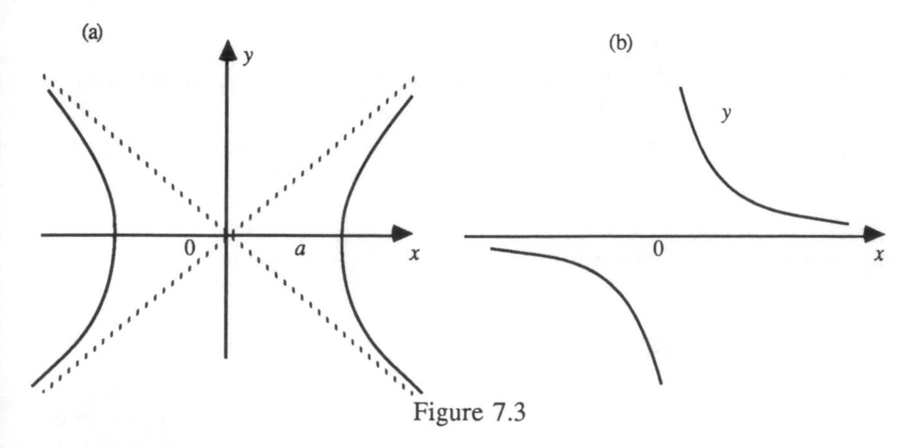

Figure 7.3

In the special case where $a = b$, a *rectangular hyperbola* (Figure 7.3(b)), the axes are rotated clockwise to coincide with the asymptotes and the equation becomes

$$xy - c^2 = 0 \tag{7.16}$$

Translation of axes

We now investigate the taking of new axes $YO'X$ parallel to the axes yOx. Let P have coordinates (x, y) in the axes yOx and coordinates (X, Y) relative to the new axes $YO'X$ (See Figure 7.4). Then

$$x = X + p, \quad y = Y + q \tag{7.17}$$

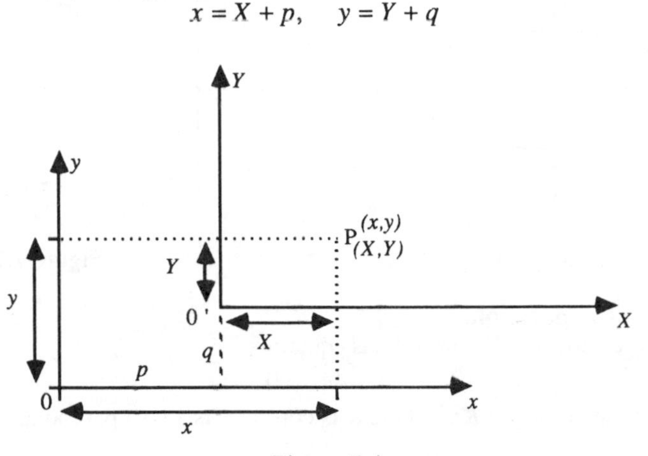

Figure 7.4

Let us apply this to a circle with centre O' at (x_0, y_0) and radius a. Take new axes $XO'Y$ through O' parallel to the axes xOy. Relative to O', the equation of the circle must be $X^2 + Y^2 = a^2$. But $x = X + x_0$ and $y = Y + y_0$, so that $X = x - x_0$ and $Y = y - y_0$. Hence relative to the original axes xOy the equation becomes $(x - x_0)^2 + (y - y_0)^2 = a^2$.

General equation of a conic

All conic sections can be represented by the general equation of second degree

$$ax^2 + 2hxy + by^2 + 2gx + 2fy + c = 0 \tag{7.18}$$

For example if $h^2 < ab$ then (7.18) represents an ellipse. Note that if the left-hand side of the equation be denoted by S then the equation

$$(1 - \lambda)S_1 + \lambda S_2 = 0 \tag{7.19}$$

represents a family of conics with $\lambda = 0$ and $\lambda = 1$ giving special cases.

Example 1

The conic sections given by

$$(1 - \lambda)(x^2 + y^2 - a^2) + \lambda(x - a)(y - a) = 0$$

represent for $0 \le \lambda \le 1$ a blending between the circular section $x^2 + y^2 - a^2 = 0$ when $\lambda = 0$ and the square section $x = a$, $y = a$ when $\lambda = 1$. Such a set of blending curves has been used in the design of aircraft fuselages.

Example 2
Find the conic which passes through the five points (0, 0), $(a, 0)$, (a, a), $(0, a)$ and $(\frac{5a}{4}, \frac{a}{4})$.

We first find two pairs of lines having the first four points as their intersections. A simple choice is $x = 0$, $x - a = 0$ and $y = 0$, $y - 1 = 0$. Refer to Figure 7.5.

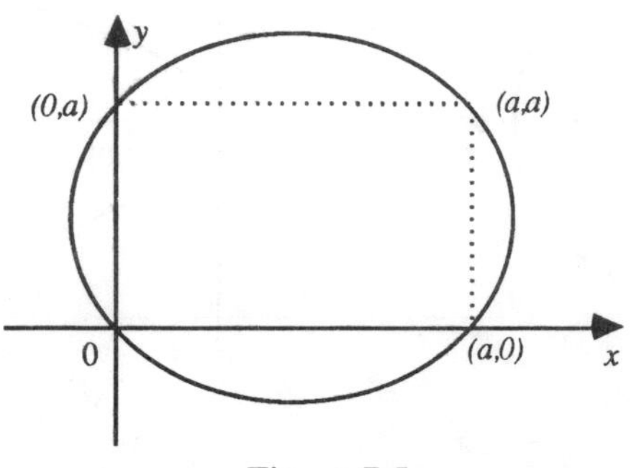

Figure 7.5

Then every conic section through these points has equation, for some λ,

$$(1 - \lambda)\, x(x - a) + \lambda y(y - a) = 0$$

If the conic is to pass through $(\frac{5a}{4}, \frac{a}{4})$ then

$$(1 - \lambda)\left[\frac{5a}{4}\right]\left[\frac{a}{4}\right] + \lambda\left[\frac{a}{4}\right]\left[-\frac{3a}{4}\right] = 0$$

or
$$5(1 - \lambda) - 3\lambda = 0$$

i.e.
$$\lambda = 5/8$$

so that the required (unique) conic is

$$3x(x - a) + 5y(y - a) = 0$$

or
$$3x^2 + 5y^2 - 3ax - 5ay = 0$$

which can easily be shown to be an ellipse.

7.2 Inequalities Involving Two Variables

Consider the equation $x^2 + y^2 = a^2$; we may write it as $f(x, y) \equiv x^2 + y^2 - a^2 = 0$. All points on the circumference satisfy $f(x, y) = 0$. It can be shown that points inside the circle satisfy $f(x, y) < 0$ and points outside the circle satisfy $f(x, y) > 0$ (see Figure 7.6(i)). This is an example of a general principle that the plane is divided into three regions by a simple curve (i.e. one which contains no loops). Consider the parabola $y^2 = 4x$ rewritten as

$f(x, y) \equiv y^2 - 4x = 0$. To find which region corresponds to $f(x, y) < 0$ and which to $f(x, y) > 0$ all we need do is take a point in either region and evaluate $f(x, y)$ there. Then, for example, $f(1, 0) = -4 < 0$ or $f(-1, 0) = 4 > 0$. The result is shown in Figure 7.6(ii).

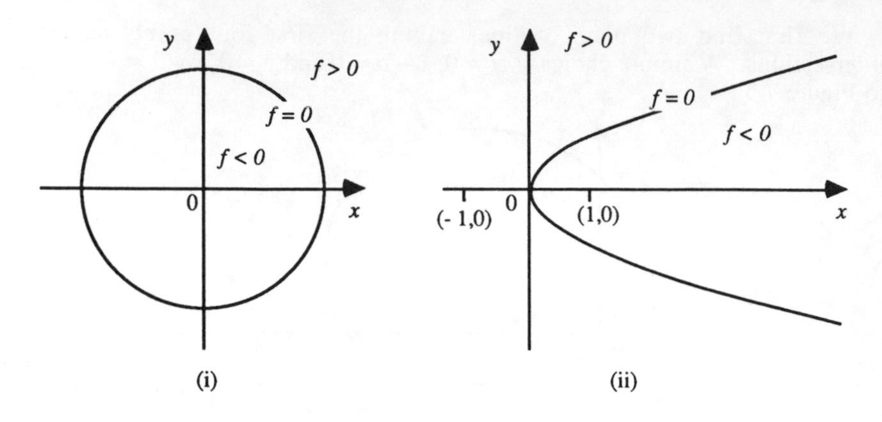

Figure 7.6

The inequalities produced by straight lines are of importance. Consider the line $x + y - 3 = 0$, then $f(x, y) = x + y - 3$ and $f(0, 0) < 0$. Hence the plane is divided into three distinct regions as shown in Figure 7.7(i).

Special cases relate to the coordinate axes. The region $x \geq 0$ lies on and to the right of the y-axis, the region $y > 0$ lies on and above the x-axis and so the shaded region in Figure 7.7(i) satisfies the simultaneous inequalities, $x \geq 0$, $y \geq 0$, $x + y - 3 \leq 0$; such a region is known as the **region of feasible solutions** to this set of inequalities.

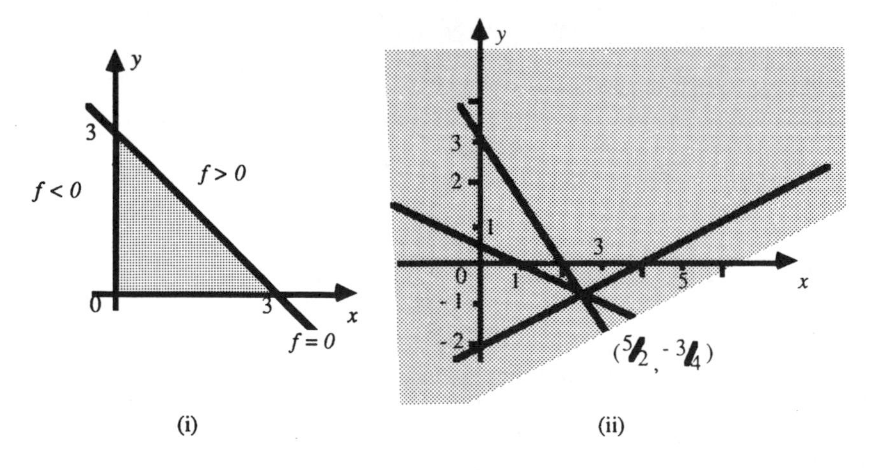

Figure 7.7

Consider the statement $3x + 2y - 6 < 2x + 4y - 2$. We can rearrange this to $3x + 2y - 6 - (2x + 4y - 2) < 0$, i.e. $x - 2y - 4 < 0$. This region is shaded in Figure 7.7(ii) and the lines $3x + 2y - 6 = 0$ and $2x + 4y - 2 = 0$ are sketched.

Note that the line $x - 2y - 4 = 0$ passes through the point of intersection of the original two lines, viz (5/2, –3/4). The values of x and y at all points in the shaded region satisfy the given inequality.

7.3 Tangents and Normals

The tangent to the curve $y = f(x)$ at the point (x_1, y_1) is given by

$$y - y_1 = f'(x_1)(x - x_1) \tag{7.20}$$

where $f'(x_1)$ is the value of df/dx at $x = x_1$.

Similarly, the normal to the curve at the same point is given by

$$y - y_1 = -(x - x_1)/f'(x_1) \tag{7.21}$$

Difficulties will arise when the tangent is parallel to the x-axis. Those of you who have not studied partial differentiation should return to this next paragraph after reading Chapter 14.

If the curve is given in the implicit form $g(x, y) = 0$ then an equation for the tangent can be written as

$$g_x(x_1, y_1).(x - x_1) + g_y(x_1, y_1)(y - y_1) = 0 \tag{7.22}$$

where $g_x(x_1, y_1)$ is the value of $\partial g/\partial x$ at (x_1, y_1).

Similarly, an equation for the normal at the same point can be written as

$$g_y(x_1, y_1).(x - x_1) - g_x(x_1, y_1)(y - y_1) = 0 \tag{7.23}$$

These latter forms avoid the difficulties mentioned.

Example

To find the tangent to the ellipse $x^2 + 4y^2 - 4 = 0$ at the point (–2, 0) we first note that $g(x, y) = x^2 + 4y^2 - 4$ so that $g_x = 2x$, $g_y = 8y$. Then $g_x(-2, 0) = -4$, $g_y(-2, 0) = 0$ and therefore the tangent is given by

$$-4(x + 2) + 0(y - 0) = 0$$

i.e. the line $x = -2$.

Similarly the normal at (–2, 0) is given by

$$0(x + 2) + 4(y - 0) = 0$$

i.e. the line $y = 0$.

In the case of the general conic $ax^2 + 2hxy + by^2 + 2gx + 2fy + c = 0$, $g_x = 2ax + 2hy + 2g$, $g_y = 2hx + 2by + 2f$.

We quote the equations for tangents to standard conics. For the ellipse the tangent at (x_1, y_1) is

$$\frac{x x_1}{a^2} + \frac{y y_1}{b^2} = 1 \tag{7.24}$$

For the parabola the equation is

$$yy_1 = 2a(x + x_1) \tag{7.25}$$

and for the hyperbola the equation is

$$\frac{x x_1}{a^2} - \frac{y y_1}{b^2} = 1 \qquad (7.26)$$

Note that another approach to finding the equation of a tangent is to consider the intersection of a straight line with a curve and determine the conditions under which there is only one point of intersection. For example, for the parabola $y^2 = 4ax$, the intersections with the line $y = mx + c$ leads to the quadratic equation $m^2x^2 + (2mc - 4a)x + c^2 = 0$. For a repeated root, $c = a/m$ and hence the equation of the tangent becomes

$$y = mx + a/m \qquad (7.27)$$

This particular form gives the equation of the tangent for any given slope m.

7.4 Simple Ideas on Curve Sketching

We speak here of *sketching* a curve as distinct from *plotting*. By this we mean that we examine the salient features of the curve and produce a sketch of its general shape. (Each point of the curve is not plotted accurately.)

We have enough results already to make a reasonable attempt at sketching the main features of some curves. As we develop further techniques we shall be able to extend the range of functions we can sketch. The idea will be to deduce the maximum amount of information with the minimum of calculation.

If $f'(x) > 0$ in some interval, then $f(x)$ is said to be **increasing** there; whereas if $f'(x) < 0$ in some interval, then $f(x)$ is said to be **decreasing** there. For example, let $f(x) = 3x^2 - 2x - 1 = (3x + 1)(x - 1)$ so that $f'(x) = 6x - 2$. Then $f'(x) > 0$ if $6x - 2 > 0$, i.e. if $x > 1/3$, and $f'(x) < 0$ if $x < 1/3$. A sketch of $f(x) = 3x^2 - 2x - 1$ is given in Figure 7.8

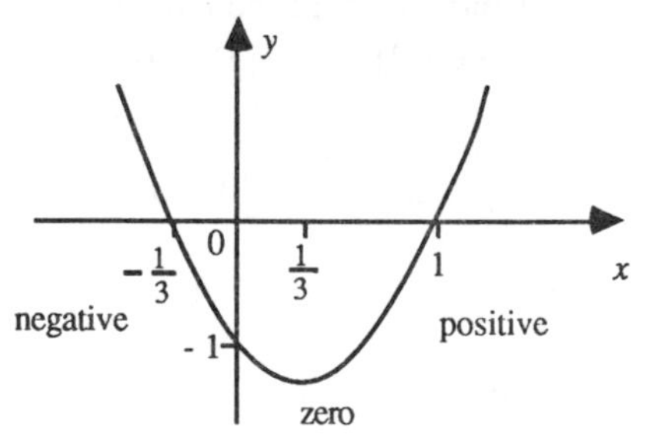

Figure 7.8

Note that $f(x) = 0$ when $x = 1$ or $-1/3$, also $f(0) = -1$; and that $f(x)$ is decreasing in the interval $(-\infty, 1/3)$ and increasing in the interval $(1/3, +\infty)$.

It is clear that where the function increases, its slope is positive and where it decreases its slope is negative. Where $f'(x) = 0$ we have a **turning point**. In the example shown $x = 1/3$ is a turning point and is a **local minimum**. We use *local* because it may not be the *absolute* minimum of the function. [If the equation is of the form $y = f(x)$, we often write dy/dx for $f'(x)$.]

We list the features to be examined when sketching a curve.

(a) *Symmetry about either axis*

If in the equation governing the curve, x can be replaced by $-x$ without altering the equation, the curve is *symmetrical about the y-axis*. You will remember that functions with this property are called **even functions**, for example Figure 7.9(i); examples of such functions are those which have an equation where x appears only as x^2, x^4 etc. If the effect of replacing x by $-x$ in a functional equation is to reverse the signs of all the terms the function is called **odd**, for example Figure 7.9(ii). We may similarly define symmetry about the x-axis; Figure 7.9(iii).

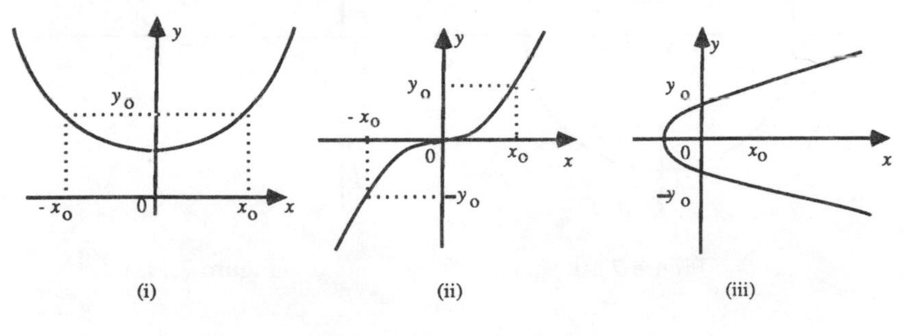

Figure 7.9

(b) *Any restrictions on the ranges of values of x and y*

For example, if $y^2 = x$ we need not consider $x < 0$.

(c) *Intersections of the curve with the axes*

These are found by putting x or y equal to zero.

(d) *The behaviour of the function for large values of x, both positive and negative*

(e) *The behaviour near the origin if this lies on the curve*

(f) *Any discontinuities in the curve*

(g) *The location and nature of any turning points*

(h) *The behaviour near discontinuities*

We are not yet in a position to deal fully with (g) and (h); however, we consider a few examples.

Example 1: $y^2 = 1 - x^2$ (Figure 7.10)

(i) The curve has symmetry about both axes

(ii) $y^2 \geq 0 \Rightarrow -1 \leq x \leq 1$, similarly $-1 \leq y \leq 1$

(iii) When $x = 0$, $y = \pm 1$ and $y = 0 \Rightarrow x = \pm 1$

This equation represents a circle, centre the origin and radius 1.

(We see from the sketch that $dy/dx = 0$ at $x = 0$ and is infinite at $x = \pm 1$.)

Example 2: $y = (x - 1)^2(x - 2)$ (Figure 7.11)

We leave you to fill in the details and explain the behaviour at $x = 1$.

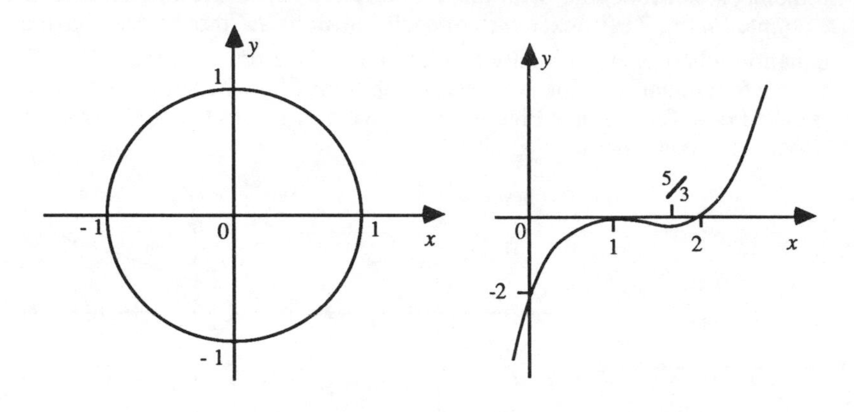

Figure 7.10

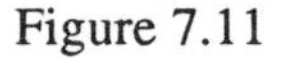

Figure 7.11

Example 3: $y = \dfrac{x-2}{(x-1)(x-3)}$ (Figure 7.12)

(i) $x = 2 \Rightarrow y = 0$; $x = 0 \Rightarrow y = -2/3$

(ii) For large $|x|$, y can be approximated by $x/x.x$, i.e. $1/x$. Hence as $x \to \infty$, $y \searrow 0$, and as $x \to -\infty$, $y \nearrow 0$

(iii) as $x \nearrow 1$, $y \to -\infty$ and as $x \searrow 1$, $y \to \infty$.

Further, as $x \nearrow 3$, $y \to -\infty$, and as $x \searrow 3$, $y \to \infty$. Note that dy/dx at $x = 2$ is -1.

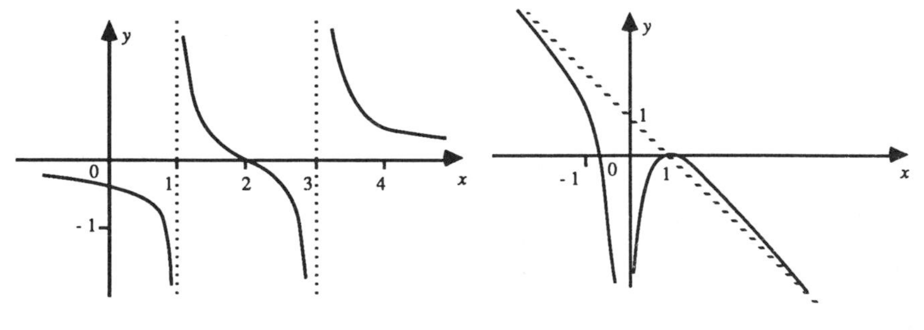

Figure 7.12

Figure 7.13

Example 4: $y = -x + 1 + \frac{1}{x} - \frac{1}{x^2}$ (Figure 7.13)

Here the important features are of an *asymptotic* nature.

For large positive x, y is large and negative; but we can go further and ignore the terms $1/x$ and $1/x^2$ to say that y is asymptotic to $(-x + 1)$ for large positive x (and also for large negative x).

Since, for large values of x the term $1/x$ dominates $-1/x^2$, we see that the curve is *above* the asymptote for large positive x and *below* it for large negative x. As x approaches zero, the dominant term is $-1/x^2$.

$$\frac{dy}{dx} = -1 - \frac{1}{x^2} + \frac{2}{x^3} = \frac{-(x^3 + x - 2)}{x^3} = \frac{-(x-1)(x^2 + x + 2)}{x^3}$$

Note (i) dy/dx is infinite at $x = 0$ and zero at $x = 1$

(ii) $y = 0$ when $x = 1$

Example 5: $y = x\sqrt{\frac{x}{x-2}}$

There is no symmetry. There is an intercept with the axes at the origin. y is not real if $0 < x < 2$ since $x/(x-2)$ is negative for this range of values of x. Since

$$y = x\left[1 - \frac{2}{x}\right]^{-1/2} \text{ then if } |x| > 2,\ y = x\left[1 + \frac{1}{x} + \frac{3}{2x^2} + 0\left[\frac{1}{x^3}\right]\right]$$

i.e.

$$y = x + 1 + \frac{3}{2x} + 0\left[\frac{1}{x^2}\right]$$

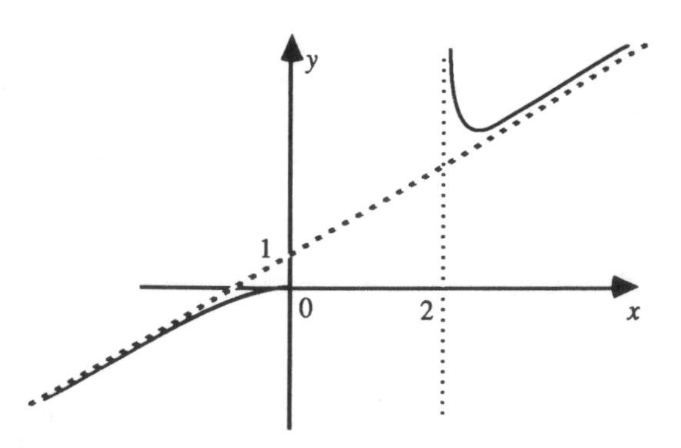

Figure 7.14

Note that $y = x + 1$ is an asymptote approached from above as $x \to \infty$ and below as $x \to -\infty$. As $x \to 2+$, $y \to \infty$ therefore $x = 2$ is an asymptote. Near the origin, $y \cong -\frac{|x^{3/2}|}{\sqrt{2}} \to 0$. Turning points occur where $\frac{\sqrt{x}\,(x-3)}{(x-2)^{3/2}} = 0$

i.e. at $(0, 0)$, $(3, \sqrt{3})$.

The curve is sketched in Figure 7.14 where the asymptotes are shown by dashed lines.

7.5 Polar Coordinates

In the *Cartesian* system, we locate a point by its distance from the two axes. In the *polar* system we locate a point by its distance r from the origin and by the angle θ between the x-axis and the line joining the point to the origin as shown in Figure 7.15

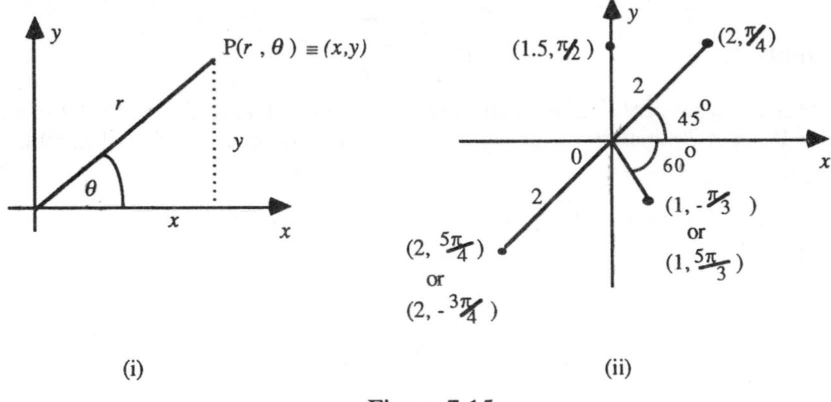

Figure 7.15

Polar coordinates are especially useful when shapes have some measure of rotational symmetry.

We see that the equations of transfer from one coordinate system to the other are

$$x = r\cos\theta, \qquad y = r\sin\theta \tag{7.28a}$$

or

$$r = \sqrt{x^2 + y^2}, \qquad \tan\theta = y/x \tag{7.28b}$$

It is important to know the values of y and x separately; for instance, the information $\tan\theta = -1$ does not allow us to distinguish between $\theta = 135°$ and $\theta = 315°$. (It is always advisable to plot the point on a diagram.) Some points are marked on Figure 7.15(ii). We shall assume that r is always positive. Sometimes angles in the third and fourth quadrants are taken negative for ease in calculation.

It is customary to describe curves in polar coordinates in the form $r = f(\theta)$. If $f(\theta) = f(-\theta)$ the curve is symmetrical about the x-axis; in particular, this will be so if $r = f(\cos\theta)$ since $\cos\theta$ is even. However, if $r = f(\sin\theta)$ there will be symmetry about the y-axis ($\theta = \pi/2$).

The equation $r = 3\sin\theta$ is defined for $0 \le \theta \le \pi$ only (since r is positive). If we attempt to convert this to cartesian coordinates we obtain

$$r^2 = 9\sin^2\theta \quad \text{or} \quad r^4 = 9r^2\sin^2\theta$$

and since $r^2 = x^2 + y^2$ and $y = r\sin\theta$ we get

$$(x^2 + y^2)^2 = 9y^2$$

That is

$$x^2 + y^2 = 3y$$

This last equation is valid for $y \ge 0$ which corresponds to $0 \le \theta \le \pi$.

We can similarly convert from an x-y equation to its polar form by putting $x = r\cos\theta$, $y = r\sin\theta$.

Example 1: $r = 3 + 2\cos\theta$

Since $r = f(\cos\theta)$ we can assume symmetry about the x-axis. We tabulate r against θ for $\theta = 0(\pi/12)\pi$.

θ	0	$\pi/12$	$\pi/6$	$\pi/4$	$\pi/3$	$5\pi/12$	$\pi/2$	$7\pi/12$	$2\pi/3$	$3\pi/4$	$5\pi/6$	$11\pi/12$	π
r (2 dp)	5.00	4.93	4.73	4.41	4.00	3.52	3.00	2.48	2.00	1.59	1.27	1.07	1.00

It should be clear from Figure 7.16(i) how the curve was obtained and you should extend the plotting of points for θ between and 2π to obtain the dashed curve.

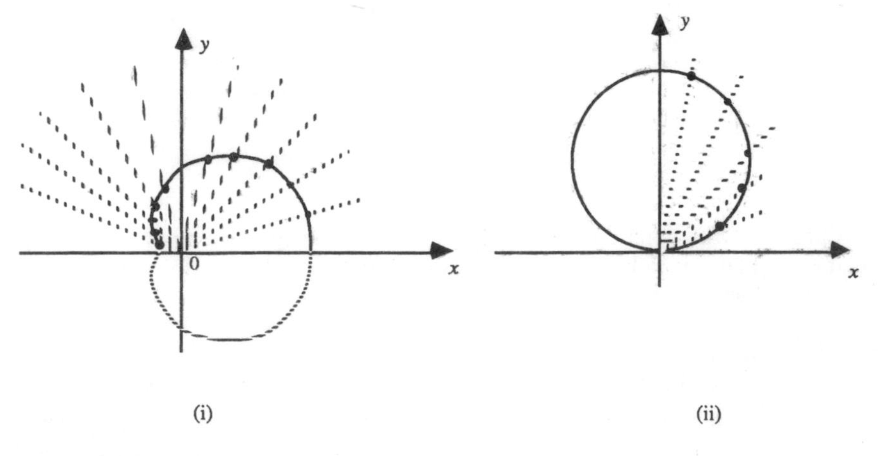

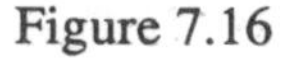
Figure 7.16

Example 2: $r = \sin \theta$

Since $r = \sin \theta$ we can assume symmetry about the y-axis and since $r \geq 0$, we have only to consider $0 < \theta < \pi$. A table of values is provided. Extend this to obtain the dashed part of the curve shown in Figure 7.16(ii).

θ	0	$\pi/12$	$\pi/6$	$\pi/4$	$\pi/3$	$5\pi/12$	$\pi/2$
r (2 d.p.)	0.00	0.26	0.50	0.71	0.87	0.97	1.00

Example 3: $r = a\theta$

This arises when a rod rotates with constant angular velocity, ω, about one end and a particle P moves along the rod with a constant velocity, u, relative to the rod. After time t, the point P will be a distance $a\theta$ from A and the angle that AP makes with the axis will be $\omega t = \theta$; Figure 7.17(i).

Hence $r = ut$, $\theta = \omega t$, and if we write the constant ratio u/ω as a then $r = a\theta$. The curve is plotted in Figure 7.17(ii). It is known as the **Archimedean spiral.**

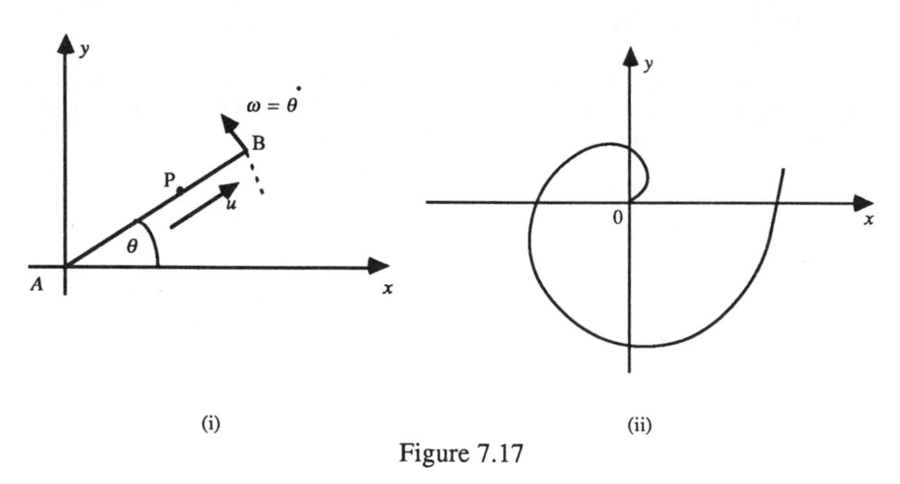

Figure 7.17

A problem with polar coordinates is that there is no simple way of connecting polar coordinates which relate to different origins and the equations of tangents and normals are not simple.

Rotation of Axes

We now turn to the problem of rotating the coordinate axes [refer to Figure 7.18(i)]. Consider a point P which has polar coordinates (r, θ) in the old axes yOx. After rotation through an angle α anti-clockwise the axes become $y'Ox'$ and the coordinates of P in this new set of axes are $(r, \theta-\alpha)$. To obtain the cartesian equivalent (x', y') we write $x' = r\cos(\theta - \alpha)$, $y' = r\sin(\theta - \alpha)$.

Now in the old axes we have $x = r \cos \theta$, $y = r \sin \theta$. Hence, since $x' = r \cos \theta \cos \alpha + r \sin \theta \sin \alpha$, we get

$$\left.\begin{aligned} x' &= x \cos \alpha + y \sin \alpha \\ y' &= -x \sin \alpha + y \cos \alpha \end{aligned}\right\} \quad (7.29a)$$

Similarly

(If α is negative we have a clockwise rotation of axes.)

We can often rotate axes to make an equation more easy to understand. In such cases we need x, y in terms of x', y', that is

$$\left.\begin{aligned} x &= x' \cos \alpha - y' \sin \alpha \\ y &= x' \sin \alpha + y' \cos \alpha \end{aligned}\right\} \quad (7.29b)$$

Why would you have expected this result?

Example

If we take the equation $\dfrac{x^2}{2} + \dfrac{y^2}{2} - xy = 2\sqrt{2}\, a(x + y)$ we can write it in the form

$$\left[\frac{x}{\sqrt{2}} - \frac{y}{\sqrt{2}}\right]^2 = 4a\left[\frac{x}{\sqrt{2}} + \frac{y}{\sqrt{2}}\right]$$

If we rotate the coordinate axes anti-clockwise through 45°, we can use equations (7.29b) with $\alpha = 45°$ to obtain $x = (x' - y')/\sqrt{2}$, $y = (x' + y')/\sqrt{2}$ and substitution gives $(y')^2 = 4ax'$, which is the equation of a parabola with axis 45° to the horizontal as shown in Figure 7.18(ii).

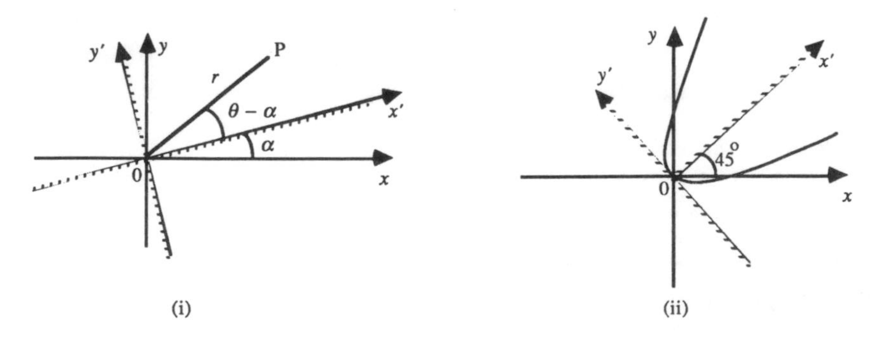

Figure 7.18

More complicated cases can arise if the rotation of axes is about a point other than the origin.

7.6 Parametric Representation of a Curve

Sometimes it is advantageous to describe a curve in terms of one subsidiary variable quantity, called a **parameter**. We illustrate by examples.

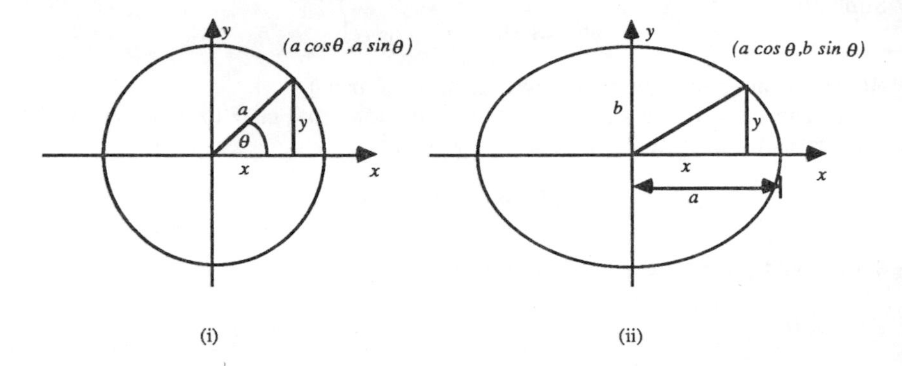

Figure 7.19

Circle and Ellipse

We can see from Figure 7.19(i) that if we make the substitutions $x = a \cos\theta$, $y = a \sin\theta$ then $x^2 + y^2 = a^2 \cos^2\theta + a^2 \sin^2\theta = a^2$, and so a point on this circle can be represented as $(a \cos\theta, a \sin\theta)$.

As θ varies from 0 to 2π, so we describe in turn all the points on the circle's circumference. Instead of the two variables x and y we now have one variable θ. An extension to the ellipse is fairly evident and is shown in Figure 7.19(ii). We write $x = a \cos\theta$ and $y = b \sin\theta$, so that $x^2/a^2 + y^2/b^2 = \cos^2\theta + \sin^2\theta = 1$. ($\theta$ here is called the *eccentric angle.*) The advantage of a parametric description is that we can plot points on the curve by evaluating $x(\theta)$ and $y(\theta)$ for successive values of θ – this is useful in numerical control of tools.

Tangent to the circle

Suppose we now seek an equation for the tangent to the circle at any point P (x_1, y_1). We saw previously that this could be written $xx_1 + yy_1 = a^2$; substituting in terms of the parameter θ_1 at P we get the equation of the tangent in the form $xa \cos\theta_1 + ya \sin\theta_1 = a^2$; which reduces to $x \cos\theta_1 + y \sin\theta_1 = a$.

Tangent to the ellipse

From the previous result for a tangent to an ellipse at the point (x_1, y_1), namely $xx_1/a^2 + yy_1/b^2 = 1$, we obtain the tangent at $(a \cos\theta_1, b \sin\theta_1)$ in the form

$$\frac{x\cos\theta_1}{a}+\frac{y\sin\theta_1}{b}=1$$

Parabola

If we now turn to the parabola, $y^2 = 4ax$, we see that a possible parametric representation is $x = at^2$, $y = 2at$. Values of t at various points on the parabola are shown in Figure 7.20(i). Each point on the parabola is associated with a unique value of t and vice-versa.

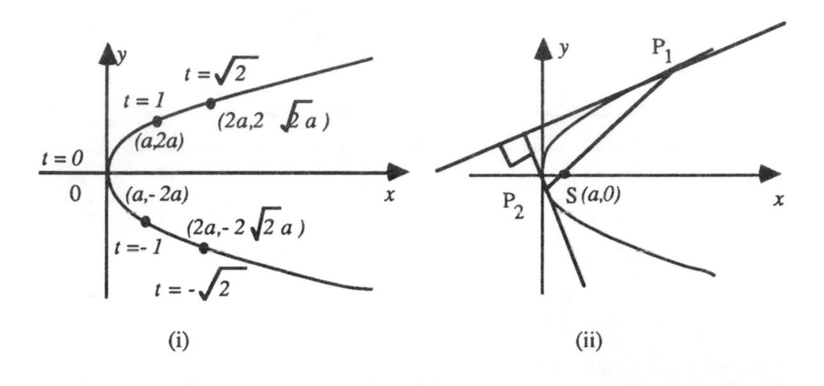

Figure 7.20

Note that x and y are each functions of t and t is a function of y, but t is not a function of x in the strict sense.

The tangent at any point (x_1, y_1) was found to be $yy_1 = 2a(x + x_1)$ and this becomes

$$y.2at_1 = 2a(x + at_1^{\,2})$$

that is,
$$t_1y = x + at_1^{\,2}$$

We can see that the tangent has slope $1/t_1$ and this is consistent with the graph. (For example, the slope decreases at t increases.) It should be clear that the gradient of the normal at t_1 is $-t_1$ and the equation of the normal at t_1 is $y - 2at_1 = -t_1(x - at_1^{\,2})$ or $y + t_1x = 2at + at_1^{\,3}$.

The focal chord is a line through the focus; its end points are denoted t_1 and t_2. The equation of the chord joining points t_1 and t_2 is

$$y - 2at_1 = \frac{2at_2 - 2at_1}{at_2^2 - at_1^2}(x - at_1^{\,2})$$

and since this is satisfied by the point $(a, 0)$ we have

$$-2at_1 = \frac{2a}{t_2 + t_1}(1 - t_1^{\,2})$$

which reduces to $t_2 t_1 = -1$.

Notice that $1/t_1 . 1/t_2 = -1$ and so the tangents at the end-points of the focal chord are perpendicular [See Figure 7.20(ii)].

Hyperbola

The equation $x^2/a^2 - y^2/b^2 = 1$ is satisfied by $x = a \cosh \theta, y = b \sinh \theta$ (see Figure 7.21).

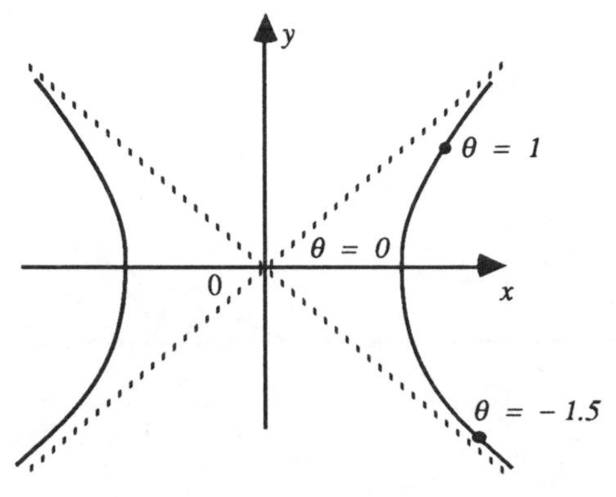

Figure 7.21

Note that this caters for the right-hand branch only so that if we wanted to cover the left-hand branch we should have to consider $x = -a \cosh \theta$, $y = b \sinh \theta$.

The tangent at θ_1 is $(x/a)\cosh \theta_1 - (y/b)\sinh \theta_1 = 1$, or $x/a - (y/b)\tanh \theta_1 = 1/\cosh \theta_1$.

Let $\theta_1 \to \infty$ then $\tanh \theta_1 \to 1$ and $\cosh \theta_1 \to \infty$, hence the tangent becomes $x/a - y/b = 0$ which is the equation for the asymptote, as found previously.

Cycloid

The path traced out by a point on the circumference of a circle which rolls without slipping on a straight line is called a cycloid [see Figure 7.22(i)].

It has parametric equations

$$x = a(\theta - \sin \theta)$$
$$y = a(1 - \cos \theta)$$

Here it is difficult to eliminate θ and find an equation connecting y and x. Indeed, even if we do find the equation, it is of very little use to us. By taking different values of θ we can easily sketch the curve as shown in Figure 7.22(ii). The values of θ are marked.

Try and see how it was obtained. Why not convince yourself that this is in the correct shape by marking a point on the rim of a circular object and seeing how it moves.

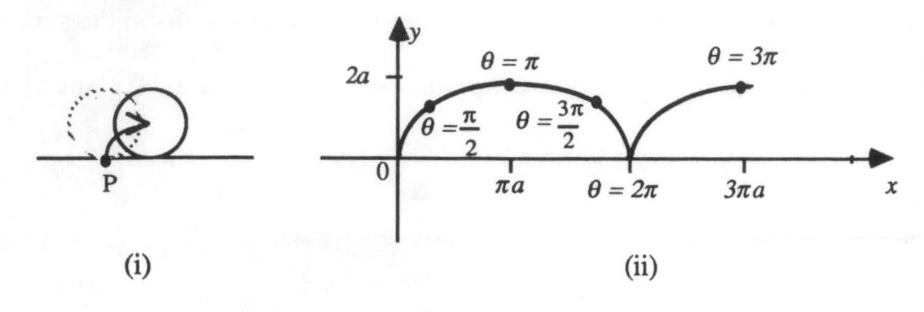

Figure 7.22

In general the tangent to the curve $x = x(t)$, $y = y(t)$ at $t = t_1$ is given by

$$x = x(t_1) + s\dot{x}\ (t_1), \quad y = y(t_1) + s\dot{y}\ (t_1)$$

where s is the parameter on the tangent and where $\dot{x}\ (t_1)$ is the value of dx/dt at $t = t_1$. (Similarly for $\dot{y}\ (t_1)$)

The normal at the same point is given by

$$x = x(t_1) + s\dot{y}\ (t_1), \quad y = y(t_1) - s\dot{x}\ (t_1)$$

7.7 An Introduction to Solid Geometry

A relationship between the coordinates (x, y, z) can determine a surface. If we substitute values of x and y into the relationship then the corresponding values of z represent points above the x-y plane on a surface; Figure 7.23 refers.

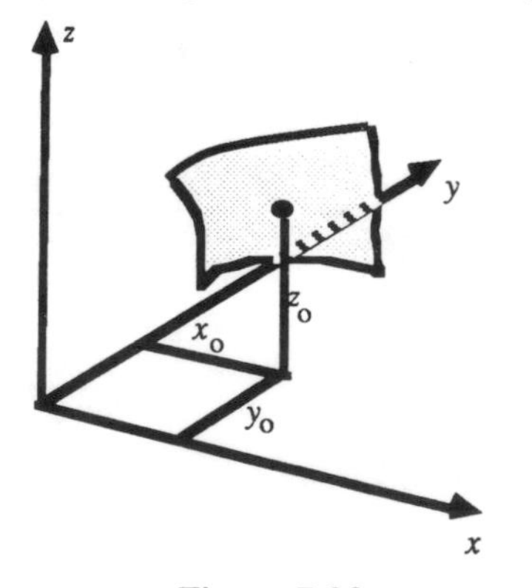

Figure 7.23

You can perhaps visualise the surface representation more clearly if you imagine that at each point (x, y) of the x-y plane, within the ranges defined, a thin matchstick is placed upright with the height of the matchstick equal to the value of the function at that point. The tops of the matchsticks form the surface representing $f(x, y)$.

One class of surfaces which is simple to sketch is the class of planes [see Figure 7.24(i) and (ii)].

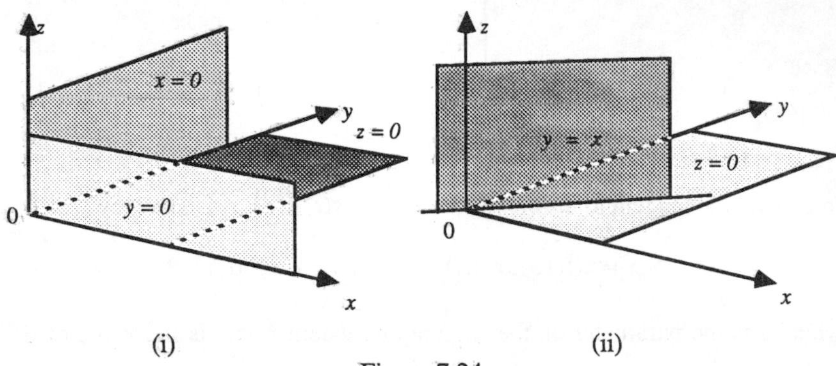

Figure 7.24

Notice that the two planes in Figure 7.24(ii) intersect in a straight line (this is generally so – see Problems 27 and 28). The specification of the plane $y = x$ must not be confused with the intersection line $y = x$ which, in three dimensions, should strictly be $y = x$, $z = 0$.

You have doubtless seen contour lines on maps which are two-dimensional representations of a three-dimensional piece of country-side. For a function of two variables the contour lines are curves joining points of equal height above sea-level and are formed by horizontal planes intersecting the surface, representing the function; see Figure 7.25(i) and (ii). The equation of such a horizontal plane is $z = c$, where c is a constant.

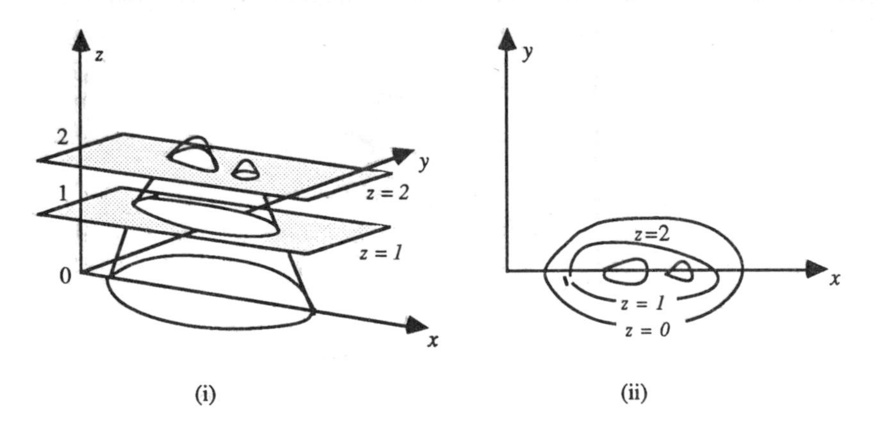

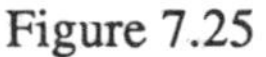

Figure 7.25

Consider the relationship $x^2 + y^2 = 1$ in three dimensions. Whatever level of

z we take the cross-section area is a circle with the given equation. The extension to three dimensions means that we have a cylindrical surface with axis the z-axis and generators vertical lines through points on the circle in the x-y plane. Part of it is shown in Figure 7.26.

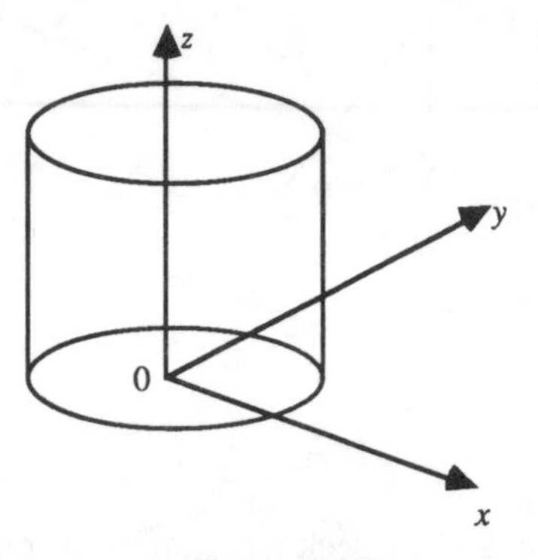

Figure 7.26

Note that in three-dimensions to specify that circle on the x-y plane we need to say $x^2 + y^2 = 1$, $z = 0$.

In the example of the cone $z = \sqrt{x^2 + y^2}$ the contour lines are concentric circles, $x^2 + y^2 = c^2$; see Figure 7.27(i) and (ii).

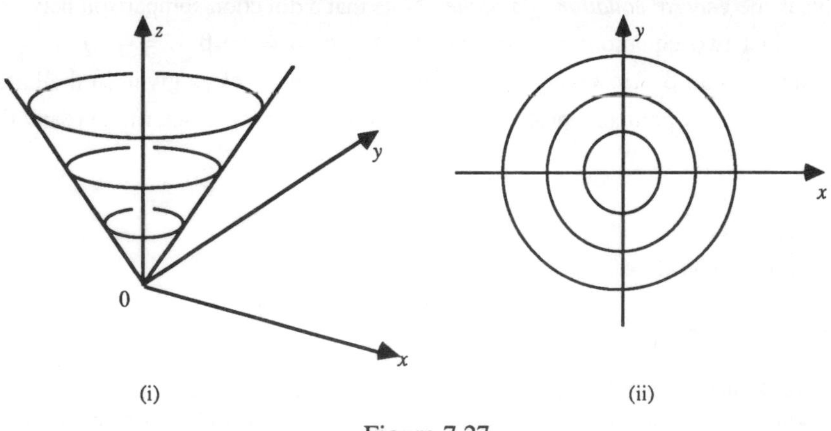

Figure 7.27

Equation of a plane

We produce the equation of a plane inclined at α to the x-axis and β to the y-axis: see Figure 7.28. From this equation you should be able to see the special cases $\alpha \to \pi/2$, $\beta \to \pi/2$.

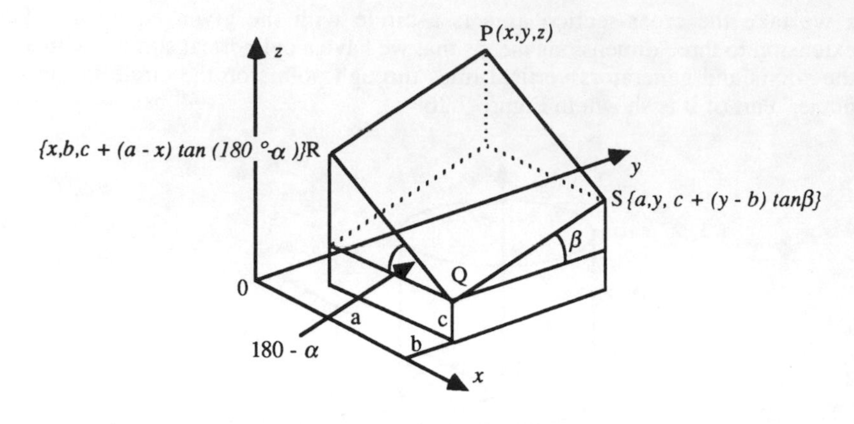

Figure 7.28

The point Q is (a, b, c) and we seek a relationship connecting the coordinates (x, y, z) of P. To get to P we first rise to R by an amount $(a - x)\tan(180° - \alpha)$ or first to S by an amount $(y - b)\tan\beta$. Similar triangles show that the total rise to P, $(z - c)$, is the sum of these and, in general, if $\alpha \neq \pi/2 \neq \beta$, we obtain the equation

$$z = c + (x - a)\tan\alpha + (y - b)\tan\beta$$

or for suitable coefficients A, B, C, D

$$Ax + By + Cz = D \tag{7.30}$$

This is the *general equation of a plane.* Note that a direction comparison between these last two equations would give $A = \tan\alpha$, $B = \tan\beta$, $C = -1$, $D = -c + a\tan\alpha + b\tan\beta$. However, the equation of a plane is often given in the latter form and in any particular problem it may not be necessary to determine α and β.

Problems

Section 7.2

1 Sketch the set of solutions of

(a) $x^2 + 2y^2 > 16$ (b) $3x^2 + 4y^2 < 25$ (c) $(x + y)^2 > 0$
(d) $x + 2y < 4$ (e) $2x - y > 3$

2 Shade the region where the three inequalities $x + y < 1$, $x + 2y < 4$, $x - 3y < 5$ are satisfied simultaneously and locate the points of intersection of pairs of lines. What happens to the region if the condition $x + 2y < 6$ is added? What happens if the condition $x + 2y < 3$ is added?

3 Sketch the region where all the inequalities $x \geq 0$, $y \geq 0$, $x + y < 3$, $x + 2y < 4$ are satisfied. Locate the vertices of this *convex region.* Suppose we add the requirement $3x + 4y > 10.1$, what do you conclude graphically? Verify this algebraically.

4 Describe the regions in Figure 7.29 by a set of simultaneous inequalities.

5 On the diagram to Figure 7.29(iii) superimpose the lines $2x + y = 0$, $2x + y = \frac{1}{2}$, $2x + y = 1$, $2x + y = 8$. What is the greatest value of c for which $2x + y = c$ has a point in common with the shaded region? What is the least value?

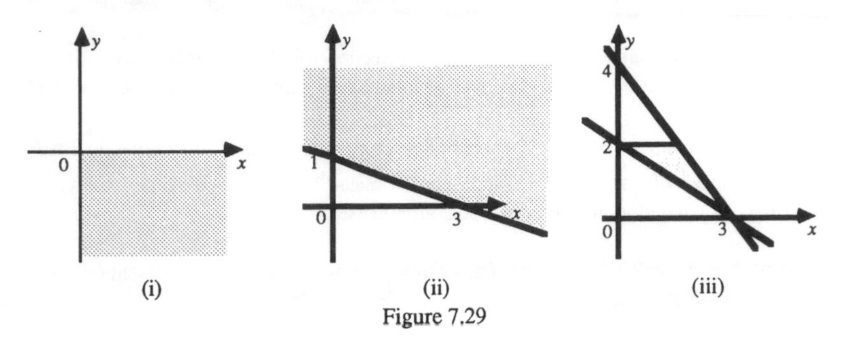

Figure 7.29

6 Repeat Problem 5 with the lines $2x + 3y = 0, 6, 9, 12$.

7 Two manufactured articles X and Y each are composed of two chemicals A and B. X comprises 2g of A and 4g of B, and Y comprises 3g of A and 3g of B. The profit on one article of X is 3 pence and on one of Y is 4 pence. If there is at most 100g of A available and 200g of B available, and if there is a demand for at least 20 articles of X and 15 articles of Y, how many of each should be manufactured if the profit is to be maximised? (Hint: let x of X be manufactured and y of Y).
The cost to a retailer of an article of X is 10p and one of Y is 20p. Assuming he must satisfy the demand, what number of each should he request to minimise his costs?

Section 7.4

8 Sketch the following curves:

(a) $y = x^2 + x + 1$ (b) $y^2 = 4 + 2x^2$ (c) $y^2 = 4 - 2x^2$
(d) $y = x^2(x - 4)$ (e) $y = (x - 1)(x - 3)^2$ (f) $y = (x - 1)/(x - 2)(x - 3)$
(g) $y = (x - 3)/(x - 1)(x - 2)$ (h) $y = (2 - x^2)/(2 + x^2)$
(i) $y = x/(1 + x^2)$ [You may assume $dy/dx = (1 - x^2)/(1 + x^2)^2$]
(j) $y = -x + 1 - 1/x$ (k) $y = -x + 1 + 1/x^2$

Section 7.5

9 Find the polar form of

(i) $y^2 = x^2(1 - x^2)$ (ii) $y^2 = x^2(y^2 + x^2)$ (iii) $(x^2 + y^2)(x - y)^2 = 4$

10 Find the Cartesian form of

(i) $r = 3\cos\theta + 4\sin\theta$ (ii) $r^2 = 2\sec 4\theta$ (iii) $r = 1 - \tan^2\theta$

11 Sketch the curves $r = 2\sin^2\theta$, $r = 3\cos\theta$, $r = -2\cos\theta$, $r = \cos^2\theta$, $r = \sin 2\theta$, $r = \cos 2\theta$, $r = \sin 3\theta$

12 Sketch the curves $r = 1 + \cos\theta$, $r = 1 + 2\cos\theta$, $r = 2 + \cos\theta$. Try to sketch $r = a + b\cos\theta$.

13 Show that
(i) $r = 2a\cos\theta$ represents a circle centre, $(a, 0)$, in polar coordinates
(ii) $r = 2a\cos(\theta - \alpha)$ represents a circle, centre (a, α).
Deduce a special case by putting $\alpha = \pi/2$. Sketch the curves on common axes.

14 Calculate algebraically the points of intersection of $r^2 = 9\cos\theta$ and $r = (1 + \cos\theta)$. Sketch the curves and locate the points of intersection graphically. What do you conclude?

15 The lemniscate $r^2 = a^2 \sin 2\theta$ is used in constructing transition curves in roads. For small values of r and θ the curve is approximated by $y = Kx^3$ for a suitable value of K. Take $a = 1$ and plot the two curves on common axes.

16 Write down the new coordinates (x', y') of a point when the axes are rotated 60° clockwise, 90° anti-clockwise.

17 Rotate the axes clockwise through 45° to obtain a more familiar form for the curve $x^2 - y^2 = 1$.

Section 7.6

18 Show that the rectangular hyperbola can be parametrised into $x = ct$, $y = c/t$. Sketch the curve and mark on it suitable values of t. Show that at the point t_1 the equation of the tangent is $t_1^2 y + x = 2ct_1$.

19 Find where the tangents to $xy = c^2$ at t_1 and t_2 intersect.

20 Trace the curve whose parametric representation is $x = a(1 + \cos\theta)$, $y = a\sin\theta$. Interpret it algebraically by first eliminating θ from the equations for x and y.

21 A rod OA of length 50 cm rotates at 7 rad/sec and a rod AB of equal length, hinged at A to the rod OA, rotates about A in the opposite sense at 1 rad/sec. B is the point (x, y). In terms of time t, obtain expressions for x and y, assuming that OAB lay originally along the x-axis.

22 Eliminate θ from the equations $x = \cos^3\theta$, $y = \sin^3\theta$. Try to sketch the curve from this new equation and from the parametric equations.

23 Let P_1 and P_2 be (x_1, y_1) and (x_2, y_2) respectively. Let $P(x, y)$ be a point on this line such that $PP_1/P_2P_1 = \mu$ (note the meaning of $\mu < 0$). Find expressions for x and y in terms of μ. Show on a diagram the points $\mu = 0$, $\mu = 1$, $\mu = -3$, $\mu = 1/2$, $\mu = 2$.

Section 7.7

24 Sketch the sets of points in 3 dimensions which satisfy
(i) $3x - y = 0$ (ii) $2x + y = 0$ (iii) $2x + y = 0, z = 0$
(iv) $x^2 + y^2 = 36$ (v) $x^2 + y^2 = 36, z = 2$

25 Find the points which lie on the intersection of
(i) $x = 2, y = 3$ (ii) $x = 0, z = 0$ (iii) $x + y = 2, x + z = 1$

26 Let two planes have equations

$$ax + by + cz = d$$
$$a'x + b'y + c'z = d'$$

Under what conditions do they intersect? Interpret geometrically.

27 Sketch the contours $z = 0(1)4$ of the surface $z = x^2 + 2y^2 - 3x + 1$ and hence describe the function.

28 Where does the plane $ax + by + cz = d$ intersect the three axes?

8

LINEAR ALGEBRA I VECTORS

Many quantities in engineering require a direction to be specified, e.g. velocity (a speed in a certain direction), force, momentum, displacement (a movement in a certain direction). Such quantities are called **vectors** as distinct from **scalar** quantities which require the specification of magnitude only.

To distinguish between vector quantities and scalars we write vectors in boldface type, such as **V**, **F** and **a**. Usually on paper or the blackboard we write $\underline{V}$, $\underline{F}$ and $\underline{a}$.

8.1 Elementary Vector Algebra

Now, just as we learn in algebra to add, subtract and multiply scalar quantities, we have to learn how to manipulate vector quantities and we shall go through the rules of addition, subtraction and multiplication to create an algebra of vectors.

Representation of a vector quantity

We can represent a vector quantity **F** graphically both in magnitude and direction by a line OP.

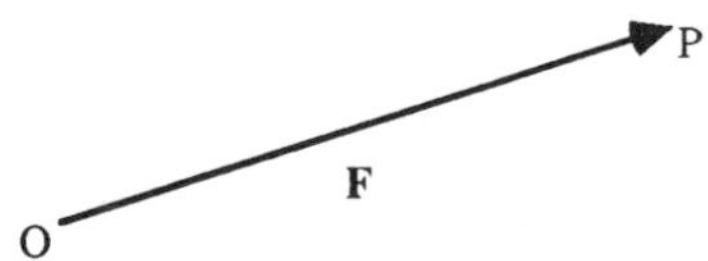

The magnitude of the vector is represented by the length of the line OP (using some convenient scale) and its direction in space is shown by an arrow on the line. Various notations are used in books when describing this vector. We could write for the vector

$$\mathbf{F} = \overline{OP} \quad \text{or} \quad \overrightarrow{OP}$$

and for its magnitude we would have

$$F = |\mathbf{F}| = |\overrightarrow{OP}| = OP$$

All these are possible notations that you will meet. Here we shall use

$$\mathbf{F} = \overrightarrow{OP} \quad \text{and } F = |\mathbf{F}| = OP$$

It is obvious from our definition that if we multiply a vector by some scalar λ, then all we are doing is to multiply the length of the vector by λ: we are not changing the direction.

For example, if **F** is the directed length $\overrightarrow{OP}$ then 3**F** would be the vector $\overrightarrow{OR}$ as follows.

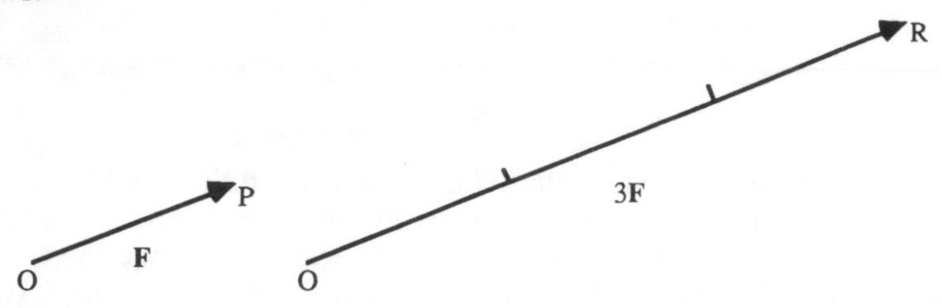

In particular if **f** is a vector of unit size (called a **unit vector**) in the direction of **F** then

$$\mathbf{F} = F\mathbf{f}$$

Sometimes a unit vector is written **a**; then $\mathbf{a} = |\mathbf{a}|\hat{\mathbf{a}} = a\hat{\mathbf{a}}$

Equality of two vectors

If two vectors are parallel, pointing in the same direction, and of equal length, we say they are *equal*. Let $\mathbf{F} = \overrightarrow{OP} = \overrightarrow{QR}$.

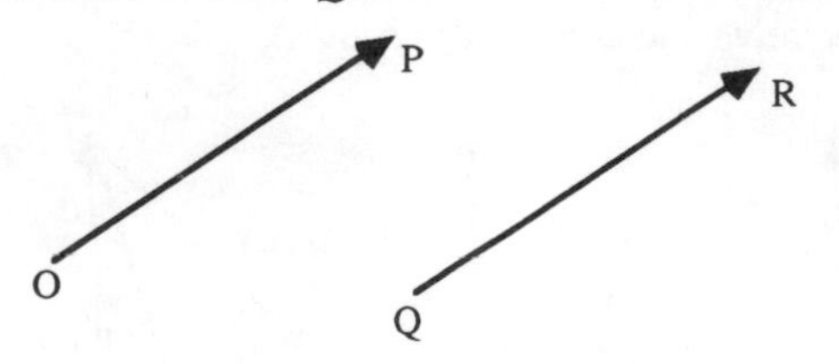

The points of application O and Q do not need to be the same point here. For example, it is common to say an aeroplane is flying due East with speed 250 m.p.h. without specifying its point in space. A vector where the point of application need not be specified is called a **free vector**.

If we have two vectors **F** and $\mathbf{F}_1$ which are parallel and of the same length but opposite in direction, then we call the second vector –**F** since it is (–1).**F** and then $\mathbf{F}_1 = -\mathbf{F}$.

Addition and subtraction of vectors

We use the parallelogram law of addition. Formally, if we want to find **E** + **F** we draw **E** and **F** with the same point of application and complete the parallelogram. The diagonal of this parallelogram is then **E** + **F** [Figure 8.1(i)]. **E** + **F** is called the resultant of **E** and **F**.

It is obvious that **F** + **E** must give the same resultant so that **E** + **F** = **F** + **E**. (Vector addition is therefore commutative.)

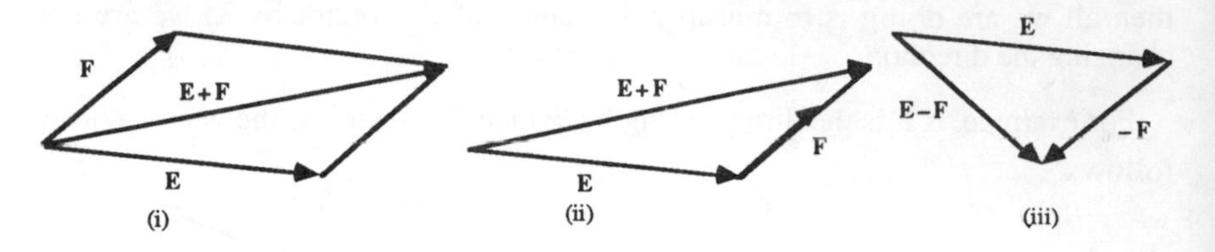

Figure 8.1

An alternative form of this is the **triangle law of addition** which is really half the above parallelogram. Here we draw the vector **E** first. At the end of **E** we place the vector **F**. Joining the point of application of **E** and the end of **F** gives the vector **E** + **F** (directions of arrows are important). See Figure 8.1(ii).

Subtraction immediately follows for we can regard **E** – **F** as meaning **E** + (–**F**) which gives Figure 8.1(iii). (Again note the direction of arrows.)

We can extend addition to more than two vectors and build up a polygon by placing the vectors successively end to end. For example, with three vectors **A**, **B**, **C** (which need not necessarily be in one plane the sum **A** + **B** + **C** is represented by the closing side of the polygon shown in Figure 8.2(i).

The order of the addition is not important and we can easily see from Figures 8.2(ii), (iii) and (iv) that

$$(\mathbf{A} + \mathbf{B}) + \mathbf{C} = \mathbf{A} + (\mathbf{B} + \mathbf{C}) = (\mathbf{A} + \mathbf{C}) + \mathbf{B}$$

(combining the *associative* and *commutative* laws).

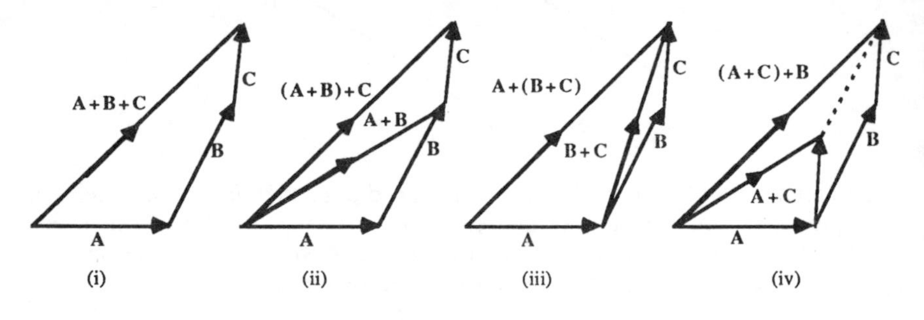

Figure 8.2

Similarly it is easy to show that $\lambda(\mathbf{A} + \mathbf{B}) = \lambda\mathbf{A} + \lambda\mathbf{B}$ and $(\lambda + \mu)\mathbf{A} = \lambda\mathbf{A} + \mu\mathbf{A}$ for scalars λ and μ.

Thus vector addition and subtraction obey all the laws of the normal algebra of real (or complex) numbers and we can treat linear vector equations in exactly the same way as in the algebra of real numbers. For example, if **A** + **B** = **C**, then **A** = **C** – **B**. We define further a **null vector** or **zero vector 0** as having no particular direction and zero length, so that we can then say in equations like the above – if **A** + **B** = **C**, then **A** + **B** – **C** = **0**. Also **A** + **0** = **0** + **A** = **A**.

A word of warning: when applying vector methods to physical problems it will be necessary at times to specify the line of action *and/or* the point of application.

For example, if a force acts on a rigid body its line of action **must** be specified since a change in the line of action will alter the torque produced on the body. Similarly when dealing with a body which changes its shape when forces are applied we must specify the points of application of these forces. Vectors which have some *fixed* point of application are called **bound vectors**.

Applications in plane geometry and mechanics

We now consider some applications in plane geometry and mechanics.

Example 1 (Refer to Figure 8.3)

Given DB = (2/5)AB and BE = (2/5)BC, *Prove* DE||AC and DE = (2/5)AC.

In vector language, $\overrightarrow{DB} = (2/5)\overrightarrow{AB}$ and $\overrightarrow{BE} = (2/5)\overrightarrow{BC}$ and we must prove that $\overrightarrow{DE}$ $= (2/5)\overrightarrow{AC}$. (Since this means that DE = (2/5)AC and is also parallel to AC.)

Now $\quad \overrightarrow{AB} + \overrightarrow{BC} = \overrightarrow{AC}, \overrightarrow{AD} = (3/5)\overrightarrow{AB}, \overrightarrow{EC} = (3/5)\overrightarrow{BC},$

but $\quad \overrightarrow{DB} + \overrightarrow{BE} = \overrightarrow{DE}$

therefore
$$\begin{aligned}\overrightarrow{DE} &= (2/5)\overrightarrow{AB} + (2/5)\overrightarrow{BC}\\ &= (2/5)(\overrightarrow{AB} + \overrightarrow{BC})\\ &= (2/5)\overrightarrow{AC}\end{aligned}$$

Hence the result is proved.

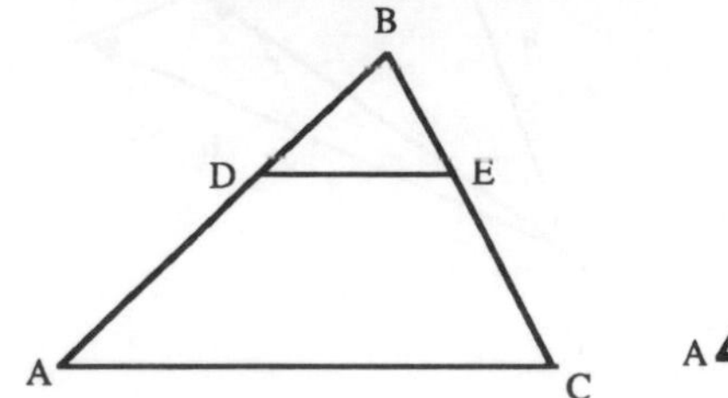

Figure 8.3

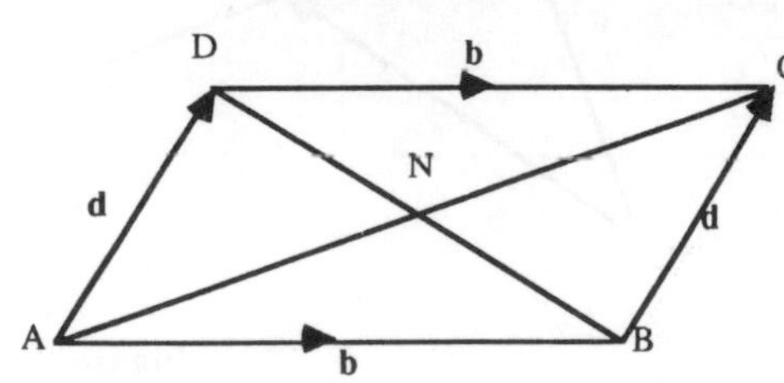

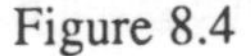

Figure 8.4

Example 2

To prove that the diagonals of a parallelogram bisect each other. Consider the parallelogram ABCD shown in Figure 8.4.

Let $\overrightarrow{AB} = \mathbf{b}$ and $\overrightarrow{AD} = \mathbf{d}$. We wish to show $\overrightarrow{BN} = \frac{1}{2}\overrightarrow{BD}$ and $\overrightarrow{AN} = \frac{1}{2}\overrightarrow{AC}$.

Now $\quad \overrightarrow{AC} = \mathbf{b} + \mathbf{d} \quad \overrightarrow{BC} = \mathbf{d} \quad \overrightarrow{DC} = \mathbf{b} \quad \overrightarrow{BD} = \mathbf{d} - \mathbf{b}$

But $\overrightarrow{BN} = \lambda\overrightarrow{BD}$ for some scalar λ, i.e. $\overrightarrow{BN} = \lambda(\mathbf{d} - \mathbf{b}) = \lambda\mathbf{d} - \lambda\mathbf{b}$

Also $\overrightarrow{AN} = \mu AC$ for some scalar μ, i.e. $\overrightarrow{AN} = \mu(\mathbf{b} + \mathbf{d}) = \mu\mathbf{b} + \mu\mathbf{d}$

Also $\overrightarrow{AB} + \overrightarrow{BN} = \overrightarrow{AN}$, therefore $\mathbf{b} + \lambda\mathbf{d} - \lambda\mathbf{b} = \mu\mathbf{b} + \mu\mathbf{d}$. Comparing both sides, $\mu = \lambda$ and $1 - \lambda = \mu = \lambda$. Solving simultaneously we obtain $\lambda = \mu = \frac{1}{2}$. That is the diagonals of a parallelogram bisect each other.

Example 3
The equation of a straight line

(i) Suppose we are given a point on the line and the direction of the line [see Figure 8.5(i)]

Take an origin O and let A be the given point such that $\overrightarrow{OA} = \mathbf{a}$ and let the unit vector $\hat{\mathbf{u}}$ describe the direction of the line. We seek an equation which is satisfied by any point P on the line.

Now $\overrightarrow{AP} = t\hat{\mathbf{u}}$, t being a scalar which can vary, and $\overrightarrow{OP} = \overrightarrow{OA} + \overrightarrow{AP}$. If $\overrightarrow{OP} = \mathbf{r}$, then

$$\mathbf{r} = \mathbf{a} + t\hat{\mathbf{u}} \tag{8.1}$$

and this is the desired equation. It gives for different values of t the value of $\mathbf{r}$ for all points on the line (e.g. $t = 0$ gives A).

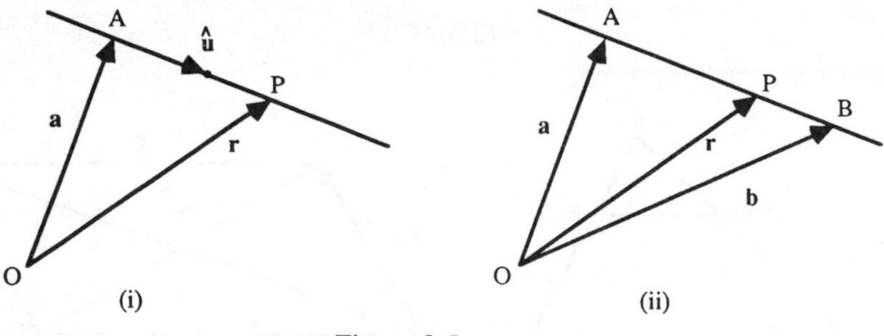

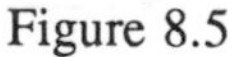
Figure 8.5

(ii) Suppose we are given two points on the line, A and B. In Figure 8.5(ii) let $\overrightarrow{OA} = \mathbf{a}$ and $\overrightarrow{OB} = \mathbf{b}$, then $\overrightarrow{AB} = \mathbf{b} - \mathbf{a}$ and if P is any point on the line then $\overrightarrow{AP} = t(\mathbf{b} - \mathbf{a})$ for some value of t. Now $\overrightarrow{OP} = \overrightarrow{OA} + \overrightarrow{AP}$, i.e. $\mathbf{r} = \mathbf{a} + t(\mathbf{b} - \mathbf{a})$, or

$$\mathbf{r} = (1 - t)\mathbf{a} + t\mathbf{b} \tag{8.2}$$

Note that $t = 0 \equiv$ A, $t = 1 \equiv$ B, $t < 0$ to left of A, $t > 0$ to right of B.

Example 4
Forces of magnitude 4 and 5 act along AB, AC respectively; find the resultant. We refer to Figure 8.6 below.

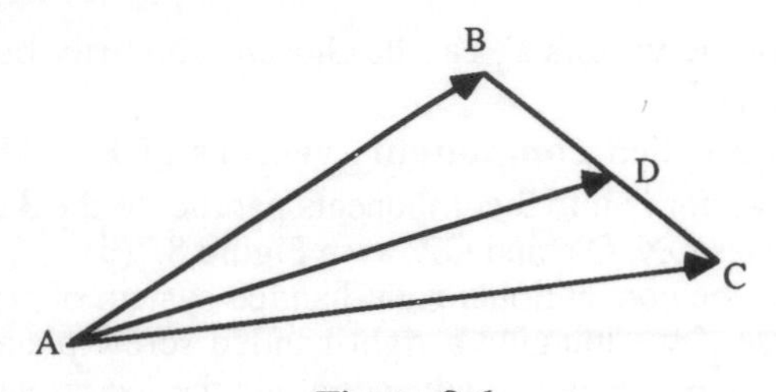

Figure 8.6

Let $\overrightarrow{AD}$ be the resultant direction. Now $4\overrightarrow{AB} = 4\,\overrightarrow{AD} + 4\overrightarrow{DB}$ and $5\overrightarrow{AC} = 5\overrightarrow{AD} + 5\overrightarrow{DC}$

therefore $$4\overrightarrow{AB} + 5\overrightarrow{AC} = 9\overrightarrow{AD} + (4\overrightarrow{DB} + 5\overrightarrow{DC})$$

Since the resultant acts along $\overrightarrow{AD}$, $4\overrightarrow{DB} + 5\overrightarrow{DC} = \mathbf{0}$. Therefore 4BD = 5DC, which locates D on the line BC.

The resultant force is of magnitude 9 acting along AD.

8.2 Vectors in Cartesian Coordinates

General components of a vector

From the definition of vector addition it is clear that any vector $\mathbf{F}$ can be written as the sum of other vectors, such as

$$\mathbf{F} = \mathbf{F}_1 + \mathbf{F}_2 + \dots + \mathbf{F}_n$$

where the n vectors $\mathbf{F}_i$ and the vector $-\mathbf{F}$ when put head to tail form a closed polygon, as in Figure 8.7(i).

Notice this need not necessarily be a plane figure.

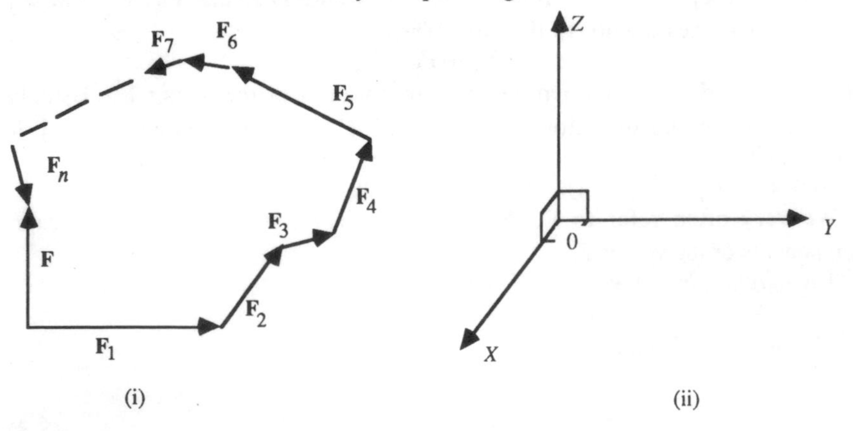

Figure 8.7

Further, $(n-1)$ of the vectors $\mathbf{F}_i$ can be chosen arbitrarily but the nth vector must close the polygon.

$\mathbf{F}_1, \mathbf{F}_2, \ldots \mathbf{F}_n$ are called **component vectors** of **F**. The most useful decomposition of a vector is into 3 components parallel to the 3 orthogonal axes of cartesian coordinates OX, OY and OZ, as in Figure 8.7(ii).

This figure shows the conventional right-handed system of coordinate axes – right-handed because if we imagine a right-handed screw placed at the origin pointing along the X-axis, this would advance along the X-axis when rotating OY towards OZ. Similarly when turning OZ to OX we would advance a screw along the Y-axis, and so on for the cyclic pattern (Figure 8.8(i)).

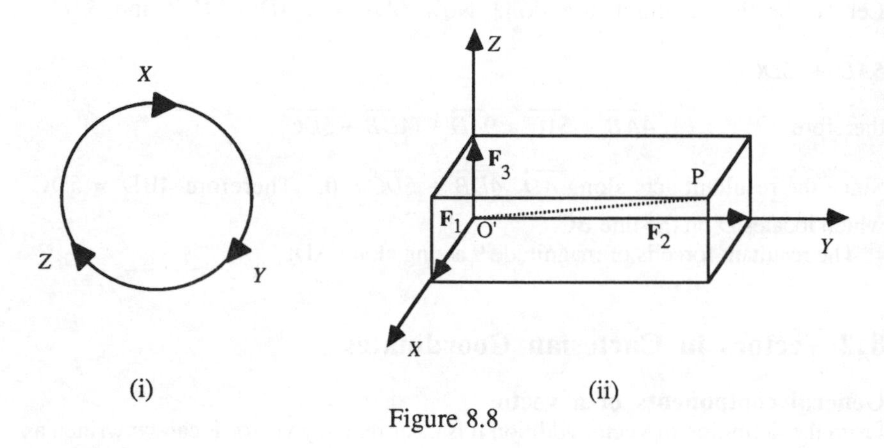

Figure 8.8

If now we take a vector **F** as $\overrightarrow{O'P}$ and draw lines through O′ parallel to the rectangular cartesian axes, we can complete the rectangular box having O′P as diagonal; this is shown in Figure 8.8(ii).

The intercepts on the lines O′X, O′Y and O′Z are $\mathbf{F}_1$, $\mathbf{F}_2$ and $\mathbf{F}_3$ respectively in the direction of the axes. Then

$$\mathbf{F} = \mathbf{F}_1 + \mathbf{F}_2 + \mathbf{F}_3$$

and $\mathbf{F}_1$, $\mathbf{F}_2$ and $\mathbf{F}_3$ are the **cartesian components** of the vector **F**. Usually unit vectors in the directions of the X, Y, Z axes are denoted by **i**, **j**, **k** respectively.

Now $\mathbf{F}_1 = F_1\mathbf{i}$, $\mathbf{F}_2 = F_2\mathbf{j}$, $\mathbf{F}_3 = F_3\mathbf{k}$, so that $\mathbf{F} = F_1\mathbf{i} + F_2\mathbf{j} + F_3\mathbf{k}$

We very often refer to F_1, F_2, F_3 as the first, second and third scalar components of the vector **F**.

The *magnitude* of the vector **F** is easily shown by Pythagoras' Theorem to be $\sqrt{F_1^2 + F_2^2 + F_3^2}$. That is

$$\text{O'P} = |\mathbf{F}| = F = \sqrt{F_1^2 + F_2^2 + F_3^2} \qquad (8.3)$$

The vectors **i**, **j** and **k** are said to be **linearly independent** (this means that no one of the quantities **i**, **j**, **k** can be expressed as a *linear combination* of the other two). Thus, for addition and subtraction of vectors we have

$$\text{if} \quad \begin{cases} \mathbf{A} = A_1\mathbf{i} + A_2\mathbf{j} + A_3\mathbf{k} \\ \mathbf{B} = B_1\mathbf{i} + B_2\mathbf{j} + B_3\mathbf{k} \end{cases}$$

$$\text{then} \quad \mathbf{A} \pm \mathbf{B} = (A_1 \pm B_1)\mathbf{i} + (A_2 \pm B_2)\mathbf{j} + (A_3 \pm B_3)\mathbf{k} \tag{8.4}$$

In other words, we may just add or subtract the respective components, which ties in with our ideas of two-dimensional vectors.

Example

If $\mathbf{A} = 3\mathbf{i} - 2\mathbf{j} + \mathbf{k}$, $\mathbf{B} = -\mathbf{i} + 2\mathbf{j} - \mathbf{k}$

then $\quad \mathbf{A} + \mathbf{B} = 2\mathbf{i}$

$\quad \mathbf{A} - \mathbf{B} = 4\mathbf{i} - 4\mathbf{j} + 2\mathbf{k} \quad \mathbf{A} + 3\mathbf{B} = 4\mathbf{j} - 2\mathbf{k}$

As an additional consequence, if two vectors are equal then the respective scalar components must be equal.

For example, if we have $\mathbf{A} = x\mathbf{i} + y\mathbf{j} + z\mathbf{k}$ and $\mathbf{A} = \mathbf{B} + \mathbf{C}$ where $\mathbf{B} = 2\mathbf{i} - \mathbf{j} + \mathbf{k}$ and $\mathbf{C} = 3\mathbf{i} + 5\mathbf{j} - \mathbf{k}$, then

$$x = 2 + 3 = 5, \quad y = -1 + 5 = 4, \quad z = 1 - 1 = 0$$

Position vector

A special case of the above component form is where we have a point P in space, with coordinates (x, y, z) referred to rectangular cartesian axes, see Figure 8.9.

Then $\overrightarrow{OP}$ is the **position vector** of the point **P**, usually denoted by **r**.

The lengths along the coordinate axes are OA = x, OB = y, OC = z so that $\mathbf{r} = x\mathbf{i} + y\mathbf{j} + z\mathbf{k}$.

The distance of this point from the origin is the length of OP and is

$$r = \sqrt{x^2 + y^2 + z^2}$$

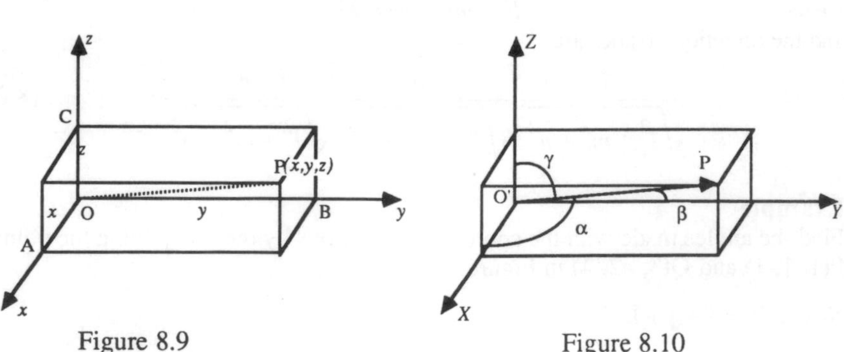

Figure 8.9 Figure 8.10

Direction cosines and direction ratios

Suppose a vector $\mathbf{F} = \overrightarrow{O'P}$, where $\mathbf{F} = F_1\mathbf{i} + F_2\mathbf{j} + F_3\mathbf{k}$.

The direction of O′P can be specified by the angles it makes with the O′X,

O′Y, O′Z directions. These are usually designated by α, β, γ as shown in Figure 8.10

Now $\quad \mathbf{F} = F_1\mathbf{i} + F_2\mathbf{j} + F_3\mathbf{k}$

$$= \text{O}'\text{P}\cos\alpha\mathbf{i} + \text{O}'\text{P}\cos\beta\mathbf{j} + \text{O}'\text{P}\cos\gamma\mathbf{k}$$

$$= \text{O}'\text{P}(\cos\alpha\mathbf{i} + \cos\beta\mathbf{j} + \cos\gamma\mathbf{k}) \tag{8.5}$$

The quantities $\cos\alpha$, $\cos\beta$, $\cos\gamma$ are called the **direction cosines** of the vector **F**.

We can see from equation (8.5) that, since O′P is the length of the vector **F**, $(\cos\alpha\mathbf{i} + \cos\beta\mathbf{j} + \cos\gamma\mathbf{k})$ must be a unit vector. That is, it has unit length so that

$$\cos^2\alpha + \cos^2\beta + \cos^2\gamma = 1$$

Notice that $\cos\alpha = F_1/F$, $\cos\beta = F_2/F$, $\cos\gamma = F_3/F$

Sometimes it is convenient to use quantities which are proportional to the direction cosines, but which do not have a sum of squares equal to 1. That is, we have three quantities (l, m, n) which are such that

$$\frac{l}{\cos\alpha} = \frac{m}{\cos\beta} = \frac{n}{\cos\gamma}$$

They are called **direction ratios** and are convenient to use since they can be taken as integers. Indeed $[F_1, F_2, F_3]$ are direction ratios of O′P. We can easily convert them into direction cosines when needed because if we let

$$\frac{l}{\cos\alpha} = \frac{m}{\cos\beta} = \frac{n}{\cos\gamma} = \lambda$$

then $\quad \cos\alpha = \frac{l}{\lambda}, \quad \cos\beta = \frac{m}{\lambda}, \quad \cos\gamma = \frac{n}{\lambda}$

But $\quad \cos^2\alpha + \cos^2\beta + \cos^2\gamma = 1$

Hence $\quad l^2 + m^2 + n^2 = \lambda^2$

and the direction cosines are

$$\frac{l}{\sqrt{l^2+m^2+n^2}}, \frac{m}{\sqrt{l^2+m^2+n^2}}, \frac{n}{\sqrt{l^2+m^2+n^2}} \tag{8.6}$$

Example

Find the angles made with the coordinate directions by the line joining the points P(1, 1, 1) and Q(3, –2, 4) in Figure 8.11.

Now $\overrightarrow{OP} = \mathbf{i} + \mathbf{j} + \mathbf{k}$

$$\overrightarrow{OQ} = 3\mathbf{i} - 2\mathbf{j} + 4\mathbf{k}$$

$$\overrightarrow{PQ} = \overrightarrow{OQ} - \overrightarrow{OP} = 2\mathbf{i} - 3\mathbf{j} + 3\mathbf{k}$$

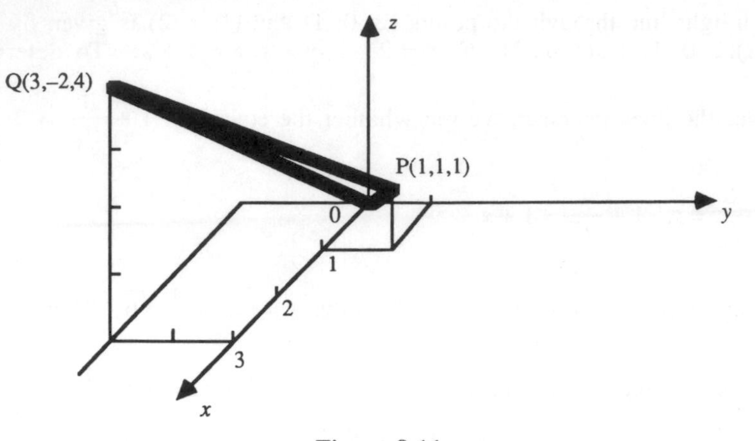

Figure 8.11

Therefore $$PQ = \sqrt{2^2 + 3^2 + 3^2} = \sqrt{22}$$

hence $$\overrightarrow{PQ} = PQ\left[\frac{2}{\sqrt{22}}\mathbf{i} - \frac{3}{\sqrt{22}}\mathbf{j} + \frac{3}{\sqrt{22}}\mathbf{k}\right]$$

The direction cosines are $\cos\alpha = \frac{2}{\sqrt{22}}$, $\cos\beta = -\frac{3}{\sqrt{22}}$, $\cos\gamma = \frac{3}{\sqrt{22}}$

Hence $\alpha = 64°17'$, $\beta = 129°46'$, $\gamma = 50°14'$

Notice that we can say $\overrightarrow{PQ} = 2\mathbf{i} - 3\mathbf{j} + 3\mathbf{k}$

Its direction ratios are [2, –3, 3]. Hence the direction cosines are

$$\left[\frac{2}{\sqrt{2^2+3^2+3^2}}, \quad \frac{-3}{\sqrt{2^2+3^2+3^2}}, \quad \frac{3}{\sqrt{2^2+3^2+3^2}}\right]$$

Example

The straight line through the point (1, 1, 1) in the direction $\left[\frac{1}{\sqrt{6}}, \frac{-2}{\sqrt{6}}, \frac{1}{\sqrt{6}}\right]$ is given by $\mathbf{r} = (1, 1, 1) + t\left[\frac{1}{\sqrt{6}}, \frac{-2}{\sqrt{6}}, \frac{1}{\sqrt{6}}\right]$. In coordinate terms this is

$$x = 1 + \frac{t}{\sqrt{6}}, y = 1 - \frac{2t}{\sqrt{6}}, z = 1 + \frac{t}{\sqrt{6}}$$

The straight line through the points (2, 0, 1) and (1, 1, 2) is given by $\mathbf{r} = (1 - s)(2, 0, 1) + s(1, 1, 2)$ or $x = 2 - s, y = s, z = 1 + s$. To determine whether the lines intersect, we see whether the equations $1 + \frac{t}{\sqrt{6}} = 2 - s$,

$$\frac{3}{2} - \frac{2t}{\sqrt{6}} = s, \quad 1 + \frac{t}{\sqrt{6}} = 1 + s.$$

From the first and last of these we find that $s = \frac{1}{2}$, $t = \frac{1}{2}\sqrt{6}$. These values also satisfy the second equation and the lines intersect at the point $(1\frac{1}{2}, \frac{1}{2}, 1\frac{1}{2})$.

8.3 Scalar Product

The work done by a force **F** in moving a body through a displacement **d** is $Fd \cos \theta$ and this is a scalar quantity: see Figure 8.12

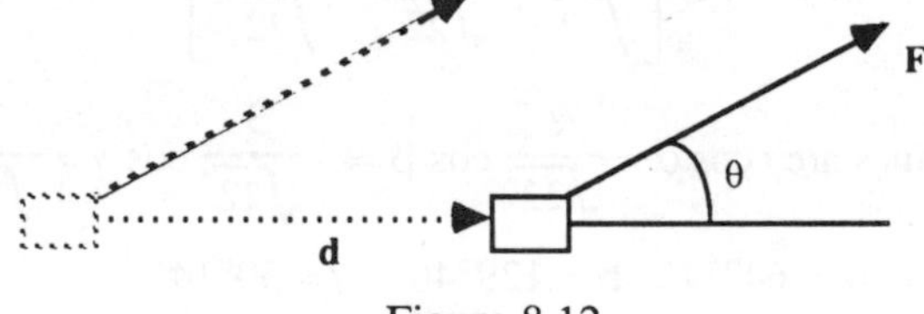

Figure 8.12

You have probably learnt that the work done is the displacement times the projection of the force along the direction of movement (the component of the force perpendicular to the direction of movement does no work).

Thus we are led to define the **scalar product** of two vectors in the following form.

Given two vectors **A** and **B**, θ being the angle between their directions ($0 \leq \theta \leq 180°$), then the *scalar product* of **A** and **B** is AB cos θ and is written **A.B** (read '**A** dot **B**').

Now **A.B** = AB cos θ is a scalar quantity. Clearly **B.A** = BA cos θ = **A.B** and the product is **commutative**.

We draw the two vectors as if they had a common point O. If they are in different planes then θ is the angle between two vectors parallel to **A** and **B** respectively drawn through any specified point.

Properties of the scalar product

1 By definition, if the two vectors are *perpendicular* then $\theta = 90°$ and $\cos 90° = 0$ so that **A.B** = 0. This is the first major difference between

vector and scalar algebra, for in scalars we have:

If $ab = 0$, then either $a = 0$ or $b = 0$

However, we have for vectors:

If $\mathbf{A}.\mathbf{B} = 0$, then either $\mathbf{A} = \mathbf{0}$ or $\mathbf{B} = \mathbf{0}$ or $\mathbf{A}$ is perpendicular to $\mathbf{B}$.

(This is consistent with the fact that work done by a force perpendicular to a displacement is zero.)

2 If $\mathbf{A}$ and $\mathbf{B}$ are parallel, then $\cos\theta = 1$ and $\mathbf{A}.\mathbf{B} = AB$. In particular if $\mathbf{B} = \mathbf{A}$, then

$$\mathbf{A}.\mathbf{A} = \mathbf{A}^2 = (\text{magnitude of } \mathbf{A})^2$$

(Note: this may be used in reverse to find the magnitude of a vector.)

3 $\mathbf{A}.\mathbf{B} = A(B\cos\theta) = (A\cos\theta)B$.

From Figure 8.13(i) we can see that this product can be interpreted as

$A \times$ component of $\mathbf{B}$ in the direction of $\mathbf{A}$, or

$B \times$ component of $\mathbf{A}$ in the direction of $\mathbf{B}$.

Note, therefore, that if $\mathbf{n}$ is a unit vector in some given direction, then $\mathbf{A}.\mathbf{n} = A.1\cos\theta = A\cos\theta =$ component of $\mathbf{A}$ in the direction of $\mathbf{n}$.

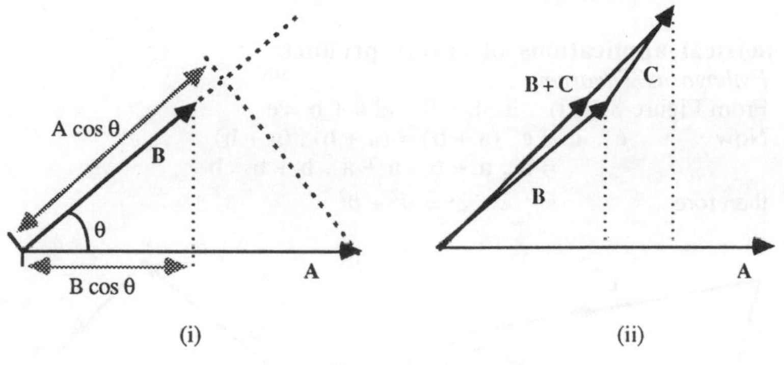

Figure 8.13

4 $\mathbf{A}.(\mathbf{B} + \mathbf{C}) = \mathbf{A}.\mathbf{B} + \mathbf{A}.\mathbf{C}$

Try to prove this by using Property 3 and considering Figure 8.13(ii)

5 Since $\mathbf{i}, \mathbf{j}, \mathbf{k}$ are *mutually orthogonal* vectors, then we have

$$\mathbf{i}.\mathbf{j} = \mathbf{j}.\mathbf{k} = \mathbf{k}.\mathbf{i} = 0$$
$$\mathbf{i}.\mathbf{i} = \mathbf{j}.\mathbf{j} = \mathbf{k}.\mathbf{k} = 1$$

Using these results we have

$$\mathbf{A}.\mathbf{B} = (A_1\mathbf{i} + A_2\mathbf{j} + A_3\mathbf{k}).(B_1\mathbf{i} + B_2\mathbf{j} + B_3\mathbf{k})$$

i.e. $$\mathbf{A}.\mathbf{B} = A_1B_1 + A_2B_2 + A_3B_3 \tag{8.7}$$

Notice that effectively only like components are multiplied. For example

$$(2\mathbf{i} - 3\mathbf{j} + \mathbf{k}).(4\mathbf{i} + 2\mathbf{j} - 2\mathbf{k}) = (2).(4) + (-3).(2) + (1).(-2) = 0$$

Neither of these vectors is zero. Hence they are perpendicular.

Notice further that $\mathbf{A}.\mathbf{A} = A^2 = A_1^2 + A_2^2 + A_3^2$ which is consistent with our finding of the magnitude of $\mathbf{A}$.

Notice that we can use the scalar product to find the angle between two vectors, for if we are given two vectors **A** and **B** then we can find **A.B**. But **A.B** = |**A**||**B**| cos θ and since we can also find |**A**| and |**B**| then cos θ is easily found giving θ the angle between the vectors.

Example

Find the angle between the vectors **A** = **i** – **j** – **k** and **B** = 2**i** + **j** + 2**k**. Now

$$\mathbf{A.B} = 1\,.\,2 - 1\,.\,1 - 1\,.\,2 = -1$$

$$|\mathbf{A}| = \sqrt{1^2 + 1^2 + 1^2} = \sqrt{3}$$

$$|\mathbf{B}| = \sqrt{2^2 + 1^2 + 2^2} = 3$$

But **A.B** = |**A**||**B**| cos θ

Hence $$\cos\theta = \frac{\mathbf{A}\,.\,\mathbf{B}}{|\mathbf{A}||\mathbf{B}|} = \frac{-1}{3\sqrt{3}}$$

Hence we can find θ as approximately 101°6′.

Geometrical applications of scalar products

(i) *Pythagoras' Theorem*

From Figure 8.14(i) **a . b** = 0 and **a + b = c.**

Now $$\mathbf{c\,.\,c} = \mathbf{c\,.\,(a + b)} = \mathbf{(a + b)\,.\,(a + b)}$$
$$= \mathbf{a\,.\,a + b\,.\,a + a\,.\,b + b\,.\,b}$$

therefore $$c^2 = a^2 + b^2$$

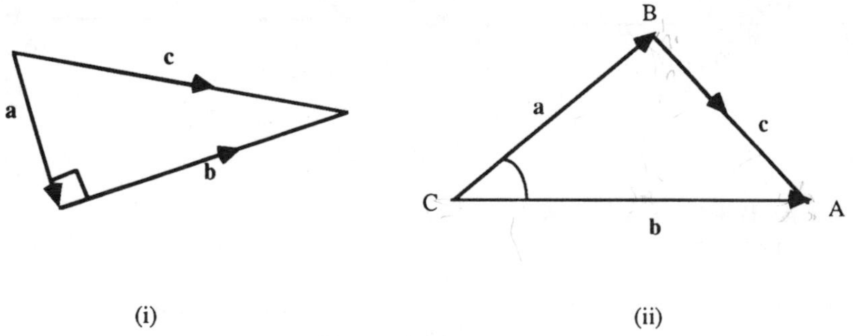

(i) (ii)

Figure 8.14

(ii) *Cosine Rule*

From Figure 8.14(ii) **b = a + c**

therefore $$\mathbf{c = b - a}$$

therefore $$\mathbf{c\,.\,c = (b - a)\,.\,(b - a)}$$

i.e. $$c^2 = \mathbf{b\,.\,b - a\,.\,b - a\,.\,b + a\,.\,a}$$
$$= b^2 + a^2 - 2ab\cos C$$

(iii) *Line perpendicular to a given line*

Let $\mathbf{s} = a\mathbf{i} + b\mathbf{j}$ be the given vector.

Let the required line pass through a point $\mathbf{A}'$: (x', y'). Let $P(x, y)$ be a point on this line. Then

$$\mathbf{A'P} = (x - x')\mathbf{i} + (y - y')\mathbf{j}$$

We require $\quad \mathbf{s} \,.\, \mathbf{A'P} = 0$

therefore $\quad (x - x')a + (y - y')b = 0$

Hence $\quad ax + by = ax' + by'$

i.e. $\quad ax + by = c$

where $c = ax' + by' =$ constant

This is the equation of the line perpendicular to $\mathbf{s} = a\mathbf{i} + b\mathbf{j}$

(iv) *Angle between two straight lines*

This angle is the same as the angle between the perpendiculars to the lines. If the lines are

$$x + 2y + 3 = 0$$
$$3x + 2y - 3 = 0$$

the perpendiculars are $\mathbf{P}_1 = \mathbf{i} + 2\mathbf{j}$; $\mathbf{P}_2 = 3\mathbf{i} + 2\mathbf{j}$ [using (iii)]. Then

$$\cos\theta = \frac{\mathbf{P}_1 \,.\, \mathbf{P}_2}{|\mathbf{P}_1| \,.\, |\mathbf{P}_2|} = \frac{7}{\sqrt{5} \,.\, \sqrt{13}}$$

(v) *Perpendicular distance from a given point to a given line*

For example, we go from $\mathbf{A} \equiv (3, 1)$ to $3x + 2y - 6 = 0$ (see Figure 8.15).

The perpendicular to the line is $\hat{\mathbf{p}} = (3\mathbf{i} + 2\mathbf{j})/|3\mathbf{i} + 2\mathbf{j}|$

therefore $\quad \hat{\mathbf{p}} = \dfrac{1}{\sqrt{13}}(3\mathbf{i} + 2\mathbf{j})$

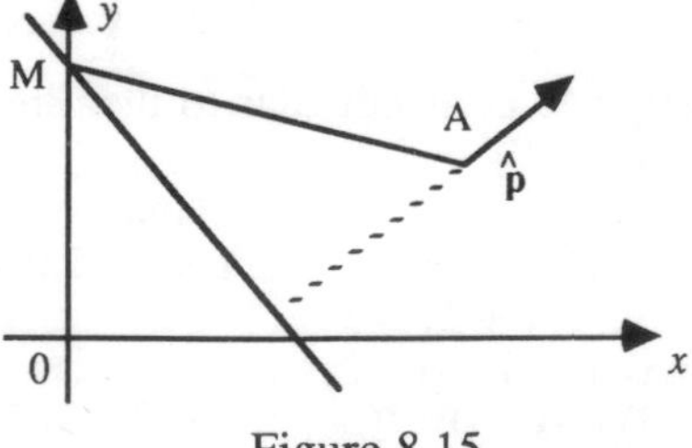

Figure 8.15

Take any point on the line, for example $M \equiv (0, 3)$, then

$$\overrightarrow{AM} = -3\mathbf{i} + 2\mathbf{j} \quad \text{and} \quad |\overrightarrow{AM} \,.\, \hat{\mathbf{p}}| = |-5(\sqrt{13}/13)| = 5/\sqrt{13}$$

is the required distance.

(vi) *Equation of a plane*

Suppose we are given a normal direction to the plane and a point B in the plane ($\overrightarrow{OB} = \mathbf{b}$) [see Figure 8.16(i)]. Let **a** be perpendicular to the plane and let P be *any* point in the plane ($\overrightarrow{OP} = \mathbf{r}$).

Now BP lies in the plane and is perpendicular to **a**.

Therefore $$(\mathbf{r} - \mathbf{b}) \cdot \mathbf{a} = 0 \tag{8.8}$$

If $\mathbf{r} = x\mathbf{i} + y\mathbf{j} + z\mathbf{k}$, $\mathbf{b} = b_1\mathbf{i} + b_2\mathbf{j} + b_3\mathbf{k}$ and $\mathbf{a} = a_1\mathbf{i} + a_2\mathbf{j} + a_3\mathbf{k}$, then we get

$$(x - b_1)a_1 + (y - b_2)a_2 + (z - b_3)a_3 = 0$$

i.e. $$a_1x + a_2y + a_3z = a_1b_1 + a_2b_2 + a_3b_3 = \text{constant}$$

Notice that the normal direction (a_1, a_2, a_3) constitutes the coefficients of x, y and z in this equation.

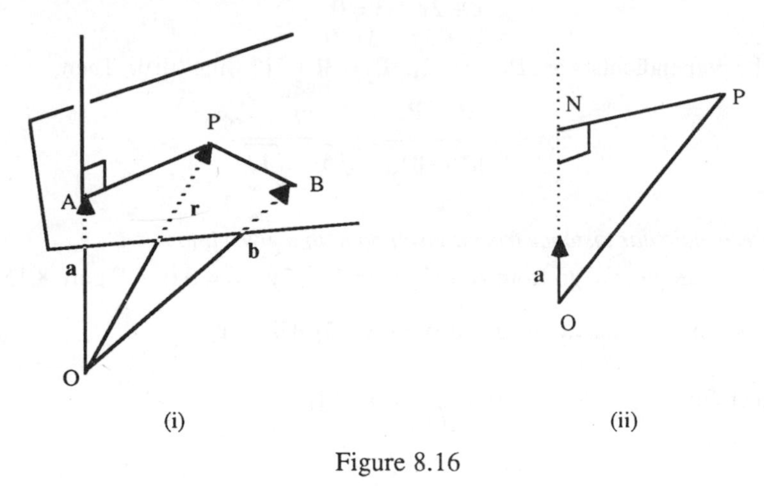

Figure 8.16

Example 1

Find the equation of the plane perpendicular to the direction $\mathbf{i} + 2\mathbf{j} - 3\mathbf{k}$ and passing through the point (1, 0, 0).

The plane must have equation $x + 2y - 3z = \text{constant}$.

Since it passes through (1, 0, 0), the constant = 1 + 2.0 – 3.0 = 1.

Hence the plane is $$x + 2y - 3z = 1$$

Alternatively its equation is $(\mathbf{r} - \mathbf{i}) \cdot (\mathbf{i} + 2\mathbf{j} - 3\mathbf{k}) = 0$

i.e. $$(x - 1) + (y - 0) \cdot 2 + (z - 0)(-3) = 0$$

as before.

Example 2

Given the plane $2x + 3y + 2z = 4$, find a unit vector perpendicular to it. A perpendicular to the plane is $\mathbf{a} = 2\mathbf{i} + 3\mathbf{j} + 2\mathbf{k}$. Now $|\mathbf{a}| = \sqrt{17}$. Hence the unit vector perpendicular to the plane is

$$\hat{\mathbf{a}} = \frac{2}{\sqrt{17}}\mathbf{i} + \frac{3}{\sqrt{17}}\mathbf{j} + \frac{2}{\sqrt{17}}\mathbf{k}$$

Let us find the perpendicular distance to the plane from O; refer to Figure 8.16(ii). Take any point **P** in the plane – say (2, 0, 0)

Then $\overrightarrow{OP} = 2\mathbf{i}$

Perpendicular distance from O = ON = $\overrightarrow{OP} \cdot \hat{\mathbf{a}} = 2\mathbf{i}.(2\mathbf{i} + 3\mathbf{j} + 2\mathbf{k})/\sqrt{17} = 4/\sqrt{17}$.

8.4 Vector Product

We now look for a product to two vectors which results in a *vector* quantity.

Given two vectors **A** and **B** with an included angle θ then the vector product of **A** and **B** is written **A** × **B** (sometimes **A**∧**B**) and is such that

(i) **A** × **B** is perpendicular to both **A** and **B**

(ii) the magnitude of **A** × **B** = *AB* sin θ

(iii) a *right-handed* rotation about the vector **A** × **B** through an angle θ would move **A** to the same direction as **B** (see Figure 8.17)

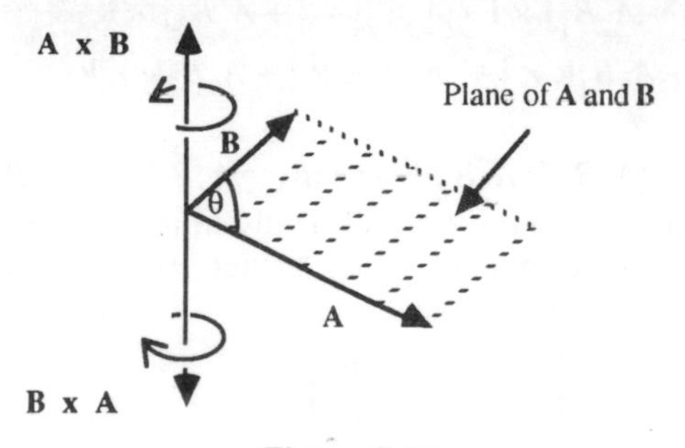

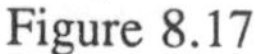
Figure 8.17

Hence $\mathbf{A} \times \mathbf{B} = AB \sin\theta\, \hat{\mathbf{n}}$ where $\hat{\mathbf{n}}$ is a unit vector in the direction of **A** × **B**.

Properties of the vector product

1 By the definition, $\mathbf{B} \times \mathbf{A} = BA \sin\theta\, (-\hat{\mathbf{n}})$ since the rotation is from **B** to **A** through an angle of θ. Hence **B** × **A** = –**A** × **B**. Thus the vector product is not commutative and this vector product does not follow the usual rule of scalar algebra.

2 If **A** and **B** are *parallel*, then $\theta = 0$ and $\sin\theta = 0 \Rightarrow \mathbf{A} \times \mathbf{B} = \mathbf{0}$.

Notice again the change from scalar algebra in that if $\mathbf{A} \times \mathbf{B} = \mathbf{0}$, then either $\mathbf{A} = \mathbf{0}$, $\mathbf{B} = \mathbf{0}$ or **A** is parallel to **B**.

3 If **A** and **B** are *perpendicular*, then $\mathbf{A} \times \mathbf{B} = AB\hat{\mathbf{n}}$.

4 The unit vectors **i**, **j**, **k** give the following results

$$\mathbf{i} \times \mathbf{i} = \mathbf{j} \times \mathbf{j} = \mathbf{k} \times \mathbf{k} = \mathbf{0}$$

$$\mathbf{i} \times \mathbf{j} = \mathbf{k}, \quad \mathbf{j} \times \mathbf{k} = \mathbf{i}, \quad \mathbf{k} \times \mathbf{i} = \mathbf{j}$$

The last products, where the unit vectors are in cyclic order, are positive, but of course, if there is any change from cyclic order, a minus sign is introduced, as follows

$$\mathbf{i} \times \mathbf{k} = -\mathbf{j}, \quad \mathbf{k} \times \mathbf{j} = -\mathbf{i}, \quad \mathbf{j} \times \mathbf{i} = -\mathbf{k}$$

5 $\mathbf{A} \times (\mathbf{B} + \mathbf{C}) = \mathbf{A} \times \mathbf{B} + \mathbf{A} \times \mathbf{C}$

The proof of this law is quite difficult and we shall not give it here.

6 If we write **A** and **B** in component form, we get

$$\mathbf{A} \times \mathbf{B} = (A_1\mathbf{i} + A_2\mathbf{j} + A_3\mathbf{k}) \times (B_1\mathbf{i} + B_2\mathbf{j} + B_3\mathbf{k})$$

Using the results of (4) and (5) we find

$$\begin{aligned} \mathbf{A} \times \mathbf{B} = & A_1B_1\mathbf{i} \times \mathbf{i} + A_1B_2\mathbf{i} \times \mathbf{j} + A_1B_3\mathbf{i} \times \mathbf{k} \\ & + A_2B_1\mathbf{j} \times \mathbf{i} + A_2B_2\mathbf{j} \times \mathbf{j} + A_2B_3\mathbf{j} \times \mathbf{k} \\ & + A_3B_1\mathbf{k} \times \mathbf{i} + A_3B_2\mathbf{k} \times \mathbf{j} + A_3B_3\mathbf{k} \times \mathbf{k} \end{aligned}$$

That is

$$\mathbf{A} \times \mathbf{B} = (A_2B_3 - A_3B_2)\mathbf{i} + (A_3B_1 - A_1B_3)\mathbf{j} + (A_1B_2 - A_2B_1)\mathbf{k}$$

This result can be written most conveniently in the form of a **determinant**. We shall not meet determinants until Chapter 10, but for completeness we quote the result in this form. It is

$$\mathbf{A} \times \mathbf{B} = \begin{vmatrix} \mathbf{i} & \mathbf{j} & \mathbf{k} \\ A_1 & A_2 & A_3 \\ B_1 & B_2 & B_3 \end{vmatrix}$$

$$\begin{aligned} &= \mathbf{i}(A_2B_3 - A_3B_2) + \mathbf{j}(A_3B_1 - A_1B_3) \\ &\quad + \mathbf{k}(A_1B_2 - A_2B_1) \end{aligned} \tag{8.9}$$

7 The magnitude of $\mathbf{A} \times \mathbf{B}$ is $AB \sin\theta$. Looking at the plane of **A** and **B** and completing the parallelogram on **A** and **B** we see from Figure 8.18 that the area of this parallelogram is $AB \sin\theta$.

Hence we have $|\mathbf{A} \times \mathbf{B}|$ = area of parallelogram with sides **A** and **B**.

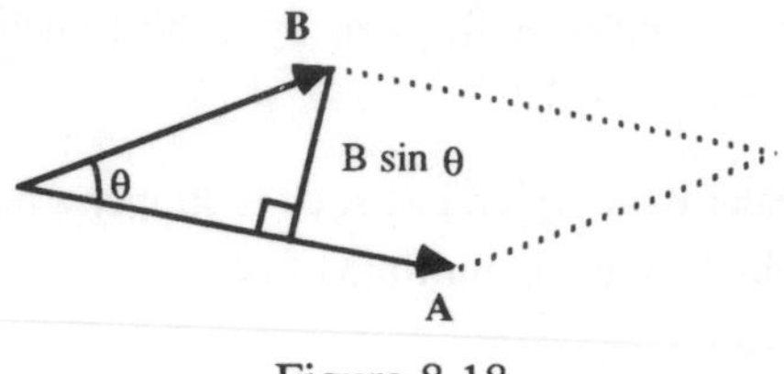

Figure 8.18

Example
We can find the length of the common perpendicular to two skew lines as follows. Refer to Figure 8.19.

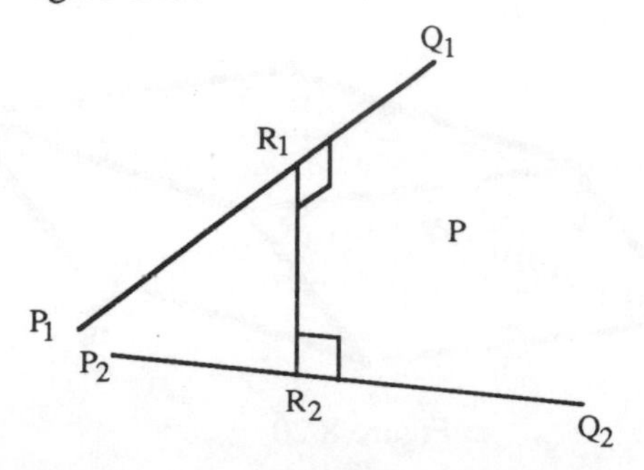

Figure 8.19

If the unit vectors along the lines P_1Q_1, P_2Q_2, R_1R_2 are $\mathbf{u}_1$, $\mathbf{u}_2$, $\mathbf{u}$ respectively and the position vectors of P_1 and P_2 are $\mathbf{r}_1$, $\mathbf{r}_2$ respectively then $\overrightarrow{OR}_1 = \mathbf{r}_1 + P_1R_1\,\mathbf{u}_1 = \mathbf{r}_2 + P_2R_2\,\mathbf{u}_2 - R_1R_2\,\mathbf{u}$. Taking scalar products with $\mathbf{u}$ we obtain

$$\mathbf{r}_1.\mathbf{u} + P_1R_1\,\mathbf{u}_1.\mathbf{u} = \mathbf{r}_2.\mathbf{u} + P_2R_2\,\mathbf{u}_2.\mathbf{u} - R_1R_2$$

But $\mathbf{u}_1.\mathbf{u} = 0 = \mathbf{u}_2.\mathbf{u}$
and hence

$$R_1R_2 = (\mathbf{r}_2 - \mathbf{r}_1).\mathbf{u}$$

Further, $\mathbf{u}$ must be proportional to $\mathbf{u}_1 \times \mathbf{u}_2$ since it is perpendicular to both $\mathbf{u}_1$ and $\mathbf{u}_2$ therefore

$$R_1R_2 = \left| \frac{(\mathbf{r}_2 - \mathbf{r}_1) \,.\, (\mathbf{u}_1 \times \mathbf{u}_2)}{|\mathbf{u}_1 \times \mathbf{u}_2|} \right|$$

8.5 Products of Three Vectors

We shall not spend very much time on such products, but they do occur in physical situations often enough to deserve some attention. For example, they occur in advanced dynamics when considering the motion of a body.

The product of two vectors $\mathbf{B}$ and $\mathbf{C}$ can be either a scalar $\mathbf{B}\,.\,\mathbf{C}$ or a vector

$\mathbf{B} \times \mathbf{C}$ and in each case can be multiplied by a third vector. Three types of product can occur.

(a) *Type* $(\mathbf{B} \cdot \mathbf{C})\mathbf{A}$

Now $\mathbf{B} \cdot \mathbf{C}$ is a scalar quantity and as such is simply a number ϕ, say. Then $(\mathbf{B} \cdot \mathbf{C})\mathbf{A} = \phi\mathbf{A}$, which is just a multiple of $\mathbf{A}$

(b) *Type* $\mathbf{A} \cdot (\mathbf{B} \times \mathbf{C})$

Since $\mathbf{B} \times \mathbf{C}$ is a vector quantity, we can perform the scalar product of this with vector $\mathbf{A}$ to generate the product $\mathbf{A} \cdot (\mathbf{B} \times \mathbf{C})$. The result is a scalar quantity and is called the scalar triple product.

Let us first of all construct the parallelopiped with three of its sides as **A**, **B**, **C** as shown in Figure 8.20

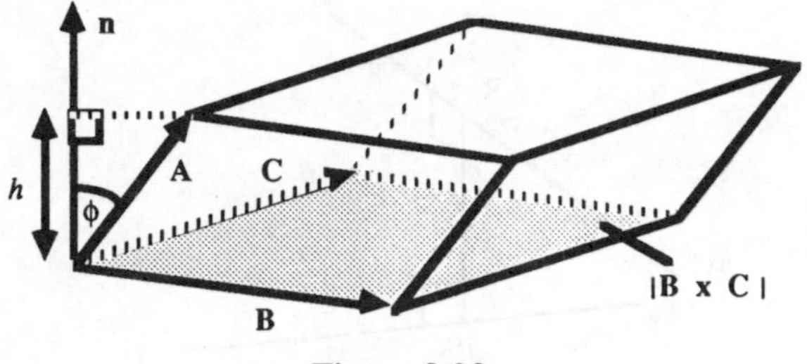

Figure 8.20

Now $\mathbf{B} \times \mathbf{C}$ is a vector in the direction of $\mathbf{n}$ with magnitude equal to the area of the shaded parallelogram. Hence

$$\begin{aligned} \mathbf{A} \cdot (\mathbf{B} \times \mathbf{C}) &= \mathbf{A} \cdot \mathbf{n} \times (\text{area of parallelogram}) \\ &= A \cos \phi \times (\text{area of parallelogram}) \\ &= h \times (\text{area of parallelogram}) \end{aligned}$$

That is, $\mathbf{A} \cdot (\mathbf{B} \times \mathbf{C}) = \text{Volume of the parallelopiped}$

Now in the above, the face constructed on **B** and **C** was taken as the base of the solid, but it is obvious that we could take any of the other faces as the base. It then follows that the expressions $\mathbf{C} \cdot (\mathbf{A} \times \mathbf{B})$ and $\mathbf{B} \cdot (\mathbf{C} \times \mathbf{A})$ also give the volume of the solid. (Notice that the cyclic order of **A**, **B**, **C** *must* be maintained to obtain the volume with positive sign.)

Hence $\mathbf{A} \cdot (\mathbf{B} \times \mathbf{C}) = \mathbf{B} \cdot (\mathbf{C} \times \mathbf{A}) = \mathbf{C} \cdot (\mathbf{A} \times \mathbf{B})$

If we interchange the last one of these expressions we get

$$\mathbf{A} \cdot (\mathbf{B} \times \mathbf{C}) = (\mathbf{A} \times \mathbf{B}) \cdot \mathbf{C}$$

The order **A**, **B**, **C** is the same, but the . and × are interchanged.

It is the practice therefore in many books to write the scalar triple product as

$$[\mathbf{A}, \mathbf{B}, \mathbf{C}]$$

since it can be any of the expressions $\mathbf{A} \cdot \mathbf{B} \times \mathbf{C}$, $\mathbf{A} \times \mathbf{B} \cdot \mathbf{C}$, together with their cyclic permutations. Notice that $(\mathbf{A} \cdot \mathbf{B}) \cdot \mathbf{C}$ has no meaning (why is this so?) and so the position of the brackets is obvious.

In component form we would get the following expression

$$\mathbf{A} \cdot (\mathbf{B} \times \mathbf{C}) = A_1(B_2C_3 - B_3C_2) + A_2(B_3C_1 - B_1C_3) + A_3(B_1C_2 - B_2C_1)$$

Try to obtain this from direct expansion of the expression on the left. Again the most convenient form is as a determinant, which we quote here:

$$\mathbf{A} \cdot (\mathbf{B} \times \mathbf{C}) = \begin{vmatrix} A_1 & A_2 & A_3 \\ B_1 & B_2 & B_3 \\ C_1 & C_2 & C_3 \end{vmatrix} \qquad (8.10)$$

Important property of the scalar triple product
The triple scalar product will vanish if
(i) any of the vectors is zero
(ii) any two of the vectors are parallel (this makes the area of a face of the solid zero)
(iii) the three vectors are coplanar (the volume of the solid is zero)
It is noted that (ii) is really a special case of (iii).

(c) *Type* $\mathbf{A} \times (\mathbf{B} \times \mathbf{C})$

Since $\mathbf{B} \times \mathbf{C}$ is a vector we can also multiply this vectorially by a third vector and obtain the expression

$$\mathbf{A} \times (\mathbf{B} \times \mathbf{C})$$

This is a vector and is called the **vector triple product**.
Looking at this expression geometrically in Figure 8.21(i), we note that $\mathbf{B} \times \mathbf{C}$ is a vector in the direction perpendicular to $\mathbf{B}$ and $\mathbf{C}$.

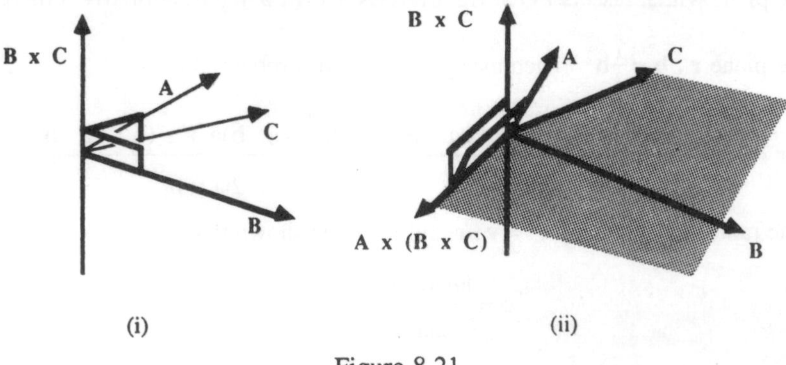

Figure 8.21

Now $\mathbf{A} \times (\mathbf{B} \times \mathbf{C})$ will be perpendicular to both $\mathbf{A}$ and $\mathbf{B} \times \mathbf{C}$. Since it is perpendicular to $\mathbf{B} \times \mathbf{C}$ it must lie in the plane of $\mathbf{B}$ and $\mathbf{C}$ [Figure 8.21(ii)]. A vector in the plane of $\mathbf{B}$ and $\mathbf{C}$ must be of the form

$$\lambda \mathbf{B} + \mu \mathbf{C} \quad \text{for some } \lambda \text{ and } \mu$$

Hence $\mathbf{A} \times (\mathbf{B} \times \mathbf{C}) = \lambda\mathbf{B} + \mu\mathbf{C}$ where the λ and μ have to be determined. We can find λ and μ by several methods – see the problem that follows here for the method of expanding into cartesian components. More elegant methods do exist, but it is not felt necessary to pursue them here. What is important, is that you should know the result, for it has great uses in problems and in future derivations. The result is

$$\mathbf{A} \times (\mathbf{B} \times \mathbf{C}) = (\mathbf{A} \cdot \mathbf{C})\mathbf{B} - (\mathbf{A} \cdot \mathbf{B})\mathbf{C} \qquad (8.11)$$

Notice that the order and brackets are important here, for example,

$$\begin{aligned}(\mathbf{A} \times \mathbf{B}) \times \mathbf{C} &= -\mathbf{C} \times (\mathbf{A} \times \mathbf{B}) \\ &= -[(\mathbf{C} \cdot \mathbf{B})\mathbf{A} - (\mathbf{C} \cdot \mathbf{A})\mathbf{B}] \\ &= (\mathbf{A} \cdot \mathbf{C})\mathbf{B} - (\mathbf{C} \cdot \mathbf{B})\mathbf{A}\end{aligned}$$

an entirely different result from equation (8.11).

Example

We seek the circle through three given points. Refer to Figure 8.22 below.

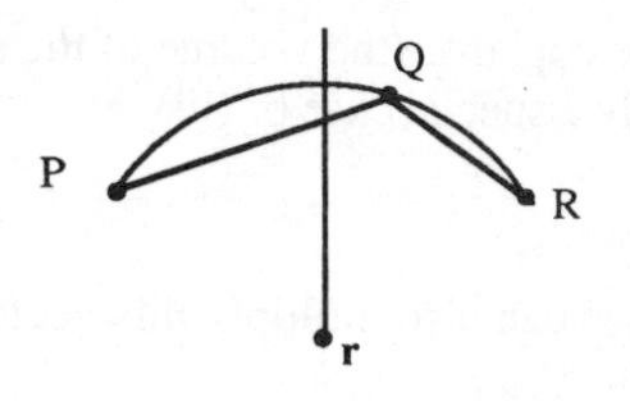

Figure 8.22

Let the points P, Q, R have position vectors $\mathbf{O}$, $\mathbf{a}$, $\mathbf{b}$ respectively. The centre of the circle with position vector $\mathbf{r}$ lies in the plane $\mathbf{r} \cdot (\mathbf{a} \wedge \mathbf{b}) = 0$. It lies on the plane which bisects PQ at right angles, i.e. $\mathbf{r} \cdot \mathbf{a} = \frac{1}{2}\mathbf{a}^2$; similarly it lies on the plane $\mathbf{r} \cdot \mathbf{b} = \frac{1}{2}\mathbf{b}^2$. Then using the result of Problem 36:

$$\mathbf{r} = \frac{a^2(\mathbf{b} \times (\mathbf{a} \times \mathbf{b})) + b^2((\mathbf{a} \times \mathbf{b}) \times \mathbf{a})}{2|\mathbf{a} \times \mathbf{b}|^2} = \frac{b^2(a^2 - \mathbf{a} \cdot \mathbf{b})\mathbf{a} + a^2(b^2 - \mathbf{a} \cdot \mathbf{b})\mathbf{b}}{2|\mathbf{a} \times \mathbf{b}|^2}$$

The radius c is given by $c^2 = \mathbf{r} \cdot \mathbf{r}$ and it can be shown that this leads to

$$c = \frac{|\mathbf{a}|\,|\mathbf{b}|\,|\mathbf{a} - \mathbf{b}|}{2|\mathbf{a} \times \mathbf{b}|}$$

Problems

Section 8.1

1 Prove that the mid-points of the sides of any quadrilateral form a parallelogram.

2 Two forces act at the corner of a quadrilateral ABDC represented by $\overrightarrow{AB}$ and $\overrightarrow{AD}$; and two at C represented by $\overrightarrow{CB}$ and $\overrightarrow{CD}$. Show that their resultant is represented by $4\overrightarrow{PQ}$ where P and Q are the mid-points of AC and BD respectively.

Section 8.2

3 If $\mathbf{r} = (2, -1, -1)$, $\mathbf{s} = (2, 1, 1)$, $\mathbf{t} = (-1, -1, 3)$

(a) find the vector $\mathbf{u}$ such that $\mathbf{r} + 2\mathbf{s} + \mathbf{u} = (-1, -1, 3)$

(b) find $|\mathbf{v}|$ if $2\mathbf{v} + 4\mathbf{r} + \mathbf{s} - 2\mathbf{t} = (1, 0, -3)$

4 Find a unit vector parallel to the resultant of the vectors $\mathbf{a} = (2, 4, -5)$ and $\mathbf{b} = (1, 2, 3)$

5 The vector $(-2\mathbf{i} + 2\sqrt{3}\mathbf{j})$ in the (x, y) plane is multiplied by the scalar λ and rotated anticlockwise through 90°. The vector $(\mathbf{i} + \mathbf{j})$ is then added and the resulting vector has magnitude 4λ. Determine the value of λ.

6 If O is the origin and $\overrightarrow{OP} = (2, 3\ -1)$, $\overrightarrow{OQ} = (4, -3, 2)$, find $\overrightarrow{PQ}$ and determine its magnitude.

7 Prove that the vectors $\mathbf{a} = (3, 1, -2)$, $\mathbf{b} = (-1, 3, 4)$, $\mathbf{c} = (4, -2, -6)$ can form the sides of a triangle, and find the lengths of the medians of this triangle.

8 Show geometrically that if $\mathbf{a}$ and $\mathbf{b}$ are non-collinear vectors in a certain plane, any other vector $\mathbf{c}$ in that plane can be represented as $(m\mathbf{a} + n\mathbf{b})$ for certain values of m and n. If $\mathbf{a} = (2\mathbf{i} - 5\mathbf{j})$, $\mathbf{b} = (-6\mathbf{i} + 4\mathbf{j})$ and $\mathbf{c} = (11\mathbf{i} - \mathbf{j})$, find m and n.

9 P, Q, R are the mid-points of the sides BC, CA, AB of the triangle ABC, and O is any other point. Show that $\overrightarrow{OP} + \overrightarrow{OQ} + \overrightarrow{OR} = \overrightarrow{OA} + \overrightarrow{OB} + \overrightarrow{OC}$.

Section 8.3

10 Determine a unit vector perpendicular to the plane containing the vectors $2\mathbf{i} + \mathbf{j} - 3\mathbf{k}$, $\mathbf{i} - 2\mathbf{j} + \mathbf{k}$. Show that the vector $\mathbf{a} = 5\mathbf{i} - 20\mathbf{j} + 35\mathbf{k}$ is perpendicular to the plane containing the vectors $\mathbf{b} = \mathbf{i} + 2\mathbf{j} + \mathbf{k}$ and $\mathbf{c} = 3\mathbf{i} - \mathbf{j} - \mathbf{k}$.

11 A force has components in the x and y directions respectively of 2 N and 5 N. Its point of application moves from (1, 5) to (6, 8). Express the work done as a dot product and evaluate it, the unit of displacement being 1 m.

12 Show that the vectors $\mathbf{a} = 3\mathbf{i} - 2\mathbf{j} + \mathbf{k}$, $\mathbf{b} = \mathbf{i} - 3\mathbf{j} + 5\mathbf{k}$, $\mathbf{c} = 2\mathbf{i} + \mathbf{j} - 4\mathbf{k}$ form a right-angled triangle.

13 Find the work done in moving an object along the vector = (3, 2, 15) if the applied force is $\mathbf{F} = (2, -1, -1)$.

14 Prove that the diagonals of a rhombus are perpendicular.

15 (a) Find the volume of a tetrahedron, three of whose edges are defined by the vectors

$$\mathbf{a} = (1, 3, -1); \quad \mathbf{b} = (1, 2, 9); \quad \mathbf{c} = (3, 1, 6)$$

(b) The points A, B, C and D are the vertices of a parallelogram and P is the intersection of the diagonals AC and BD.

Let $\mathbf{AB} = \mathbf{p}$; $\mathbf{AD} = \mathbf{q}$; $\dfrac{AP}{AC} = \lambda$ and $\dfrac{BP}{BD} = \mu$

(i) Show that $\mathbf{AP} = \lambda(\mathbf{p} + \mathbf{q})$ and $\mathbf{BP} = \mu(\mathbf{q} - \mathbf{p})$

(ii) Prove that the diagonals bisect each other.

(c) After finding the scalar product with **u**, solve the vector equation $\alpha\mathbf{x} + (\mathbf{x}.\mathbf{u})\mathbf{v} = \mathbf{w}$, in which $\alpha \neq 0$ and **u**, **v** and **w** are given vectors.

(d) The plane $3x - 2y + z = 4$ and the point (1, 2, 3) are given.

(i) Find a unit vector normal to the plane.

(ii) Find the vector equation of the line normal to the plane which passes through the given point.

(iii) Find the Cartesian coordinates of the point where the normal line in (ii) intersects the plane. (EC)

16 (a) Explain carefully the meaning of each of the three symbols **a**, a and $\hat{\mathbf{a}}$ in the expression $\mathbf{a} = a\hat{\mathbf{a}}$.

(b) A rectangular structure with six plane faces has dimensions as shown in Figure 8.23. Using vectors find,

(i) the angle between the face diagonals OG and CD

(ii) the area of the triangle with vertices C, D and E

(iii) the shortest distance from F to the diagonal BD.

(c) Derive the vector equation $(\mathbf{r} - \mathbf{b}) \cdot \mathbf{a} = 0$ for a plane in which **b** is the position vector of a given point on the plane and **a** is any vector perpendicular to the plane. Hence find the equation of the plane passing through the point (1, 0, 0) perpendicular to the direction $\mathbf{i} + 2\mathbf{j} - 3\mathbf{k}$.

Where does this plane cross the y-axis? (EC)

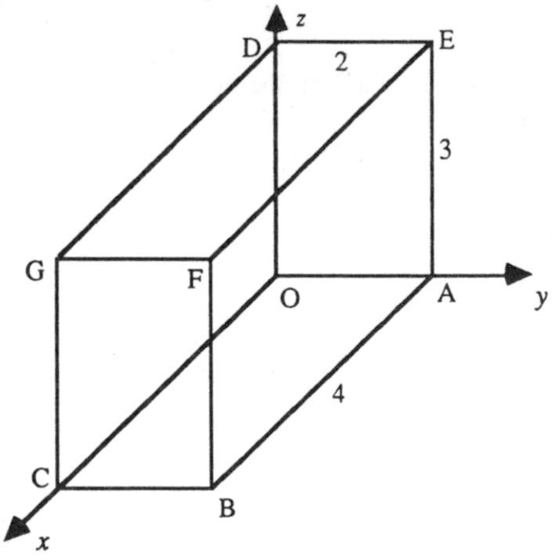

Figure 8.23

Section 8.4

17 Given $\mathbf{A} = 2\mathbf{i} - \mathbf{j} + \mathbf{k}$, $\mathbf{B} = 3\mathbf{i} + 5\mathbf{j} - \mathbf{k}$, find $\mathbf{A} \times \mathbf{B}$.

18 Find the area of the triangle with vertices at the points P(1, 3, 2), Q(2, –1, 1), R(–1, 2, 3).

19 Show $(\mathbf{A} - \mathbf{B}) \times (\mathbf{A} + \mathbf{B}) = 2(\mathbf{A} \times \mathbf{B})$.

20 Prove the sine rule for a triangle.

21 If $\mathbf{a} = (12, 1, -13)$ and $\mathbf{c} = (8, -5, 7)$ find $\mathbf{b}$ so that $\mathbf{a} \times \mathbf{b} = \mathbf{c}$ and $\mathbf{a} . \mathbf{b} = 2$.

22 The magnetic induction $\mathbf{B}$ is defined by the *Lorentz force equation* $\mathbf{F} = q(\mathbf{V} \times \mathbf{B})$ where $\mathbf{F}$ is the force on a charge q moving with velocity $\mathbf{V}$. In three experiments it was found that

when $\mathbf{V} = \mathbf{i}$, $\mathbf{F}/q = -\mathbf{j} - \mathbf{k}$

when $\mathbf{V} = \mathbf{j}$, $\mathbf{F}/q = \mathbf{i} - 2\mathbf{k}$

when $\mathbf{V} = \mathbf{k}$, $\mathbf{F}/q = \mathbf{i} + 2\mathbf{j}$

Using these results calculate $\mathbf{B}$.

23 (a) If $\mathbf{p}$ and $\mathbf{q}$ are vectors with magnitudes p and q respectively, prove that

$$(\mathbf{p} + \mathbf{q}) . (\mathbf{p} - \mathbf{q}) \equiv p^2 - q^2 \quad \text{and} \quad (\mathbf{p} + \mathbf{q}) \times (\mathbf{p} - \mathbf{q}) \equiv 2\mathbf{q} \times \mathbf{p}$$

Further if $\mathbf{p}$ and $\mathbf{q}$ are adjacent sides of a rectangle, draw a diagram showing the vectors $\mathbf{p}$, $\mathbf{q}$, $(\mathbf{p} + \mathbf{q})$ and $(\mathbf{p} - \mathbf{q})$.

What does $|\mathbf{q} \times \mathbf{p}|$ represent? Give a geometrical interpretation of the second of the identities.

(b) Verify that the two lines

$$\mathbf{r}_1 = (-1, 2, 3) + \lambda_1(-2, 1, 4)$$

$$\text{and } \mathbf{r}_2 = (3, 0, -5) + \lambda_2(1, 2, 3)$$

intersect each other, and find the acute angle between them. (EC)

24 (a) Prove that the free vectors $\mathbf{u} = 2\mathbf{i} + \mathbf{j} + 2\mathbf{k}$, $\mathbf{v} = \mathbf{i} + 3\mathbf{j} - 7\mathbf{k}$, $\mathbf{w} = 3\mathbf{i} + 4\mathbf{j} - 5\mathbf{k}$ can form a right-angled triangle.

(b) If $\mathbf{p} = (12, 1, -13)$ and $\mathbf{q} = (8, -5, 7)$, find the vector $\mathbf{s}$ such that $\mathbf{p} \times \mathbf{s} = \mathbf{q}$ and $\mathbf{p} . \mathbf{s} = 2$.

(c) Show that the perpendicular distance from the origin to the line $\mathbf{r}(t) = \mathbf{a} + \mathbf{b}t$ is given by

$$\left| \mathbf{a} - \frac{(\mathbf{a} . \mathbf{b})\mathbf{b}}{b^2} \right|$$

Find the perpendicular distance from the origin to $\mathbf{r} = 3\mathbf{j} + 4\mathbf{k} + t(5\mathbf{i} + 5\mathbf{j})$. (EC)

25 (a) Given the vectors $\mathbf{a} = 3\mathbf{i} + 4\mathbf{j} - \mathbf{k}$, $\mathbf{b} = 2\mathbf{i} - \mathbf{j} + 3\mathbf{k}$, $\mathbf{c} = \mathbf{i} + 5\mathbf{j} - 4\mathbf{k}$, show that they can form the sides of a triangle. Use vector methods to find

(i) the area of this triangle (ii) its angles (iii) the length of its sides.

(b) For the vectors $\mathbf{a}$, $\mathbf{b}$, $\mathbf{c}$ in section (a) above verify the vector identity $\mathbf{a} \wedge (\mathbf{b} \wedge \mathbf{c}) = (\mathbf{a} . \mathbf{c})\mathbf{b} - (\mathbf{a} . \mathbf{b})\mathbf{c}$.

(c) Write down the equation of the line through the points with position vectors $\mathbf{f} = (1, -2, -1)$ and $\mathbf{g} = (2, 3, 1)$. Where does this line cut the x-y plane? (EC)

26 (a) The position vectors, measured relative to an origin O, of four points A, B, C and D are given by

$$\mathbf{4i + 5j + k} \quad \mathbf{-3i + k} \quad \mathbf{3i + 9j + 4k} \quad \mathbf{-4i + 4j + 4k}$$

respectively.

(i) Write down the vectors AB, AC and AD in terms of **i, j** and **k**

(ii) Prove that the points A, B, C and D lie in the same plane.

(b) The four points O, P, Q and R are the vertices of a structure in the form of a tetrahedron, see Figure 8.24.

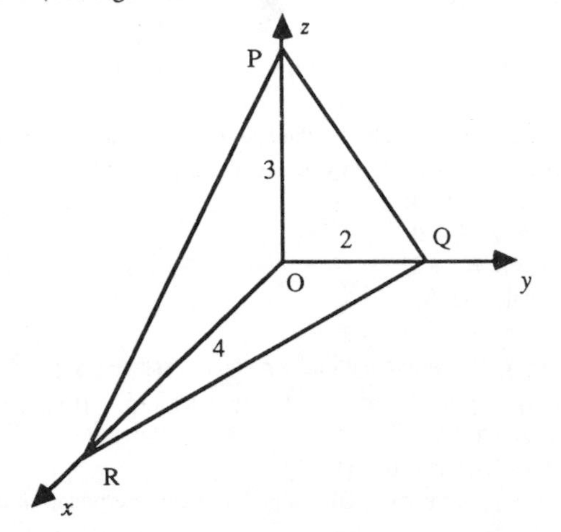

Figure 8.24

(i) Find, using vectors, the acute angle between the two edges PQ and QR

(ii) Find a unit vector perpendicular to the face PQR, and hence find the angle between the two faces PQR and OPQ

(iii) Find, using vectors, the area of the sloping face PQR.

(c) The vector equation $\alpha\mathbf{x} + \mathbf{x} \times \mathbf{a} = \mathbf{b}$ can be shown to have a solution for $\mathbf{x}$ given by

$$(\alpha^2 + a^2)\mathbf{x} = \alpha^{-1}(\mathbf{a.b})\mathbf{a} + \alpha\mathbf{b} + \mathbf{a} \times \mathbf{b}$$

Find the solution in Cartesian form for the equation

$$2\mathbf{x} + \mathbf{x} \times (2\mathbf{i} + 3\mathbf{j}) = 4\mathbf{k}$$

(EC)

Section 8.5

27 $\mathbf{A} = \mathbf{i} + 2\mathbf{j} - \mathbf{k}$, $\mathbf{B} = 2\mathbf{i} + \mathbf{j} + 3\mathbf{k}$, $\mathbf{C} = \mathbf{i} - \mathbf{j} - 2\mathbf{k}$. Find $\mathbf{A} \cdot (\mathbf{B} \times \mathbf{C})$.

28 Show that $(\mathbf{A} \times \mathbf{B}) \cdot (\mathbf{C} \times \mathbf{D}) = (\mathbf{A} \cdot \mathbf{C})(\mathbf{B} \cdot \mathbf{D}) - (\mathbf{A} \cdot \mathbf{D})(\mathbf{B} \cdot \mathbf{C})$.

29 Prove the vectors $\mathbf{A} = 2\mathbf{i} + \mathbf{j} + 2\mathbf{k}$, $\mathbf{B} = \mathbf{i} + 3\mathbf{j} - 7\mathbf{k}$, $\mathbf{C} = 3\mathbf{i} + 4\mathbf{j} - 5\mathbf{k}$ are coplanar and form a right-angled triangle.

30 $\mathbf{A} = \lambda\mathbf{B} + \mu\mathbf{C}$, show $\mathbf{A} \cdot (\mathbf{B} \times \mathbf{C}) = 0$ and vice-versa.

31 If $\mathbf{A} = \mathbf{i} + \mathbf{j} + \mathbf{k}$, $\mathbf{B} = 2\mathbf{i} + \mathbf{j} - \mathbf{k}$, $\mathbf{C} = 3\mathbf{i} - 2\mathbf{j} + 2\mathbf{k}$, verify equation (8.11) by multiplying out the left- and right-hand side and comparing results.

32 Choose axes so that $\mathbf{B} = B_1\mathbf{i}$, $\mathbf{C} = C_1\mathbf{i} + C_2\mathbf{j}$, $\mathbf{A} = A_1\mathbf{i} + A_2\mathbf{j} + A_3\mathbf{k}$. Expand $\mathbf{A} \times (\mathbf{B} \times \mathbf{C})$ and collect the terms, remembering that the result required is $\lambda\mathbf{B} + \mu\mathbf{C}$ or here $\lambda B_1\mathbf{i} + \mu(C_1\mathbf{i} + C_2\mathbf{j})$. Hence show $\mathbf{A} \times (\mathbf{B} \times \mathbf{C}) = (\mathbf{A} \, . \, \mathbf{C})\mathbf{B} - (\mathbf{A} \, . \, \mathbf{B})\mathbf{C}$.

33 If $\mathbf{a} = (3, -1, 2)$, $\mathbf{b} = (2, 1, -1)$, $\mathbf{c} = (1, -2, 2)$ find $(\mathbf{a} \times \mathbf{b}) \times \mathbf{c}$ and $\mathbf{a} \times (\mathbf{b} \times \mathbf{c})$.

34 An electric charge q_1 moving with velocity $\mathbf{v}_1$ produces a magnetic induction $\mathbf{B}$ given by

$$\mathbf{B} = \frac{\mu}{4\pi} q_1 \frac{\mathbf{v}_1 \times \mathbf{r}}{r^2} \qquad \textit{(Biot-Savart Law)}$$

The magnetic force exerted on a second charge q_2, moving with velocity $\mathbf{v}_2$ is given by

$$\mathbf{F} = q_2 \mathbf{v}_2 \times \mathbf{B}$$

Show that

$$\mathbf{F} = \frac{\mu}{4\pi r^2} q_1 q_2 \, \mathbf{v}_2 \times (\mathbf{v}_1 \times \mathbf{r})$$

and find $\mathbf{F}$ if $\mathbf{v}_1 = (\mathbf{i} + \mathbf{j} - \mathbf{k})$, $\mathbf{v}_2 = -3\mathbf{i} + 2\mathbf{j}$, $\mathbf{r} = 2\mathbf{i} - \mathbf{j} + 3\mathbf{k}$, $q_1 = q_2 = 1$.

35 Show that $(\mathbf{A} \times \mathbf{B}) \times (\mathbf{C} \times \mathbf{D}) = (\mathbf{A} \, . \, \mathbf{B} \times \mathbf{D})\mathbf{C} - (\mathbf{A} \, . \, \mathbf{B} \times \mathbf{C})\mathbf{D}$

36 Show that the three planes $\mathbf{r} \, . \, \mathbf{u}_1 = \mathbf{a}_1$, $\mathbf{r} \, . \, \mathbf{u}_2 = \mathbf{a}_2$, $\mathbf{r} \, . \, \mathbf{u}_3 = \mathbf{a}_3$ intersect at

$$\mathbf{r} = \frac{\mathbf{a}_1(\mathbf{u}_2 \times \mathbf{u}_3) + \mathbf{a}_2(\mathbf{u}_3 \times \mathbf{u}_1) + \mathbf{a}_3(\mathbf{u}_1 \times \mathbf{u}_2)}{\mathbf{u}_1 \, . \, (\mathbf{u}_2 \times \mathbf{u}_3)}$$

9

LINEAR ALGEBRA II LINEAR EQUATIONS

9.1 Introduction

Many mathematical models in engineering and other disciplines are formulated as systems of linear equations. As an example, consider the electrical circuit shown in Figure 9.1. We wish to evaluate the currents between the nodes A, B, C, D, E, F, G, H.

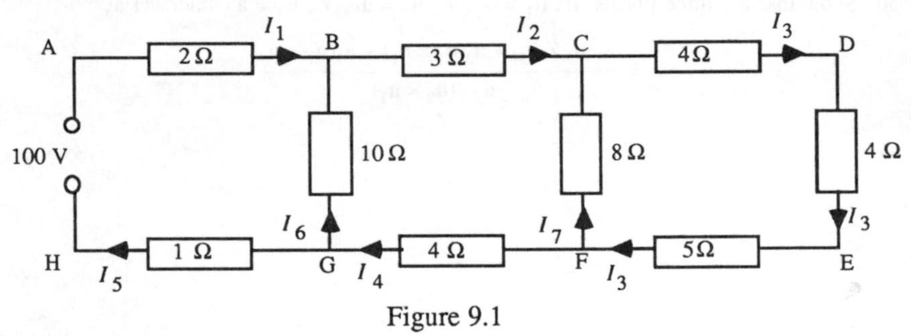

Figure 9.1

Kirchhoff's law of currents states that the net flow of current through each junction of a circuit is zero. Hence, at node B

$$I_1 - I_2 + I_6 = 0 \tag{9.1}$$

Applying the law at the nodes C, F and G in turn:

$$I_2 - I_3 + I_7 = 0 \tag{9.2}$$

$$I_3 - I_4 - I_7 = 0 \tag{9.3}$$

$$I_4 - I_5 - I_6 = 0 \tag{9.4}$$

We need three more equations to solve the system. Kirchhoff's law of potentials states that the net voltage drop around each closed loop of the circuit is zero. Applying this law to the loops ABGHA, BCFGB, CDEFC and ABCDEFGHA in turn, we obtain the equations

$$2I_1 - 10I_6 + I_5 = 100 \tag{9.5}$$

$$3I_2 - 8I_7 + 4I_4 + 10I_6 = 0 \tag{9.6}$$

$$4I_3 + 4I_3 + 5I_3 + 8I_7 = 0 \tag{9.7}$$

$$2I_1 + 3I_2 + 4I_3 + 4I_3 + 5I_3 + 4I_4 + I_5 = 100 \tag{9.8}$$

Equation (9.8) can, in fact, be obtained by adding together (9.5), (9.6) and (9.7); this implies that (9.8) tells us nothing new about the circuit and can be discarded. It is not immediately obvious that all the equations that remain are independent from each other; this may emerge during the solution process.

Rewriting the system of equations, we obtain

$$\left.\begin{array}{l} 2I_1 + I_5 - 10I_6 = 100 \\ I_1 - I_2 + I_6 = 0 \\ I_2 - I_3 + I_7 = 0 \\ I_3 - I_4 - I_7 = 0 \\ I_4 - I_5 - I_6 = 0 \\ 3I_2 + 4I_4 + 10I_6 - 8I_7 = 0 \\ 13I_3 + 8I_7 = 0 \end{array}\right\} \qquad (9.9)$$

In general, a system of linear equations is a set of equations of the form

$$\left.\begin{array}{l} a_{11}x_1 + a_{12}x_2 + \ldots + a_{1n}x_n = b_1 \\ a_{21}x_1 + a_{22}x_2 + \ldots + a_{2n}x_n = b_2 \\ \vdots \\ a_{m1}x_1 + a_{m2}x_2 + \ldots + a_{mn}x_n = b_n \end{array}\right\} \qquad (9.10)$$

This chapter is concerned with methods of solving this system for $x_1, x_2, \ldots, x_n$ given the coefficients a_{ij} $(i = 1, 2, \ldots, n; j = 1, 2, \ldots, n)$ and the constants b_i $(i = 1, 2, \ldots, n)$.

First we consider the solution of two equations in two unknowns. These will have the form

$$\left.\begin{array}{l} a_{11}x_1 + a_{12}x_2 = b_1 \\ a_{21}x_1 + a_{22}x_2 = b_2 \end{array}\right\} \qquad (9.11)$$

We know from earlier work that each equation represents a straight line in the $x_1 - x_2$ plane. The solution to these equations is represented by the point of intersection of the two lines. Three cases arise in theory:

(i) Unique solution, e.g. $2x_1 + x_2 = 1$, $x_1 - 2x_2 = 0.1$

(ii) No solution, e.g. $2x_1 - x_2 = -1$, $2x_1 - x_2 = -2$ (lines parallel)

(iii) Infinite number of solutions, e.g. $x_1 - x_2 = 1$, $3x_1 - 3x_2 = 3$ (lines coincident)

We say that cases (i) and (iii) are **consistent**.

From a practical point of view, there is a fourth possibility. This occurs when the straight lines are almost parallel. Take, for example,

$$\left.\begin{aligned} x_1 + 3x_2 &= 4 \\ 30x_1 + 90.1x_2 &= 120.1 \end{aligned}\right\}$$

These have a solution $x_1 = 1, x_2 = 1$; however, a change of +(1/9)% in the coefficient of x_2 in the first equation yields a pair of equations which have no solution but, if this is accompanied by a change of +(1/12)% in the constant term, yields a pair of equations which are identical. These changes in the coefficients could easily be caused by inexact data. Alternatively, if the data is exact, the lack of precision in subsequent arithmetic could greatly affect the end result. We say that a set of equations so finely balanced is **ill-conditioned** (or **unstable**) and needs special attention.

Consider the general equations (9.11); if we multiply the first one by a_{22} and subtract a_{12} times the second we obtain

$$(a_{11}a_{22} - a_{21}a_{12})x_1 = b_1a_{22} - b_2a_{12}$$

Similarly, we find that

$$(a_{11}a_{22} - a_{21}a_{12})x_2 = b_2a_{11} - b_1a_{21}$$

Therefore, if $a_{11}a_{22} - a_{21}a_{12} \neq 0$ we can write down a unique solution. If $a_{11}a_{22} - a_{21}a_{12} = 0$ then, generally speaking, we have $a_{11}/a_{12} = a_{21}/a_{22}$ so that the slopes of the straight lines are equal (and if this ratio equals b_1/b_2 the lines are coincident). We say **generally** since there are special cases such as $a_{11} = 0 = a_{21}$ etc. (What do these relate to geometrically?)

For short-hand, we write $a_{11}a_{22} - a_{21}a_{22} \equiv \begin{vmatrix} a_{11} & a_{12} \\ a_{21} & a_{22} \end{vmatrix}$ and this is an example of a **determinant**, considered in more detail in Section 10.3.

For illustrative purposes we shall use a system of three equations in three unknowns. This is because the system of two equations in two unknowns has special features which systems of higher order do not; conversely, the only real additional feature in a system of an order greater than three is the increase in computation required for its solution (with a consequent potential increase in round-off error).

It is noted that, just as with two equations in two unknowns, cases can arise where we get

(i) a unique solution
(ii) no solution, or
(iii) an infinite number of solutions.

We shall consider cases (ii) and (iii) later in more detail with the reasons for their occurrence, *but at the moment* let us assume that the equations have a *unique* answer and hence proceed on to develop for these a numerical method of solution.

9.2 Gauss Elimination

The basic method of Gauss reduces the equations to the triangular form

$$\begin{aligned} a_{11}x_1 + a_{12}x_2 + a_{13}x_3 &= b_1 \\ A_{22}x_2 + A_{23}x_3 &= B_2 \\ A_{33}x_3 &= B_3 \end{aligned}$$

(Note that the operations involved will change the coefficients in the second and third lines.)

The third equation then gives $x_3 = B_3/A_{33}$, assuming $A_{33} \neq 0$.
Back substitution in the second equation now gives x_2, assuming $A_{22} \neq 0$.
Finally, back substitution in the first equation gives x_1, assuming $a_{11} \neq 0$.

Let us work through an example.

Example 1

$$\left.\begin{aligned} 2x_1 - 4x_2 + x_3 &= -9 &\quad (9.12a) \\ 4x_1 + 2x_2 - 3x_3 &= 17 &\quad (9.12b) \\ x_1 - x_2 + 5x_3 &= -11 &\quad (9.12c) \end{aligned}\right\}$$

First of all we eliminate x_1 from equations (9.12b) and (9.12c) by

(i) subtracting $2 \times$ (9.12a) from (9.12b) and

(ii) subtracting $\frac{1}{2}\times$ (9.12a) from (9.12c)

This gives

$$\begin{aligned} 2x_1 - 4x_2 + x_3 &= -9 &\quad (9.12a) \\ 10x_2 - 5x_3 &= 35 &\quad (9.12d) \\ x_2 + (9/2)x_3 &= -13/2 &\quad (9.12e) \end{aligned}$$

For convenience divide equation (9.12d) by the common factor 5, to obtain

$$\begin{aligned} 2x_1 - 4x_2 + x_3 &= -9 &\quad (9.12a) \\ 2x_2 - x_3 &= 7 &\quad (9.12f) \\ x_2 + (9/2)x_3 &= -13/2 &\quad (9.12e) \end{aligned}$$

We now eliminate x_2 from (9.12e) by taking (9.12e) $-\frac{1}{2}\times$ (9.12f) giving

$$\begin{aligned} 2x_1 - 4x_2 + x_3 &= -9 &\quad (9.12a) \\ 2x_2 - x_3 &= 7 &\quad (9.12f) \\ 5x_3 &= -10 &\quad (9.12g) \end{aligned}$$

Then we begin back substituting
(9.12g) $\Rightarrow x_3 = -2$
(9.12b) $\Rightarrow x_2 = 7/2 + x_3/2 = 5/2$
(9.12a) $\Rightarrow x_1 = -9/2 + 2x_2 - x_3/2 = -9/2 + 5 + 1 = 3/2$

Interpreting our result geometrically, we can say that the three planes represented by equation (9.12) meet at the point (3/2, 5/2, –2).

You will notice that this method relies on the coefficients only and so we can isolate them for purposes of calculation. (It would be worth having a check of some kind; one of the simplest is a row sum check in which you could sum all the constants in each equation and whatever operations are done on the equations should also check with the same operation on the row sums.)

Let us do the above example again, isolating the coefficients. We have

$$\begin{bmatrix} 2 & -4 & 1 \\ 4 & 2 & -3 \\ 1 & -1 & 5 \end{bmatrix}\begin{bmatrix} -9 \\ 17 \\ -11 \end{bmatrix} \quad \begin{matrix} \\ \text{Row 2} - (2 \times \text{Row 1}) \\ \text{Row 3} - (\tfrac{1}{2} \times \text{Row 1}) \end{matrix}$$

$$\begin{bmatrix} 2 & -4 & 1 \\ 0 & 10 & -5 \\ 0 & 1 & 9/2 \end{bmatrix}\begin{bmatrix} -9 \\ 35 \\ -13/2 \end{bmatrix} \quad \begin{matrix} \\ \text{Row 2} \div 5 \\ \\ \end{matrix}$$

$$\begin{bmatrix} 2 & -4 & 1 \\ 0 & 2 & -1 \\ 0 & 1 & 9/2 \end{bmatrix}\begin{bmatrix} -9 \\ 7 \\ -13/2 \end{bmatrix} \quad \begin{matrix} \\ \\ \text{Row 3} - (\tfrac{1}{2} \times \text{Row 2}) \end{matrix}$$

$$\begin{bmatrix} 2 & -4 & 1 \\ 0 & 2 & -1 \\ 0 & 0 & 5 \end{bmatrix}\begin{bmatrix} -9 \\ 7 \\ -10 \end{bmatrix}$$

A possible check is to substitute the values obtained for x_1, x_2, x_3 into the second and third equations of the original set (why not the first?).

Hence, substitution in equation (9.12b) gives

$$(4 \times 3/2) + (2 \times 5/2) - (3 \times -2) = 17$$

and in equation (9.12c) produces

$$(1 \times 3/2) - (1 \times 5/2) - (5 \times -2) = -11$$

Example 2

Let us work through one more example using this method. Solve

$$\left.\begin{aligned} 3x + 2y + z &= 4 \\ x - y + z &= 2 \\ -2x + 2z &= 5 \end{aligned}\right\} \qquad (9.13)$$

We first write the system in a form which isolates the coefficients as follows:

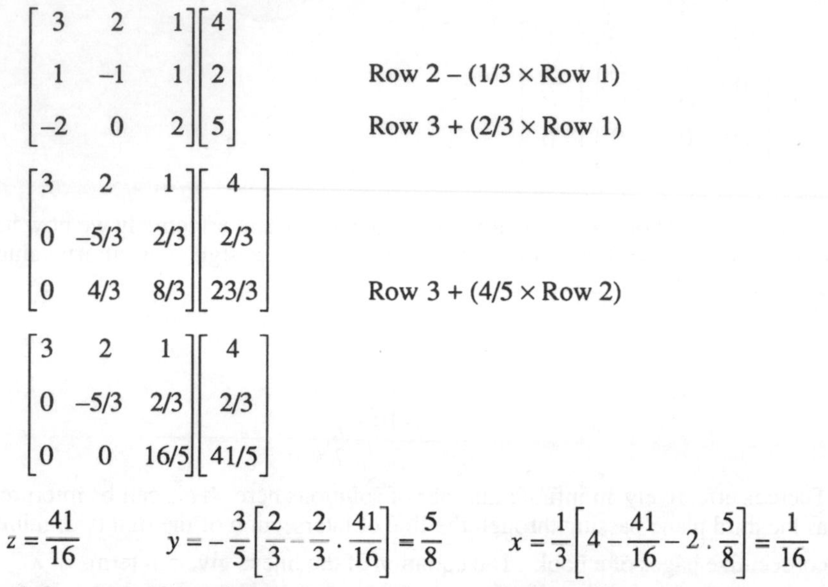

$$\begin{bmatrix} 3 & 2 & 1 \\ 1 & -1 & 1 \\ -2 & 0 & 2 \end{bmatrix}\begin{bmatrix} 4 \\ 2 \\ 5 \end{bmatrix} \qquad \begin{matrix} \\ \text{Row } 2 - (1/3 \times \text{Row } 1) \\ \text{Row } 3 + (2/3 \times \text{Row } 1) \end{matrix}$$

$$\begin{bmatrix} 3 & 2 & 1 \\ 0 & -5/3 & 2/3 \\ 0 & 4/3 & 8/3 \end{bmatrix}\begin{bmatrix} 4 \\ 2/3 \\ 23/3 \end{bmatrix} \qquad \begin{matrix} \\ \\ \text{Row } 3 + (4/5 \times \text{Row } 2) \end{matrix}$$

$$\begin{bmatrix} 3 & 2 & 1 \\ 0 & -5/3 & 2/3 \\ 0 & 0 & 16/5 \end{bmatrix}\begin{bmatrix} 4 \\ 2/3 \\ 41/5 \end{bmatrix}$$

$$z = \frac{41}{16} \qquad y = -\frac{3}{5}\left[\frac{2}{3} - \frac{2}{3}\cdot\frac{41}{16}\right] = \frac{5}{8} \qquad x = \frac{1}{3}\left[4 - \frac{41}{16} - 2\cdot\frac{5}{8}\right] = \frac{1}{16}$$

Check for yourself that these values satisfy the original equations (9.13).

Inconsistencies

There are *three* types of snags which can occur.

Case 1

If we consider the solution of the equations

$$\begin{aligned} 13x_1 + 4x_2 - x_3 &= -14 \\ 4x_1 + 5x_2 - 3x_3 &= 0 \\ 3x_1 + 2x_2 - x_3 &= -2 \end{aligned}$$

or

$$\begin{bmatrix} 13 & 4 & -1 \\ 4 & 5 & -3 \\ 3 & 2 & -1 \end{bmatrix}\begin{bmatrix} -14 \\ 0 \\ -2 \end{bmatrix} \qquad \begin{matrix} \\ \text{Row } 2 - \left[\frac{4}{13} \times \text{Row } 1\right] \\ \text{Row } 3 - \left[\frac{3}{13} \times \text{Row } 1\right] \end{matrix}$$

so that

$$\begin{bmatrix} 13 & 4 & -1 \\ 0 & \frac{49}{13} & -\frac{35}{13} \\ 0 & \frac{14}{13} & -\frac{10}{13} \end{bmatrix}\begin{bmatrix} -14 \\ \frac{56}{13} \\ \frac{16}{13} \end{bmatrix} \qquad \begin{matrix} \\ \\ \text{Row } 3 - \left[\frac{14}{49} \times \text{Row } 2\right] \end{matrix}$$

and finally we obtain

$$\begin{bmatrix} 13 & 4 & -1 \\ 0 & \frac{49}{13} & -\frac{35}{13} \\ 0 & 0 & 0 \end{bmatrix} \begin{bmatrix} -14 \\ \frac{56}{13} \\ 0 \end{bmatrix}$$

The last equation has been eliminated completely and effectively we now have only 2 equations in 3 unknowns. We can, of course, assign an arbitrary value to $x_3 = \lambda$ (say) and we get

$$x_2 = \frac{8}{7} + \frac{5}{7}\lambda$$

$$x_1 = \frac{-10}{7} - \frac{\lambda}{7}$$

There is effectively an infinite number of solutions here. This can be interpreted as the third plane passing through the line of intersection of the first two; think of consecutive pages of a book. The equation of the line is given in terms of λ.

Case 2

Consider now the solution of the equations

$$\begin{aligned} 3x_1 + 2x_2 + x_3 &= 4 \\ x_1 - x_2 + x_3 &= 2 \\ 9x_1 + x_2 + 5x_3 &= 10 \end{aligned}$$

We obtain successively

$$\begin{bmatrix} 3 & 2 & 1 \\ 1 & -1 & 1 \\ 9 & 1 & 5 \end{bmatrix} \begin{bmatrix} 4 \\ 2 \\ 10 \end{bmatrix} \qquad \begin{matrix} \\ \text{Row } 2 - [\frac{1}{3} \times \text{Row } 1] \\ \text{Row } 3 - (3 \times \text{Row } 1) \end{matrix}$$

$$\begin{bmatrix} 3 & 2 & 1 \\ 0 & -\frac{5}{3} & \frac{2}{3} \\ 0 & -5 & 2 \end{bmatrix} \begin{bmatrix} 4 \\ \frac{2}{3} \\ -2 \end{bmatrix} \qquad \begin{matrix} \\ \\ \text{Row } 3 - (3 \times \text{Row } 2) \end{matrix}$$

$$\begin{bmatrix} 3 & 2 & 1 \\ 0 & -\frac{5}{3} & \frac{2}{3} \\ 0 & 0 & 0 \end{bmatrix} \begin{bmatrix} 4 \\ \frac{2}{3} \\ -4 \end{bmatrix}$$

The last equation is **false** and thus our elimination procedure has led to a false result. Hence the equations have **no** solution. The third plane does not meet the line of intersection of the first two.

The two examples above illustrate that sometimes we shall get no answer at all or perhaps an infinite number of solutions. Can we recognise when these situations will occur? In fact we can. You will remember that with two equations in two unknowns it was important that the *determinant* of the l.h.s. coefficients should not be zero.

For three equations in three unknowns the same criterion applies; that is, for this special equation (9.10), the determinant

$$\begin{vmatrix} a_{11} & a_{12} & a_{13} \\ a_{21} & a_{22} & a_{23} \\ a_{31} & a_{32} & a_{33} \end{vmatrix} \neq 0$$

We do not yet know how to evaluate such determinants but we shall deal with this in the next chapter. However, at the moment we state that

$$\begin{vmatrix} a_{11} & a_{12} & a_{13} \\ a_{21} & a_{22} & a_{23} \\ a_{31} & a_{32} & a_{33} \end{vmatrix} = a_{11}a_{22}a_{33} + a_{12}a_{23}a_{31} + a_{13}a_{21}a_{32} - a_{11}a_{23}a_{32} - a_{12}a_{21}a_{33} - a_{13}a_{31}a_{22}$$

Looking at the two examples we have just done and considering the determinant of the left-hand side we find in the example of Case 1

$$\begin{vmatrix} 13 & 4 & -1 \\ 4 & 5 & -3 \\ 3 & 2 & -1 \end{vmatrix} = -65 - 36 - 8 + 78 + 16 + 15 = 0$$

and for the example in Case 2

$$\begin{vmatrix} 3 & 2 & 1 \\ 1 & -1 & 1 \\ 9 & 1 & 5 \end{vmatrix} = -15 + 18 + 1 - 3 - 10 + 9 = 0$$

It should be fairly obvious how to extend this check to $4 \times 4, \ldots, n \times n$ simultaneous linear equations.

We have then a straightforward check as to when snags discussed in Cases 1 and 2 will arise. *This is when the determinant of the coefficients of the equations is zero.*

Case 3

The third possibility is that of ill-conditioning which occurs when the determinant of coefficients is relatively small. You will remember the example of ill-conditioned equations in Section 9.1

$$\left.\begin{aligned} x_1 + 3x_2 &= 4 \\ 30x_1 + 90.1x_2 &= 120.1 \end{aligned}\right\}$$

The determinant of coefficients is

$$\begin{vmatrix} 1 & 3 \\ 30 & 90.1 \end{vmatrix} = 0.1$$

which is relatively small compared with the exact answers of $x_1 = 1, x_2 = 1$.

Example

A well-known ill-conditioned set of 3 equations is

$$\left.\begin{aligned} \frac{1}{3}x_1 + \frac{1}{4}x_2 + \frac{1}{5}x_3 &= 0 \\ \frac{1}{2}x_1 + \frac{1}{3}x_2 + \frac{1}{4}x_3 &= 0 \\ x_1 + \frac{1}{2}x_2 + \frac{1}{3}x_3 &= 1 \end{aligned}\right\}$$

The determinant of the coefficients is

$$\begin{vmatrix} \frac{1}{3} & \frac{1}{4} & \frac{1}{5} \\ \frac{1}{2} & \frac{1}{3} & \frac{1}{4} \\ 1 & \frac{1}{2} & \frac{1}{3} \end{vmatrix} = -\frac{1}{27}$$

and the exact answers to this set of equations are

$$x_1 = 9 \qquad x_2 = -36 \qquad x_3 = 30$$

If we approximate 1/3 by 0.33 and solve the equations using Gauss elimination, we find

$$x_1 = -13.67 \qquad x_2 = 17 \qquad x_3 = -18.69$$

Thus for a small change in the coefficients, we produce a very large change in the solution.

It is worth emphasising at this stage that we have to be careful when considering approximate solutions. If we have an exact solution, each equation is exactly in balance; however, if the solution process leads to an approximate solution how can we be satisfied that the approximation is satisfactory? The test available is to substitute the supposed solution into the equations to see how closely they balance. The danger inherent in that test is illustrated by these ill-conditioned systems – approximate balancing of the equations does not necessarily mean a good approximation to the true solution. Any round-off introduced during the process of solution means that we are not solving exactly the original system and this may lead to a quite different solution from the true one.

9.3 Refinement of the Solution

The method of Gauss Elimination should cause no difficulties when the values of the unknowns are integers or simple decimals, but real-life problems seldom have such easy answers. Furthermore, we may frequently have to solve sets of 50 or more equations; here, the problems of round-off error may well become severe. The strategy of partial pivoting aims to reduce the build-up of this error. To illustrate the ideas involved, we consider the system

$$\begin{bmatrix} 2 & -1 & -3 \\ 4 & 5 & -1 \\ 3 & 2 & 2 \end{bmatrix} \begin{bmatrix} 3 \\ 7 \\ 17 \end{bmatrix} \tag{9.14}$$

First we identify the coefficient in column 1 with the largest magnitude: this is circled above. It this is not in the first row then the pivot row is interchanged with the first row, as is required in our example:

$$\begin{bmatrix} 4 & 5 & -1 \\ 2 & -1 & -3 \\ 3 & 2 & 2 \end{bmatrix} \begin{bmatrix} 7 \\ 3 \\ 17 \end{bmatrix} \quad \begin{matrix} \\ R2-(2/4)R1 \\ R3-(3/4)R1 \end{matrix}$$

(We see that the multiples of rows being subtracted are in magnitude < 1.)

$$\begin{bmatrix} 4 & 5 & -1 \\ 0 & -7/2 & -5/2 \\ 0 & -7/4 & 11/4 \end{bmatrix} \begin{bmatrix} 7 \\ -1/2 \\ 47/4 \end{bmatrix} \qquad R3-\left(\frac{-7}{4}\right)\Big/\left(\frac{-7}{2}\right) R2, \quad \text{i.e. } R3-(\tfrac{1}{2})R2$$

(No need for an interchange here.)

$$\begin{bmatrix} 4 & 5 & -1 \\ 0 & -7/2 & -5/2 \\ 0 & 0 & 16/4 \end{bmatrix} \begin{bmatrix} 7 \\ -1/2 \\ 48/4 \end{bmatrix}$$

Hence $x_3 = 48/4 \div 16/4 = 3$

$x_2 = (-1/2 + 5/2\,(3)) \div (-7/2) = -2$

$x_1 = (7 - 5\,(-2) + 1\,(3)) \div 4 = 5$

At each stage of the elimination process we look among the rows from which the next variable is to be eliminated to find the pivot row. This is the row which contains the coefficient of largest modulus in the column relating to the variable under consideration. Such a process is called **partial pivoting** or maximal column pivoting.

We were working with exact arithmetic so that the solution obtained was identical to that which could be obtained without pivoting. To see the effects of pivoting, consider the following example which was solved on a Prime computer.

$$\begin{aligned}
5x_1 + 2\times10^6x_2 + 10^5x_3 + 3x_4 &= 7 \\
2\times10^6x_1 + 10^5x_2 + 9x_3 + 4x_4 &= 3 \\
3\times10^6x_1 + 8x_2 + 10^6x_3 + 2x_4 &= 7 \\
20x_1 + 10^6x_2 + 30x_3 + 40x_4 &= 8
\end{aligned}$$

Without pivoting the solution was

(x_1, x_2, x_3, x_4)

$= (1.335144 \times 10^{-6},\ 3.108558 \times 10^{-6},\ 4.160305 \times 10^{-6},\ 1.222820 \times 10^{-1})$

whereas with pivoting we obtained

$(1.099991 \times 10^{-6},\ 3.145079 \times 10^{-6},\ 3.457263 \times 10^{-6},\ 1.213700 \times 10^{-1})$

These solutions were substituted into the equations and the values of the left-hand sides subtracted from the given right-hand sides to obtain respectively the error vectors

$(0, -0.470309,\ -1.41032,\ -1.907349 \times 10^{-6})$ and

$(2.861023 \times 10^{-6},\ 9.536743 \times 10^{-7},\ 9.536743 \times 10^{-7},\ 9.536743 \times 10^{-7})$

The improvement obtained by pivoting is apparent.

We should beware of placing too much credence in the second of these vectors; the coincidence in the last three components is a signal that we are down at the limit of machine accuracy.

A program for Gauss elimination with partial pivoting is shown on the following page.

Other pivoting Strategies

Although partial pivoting ensures that the multiple, p, of one row that is subtracted from another row is never more than 1 in magnitude, there are systems of equations for which such a strategy by itself may not be sufficient. Consider the system

$$\begin{aligned}
2 \times 10^{-10} x_1 + x_2 &= 1 \\
x_1 + x_2 &= 2
\end{aligned} \tag{9.15}$$

Using Gauss elimination in a Fortran 77 program on a Prime computer we obtained the 'solution' $x_2 = 1, x_1 = 0$, which is clearly nonsense. The main reason for this is the inability of the computer to store a number such as $1 - \frac{1}{2}\times 10^{10}$ any differently from $-\frac{1}{2}\times 10^{10}$. When, however, we employed the partial pivoting algorithm we obtained the result $x_2 = 1, x_1 = 1$, which is as good an approximation as we can expect. However, suppose that the system had been presented to us in the form

$$\begin{aligned}
2x_1 + 10^{10} x_2 &= 10^{10} \\
x_1 + x_2 &= 2
\end{aligned} \tag{9.16}$$

Since no interchange of rows is needed both straightforward Gauss elimination and elimination with pivoting gave the result $x_2 = 1, x_1 = 0$.

```
>
  10  CLS
  20  DIM A(10,11),X(10),P(10),B(11)
  30  PRINTTAB(20,1)"Gauss Elimination with
      Partial Pivoting"
  40  INPUTTAB(0,3)"Enter number of equations ",N
  50  FOR row =1 TO N
  60    FOR col =1 TO N
  70      PRINT"Enter coefficient for row ";row;
          " column ";col;
  80      INPUTA(row,col)
  90      NEXT col
 100    PRINTTAB(12)"Enter constant for row ";row;
 110    INPUTA(row,N+1)
 120    NEXT row
 130  FOR L = 1 TO N-1
 140    A1 = ABS(A(L,L))
 150    L2 = L
 160    FOR I = L+1 TO N
 170      IF A1 > ABS(A(I,L)) THEN 200
 180      A1 = A(I,L)
 190      L2 = I
 200      NEXT I
 210    IF L2 = L THEN 270
 220    FOR J = L TO N+1
 230      W = A(L,J)
 240      A(L,J) = A(L2,J)
 250      A(L2,J) = W
 260      NEXT J
 270    FOR I = L+1 TO N
 280      P(I) = -1*A(I,L)/A(L,L)
 290      FOR J = L TO N+1
 300        A(I,J) = A(I,J)+P(I)*A(L,J)
 310        NEXT J
 320      NEXT I
 330    NEXT L
 340  X(N) = A(N,N+1)/A(N,N)
 350  FOR I = 1 TO N-1
 360    N3 = N-I
 370    X(N3) = A(N3,N+1)
 380    N4 = N3+1
 390    FOR J = N4 TO N
 400      X(N3) = X(N3)-A(N3,J)*X(J)
 410      NEXT J
 420    X(N3) = X(N3)/A(N3,N3)
 430    NEXT I
 440  FOR I = 1 TO N
 450    PRINT"X ";I;" = ";X(I)
 460    NEXT I
 470  END
```

Therefore, we need a more comprehensive pivoting strategy to cater for this case. The technique of **scaled partial pivoting** is an endeavour to overcome the difficulty above. In essence, it proceeds thus: in searching for a pivot row, each candidate for the pivot is divided by the largest element in its row (on the left-hand side); then the scaled value of greatest magnitude is chosen as the pivot. It is clear that the system (9.16) would lead to scaled values of 2×10^{-10} and 1 respectively and hence the equations would be interchanged to give

$$\begin{aligned} x_1 + \quad x_2 &= 2 \\ 2x_1 + 10^{10}x_2 &= 10^{10} \end{aligned}$$

Subtracting twice the first equation from the second gives the system

$$\begin{aligned} x_1 + \quad x_2 &= 2 \\ 10^{10}x_2 &= 10^{10} \end{aligned}$$

From this we obtain $x_2 = 1, x_1 = 1$. This process can be seen as a means of 'balancing' the equations in the system.

We might mention that partial pivoting is usually safe; however, the example we have illustrated is perhaps an extreme case but for the sake of peace of mind we can employ the scaled partial pivoting algorithm as a matter of course.

A further strategy, known as **complete scaled pivoting** allows the possibility of column interchanges in its search for a pivot. The amount of extra computation required in the extra searching and the interchanges is seldom justified and the algorithm would be best saved for especially unyielding systems of equations.

Residuals

Suppose that the system

$$\begin{aligned} 2x_1 - \quad x_2 - 3x_3 &= 3 \\ 4x_1 + 5x_2 - \quad x_3 &= 7 \\ 3x_1 + 2x_2 + 2x_3 &= 17 \end{aligned} \tag{9.14}$$

has been 'solved' to give the approximate values $(x_1, x_2, x_3) = (5.0001, -1.9999, 2.9998)$. When substituting back into the equations to check, we obtain left-hand sides of (3.0007, 7.0011, 17.0001).

If we subtract from this vector the vector of the right-hand sides, we obtain the **residual vector** (0.0007, 0.0011, 0.0001). Ideally, this should be the zero vector and will be so when the calculations for x_1, x_2, x_3 are exact. We write the exact solution to the system as

$$X_1 = 5, \quad X_2 = -2, \quad X_3 = 3$$

If we write $x_1 = 5.0001 = X_1 + \varepsilon_1$ etc, then the **error vector** is

$$(\varepsilon_1, \varepsilon_2, \varepsilon_3) = (0.0001,\ 0.0001,\ -0.0002)$$

We again raise the question: should we be satisfied with a small error vector or a small residual vector; and how do we measure 'small'?

When we substitute the approximate solution into the first equation of (9.14) we find that

$$2(X_1 + \varepsilon_1) - (X_2 + \varepsilon_2) - 3(X_3 + \varepsilon_3) = 3 + 0.0007$$

and since $\quad 2X_1 - X_2 - 3X_3 = 3 \quad$ (exactly)

then it follows that

$$2\varepsilon_1 - \varepsilon_2 - 3\varepsilon_3 = 0.0007$$

Similarly,

$$4\varepsilon_1 + 5\varepsilon_2 - \varepsilon_3 = 0.0011$$

$$3\varepsilon_1 + 2\varepsilon_2 + 2\varepsilon_3 = 0.0001$$

This system of equations can, in principle, be solved to give ε_1, ε_2 and ε_3 and hence the approximate values x_1, x_2, x_3 can be corrected to give the exact solutions X_1, X_2, X_3. Furthermore, since the coefficients on the left-hand sides are the same as those of the original system, the same row operations in the same order as were used to solve the original system can be used to solve this second system. However, since the first solution process was inexact, there is no reason to suppose that the second process will be any better. In other words, we shall find only approximate values for the errors $\varepsilon_1, \varepsilon_2, \varepsilon_3$. These may be good enough for our purposes: if not, we shall have to use the method again to 'correct the corrections'.

9.4 Gauss-Jordan Method

This is a slight variation of the Gauss elimination procedure which allows the values of $x_1, x_2, \ldots$ to be read off directly.

The elimination procedure is extended to reduce the array of coefficients, for example, in a 3×3 system of equations, to the form

$$\begin{bmatrix} a_{11} & 0 & 0 \\ 0 & a'_{22} & 0 \\ 0 & 0 & a'_{33} \end{bmatrix}$$

Sometimes this is further extended to give

$$\begin{bmatrix} 1 & 0 & 0 \\ 0 & 1 & 0 \\ 0 & 0 & 1 \end{bmatrix}$$

Returning to

$$\left.\begin{aligned} 2x_1 - 4x_2 + x_3 &= -9 \\ 4x_1 + 2x_2 - 3x_3 &= 17 \\ x_1 - x_2 + 5x_3 &= -11 \end{aligned}\right\} \qquad (9.12)$$

the procedure would be as before to write

$$\begin{bmatrix} 2 & -4 & 1 \\ 4 & 2 & -3 \\ 1 & -1 & 5 \end{bmatrix}\begin{bmatrix} -9 \\ 17 \\ -11 \end{bmatrix}$$

and then eliminating to obtain

$$\begin{bmatrix} 2 & -4 & 1 \\ 0 & 2 & -1 \\ 0 & 0 & 5 \end{bmatrix}\begin{bmatrix} -9 \\ 7 \\ -10 \end{bmatrix}$$

Gauss-Jordan then extends this via

Row 1 – (1/5 × Row 3)

Row 2 + (1/5 × Row 3)

to obtain

$$\begin{bmatrix} 2 & -4 & 0 \\ 0 & 2 & 0 \\ 0 & 0 & 5 \end{bmatrix}\begin{bmatrix} -7 \\ 5 \\ -10 \end{bmatrix} \qquad \text{Row 1} + (2 \times \text{Row 2})$$

which gives

$$\begin{bmatrix} 2 & 0 & 0 \\ 0 & 2 & 0 \\ 0 & 0 & 5 \end{bmatrix}\begin{bmatrix} 3 \\ 5 \\ -10 \end{bmatrix} \qquad \begin{matrix} \text{Row 1} \div 2 \\ \text{Row 2} \div 2 \\ \text{Row 3} \div 5 \end{matrix}$$

$$\text{or} \quad \begin{bmatrix} 1 & 0 & 0 \\ 0 & 1 & 0 \\ 0 & 0 & 1 \end{bmatrix}\begin{bmatrix} 3/2 \\ 5/2 \\ -2 \end{bmatrix}$$

giving directly

$x_1 = 3/2, \quad x_2 = 5/2, \quad x_3 = -2$

An application of this method is found in Section 10.2.

Efficiency of Methods of Solution of Simultaneous Linear Equations

It is of interest to compare the methods outlined above for solving simultaneous linear equations.

For n simultaneous equations, using the direct method of solution involving determinants requires ~(1.72 × n × n!) multiplications and divisions (additions being regarded as taking negligible time compared with multiplication and division). For two equations the number is exactly 8, and for 100 equations it is

$\sim 1.6 \times 10^{160}$, which should take a fast digital computer several centuries.

The Gauss elimination method requires $[(n^3/3) + n^2 - (n/3)]$ multiplications and divisions; for two equations this becomes 6 operations and for 100 equations 343 333 operations – a considerable saving, relative to the determinant method.

The Gauss-Jordan variation requires $\frac{1}{2}n^3 + n^2 - \frac{1}{2}n$ multiplications and divisions.

An alternative approach is provided in Section 9.5.

9.5 Iterative Methods

In this section we consider two methods of solution which rely on an iterative approach.

The system

$$\begin{aligned} 5x_1 + x_2 + 2x_3 &= 13 \\ x_1 + 5x_2 - 3x_3 &= -9 \\ 2x_1 + 2x_2 - 5x_3 &= -8 \end{aligned} \tag{9.17}$$

can be re-written as

$$\begin{aligned} x_1 &= 2.6 - 0.2x_2 - 0.4x_3 \\ x_2 &= -1.8 - 0.2x_1 + 0.6x_3 \\ x_3 &= 1.6 + 0.4x_1 + 0.4x_2 \end{aligned}$$

If we 'guess' that the solution is $x_1 = 2$, $x_2 = -1$, $x_3 = 2$ then we can check by substituting these values into the right-hand sides: we obtain 2, –1, 2 respectively, which shows that our guess was correct.

However, a different guess of $x_1 = x_2 = x_3 = 0$ gives, on substitution 2.6, –1.8, 1.6 respectively, showing that this guess is wrong. Instead of giving up, suppose that we try $x_1 = 2.6$, $x_2 = -1.8$, $x_3 = 1.6$ as a second guess. This time, on substituting we obtain 2.32, –1.36, 1.92 and these values in turn lead to 2.104, –1.108, 1.984. It should be clear that the successive application of this procedure is taking us closer to the exact solution; it should also be evident that we generate one extra decimal place in our calculated values on each application. Further calculations are shown in Table 9.1; results are recorded to 4 d.p.

Table 9.1

Iteration	0	1	2	3	4	5	6
x_1	0	2.6	2.32	2.104	2.0280	2.0067	2.0017
x_2	0	–1.8	–1.36	–1.108	–1.0304	–1.0066	–1.0019
x_3	0	1.6	1.92	1.984	1.9984	1.9990	2.0001

This process is known as the **Jacobi method** and can be described formally as follows

$$\begin{aligned} x_1^{(r+1)} &= 2.6 - 0.2x_2^{(r)} - 0.4x_3^{(r)} \\ x_2^{(r+1)} &= -1.8 - 0.2x_1^{(r)} + 0.6x_3^{(r)} \\ x_3^{(r+1)} &= 1.6 + 0.4x_1^{(r)} + 0.4x_2^{(r)} \end{aligned} \tag{9.18}$$

where $x_1^{(r)}$ is the rth approximation to x_1, etc.

The **Gauss-Seidel method** is a sensible modification of the above procedure. If the Jacobi method produces successively better approximations to x_1, x_2, x_3 then it would seem reasonable to use the latest values available for each variable as soon as they become available. In our current example

$$\begin{aligned} x_1^{(r+1)} &= 2.6 - 0.2x_2^{(r)} - 0.4x_3^{(r)} \\ x_2^{(r+1)} &= -1.8 - 0.2x_1^{(r+1)} + 0.6x_3^{(r)} \\ x_3^{(r+1)} &= 1.6 + 0.4x_1^{(r+1)} + 0.4x_2^{(r+1)} \end{aligned} \tag{9.19}$$

Comparison with the equations for the Jacobi method will be beneficial to the reader.

Hence starting from $x_1^{(0)} = x_2^{(0)} = x_3^{(0)} = 0$ the first equation yields

$$x_1^{(1)} = 2.6$$

$$x_2^{(1)} = -1.8 - 0.2 \times 2.6 + 0 = -2.32$$

$$x_3^{(1)} = 1.6 + 0.4 \times 2.6 + 0.4 \times (-2.32) = 1.712 \quad \text{etc}$$

Table 9.2 shows the application of these formulae; note the increased speed of convergence to the exact solution.

Table 9.2

Iteration	0	1	2	3	4	5	6
x_1	0	2.6	2.3792	2.0288	1.9861	1.9965	2.0001
x_2	0	−2.32	−1.2486	−0.9744	−0.9841	−0.9988	−1.0006
x_3	0	1.712	2.0522	2.0217	2.0008	1.9991	1.9998

What factors affect the speed of convergence? Since the vector of approximate solutions $(x_1^{(r)}, x_2^{(r)}, x_3^{(r)})$ is undergoing a rotation on each cycle of operations, then the closer the initial orientation of the vector is to its final direction (as the exact solution vector), the more quickly will convergence occur. But will convergence always occur?

It can be shown that if the matrix of coefficients of the l.h. sides of (9.17), **A**, is diagonally dominant,

i.e. $$|a_{rr}| > \sum_{\substack{j=1 \\ j \neq r}}^{n} |a_{rj}| \text{ for } r = 1 \text{ to } n$$

then the Gauss-Seidel method converges.In our example the criterion becomes

$$|a_{11}| > |a_{12}| + |a_{13}|, \quad |a_{22}| > |a_{21}| + |a_{23}| \quad \text{and} \quad |a_{33}| > |a_{31}| + |a_{32}|$$

```
>
 10  CLS
 20  PRINTTAB(20,1)"Gauss-Seidel Iteration"
 30  DIM A(3,3),B(3),X(3),OLDX(3)
 40  INPUTTAB(0,2)"Enter specified tolerance ",tol
 50  INPUT"Enter maximum number of iterations ",itmax
 60  PRINTTAB(0,6)"Enter matrix coefficients"
 70  FOR row = 1 TO 3
 80    FOR col = 1 TO 3
 90      PRINT"Enter the value for row ";row;
         " column ";col;
100      INPUT A(row,col)
110      NEXT col
120    PRINTTAB(13)"Enter row ";row;" constant ";
130    INPUT B(row)
140    PRINTTAB(7)"Enter initial guess for X";row;" ";
150    INPUTOLDX(row)
160    PRINT
170    NEXT row
180  itcoun=0
190  REPEAT
200    ON ERROR PRINT"NUMBER TOO BIG":END
210    X(1)=(B(1)-A(1,3)*OLDX(3)-A(1,2)*OLDX(2))/A(1,1)
220    X(2)=(B(2)-A(2,3)*OLDX(3)-A(2,1)*X(1))/A(2,2)
230    X(3)=(B(3)-A(3,2)*X(2)-A(3,1)*X(1))/A(3,3)
240    itcoun=itcoun+1
250    gap1=ABS(X(1)-OLDX(1))
260    IF gap1>= tol THEN 310
270    gap2=ABS(X(2)-OLDX(2))
280    IF gap2>= tol THEN 310
290    gap3=ABS(X(3)-OLDX(3))
300    IF gap3< tol THEN 360
310    IF itcoun>= itmax THEN PRINT"No solution
       after ";itmax;" iterations":END
320    FOR row = 1 TO 3
330      OLDX(row)=X(row)
340      NEXT row
350    UNTIL FALSE
360  FOR row = 1 TO 3
370    PRINT"X";row;" value is ";X(row)
380    NEXT row
390  PRINT"After ";itcoun;" iterations"
400  END
```

Substituting the actual values gives

$$|5| > |1| + |2|, \quad |5| > |1| + |-2|, \quad |-5| > |2| + |2|$$

and the criterion is satisfied.

Notice that if the largest sized coefficient in each left-hand side is underlined then the underlined coefficients lie on the leading diagonal of **A**. This is so for our example.

The system

$$\begin{aligned} x_1 + 5x_2 - 3x_3 &= -9 \\ 2x_1 + 2x_2 - 5x_3 &= -8 \\ 5x_1 + x_2 + 2x_3 &= 13 \end{aligned} \tag{9.20}$$

as it stands is not suitable for applying the Gauss-Seidel method. Table 9.3 shows the first few applications.

Table 9.3

Iteration	0	1	2	3	4
x_1	0	–9	45.5	–439.625	2355.03125
x_2	0	5	16.75	146.5625	221.671875
x_3	0	26.5	–115.625	1032.28125	–5991.91406

The system can be reorganised to make Gauss-Seidel a suitable method: if the last equation is moved to the first place, convergence should occur.

A moment's thought will show that if the coefficient matrix contains several zero entries then diagonal dominance, if it occurs, will be enhanced and this will indicate a high rate of convergence. Such matrices arise, for example, in the solution of partial differential equations.

A possible criterion for stopping the Gauss-Seidel process is to continue until the numbers

$$|x_i^{(r+1)} - x_i^{(r)}|$$

are all less than a pre-assigned value.

The program on the previous page will solve a system of three equations in three unknowns by the Gauss-Seidel method.

Problems

Section 9.2

1 Solve the following simultaneous equations using Gauss' method working with exact arithmetic (fractions or whole numbers) and using a row sum check.

(a) $\begin{cases} 5x - y + 2z - 3 = 0 \\ 2x + 4y + z - 8 = 0 \\ x + 3y - 3z - 2 = 0 \end{cases}$ (b) $\begin{cases} 25x_1 + 36x_2 + 36x_3 = 12 \\ 4x_1 + 81x_2 + 4x_3 = 27 \\ 9x_1 + 9x_2 + 64x_3 = 3 \end{cases}$

(c) For the circuit shown in Figure 9.2, Kirchhoff's Laws produce the equations

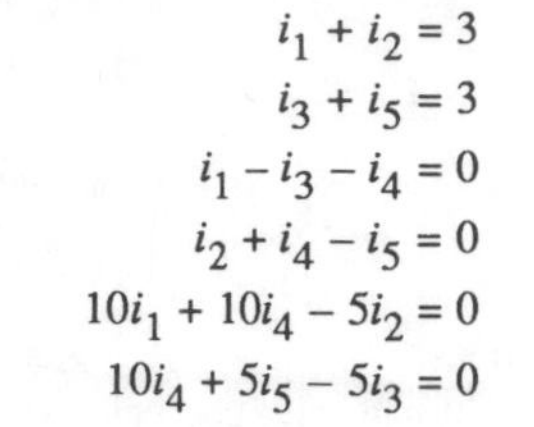

$$\begin{aligned} i_1 + i_2 &= 3 \\ i_3 + i_5 &= 3 \\ i_1 - i_3 - i_4 &= 0 \\ i_2 + i_4 - i_5 &= 0 \\ 10i_1 + 10i_4 - 5i_2 &= 0 \\ 10i_4 + 5i_5 - 5i_3 &= 0 \end{aligned}$$

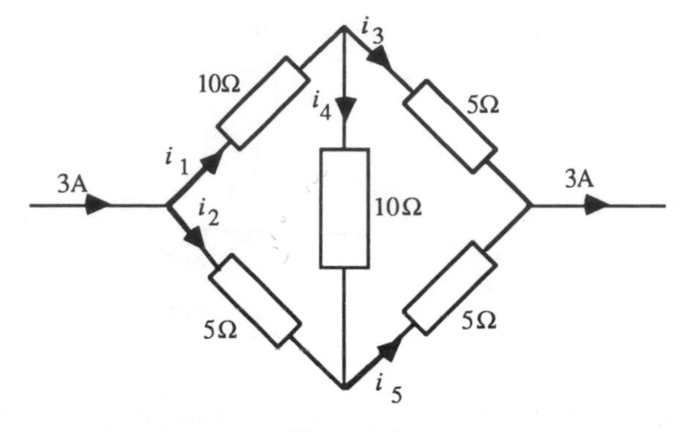

Figure 9.2

Show that the fourth equation can be removed without altering the problem and solve the resulting set of equations.

(d) *(Beware!)*
$$\begin{cases} 20x - 12y - 14z = 20 \\ 8x + 4y - 6z = 16 \\ 3x - 4y - 2z = 2 \end{cases}$$

(e)
$$\begin{cases} 3x - 4y + 7z = 16 \\ 2x + 3y + z = 11 \\ x - 5z = 14 \end{cases}$$

(f)
$$\begin{cases} 2x + 3y + 4z = 3 \\ 6x - 3y + 8z = 4 \\ -2x + 6y - 12z = -2 \end{cases}$$

2 Solve by Gauss elimination method, where $i^2 = -1$

(a) $(2 + i)x - iy = i - 2$
$3ix + (1 + i)y = -1$

(b) $(3 - 2i)x + (1 - 2i)y = 1 + 2i$
$(6 - 4i)x + (6 - 3i)y = -5 - 2i$

3 Show, using Gaussian elimination, that each of the sets of equations

$$\begin{cases} 2x + 4y = 3 \\ 3x + 6y = 4.5 \end{cases}$$

$$\begin{cases} 5x - 3y + 4z = 1 \\ 18x - 3y + 13z = 6 \\ 8x + 3y + 5z = 4 \end{cases}$$

$$\begin{cases} x + y + 2z = 3 \\ 4x - y - 2z = 4 \\ 2x - 3y - 6z = -2 \end{cases} \qquad \begin{cases} 2x - 3y - z = 2 \\ 8x - 23y - 7z = 3 \\ -x + 7y + 2z = 3/2 \end{cases}$$

has an infinite number of solutions.
Evaluate the determinant of the left-hand side coefficients in each case.

4 Show that each of the sets of equations

$$\begin{cases} 2x + 4y = 3 \\ 3x + 6y = 4.9 \end{cases} \qquad \begin{cases} 7x + y + 3z = 14 \\ 2x - 2y + 2z = 3 \\ 3x + 5y - z = 9 \end{cases}$$

$$\begin{cases} 2x - 3y - z = 2 \\ 8x - 23y - 7z = 4 \\ -2x + 14y + 4z = 6 \end{cases} \qquad \begin{cases} x + y + z = 1 \\ 2x - 2y + z = 4 \\ 3x - y + 2z = 6 \end{cases}$$

has no solution. Check that the determinant of l.h.s. coefficients is zero in each case.

5 By Gauss elimination using exact arithmetic, solve the simultaneous equations

$$x_1 + x_2 + x_3 = 1$$
$$2x_1 - 2x_2 + 3x_3 = q$$
$$3x_1 - x_2 + px_3 = 2$$

For what values of p and q do these equations
(i) have no solutions
(ii) have an infinite number of solutions
(iii) have a unique solution?
Write down, in terms of p and q, the solutions in cases (ii) and (iii). (EC)

6 (a) Calculate the value of m from the following three simultaneous equations

$$3l - 2m + 4n + 2 = 0$$
$$2l - 3m + 2n = 0$$
$$-2l + 2m + 5 = 0$$

(b) (i) Assuming that a, b and c are not all zero, establish the conditions upon k such that the following set of simultaneous equations has a unique solution for x, y and z. Do *not* solve the equations.

$$2x - 3y + kz = a$$
$$-10x + y - z = b$$
$$kx - 2y + kz = c$$

(ii) If in the equations in (i) above a, b and c are all zero, what values of k will lead to a non-trivial consistent solution for x, y and z? Further, with k chosen to be one of the values, solve the equations for x, y and z. (EC)

7 Solve the equations

$$\begin{cases} x_1 - 2x_2 + x_3 = -2 \\ 3x_1 - 2x_2 + 4x_3 = 36 \\ 8x_1 - 20x_2 + 9x_3 = -38 \end{cases}$$

using exact arithmetic.
Solve the equations again when they are in the form

$$\begin{cases} x_1 - 2x_2 + x_3 = -2 \\ x_1 - 0.67x_2 + 1.33x_3 = 12 \\ x_1 - 2.50x_2 + 1.13x_3 = -4.75 \end{cases}$$

working to 2 d.p.
Give reasons for the inconsistencies.

8 Solve, using Gaussian elimination, first using exact arithmetic and then working to 2 d.p.

$$\begin{cases} 5x_1 + 7x_2 + 6x_3 + 5x_4 = 23 \\ 7x_1 + 10x_2 + 8x_3 + 7x_4 = 32 \\ 6x_1 + 8x_2 + 10x_3 + 9x_4 = 33 \\ 5x_1 + 7x_2 + 9x_3 + 10x_4 = 31 \end{cases}$$

9 For the equations of Problem 8, try the effect of substituting for x_1, x_2, x_3 and x_4
(i) 14.6, –7.2, –2.5, 3.1 (ii) 2.36, 0.18, 0.65, 1.21
Comment on your results.

Section 9.3

10 Solve to 2 d.p. using Gaussian elimination with and without pivoting; work to 3 d.p.

(a) $$\begin{cases} -i_1 - 3i_2 + 7i_3 = 2 \\ 4i_1 - 2i_2 - i_3 = 10 \\ -2i_1 + 14i_2 - 3i_3 = 5 \end{cases}$$

(b) $$\begin{cases} 0.19x + 0.22y + 0.42z = 0.25 \\ 0.27x + 0.34y + 0.56z = 0.18 \\ 0.52x + 0.41y + 0.17z = 0.69 \end{cases}$$

(c) $$\begin{cases} 0.22x + 0.40y + 0.39z = 0.15 \\ 0.41x + 0.37y + 0.70z = 0.40 \\ 0.36x + 0.37y + 0.19z = 0.42 \end{cases}$$

(d) $$\begin{cases} 0.732x + 1.013y - 5.421z = 4.256 \\ 3.491x + 0.782y + 2.203z = 7.113 \\ 0.961x + 4.265y - 1.523z = 3.727 \end{cases}$$

11 Solve the simultaneous equations

$$\begin{aligned} 2x_1 - 4x_2 + x_3 &= -9 \\ 4x_1 + 2x_2 - 3x_3 &= 17 \\ x_1 - x_2 + 5x_3 &= -11 \end{aligned}$$

via Gaussian elimination, with or without pivoting, to solve them, indicating clearly the individual elementary row operation at each stage. (EC)

12 (a) Solve the following system of simultaneous linear equations using Gaussian elimination with an augmented matrix

$$x - y + z = 2$$
$$-2x + 2z = 5$$
$$3x + 2y + z = 4$$

Explain each elementary row operation using appropriate symbols.

(b) The equations below are given

$$2x - 3y - z = 2$$
$$8x - 23y - 7z = 4$$
$$-2x + 14y + 4z = p$$

(i) Using determinants show that if $p = 2$, the equations have an infinite number of solutions.

(ii) If $p = 6$, what can be said about the solution of the new system? (EC)

13 Find the residuals in Problems 10(a) and (b). Using these, improve the accuracy of the solutions by one application of the process given in the above theory.

Section 9.4

14 Solve the simultaneous equations given in Problems 1 and 10 using Gauss-Jordan elimination.

Section 9.5

15 Consider each of the following systems of three simultaneous equations

$$\text{(i)} \quad \begin{cases} 3x_1 - 3x_2 + 7x_3 = 18 \\ x_1 + 6x_2 - x_3 = 10 \\ 10x_1 - 2x_2 + 7x_3 = 27 \end{cases} \qquad \text{(ii)} \quad \begin{cases} 4x_1 + x_2 + 2x_3 = 16 \\ x_1 + 3x_2 + x_3 = 10 \\ x_1 + 2x_2 + 5x_3 = 12 \end{cases}$$

(a) Without rearranging the equations, try to find the solutions iteratively using both the Jacobi and the Gauss-Seidel methods starting with values (0, 0, 0), (1, 1, 1) and (1.01, 2.01, 3.01) for (x_1, x_2, x_3).

(b) Rearrange the equations if necessary to satisfy the convergence criteria on page 248 and repeat (a).

(c) Check your solutions in the original equations.

10

LINEAR ALGEBRA III MATRICES

10.1 Matrix Algebra

We saw with the solution of linear equations that the whole calculation could proceed using only arrays of coefficients. Such arrays are called **matrices** and are simply a convenient way of storing information. They have their own algebra for addition and multiplication with which we have to become familiar.

It must be emphasised at this point that there is a fundamental difference between matrices and determinants. A matrix is an *array* of elements, each of which has its own distinct position in the array. A determinant is a *number* which is produced by combining these elements in a prescribed manner.

In our example on Gauss elimination (Chapter 9) we met the matrices

$$\begin{bmatrix} 2 & -4 & 1 \\ 4 & 2 & -3 \\ 1 & -1 & 5 \end{bmatrix}$$

which is an example of a 3×3 matrix since it has 3 rows and 3 columns, and

$$\begin{bmatrix} -9 \\ 17 \\ -11 \end{bmatrix}$$

which is an example of a 3×1 matrix as it has 3 rows and 1 column.

A matrix $\mathbf{A}$ is of order $m \times n$ if it has m rows and n columns, written as

$$\mathbf{A} = \begin{bmatrix} a_{11} & a_{12} & a_{13} & \cdots\cdots & a_{1n} \\ a_{21} & a_{22} & a_{23} & \cdots\cdots & a_{2n} \\ \vdots & & & & \\ a_{m1} & a_{m2} & a_{m3} & \cdots\cdots & a_{mn} \end{bmatrix}$$

If $m = n$, the matrix is **square**.
If $m = 1$, the matrix is called a **row vector**, e.g. $(1, -1, 2, 4)$

If $n = 1$, the matrix is called a **column vector**, e.g. $\begin{bmatrix} 1 \\ 0 \\ -1 \end{bmatrix}$

If $m = n = 1$, the matrix is a single number, e.g. (3)
The $a_{11}, a_{12}, \ldots\ldots, a_{mn}$ are called **elements**.
A matrix having all its elements **zero** is called the **null** matrix. There are many such null matrices, e.g.

$$(0) \qquad \begin{bmatrix} 0 & 0 \\ 0 & 0 \end{bmatrix} \qquad (0 \quad 0 \quad 0 \quad 0)$$

A square matrix whose only non-zero elements are on the main diagonal (from the top left to the bottom right) is called a **diagonal** matrix. For example

$$\begin{bmatrix} -3 & 0 & 0 & 0 \\ 0 & 2 & 0 & 0 \\ 0 & 0 & 0 & 0 \\ 0 & 0 & 0 & -5 \end{bmatrix}$$

Transpose of a matrix A

This results if we interchange columns and rows of matrix **A** and is written $\mathbf{A}^{\mathrm{T}}$. Examples:

$$\mathbf{A} = (1, 1, -1, 0), \quad \mathbf{A}^{\mathrm{T}} = \begin{bmatrix} 1 \\ 1 \\ -1 \\ 0 \end{bmatrix}; \quad \mathbf{A} = \begin{bmatrix} 1 & 2 & 3 \\ 0 & -1 & 4 \end{bmatrix}, \quad \mathbf{A}^{\mathrm{T}} = \begin{bmatrix} 1 & 0 \\ 2 & -1 \\ 3 & 4 \end{bmatrix}$$

Equality

Two matrices are **equal** if they are the same shape (i.e. have the same number of rows and columns) and if the corresponding elements are equal.

Addition

If matrices **A** and **B** are of the same shape, they can be added by adding corresponding elements and forming a matrix of the same shape as the original ones. For example,

$$\begin{bmatrix} 1 & 2 & 0 \\ 4 & -1 & 8 \end{bmatrix} + \begin{bmatrix} 0 & 1 & 4 \\ -1 & 1 & 4 \end{bmatrix} = \begin{bmatrix} 1 & 3 & 4 \\ 3 & 0 & 12 \end{bmatrix}$$

The resulting matrix is written $\mathbf{A} + \mathbf{B}$; observe that $\mathbf{A} + \mathbf{B} \equiv \mathbf{B} + \mathbf{A}$.

We cannot add $\begin{bmatrix} 1 & 2 \\ 3 & 4 \end{bmatrix}$ and $\begin{bmatrix} 1 & 1 & 1 \\ 1 & 1 & 1 \end{bmatrix}$; why?

Subtraction

Subtraction is obtained in the same way by subtraction of corresponding elements, e.g.

$$\begin{bmatrix} 1 & 2 & 0 \\ 4 & -1 & 8 \end{bmatrix} - \begin{bmatrix} 0 & 1 & 4 \\ -1 & 1 & 4 \end{bmatrix} = \begin{bmatrix} 1 & 1 & -4 \\ 5 & -2 & 4 \end{bmatrix}$$

Multiplication

(a) *Multiplication of a matrix by a scalar*

$$\text{If } \mathbf{A} \text{ is a matrix} = \begin{bmatrix} a_{11} & a_{12} & a_{13} & \cdots\cdots & a_{1n} \\ a_{21} & a_{22} & a_{23} & \cdots\cdots & a_{2n} \\ \vdots & & & & \\ \vdots & & & & \\ a_{m1} & a_{m2} & a_{m3} & \cdots\cdots & a_{mn} \end{bmatrix}$$

$$\text{then } k\mathbf{A} = \begin{bmatrix} ka_{11} & ka_{12} \cdots & ka_{1n} \\ ka_{21} & ka_{22} \cdots & ka_{2n} \\ \vdots & & \\ \vdots & & \\ ka_{m1} & ka_{m2} \cdots & ka_{mn} \end{bmatrix}$$

That is, every element is multiplied by the scalar k, e.g.

$$3\begin{bmatrix} 1 & 2 \\ 3 & 4 \end{bmatrix} = \begin{bmatrix} 3 & 6 \\ 9 & 12 \end{bmatrix}$$

(b) *Multiplication of two matrices*

Two matrices **A** and **B** can be multiplied in the order $\mathbf{A} \times \mathbf{B}$ if the number of columns in **A** = number of rows in **B**. If this condition does not hold, multiplication cannot be defined.

If **A** is an $m \times n$ and **B** is an $n \times p$ matrix, then the product **C** is an $m \times p$ matrix

$$\underset{m\times n}{\mathbf{A}} \times \underset{n\times p}{\mathbf{B}} = \underset{m\times p}{\mathbf{C}}$$

We say that **A** is post-multiplied by **B** or that **B** is pre-multiplied by **A**.

The element of **C** in the kl position is obtained by multiplying and summing the corresponding elements of the kth row of **A** with the lth column of **B**. That is,

$$\text{if } \mathbf{A} = \begin{bmatrix} a_{11} & a_{12} & \cdots & a_{1n} \\ a_{21} & a_{22} & \cdots & a_{2n} \\ \vdots & & & \\ a_{k1} & a_{k2} & \cdots & a_{kn} \\ \vdots & & & \\ a_{m1} & a_{m2} & \cdots & a_{mn} \end{bmatrix}$$

$$\text{and } \mathbf{B} = \begin{bmatrix} b_{11} & b_{12} & \cdots & b_{1l} & \cdots & b_{1p} \\ b_{21} & b_{22} & \cdots & b_{2l} & \cdots & b_{2p} \\ \vdots & & & & & \\ b_{n1} & b_{n2} & \cdots & b_{nl} & \cdots & b_{np} \end{bmatrix}$$

then **C** =

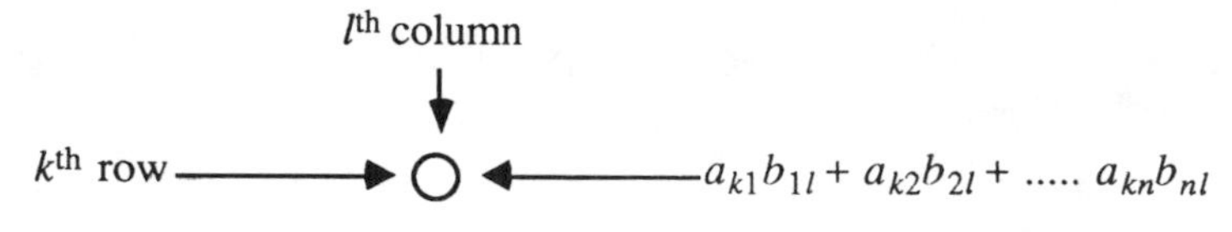

The process will be clearer by studying two examples.

Example 1

$$\begin{bmatrix} 1 & -1 \\ 2 & 0 \end{bmatrix} \begin{bmatrix} 2 & -1 \\ 1 & 3 \end{bmatrix} = \begin{bmatrix} 1 \times 2 + (-1) \times 1 & 1 \times (-1) + (-1) \times 3 \\ 2 \times 2 + 0 \times 1 & 2 \times (-1) + 0 \times 3 \end{bmatrix}$$

$$= \begin{bmatrix} 1 & -4 \\ 4 & -2 \end{bmatrix}$$

Example 2

$$\underset{(3 \times 3)}{\begin{bmatrix} 1 & -1 & 2 \\ 0 & 3 & 4 \\ -2 & 5 & -1 \end{bmatrix}} \underset{(3 \times 2)}{\begin{bmatrix} 2 & 0 \\ -1 & 3 \\ 1 & 1 \end{bmatrix}} = \begin{bmatrix} 1{\times}2+(-1){\times}(-1)+2{\times}1 & 1{\times}0+(-1){\times}3+2{\times}1 \\ 0{\times}2+3{\times}(-1)+4{\times}1 & 0{\times}0+3{\times}3+4{\times}1 \\ -2{\times}2+5{\times}(-1)+(-1){\times}1 & -2{\times}0+5{\times}3+(-1){\times}1 \end{bmatrix}$$

$$= \begin{bmatrix} 5 & -1 \\ 1 & 13 \\ -10 & 14 \end{bmatrix}$$
$$(3 \times 2)$$

Notice that in the reverse order the product

$$\begin{bmatrix} 2 & 0 \\ -1 & 3 \\ 1 & 1 \end{bmatrix} \begin{bmatrix} 1 & -1 & 2 \\ 0 & 3 & 4 \\ -2 & 5 & -1 \end{bmatrix}$$

is not defined since it is a 3×2 times a 3×3 matrix. If the matrices are both square, then we can form $\mathbf{A} \times \mathbf{B}$ and $\mathbf{B} \times \mathbf{A}$ but the answers are not necessarily the same, e.g.

$$\begin{bmatrix} 1 & 3 \\ 2 & 4 \end{bmatrix} \begin{bmatrix} 2 & 5 \\ 1 & 4 \end{bmatrix} = \begin{bmatrix} 5 & 17 \\ 8 & 26 \end{bmatrix}$$

and
$$\begin{bmatrix} 2 & 5 \\ 1 & 4 \end{bmatrix} \begin{bmatrix} 1 & 3 \\ 2 & 4 \end{bmatrix} = \begin{bmatrix} 12 & 26 \\ 9 & 19 \end{bmatrix}$$

That is, matrix multiplication is **not commutative**.
As in Example 2, note that

$$\begin{bmatrix} 1 & 2 \\ 2 & 2 \end{bmatrix} \begin{bmatrix} 1 & 1 & 1 \\ 1 & 1 & 1 \end{bmatrix} = \begin{bmatrix} 3 & 3 & 3 \\ 4 & 4 & 4 \end{bmatrix}$$

but $\begin{bmatrix} 1 & 1 & 1 \\ 1 & 1 & 1 \end{bmatrix} \begin{bmatrix} 1 & 2 \\ 2 & 2 \end{bmatrix}$ is *not* defined.

A further difference from normal algebra is that we can have $\mathbf{A} \,.\, \mathbf{B} = \mathbf{0}$ = null matrix, without either $\mathbf{A} = \mathbf{0}$ or $\mathbf{B} = \mathbf{0}$; i.e. neither $\mathbf{A}$ nor $\mathbf{B}$ is a null matrix, e.g.

$$\begin{bmatrix} 2 & 1 & 4 \\ 4 & 2 & 8 \end{bmatrix} \begin{bmatrix} 3 & 2 \\ 2 & 4 \\ -2 & -2 \end{bmatrix} = \begin{bmatrix} 0 & 0 \\ 0 & 0 \end{bmatrix}$$

However, with matrices of the correct shape for which multiplication can be defined,

$$\mathbf{A} \,.\, \mathbf{0} \equiv \mathbf{0}$$

Unit matrix

The special square matrices of the form

$$[1] \begin{bmatrix} 1 & 0 \\ 0 & 1 \end{bmatrix} \begin{bmatrix} 1 & 0 & 0 \\ 0 & 1 & 0 \\ 0 & 0 & 1 \end{bmatrix} \begin{bmatrix} 1 & 0 & 0 & 0 \\ 0 & 1 & 0 & 0 \\ 0 & 0 & 1 & 0 \\ 0 & 0 & 0 & 1 \end{bmatrix}$$

and so on which have ones on the main diagonal and zeros everywhere else are called **unit matrices**. Any unit matrix is usually denoted by **I**. A unit matrix plays the same role in matrix algebra as "1" in ordinary arithmetic since, for appropriately shaped matrices

$$\mathbf{AI} = \mathbf{IA} = \mathbf{A} \qquad \mathbf{BI} = \mathbf{B}$$

e.g.

$$\begin{bmatrix} 1 & 1 & 1 \\ 2 & 0 & 4 \\ -1 & 4 & 7 \end{bmatrix} \begin{bmatrix} 1 & 0 & 0 \\ 0 & 1 & 0 \\ 0 & 0 & 1 \end{bmatrix} = \begin{bmatrix} 1 & 1 & 1 \\ 2 & 0 & 4 \\ -1 & 4 & 7 \end{bmatrix} = \begin{bmatrix} 1 & 0 & 0 \\ 0 & 1 & 0 \\ 0 & 0 & 1 \end{bmatrix} \begin{bmatrix} 1 & 1 & 1 \\ 2 & 0 & 4 \\ -1 & 4 & 7 \end{bmatrix}$$

$$\begin{bmatrix} 1 & 2 & 3 \\ -1 & 1 & 4 \end{bmatrix} \begin{bmatrix} 1 & 0 & 0 \\ 0 & 1 & 0 \\ 0 & 0 & 1 \end{bmatrix} = \begin{bmatrix} 1 & 2 & 3 \\ -1 & 1 & 4 \end{bmatrix}$$

IB is not defined in this case unless we choose **I** to be 2×2 when the result is

$$\begin{bmatrix} 1 & 0 \\ 0 & 1 \end{bmatrix} \begin{bmatrix} 1 & 2 & 3 \\ -1 & 1 & 4 \end{bmatrix} = \begin{bmatrix} 1 & 2 & 3 \\ -1 & 1 & 4 \end{bmatrix}$$

Also,

$$\begin{bmatrix} 1 & 2 \end{bmatrix} \begin{bmatrix} 1 & 0 \\ 0 & 1 \end{bmatrix} = \begin{bmatrix} 1 & 2 \end{bmatrix}$$

10.2 Matrix Notation for Simultaneous Equations

For simultaneous equations we now have a very simple way of writing down the system.

Take the first example we considered,

$$\begin{cases} 3x - y = 2 \\ x + y = 4 \end{cases}$$

We can write these equations as

$$\begin{bmatrix} 3 & -1 \\ 1 & 1 \end{bmatrix}\begin{bmatrix} x \\ y \end{bmatrix} = \begin{bmatrix} 2 \\ 4 \end{bmatrix}$$

Our next worked example was the solution of the equations

$$\begin{cases} 2x_1 - 4x_2 + x_3 = -9 \\ 4x_1 + 2x_2 - 3x_3 = 17 \\ x_1 - x_2 + 5x_3 = -11 \end{cases}$$

We could write this system of equations in the form

$$\begin{bmatrix} 2 & -4 & 1 \\ 4 & 2 & -3 \\ 1 & -1 & 5 \end{bmatrix}\begin{bmatrix} x_1 \\ x_2 \\ x_3 \end{bmatrix} = \begin{bmatrix} -9 \\ 17 \\ -11 \end{bmatrix}$$

Furthermore when using Gauss elimination we have been working with this matrix notation. We could write the next step in the process as

$$\begin{bmatrix} 2 & -4 & 1 \\ 0 & 10 & -5 \\ 0 & 1 & \frac{9}{2} \end{bmatrix}\begin{bmatrix} x_1 \\ x_2 \\ x_3 \end{bmatrix} = \begin{bmatrix} -9 \\ 35 \\ -\frac{13}{2} \end{bmatrix}$$

and the next as

$$\begin{bmatrix} 2 & -4 & 1 \\ 0 & 2 & -1 \\ 0 & 1 & \frac{9}{2} \end{bmatrix}\begin{bmatrix} x_1 \\ x_2 \\ x_3 \end{bmatrix} = \begin{bmatrix} -9 \\ 7 \\ -\frac{13}{2} \end{bmatrix}$$

with finally

$$\begin{bmatrix} 2 & -4 & 1 \\ 0 & 2 & -1 \\ 0 & 0 & 5 \end{bmatrix}\begin{bmatrix} x_1 \\ x_2 \\ x_3 \end{bmatrix} = \begin{bmatrix} -9 \\ 7 \\ -10 \end{bmatrix}$$

Notice that when we use Gauss-Jordan elimination the aim is to produce a unit matrix on the left-hand side. Taking the example on page 235 further, we had reached the stage shown on the next page.

$$\begin{bmatrix} 2 & -4 & 1 \\ 0 & 2 & -1 \\ 0 & 0 & 5 \end{bmatrix}\begin{bmatrix} x_1 \\ x_2 \\ x_3 \end{bmatrix} = \begin{bmatrix} -9 \\ 7 \\ -10 \end{bmatrix}$$

and you may remember that the Gauss-Jordan method then gave

$$\begin{bmatrix} 2 & -4 & 0 \\ 0 & 2 & 0 \\ 0 & 0 & 5 \end{bmatrix}\begin{bmatrix} x_1 \\ x_2 \\ x_3 \end{bmatrix} = \begin{bmatrix} -7 \\ 5 \\ -10 \end{bmatrix}$$

then

$$\begin{bmatrix} 2 & 0 & 0 \\ 0 & 2 & 0 \\ 0 & 0 & 5 \end{bmatrix}\begin{bmatrix} x_1 \\ x_2 \\ x_3 \end{bmatrix} = \begin{bmatrix} 3 \\ 5 \\ -10 \end{bmatrix}$$

and finally

$$\begin{bmatrix} 1 & 0 & 0 \\ 0 & 1 & 0 \\ 0 & 0 & 1 \end{bmatrix}\begin{bmatrix} x_1 \\ x_2 \\ x_3 \end{bmatrix} = \begin{bmatrix} \frac{3}{2} \\ \frac{5}{2} \\ -2 \end{bmatrix}$$

But the left-hand side is $\mathbf{I}\begin{bmatrix} x_1 \\ x_2 \\ x_3 \end{bmatrix}$ and this is just $\begin{bmatrix} x_1 \\ x_2 \\ x_3 \end{bmatrix}$

Therefore

$$\begin{bmatrix} x_1 \\ x_2 \\ x_3 \end{bmatrix} = \begin{bmatrix} \frac{3}{2} \\ \frac{5}{2} \\ -2 \end{bmatrix}$$

Multiplication by special matrices

Special multiplications involving the columns of the unit matrix are most important. Consider for example

$$\begin{bmatrix} a_{11} & a_{12} & a_{13} \\ a_{21} & a_{22} & a_{23} \\ a_{31} & a_{32} & a_{33} \end{bmatrix} \begin{bmatrix} 1 \\ 0 \\ 0 \end{bmatrix}$$

Performing the multiplication we just get

$$\begin{bmatrix} a_{11} \\ a_{21} \\ a_{31} \end{bmatrix}$$

that is, the first column of the matrix **A**.

Similarly,

$$\begin{bmatrix} a_{11} & a_{12} & a_{13} \\ a_{21} & a_{22} & a_{23} \\ a_{31} & a_{32} & a_{33} \end{bmatrix} \begin{bmatrix} 0 \\ 1 \\ 0 \end{bmatrix} = \begin{bmatrix} a_{12} \\ a_{22} \\ a_{32} \end{bmatrix} = \text{second column of matrix } \mathbf{A}$$

and

$$\begin{bmatrix} a_{11} & a_{12} & a_{13} \\ a_{21} & a_{22} & a_{23} \\ a_{31} & a_{32} & a_{33} \end{bmatrix} \begin{bmatrix} 0 \\ 0 \\ 1 \end{bmatrix} = \begin{bmatrix} a_{13} \\ a_{23} \\ a_{33} \end{bmatrix} = \text{third column of matrix } \mathbf{A}$$

Although we have only illustrated work with columns of the 3×3 unit matrix it is obvious that it will be exactly the same for unit matrices of other sizes.

10.3 Determinants

We have already met determinants of order two of the form

$$\begin{vmatrix} a_{11} & a_{12} \\ a_{21} & a_{22} \end{vmatrix}$$

The order of a determinant is the number of rows (or columns) it has. The value of this determinant is $a_{11}a_{22} - a_{12}a_{21}$. Here again we emphasise this is simply a *number* in contrast to the corresponding matrix which is an array.

We mentioned briefly in the solution of linear equations the third order determinant which follows:

$$D = \begin{vmatrix} a_{11} & a_{12} & a_{13} \\ a_{21} & a_{22} & a_{23} \\ a_{31} & a_{32} & a_{33} \end{vmatrix}$$

If the corresponding matrix

$$\begin{bmatrix} a_{11} & a_{12} & a_{13} \\ a_{21} & a_{22} & a_{23} \\ a_{31} & a_{32} & a_{33} \end{bmatrix}$$

is written **A**, this determinant is written |**A**|, pronounced 'det **A**'.

We wish to study this and higher order determinants a little more fully.

Formally, the value of **A** is given by

$$|\mathbf{A}| = a_{11}\begin{vmatrix} a_{22} & a_{23} \\ a_{32} & a_{33} \end{vmatrix} - a_{12}\begin{vmatrix} a_{21} & a_{23} \\ a_{31} & a_{33} \end{vmatrix} + a_{13}\begin{vmatrix} a_{21} & a_{22} \\ a_{31} & a_{32} \end{vmatrix} \qquad (10.1)$$

which is called *expanding along the top row.* Notice that each element in the top row is multiplied by the determinant of the elements left when the column and row through that element are struck out. For example, looking at the element a_{12}

$$\begin{vmatrix} a_{11} & \cdots\cdots & a_{12} & \cdots\cdots & a_{13} \\ & & \vdots & & \\ a_{21} & & a_{22} & & a_{23} \\ & & \vdots & & \\ a_{31} & & a_{32} & & a_{33} \end{vmatrix}$$

the determinant of elements left over is

$$\begin{vmatrix} a_{21} & a_{23} \\ a_{31} & a_{33} \end{vmatrix}$$

These determinants are called **minors**.

So

$$\begin{vmatrix} a_{21} & a_{23} \\ a_{31} & a_{33} \end{vmatrix}$$

is the **minor** of a_{12}.

You will notice in relation to (10.1) that a minus sign is allocated to a_{12}; this arises purely from position according to the following pattern of signs

$$\begin{matrix} + & - & + \\ - & + & - \\ + & - & + \end{matrix}$$

For any size determinant we always start with a + in the top left position and from this position work both horizontally and vertically with alternating signs

$$\begin{matrix} + & \rightarrow & - & \rightarrow & + \\ \downarrow & & & & \\ - & & & & \\ \downarrow & & & & \\ + & & & & \end{matrix}$$

If we write down a minor and multiply this by its sign from the position pattern we then call it a **cofactor**. So

$$-\begin{vmatrix} a_{21} & a_{23} \\ a_{31} & a_{33} \end{vmatrix}$$

is the **cofactor** of a_{12}, and is written $\mathbf{A}_{12}$. In general, the co-factor of a_{ij} is written $\mathbf{A}_{ij}$.

The rules for a larger order determinant are exactly the same, e.g.

$$\begin{vmatrix} a_{11} & a_{12} & a_{13} & a_{14} \\ a_{21} & a_{22} & a_{23} & a_{24} \\ a_{31} & a_{32} & a_{33} & a_{34} \\ a_{41} & a_{42} & a_{43} & a_{44} \end{vmatrix} = a_{11}\begin{vmatrix} a_{22} & a_{23} & a_{24} \\ a_{32} & a_{33} & a_{34} \\ a_{42} & a_{43} & a_{44} \end{vmatrix} - a_{12}\begin{vmatrix} a_{21} & a_{23} & a_{24} \\ a_{31} & a_{33} & a_{34} \\ a_{41} & a_{43} & a_{44} \end{vmatrix}$$

$$+ a_{13}\begin{vmatrix} a_{21} & a_{22} & a_{24} \\ a_{31} & a_{32} & a_{34} \\ a_{41} & a_{42} & a_{44} \end{vmatrix} - a_{14}\begin{vmatrix} a_{21} & a_{22} & a_{23} \\ a_{31} & a_{32} & a_{33} \\ a_{41} & a_{42} & a_{43} \end{vmatrix}$$

Notice the sign pattern is

$$\begin{matrix} + & - & + & - \\ - & + & - & + \\ + & - & + & - \\ - & + & - & + \end{matrix}$$

We then expand the 3rd order determinants as before.

In the two following examples, we shall evaluate two determinants.

Example 1

$$\begin{vmatrix} 1 & -1 & 3 \\ 0 & 2 & 5 \\ -2 & 1 & 6 \end{vmatrix} = 1\begin{vmatrix} 2 & 5 \\ 1 & 6 \end{vmatrix} - (-1)\begin{vmatrix} 0 & 5 \\ -2 & 6 \end{vmatrix} + 3\begin{vmatrix} 0 & 2 \\ -2 & 1 \end{vmatrix}$$

$$= 7 + 10 + 12 = 29$$

Example 2

$$\begin{vmatrix} 1 & 2 & 1 & 2 \\ 2 & 1 & 2 & 1 \\ 0 & 1 & 1 & 0 \\ 0 & 4 & -1 & 3 \end{vmatrix} = 1\begin{vmatrix} 1 & 2 & 1 \\ 1 & 1 & 0 \\ 4 & -1 & 3 \end{vmatrix} - 2\begin{vmatrix} 2 & 2 & 1 \\ 0 & 1 & 0 \\ 0 & -1 & 3 \end{vmatrix} + 1\begin{vmatrix} 2 & 1 & 1 \\ 0 & 1 & 0 \\ 0 & 4 & 3 \end{vmatrix}$$

$$- 2\begin{vmatrix} 2 & 1 & 2 \\ 0 & 1 & 1 \\ 0 & 4 & -1 \end{vmatrix}$$

$$= 1\left\{1\begin{vmatrix} 1 & 0 \\ -1 & 3 \end{vmatrix} - 2\begin{vmatrix} 1 & 0 \\ 4 & 3 \end{vmatrix} + 1\begin{vmatrix} 1 & 1 \\ 4 & -1 \end{vmatrix}\right\}$$

$$-2\left\{2\begin{vmatrix} 1 & 0 \\ -1 & 3 \end{vmatrix} - 2\begin{vmatrix} 0 & 0 \\ 0 & 3 \end{vmatrix} + 1\begin{vmatrix} 0 & 1 \\ 0 & -1 \end{vmatrix}\right\}$$

$$+1\left\{2\begin{vmatrix} 1 & 0 \\ 4 & 3 \end{vmatrix} - 1\begin{vmatrix} 0 & 0 \\ 0 & 3 \end{vmatrix} + 1\begin{vmatrix} 0 & 1 \\ 0 & 4 \end{vmatrix}\right\}$$

$$-2\left\{2\begin{vmatrix} 1 & 1 \\ 4 & -1 \end{vmatrix} - 1\begin{vmatrix} 0 & 1 \\ 0 & -1 \end{vmatrix} + 2\begin{vmatrix} 0 & 1 \\ 0 & 4 \end{vmatrix}\right\}$$

$$= 3 - 6 - 5 - 12 + 6 + 20$$
$$= 6$$

Properties of Determinants (no proofs given)

(i) If two rows (or two columns) are interchanged, $|\mathbf{A}| \to -|\mathbf{A}|$, i.e. the value of the determinant is multiplied by -1

(ii) If the rows and columns are interchanged (transposed) the value is not changed; $|\mathbf{A}^T| = |\mathbf{A}|$

(iii) If two rows or columns are identical, $|\mathbf{A}| = 0$

(iv) If a row or column has a common factor we proceed thus:

$$\begin{vmatrix} a_{11} & a_{12} & a_{13} \\ ka_{21} & ka_{22} & ka_{23} \\ a_{31} & a_{32} & a_{33} \end{vmatrix} = k \begin{vmatrix} a_{11} & a_{12} & a_{13} \\ a_{21} & a_{22} & a_{23} \\ a_{31} & a_{32} & a_{33} \end{vmatrix} = k|\mathbf{A}|$$

(v)
$$\begin{vmatrix} a_{11} + ka_{12} + la_{13} & a_{12} + na_{13} & a_{13} \\ a_{21} + ka_{22} + la_{23} & a_{22} + na_{23} & a_{23} \\ a_{31} + ka_{32} + la_{33} & a_{32} + na_{33} & a_{33} \end{vmatrix} = |\mathbf{A}|$$

This means that we can add multiples of corresponding elements of one column to another. (Similarly for rows.)

(vi)
$$\begin{vmatrix} a_{11} + \alpha & a_{12} & a_{13} \\ a_{21} + \beta & a_{22} & a_{23} \\ a_{31} + \gamma & a_{32} & a_{33} \end{vmatrix} = \begin{vmatrix} a_{11} & a_{12} & a_{13} \\ a_{21} & a_{22} & a_{23} \\ a_{31} & a_{32} & a_{33} \end{vmatrix} + \begin{vmatrix} \alpha & a_{12} & a_{13} \\ \beta & a_{22} & a_{23} \\ \gamma & a_{32} & a_{33} \end{vmatrix}$$

These properties hold for determinants of any order. Usually, what is true for columns is true for rows and vice-versa. Property (i) implies that we can expand a determinant along any row and together with property (ii) it implies that we can expand along any row or column as long as we take account of the sign positions.

We have for instance

$$\begin{vmatrix} a_{11} & a_{12} & a_{13} \\ a_{21} & a_{22} & a_{23} \\ a_{31} & a_{32} & a_{33} \end{vmatrix} = -a_{21} \begin{vmatrix} a_{12} & a_{13} \\ a_{32} & a_{33} \end{vmatrix} + a_{22} \begin{vmatrix} a_{11} & a_{13} \\ a_{31} & a_{33} \end{vmatrix}$$

$$- a_{23} \begin{vmatrix} a_{11} & a_{12} \\ a_{31} & a_{32} \end{vmatrix} \qquad \text{(Expand along second row)}$$

$$= a_{13} \begin{vmatrix} a_{21} & a_{22} \\ a_{31} & a_{32} \end{vmatrix} - a_{23} \begin{vmatrix} a_{11} & a_{12} \\ a_{31} & a_{32} \end{vmatrix} + a_{33} \begin{vmatrix} a_{11} & a_{12} \\ a_{21} & a_{22} \end{vmatrix}$$

(Expand along 3rd column.)

Property (v) is probably the most useful since it allows us to simplify a determinant by introducing zeros. We shall take one of our previous examples.

Example 3

$$\begin{vmatrix} 1 & -1 & 3 \\ 0 & 2 & 5 \\ -2 & 1 & 6 \end{vmatrix}$$

Property (v) says that we can add any multiple of one row (or column) to another and not change the value of the determinant. We use this as follows

$$D = \begin{vmatrix} 1 & -1 & 3 \\ 0 & 2 & 5 \\ -2 & 1 & 6 \end{vmatrix} = \begin{vmatrix} 1 & 0 & 3 \\ 0 & 2 & 5 \\ -2 & -1 & 6 \end{vmatrix} \quad \text{add column 1 to column 2}$$

$$= \begin{vmatrix} 1 & 0 & 0 \\ 0 & 2 & 5 \\ -2 & -1 & 12 \end{vmatrix} \quad \text{add column } 1 \times (-3) \text{ to column 3}$$

Notice that we could have saved writing by doing both these operations together.

Now if we expand along the top row, we get

$$D = 1\begin{vmatrix} 2 & 5 \\ -1 & 12 \end{vmatrix} - 0\begin{vmatrix} 0 & 5 \\ -2 & 12 \end{vmatrix} + 0\begin{vmatrix} 0 & 2 \\ -2 & -1 \end{vmatrix}$$

$$= \begin{vmatrix} 2 & 5 \\ -1 & 12 \end{vmatrix}$$

We have reduced the 3rd order determinant to a second order determinant giving

$$D = 12 \times 2 - 5 \times (-1) = 29$$

We could, of course, have obtained zeros in any other row or column. It would have been easier to work on the first column since there is already one zero there as follows.

$$D = \begin{vmatrix} 1 & -1 & 3 \\ 0 & 2 & 5 \\ -2 & 1 & 6 \end{vmatrix} = \begin{vmatrix} 1 & -1 & 3 \\ 0 & 2 & 5 \\ 0 & -1 & 12 \end{vmatrix} \quad \text{Add } (2 \times \text{Row 1}) \text{ to Row 3}$$

If we now expand down the first column we get

$$D = 1\begin{vmatrix} 2 & 5 \\ -1 & 12 \end{vmatrix} - 0\begin{vmatrix} \cdot & \cdot \\ \cdot & \cdot \end{vmatrix} + 0\begin{vmatrix} \cdot & \cdot \\ \cdot & \cdot \end{vmatrix}$$

giving the same result as before.

Another useful device is to reduce the determinant to one with zeros everywhere below the main diagonal (**upper triangular**) or to one with zeros everywhere above the main diagonal (**lower triangular**).

$$\begin{vmatrix} 1 & -1 & 3 \\ 0 & 2 & 5 \\ -2 & 1 & 6 \end{vmatrix} = \begin{vmatrix} 1 & -1 & 3 \\ 0 & 2 & 5 \\ 0 & -1 & 12 \end{vmatrix} \qquad \text{Add } (2 \times \text{Row 1}) \text{ to Row 3}$$

$$= \begin{vmatrix} 1 & -1 & 3 \\ 0 & 2 & 5 \\ 0 & 0 & \frac{29}{2} \end{vmatrix} \qquad \text{Add } (\tfrac{1}{2} \times \text{Row 2}) \text{ to Row 3}$$

The value of such a determinant is simply the product of its diagonal elements, i.e. $1 \times 2 \times 29/2 = 29$, as before. Try reducing the original determinant to a lower triangular form and check the value thus obtained.

Let us now take an example of a fourth order determinant.

Example 4

$$D = \begin{vmatrix} -3 & 2 & 2 & 5 \\ 3 & -1 & 1 & 4 \\ 6 & -2 & 3 & 7 \\ 8 & -3 & -4 & 2 \end{vmatrix}$$

It will be easiest to avoid fractions by fixing our attention on the "1" present in row 2, column 3, and producing zeros in the row or column in which it lies. We get

$$D = \begin{vmatrix} -3 & 2 & 2 & 5 \\ 3 & -1 & \textcircled{1} & 4 \\ 6 & -2 & 3 & 7 \\ 8 & -3 & -4 & 2 \end{vmatrix}$$

$$= \begin{vmatrix} -9 & 4 & 0 & -3 \\ 3 & -1 & \textcircled{1} & 4 \\ -3 & 1 & 0 & -5 \\ 20 & -7 & 0 & 18 \end{vmatrix} \qquad \begin{matrix} \text{Add } (-2 \times \text{Row 2}) \text{ to Row 1} \\ \\ \text{Add } (-3 \times \text{Row 2}) \text{ to Row 3} \\ \text{Add } (4 \times \text{Row 2}) \text{ to Row 4} \end{matrix}$$

Now if we expand down the 3rd column we effectively reduce the determinant to one of 3rd order. Remembering the position signs we get

$$D = -1\begin{vmatrix} -9 & 4 & -3 \\ -3 & 1 & -5 \\ 20 & -7 & 18 \end{vmatrix}$$

There is a "1" here again so we work to manufacture zeros in the row or column containing this. We proceed:

$$D = -1\begin{vmatrix} 3 & 4 & 17 \\ 0 & 1 & 0 \\ -1 & -7 & -17 \end{vmatrix} \quad \begin{array}{l} \text{Add } (2 \times \text{Column 2}) \text{ to Column 1} \\ \\ \text{Add } (5 \times \text{Column 2}) \text{ to Column 3} \end{array}$$

$$= -1\begin{vmatrix} 3 & 17 \\ -1 & -17 \end{vmatrix} \quad \text{(Expand along 2}^{\text{nd}}\text{ Row)}$$

$$= 34$$

If there is not a "1" present, then we can manufacture one by dividing a whole row or column by one of its elements, taking this outside as a factor. Thus

$$\begin{vmatrix} 3 & 6 & -5 & 4 \\ 5 & 5 & 3 & 5 \\ 9 & -7 & 8 & 4 \\ 8 & -11 & 14 & 19 \end{vmatrix} = 3\begin{vmatrix} 1 & 6 & -5 & 4 \\ \frac{5}{3} & 5 & 3 & 5 \\ 3 & -7 & 8 & 4 \\ \frac{8}{3} & -11 & 14 & 19 \end{vmatrix} \quad \text{and so on.}$$

10.4 Inverse of a Matrix

Let the matrix **A** be square. If we can find a matrix **B** which is such that **AB** = **I** = **BA**, then **B** is said to be the **inverse** of **A** and it is written $\mathbf{A}^{-1}$. (Notice **A** must be square). As an example,

$$\mathbf{A} = \begin{bmatrix} 2 & 3 & -1 \\ 0 & 1 & 1 \\ -1 & 2 & 1 \end{bmatrix} \qquad \mathbf{B} = \frac{1}{6}\begin{bmatrix} 1 & 5 & -4 \\ 1 & -1 & 2 \\ -1 & 7 & -2 \end{bmatrix}$$

$$\mathbf{AB} = \frac{1}{6}\begin{bmatrix} 6 & 0 & 0 \\ 0 & 6 & 0 \\ 0 & 0 & 6 \end{bmatrix} = \begin{bmatrix} 1 & 0 & 0 \\ 0 & 1 & 0 \\ 0 & 0 & 1 \end{bmatrix} = \mathbf{I}$$

and

$$\mathbf{BA} = \frac{1}{6}\begin{bmatrix} 6 & 0 & 0 \\ 0 & 6 & 0 \\ 0 & 0 & 6 \end{bmatrix} = \mathbf{I}$$

Thus $\mathbf{B} = \mathbf{A}^{-1}$

If the inverse exists, then it is unique.

Suppose there are two matrices **B** and **C** which are such that

$$\mathbf{AB} = \mathbf{I} = \mathbf{BA}$$

and $$\mathbf{AC} = \mathbf{I} = \mathbf{CA}$$

Then $$\mathbf{AB} - \mathbf{AC} = \mathbf{0}$$

Premultiplication by **B** gives $\mathbf{BA}(\mathbf{B} - \mathbf{C}) = \mathbf{B0} = \mathbf{0}$

That is, $$\mathbf{I}(\mathbf{B} - \mathbf{C}) = \mathbf{0}$$

or $$\mathbf{B} - \mathbf{C} = \mathbf{0}$$

(since multiplying by **I** leaves the matrix unchanged)

Hence $$\mathbf{B} = \mathbf{C}$$

To find the inverse of a matrix

Consider the set of simultaneous equations

$$\begin{bmatrix} a_{11} & a_{12} & a_{13} \\ a_{21} & a_{22} & a_{23} \\ a_{31} & a_{32} & a_{33} \end{bmatrix}\begin{bmatrix} x_1 \\ x_2 \\ x_3 \end{bmatrix} = \begin{bmatrix} b_1 \\ b_2 \\ b_3 \end{bmatrix}$$

This can be represented in matrix notation as

$$\mathbf{AX} = \mathbf{B}$$

First of all, if we use Gauss-Jordan elimination we transform this to

$$\mathbf{IX} = \mathbf{B}'$$

Looking at the equation $\mathbf{AX} = \mathbf{B}$ from a different viewpoint, if we pre-multiply both sides of the equation by $\mathbf{A}^{-1}$ we get

$$\mathbf{A}^{-1}\mathbf{AX} = \mathbf{A}^{-1}\mathbf{B}$$

or $$\mathbf{IX} = \mathbf{A}^{-1}\mathbf{B}$$

Thus $$\mathbf{B}' = \mathbf{A}^{-1}\mathbf{B}$$

By picking **B** as the special column $\begin{bmatrix} 1 \\ 0 \\ 0 \end{bmatrix}$

we know that $\mathbf{A}^{-1}\mathbf{B}$ will be the first column of $\mathbf{A}^{-1}$.

Similarly by choosing **B** as $\begin{bmatrix} 0 \\ 1 \\ 0 \end{bmatrix}$

we shall obtain the second column of $\mathbf{A}^{-1}$ and finally if $\mathbf{B}$ is chosen as $\begin{bmatrix} 0 \\ 0 \\ 1 \end{bmatrix}$

we get the third column of $\mathbf{A}^{-1}$.

We can thus find all the columns of $\mathbf{A}^{-1}$.

Example

Let us take an example: find the inverse of $\mathbf{A} = \begin{bmatrix} 2 & 3 & -1 \\ 0 & 1 & 1 \\ -1 & 2 & 1 \end{bmatrix}$

Consider the equation $\begin{bmatrix} 2 & 3 & -1 \\ 0 & 1 & 1 \\ -1 & 2 & 1 \end{bmatrix}\begin{bmatrix} x_1 \\ x_2 \\ x_3 \end{bmatrix} = \begin{bmatrix} 1 \\ 0 \\ 0 \end{bmatrix}$

We get

$$\begin{bmatrix} 2 & 3 & -1 \\ 0 & 1 & 1 \\ 0 & \frac{7}{2} & \frac{1}{2} \end{bmatrix}\begin{bmatrix} x_1 \\ x_2 \\ x_3 \end{bmatrix} = \begin{bmatrix} 1 \\ 0 \\ \frac{1}{2} \end{bmatrix} \qquad \text{Row 3} + (\tfrac{1}{2} \times \text{Row 1})$$

$$\begin{bmatrix} 2 & 3 & -1 \\ 0 & 1 & 1 \\ 0 & 0 & -3 \end{bmatrix}\begin{bmatrix} x_1 \\ x_2 \\ x_3 \end{bmatrix} = \begin{bmatrix} 1 \\ 0 \\ \frac{1}{2} \end{bmatrix} \qquad \text{Row 3} - (7/2 \times \text{Row 2})$$

$$\begin{bmatrix} 2 & 3 & 0 \\ 0 & 1 & 0 \\ 0 & 0 & -3 \end{bmatrix}\begin{bmatrix} x_1 \\ x_2 \\ x_3 \end{bmatrix} = \begin{bmatrix} \frac{5}{6} \\ \frac{1}{6} \\ \frac{1}{2} \end{bmatrix} \qquad \begin{array}{l} \text{Row 1} - (1/3 \times \text{Row 3}) \\ \text{Row 2} + (1/3 \times \text{Row 3}) \end{array}$$

$$\begin{bmatrix} 2 & 0 & 0 \\ 0 & 1 & 0 \\ 0 & 0 & -3 \end{bmatrix}\begin{bmatrix} x_1 \\ x_2 \\ x_3 \end{bmatrix} = \begin{bmatrix} \frac{2}{6} \\ \frac{1}{6} \\ \frac{1}{2} \end{bmatrix} \qquad \text{Row 1} - (3 \times \text{Row 2})$$

and finally

$$\begin{bmatrix} 1 & 0 & 0 \\ 0 & 1 & 0 \\ 0 & 0 & 1 \end{bmatrix} \begin{bmatrix} x_1 \\ x_2 \\ x_3 \end{bmatrix} = \begin{bmatrix} \frac{1}{6} \\ \frac{1}{6} \\ -\frac{1}{6} \end{bmatrix} \qquad \begin{matrix} \text{Row } 1 \div 2 \\ \\ \text{Row } 3 \div (-3) \end{matrix}$$

Thus the first column of $\mathbf{A}^{-1}$ is $\begin{bmatrix} \frac{1}{6} \\ \frac{1}{6} \\ -\frac{1}{6} \end{bmatrix}$

We now consider

$$\begin{bmatrix} 2 & 3 & -1 \\ 0 & 1 & 1 \\ -1 & 2 & 1 \end{bmatrix} \begin{bmatrix} x_1 \\ x_2 \\ x_3 \end{bmatrix} = \begin{bmatrix} 0 \\ 1 \\ 0 \end{bmatrix}$$

We shall perform exactly the same operations to reduce the l.h.s. to **IX**. Hence we get for r.h.s.

$$\begin{bmatrix} 0 \\ 1 \\ 0 \end{bmatrix} \quad \begin{bmatrix} 0 \\ 1 \\ 0 \end{bmatrix} \quad \begin{bmatrix} 0 \\ 1 \\ -\frac{7}{2} \end{bmatrix} \quad \begin{bmatrix} \frac{7}{6} \\ -\frac{1}{6} \\ -\frac{7}{2} \end{bmatrix}$$

Row 3 + ($\frac{1}{2}$Row 1) Row 3 – (7/2 × Row 2) Row 1 – (1/3 × Row 3)

Row 2 + (1/3 × Row 3)

$$\begin{bmatrix} \frac{10}{6} \\ -\frac{1}{6} \\ -\frac{7}{2} \end{bmatrix} \quad \begin{bmatrix} \frac{5}{6} \\ -\frac{1}{6} \\ \frac{7}{6} \end{bmatrix}$$

Row 1 – (3 × Row 2) Row 1 ÷ 2

Row 3 ÷ (–3)

This is the 2nd column of $\mathbf{A}^{-1}$

Similarly we find that the elements of the third column are $-\frac{4}{6}, \frac{2}{6}, -\frac{2}{6}$.

Hence

$$\mathbf{A}^{-1} = \frac{1}{6}\begin{bmatrix} 1 & 5 & -4 \\ 1 & -1 & 2 \\ -1 & 7 & -2 \end{bmatrix}$$

Now since we are performing the same operations each time with different right-hand sides we could set up the system in tabloid form dealing with all the right-hand sides simultaneously as follows.

$$\begin{bmatrix} 2 & 3 & -1 \\ 0 & 1 & 1 \\ -1 & 2 & 1 \end{bmatrix}\begin{bmatrix} 1 & 0 & 0 \\ 0 & 1 & 0 \\ 0 & 0 & 1 \end{bmatrix}$$

$$\begin{bmatrix} 2 & 3 & -1 \\ 0 & 1 & 1 \\ 0 & \frac{7}{2} & \frac{1}{2} \end{bmatrix}\begin{bmatrix} 1 & 0 & 0 \\ 0 & 1 & 0 \\ \frac{1}{2} & 0 & 1 \end{bmatrix} \qquad \text{Row 3} + (\tfrac{1}{2} \times \text{Row 1})$$

$$\begin{bmatrix} 2 & 3 & -1 \\ 0 & 1 & 1 \\ 0 & 0 & -3 \end{bmatrix}\begin{bmatrix} 1 & 0 & 0 \\ 0 & 1 & 0 \\ \frac{1}{2} & -\frac{7}{2} & 1 \end{bmatrix} \qquad \text{Row 3} - (7/2 \times \text{Row 2})$$

$$\begin{bmatrix} 2 & 3 & 0 \\ 0 & 1 & 0 \\ 0 & 0 & -3 \end{bmatrix}\begin{bmatrix} \frac{5}{6} & \frac{7}{6} & -\frac{1}{3} \\ \frac{1}{6} & -\frac{1}{6} & \frac{1}{3} \\ \frac{1}{2} & -\frac{7}{2} & 1 \end{bmatrix} \qquad \begin{matrix} \text{Row 1} - (1/3 \times \text{Row 3}) \\ \text{Row 2} + (1/3 \times \text{Row 3}) \\ \end{matrix}$$

$$\begin{bmatrix} 2 & 0 & 0 \\ 0 & 1 & 0 \\ 0 & 0 & -3 \end{bmatrix}\begin{bmatrix} \frac{2}{6} & \frac{10}{6} & -\frac{8}{6} \\ \frac{1}{6} & -\frac{1}{6} & \frac{1}{3} \\ \frac{1}{2} & -\frac{7}{2} & 1 \end{bmatrix} \qquad \text{Row 1} - (3 \times \text{Row 2})$$

$$\begin{bmatrix} 1 & 0 & 0 \\ 0 & 1 & 0 \\ 0 & 0 & 1 \end{bmatrix} \begin{bmatrix} \frac{1}{6} & \frac{5}{6} & -\frac{4}{6} \\ \frac{1}{6} & -\frac{1}{6} & \frac{2}{6} \\ -\frac{1}{6} & \frac{7}{6} & -\frac{2}{6} \end{bmatrix} \quad \begin{matrix} \text{Row } 1 \div 2 \\ \\ \text{Row } 3 \div (-3) \end{matrix}$$

$$\uparrow$$
$$\mathbf{A}^{-1}$$

Notice that $\mathbf{A}^{-1}$ has become $\mathbf{I}$ and $\mathbf{I}$ has become $\mathbf{A}^{-1}$. We have assumed that $\mathbf{A}^{-1}$ does in fact exist – there are, of course, cases when we cannot find $\mathbf{A}^{-1}$. There is a direct way of finding $\mathbf{A}^{-1}$ if it exists which we show you in the next section.

```
>
   10   CLS:@%=10
   20   PRINTTAB(20,2)"3x3 Matrix Inverse"
   30   DIM A(3,3),B(3,3)
   40   FOR row= 1 TO 3
   50     FOR col= 1 TO 3
   60       PRINT"Enter the value for row ";row;
            " column ";col;" ";
   70       INPUT A(col,row)
   80       NEXT col
   90     NEXT row
  100   X=A(1,1)*(A(2,2)*A(3,3)-A(2,3)*A(3,2))
  110   Y=A(2,1)*(A(1,2)*A(3,3)-A(1,3)*A(3,2))
  120   Z=A(3,1)*(A(1,2)*A(2,3)-A(1,3)*A(2,2))
  130   LET D=X-Y+Z
  140   B(1,1)=A(2,2)*A(3,3)-A(2,3)*A(3,2)
  150   B(1,2)=(A(1,2)*A(3,3)-A(1,3)*A(3,2))*-1
  160   B(1,3)=A(1,2)*A(2,3)-A(1,3)*A(2,2)
  170   B(2,1)=(A(2,1)*A(3,3)-A(2,3)*A(3,1))*-1
  180   B(2,2)=A(1,1)*A(3,3)-A(1,3)*A(3,1)
  190   B(2,3)=(A(1,1)*A(2,3)-A(1,3)*A(2,1))*-1
  200   B(3,1)=A(2,1)*A(3,2)-A(2,2)*A(3,1)
  210   B(3,2)=(A(1,1)*A(3,2)-A(1,2)*A(3,1))*-1
  220   B(3,3)=A(1,1)*A(2,2)-A(1,2)*A(2,1)
  230   PRINTTAB(0,15)"The inverse of the matrix is "
  235   @%=&20309
  240   FOR row= 1 TO 3
  250     PRINT B(1,row)/D,B(2,row)/D,B(3,row)/D
  260     NEXT row
```

The computer program for Gauss-Jordan elimination can be modified to invert a matrix. There is no column of constants to be catered for, but there are two arrays **A** and **B**. The necessary operations on **A** are repeated on **B**. The program is shown above.

10.5 Formal Evaluation of the Inverse of a Matrix

We have seen how to find the inverse of a matrix by the use of elementary operations. As mentioned then, there is an *alternative method of finding the inverse*.

This procedure is as follows.

(i) Take the matrix **A** and replace each element by its cofactor

(ii) Transpose the result to form the **adjoint** matrix **adj A**

(iii) Find the determinant of $\mathbf{A} = |\mathbf{A}|$

(iv) Then the inverse matrix $\mathbf{A}^{-1} = \mathbf{adj\ A}/|\mathbf{A}|$, provided $|\mathbf{A}| \neq 0$

Example 1

$$\mathbf{A} = \begin{bmatrix} 3 & -1 \\ 2 & -2 \end{bmatrix}$$

Replacing each element by its cofactor, we get (Rule of signs $\begin{smallmatrix} + & - \\ - & + \end{smallmatrix}$)

$$\mathbf{adj\ A} = \begin{bmatrix} -2 & -2 \\ 1 & 3 \end{bmatrix}^T = \begin{bmatrix} -2 & 1 \\ -2 & 3 \end{bmatrix}$$

$$|\mathbf{A}| = -4$$

therefore

$$\mathbf{A}^{-1} = -\frac{1}{4}\begin{bmatrix} -2 & 1 \\ -2 & 3 \end{bmatrix}$$

Check

$$\mathbf{AA}^{-1} = -\frac{1}{4}\begin{bmatrix} 3 & -1 \\ 2 & -2 \end{bmatrix}\begin{bmatrix} -2 & 1 \\ -2 & 3 \end{bmatrix} = -\frac{1}{4}\begin{bmatrix} -4 & 0 \\ 0 & -4 \end{bmatrix} = \mathbf{I}$$

and

$$\mathbf{A}^{-1}\mathbf{A} = -\frac{1}{4}\begin{bmatrix} -2 & 1 \\ -2 & 3 \end{bmatrix}\begin{bmatrix} 3 & -1 \\ 2 & -2 \end{bmatrix} = -\frac{1}{4}\begin{bmatrix} -4 & 0 \\ 0 & -4 \end{bmatrix} = \mathbf{I}$$

Example 2

$$\mathbf{A} = \begin{bmatrix} 3 & 1 & 2 \\ 5 & -4 & 1 \\ -1 & 2 & 1 \end{bmatrix}$$

Replacing each element by its cofactor and, remembering the rule of signs

$$\begin{matrix} + & - & + \\ - & + & - \\ + & - & + \end{matrix}$$

we have

$$\begin{bmatrix} \begin{vmatrix} -4 & 1 \\ 2 & 1 \end{vmatrix} & -\begin{vmatrix} 5 & 1 \\ -1 & 1 \end{vmatrix} & \begin{vmatrix} 5 & -4 \\ -1 & 2 \end{vmatrix} \\ -\begin{vmatrix} 1 & 2 \\ 2 & 1 \end{vmatrix} & \begin{vmatrix} 3 & 2 \\ -1 & 1 \end{vmatrix} & -\begin{vmatrix} 3 & 1 \\ -1 & 2 \end{vmatrix} \\ \begin{vmatrix} 1 & 2 \\ -4 & 1 \end{vmatrix} & -\begin{vmatrix} 3 & 2 \\ 5 & 1 \end{vmatrix} & \begin{vmatrix} 3 & 1 \\ 5 & -4 \end{vmatrix} \end{bmatrix}$$

$$= \begin{bmatrix} -6 & -6 & 6 \\ 3 & 5 & -7 \\ 9 & 7 & 17 \end{bmatrix}$$

Thus $$\mathbf{adj\ A} = \begin{bmatrix} -6 & -6 & 6 \\ 3 & 5 & -7 \\ 9 & 7 & -17 \end{bmatrix}^{\mathrm{T}} = \begin{bmatrix} -6 & 3 & 9 \\ -6 & 5 & 7 \\ 6 & -7 & -17 \end{bmatrix}$$

$$|\mathbf{A}| = \begin{vmatrix} 3 & 1 & 2 \\ 5 & -4 & 1 \\ -1 & 2 & 1 \end{vmatrix} = \begin{vmatrix} 5 & -3 & 0 \\ 6 & -6 & 0 \\ -1 & 2 & 1 \end{vmatrix} \begin{matrix} \text{Row 1} - (2 \times \text{Row 3}) \\ \text{Row 2} - \text{Row 3} \\ \end{matrix}$$

$$= \begin{vmatrix} 5 & -3 \\ 6 & -6 \end{vmatrix}$$

$$= -12$$

A quicker method of finding $|\mathbf{A}|$ in these problems is to remember that $|\mathbf{A}| = a_{11}\mathbf{A}_{11} + a_{12}\mathbf{A}_{12} + a_{13}\mathbf{A}_{13}$ and notice that $\mathbf{A}_{11}, \mathbf{A}_{12}, \mathbf{A}_{13}$ have already been evaluated. Therefore

$$\begin{aligned}|\mathbf{A}| &= 3 \times (-6) + 1 \times (-6) + 2 \times 6 \\ &= -12, \text{ as before}\end{aligned}$$

Thus

$$\mathbf{A}^{-1} = \frac{-1}{12}\begin{bmatrix} -6 & 3 & 9 \\ -6 & 5 & 7 \\ 6 & -7 & -17 \end{bmatrix}$$

Check

$$\begin{aligned}\mathbf{AA}^{-1} &= \frac{-1}{12}\begin{bmatrix} 3 & 1 & 2 \\ 5 & -4 & 1 \\ -1 & 2 & 1 \end{bmatrix}\begin{bmatrix} -6 & 3 & 9 \\ -6 & 5 & 7 \\ 6 & -7 & -17 \end{bmatrix} \\ &= \frac{-1}{12}\begin{bmatrix} -12 & 0 & 0 \\ 0 & -12 & 0 \\ 0 & 0 & -12 \end{bmatrix} \\ &= \mathbf{I}\end{aligned}$$

Similarly $\mathbf{A}^{-1}\mathbf{A} = \mathbf{I}$. Remember that *both* $\mathbf{AA}^{-1}$ *and* $\mathbf{A}^{-1}\mathbf{A}$ must equal $\mathbf{I}$.

Singular Matrix

Since, in finding $\mathbf{A}^{-1}$ we divide **adj A** by $|\mathbf{A}|$, we shall not be able to find $\mathbf{A}^{-1}$ if $|\mathbf{A}| = 0$. Such a matrix is called a **singular matrix**. Hence a *non-singular* matrix *has* an inverse, a *singular* matrix does *not*.

Solution of simultaneous equations using the inverse method

We have considered the general set of equations

$$\begin{cases} a_{11}x_1 + a_{12}x_2 + a_{13}x_3 + \ldots\ldots + a_{1n}x_n = b_1 \\ a_{21}x_1 + a_{22}x_2 + \ldots\ldots\ldots\ldots + a_{2n}x_n = b_2 \\ \vdots \qquad\qquad\qquad\qquad\qquad \vdots \qquad \vdots \\ a_{n1}x_1 + a_{n2}x_2 + \ldots\ldots\ldots\ldots + a_{nn}x_n = b_n \end{cases}$$

We know that we can write this system of equations in matrix form as

$$\mathbf{AX} = \mathbf{B}$$

where

$$\mathbf{A} = \begin{bmatrix} a_{11} & \dots a_{1n} \\ \vdots & \\ a_{n1} & \dots a_{nn} \end{bmatrix}, \qquad \mathbf{X} = \begin{bmatrix} x_1 \\ \vdots \\ x_n \end{bmatrix}, \qquad \mathbf{B} = \begin{bmatrix} b_1 \\ b_2 \\ \vdots \\ b_n \end{bmatrix}$$

If we now pre-multiply both sides of this matrix equation by $\mathbf{A}^{-1}$, then

$$\mathbf{A}^{-1}\mathbf{AX} = \mathbf{A}^{-1}\mathbf{B}$$

That is, $$\mathbf{IX} = \mathbf{A}^{-1}\mathbf{B}$$

or $$\mathbf{X} = \mathbf{A}^{-1}\mathbf{B}$$

Thus we can solve the equations by multiplying the vector of the right-hand sides by $\mathbf{A}^{-1}$.

10.6 Geometrical Transformations

In computer graphics it is necessary to carry out geometrical transformations on objects and to describe these transformations mathematically. We collect some results below.

If $\mathbf{r}$ represents a point before a transformation and $\mathbf{r}'$ represents the point afterwards then a translation can be represented by

$$\mathbf{r}' = \mathbf{r} + \mathbf{t}$$

A rotation about the z-axis through an angle θ can be written as

$$\mathbf{r}' = \mathbf{Ar} \quad \text{where } \mathbf{A} = \begin{bmatrix} \cos\theta & -\sin\theta & 0 \\ \sin\theta & \cos\theta & 0 \\ 0 & 0 & 1 \end{bmatrix} \quad \text{where } \mathbf{r} = \begin{bmatrix} x \\ y \\ z \end{bmatrix}$$

The matrix $\mathbf{A} = \begin{bmatrix} \lambda & 0 & 0 \\ 0 & \mu & 0 \\ 0 & 0 & \upsilon \end{bmatrix}$ represents a scaling of λ, μ, υ in the x, y, z axes respectively. Note that if $\lambda = \mu = \upsilon$ we have an enlargement.

The matrix $\mathbf{A} = \begin{bmatrix} 1 & \alpha & 0 \\ 0 & 1 & 0 \\ 0 & 0 & 1 \end{bmatrix}$ represents a shear along planes parallel to the x–y plane and the direction of shear is parallel to the x-axis.

The matrix $\mathbf{A} = \begin{bmatrix} -1 & 0 & 0 \\ 0 & 1 & 0 \\ 0 & 0 & 1 \end{bmatrix}$ represents a reflection in the y-z plane.

Problems

Section 10.1

1 Find, where they exist, the matrices [A + B], [A – B], [AB], [BA], [A + B][A – B] and $[\mathbf{A}^2 - \mathbf{B}^2]$, if:

(i) $\mathbf{A} = \begin{bmatrix} 4 & -5 \\ 2 & 0 \\ -6 & 3 \end{bmatrix}$ $\quad \mathbf{B} = \begin{bmatrix} -1 & 0 \\ 0 & 7 \\ 2 & -5 \end{bmatrix}$

(ii) $\mathbf{A} = \begin{bmatrix} 2 & 3 & 4 \\ 1 & 5 & 6 \end{bmatrix}$ $\quad \mathbf{B} = \begin{bmatrix} 1 \\ 2 \\ 3 \end{bmatrix}$

(iii) $\mathbf{A} = \begin{bmatrix} 3 & -4 \\ 1 & 5 \\ -2 & 2 \end{bmatrix}$ $\quad \mathbf{B} = \begin{bmatrix} 1 & 2 & 1 \\ 4 & 0 & 2 \end{bmatrix}$

(iv) $\mathbf{A} = \begin{bmatrix} 1 & 2 & 3 \\ -4 & 0 & 6 \end{bmatrix}$ $\quad \mathbf{B} = \begin{bmatrix} 2 & 6 \\ 8 & 4 \\ -1 & 0 \\ 0 & 0 \end{bmatrix}$

(v) $\mathbf{A} = \begin{bmatrix} 2 & -1 & 1 \\ 0 & 1 & 2 \\ 1 & 0 & 1 \end{bmatrix}$ $\quad \mathbf{B} = \begin{bmatrix} 0 & 3 & -5 \\ 2 & 0 & 6 \\ 1 & -1 & 4 \end{bmatrix}$

2 For the matrices A, B and C below verify that [AB]C = A[BC].

$\mathbf{A} = \begin{bmatrix} 1 & 2 \\ -5 & 4 \\ 3 & -8 \end{bmatrix}$ $\quad \mathbf{B} = \begin{bmatrix} 1 & 0 & -1 & 4 \\ -5 & 6 & 7 & 0 \end{bmatrix}$ $\quad \mathbf{C} = \begin{bmatrix} 1 \\ 2 \\ 1 \\ 2 \end{bmatrix}$

This is true in general and means that matrix multiplication is *associative*.

3 For 2×2 and 3×3 matrices, show that $\mathbf{A}\mathbf{A}^{\mathrm{T}}$ is a symmetric matrix – that is, a matrix of the general form

$$\begin{bmatrix} a_{11} & a_{12} & a_{13} & \cdots\cdots \\ a_{12} & a_{22} & a_{23} & \cdots\cdots \\ a_{13} & a_{23} & a_{33} & \cdots\cdots \\ \cdot & \cdot & \cdot & \cdots\cdots \\ \cdot & \cdot & \cdot & \cdots\cdots \end{bmatrix} \qquad (\text{or } a_{ij} = a_{ji})$$

Verify this for the matrix $\mathbf{A} = \begin{bmatrix} 1 & 0 & 7 \\ -3 & 5 & -4 \\ 0 & 2 & 1 \end{bmatrix}$

Section 10.2

4 Show that $\begin{bmatrix} 1 & 3 & 3 \\ 1 & 4 & 3 \\ 1 & 3 & 4 \end{bmatrix}\begin{bmatrix} 7 & -3 & -3 \\ -1 & 1 & 0 \\ -1 & 0 & 1 \end{bmatrix} = \mathbf{I}$, the unit matrix

5 Show that if $\mathbf{A} = \begin{bmatrix} \frac{3}{13} & \frac{4}{13} & \frac{12}{13} \\ \frac{4}{5} & -\frac{3}{5} & 0 \\ \frac{36}{65} & \frac{48}{65} & -\frac{25}{65} \end{bmatrix}$ then $\mathbf{A}\mathbf{A}^{\mathrm{T}} = \mathbf{I}$

6 Show that pre-multiplying the matrix $\mathbf{A} = \begin{bmatrix} 1 & 2 & 3 \\ 4 & 5 & 6 \\ 7 & 8 & 9 \end{bmatrix}$

by the matrix $\begin{bmatrix} 1 & 0 & 0 \\ 0 & 0 & 1 \\ 0 & 1 & 0 \end{bmatrix}$

interchanges rows two and three and that post-multiplying by

$$\begin{bmatrix} 0 & 0 & 1 \\ 0 & 1 & 0 \\ 1 & 0 & 0 \end{bmatrix}$$

interchanges rows one and three.

7 Show that the effect of post-multiplying a matrix with three columns by the matrix

$$\begin{bmatrix} 1 & 0 & 0 \\ 0 & 1 & 0 \\ 2 & 0 & 1 \end{bmatrix}$$

is to replace the first column by (the first column + twice the third column).

Generally speaking, the rule is: do to the unit matrix that which you want to do to the given matrix, post-multiplying if the effect is to be on columns, pre-multiplying if the effect is to be on rows.

8 Find matrices which will perform each of the following operations on a 3×3 matrix and verify your choice by application to matrix **A** of Problem 6:
(a) add three times row 3 to row 1 (b) halve column 2
(c) subtract row 3 from row 2.

Section 10.3

9 Check properties (i), (ii), (iii) of determinants (see page 266) for a general 3×3 determinant.

10 Evaluate the following determinants:

(i) $\begin{vmatrix} 3 & 1 & -2 \\ 8 & -5 & 7 \\ 4 & 0 & 1 \end{vmatrix}$ (ii) $\begin{vmatrix} 2.7 & 3.6 & 4.1 \\ 2.5 & 3.9 & 4.5 \\ 2.9 & 3.3 & 4.2 \end{vmatrix}$

(iii) $\begin{vmatrix} 2 & 1+i & 4 \\ 1-i & 3 & 2-i \\ 4 & 2+i & 1 \end{vmatrix}$ (iv) $\begin{vmatrix} 2 & 4 & 5 & 7 \\ 3 & 5 & 9 & 2 \\ 4 & 1 & 7 & 3 \\ 2 & 11 & 7 & 13 \end{vmatrix}$

$(i^2 = -1)$

11 Show that $\begin{vmatrix} 1 & a & b \\ a & 1 & b \\ a & b & 1 \end{vmatrix} = (a-1)(b-1)(a+b+1)$

12 Show that $\begin{vmatrix} 1 & a & bc \\ 1 & b & ca \\ 1 & c & ab \end{vmatrix} = (a-b)(b-c)(c-a)$

13 For the determinant

$$D = \begin{vmatrix} 4 & 1 & 1 \\ 1 & 2 & 3 \\ 3 & 1 & 2 \end{vmatrix}$$

find the cofactor of each element and show that

$$D = a_{11}A_{11} + a_{12}A_{12} + a_{13}A_{13}$$
$$D = a_{11}A_{11} + a_{21}A_{21} + a_{31}A_{31}$$
$$D = a_{31}A_{31} + a_{32}A_{32} + a_{33}A_{33}$$

What other expressions can you find which also equal D?

Section 10.4

14 Verify that **PQ** = **R** where

$$\mathbf{P} = \begin{bmatrix} 1 & 0 & 0 \\ 2 & 1 & 0 \\ 3 & 2 & 1 \end{bmatrix}; \mathbf{Q} = \begin{bmatrix} 2 & 3 & 4 \\ 0 & 4 & 1 \\ 0 & 0 & 6 \end{bmatrix}; \mathbf{R} = \begin{bmatrix} 2 & 3 & 4 \\ 4 & 10 & 9 \\ 6 & 17 & 20 \end{bmatrix}$$

All working must be clearly shown.
Find det **P** and det **Q** and explain why det **R** = 48. (EC)

15 (i) Show that for an $n \times n$ matrix **A**, $\det(2\mathbf{A}) = 2^n \det \mathbf{A}$
(ii) For any non-singular matrix **A** the adjoint, adj **A**, satisfies **A** adj **A** = (det **A**)**I**. By replacing **A** by 2**A** show that $\operatorname{adj}(2\mathbf{A}) = (2^{n-1} \det \mathbf{A})\mathbf{A}^{-1}$ (EC)

16 Find the inverses where possible of the following matrices

(i) $\begin{bmatrix} -2 & 3 \\ 1 & 4 \end{bmatrix}$ (ii) $\begin{bmatrix} 2 & 3 \\ 1.8 & 2.7 \end{bmatrix}$ (iii) $\begin{bmatrix} 1 & 2 & 3 \\ 2 & -2 & -4 \\ 3 & -4 & -3 \end{bmatrix}$

(iv) $\begin{bmatrix} 1 & 3 & 3 \\ 1 & 4 & 3 \\ 1 & 3 & 4 \end{bmatrix}$ (v) $\begin{bmatrix} 1 & 3 & 3 \\ 1 & 4 & 3 \\ 1 & 5 & 3 \end{bmatrix}$ (vi) $\begin{bmatrix} 1 & -1 & 2 & 1 \\ -1 & 0 & 3 & 2 \\ 2 & 1 & 0 & -1 \\ 2 & -2 & 1 & 3 \end{bmatrix}$

17 You are given that **A** and **B** are non-singular matrices, **I** is a unit matrix, and the relation

$$(\mathbf{AB})^{-1}(\mathbf{A} + \mathbf{I}) = 2\mathbf{B}.$$

Solve this equation for **A** in terms of **B** and **I**. (EC)

18 Three vectors **y**, **x** and **u** and two matrices **A** and **B** are connected by the equation **y** = **Ax** + **Bu**. It is required to write $\mathbf{y} = \mathbf{A}_c\mathbf{x}$ by putting **u** = –**Kx** in which the matrix

K is to be determined. Show that $\mathbf{K} = \mathbf{B}^{-1}(\mathbf{A} - \mathbf{A}_c)$. (EC)

19 Find the transpose and inverse of the matrix $\mathbf{A} = \begin{bmatrix} \alpha & \beta \\ -\beta & \alpha \end{bmatrix}$

For what values of α and β does $\mathbf{A}^T = \mathbf{A}^{-1}$?
Put $\alpha = \cos\theta$, $\beta = \sin\theta$ and interpret geometrically the equation

$$\begin{bmatrix} x' \\ y' \end{bmatrix} = \mathbf{A}\begin{bmatrix} x \\ y \end{bmatrix}$$ (EC)

20 A certain engineering problem produced the following four circuit relations

$$\mathbf{i} = \mathbf{C}\mathbf{u}; \quad \mathbf{u} = \mathbf{Z}^{-1}\mathbf{v}; \quad \mathbf{v} = \mathbf{C}^{\mathrm{T}}\mathbf{e}; \quad \mathbf{Z} = \mathbf{C}^{\mathrm{T}}\mathbf{W}\,\mathbf{C}$$

in which **i**, **u**, **v** and **e** are column vectors, and C, **Z** and **W** are non-singular matrices. Using the fact that for any non-singular matrices **P**, **Q** and **R**, $(\mathbf{PQR})^{-1} = \mathbf{R}^{-1}\mathbf{Q}^{-1}\mathbf{P}^{-1}$ and $(\mathbf{P}^{\mathrm{T}})^{-1} = (\mathbf{P}^{-1})^{\mathrm{T}}$, show that the four circuit relations condense to $\mathbf{i} = \mathbf{W}^{-1}\mathbf{e}$. (EC)

21 Three matrices **A**, **B** and **C** are given by

$$\mathbf{A} = \begin{bmatrix} 2 & 5 \\ -4 & 1 \end{bmatrix} \qquad \mathbf{B} = \begin{bmatrix} 1 & 3 \\ -2 & 6 \end{bmatrix} \qquad \mathrm{C} = \begin{bmatrix} 2 & 1 \\ -3 & -1 \end{bmatrix}$$

(i) Find $\mathbf{A}^{-1}$ and $\mathbf{B}^{-1}$

(ii) By pre-multiplying and post-multiplying with suitable matrices solve the equation **AXB** = **C** for the matrix **X**. (EC)

22 The matrix **A** is given by $\mathbf{A} = \begin{bmatrix} a & 0 & 0 \\ b & a & b \\ 0 & 0 & a \end{bmatrix}$

Find $\mathbf{A}^{-1}$ and determine if there are any values of a and b such that $\mathbf{A}^{-1} = \mathbf{A}$ (EC)

23 (i) If $\mathbf{F} = \mathbf{GDG}^{-1}$, show that $\mathbf{F}^2 = \mathbf{GD}^2\mathbf{G}^{-1}$ and $\mathbf{F}^3 = \mathbf{GD}^3\mathbf{G}^{-1}$

(ii) If $\mathbf{F} = \begin{bmatrix} 5 & -2 \\ 7 & -4 \end{bmatrix}$ and $\mathbf{G} = \begin{bmatrix} 2 & 1 \\ 7 & 1 \end{bmatrix}$ show that the diagonal matrix **D** given by

$\mathbf{G}^{-1}\mathbf{FG}$ is $\mathbf{D} = \begin{bmatrix} -2 & 0 \\ 0 & 3 \end{bmatrix}$ and *deduce* $\mathbf{F}^4$ (EC)

Section 10.5

24 Find the inverse of the matrix $\begin{bmatrix} 5 & -1 & 2 \\ 2 & 4 & 1 \\ 1 & 3 & -3 \end{bmatrix}$ and hence solve the simultaneous

equations $\begin{cases} 5x - y + 2z = 3 \\ 2x + 4y + z = 8 \\ x + 3y - 3z = 2 \end{cases}$

25 Find the inverse of

$$\mathbf{A} = \begin{bmatrix} 1 & 3 & 3 \\ 1 & 4 & 3 \\ 1 & 3 & 4 \end{bmatrix}$$

and check that $\mathbf{AA}^{-1} = \mathbf{A}^{-1}\mathbf{A} = \mathbf{I}$

26 Use the formal method of finding an inverse matrix to check your answers obtained to Problems 16 (i) and (iii) on page 283 and use these to solve the equations

$$\begin{cases} -2x_1 + 3x_2 = 1 \\ x_1 + 4x_2 = 3 \end{cases} \qquad \begin{cases} x_1 + 2x_2 + 3x_3 = 12 \\ 2x_1 - 2x_2 - 4x_3 = -14 \\ 3x_1 - 4x_2 - 3x_3 = -4 \end{cases}$$

27 If a set of three-dimensional orthogonal coordinate axes is rotated about the z-axis through an angle α, so that a fixed point P has coordinates (x, y, z) in one system and (x', y', z') in the second system, show that

$$x = x' \cos \alpha - y' \sin \alpha$$
$$y = x' \sin \alpha + y' \cos \alpha$$
$$z = z'$$

and write these relations in the form of a transformation $\mathbf{X} = \mathbf{AX}'$ where $\mathbf{X} = \begin{bmatrix} x \\ y \\ z \end{bmatrix}$,

$\mathbf{X}' = \begin{bmatrix} x' \\ y' \\ z' \end{bmatrix}$ and $\mathbf{A}$ is a matrix.

Find the inverse matrix $\mathbf{A}^{-1}$ (EC)

28 In two-dimensional geometrical transformations the matrix $\mathbf{Rot}(z, \theta)$ occurs where

$$\mathbf{Rot}(z, \theta) = \begin{bmatrix} \cos\theta & -\sin\theta & 0 \\ \sin\theta & \cos\theta & 0 \\ 0 & 0 & 1 \end{bmatrix}$$

Find the product $\mathbf{Rot}(z, \phi).\mathbf{Rot}(z, \theta)$, and show that this product is the same as $\mathbf{Rot}(z, \phi+\theta)$ (EC)

11

COMPLEX NUMBERS

11.1 The Idea of a Complex Number

When mathematicians encounter equations whose solutions cannot be expressed in terms of known functions they often define the solution to be a new function and set about determining its main features. In the past the failure of the existing rational numbers to provide a solution to $x^2 = 2$ led to the extension of the number system to the real numbers which incorporated the irrationals. A similar failure of the reals to help solve equations such as $x^2 = -1$, $(x - 3)^4 = -1$ led to the extension of the reals to the **complex numbers**. We shall find that, by introducing these so-called complex numbers, we can always solve a polynomial equation of the nth degree to give n roots. This is so even if the coefficients of the polynomial are themselves complex numbers (whereas an equation such as $x^2 - 2 = 0$ with rational coefficients cannot be solved in terms of rationals). In this sense the complex numbers are *complete.*

Since we wish to be able to solve an equation such as $x^2 = -1$ which gives $x = \pm\sqrt{-1}$, we introduce a quantity i which obeys the rule $\mathrm{i}^2 = -1$. Then for this equation $x = \pm\mathrm{i}$.

We now define a complex number to be of the form $a + \mathrm{i}b$ where a and b are real numbers. For example, $3 + \mathrm{i}2$, $\sqrt{3} + \mathrm{i}\pi$, $0 - \mathrm{i}\pi$, $\sqrt{2} - \mathrm{i}2.53$ are all complex numbers; the real numbers themselves can be regarded as special cases of the complex numbers for we can say for example $2 = 2 + \mathrm{i}0$, $\sqrt{2} = \sqrt{2} + \mathrm{i}0$. (Many engineers and books use j instead of i.)

You may rightly point out that all this is rather arbitrary and remote from your engineering subjects. However, for reasons which will become clearer as you read through this book, complex numbers and associated quantities act as a useful vehicle in the course of many calculations and save a great deal of work.

Having introduced these numbers we must develop rules for the arithmetic operations involving them, but we must do so in a way which is consistent with the special case of real numbers [$b = 0$].

We usually write $z \equiv x + \mathrm{i}y$ for a general complex number and call x the **real part** of z, written $\Re(z)$, and y the **imaginary part** of z, written $\Im(z)$.

Numbers of the form ib are called purely imaginary. Hence the real part of $-7 + 3i$ is -7 and the imaginary part is 3; $i\pi$ is a *purely imaginary* number.

Since a complex number comprises two real numbers which are independent, we may use the notation $z \equiv (x, y)$ and this suggests an analogy to coordinate geometry. We visualise a complex number as a point on a plane or as the line joining this end-point to the coordinate origin. The points $z_1 = 3 + i \equiv (3, 1)$, $z_2 = 1 - i \equiv (1, -1)$, $z_3 = -2 + 4i \equiv (-2, 4)$, $z_4 = -1 - 2i \equiv (-1, -2)$, $z_5 = -3$, $z_6 = -2i$, are shown in Figure 11.1 which is an example of an **Argand diagram**. Note that there is a *one-to-one correspondence* between the complex numbers and points in the plane.

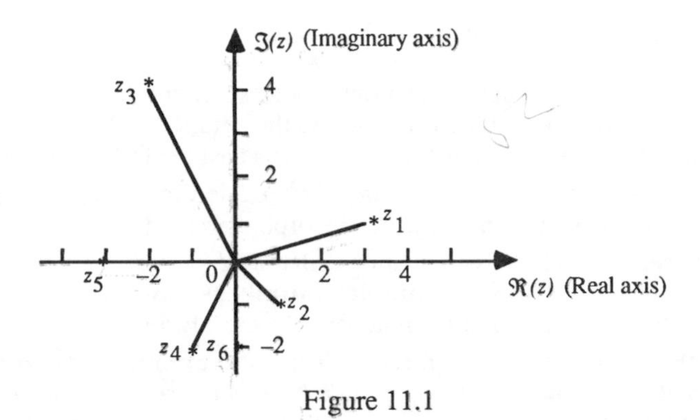

Figure 11.1

This diagrammatic representation suggests the use of polar coordinates (r, θ) as an alternative to (x, y) coordinates. r is called the **modulus** of the complex number and is sometimes written $|z|$; it corresponds to the distance of the point z from the origin and this is really what is meant by the real equivalent $|x|$. θ is called the **argument** or amplitude of a complex number and written $\arg(z)$. Notice that for real positive numbers $\theta = 0$, for real negative numbers $\theta = \pi$ and a purely imaginary number will have an argument $\theta = \pi/2$ or $\theta = 3\pi/2$. An alternative notation to (r, θ) is $r\underline{/\theta}$. See Figure 11.2(i).

We use the relations $x = r \cos \theta$, $y = r \sin \theta$ to convert to polars and $r = \sqrt{x^2 + y^2}$, $\tan \theta = y/x$ to convert from polars. One danger is that in the range [0, 360°] there are two values of θ for each value of $\tan \theta$ and to avoid ambiguity it is safer to plot the desired point on an Argand diagram. (The convention is usually used that θ is taken to be in the range $-\pi < \theta \leq \pi$)

Let us take a few examples (i) $z = -3$, (ii) $z = i\pi/2$, (iii) $z = -3i$, (iv) $z = -1/\sqrt{2} - i/\sqrt{2}$. We plot these first on an Argand diagram [Figure 11.2(ii)]. The equivalent polar forms are then (3, 180°), (π/2, 90°), (3, −90°), (1, −135°).

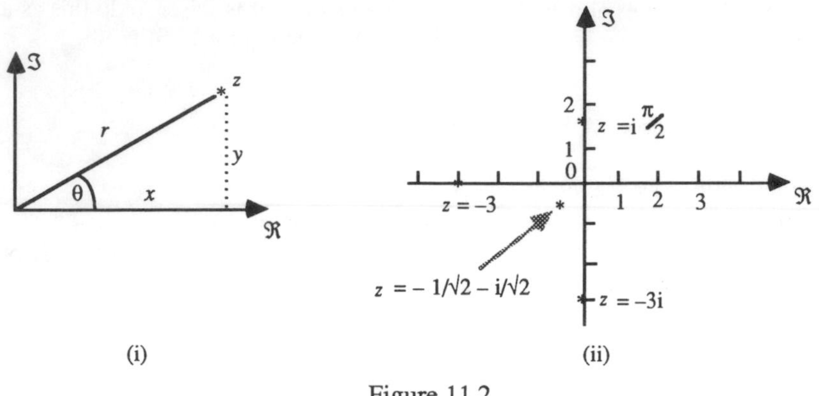

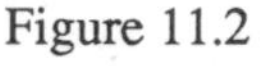

Figure 11.2

Example 1

Plot $2\underline{/45°}$ on an Argand diagram and find the equivalent cartesian form. The diagram is Figure 11.3(i). $x = 2\cos 45° = \sqrt{2}$, $y = 2\sin 45° = \sqrt{2}$, and the equivalent form is $\sqrt{2} + \sqrt{2}\mathrm{i}$.

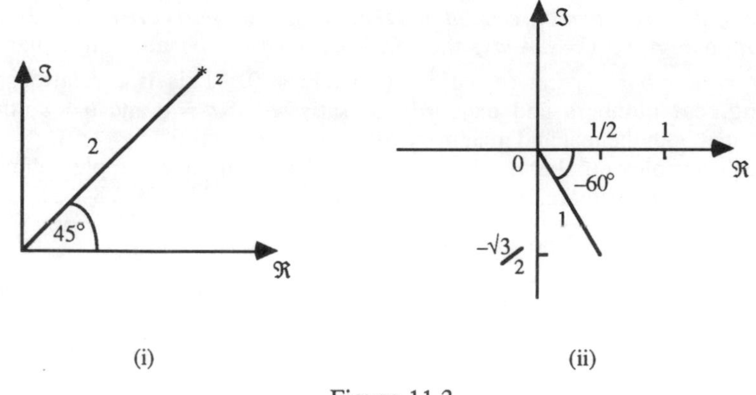

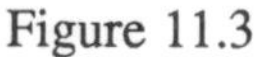

Figure 11.3

Example 2

Determine the polar form of $1/2 - (\sqrt{3}/2)\mathrm{i}$ and plot on an Argand diagram.

$$r = \sqrt{\left[\frac{1}{2}\right]^2 + \left[\frac{-\sqrt{3}}{2}\right]^2} = \sqrt{\frac{1}{4} + \frac{3}{4}} = 1; \qquad \tan\theta = -\frac{\sqrt{3}}{2}\Big/\frac{1}{2} = -\sqrt{3}$$

Now there are two angles 120° and –60° which have tangent of $-\sqrt{3}$ and if we note that $r\cos\theta = 1/2 > 0$ and $r\sin\theta = -\sqrt{3}/2 < 0$ we can see that θ lies in the fourth quadrant, selecting –60° as the required angle. A plot on an Argand diagram would have established this straight away and it is wiser to do this as a first step. Then the polar form required is $1\angle -60°$. See Figure 11.3(ii).

11.2 Complex Arithmetic

We define addition as one might expect; an example first

$$(3-2i)+(1+4i)=(3+1)+(-2+4)i=4+2i$$

Note that we are defining the + between the parentheses on the left-hand side of the equation. Subtraction follows similarly:

$$(3-2i)-(1+4i)=(3-1)+(-2-4)i=2-6i$$

In the general case for two complex numbers $(a+bi)$ and $(c+di)$

$$(a+bi)+(c+di)=(a+c)+(b+d)i \tag{11.1a}$$

$$(a+bi)-(c+di)=(a-c)+(b-d)i \tag{11.1b}$$

Multiplication is carried out as in ordinary algebra except that wherever i^2 occurs, it is replaced by -1. For example,

$$(3-2i).(1+4i)=3-2i+12i-8i^2=3-2i+12i+8=11+10i$$

More generally,

$$(a+bi).(c+di)=(ac-bd)+(bc+ad)i \tag{11.2}$$

Before turning to division we need two further concepts. Two complex numbers are **equal** *if their real parts are equal and their imaginary parts are equal.* To see this, suppose $a+bi=c+di$; then $(a-c)=(d-b)i$, and on squaring $(a-c)^2=(d-b)^2.-1$, i.e. $(a-c)^2+(d-b)^2=0$. This is a relationship involving real numbers and can *only* be satisfied if $a=c$ and $b=d$, thus showing that both the real and imaginary parts must be equal.

To each complex number $a+bi$ there corresponds a unique complex number $a-bi$ called the **conjugate**. For example, the conjugate of $3-2i$ is $3+2i$.

The conjugate of z is denoted by $\bar{z}$.

Observe that if z is real, $\bar{z}=z$ and if z is *purely imaginary*, $\bar{z}=-z$. Now it can be shown that $z.\bar{z}$ is *always* real; for if $z=x+iy$, then $\bar{z}=x-iy$ and hence $z\bar{z}=(x+iy)(x-iy)=x^2-i^2y^2=x^2+y^2$. This property allows us to carry out division of two complex numbers – we *multiply top and bottom by the conjugate of the denominator*, e.g.

$$\frac{3-2i}{1+4i}=\frac{3-2i}{1+4i}\cdot\frac{1-4i}{1-4i}=\frac{3-2i-12i+8i^2}{1-16i^2}=\frac{-5-14i}{17}=\frac{-5}{17}-\frac{14}{17}i$$

In algebraic terms,

$$\frac{z_1}{z_2}=\frac{z_1}{z_2}\cdot\frac{\bar{z}_2}{\bar{z}_2}=\frac{z_1\bar{z}_2}{z_2\bar{z}_2} \tag{11.3}$$

and the denominator is real.

Note, in particular, that $1/i = -i$.

We have said that the complex numbers include the reals as a special case; you should verify that the arithmetic operations defined above are valid for real numbers.

Multiplication and Division in Polar Form

We now turn to the multiplication and division of two complex numbers in polar form. First we observe that $z = (r, \theta) \equiv r \cos \theta + ir \sin \theta$.

Then if $z_1 = (r_1, \theta_1)$ and $z_2 = (r_2, \theta_2)$,

$$\begin{aligned} z_1 z_2 &= (r_1 \cos \theta_1 + ir_1 \sin \theta_1).(r_2 \cos \theta_2 + ir_2 \sin \theta_2) \\ &= r_1 r_2 \cos \theta_1 \cos \theta_2 + i^2 r_1 r_2 \sin \theta_1 \sin \theta_2 + \\ &\qquad i[r_1 r_2 \cos \theta_1 \sin \theta_2 + r_1 r_2 \sin \theta_1 \cos \theta_2] \\ &= r_1 r_2 [\cos (\theta_1 + \theta_2) + i \sin (\theta_1 + \theta_2)] \end{aligned} \qquad (11.4)$$

In other words, *to multiply two complex numbers, we* **multiply** *their* **moduli** *and* **add** *their* **arguments**. In shorthand $r_1 \underline{/\theta_1} \times r_2 \underline{/\theta_2}$ gives $r_1 r_2 \underline{/(\theta_1 + \theta_2)}$.

Example

Consider $(1 + i).[1/2 + (\sqrt{3}/2)i] = 1/2 + (1/2)i + (\sqrt{3}/2)i + (\sqrt{3}/2)i^2 = (-\sqrt{3} + 1)/2 + i[(1 + \sqrt{3})/2]$.

We can do the same problem in polar form using the results just obtained; for $1 + i = \sqrt{2}\underline{/45^\circ}$ and $1/2 + (\sqrt{3}/2)i = 1\underline{/60^\circ}$.

Hence their product is $\sqrt{2}\underline{/45^\circ + 60} = \sqrt{2}\underline{/105^\circ}$, which can be written

$$\sqrt{2} \cos 105^\circ + i\sqrt{2} \sin 105^\circ = -\sqrt{2} \cos 75^\circ + i\sqrt{2} \sin 75^\circ$$

Convince yourself that this is the same as the result obtained when doing the problem in cartesian coordinates.

Division in polars follows a similar pattern:

$$\frac{z_1}{z_2} = \frac{r_1}{r_2}[\cos(\theta_1 - \theta_2) + i \sin(\theta_1 - \theta_2) = \frac{r_1}{r_2}\underline{/(\theta_1 - \theta_2)} \qquad (11.5)$$

Hence in *division* we **divide** *the* **moduli**, *and* **subtract** *the* **arguments**. For example

$$\frac{1 + i}{1/2 + (\sqrt{3}/2)i} = \frac{\sqrt{2}\,\underline{/45^\circ}}{1\underline{/60^\circ}} = \frac{\sqrt{2}}{1}\underline{/45^\circ - 60^\circ} = \sqrt{2}\underline{/-15^\circ}$$

These rules can obviously be combined.

For example

$$\frac{(3\angle 60^\circ)\times(2\angle 40^\circ)}{4\angle 30^\circ}=\frac{3}{2}\angle 70^\circ$$

Two observations are worthy of attention:

1 $1/z_1=\frac{1}{r_1}\angle -\theta_1$, $z_1^2=r_1^2\angle 2\theta_1$, $z_1^3=r_1^3\angle 3\theta_1$ etc

2 It is generally true that multiplication and division are best carried out in polars.

Complex numbers as vectors

At this stage it will be worthwhile examining the consequences of the rule for addition and we represent this graphically in Figure 11.4

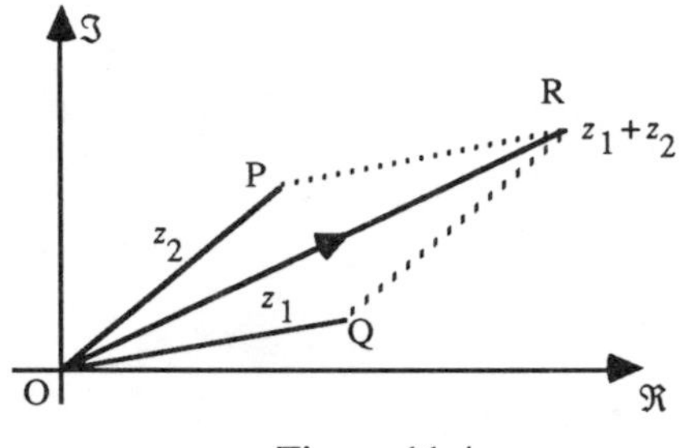

Figure 11.4

If OQ and OP represent the complex numbers z_1 and z_2 respectively, and OPRQ is a parallelogram, then it is not too hard to show by coordinate geometry that R has coordinates of $(z_1 + z_2)$ and hence OR represents the sum $z_1 + z_2$.

We may regard a complex number as an ordered pair of real numbers (x, y) and define addition as $(x_1, y_1) + (x_2, y_2) = [(x_1 + x_2), (y_1 + y_2)]$.

Entities which obey this rule are called *two-dimensional vectors* and these are dealt with in the next section. It is important to bear in mind that complex numbers can be regarded as vectors for some purposes.

11.3 De Moivre's Theorem and Complex Roots

An important consequence of the polar rule for multiplication is *de Moivre's Theorem.* This states that if p/q be a rational number, one value of

$$(\cos\theta + i\sin\theta)^{p/q} \quad \text{is} \quad (\cos\frac{p}{q}\theta + i\sin\frac{p}{q}\theta)$$

We shall prove this only for the special case p/q is a positive integer, n, using induction.

Clearly the result holds for $n = 1$.

Let us assume that $(\cos\theta + i\sin\theta)^m = \cos m\theta + i\sin m\theta$ for a positive integer m.

Multiplying both sides by $\cos\theta + i\sin\theta$ we get

$$
\begin{aligned}
(\cos\theta + i\sin\theta)^{m+1} &= (\cos m\theta + i\sin m\theta)(\cos\theta + i\sin\theta) \\
&= \cos m\theta\cos\theta + i\sin m\theta\cos\theta + i\cos m\theta\sin\theta \\
&\quad - \sin m\theta\sin\theta \\
&= \cos(m+1)\theta + i\sin(m+1)\theta
\end{aligned}
$$

Hence if the result is true for a positive integer m, it is also true for $m + 1$. But it is true for $m = 1$.

Hence, by induction, the result is true for all integers.

Example

We can use this result to evaluate $(1 - i)^8$. We know that $1 - i \equiv \sqrt{2}\underline{/-45^\circ}$. Therefore

$$(1 - i)^8 = (\sqrt{2})^8\,\underline{/-360^\circ} = 16\underline{/0^\circ} = 16$$

We note that $(\cos\theta + i\sin\theta)^{-r}$, can be expressed as

$$\cos(-r)\theta + i\sin(-r)\theta = \cos r\theta - i\sin r\theta$$

where r is any real rational number.
In particular,

$$(\cos\theta + i\sin\theta)^{-1} = \cos\theta - i\sin\theta$$

Hence, if $\cos\theta + i\sin\theta = z$, then $(\cos\theta + i\sin\theta)^{-1} = 1/z = \cos\theta - i\sin\theta$,

and further, by addition, $\cos\theta = \frac{1}{2}[z + (1/z)]$, and subtracting, $\sin\theta = (1/2i)[z - (1/z)]$.

Let us now investigate the effect of allowing n to increase without limit in the expression $[\cos(\theta/n) + i\sin(\theta/n)]^n$. By de Moivre's theorem we know that the expression takes the value $(\cos\theta + i\sin\theta)$ for any value of n and hence the limiting value must also be $(\cos\theta + i\sin\theta)$.

But we shall find later that $\cos\phi = 1 - (\phi^2/2!) + (\phi^4/4!) - \ldots$ and $\sin\phi = \phi - (\phi^3/3!) + (\phi^5/5!) - \ldots$. Hence

$$\cos(\theta/n) = 1 + 0(1/n^2) \qquad \text{and} \qquad \sin(\theta/n) = \theta/n + 0(1/n^3)$$

where $0(1/n^2)$ means that the next term is of order $1/n^2$, similarly $0(1/n^3)$

Hence

$$\left[\cos\frac{\theta}{n} + i\sin\frac{\theta}{n}\right]^n = \left[1 + i\frac{\theta}{n} + 0\left[\frac{1}{n^2}\right]\right]^n$$

By treating $i\theta$ in the same way that we treated x in the definition of e^x, it may be shown that

$$\lim_{n\to\infty}\left[1 + i\frac{\theta}{n} + 0\left[\frac{1}{n^2}\right]\right]^n = \lim_{n\to\infty}\left[1 + i\frac{\theta}{n}\right]^n = e^{i\theta}$$

Hence

$$\cos\theta + i\sin\theta \equiv e^{i\theta} \tag{11.6}$$

which can be regarded as a definition of $e^{i\theta}$ as an alternative. The two expressions on either side of the identity sign behave identically for all values of θ.

Using $\cos\theta + i\sin\theta \equiv e^{i\theta}$, de Moivre's theorem now becomes $(e^{i\theta})^{p/q} = e^{i(p\theta/q)}$.

Further, since $x + iy = r\cos\theta + ir\sin\theta = r(\cos\theta + i\sin\theta)$, we can write

$$z = re^{i\theta} \tag{11.7}$$

and this is one of the most useful forms of a complex number. It is called the **exponential form**. Also we have

$$e^{-i\theta} = \cos\theta - i\sin\theta \tag{11.8}$$

and so

$$\left.\begin{aligned}\cos\theta &= \frac{1}{2}(e^{i\theta} + e^{-i\theta})\\ \sin\theta &= \frac{1}{2i}(e^{i\theta} - e^{-i\theta})\end{aligned}\right\} \tag{11.9}$$

We can obtain expressions for $\cos n\theta$ and $\sin n\theta$ by using these results, e.g.

$$\begin{aligned}\cos 3\theta + i\sin 3\theta &= (\cos\theta + i\sin\theta)^3\\ &= \cos^3\theta + 3\cos^2\theta\, i\sin\theta + 3\cos\theta\, i^2\sin^2\theta + i^3\sin^3\theta\\ &= \cos^3\theta - 3\cos\theta\sin^2\theta + i(3\cos^2\theta\sin\theta - \sin^3\theta)\end{aligned}$$

Hence picking out real and imaginary parts

$$\begin{aligned}\cos 3\theta = \cos^3\theta - 3\cos\theta\sin^2\theta &= \cos^3\theta - 3\cos\theta(1 - \cos^2\theta)\\ &= 4\cos^3\theta - 3\cos\theta\\ \sin 3\theta = 3\cos^2\theta\sin\theta - \sin^3\theta &= 3\sin\theta(1 - \sin^2\theta) - \sin^3\theta\\ &= 3\sin\theta - 4\sin^3\theta\end{aligned}$$

Likewise, we can find expressions for $\cos^n\theta$, $\sin^n\theta$. For example,

$$\begin{aligned}\cos^4\theta &= \frac{1}{16}(e^{i\theta} + e^{-i\theta})^4\\ &= \frac{1}{16}(e^{4i\theta} + 4e^{3i\theta}e^{-i\theta} + 6e^{2i\theta}e^{-2i\theta} + 4e^{i\theta}e^{-3i\theta} + e^{-4i\theta})\end{aligned}$$

$$= \frac{1}{16}(e^{4i\theta} + e^{-4i\theta}) + \frac{4}{16}(e^{2i\theta} + e^{-2i\theta}) + \frac{6}{16}$$

Now $\qquad \cos 4\theta + i \sin 4\theta = e^{4i\theta}$

and $\qquad \cos 4\theta - i \sin 4\theta = e^{-4i\theta}$

Hence $\qquad e^{4i\theta} + e^{-4i\theta} = 2 \cos 4\theta$

and so

$$\cos^4\theta = \frac{1}{16}.2 \cos 4\theta + \frac{4}{16}.2 \cos 2\theta + \frac{6}{16}$$

$$= \frac{1}{8}\cos 4\theta + \frac{1}{2}\cos 2\theta + \frac{3}{8}$$

Roots of a complex number

Let us consider the problem of finding the cube roots of 1. Although we might expect three roots, we know there is only one real root. Let z be any root so that $z^3 = 1$. If z be (r, θ), then $z^3 = (r^3, 3\theta)$ and $1 \equiv (1, 0)$. Hence

$$r^3(\cos 3\theta + i \sin 3\theta) = 1(\cos 0° + i \sin 0°) \qquad (11.10)$$

From this we see that $r^3 = 1$ and since r is real, $r = 1$.

Hence the root lies on the unit circle $|z| = 1$; $3\theta = 0$ and so the root z_1 is $(1, 0)$, which is the real root we knew already. But $\sin(360° + \phi) \equiv \sin \phi$ and $\cos(360° + \phi) \equiv \cos \phi$ and the number 1 may also be represented $(1, 360°)$; hence equation (11.10) may be replaced by

$$r^3(\cos 3\theta + i \sin 3\theta) = 1(\cos 360° + i \sin 360°)$$

which gives $r = 1$ and $\theta = 120°$. We claim that a second root is

$$z_2 = \cos 120° + i \sin 120° = -\frac{1}{2} + \frac{\sqrt{3}}{2}i$$

You should check formally that $z_2^3 = 1$.

(11.10) may also be replaced by

$$r^3(\cos 3\theta + i \sin 3\theta) = 1(\cos 720° + i \sin 720°)$$

whence $r = 1$, $\theta = 240°$ and $z_3 = -\frac{1}{2} - \frac{\sqrt{3}}{2}i$.

Let us plot these three roots on an Argand diagram (Figure 11.5).

There are several features worthy of note.

1 The roots are equally spaced around the unit circle $|z| = 1$.

2 There are no more roots to be found; if we replaced (11.10) by $r^3(\cos 3\theta + i \sin 3\theta) = 1(\cos 1080° + i \sin 1080°)$ we would obtain $\theta = 360°$ which

coincides with z_1 and we would merely restart the cycle.

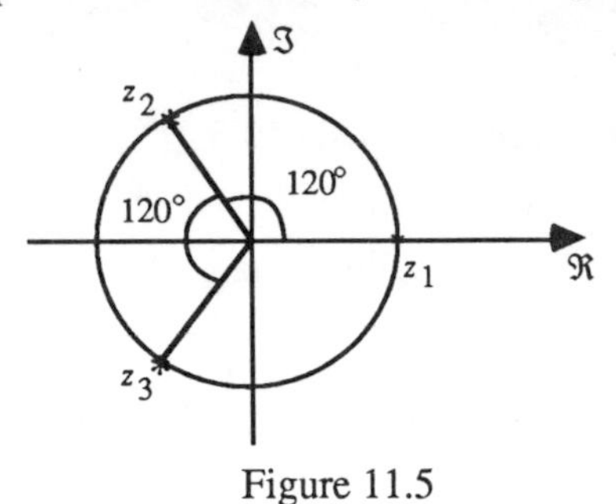

Figure 11.5

3 $\bar{z}_3 = z_2$ and $\bar{z}_2 = z_3$.

4 Since $z_3 = (1, 240°)$ and $z_2 = (1, 120°)$, we have $z_3 = z_2^2$.

We label the roots, 1, ω, ω^2 therefore, and the equation $z^3 - 1 = 0$ can be written as $(z - 1)(z - \omega)(z - \omega^2) = 0$. The left-hand side can be multiplied out to give

$$z^3 - z^2(1 + \omega + \omega^2) + z(\omega + \omega^2 + \omega^3) - \omega^3 = 0$$

Since $z_2 = \omega$ is a root of $z^3 = 1$, $\omega^3 = 1$ and as there is no z^2 term in the equation $z^3 - 1 = 0$, the sum of the roots $= 1 + \omega + \omega^2 = 0$. This also leads to a vanishing of the z term and the equation reduces to $z^3 = 1$.

It is useful to observe that we may cast the equation as

$$(z - 1)[z^2 - z(\omega + \omega^2) + \omega^3] = 0$$

i.e. $(z - 1)(z^2 + z + 1) = 0$ and we have the left-hand side as the product of real linear and real quadratic factors.

In fact there are two general rules which apply to polynomial equations

$$z^n + a_{n-1}z^{n-1} + \ldots + a_1z + a_0 = 0$$

where a_r are *real* coefficients: any complex roots occur in *conjugate pairs* (z and $\bar{z}$) and the left-hand side can be decomposed into a product of real linear and real quadratic factors. You should show that the second result follows from the first.

Example

Let us find the fifth roots of 1 + i. First, we write 1 + i as $\sqrt{2}[(1/\sqrt{2}) + i/\sqrt{2}]$ and hence in polar form as $\sqrt{2}\underline{/45°}$.

De Moivre's theorem in its general form states that if $z = (r, \theta)$

$$z^{p/q} = r^{p/q}\left[\cos\frac{p(\theta + 2k\pi)}{q} + i\sin\frac{p(\theta + 2k\pi)}{q}\right]$$

where $k = 0, 1, 2, \ldots, (q - 1)$.

This may be remembered more easily by noting that all the roots are equally spaced around the circle of radius $r^{p/q}$. The radius of the circle we seek is $(\sqrt{2})^{1/5}$, i.e. $2^{1/10} \cong 1.072$. One root, called the *principal root*, has argument $(45°/5)$, i.e. $9°$. Since the others are equally spaced, they must be $(360/5)°$, i.e. $72°$ apart; i.e. they have arguments $81°$, $153°$, $225°$, $297°$. They are plotted on an Argand diagram, Figure 11.6

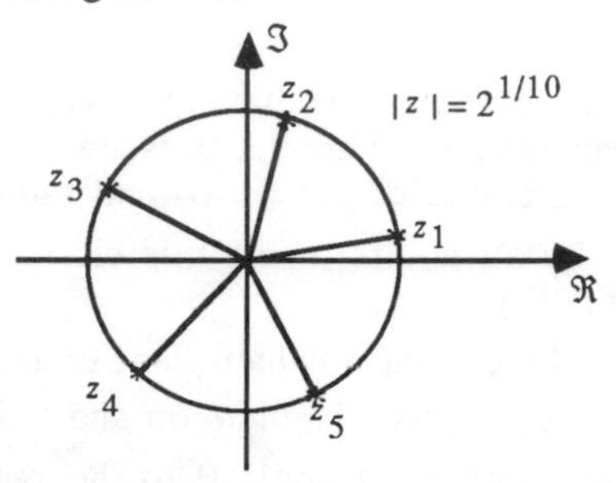

Figure 11.6

In this case no root is the conjugate of one of the others. Why not?

Summary of the method for finding the roots of a complex number

If we wish to find $\sqrt[n]{a + ib}$

(i) plot $a + ib$ on an Argand diagram
(ii) find $|a + ib| = r$ and $\arg(a + ib) = \theta$
(iii) find $\sqrt[n]{r}$ and draw a circle centre the origin of this radius
(iv) the first root lies on this circle with argument θ/n
(v) find $360°/n$ and, starting at the first root, plot out the remaining roots on the circle by consistently increasing the argument by $360°/n$.

11.4 Regions in the Complex Plane

Regions in the z-plane may be described in terms of complex numbers. For example, the equation $|z| = 2$ states that the length of the line joining the origin to z is always equal to 2. Hence the point represented by z must move on a circle, centre the origin, radius 2 [Figure 11.7(i)].

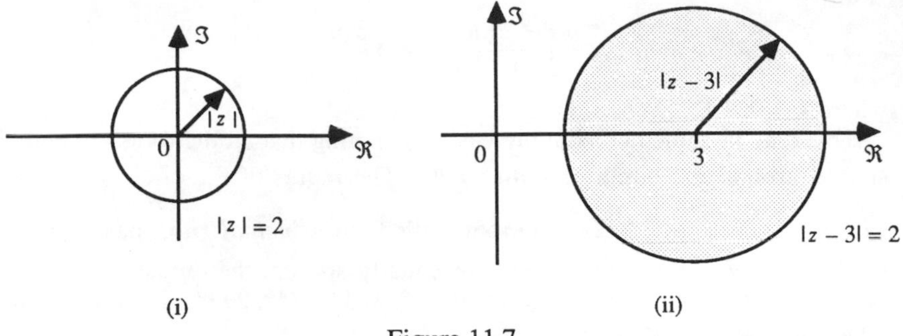

Figure 11.7

Similarly the equation $|z-3| = 2$ can be thought of as $|z-(3+i0)| = 2$ and says that the distance between the point z and (3, 0) is always equal to 2. Hence the locus of the point z is a circle, centre (3, 0) and radius 2. The inequality $|z-3| < 2$ is satisfied by all points inside this circle and so represents the shaded area in Figure 11.7(ii).

The inequality $\Re(z) > 4$ represents the half-plane to the right of the line $x = 4$; the inequality $\Im(z) \leq 3$ represents all points on and below the line $y = 3$. A straight line from the point z_0 at angle θ to the x-axis is described by $\arg(z-z_0) = \theta$ [Figure 11.8(i)]. To continue the line of the other side of z_0 would require $\arg(z-z_0) = \theta + 180°$.

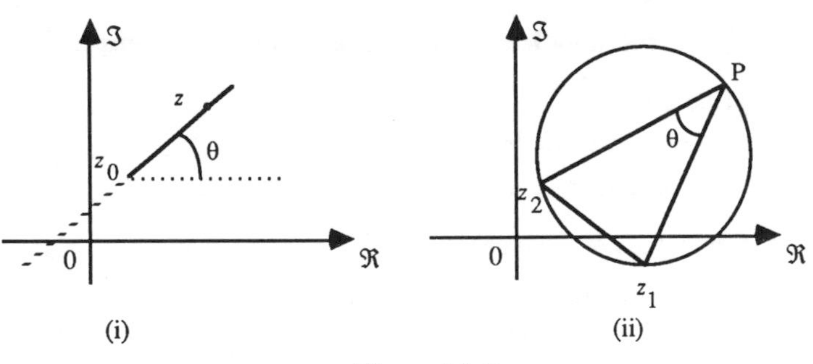

Figure 11.8

Consider the arc of a circle as shown in Figure 11.8(ii). The arc stands on the chord with end-points z_1 and z_2 and an angle θ is subtended at the circumference. If P, represented by z, is any point on the arc then we have $\arg[(z-z_1)/(z-z_2)] = \theta = \text{constant}$, and this is the equation of the arc.

Note that the left-hand side is $\arg(z-z_1) - \arg(z-z_2)$ [Problem 19, page 302]. Try and interpret this from the diagram and convince yourself of the suitability of the polar form by writing the cartesian equivalent.

Finally, we see how to unravel an inequality by algebra and compare with a geometrical interpretation. Consider $|z - 1| \leq |z + 1|$. To see what domain in the complex plane this represents, we note that both sides of the inequality are positive and so we may square them to obtain

$$|z - 1|^2 \leq |z + 1|^2$$

i.e. $(z - 1)(\overline{z - 1}) \leq (z + 1)(\overline{z + 1})$ (using $|z|^2 = z\bar{z}$)

i.e. $(z - 1)(\bar{z} - 1) \leq (z + 1)(\bar{z} + 1)$ (using $\overline{z_1 + z_2} = \bar{z}_1 + \bar{z}_2$)

i.e. $z\bar{z} - \bar{z} - z + 1 \leq z\bar{z} + \bar{z} + z + 1$

i .e. $-\bar{z} - z \leq \bar{z} + z$

i.e. $2(z + \bar{z}) \geq 0$

therefore $z + \bar{z} \geq 0$

but $z + \bar{z} = 2\Re(z)$

hence $2\Re(z) \geq 0$

This represents all points on and to the right of the line $x = 0$

Now the *geometrical* interpretation of the original inequality is that z moves in such a way that its distance from the point 1 is less than or, at worst, equal to its distance from the point -1; which leads to the same region of the plane. Further $|z - 1| = |z + 1|$ gives the perpendicular bisector of the line segment joining $+1$ and -1 (that is, in the y-axis).

Note that we could have completed this problem by an alternative method. In this alternative method we write $z = x + iy$ and reduce the equation to x, y form. Thus

$$|z - 1| \leq |z + 1|$$

becomes $|x - 1 + iy| \leq |x + 1 + iy|$

i.e. $\sqrt{(x - 1)^2 + y^2} \leq \sqrt{(x + 1)^2 + y^2}$

i.e. $(x - 1)^2 + y^2 \leq (x + 1)^2 + y^2$

i.e. $x^2 - 2x + 1 \leq x^2 + 2x + 1$

i.e. $0 \leq 4x$

Hence $x \geq 0$, as before.

This latter method is often the only one that can be used easily.

11.5 Application – Linear AC Circuits

A periodic disturbance may be represented via complex variables. If a real quantity x varies harmonically with time as $x = a \cos(\omega t + \phi)$, a is called the *amplitude*, $\omega t + \phi$ the *phase*, $2\pi/\omega$ the *period* and $\omega/2\pi$ the *frequency* of the disturbance. We can **represent** the disturbance by $z = ae^{i(\omega t+\phi)}$ $=a[\cos(\omega t + \phi) + i\sin(\omega t + \phi)]$; the *real* part of this complex quantity is the *disturbance* x. In applications to linear AC circuits, we shall need to attach a meaning to dz/dt. We can find dx/dt as $-\omega a \sin(\omega t + \phi)$; since we want this to be the real part of some complex expression we note that $i\omega ae^{i\omega t}$ = $i\omega a(\cos \omega t + i \sin \omega t) = i\omega a \cos \omega t - \omega a \sin \omega t$ and hence dx/dt is the real part of $i\omega ae^{i(\omega t+\phi)}$.

If you have already done differentiation you will appreciate that this is consistent with saying that $dz/dt = i\omega ae^{i(\omega t+\phi)}$ which follows from the usual rules for differentiation.

Consider the circuit shown in Figure 11.9(i), where the applied e.m.f. $E = E_0 \cos \omega t$. When the circuit settles down to a steady oscillatory behaviour after the effective decay of any transient behaviour we can assume the alternating current to be of the same frequency as the applied e.m.f. but not necessarily in phase with it, i.e. $I = I_0 \cos(\omega t + \phi)$. It is customary in such problems to write the current representation as i and hence $\sqrt{-1}$ must be given the alternative symbol, j.

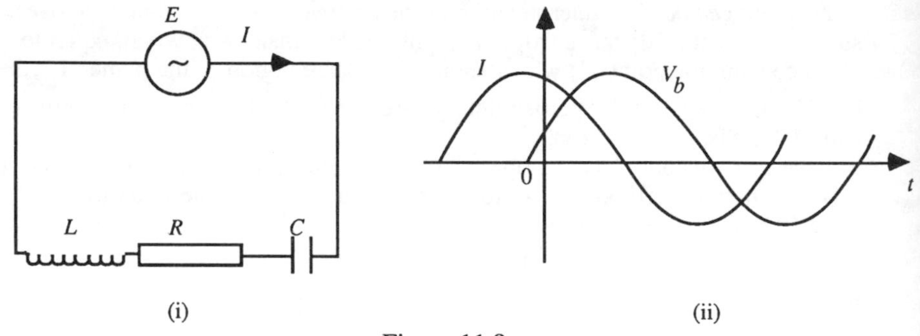

Figure 11.9

When a current I passes through an inductance coil of self-inductance L the back e.m.f. developed is $L(dI/dt) = V_b$.

If

$$I = I_0 \cos(\omega t + \phi), \qquad V_b = -\omega L I_0 \sin(\omega t + \phi)$$

$$= \omega L I_0 \cos(\omega t + \phi + \tfrac{\pi}{2})$$

In other words, the back e.m.f. is ahead of the current by 90°. See Figure 11.9(ii).

If we *represent* the current by $i = I_0\, e^{j(\omega t+\phi)}$ then the *representation* of the back e.m.f. is

$$v_b = \omega L I_0\, e^{j(\omega t+\phi+\pi/2)} = \omega L I_0\, e^{j(\omega t+\phi)}\, e^{j\pi/2}$$
$$= j\omega L i \qquad (\text{since } e^{j\pi/2} \equiv j)$$

Similarly, it can be shown that the representations v_R, v_C of the e.m.f.'s in the resistance and in the condenser are given by $v_R = Ri$ and $v_C = i/j\omega C = -ji/\omega C$. Now ωL and $1/\omega C$ are the *reactances* of the inductance and the capacitance; they play the same role as the resistance.

The *representation* of the total e.m.f. on the circuit,

$$e = i\left(j\omega L + R + \frac{-j}{\omega C}\right) \tag{11.11}$$

e/i is called the *complex impedance* of the circuit, often denoted by z.

Then $|z| = [R^2 + (\omega L - 1/\omega C)^2]^{1/2}$. If we let $\arg z = -\phi$ then

$$\tan\phi = (1/\omega C - \omega L)/R = (1 - \omega^2 LC)/R\omega C$$

Transforming back to real variables, and noting that $z = |z|e^{-j\phi}$

$$I = \Re(i) = \Re\{e/z\} = \Re\{E_0\, e^{j\omega t}\, e^{j\phi} / [R^2 + (\omega L - \frac{1}{\omega C})^2]^{1/2}\}$$

$$= \frac{E_0 \cos(\omega t + \phi)}{[R^2 + (\omega L - 1/\omega C)^2]^{1/2}}$$

Note that equation (11.11) contains all the information about the solution and can yield this information after complex algebra.

We see that the concept of complex impedance is one which facilitates solution to a problem.

Problems

Section 11.1

1 Mark on an Argand diagram the points representing the following numbers: $3 + 4i$, $-2 + 3i$, $-3 - 2i$, $2 - 4i$

2 Express the following complex numbers in polar form: 3, $2i$, -1, $-2i$, $5 + 12i$, $-\sqrt{3}+i$, $-6 - 8i$, $\sqrt{2} - i\sqrt{2}$, $2 + 2i\sqrt{3}$, $-3 + 2i$, $-1 - i$, $1 - 2i$

3 Express the following complex numbers in cartesian form: $\sqrt{2}\underline{/\pi/4}$, $2\underline{/140^\circ}$, $2\underline{/-5\pi/6}$, $5\underline{/-55^\circ}$

Section 11.2

4 Simplify, giving answer in cartesian form: $(5 - 3i)(2 + i) - \dfrac{4(3 - i)}{1 - i}$

5 (i) Find the roots of the equation $x^2 + 4x + 13 = 0$ and express them in the polar form.

(ii) If $b^2 < ac$, show that the roots of the equation $ax^2 + 2bx + c = 0$ can be expressed as $\sqrt{c/a}\ \angle \pm\cos^{-1}(-b/\sqrt{ac})$

6 (i) If $z = (2 - i)/(1 + i) - 2(3 + 4i)/(u + i)$ where u is a real number, find the values of u which make the complex number z lie on the line $x = y$ in the Argand diagram.

(ii) Find the complex number z such that $\arg z = \pi/4$ and $|z - 3 + 2i| = |z + 3i|$

(iii) If x and y are real numbers and

$$\frac{2(x + iy)}{1 - i} + \frac{2(x - iy)}{i} = \frac{5(1 + i)}{2 - i}$$

find x and y.

7 Show that $z\bar{z}$ is real and equal to $|z|^2$.

8 Show that $\Re(z) = \frac{1}{2}(z + \bar{z})$ and find an expression for $\Im(z)$.

9 Show that $1/i = -i$, $1/(1 + i) = \frac{1}{2}(1 - i)$

10 Evaluate $(3 + i)(3 - i)$, $(1 + 4i)(2 - 6i)$, $(5i + 6)/7i$, $(3 + 4i)/(4 + 3i)$

11 Find the modulus and argument of $-1 + \sqrt{3}i$, $3 + 4i$, $4 - 2i$; write each number in polar form.

12 Plot $z, \bar{z}$ on an Argand diagram and express $\arg \bar{z}$ in terms of $\arg z$.

13 Given that $|z - 3| = 4$ and $\arg z = \pi/4$, find z.

14 (i) Find the modulus and argument of each of the complex numbers which satisfy the equation $z^2 + 2z + 5 = 3i$

(ii) What may one deduce if

(a) the quotient of two complex numbers is real

(b) the product of two complex numbers is real and their difference is imaginary?

15 Given a complex number z, draw in an Argand diagram the positions of the points

(i) iz (ii) $\bar{z}$ (iii) $1/z$ (iv) z^3

Assume $|z| > 1$.

16 Prove the triangle inequality $|z_1 + z_2| \le |z_1| + |z_2|$

17 Evaluate $(2 + i)^3 - (2 - i)^3$, $(2 + i)^{-2} + (2 - i)^{-2}$

18 Show that $\overline{(\bar{z})} = z$, $\overline{\left[\frac{1}{z}\right]} = 1/\bar{z}$, $(\bar{z})^n = (\overline{z^n})$ by induction and the result $\overline{z_1 z_2} = \bar{z}_1 \bar{z}_2$

19 Show that $\arg(z_1/z_2) = \arg z_1 - \arg z_2$

20 Simplify, giving answer in cartesian form:

(i) $\dfrac{(\sqrt{2}\,\angle 3\pi/4)^2\,(2\angle{-2\pi/3})^2}{(2\angle{-\pi/6})}$ (ii) $\dfrac{(3\angle\pi/3)^3(2\angle{-\pi/4})^5}{(4\angle\pi/2)}$

21 Simplify, giving answer in polar form:

(i) $\dfrac{(3\angle 15^\circ)^2\times(2\angle 40^\circ)^3}{6\angle 25^\circ\times 4\angle 35^\circ}$ (ii) $2\angle\pi/3 + 2\angle 5\pi/6$ (iii) $\dfrac{(-1+\mathrm{i})^2(-1-\mathrm{i}\sqrt{3})^2}{\sqrt{3}-\mathrm{i}}$

Section 11.3

22 Use de Moivre's theorem to obtain $\cos 5\theta$, $\sin 5\theta$ as polynomials in $\cos\theta$, $\sin\theta$ respectively. Hence obtain surd expressions for $\cos 18^\circ$ and $\sin 36^\circ$.

23 Use the definitions of $\cos\theta$, $\sin\theta$ in terms of complex exponentials to show that

$$\cos^4\theta\,\sin^2\theta = \frac{1}{32}\,(2 + \cos 2\theta - 2\cos 4\theta - \cos 6\theta)$$

24 If $z = \cos\theta + \mathrm{i}\sin\theta$, show that for positive integers n, $z^n + (1/z^n) = 2\cos n\theta$ and hence obtain the identity $4\cos^3\theta - 3\cos\theta = \cos 3\theta$

25 (i) If a and b are real and $(a + \mathrm{i}b)\mathrm{e}^{\mathrm{i}(5\pi/6)} + \sqrt{2}\mathrm{e}^{-\mathrm{i}(\pi/4)} = 2 - \mathrm{i}$, find a, b

(ii) If $3\mathrm{e}^{\mathrm{i}\theta} - \mathrm{e}^{-\mathrm{i}\theta} = \sqrt{2} + \mathrm{i}b$, where θ and b are real, show that $\theta = 2k\pi \pm \pi/4$ (k an integer), $b = \pm 2\sqrt{2}$

26 Solve the equation $\sin z = 2$

27 Using de Moivre's theorem, or otherwise, show that

$$\frac{\sin 5\theta}{\sin\theta} = 16\cos^4\theta - 12\cos^2\theta + 1$$

For what values of θ is this result not true?

28 (i) Show, using de Moivre's theorem that

$$\cos 4\theta = \cos^4\theta - 6\cos^2\theta\,\sin^2\theta + \sin^4\theta$$

and

$$\sin 4\theta = 4\sin\theta\cos\theta(\cos^2\theta - \sin^2\theta)$$

If, further, we define $T_n(x) = \cos n\theta$, where $x = \cos\theta$, obtain an expression for $T_4(x)$ in terms of x only, and derive the inverse relation $8x^4 = T_4(x) + 4T_2(x) + 3T_0(x)$

(ii) Draw the curve represented by $|z| = a$ on an Argand diagram. If $Z = z\mathrm{e}^{\mathrm{i}\theta}$, what is the locus of the point Z as z moves on the above curve?
If also $W = z + (a^2/z)$, show that the point W will describe a straight line segment and indicate this on the Argand diagram.

29 $\cos\theta + \mathrm{i}\sin\theta$ is sometimes written cis θ. Find cis |30°|.

30 Using the exponential form of z, produce a suitable interpretation of log z; what difficulties does your definition give?

31 Demonstrate that $e^{i\pi} = -1$. What is a suitable value for $e^{i\pi/2}$?

32 Show that $\sin(iz) = i \sinh z$, $\cos(iz) = \cosh z$. Find $\cosh(2 + i)$.

33 The current entering a telephone line is the real part of

$$\frac{\cos \omega t + i \sin \omega t}{\cosh(s + is)}$$

Express this in the form $A \sin(\omega t + \alpha)$.

34 By taking logarithms show that a suitable value for i^i is $e^{-\pi/2}$.

35 (i) Determine the cube roots of $\sqrt{3} - i$, and express them in cartesian form.

(ii) Determine the fourth roots of $z = \dfrac{(-2 + i)^3(-2 - 3i)^3}{(4 - 3i)^2}$ expressing them in the cartesian form.

36 (i) Find the complex number z which satisfies the equation

$$(1 + 2i)z + (3 + i)\bar{z} = \frac{5(1 - i)}{2 + i}$$

(ii) If $x = 1 + i$ is a root of the quartic equation $x^4 + ax^3 + bx^2 + 8 = 0$, where a and b are real coefficients, find the other roots and determine the values of a and b.

37 Solve the equations (i) $z + z^5 = 0$
(ii) $z^6 - z^3 - 12 = 0$

38 (i) Find the cube roots of $\sqrt{3} + i$. (ii) Find the fourth roots of $3 + 4i$.

39 (i) Show that the cube roots of a complex number may be written in the form z_1, ωz_1, $\omega^2 z_1$ (where $\omega = e^{2\pi i/3}$).

(ii) Show that the fourth roots of a complex number may be written as $\pm z_2$, $\pm i z_2$.

40 Verify directly that $1 + \omega + \omega^2 = 0$.

41 Using the fact that the nth roots of 1 are spaced equally round the unit circle with one root of $z = 1$, verify that the fourth roots of 1 are 1, i, –1, –i. Find the fifth roots of 1. Label the first root anti-clockwise from $z = 1$ as γ. Do the other roots have a simple relationship to γ? Are there any relations akin to $1 + \omega + \omega^2 = 0$? Can you say anything in general about the nature of the nth roots of 1? Consider the cases n even and n odd.

Section 11.4

42 (i) If $z_2 = \rho i z_1$ with ρ real and positive prove algebraically that $|z_1|^2 + |z_2|^2 = |z_1 - z_2|^2$ and interpret this result geometrically in an Argand diagram.

(ii) If the point P represents the complex number z in an Argand diagram and $\arg[(z-3)/(z-1)] = \pm\pi/2$, show that P lies on the circle of unit radius with centre at the point (2, 0).

43 (i) The centre of an equilateral triangle is represented by $-1 + i$ and one vertex by $3 + 5i$. Find the complex numbers representing the other vertices.

(ii) If the complex number z is such that the expression $(z - 1)/(z + i)$ is wholly imaginary, show that the locus of z in the Argand diagram is a circle with centre at the point $(1/2) - (1/2)i$ and radius $(1/2)\sqrt{2}$.

44 Plot on an Argand diagram $7 + 3i$, $6i$, $-3 - i$ and show that they are vertices of a square; find the fourth vertex.

45 Sketch the domains

(i) $|z| \leq 4$ (ii) $|\arg z| < \pi/6$

(iii) $(|z-1|/|z-2|) < 1$ (iv) $0 < \arg[(z-2)/(z-1)] < \pi/2$

46 Evaluate iz in polar form and show that the effect of multiplying a complex number by i is to rotate the line joining it to the origin by 90° anticlockwise. Hence prove $i^2 = -1$.

47 If $z(\equiv x + iy)$ is a complex number and $w = (z-2)/(z-i)$

(a) show that when the point in the Argand diagram represented by w moves along the real axis, z traces a straight line through (2, 0) and (0, 1)

(b) determine $|z - 1 - \frac{1}{2}i|$ when w lies on the imaginary axis. What is the locus of z as w moves along the imaginary axis?

48 Determine the region in the Argand plane defined by $|z - 1| + |z - i| < 4$

49 Show that the equation of any circle in the complex plane may be written in the form $z\bar{z} + \alpha z + \bar{\alpha}\bar{z} + a = 0$ where a is a real constant and α is (in general) a complex constant. Express in this form the equation of the circle passing through the points $1 - i$, $2i$, $1 + i$ and find its radius and the position of its centre.

50 (cos 60° + i sin 60°) is represented by a line segment OP. What number is represented by OP turned through 20° and halved in length?

51 Determine the locus $|z + 3i|^2 - |z - 3i|^2 = 12$

Section 11.5

52 Develop the solution for the circuit of Figure 11.10(i). You will find it easier to work in terms of the *complex admittance*, which is the reciprocal of the complex impedance.

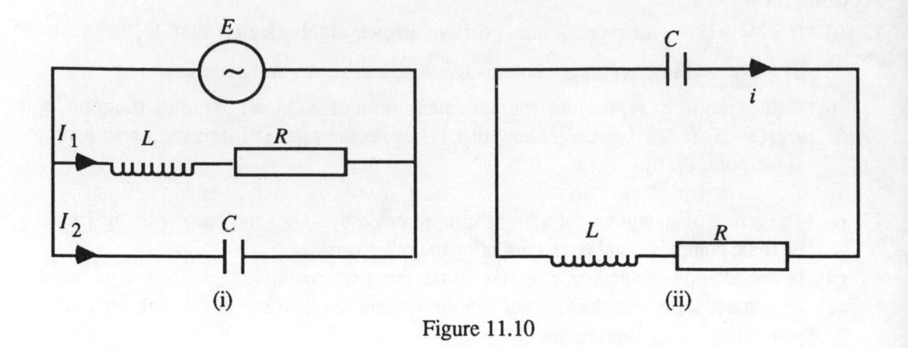

Figure 11.10

53 For a circuit shown in Figure 11.10(ii) with no e.m.f. generator, allow the current to be represented by $i = I_0\ e^{j\omega t}$. Find an equation for ω by equating the total e.m.f. to zero. Write down the solution for i. The effect of the factor $e^{-Rt/2L}$ denotes attenuation; explain what happens if $R = 0$. Distinguish between the cases $4C - R^2 > 0, = 0, < 0$.

12

DIFFERENTIATION

Consider the mechanism shown in Figure 12.1(i). It is a piston connected to a crank OQ by a connecting rod PQ; when the crank rotates about O with constant angular velocity ω the piston moves in a straight line. We wish to discover how the piston moves. To do this we must first find a relationship between the linear displacement of the piston and the angle turned through by the crank. We use the schematic diagram of Figure 12.1(ii).

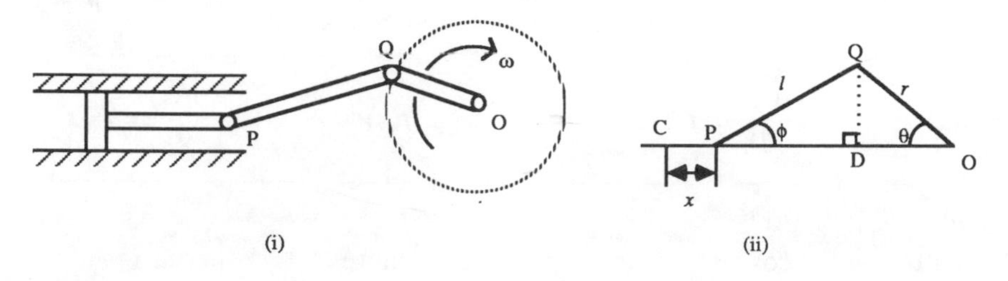

Figure 12.1

When PQO is a straight linc ($\theta = \phi = 0$) then the piston is as far to the left as it will go and P is in the position C. This is called the *dead centre* position. The length PQO is then $(l + r)$ and, at that stage, $x = 0$; hence $CO = l + r$. As θ increases, so the displacement x increases until θ reaches π when x has a maximum of $2r$; thereafter x decreases to zero when $\theta = 2\pi$, whence the motion repeats every 2π. In Figure 12.1(ii), the distance PO is $l \cos \phi + r \cos \theta$ and so $x = l + r - (l \cos \phi + r \cos \theta)$. However from triangles PQD and OQD we can see that $l \sin \phi = r \sin \theta$ therefore

$$\cos \phi = \sqrt{1 - \frac{r^2}{l^2} \sin^2 \theta}$$

It is convenient at this point to let the ratio l/r be denoted by m; usually, in practice, $l >> r$. Then we have the equation

$$x = r(1 - \cos \theta + m - \sqrt{m^2 - \sin^2 \theta}) \tag{12.1}$$

If we wish to determine the velocity dx/dt and acceleration d^2x/dt^2 of the piston for the purpose of estimating the force on it we must clearly have the expertise to differentiate the right-hand side of equation (12.1). For this purpose, we need some general techniques of differentiation.

In Chapter 3 we discussed the concept of differentiation but we only learned how to find the derivatives of simple expressions. Now we shall study further techniques of differentiation, returning to the piston problem on pages 311 and 316.

12.1 Techniques of Differentiation

First, we provide a table of standard derivatives, Table 12.1; you should familiarise yourselves with the results therein. They can all be derived from first principles, but in the problems at the end of this chapter some other ways of obtaining them are indicated.

Table 12.1
Table of standard derivatives

$f(x)$	$\frac{df(x)}{dx} \equiv f'(x)$	$f(x)$	$\frac{df(x)}{dx} \equiv f'(x)$
x^n	nx^{n-1}		
$\sin x$	$\cos x$	$\sinh x$	$\cosh x$
$\cos x$	$-\sin x$	$\cosh x$	$\sinh x$
$\tan x$	$\sec^2 x$	$\tanh x$	$\text{sech}^2 x$
$\text{cosec}\, x$	$-\text{cosec}\, x \cot x$	$\ln x$	$\frac{1}{x}$
$\sec x$	$\sec x \tan x$	$\ln[f(x)]$	$f'(x)/f(x)$
$\cot x$	$-\text{cosec}^2 x$	e^x	e^x
$\sin^{-1} x$	$\frac{1}{\sqrt{1-x^2}}$	$\sinh^{-1} x$	$\frac{1}{\sqrt{x^2+1}}$
$\cos^{-1} x$	$\frac{-1}{\sqrt{1-x^2}}$	$\cosh^{-1} x$	$\frac{1}{\sqrt{x^2-1}}$
$\tan^{-1} x$	$\frac{1}{1+x^2}$		

In the case of inverse functions the range is restricted.

Rules for combinations of functions

These rules come directly from the corresponding rules for limits of sequences; they should really be applied at a point x_0, but for practical purposes we need to assume that the functions are *differentiable in some interval of x*. Also we should confine the term *derivative* to a point, e.g. $f'(x_0)$ and speak of the *derived function* $f'(x)$ but we shall sometimes, for brevity, call $f'(x)$ the *derivative of f(x)*.

For these rules we assume that $f(x)$ and $g(x)$ are differentiable functions in some interval I and that α is some number. Then

(i)
$$\frac{d}{dx}[f(x) \pm g(x)] = f'(x) \pm g'(x) \qquad (12.2)$$

e.g.

$$y = x^2 + \sin x; \quad \frac{dy}{dx} = \frac{d}{dx}(x^2) + \frac{d}{dx}(\sin x) = 2x + \cos x$$

(ii)
$$\frac{d}{dx}[\alpha f(x)] = \alpha f'(x) \qquad (12.3)$$

e.g.

$$y = \sqrt{\pi}\cos x; \quad \frac{dy}{dx} = \sqrt{\pi}\frac{d}{dx}(\cos x) = -\sqrt{\pi}\sin x$$

(iii)
$$\frac{d}{dx}[f(x).g(x)] = f'(x).g(x) + f(x).g'(x) \qquad (12.4)$$

e.g.

$$y = e^x \tan x; \quad \frac{dy}{dx} = \frac{d}{dx}(e^x)\tan x + e^x\frac{d}{dx}(\tan x) = e^x(\tan x + \sec^2 x)$$

(iv)
$$\frac{d}{dx}\left[\frac{f(x)}{g(x)}\right] = \frac{f'(x)g(x) - f(x)g'(x)}{[g(x)]^2} \qquad (12.5)$$

provided $g(x) \neq 0$ in I,

e.g.

$$\frac{d}{dx}[(\ln x)/x] = [(1/x).x - \ln x.1]/x^2 = (1 - \ln x)/x^2$$

provided I does not include $x = 0$.

Try to obtain the special case of this **Quotient Rule** when $f(x) \equiv 1$.

Example

Let us demonstrate these rules for the function $h(x) = \dfrac{(x+2)(x^2 - 2x + 1)}{(4x - 4)}$

By rules (12.2) and (12.3) the derivative of $(x + 2)$ is 1,

and that of $(x^2 - 2x + 1)$ is $(2x - 2)$ and that of $(4x - 4)$ is 4.
Then, by rule (12.4) the derivative of the numerator is

$$1.(x^2 - 2x + 1) + (x + 2)(2x - 2)$$

that is
$$(3x^2 - 3)$$

Finally, using rule (12.5) we obtain

$$h'(x) = \frac{(3x^2 - 3)(4x - 4) - (x + 2)(x^2 - 2x + 1).4}{(4x - 4)^2}$$

$$= \frac{12(x^2 - 1)(x - 1) - 4(x + 2)(x - 1)^2}{16(x - 1)^2}$$

$$= \frac{3(x + 1) - (x + 2)}{4}$$

$$= \frac{2x + 1}{4}$$

Of course, this result could have been, and should have been, obtained more easily had we noticed that $h(x) = \frac{1}{4}(x + 2)(x - 1) = \frac{1}{4}(x^2 + x - 2)$. (For what value of x is our result invalid?)

Chain Rule

When we are dealing with the composition of two functions the chain rule is applied. This rule is as follows: *if y is a differentiable function of u and u is a differentiable function of x, then y is a differentiable function of x and*

$$\frac{\mathrm{d}y}{\mathrm{d}x} = \frac{\mathrm{d}y}{\mathrm{d}u} \cdot \frac{\mathrm{d}u}{\mathrm{d}x} \qquad (12.6)$$

Example 1

Consider $y = (2x^2 + 3)^3 + 2$

If we put $u = 2x^2 + 3$ then $y = u^3 + 2$ so that $\mathrm{d}u/\mathrm{d}x = 4x$ and $\mathrm{d}y/\mathrm{d}u = 3u^2$. The rule tells us that

$$\frac{\mathrm{d}y}{\mathrm{d}x} = 3u^2.4x = 3(2x^2 + 3)^2\, 4x = 12x(2x^2 + 3)^2$$

You should first see how this result could have been obtained in one step and then check the answer by expanding the original expression for y and differentiating.

Example 2

We now consider $h(x) = \sin^2(3x^3 + 1)$. We write $u = 3x^3 + 1$ and note that $\sin^2 u \equiv (\sin u)^2$. Then $\mathrm{d}h/\mathrm{d}u = 2 \sin u.\cos u$ by one application of the rule and by another $dh/dx = 2 \sin u \cos u.9x^2 = (\sin 2u).9x^2 = [\sin 2(3x^3 + 1)].9x^2$.

Example 3
Finally, we return to the motion of the piston discussed at the beginning of this chapter. We have equation (12.1), i.e. $x = r(1 - \cos\theta + m - \sqrt{m^2 - \sin^2\theta})$ so that x is a function of θ where θ itself is a function of time, t.

Then by the chain rule $dx/dt = dx/d\theta.d\theta/dt$, where $d\theta/dt$ is the angular velocity which has constant magnitude, ω. The tricky part in finding $dx/d\theta$ is to differentiate $\sqrt{m^2 - \sin^2\theta} \equiv (m^2 - \sin^2\theta)^{1/2}$; applying the chain rule, the derivative of this is

$$\frac{1}{2}.\frac{-2\sin\theta.\cos\theta}{(m^2 - \sin^2\theta)^{1/2}} \equiv \frac{-\sin 2\theta}{2\sqrt{m^2 - \sin^2\theta}}$$

Hence the velocity of the piston

$$v \equiv \frac{dx}{dt} = r\left[\sin\theta + \frac{\frac{1}{2}\sin 2\theta}{\sqrt{m^2 - \sin^2\theta}}\right].\omega \tag{12.7}$$

Try to sketch a graph of v against θ; when $\theta = \pi/2$, $v = \omega r$ and you could show that this is the maximum value of v. (When is v a minimum?) It should be observed that ω is measured in radians/sec; a dimensional check on equation (12.7) will verify this. We consider this problem further on page 316.

Note that the chain rule can be extended to more than one stage: see Problem 9.

Inverse Functions
In the special case of the chain rule where y is a function of u and u is a function of y we have $1 = (dy/du).(du/dy)$ and so

$$\frac{du}{dy} = 1\bigg/\frac{dy}{du} \tag{12.8}$$

This is **not** a case of treating dy/du as a fraction; nor was this so for the general chain rule.

Examples

(i) If $y = u^3$, $u = y^{1/3}$ and u is a function of y so that

$$\frac{du}{dy} = \frac{1}{3u^2} = \frac{1}{3y^{2/3}}$$

as we might expect.

(ii) If $y = \sin 3u$, $u = (1/3)\sin^{-1}y$ and if we restrict the function y to the domain $-\pi/6 \le u \le \pi/6$, u is a function of y with

$$du/dy = 1/(3 \cos 3u) = 1/(3\sqrt{1-y^2})$$

(iii) If $y = u^2$, $u = y^{1/2}$ and $1/(dy/du) = 1/2u$.

Logarithmic Differentiation

It can be shown that the definition of the exponential function e^x given in Section 2.3 is equivalent to the definition that e^x satisfies the identity $f'(x) = f(x)$ with $f(0) = 1$. Since the inverse of the function $x = e^u$ is the function $u = \ln x$ it follows that

$$\frac{d}{dx}(\ln x) = \frac{1}{e^u} = \frac{1}{x}$$

This, together with the rules for logarithms, allows us to differentiate products and quotients of functions more easily. We need the two additional results

$$\frac{d}{dx}[\ln f(x)] = \frac{f'(x)}{f(x)} \quad \text{and} \quad \frac{d}{dx}\ln y = \left[\frac{d}{dy}\ln y\right]\frac{dy}{dx} = \frac{1}{y}\frac{dy}{dx}$$

both of which follow from the chain rule, as you can verify.

Example

$$y = \frac{(2x+3)^2(x-4)^3}{(7x-2)^2(x^2+2x+2)}$$

Now,

$$\begin{aligned} \ln y &= \ln(2x+3)^2 + \ln(x-4)^3 - \ln(7x-2)^2 - \ln(x^2+2x+2) \\ &= 2\ln(2x+3) + 3\ln(x-4) - 2\ln(7x-2) - \ln(x^2+2x+2) \end{aligned}$$

Differentiating both sides of this equation with respect to x we obtain

$$\frac{1}{y}\frac{dy}{dx} = 2.\frac{2}{2x+3} + \frac{3}{x-4} - 2.\frac{7}{7x-2} - \frac{2x+2}{x^2+2x+2}$$

Hence we may obtain dy/dx. Check the example on page 309 by this method.

Implicit Differentiation

Often we have an equation of the form $f(x, y) = 0$ where it would be difficult or impossible to write it as $y = g(x)$ or where perhaps so doing would lead to a complicated expression. In such cases, it is better to differentiate the equation as it stands.

Example 1

$$y^2 + 2ay + xy + \sin y = 4$$

Differentiating with respect to x, we obtain the following equation

$$2y\frac{dy}{dx} + 2a\frac{dy}{dx} + 1.y + x\frac{dy}{dx} + \cos y\frac{dy}{dx} = 0$$

therefore
$$\frac{dy}{dx} = \frac{-y}{2y + 2a + x + \cos y}$$

Example 2

The circle $x^2 + y^2 = a^2$ gives on differentiating w.r.t. x $\quad 2x + 2y\dfrac{dy}{dx} = 0$

Therefore,
$$\frac{dy}{dx} = -\frac{x}{y}$$

This result gives the gradient of the tangent to the circle at any point. Take a point on the circle from each of the four quadrants and check that the slope has the right sign; qualitatively check the slope as x moves from 0 to a in the first quadrant.

The slope at the point (x_1, y_1) is $(-x_1/y_1)$ and the equation of the tangent to the circle at that point is $y - y_1 = (-x_1/y_1)(x - x_1)$, that is,

$$yy_1 - y_1^2 + x_1x - x_1^2 = 0$$

or
$$yy_1 + xx_1 = x_1^2 + y_1^2 = a^2$$

which is the standard equation.

Parametric Differentiation

When a curve is given in parametric form, e.g. $x = x(t)$, $y = y(t)$, it is often not wise to eliminate the parameter t (even if this is possible) before differentiating to find dy/dx. We consider first the parabola $y^2 = 4ax$, which in parametric form is $x = at^2$, $y = 2at$. Now it is straightforward to find $2y(dy/dx) = 4a$ and hence $dy/dx = 2a/y$.

But $dx/dt = 2at$ and $dy/dt = 2a$, therefore

$$\frac{dy}{dx} = \frac{dy}{dt} \times \frac{dt}{dx} = \frac{dy}{dt}\bigg/\frac{dx}{dt} = \frac{2a}{2at} = \frac{1}{t}$$

we see that we get the same result as before; however, t is **not** a function of x and we need care in interpreting this result. In this case there should be little difficulty.

Example

The equations for a cycloid are $x = a(\theta - \sin\theta)$, $y = a(1 - \cos\theta)$. Here we have

$$\frac{dx}{d\theta} = a(1 - \cos\theta), \quad \frac{dy}{d\theta} = +a\sin\theta \quad \text{and} \quad \frac{dy}{dx} = \frac{\sin\theta}{1 - \cos\theta} = \cot\frac{\theta}{2}$$

It would not be easy to express y as a function of x and less easy to differentiate the resulting expression. Is our result for dy/dx subject to misinterpretation? Compare the slope from Figure 12.2 with the formula for dy/dx. In this example the parameter θ has a physical significance. A **cycloid** is the path traced out by a point on the circumference of a wheel of radius a if the point is initially in contact with the ground; θ is the angle turned through by the wheel.

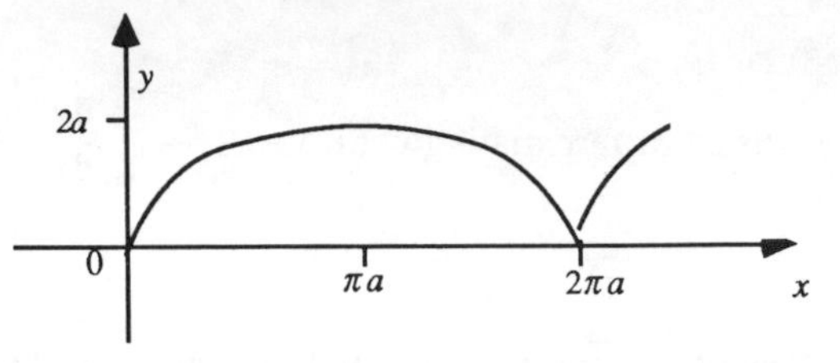

Figure 12.2

12.2 Maximum and Minimum Values of a Function

We gave the definitions of *increasing* and *decreasing* regions of a function and also of a *local minimum* in Section 7.4. For completeness we state the following rules:

(i) If $f'(x)$ is zero at a point x_0, then x_0 is said to be a **turning point** of $f(x)$

(iia) If $f'(x_0-) < 0$ and $f'(x_0+) > 0$ the turning point is a **local minimum** [e.g. $f(x) = x^2$ at $x_0 = 0$]
[Remember that $f'(x_0-)$ means the derivative of $f(x)$ at the point $(x_0 - \varepsilon)$ where ε is a small positive number.]

(iib) If $f'(x_0-) > 0$ and $f'(x_0+) < 0$ the turning point is a **local maximum** [e.g. $f(x) = 1 - x^2$ at $x_0 = 0$]

(iic) If $f'(x_0-)$ and $f'(x_0+)$ have the same sign then the turning point is a **Point of Inflection** [e.g. $f(x) = x^3$ at $x_0 = 0$]

We shall return to the case of points of inflection in detail in the next section.

Example
Consider $f(x) = x^3 - 12x + 4$. We can easily find $f'(x) = 3x^2 - 12$. It should be obvious that $f'(x) = 0$ when $x = \pm 2$, so these are the turning points. Since $f(x)$ is *continuous* we can take x_0- and x_0+ at convenient points between these two values of x without having to worry about being too close to

± 2 (we shall not lose local effects here). Now $f'(-3) > 0$ and $f'(-1) < 0$ hence $x = -2$ is a *local maximum*; likewise $f'(1) < 0$ and $f'(3) > 0$ so that $x = 2$ is a *local minimum*.

The values of $f(x)$ at the turning points are called **stationary values** and in this example they are $f(-2) = 20$ and $f(2) = -12$. A sketch of $f(x)$ is shown in Figure 12.3. What other information was needed for the sketch?

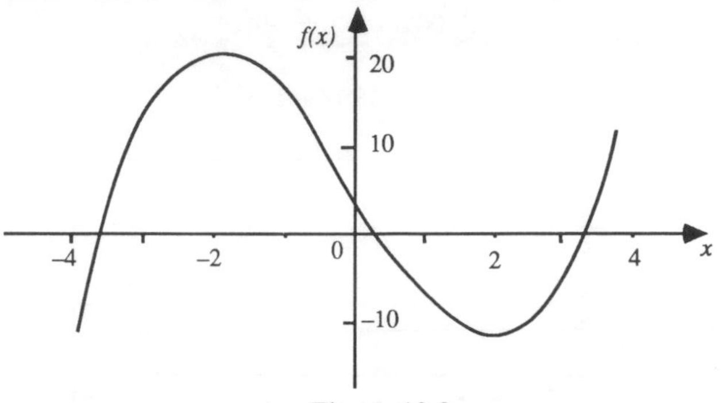

Figure 12.3

It should be borne in mind that we have so far mentioned *local* maxima and *local* minima; these are classed as *local* **extrema** and may not coincide with the *overall* extreme values of the function. In the example above, if we seek the greatest and least values of $f(x)$ in the domain $[-3, 4]$ we find these occur at $x = 4$ and $x = -2$ respectively; in this case, the local maximum is achieved at an end-point of that interval. It is a general result that if an overall maximum (or minimum) does not coincide with its local counterparts then it will occur at an end-point of the interval in question, if this is finite. (Why should this be so? What happens if the interval is not finite? Try to find some relevant examples.)

In general, to find overall extreme values, it is advisable to evaluate $f(x)$ at points where $f'(x) = 0$ and at the end-points of the relevant interval and also at points where $f'(x)$ does not exist (remember $f(x) = |x|$?); then we sift through all these values of $f(x)$ to find the greatest and the least ones.

Example 1

Find the greatest value of $\dfrac{1}{13 - 6x + x^2}$

We can rearrange the expression to $\dfrac{1}{4 + (x-3)^2}$

and it will take its greatest value when the denominator is least; this being the sum of two positive quantities is least when $(x - 3)^2 = 0$, i.e. when $x = 3$. The value of the original expression is then $1/4$.

A second situation is where the determination of the nature of any turning point can be achieved by physical reasoning.

Example 2

Find the maximum volume of a lidless rectangular box cut from a sheet of metal measuring 80 cm by 40 cm (see Figure 12.4).

If the depth of the box is x cm, then the cross-sectional area of its base is $(40 - 2x)(80 - 2x)\text{cm}^2$. The volume of the box is therefore

$$V(x) = (40 - 2x)(80 - 2x)x$$
$$= 3200x - 240x^2 + 4x^3$$

Hence

$$V'(x) = 3200 - 480x + 12x^2$$

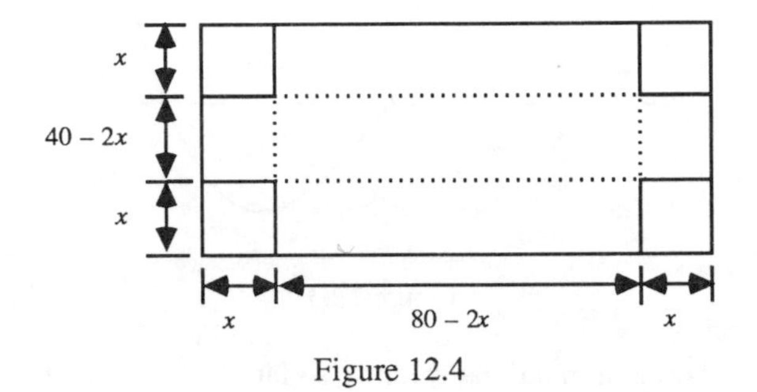

Figure 12.4

Now $V'(x) = 0$ when $4(3x^2 - 120x + 800) = 0$, i.e. when $x = 20[1 \pm (1/\sqrt{3})]$. On physical grounds we reject the + sign since this would mean $40 - 2x < 0$ and we note that this is impossible by reference to Figure 12.4. It is easy to see that x is confined to the range $0 \leq x \leq 20$ and, since $x = 0$ or $x = 20$ give zero volume, it follows that $x = 20\,[1 - (1/\sqrt{3})]$ will provide the maximum volume.

Example 3

Let us return to equation (12.7): $v = \omega r(\sin\theta + \frac{1}{2}\sin 2\theta/\sqrt{m^2 - \sin^2\theta})$; we see that when $\theta = 0$ or π or 2π, $v = 0$ and hence x has local extreme values; check that these correspond to extreme positions of the piston. In fact, $v = 0$ when

$$\sin\theta + \frac{\frac{1}{2}\sin 2\theta}{\sqrt{m^2 - \sin^2\theta}} = 0, \quad \text{i.e. } \sqrt{m^2 - \sin^2\theta}\,.\sin\theta + \sin\theta\cos\theta = 0$$

which reduces to $\sin\theta[\sqrt{m^2 - \sin^2\theta} + \cos\theta] = 0$

Thus *either* $\sin\theta = 0$

or $\sqrt{m^2 - \sin^2\theta} = -\cos\theta$

i.e. $m^2 - \sin^2\theta = \cos^2\theta$

i.e. $$m^2 = \cos^2\theta + \sin^2\theta$$

i.e. $$m^2 = 1$$

If $m = -1$ then, remembering that $m = l/r$, we find that $l = -r$, which is nonsense since l and r are lengths, and must be positive. If $m = 1$, then $l = r$. *However*, it can be reasoned on physical grounds that m must be at least 2. The case $m = -1$ has arisen because at one stage we squared an equation in order to obtain a solution for m. This is a spurious root arising from algebraic manipulation. On the other hand, the case $m = 1$ is a genuine mathematical solution but in the context of this particular problem is physically inadmissible. Thus this example highlights the danger of not referring back to the original physical problem. We are left with the only solution $\sin\theta = 0$ which yields $\theta = 0$, π, 2π as claimed earlier.

Curve Sketching

We are now in a position to add more detail to the sketching of curves. Consider, for example, $y^2 = x^5$. We have symmetry about the x-axis; x cannot be negative; the only intersection with either axis is at the origin; for large values of x, y is larger than x. Now we find that

$$\frac{dy}{dx} = \pm\frac{5}{2}x^{3/2}$$

which means that as x increases, the slope increases in magnitude from its value of zero at the origin. The feature there is called a **cusp**, as shown in Figure 12.5.

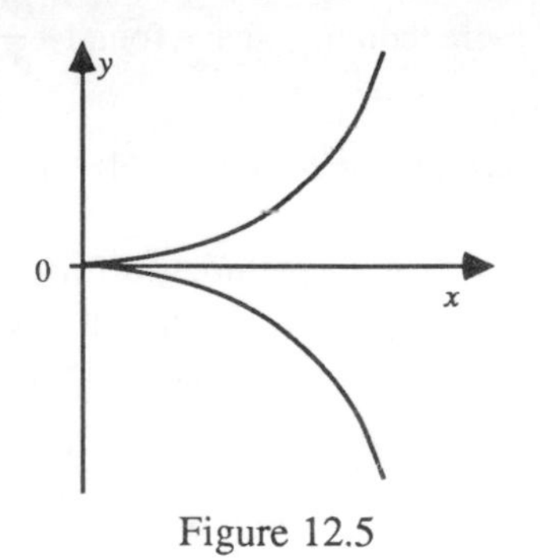

Figure 12.5

12.3 Higher Derivatives

We have seen that a tabulated function has second, third and higher differences, so a function given by the rule $f(x)$ has higher derivatives $f''(x), f'''(x)$, etc. *if they exist.* The nth order derivative is denoted $f^{(n)}(x)$ or d^ny/dx^n.

Some functions are *infinitely differentiable*, i.e. they possess derivatives of all orders. Examples are

(i) finite polynomials which at some stage have a zero derivative and further differentiation yields zero on each subsequent occasion;

(ii) $\sin x$ for which the derivatives have a cyclic pattern: $\cos x$, $-\sin x$, $-\cos x$, $\sin x$, ...

There are examples where each time we differentiate a function it becomes more badly behaved and at some stage we may obtain a non-differentiable function: for example $f(x) = |x|$ gives $f'(x)$ which is not defined at $x = 0$.

Many of the techniques we have developed so far carry through for higher derivatives.

Examples

(i) We found on differentiating implicitly the equation $x^2 + y^2 = a^2$ that $2x + 2y(\mathrm{d}y/\mathrm{d}x) = 0$. If we repeat the process we obtain

$$2 + 2\left[\frac{\mathrm{d}y}{\mathrm{d}x}\right]\frac{\mathrm{d}y}{\mathrm{d}x} + 2y\frac{\mathrm{d}^2y}{\mathrm{d}x^2} = 0$$

On substituting for $\mathrm{d}y/\mathrm{d}x$, $2 + 2\left[\frac{-x}{y}\right]^2 + 2y\frac{\mathrm{d}^2y}{\mathrm{d}x^2} = 0$

whence $$\frac{\mathrm{d}^2y}{\mathrm{d}x^2} = -\frac{1 + (x^2/y^2)}{y} = -\frac{a^2}{y^3}$$

It would have been more tedious to have found $\frac{\mathrm{d}y}{\mathrm{d}x} = \frac{-x}{\sqrt{a^2 - x^2}}$

and differentiated this equation to find $\mathrm{d}^2y/\mathrm{d}x^2$.

(ii) For the parabola $y^2 = 4ax$, in parametric form we had $\mathrm{d}y/\mathrm{d}x = 1/t$. In general, if we seek $\mathrm{d}^2y/\mathrm{d}x^2$ we note that this is $(\mathrm{d}/\mathrm{d}x)(\mathrm{d}y/\mathrm{d}x)$ but since $\mathrm{d}y/\mathrm{d}x$ is a function of t, we apply the chain rule so that

$$\frac{\mathrm{d}^2y}{\mathrm{d}x^2} = \frac{\mathrm{d}}{\mathrm{d}t}\left[\frac{\mathrm{d}y}{\mathrm{d}x}\right]\cdot\frac{\mathrm{d}t}{\mathrm{d}x} \tag{12.9}$$

Hence

$$\frac{\mathrm{d}^2y}{\mathrm{d}x^2} = \frac{\mathrm{d}}{\mathrm{d}t}\left[\frac{1}{t}\right]\cdot\frac{\mathrm{d}t}{\mathrm{d}x} = \left[\frac{-1}{t^2}\right]\cdot\frac{1}{2at} = -\frac{1}{2at^3}$$

[A common error is to say that

$$\frac{\mathrm{d}^2y}{\mathrm{d}x^2} = \frac{\mathrm{d}^2y}{\mathrm{d}t^2}\bigg/\frac{\mathrm{d}^2x}{\mathrm{d}t^2}$$

This is **absolutely wrong** – we emphasise that d^2y/dx^2 means *differentiate y w.r.t. x twice.*]

Again we must take care in interpreting the result.

(iii) What interpretation can we put on the second derivative? A physical example is where x represents displacement and t time; then d^2x/dt^2 is the rate of change of dx/dt with time, i.e. it is an acceleration. With the piston example we may differentiate equation (12.7) to obtain

$$\frac{d^2x}{dt^2} = \omega^2 r\left[\cos\theta + \frac{\sin^4\theta + m^2\cos 2\theta}{(m^2 - \sin^2\theta)^{3/2}}\right] \qquad (12.10)$$

Check this for yourself. Note that the knowledge of this acceleration would allow us to evaluate the forces involved by using Newton's Laws.

Geometrical Interpretation of the Second Derivative

Geometrically d^2y/dx^2 represents the rate of change of slope of $y = f(x)$ with x [since $d^2y/dx^2 = d/dx(dy/dx)$].

To consider the applications of this idea see Figure 12.6.

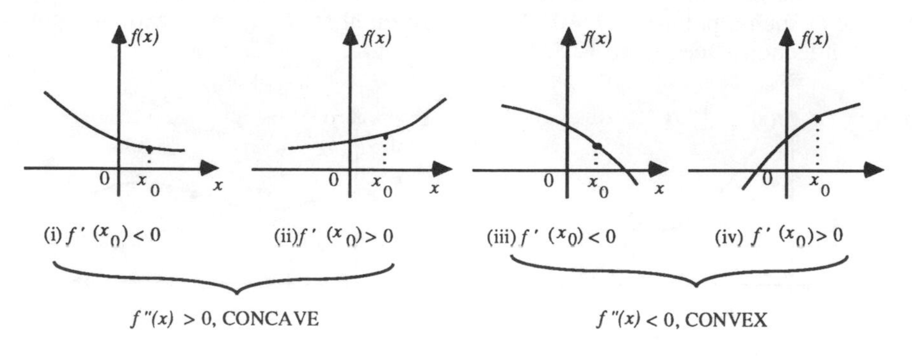

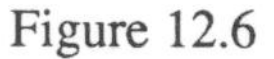
Figure 12.6

We see that $f''(x)$ gives the *local* sense of *concavity* of a function. By matching (i) and (ii) we see that a local minimum, $f''(x_0) \geq 0$ and by matching (iii) and (iv) at a local maximum, $f''(x_0) \leq 0$.

These provide alternative rules for determination of a turning point.

But in the case (iic) on page 314 we had a point where $f'(x)$ kept the same sign as it passed through the turning point. To see what implication this has on $f''(x_0)$ consider the following examples.

Example 1

Consider $f(x) = x^3$. Now $f'(x) = 3x^2$, $f''(x) = 6x$. At $x = 0$, $f'(x) = 0$; hence we have a turning point. If we now look at the expression for $f''(x)$ we find this is also zero at $x = 0$. The graph of $f(x) = x^3$, Figure 12.7(i), shows that $f'(x) = 0$ at the origin. Comparing with the diagrams of Figure 12.6 we see that the graph is made up from the two parts shown in Figure 12.7(ii). As we pass through the origin, the sense of concavity changes and $f''(x)$ changes from negative to positive. Thus, at the origin we must have $f''(x) = 0$ as we have found algebraically.

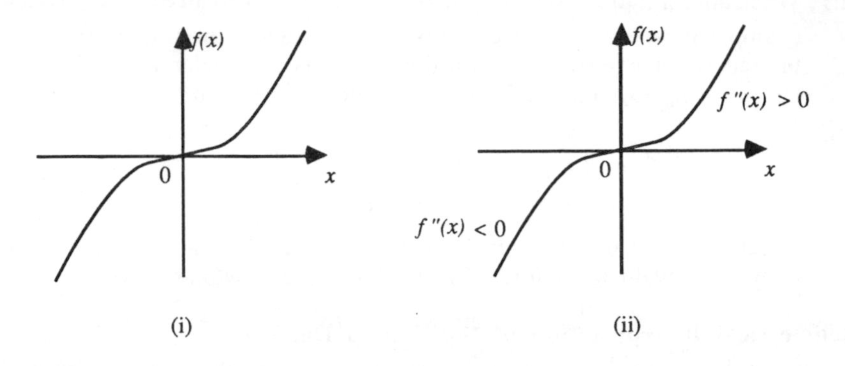

Figure 12.7

We define a *point of inflection* as being a point *where the concavity changes* and at such a point $f''(x) = 0$. It is not essential for $f'(x)$ to be zero at a point of inflection – see Figure 12.8.

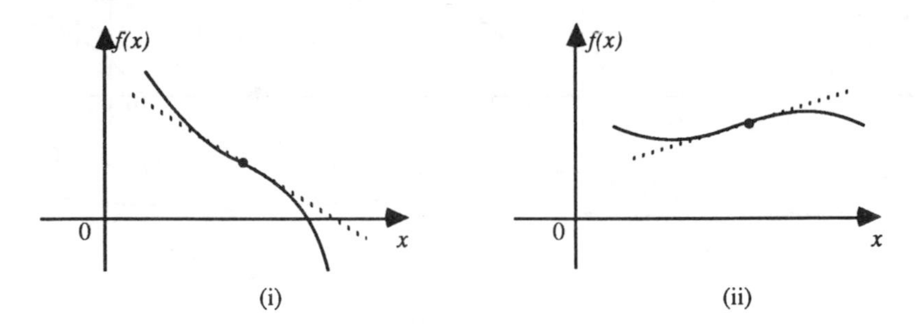

Figure 12.8

It is to be noted that where $f'(x) = 0$ at the same point, trouble can occur in identifying the nature of the turning point.

Example 2

Consider $f(x) = x^4$; then $f'(x) = 4x^3$ and $f''(x) = 12x^2$. Now both $f'(x)$ and $f''(x)$ vanish at $x = 0$, yet from Figure 12.9 we can see that, in fact, there is a local minimum there.

For the time being, we advise that where both first and second derivatives vanish at a value of x it is best to examine $f'(x)$ on either side of this value; we return with a better explanation of the situation in Section 16.5.

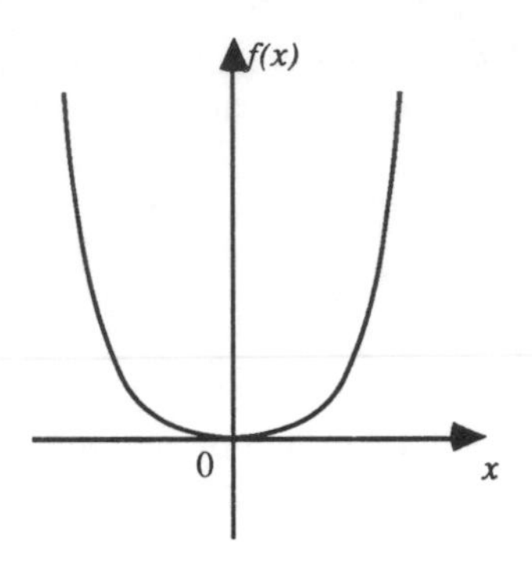

Figure 12.9

Example 3
Figure 12.10 shows an encastre beam which is fixed into a supporting structure so that the slope at each end is zero. We denote distance from the left-hand wall by x and the downwards deflection of that point from the level by y.

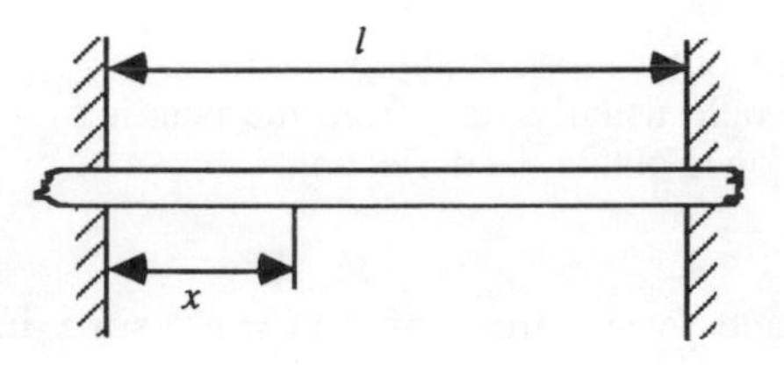

Figure 12.10

A uniformly distributed load w/unit length is applied to the beam. It can be shown that the equation for the deflection, *assumed small*, is

$$EIy = \frac{w}{24}(x^4 - 2lx^3 + l^2x^2)$$

where EI is the (constant) flexural rigidity of the beam. Check that this formula gives $y(0) = y(l) = 0$. On differentiating, we find that

$$EI\frac{\mathrm{d}y}{\mathrm{d}x} = \frac{w}{24}(4x^3 - 6lx^2 + 2l^2x) = \frac{2w}{24}x(2x^2 - 3lx + l^2)$$

Hence $\mathrm{d}y/\mathrm{d}x = 0$ at $x = 0$ or where $2x^2 - 3lx + l^2 = 0$, i.e. $x = 0$ or $(x - l)(2x - l) = 0$, i.e. $\mathrm{d}y/\mathrm{d}x = 0$ at $x = 0$, $x = l$ or $x = l/2$.

A second differentiation yields

$$-M = EI\frac{\mathrm{d}^2y}{\mathrm{d}x^2} = \frac{w}{12}(6x^2 - 6lx + l^2)$$

which at $x = l/2$ implies $d^2y/dx^2 > 0$ and hence a local minimum; this is where the deflection (measured downwards) is greatest, as we might have expected from symmetry.

Now $d^2y/dx^2 = 0$ when $6x^2 - 6lx + l^2 = 0$, that is,

$$x = \frac{1}{6}(3 \pm \sqrt{3})$$

$$\cong 0.79l \text{ and } 0.21l$$

These values give the points of inflection and are points where, *under the assumption of small deflection*, the bending moment, M, is zero. (Note they are symmetrically placed about $x = l/2$.) A further differentiation will give

$$-F = EI\frac{d^3y}{dx^3} = \frac{w}{2}(2x - l)$$

where F is the shear force. Finally,

$$EI\frac{d^4y}{dx^4} = w$$

which is a statement about the load at any point.

In practice, we would usually start from the bending moment equation and undo the differentiation to obtain the deflection.

Example 4

Find the greatest and least values of $f(x) = e^{-x} \sin x$ in $[0, \infty]$. We find $f'(x) = e^{-x}(-\sin x + \cos x)$ and this vanishes where $\tan x = +1$, that is when

$$x = (4n + 1)\frac{\pi}{4}; \quad n = 0, 1, 2, ...$$

Further,

$$f''(x) = e^{-x}(\sin x - 2\cos x - \sin x) = -2\cos x\, e^{-x};$$

at $x = \pi/4, 9\pi/4, 17\pi/4...$ $f''(x) < 0$, indicating a local maximum, whereas at $x = 5\pi/4, 13\pi/4, 21\pi/4, \ldots f''(x) > 0$, indicating a local minimum.

However, the absolute values of these stationary points are decaying with increasing x. The first local maximum of $f(x)$, viz $(1/\sqrt{2})e^{-\pi/4}$, is the greatest value we seek and the first local minimum, $(-1/\sqrt{2})e^{-5\pi/4}$ is the least value.

12.4 Curvature

The curvature of an arc AB of the curve $y = f(x)$ is the angle through which the tangent moves as its point of contact traverses the arc from A to B. The angle is shown as $\delta\psi$ and if the length of the arc AB is δs (both assumed small) we have an average rate of curvature of $\delta\psi/\delta s$. (See Figure 12.11)

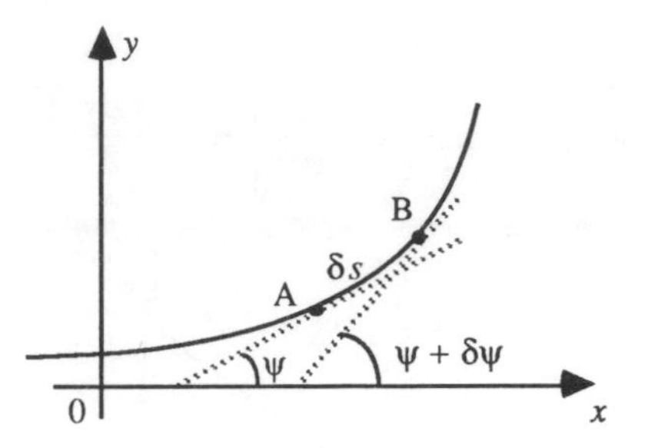

Figure 12.11

We define **curvature at a point** as

$$\kappa = \frac{d\psi}{ds} \tag{12.11}$$

Clearly the larger κ, the more rapid the *bending* of the curve.

Related to this concept is that of **radius of curvature**. The radius of curvature at a point is given by

$$\rho = \frac{1}{\kappa} = \frac{ds}{d\psi} \tag{12.12}$$

The basic idea stems from that of a circle whose radius of curvature is constant (and equal to its radius). Figure 12.12 shows us that $\delta s \cong r\delta\psi$ (exact equality occurs for a circle) and hence $r \cong \delta s/\delta\psi$, which gives us ρ.

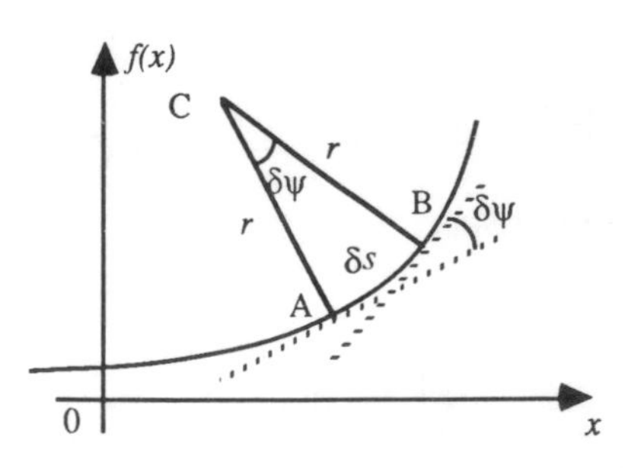

Figure 12.12

It follows that the *larger* ρ is the *flatter* the curve and the *smaller* ρ is the *sharper* the bending.

Locally, the arc AB can be taken to be the arc of a circle of radius r.

C is called the local **centre of curvature**.

To derive the formula for ρ in Cartesian coordinates we refer to Figure 12.13. If the arc AB is *small*, it can be approximated by a straight line and we have, in the limit, the relationships

$$\frac{dy}{dx} = \tan\psi \quad \text{and} \quad \left[\frac{ds}{dx}\right]^2 = 1 + \left[\frac{dy}{dx}\right]^2 \tag{12.13}$$

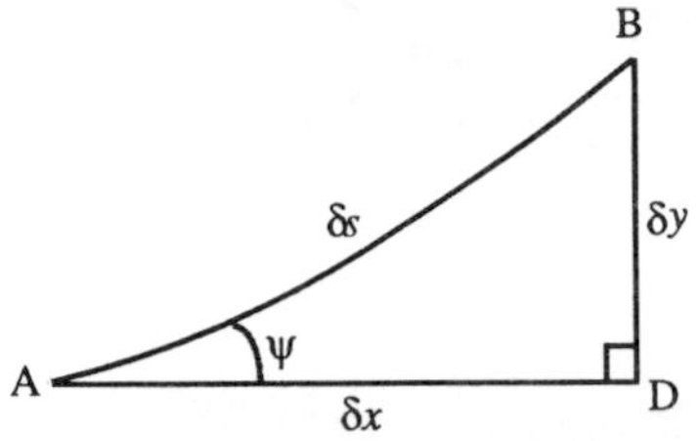

Figure 12.13

We attempt to find $d\psi/ds$ and to do this we begin with

$$\frac{d}{ds}\left[\frac{dy}{dx}\right] = \frac{d}{ds}(\tan\psi) = \frac{d}{d\psi}(\tan\psi).\frac{d\psi}{ds} = \sec^2\psi\,\frac{d\psi}{ds}$$

$$= (1+\tan^2\psi)\,\frac{d\psi}{ds} = \left\{1+\left[\frac{dy}{dx}\right]^2\right\}\frac{d\psi}{ds}$$

But

$$\frac{d}{ds}\left[\frac{dy}{dx}\right] = \frac{d}{dx}\left[\frac{dy}{dx}\right]\frac{dx}{ds} = \frac{d^2y}{dx^2}\cdot\cos\psi = \frac{d^2y}{dx^2}\,\frac{1}{\sqrt{1+\tan^2\psi}}$$

Hence, on making the comparison,

$$\rho = \frac{\left\{1+\left[\dfrac{dy}{dx}\right]^2\right\}^{3/2}}{\dfrac{d^2y}{dx^2}} \tag{12.14}$$

[Notice that if dy/dx is small then ρ is approximately $1/(d^2y/dx^2)$, hence the approximation $M = -EI(d^2y/dx^2)$ for small deflections of beams.]

Example 1

Find the radius of curvature of the curve $y = \ln x$.
We have

$$\frac{dy}{dx} = \frac{1}{x} \qquad \text{and} \quad \frac{d^2y}{dx^2} = -\frac{1}{x^2}$$

Hence

$$\rho = \frac{\left\{1 + \dfrac{1}{x^2}\right\}^{3/2}}{-\dfrac{1}{x^2}} = \frac{-(x^2+1)^{3/2}}{x}$$

Check this result against the graph of $\ln x$ on page 49.

Example 2

Find the radius of curvature of the hyperbola $xy = 1$ at the point $(1, 1)$.

Since $\quad y = \dfrac{1}{x}, \quad \rho = \dfrac{\left\{1 + \left[\dfrac{-1}{x^2}\right]^2\right\}^{3/2}}{2/x^3} = \dfrac{x^3}{2}\left\{1 + \dfrac{1}{x^4}\right\}^{3/2}$

At $(1, 1)$ $\rho = \frac{1}{2}\{2\}^{3/2} = \sqrt{2}$

Observe that at the point $(-1, -1)$ $\rho = -\sqrt{2}$ and the negative sign merely indicates the sense of bending. For large x, ρ is approximately $\frac{1}{2}(x^3)$ and the curve flattens out as $x \to \infty$; for small x, ρ is approximately $x^3/2x^6 = 1/2x^3$ and as $x \to 0$ the curve again flattens out.

If the centre of curvature C has coordinates (x_c, y_c) then it can be shown (see Problem 57) that

$$x_c = x - \rho \sin \psi$$
$$y_c = y + \rho \cos \psi$$

or

$$x_c = x - \frac{\left\{1 + \left[\dfrac{dy}{dx}\right]^2\right\}\dfrac{dy}{dx}}{\dfrac{d^2y}{dx^2}} \qquad y_c = y + \frac{\left\{1 + \left[\dfrac{dy}{dx}\right]^2\right\}}{\dfrac{d^2y}{dx^2}} \tag{12.15}$$

Hence for the hyperbola $xy = 1$

$$x_c = x + \frac{x}{2}\left[1 + \frac{1}{x^4}\right]; \qquad y_c = \frac{1}{x} + \frac{x^3}{2}\left[1 + \frac{1}{x^4}\right]$$

i.e.
$$x_c = \frac{1}{2}\left[3x + \frac{1}{x^3}\right]; \qquad y_c = \frac{1}{2}\left[x^3 + \frac{3}{x}\right]$$

As we move along a curve the centre of curvature will trace out a locus called the **evolute** of the original curve. The evolute of the hyperbola $xy = 1$ is shown in Figure 12.14. For the circle $x^2 + y^2 = a^2$ you should be able to show that $(x_c, y_c) \equiv (0, 0)$ and therefore the evolute here reduces to a *single point.*

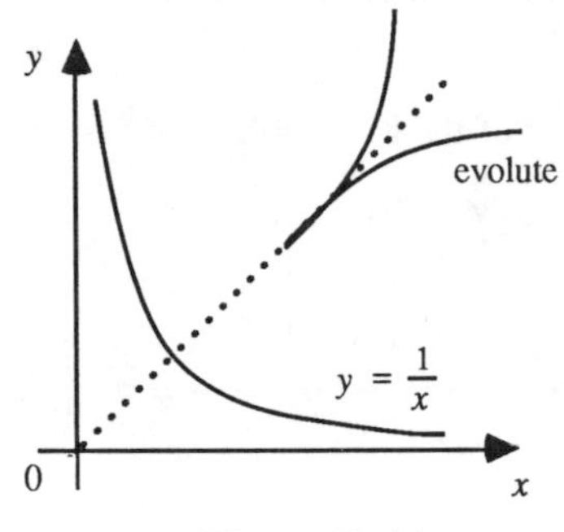

Figure 12.14

Problems

Section 12.1

1 Differentiate the following functions with respect to x (use the product, quotient and chain rules as appropriate)

(i) $x^3 \sin x$ (ii) $e^x \cos x$ (iii) $x^4 \ln x$
(iv) $x^2/(1 + x)$ (v) $(\sec x)/x$ (vi) $(3 - 2x)^5$
(vii) $\sqrt{1 - x^2}$ (viii) $\sqrt{1 - x^2}\, \sin^{-1} x$ (ix) $\sin 3x$
(x) $\sin^2 4x$ (xi) $\cos(\sin 3x)$ (xii) e^{3x+2}
(xiii) $x^2 \tan^2 2x$ (xiv) $\ln(1 - x + x^2)$ (xv) $\ln[(1 + x)/(1 - x)]$
(xvi) $\ln(\cot x)$ (xvii) $e^{2x} \ln(1 + x^2)$

2 The distance of a moving particle from its starting point is $s = 3 + 8t - 7t^2$ (cm) where t is the time in seconds. Find the speed and acceleration of the particle after 3 seconds.

3 The velocity of a body moving through a resisting medium is $20(1 - e^{-0.02t})$; find the acceleration when $t = 5$.

4 A spring moves according to $x = e^{-0.2t}(0.6 \cos 4t - 0.4 \sin 4t)$ (Damped Motion). Sketch $x(t)$ and find the velocity at any time.

5 Show that $y = (\text{cosec } x)/[\sinh(1/x)]$ satisfies the relation

$$\frac{dy}{dx} = y\left[-\cot x + \frac{1}{x^2}\coth\left(\frac{1}{x}\right)\right] \qquad \text{(EC)}$$

6 A spherical bubble expands. The rate at which its radius is increasing is 3 cm/sec. Find the rate of increase of its volume when the radius is 25 cm.

7 The ends of a rod PQ of length a are constrained to move as shown in Figure 12.15. Q is made to move such that $x = 4 \sin 3t$.

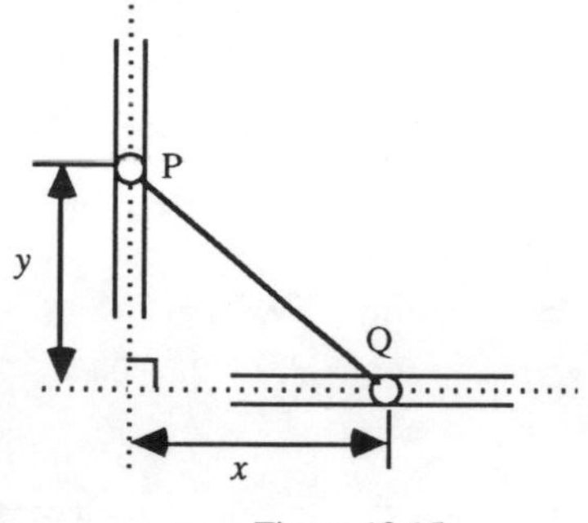

Figure 12.15

Find the speed of the end P at any instant.

8 Soil falls into a conical heap such that $h = (2/3)r$. Given that the soil falls on to the heap at the rate of 10 ml/sec, find the rate at which h increases.

9 The chain rule can be extended to the result

$$\frac{dy}{dx} = \frac{dy}{du} \cdot \frac{du}{dv} \cdot \frac{dv}{dx}$$

Check this for $y = \sqrt{u}$, $u = v(1 + v)$, $v = x^3$, by finding dy/dx by two methods.

10 Find dy/dx for each of the following functions

(i) $y = \sin^{-1}(\cos x)$, $0 < x < \pi$

(ii) $y = \tan^{-1}(\ln x)$

(iii) $y = \sin[(x + y)^2]$

(iv) $y = (1 + x)(25 - x^2)^{1/2}$

(v) $y = [x + (1/x)]^x$

(vi) $x = a \sin t - b \sin(at/b)$, $y = a \cos t - b \cos(at/b)$

(vii) $x = \cos 2\theta$, $y = 2\theta + \sin 2\theta$

(viii) $x = 2y - \tan^{-1}y$

(ix) $x^2 + y^2 + 3x + 2y - 5 = 0$ (x) $y = (2 - x)\sqrt{\dfrac{3 - x}{1 + x}}$

(xi) $y = \sin^{-1}\left[\dfrac{1 + 2\sin x}{2 + \sin x}\right]$ (xii) $y = \sin^{-1}(2x\sqrt{1 - x^2})$

11 By considering $\sin y = x$ and using implicit differentiation, derive the standard result

$$\frac{d}{dx}(\sin^{-1} x) = \frac{1}{\sqrt{1 - x^2}}$$

12 Show that

$$\frac{d}{dx}\left[\frac{1}{a}\tan^{-1}\frac{x}{a}\right] = \frac{1}{x^2 + a^2}$$

13 Find

(i) $\dfrac{d}{dx}\left[\sqrt{\dfrac{2x + 3}{2x - 3}}\right]$ (ii) $\dfrac{d}{dx}[e^x(1 + x)^2/\sqrt{(1 - x)^3}]$

14 Find dy/dx if

(i) $x^2 + 3xy + y^2 = 2$ (ii) $x^3/y^2 = 5$ (iii) $\sin(x + y) = 3xy$

(iv) $\ln y = \sin x + \cos y$ (v) $e^{x-y} - 2x = 0$

15 Find dy/dx if

(i) $x = \sin t, y = \sin \lambda t$ (ii) $x = a\cos\theta, y = b\sin\theta$

(iii) $x = t^3, y = 2t^2 - 1$ (iv) $x = a\cos 2\theta, y = \cos\theta + 1$

16 Using logarithmic differentiation, show that

(i) $\dfrac{d}{dx}(10^x) = 10^x.\ln 10$ (ii) $\dfrac{d}{dx}(x^x) = (1 + \ln x)x^x$

17 Differentiate with respect to x

(i) $\tan^{-1}\dfrac{4\sin x}{3 + 5\cos x}$ (ii) $\ln\left[\dfrac{x^2 - 1}{x^2 + 1}\right]$

(iii) $e^{-2x}(2x^2 + 2x - 1)$ (iv) $\tan^{-1}\left[\dfrac{2x}{1 - x^2}\right]$

expressing each answer in its simplest form. (LU)

18 Find the values when $x = 1$ of the differential coefficients with respect to x of the following functions

(i) $x^2 e^x \sin(\pi x)$ (ii) $\tan^{-1}(2x)$ (iii) $\ln\left[\dfrac{3x^2-2}{2x^2+3}\right]$ (LU)

19 Find the value, when $x = 1$, of the derivative with respect to x of each of the following expressions

(i) $e^{-2x}\cos \pi x$ (ii) $\ln\left[\dfrac{x^2}{1+x^3}\right]$

(iii) $\cot^{-1} x + \cot^{-1}(1/x)$ (iv) x^x (LU)

20 (i) If $y = \left[\dfrac{1+x}{1-x}\right]^k$ prove that $(1-x^2)\dfrac{dy}{dx} = 2ky$

(ii) If $y = \sqrt{ax - x^2} - a\tan^{-1}\sqrt{\dfrac{a-x}{x}}$ prove that $x\left[\dfrac{dy}{dx}\right]^2 = a - x$

(iii) If $y = \ln(\sec\theta + \tan\theta) - (2 - \sec\theta)\tan\theta$, where $0 < \theta < \frac{1}{2}\pi$, show that $dy/d\theta$ is positive and deduce that y is always positive in this range. (LU)

21 (i) If $y = \dfrac{\sin^{-1} x}{\sqrt{(1-x^2}}$ prove that $(1-x^2)\dfrac{dy}{dx} = xy + 1$

(ii) Find dy/dx in its simplest form if $y = \dfrac{(2x-1)(3-x)^3}{(2-x)^2}$ (LU)

22 If $y = \tan^{-1}\left[\dfrac{ae\sin x}{b + a\cos x}\right]$ where $a^2e^2 = a^2 - b^2$, show that $\dfrac{dy}{dx} = \dfrac{ae}{a + b\cos x}$ (LU)

23 (i) Find the value when $x = \frac{1}{2}$ of the differential coefficient with respect to x of

(a) $x\sin^{-1} x + \sqrt{(1-x^2)}$ (b) $\cosh^{-1}(1/x)$ (c) x^{2x}

(ii) If $f(x) = x - \ln(1 + x)$, find $f'(x)$ and hence show that $f(x) > 0$ for $x > 0$

(LU)

24 Given that the implicit parametric equations of a particular curve representing a particular streamline in a two-dimensional flow pattern are

$$\cos ax + \sin b\theta = x\theta, \quad \sin ay + \cos b\theta = y\theta$$

find dy/dx in terms of x, y and θ. (EC)

Section 12.2

25 (i) If $y = A \ln x + B$, prove that $x\dfrac{d^2y}{dx^2} + \dfrac{dy}{dx} = 0$

(ii) If $pV^{\gamma} = c$, show $V^2\dfrac{d^2p}{dV^2} = \gamma(\gamma + 1)p$ and $p^2\dfrac{d^2V}{dp^2} = \dfrac{\gamma + 1}{\gamma^2}V$

26 Find d^2y/dx^2 for each of the parts of Questions 14 and 15 of Section 12.1.

27 The profit P in a certain manufacturing process is given by

$$P = px - [F + V(x/k)^{3/2}]$$

where x is the number produced, p is the sale price and V, F and k are constants of cost. Find the maximum profit.

28 The potential energy of one atom due to another is given by

$$V = \frac{A}{r^{12}} - \frac{B}{r^6} \qquad (A, B > 0)$$

where r is the separation of the atoms.
Find where $dV/dr = 0$ (the equilibrium separation) and show that $d^2V/dr^2 > 0$ at this value of r.

29 If $y = \sin(2\sin^{-1}x)$, show that $(1 - x^2)\dfrac{d^2y}{dx^2} - x\dfrac{dy}{dx} + 4y = 0$

30 Prove that $\dfrac{d^2x}{dy^2} + \left[\dfrac{dx}{dy}\right]^3 \dfrac{d^2y}{dx^2} = 0$ (EC)

31 (i) If $x = a\left[t + \dfrac{1}{t}\right]$ and $y = a\left[t - \dfrac{1}{t}\right]$ prove that $y^3\dfrac{d^2y}{dx^2} + 4a^2 = 0$

(ii) When $y = \dfrac{x}{x + \sqrt{1 + x^2}}$ show that $\sqrt{1 + x^2}\dfrac{dy}{dx} = \dfrac{y^2}{x^2}$ (LU)

32 (i) Given that $x = 1/(1 + t^2)$, $y = t^3/(1 + t^2)$, obtain expressions for dy/dx and d^2y/dx^2 in terms of t.

(ii) By considering the stationary value of $f(x) = e^{ax} - x$, where a is real and positive, prove that the equation $f(x) = 0$ has two and only two real roots when $a < e^{-1}$. Show that, when a is small, the smaller of these roots is

$1 + a + \dfrac{3}{2}a^2 + O(a^3)$ (LU)

33 If $y = (1 - x^2)^{1/2} \sin^{-1} x$, prove that $(1 - x^2)\dfrac{d^2y}{dx^2} - x\dfrac{dy}{dx} + 2x + y = 0$ (LU)

34 *Leibnitz' Theorem*

Let u and v be functions of x which possess derivatives of any order. Starting from

$\dfrac{d}{dx}(uv) = u\dfrac{dv}{dx} + v\dfrac{du}{dx}$ show that $\dfrac{d^2}{dx^2}(uv) = u\dfrac{d^2v}{dx^2} + 2\dfrac{du}{dx}\cdot\dfrac{dv}{dx} + \dfrac{d^2u}{dx^2}v$ and deduce

that $\dfrac{d^3}{dx^3}(uv) = u\dfrac{d^3v}{dx^3} + 3\dfrac{du}{dx}\cdot\dfrac{d^2v}{dx^2} + 3\dfrac{d^2u}{dx^2}\dfrac{dv}{dx} + \dfrac{d^3u}{dx^3}v$

By induction show that $\dfrac{d^n}{dx^n}(uv) = \sum_{r=0}^{n} {}^nC_r \dfrac{d^ru}{dx^r}\dfrac{d^{n-r}v}{dx^{n-r}}$ where $\dfrac{d^0u}{dx^0} \equiv u$, $\dfrac{d^0v}{dx^0} \equiv v$

35 Using Leibnitz' Theorem for the nth derivative of the product of two functions, show that

$$\frac{d^n}{dx^n}(x \ln x) = (-1)^n \frac{(n-2)!}{x^{n-1}}$$ (EC)

36 In Mechanics, a point of equilibrium is said to be *stable* if the potential energy V is a minimum at that point and *unstable* if V is a maximum there; otherwise it is in *neutral equilibrium.* Investigate the stability of the points of equilibrium of the motion described by

$$V(x) = \begin{cases} c & \text{for } |x| > a \\ c - \lambda \exp\{-x^2/(a^2 - x^2)\} & \text{for } |x| \le a \end{cases}$$ (EC)

37 (i) Find the sixth derivative with respect to x of $x^2 \ln x$

(ii) If $y = \exp\{x + \sqrt{(1 + x^2)}\}$, find d^2y/dx^2 when $x = 0$ (LU)

38 State Leibnitz' Theorem on the nth differential coefficient of a product.

If $y = \sin x/(1 - x^2)$, show that

(i) $(1 - x^2)\dfrac{d^2y}{dx^2} - 4x\dfrac{dy}{dx} - (1 + x^2)y = 0$

(ii) $y_{n+2} - (n^2 + 3n + 1)y_n - n(n-1)y_{n-2} = 0$

where y_n is the value of d^ny/dx^n when $x = 0$. (LU)

39 Show that $(x^2 - 1)^n$ satisfies the equation $(x^2 - 1)\dfrac{dy}{dx} - 2nxy = 0$ and hence deduce that

$$(x^2-1)\frac{d^{n+2}}{dx^{n+2}}(x^2-1)^n + 2x\frac{d^{n+1}}{dx^{n+1}}(x^2-1)^n - n(n+1)\frac{d^n}{dx^n}(x^2-1)^n = 0$$ (LU)

40 If $f(x) = \sin(k\sin^{-1}x)$ show that

$$(1-x^2)f'' - xf' + k^2 f = 0$$

and

$$(1-x^2)f^{(n+2)} - (2n+1)xf^{(n+1)} + (k^2-n^2)f^{(n)} = 0$$ (LU)

41 (a) The equation of a curve in implicit form is $x^4 - x^3y + y^4 = 8$. Find dy/dx in terms of x and y, and derive an equation involving d^2y/dx^2.

(b) If $x = t^2$ and $y = 2t$ find dy/dx and d^2y/dx^2 in terms of t. (EC)

Section 12.3

42 (i) Show that the greatest and least values of $f(x) = x^2 - 3x + 2$ in the interval [0, 4] are 6 and $-\frac{1}{4}$ respectively. Repeat for the intervals [–3, 3], [–3, 0] and [1, 2]

(ii) Repeat the above for $f(x) = 2 + x - x^2$

43 Examine the following for maximum and minimum values:

(i) $(x-4)^2(x+3)^3$ (ii) $x^4 - 8x^3 + 18x^2 - 14$

(iii) $x^2 + (54/x)$ (iv) $x^{2/3}(6x - x^2 - 8)^{1/3}$

(v) $\theta\sin\theta + (1+h)\cos\theta$ (consider the cases $h < 1$ and $h \geq 1$ separately)

44 The velocity of a signal at a distance r along a cable is $v = kr \ln 1/r$ where $0 < r < 1$; at what distance is the velocity greatest? Sketch the graph of $v(r)$.

45 The velocity of a wave in a deep channel is $v = c\sqrt{\frac{\lambda}{b} + \frac{b}{\lambda}}$ where λ is the wavelength, b and c constants. What is the least velocity?

46 The transmitted power of a belt can be represented as $H = Cv(T_0 - wv^2/g)$ where T_0 is the initial tension in the belt, w its weight/unit length, v its running speed and C a constant. Find the speed at which maximum power is transmitted.

47 In an LRC circuit the impedance is given by $Z = \sqrt{R^2 + \left[\omega L - \frac{1}{\omega C}\right]^2}$. In a particular situation R, L and C have given positive values and ω is varied. For what value of ω will Z have a maximum or minimum value? Using a sketch of Z against ω, or otherwise, explain whether this point gives a maximum or a minimum value of Z. (EC)

48 The torque exerted by an induction motor is given by $T = k^2R/(R^2 + k^2)$ where R is the rotor resistance, and k is a constant. Find the maximum torque.

49 A high-frequency coaxial feeder has internal radius r_1 and external radius r_2. To determine its optimum dimensions we must find the value of r_1, given r_2, for which attenuation is least; this requires the least value of $\left[\frac{1}{r_1} + \frac{1}{r_2}\right] \Big/ \ln \frac{r_2}{r_1}$.

Write $x = r_2/r_1$ and find an equation for x. Show that $3 < x < 4$ and use the Newton-Raphson method to obtain x to 3 d.p.

50 Sketch the curves
(i) $y^2(2 + x) = x^2(2 - x)$ (ii) $x^3 + y^3 - 9x^2 = 0$

51 Show that the equation $x^4 - x^3 - 6x^2 + 4x + 8 = 0$ has two coincident roots. Hence find the four roots of the equation.

52 Sketch and discuss the curve $y^4 = x^2y - 4$, showing clearly the salient features, including any asymptotes and extrema. (EC)

53 Sketch and discuss the curve $y^2(x + 1) = x^2(1 - x)$. Include in your discussion the maximum and minimum points and the asymptotes. (EC)

54 (i) If $f(x) = \ln(1 - x) + x/(1 - x)$, find $f'(x)$ and deduce that $f(x) > 0$ for $0 < x < 1$
(ii) Find the maximum and minimum values of the function $g(x) = (x^2 + 2x - 1)e^{-2x}$. Show that the curve $y = g(x)$ has points of inflection and sketch the curve. (LU)

55 Find the stationary values of $y = x + 5 + \frac{3}{x} - \frac{1}{x^2}$ and determine its finite maximum and minimum values, if any. Show that dy/dx is not negative if x is positive or less than -2. Prove that $\lim_{x\to\infty} \frac{dy}{dx} = 1$ and give a careful sketch of the curve represented by the given equation. (LU)

56 (a) Find the maximum and/or minimum values of $y = 3\exp(-t) - 4\exp(-3t)$ for $t \geq 0$. Also determine where it is zero and find any points of inflexion. Hence sketch the graph of y for $t \geq 0$.
(b) Find any local maxima, minima and points of inflexion of the function $y = x^2 \exp(-x)$. Hence sketch the curve for $-\infty < x < +\infty$. (EC)

Section 12.4

57 By considering Figure 12.16, show that for the centre of curvature C

$$x_c = x - \rho \sin \psi, \qquad y_c = y + \rho \cos \psi$$

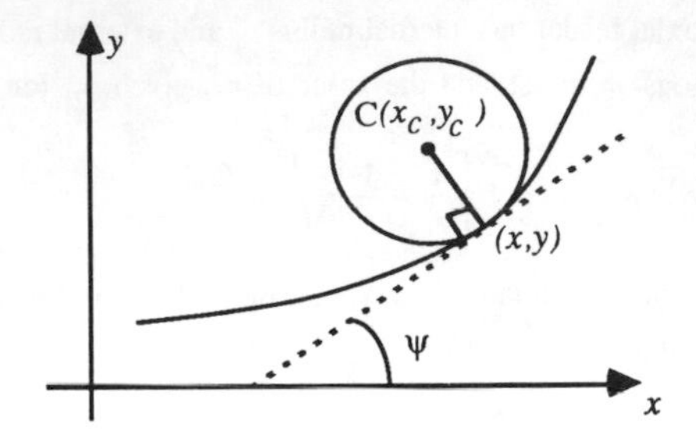

Figure 12.16

Given that $\tan\psi = dy/dx$ and $\rho = \dfrac{\left\{1+\left[\dfrac{dy}{dx}\right]^2\right\}^{3/2}}{\dfrac{d^2y}{dx^2}}$

show that $x_c = x - \left\{1+\left[\dfrac{dy}{dx}\right]^2\right\}\dfrac{dy}{dx}\bigg/\dfrac{d^2y}{dx^2}$ and $y_c = y + \left\{1+\left[\dfrac{dy}{dx}\right]^2\right\}\bigg/\dfrac{d^2y}{dx^2}$

58 Find the radius of curvature ρ of the catenary $y = c\cosh(x/c)$ and show that $\rho = y^2/c$.

59 Find the radius of curvature of

(a) $y = \dfrac{1}{\sqrt{3}}(2x^2 - x)$ at the point $\left[1, \dfrac{1}{\sqrt{3}}\right]$

(b) $y = x^3$ at the point (1, 1) and at the origin

(c) $y = \dfrac{1}{1-x}$ at (2, –1)

(d) $\dfrac{x^2}{25} + \dfrac{y^2}{12} = 1$ at (5/2, 3)

60 Obtain the radius of curvature at the point (x_1, y_1) on the curve $y = 1 + 2x - 4x^2$. Sketch the curve. Show that the circle of curvature at (1/4, 5/4) is given by the equation $16(x^2 + y^2) - 8x - 36y + 21 = 0$ (LU)

61 A curve is given in terms of the parameter θ by the equations

$$x = a(\cos\theta + \theta\sin\theta), \qquad y = a(\sin\theta - \theta\cos\theta)$$

Show that the radius of curvature at any point is $a\theta$. Show also that the locus of the centre of curvature is a circle with centre at the origin and radius a. (LU)

62 Given the function $y = k\ln(x/a)$, $k > 0$, $a > 0$, find the curvature of its graph and the coordinates of the centre of curvature, for the point where the graph meets the x-axis. (LU)

63 Find the maximum and minimum values of $x^2(x-2)^3$. Find also the radius of curvature of the curve $y = x^2(x-2)^3$ at the point (1, –1). (LU)

13

NON-LINEAR EQUATIONS

A slender column of length ℓ is built in at its lower end A and pinned and laterally supported at its upper end B, as shown in Figure 13.1. It is subjected to an increasing axial load P. In Section 23.5 we show how to set up and solve the equation which describes the deflected profile $y(x)$. Initially, the column will remain straight; however, when the load P reaches a critical value P_{cr} the column will buckle into a shape similar to that shown in Figure 13.1. To find this load it is necessary to solve the equation

$$\tan u = u \qquad (13.1)$$

where
$$u = \sqrt{\frac{P}{EI}}\, l$$

and EI is the flexural rigidity of the column.

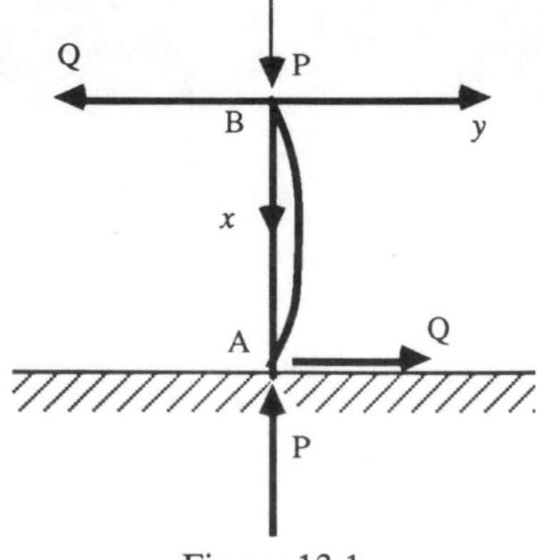

Figure 13.1

How can we solve equation (13.1) if no formula exists as in the case of a quadratic equation? On page 68 we met the idea of iterative solutions of equations. There we considered making successive point estimates x_0, x_1, x_2,..... of the roots of an equation $f(x) = 0$, or rather $x = F(x)$. An alternative approach is to 'trap' the root we seek in an interval (x_L, x_R) and progressively narrow this interval. First, we need to have some idea of the location of the roots of the equation under consideration and this is dealt with in Section 13.1. In the next three sections we examine some iterative techniques and in Section 13.5 we look at the special case of polynomial equations.

We should perhaps ask how we decide when the approximation is sufficiently accurate. Consider Figure 13.2(i); here we see a sketch of the function $y(x) = (x - 2)^{10}$. It is clear that the only real root of $f(x) = 0$ is $x = 2$; but when $x = 2.2$ $f(x)$ is only about 10^{-7}, so that for $1.8 < x < 2.2$, $|f(x)| < 10^{-7}$ and a test for convergence which asked for $|f(x)| < 10^{-7}$ would not pin-point the root to an acceptable level of accuracy. Conversely, Figure 13.2(ii) depicts $f(x) = (x-2)^{0.1}$; in order for $|f(x)| < 10^{-7}$, x must lie in the interval $(2 - 10^{-70}, 2 + 10^{-70})$ and this is almost certainly too precise for our needs. However, if $x = 2 + 10^{-7}$, $f(x) = 0.1995$ and this value is not one we would regard as "small". Therefore the question arises: should we base a convergence criterion on the magnitude of $|f(x)|$ or should we incorporate an element which looks at successive estimates of the root itself? Then there is the consideration of the value of the root: convergence to within 10^{-7} in absolute value is probably acceptable when the root is of order of magnitude 1, but if the root lies near 10^{30} such accuracy may be unobtainable. Most practical computer routines will quit when either x is obtained to a pre-determined tolerance or $|f(x)|$ is so small that to machine accuracy it registers as zero. (If the root is near zero then it is advisable to consider carefully the pre-determined tolerance you will choose.) Some routines also include a maximum number of iterations as a safety measure. The above considerations show that the stopping criterion of a root-finding routine is important. For hand calculation it helps to have some knowledge of the behaviour of $f(x)$ near the root.

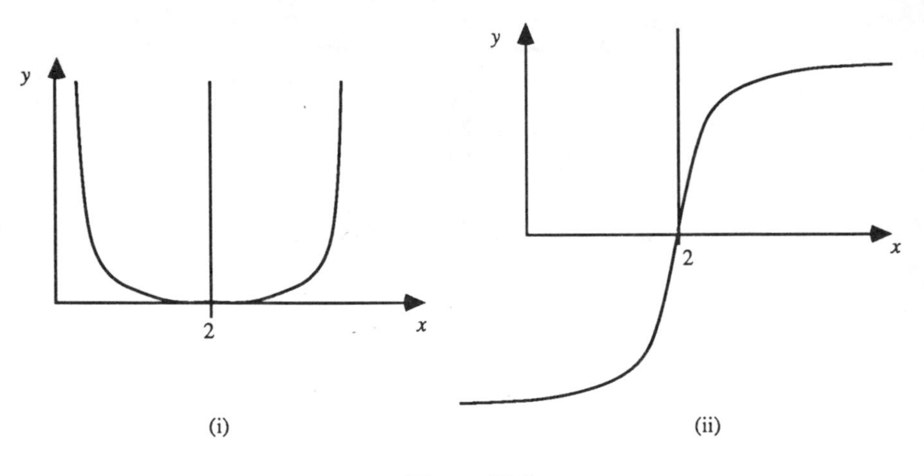

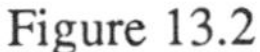
Figure 13.2

13.1 Location of Roots

Often the first step in locating the roots of $f(x) = 0$ is to sketch a graph of the function $y = f(x)$; this will give us some idea as to the *nature*, *number* and *approximate location* of the roots. Sometimes, however, a little forethought will simplify the sketching. Consider the equation $x^3 - x - 1 = 0$. It would be tedious to sketch the cubic function directly, but by rewriting the equation as $x^3 = x + 1$ and sketching the graphs of $y = x^3$ and $y = x + 1$ on the same axes, Figure 13.3(i), we can see at once that there is one positive real root. Similarly, we can rewrite the equation $x^3 - x + 1 = 0$ as $x^3 = x - 1$ and via Figure 13.3(ii) we see that there is one negative real root. If the sketches were reasonably accurate we would obtain first approximations to the roots of 1.25 and –1.25 respectively. Alternatively, we may tabulate each function near its zero (a zero of $f(x)$ is a root of $f(x) = 0$). In Table 13.1 we consider the equation $x^3 - x - 1 = 0$.

Table 13.1

x	0	1	2	1.5
x^3	0	1	8	3.375
$x + 1$	1	2	3	2.5
$x^3 - x - 1$	–1	–1	5	0.875

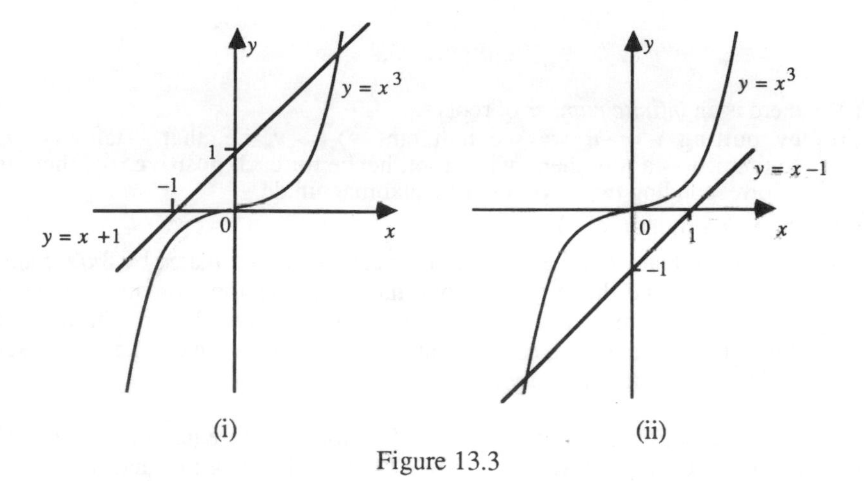

Figure 13.3

Notice that the function $f(x) = x^3 - x - 1$ changes sign between $x = 1$ and $x = 1.5$, in other words $x^3 < x + 1$ at $x = 1$, $x^3 > x + 1$ at $x = 1.5$. We can conclude that there is a root of $f(x) = 0$ in the interval $(1, 1.5)$. This follows from the **Intermediate Value Theorem** which states that if $f(x)$ is a real-valued function which is continuous on the interval $[a, b]$ and $f(a) \neq f(b)$ then for any value y_1 between $f(a)$ and $f(b)$ there is at least one value x_1 in $[a, b]$ for which $f(x_1) = y_1$. In our example, $f(x)$ is continuous everywhere and with $a = 1$, $b = 1.5$ it is clear that the theorem predicts at least one value of x in $(1, 1.5)$ for which $f(x) = 0$.

We must emphasise the importance of the requirement for continuity: although the function $f(x) = \dfrac{1}{x-2}$ is negative at $x = 1$ and positive at $x = 3$ it would be quite wrong to assume a root of $f(x) = 0$ in the interval (1, 3).

Example
The equation $\tan u = u$ is already in a form which suggests the super-position of two graphs. Remembering that the slope of $\tan u$ at $u = 0$ is 1 we can refer to Figure 13.4. Four points can be made:

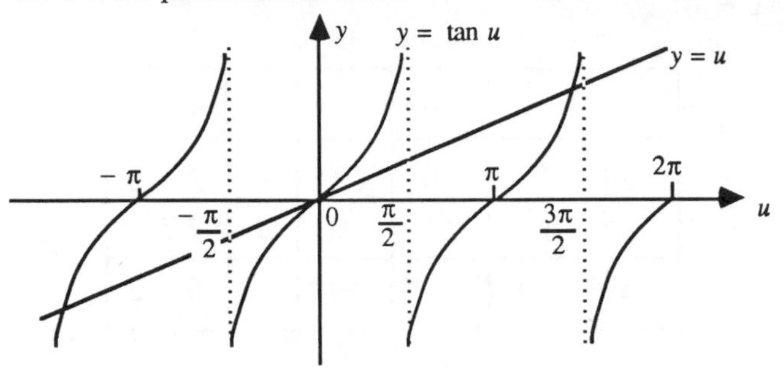

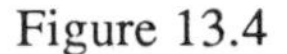
Figure 13.4

(i) there is an *infinite number* of roots
(ii) by putting $v = -u$ we see that $\tan(-v) = -v$, i.e. that $-\tan v = -v$, so that if v is a root then $-v$ is a root; hence for each **positive** root there is a corresponding **negative** root of equal magnitude
(iii) there is a root at $u = 0$
(iv) apart from that root, the positive roots can be approximated by $3\pi/2$, $5\pi/2$, $7\pi/2$, ... etc and the larger the root, the better the approximation. (These results follow since each branch of the curve is cut by the straight line and the further the branch is to the right, the higher up is the cut, i.e. the closer to the point of discontinuity.)

Next, we mention a useful technique for polynomial equations, with real coefficients known as **Descartes' rule of signs**. Bearing in mind that if the polynomial $p(x)$ has real coefficients then any complex roots must occur in conjugate pairs (i.e. if $x = a + ib$ is a root so is $x = a - ib$) then the equation $x^5 - x^3 + 2x^2 - 7x + 6 = 0$ has either 5 real roots or 3 real roots and 2 complex conjugate roots or 1 real root and 2 pairs of complex roots. Similarly, the equation $x^6 - x^4 + 2x^2 - 7x + 6 = 0$ has either 6 real roots or 4 real roots and a pair of complex roots or 2 real roots and 2 pairs of complex roots or no real roots and 3 pairs of complex roots.

Descartes' rule states that if $p(x)$ is written in the order of descending powers of x then the number of positive roots of $p(x) = 0$ is equal either to the number of sign changes in the coefficients, n, or to $(n - 2)$ or to $(n - 4)$ ....., this sequence of possibilities terminating at 1 or at 0. Similarly, information about the number of negative roots is found by examining the number of sign changes in the ordered coefficients of $p(-x)$.

Examples

(i) $p(x) = x^3 + 3x - 3$. The coefficients, including that of x^2, are 1, 0, 3, –3 and the one sign change (between 3 and –3) implies one positive root. Since $f(0) < 0 < f(1)$ we have a rough location for the root. Now $p(-x) = -x^3 - 3x - 3$ so that there are no sign changes in the coefficients –1, 0, –3, –3 and hence no negative roots. The remaining two roots are therefore complex.

(ii) $p(x) = x^3 - 2x^2 - 2x + 4$. $p(x)$ has two sign changes whereas $p(-x) = -x^3 - 2x^2 + 2x + 4$ has only one. Hence the possibilities are two positive roots and one negative root or no positive roots and one negative root.

(iii) $p(x) = x^4 - 3x^2 + 6x + 2$. There are at most two positive roots and two negative roots. Now $p(0) = 2$ and $p(-1) = -6$ so that we have located one of the negative roots. Further $p(-2) = -6$ and $p(-3) = 38$ so that we have found the location of the other negative root. However, $p(1) = 6$, $p(2) = 18$ and we wonder whether we have missed the two positive roots by using too wide a spacing in the x values, whether we have not gone far enough or whether there are no positive roots to be found.

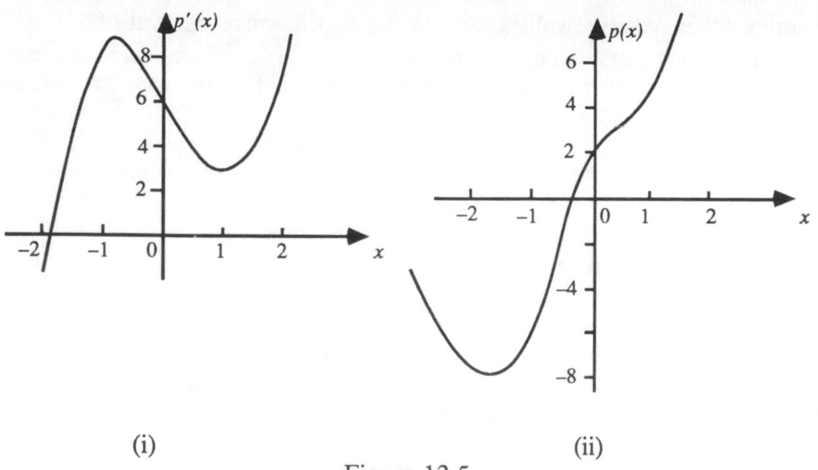

Figure 13.5

A sketch of $p'(x) = 4x^3 - 6x + 6$ is shown in Figure 13.5(i); from this we see that the slope of $p(x)$ is positive after $x \cong -1.6$ and therefore $p(x)$ is increasing from that point; since $p(0) = 2$ there are no positive roots, as indicated in Figure 13.5(ii).

Finally in this section we discuss the problem of bracketing a root. The aim is to find an interval (a, b) such that $f(a)$ and $f(b)$ have opposite signs. Provided that the function $f(x)$ is continuous then we know that there is at least one zero of the function in (a, b). Suppose however that $f(x)$ has a double root at $x = c$ (see Figure 13.6(i)); then no matter how narrow the interval (a, b) which contains the point c it will not be possible for the product $f(a).f(b)$ to be negative, i.e. for $f(a)$ and $f(b)$ to have opposite signs. How do we distinguish this case from that of Figure 13.6(ii) where there are no roots in (a, b)? In the latter case, we clearly need to know more about the behaviour of $f(x)$.

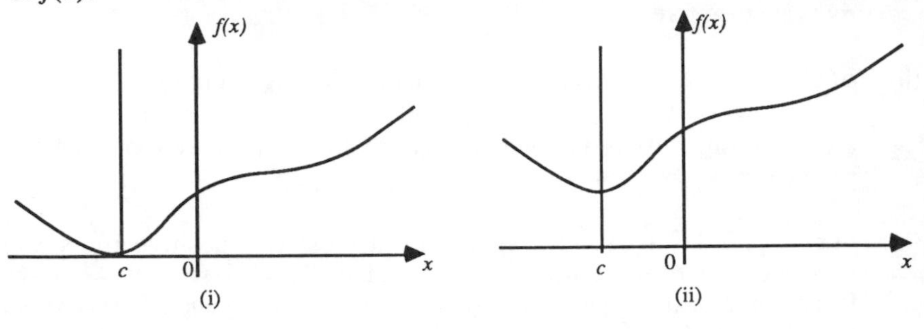

Figure 13.6

An extreme example of another problem that might arise is the function $f(x) = 6x^2 + 1 + 0.01 \ln|1-x|$. As can be verified fairly readily, this function takes negative values only in the approximate interval of 1 ± 10^{-304}. Since most computers will not be able to distinguish 10^{-304} from zero it is easy to see the difficulty, or rather the impossibility of finding the root by a standard numerical routine. Of course, you could argue that such a contrived function is virtually certain not to occur in practice, so why worry? The point is that extreme examples cause us to think carefully about the robustness of a numerical routine.

13.2 Interval Reduction Methods

In this section we look at two methods which reduce progressively the interval in which a root is known to lie. Each of the methods is known as a two-point method, since at each stage it requires two previous approximations to the root in order to obtain the next approximation.

Bisection Method

The function $f(x)$ is evaluated at the mid-point of the interval of search; one of the sub-intervals is chosen for the continuation of the search. Hence, at each stage, the interval of uncertainty is halved. A possible algorithm for the method is as follows:

STEP 1 Select x_L and x_R so that $f(x_L).f(x_R) < 0$

STEP 2 Evaluate $x_M = \frac{1}{2}(x_L + x_R)$, and $f(x_M)$. If $(x_R - x_L)$ is sufficiently small and if $|f(x_M)|$ is sufficiently small, then quote x_M as the root and stop; if not, go to Step 3

STEP 3 If $f(x_L).f(x_M) < 0$ then the root lies in the left half-interval; put $x_R = x_M$ and go to Step 2.
If $f(x_L).f(x_M) > 0$ then the root lies in the right half-interval; put $x_L = x_M$ and go to Step 2.

A possible Basic program is:

```
>
 10  CLS:@%=10
 20  PRINTTAB(10,2)"Successive Bisection"
 30  INPUTTAB(0,4)"Enter X1 and X2 ",X1,X2
 40  INPUT"Enter E1 and E2 ",E1,E2
 50  INPUT"Enter N ",N
 60  LET A=FNA(X1)
 70  LET B=FNA(X2)
 80  IF A*B>=0 THEN PRINTTAB(0,10)"No root in this
     interval ":END
 90  LET I=0
100  REPEAT
110    LET P=(X1+X2)/2
120    LET M=FNA(P)
130    IF ABS(X1-X2)<E1 AND ABS(M)<E2 THEN GOTO 180
140    IF M*A>0 THEN X1=P:A=M
150    IF M*A<=0 THEN X2=P:B=M
160    LET I=I+1
170    UNTIL I>N
180  PRINTTAB(0,10)"X1 = ";X1
190  PRINT"X2 = ";X2
200  DEF FNA(X)=X-COS(X)
210  END
```

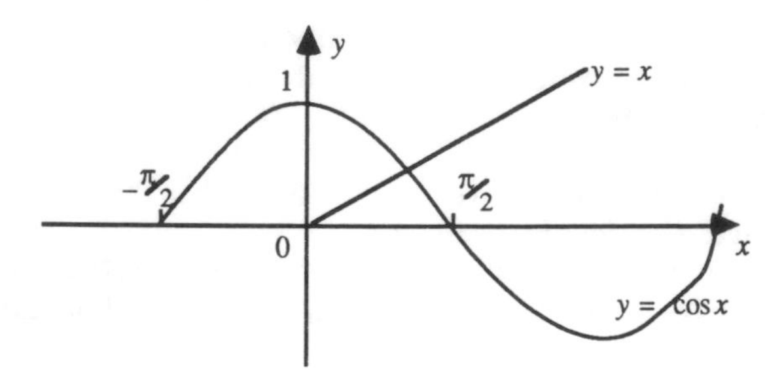

Figure 13.7

As an example of the application of the method we consider the equation $x - \cos x = 0$. From a sketch (Figure 13.7) we see that there is a root in $(0, \pi/2)$.

A simple arithmetic check shows that $f(0) = -1$, $f(\pi/2) = \pi/2$, $f(1) = 1 - 0.540302 = 0.459698$. Hence the root is in the interval (0, 1) and this is the starting point for our search. Therefore $x_L = 0$, $x_R = 1$ and $f(x_L).f(x_R) < 0$. Then $x_M = \frac{1}{2}(0 + 1) = 0.5$ and $f(x_M) = -0.37748$. Clearly the root lies in the interval (0.5, 1) and we continue our search by relabelling 0.5 as x_L before calculating $x_M = \frac{1}{2}(0.5 + 1)$. The first five iterations and the twenty-fourth are shown in Table 13.2.

Table 13.2

x_L	x_R	x_M	$f(x_L)$	$f(x_R)$	$f(x_M)$
0	1	0.5	–1	0.459698	–0.37758
0.5	1	0.75	–0.37758	0.459698	0.0183111
0.5	0.75	0.625	–0.37758	0.0183111	–0.185963
0.625	0.75	0.6875	–0.185963	0.0183111	–0.0853349
0.6875	0.75	0.71875	–0.0853349	0.0183111	–0.387937
0.739085078	0.739085198	0.739085138			7.91624×10^{-9}

At a stage where we have x_L and x_R the true root may be anywhere in the interval (x_L, x_R). The discrepancy between the true value and x_M is less than or equal to $\frac{1}{2}(x_R - x_L)$. Therefore, at stage 2 we could say that x lies in 0.75 ± 0.03125. Notice that $0.03125 = 0.71875 - 0.6875$, i.e. the difference between the current and previous values of x_M; this is generally true.

False Position Method

It is clear that the Bisection Method will continue to narrow the interval containing the root until it is as small as required. If we want to quote the root correct to say 3 d.p. then we require the $\pm$ part to be smaller than 0.0005 in size and hence we must ensure that $x_R - x_L < 0.001$. The only obvious difficulty on the horizon is that we might ask for more decimal places of accuracy than the computer or calculator can provide.

However, it might be argued that we could do better than throw away half the interval at each stage. Consider Figure 13.8.

The root is very close to x_L and if we could retain only that sub-interval (x_L, x_M) we would make a considerable reduction in the interval of search.

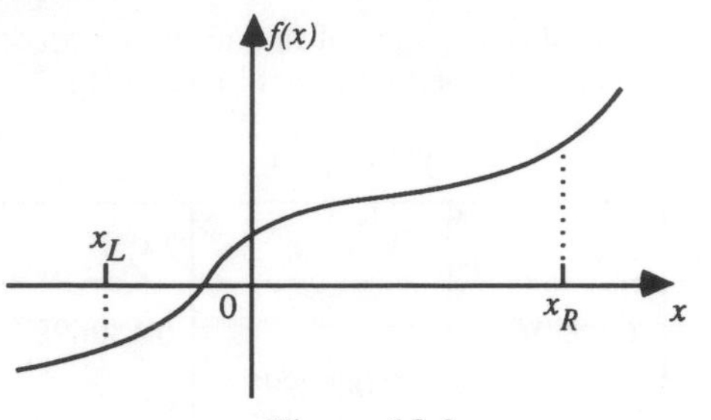

Figure 13.8

The position of x_M can be obtained by linearly interpolating the function $f(x)$ between x_L and x_R. From similar triangles it is easy to show that $\dfrac{x_R - x_M}{x_M - x_L} = -\dfrac{f(x_R)}{f(x_L)}$, the minus sign on the right-hand side being necessary since the values $f(x_R)$ and $f(x_L)$ must, by definition, be of opposite sign. Rearranging produces the formula

$$x_M = \frac{x_R\, f(x_L) - x_L\, f(x_R)}{f(x_L) - f(x_R)} \tag{13.2}$$

The strategy for the False Position Method follows precisely that for the Bisection Method except that we use formula (13.2) to evaluate x_M. In practice, as we approach the root $|f(x_L)|$ and $|f(x_R)|$ both become small and we could lose precision as a result of the subtraction on the denominator. A preferred rearrangement is:

$$x_M = x_R - \left\{ \frac{f(x_R)\,(x_L - x_R)}{f(x_L) - f(x_R)} \right\} \tag{13.3}$$

We apply the method to the example $x - \cos x = 0$.

Again we use $x_L = 0$, $x_R = 1$ but this time we **need** the values $f(x_L) = -1$, $f(x_R) = 0.459698$.

Then

$$x_M = 1 - \left\{ \frac{0.459698\ (0 - 1)}{-1 - 0.459698} \right\} = 0.68507 \qquad \text{(5 d.p.)}$$

Also $f(x_M) = -0.0892993$; therefore, we ignore (x_L, x_M) and relabel x_L as 0.68507. Once more we calculate x_M:

$$x_M = 1 - \left\{ \frac{0.459698\ (0.68507 - 1)}{-0.0892993 - 0.459698} \right\} = 0.73630.$$

Table 13.3 shows the first five stages of the iterative procedure. Compare the speed at which the root is approached with that in Table 13.2.

Table 13.3

x_L	x_R	x_M	$f(x_L)$	$f(x_R)$	$f(x_M)$
0	1	0.68507	–1	0.459698	–0.0892993
0.68507	1	0.73630	–0.0892993	"	–0.00466
0.73630	1	0.738945	–0.00466	"	–0.000234
0.738945	1	0.739078	–0.000234	"	–0.00001172
0.739078	1	0.739084	–0.00001172	"	-0.5865×10^{-6}

Note that if $f(x)$ is entirely convex in the interval (x_L, x_R) then from that stage one end-point is fixed and the other moves in towards the root. This implies a somewhat slow rate of convergence (but still faster than bisection). An example where false position is slower than bisection is provided by the equation $x^8 - 1 = 0$ with an initial interval (0, 15). Tables 13.4 and 13.5 show the first four stages and the twentieth stage for the false position method and the bisection method respectively.

Table 13.4

x_L	x_R	x_M	$f(x_L)$	$f(x_R)$	$f(x_M)$
0	1.5	0.058528	–1	24.6289	–1
0.058528	1.5	0.114772	–1	"	–0.9999997
0.114772	1.5	0.1688	0.9999997	"	–0.9999993
0.1688	1.5	0.220761	–0.185963	"	–0.999994358
"	"	"			
0.785715	1.5	0.80967315	–0.085475	"	–0.8152953

Table 13.5

x_L	x_R	x_M	$f(x_L)$	$f(x_R)$	$f(x_M)$
0	1.5	0.75	–1	24.6289	–0.899887
0.75	1.5	1.125	–0.899887	24.6289	–0.56578451
0.75	1.125	0.9375	–0.899887	0.56578451	–0.4032805
0.9375	1.125	0.984375	–0.4032805	0.56578451	0.279121187
"					
0.999999046	1.000001911	1.00000048	-7.6×10^{-6}	1.5×10^{-5}	2.8147×10^{-9}

13.3 Fixed-Point Iteration

If we rearrange the equation $x - \cos x = 0$ to the form $x = \cos x$ then the root we seek is called a **fixed point** of the function $\cos x$, since it is unaffected on applying the cosine mapping.

In general, the function $g(x)$ has at least one fixed point in $[a, b]$ if $f(x)$ is continuous on $[a, b]$ and if the values $g(x)$ satisfy $a \le g(x) \le b$ whenever $a \le x \le b$. This can be proved by considering the function $h(x) = x - g(x)$. It is simple to show that $h(x)$ is continuous on $[a, b]$. If we assume that neither a nor b is a fixed point then $g(a) > a$ and $g(b) < b$ so that $h(a) < 0$ and $h(b) > 0$. By the Intermediate Value Theorem, there is at least one point ξ in (a, b) where $h(\xi) = 0$, i.e. $\xi = g(\xi)$.

Further, if $|g'(x)| \le k < 1$ for every x in (a, b) then the fixed point is unique.

This follows from the following argument. Let ξ_1 and ξ_2 be two fixed points which lie in (a, b); then

$$\begin{aligned} |\xi_1 - \xi_2| &= |g(\xi_1) - g(\xi_2)| \\ &= |g'(\xi)|.|\xi_1 - \xi_2| \quad \text{by the Mean Value Theorem} \\ &\le k.|\xi_1 - \xi_2| < |\xi_1 - \xi_2|. \end{aligned}$$

Following through the chain of inequalities we see that $|\xi_1 - \xi_2| < |\xi_1 - \xi_2|$ which can never occur. Therefore $\xi_1 = \xi_2$, establishing uniqueness.

To see the implication of this, consider first our example with $g(x) = \cos x$ and $(a, b) = (0, 1)$. For all x in $(0, 1)$, $\cos x$ also lies in $(0, 1)$ and $|g'(x)| = |\sin x| \le \sin 1 < 1$ hence there is a unique fixed point.

However, for the example $g(x) = e^{-2x}$ and $(a, b) = (0, 1)$ it is straightforward to show graphically that there is a unique fixed point of approximate value 0.36. However $|y'(x)| = |-2e^{-2x}| \ge 1$ for some values of x in $(0, 1)$.

The condition $|g'(x)| \le k < 1$ is therefore a **sufficient** condition for uniqueness, but not a necessary condition. To find the fixed point for $g(x) = \cos x$ we make an initial guess $x_0 = 0.5$ and generate $x_1 = \cos x_0 = 0.87758$. Then we can continue the process to obtain $x_2 = \cos x_1 = 0.63901$, $x_3 = \cos x_2 = 0.80269$, and similarly $x_4 = \cos x_3 = 0.694778$. (You can easily check these values by putting a pocket calculator into radian mode, entering 0.5 and repeatedly pressing the 'cosine' button.) The values of x_{20} and x_{21} are, respectively, 0.73904958 and 0.73910908, and these agree with the required value to 3 d.p. The process of using the iterative formula

$$x_{r+1} = g(x_r)$$

repeatedly to obtain better approximations to the fixed point is sometimes known as *basic iteration.*

An algorithm for fixed-point iteration follows.

STEP 1 Put a counter, i, equal to 1 and input the maximum permitted number of iterations, n

STEP 2 Input an initial approximation x_0

STEP 3 While $i \leq n$ carry out Steps 4 to 6

STEP 4 Calculate $x_1 = g(x_0)$

STEP 5 If $|x_1 - x_0|$ is sufficiently small then output x_1 and stop; if not, go to Step 6

STEP 6 Increase i by 1 and put $x_0 = x_1$

STEP 7 Output a message to indicate that the method has not converged after n iterations.

A suitable basic program is:

```
>
 10   CLS
 20   PRINTTAB(10,2)"Fixed Point Iteration"
 30   INPUTTAB(0,4)"Enter accuracy E ",E
 40   INPUT"Enter number of iterations N ",N
 50   INPUT"Enter initial guess X0 ",X0
 60   LET XNEW=X0
 70   LET I=0
 80   REPEAT
 90     LET XOLD=XNEW
100     LET XNEW=FNA(XOLD)
110     IF ABS(XOLD-XNEW)<E THEN PRINTTAB(0,10)
        "The root is ";XNEW:END
120     LET I=I+1
130     UNTIL I>N
140   PRINTTAB(0,10)"Accuracy not achieved"
150   DEF FNA(X)=COS(X)
```

Note that Step 5 could be modified to allow the condition $|g(x_1)|$ sufficiently small as an alternative or an additional criterion for convergence. What implications does this have for the order in which the calculations are carried out?

We can apply the Mean Value Theorem to establish a criterion for convergence. Let the fixed point be x. Then

$$\begin{aligned} x_{r+1} - x &= g(x_r) - g(x) \\ &= g'(\xi)(x_r - x) \qquad \text{for at least one } \xi \text{ in } (x_r, x) \end{aligned}$$

Now let $e_i = x_r - x$ and $e_{i+1} = x_{r+1} - x$.

Then $\quad e_{i+1} \quad = g'(\xi)e_i$

so that $\quad |e_{i+1}| \quad = |g'(\xi)|.|e_i| \qquad (13.4)$

For the process to converge we require that (usually)

$$|e_{i+1}| < |e_i|$$

so that $|g'(\xi)| \leq k < 1$.

As before this is a sufficient condition, but not necessary. Note that if $g'(\xi) < 0$ then e_i and e_{i+1} are of opposite signs so that the iterates are alternatively above and below the correct value of the fixed point.

Example

Consider the equation $x^3 + 5x^2 - 12 = 0$. In seeking to rearrange this as $x = g(x)$ we have many different choices. Here are some of them:

(a) $x = \left[\dfrac{12 - x^3}{5}\right]^{1/2}$ (b) $x = \left[12 - 5x^2\right]^{1/3}$

(c) $x = \left[\dfrac{12}{x} - 5x\right]^{1/2}$ (d) $x = x - x^3 - 5x^2 + 12$

(e) $x = \left[\dfrac{12}{5 + x}\right]^{1/2}$ (f) $x = x - \dfrac{(x^3 + 5x^2 - 12)}{(3x^2 + 10x)}$

Check that each of these equations is, indeed, a rearrangement of the original equation. We programmed a BBC micro to carry out the iterative process using each rearrangement. The equation has one positive root (as Descartes' rule of signs will readily show) and this occurs in (1, 2) since $x^3 + 5x^2 - 12$ takes opposite signs at $x = 1$ and $x = 2$; we use an initial approximation of $x_0 = 1.5$. Table 13.6 summarises the results of the computations.

Table 13.6

n	(a)	(b)	(c)	(d)	(e)	(f)
1	1.31339255	0.908560296	0.707106781	-1.125	1.35873244	1.37931034
2	1.39530626	1.989325709	3.6653825	5.9707313	1.37374254	1.37230446
3	1.36260829	-1.982097214	Attempt to take square root of negative number	-373.127138	1.37212401	1.37228132
4	1.37622986			51251720.3	1.37229826	1.37228132
5	1.37065012				1.3722795	1.37228132
15	1.372281095					

The results are dramatically different: (d), (c) and (b) go haywire; (a) converges slowly; (e) converges quite quickly and (f) converges very quickly indeed.

If we have an equation in the form $f(x) = 0$ it would be useful to know whether a particular rearrangement $x = g(x)$ will lead to a convergent iteration formula $x_{n+1} = g(x_n)$ and which rearrangement leads to the most rapid convergence. The key is provided by the behaviour of $g'(x)$. It will be easier to see what happens by considering a simpler example. The equation $x - x^2 = 0$ has roots $x = 0$ and $x = 1$.

The rearrangements we consider are:

(a) $x = x^2$ and (b) $x = \sqrt{x}$

In each case we start with initial guesses 0.2, 0.8 and 1.2. Table 13.7 summarises the results.

For rearrangement (a), whereas we locate the root at 0 easily enough, it seems that we are unable to find the root at 1; it is almost as though we are repelled from it. On the other hand the reverse is true for rearrangement (b). Why should this be?

Table 13.7

n	(a)			(b)		
0	0.2	0.8	1.2	0.2	0.8	1.2
1	0.04	0.64	1.44	0.4472	0.8944	1.0954
2	0.0016	0.4096	2.0736	0.6687	0.9457	1.0466
3	0.000256	0.1678	4.2998	0.8178	0.9725	1.0231
	↓	↓		↓	↓	↓
	0	0		1	1	1

In the case of rearrangement (a), $g(x) = x^2$ so that $g'(x) = 2x$; as we approach $x = 1$ the value of $g'(x)$ is approximately 2 and from (13.4) we see that the error after each iteration is approximately twice as large as that before.

With rearrangement (b), $g'(x) = \dfrac{1}{2x^{1/2}}$ which has value $\frac{1}{2}$ at $x = 1$ and gets increasingly larger as x approaches 0.

If we examine the errors associated with the sequence of guesses (0.8, 0.8944, 0.9457, 0.9725) we see that these are respectively {0.2, 0.1056, 0.0543, 0.0275} so that the errors are approximately halved at each step. The fact that $g'(x)$ is positive in all cases indicates that the approach to the root is always from the same side. Referring back to Table 13.6 we see that there is an advantage in the successive guesses being alternatively above and below the true value: in case (a) for example with $n = 4$ and $n = 5$ we can say that the root lies between the successive guesses 1.37622986 and 1.37065012; in general, the root is bracketed by each pair of successive guesses. Figure 13.9 shows the four situations that arise depending on the nature of $g'(x)$ near the root.

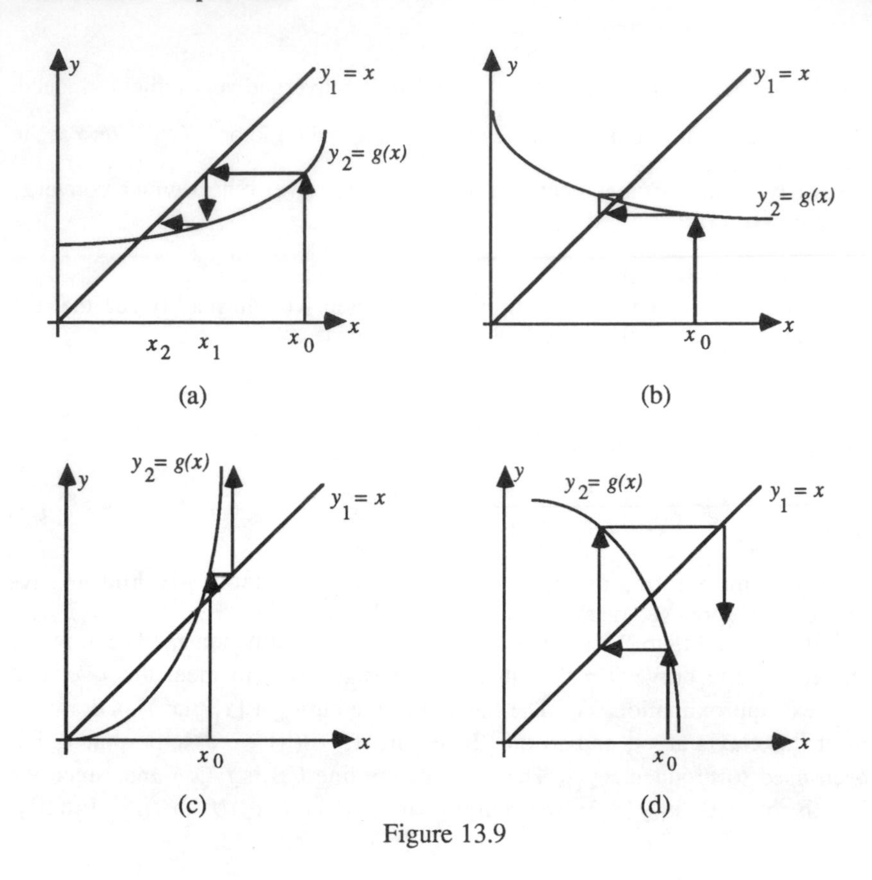

Figure 13.9

13.4 Newton-Raphson and Secant Methods

In the last section we took the equation

$$x^3 + 5x^2 - 12 = 0$$

and rearranged it in the form $x = g(x)$ in six ways. The last of these, probably the least obvious, was

$$x = x - \frac{(x^3 + 5x^2 - 12)}{(3x^2 + 10x)}$$

and this led to the most rapidly converging iterative formula of the six. In equation (3.2) we used the iterative formula

$$x_{r+1} = \frac{x_r^2 + A}{2x_r} \qquad \text{or} \qquad x_{r+1} = x_r - \frac{(x_r^2 - A)}{2x_r}$$

in order to approximate $\sqrt{A}$ and we found that it converged very rapidly. Indeed, you can show that in both cases the approximate scale factor $|g'(x)|$ is zero at the root in question.

Clearly, such rearrangements are to be preferred over others which converge more slowly.

Newton-Raphson Method

If we consider the general equation $f(x) = 0$ then we can readily see that the examples above are both related to the rearrangement

$$x = x - \frac{f(x)}{f'(x)}$$

and the companion iterative formula

$$x_{r+1} = x_r - \frac{f(x_r)}{f'(x_r)} \tag{13.5}$$

This formula forms the basis of the **Newton-Raphson Method** and we shall derive it both geometrically and analytically.

Referring to Figure 13.10(i), we start with an approximation x_0. We draw the tangent to the curve $y = f(x)$ at the point $[x_0, f(x_0)]$ to meet the x-axis at the next approximation, x_1. The tangent to the curve at $[x_1, f(x_1)]$ is drawn to meet the x-axis at x_2, and so on. In Figure 13.10(ii) we assume that x_r has been used to produce x_{r+1}. The slope of the line CB is $f'(x_r)$ and, since the length of AB is $f(x_r)$, the length of CA is $f(x_r)/f'(x_r)$. Finally, OC = OA – CA i.e. $x_{r+1} = x_r - f(x_r)/f'(x_r)$, as required.

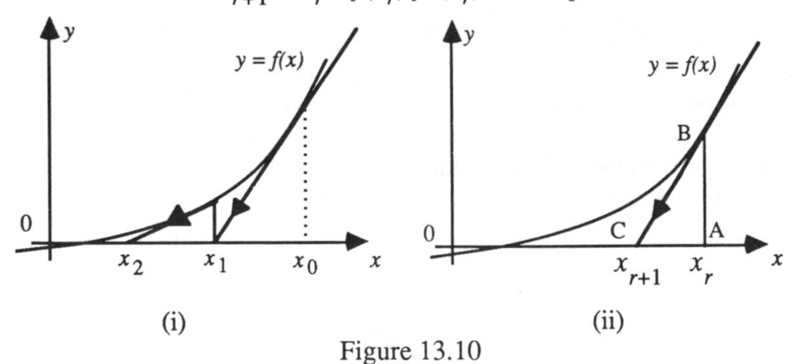

Figure 13.10

The analytical derivation starts from a result which appears in Section 16.5:

If $f(x)$ and its first derivative $f'(x)$ are both continuous in $a \le x \le a + h$ and if $f''(x)$ exists in $a < x < a + h$ then

$$f(a + h) = f(a) + hf'(a) + \frac{h^2}{2} f''(a + \theta h) \tag{13.6}$$

where $0 < \theta < 1$.

First, we put $a = x_r,\ a + h = x_{r+1}$ to obtain

$$f(x_{r+1}) \quad = f(x_r) + (x_{r+1} - x_r) f'(x_r) + \frac{(x_{r+1} - x_r)^2}{2} f''(\xi)$$

where ξ lies between x_r and x_{r+1}.

If $x_{r+1} - x_r$ is small then we can ignore the last term on the right-hand side; further, if x_{r+1} is chosen to be where $f(x_{r+1}) = 0$ then we obtain

$$0 \simeq f(x_r) + (x_{r+1} - x_r) f'(x_r) \tag{13.7}$$

from which we can readily obtain (13.5).

Because we have ignored a term, then x_{r+1} can only be but an approximation to the required root.

The algorithm for the Newton-Raphson method is very similar to that given for fixed-point iteration. Step 4 is replaced by

STEP 4 Calculate $x_1 = x_0 - f(x_0)/f'(x_0)$.

The changes to the Basic program are simple, but for completeness we reproduce the program in full.

```
>
  10  CLS:@%=10
  20  PRINTTAB(10,2)"Newton-Raphson"
  30  INPUTTAB(0,4)"Enter accuracy E ",E
  40  INPUT"Enter number of iterations N ",N
  50  INPUT"Enter initial guess X0 ",X0
  60  LET XNEW=X0
  70  LET I=0
  80  REPEAT
  90    LET XOLD=XNEW
 100    LET XNEW=XOLD-(XOLD-FNA(XOLD))/(FNB(XOLD))
 110    IF ABS(XOLD-XNEW)<E THEN PRINTTAB(0,10)
        "The root is ";XNEW
 120    IF ABS(XOLD-XNEW)<E THEN PRINT"After ";I;
        " iterations":END
 130    LET I=I+1
 140    UNTIL I>N
 150  PRINTTAB(0,10)"Accuracy not achieved"
 160  DEF FNA(X)=(COS(X))
 170  DEF FNB(X)=(1+SIN(X))
 180  END
```

Returning to (13.6) we put $a = x_r$ and $a + h$ is the required root x ($f(a + h)$ is then zero) so that $h = x - x_r$ and

$$0 = f(x_r) + (x - x_r)f'(x_r) + \frac{(x - x_r)^2}{2}f''(\xi) \tag{13.8}$$

where ξ lies between x and x_r.

If the error in the approximation x_r is given by $e_r = x_r - x$ then using (13.5) and (13.8)

$$e_{r+1} = x_{r+1} - x \qquad = x_r - \frac{f(x_r)}{f'(x_r)} - x$$

$$= -\frac{f(x_r)}{f'(x_r)} + \left\{\frac{f(x_r)}{f'(x_r)} + \frac{e_r^2}{2}\frac{f''(\xi)}{f'(x_r)}\right\}$$

$$= \left[\frac{f''(\xi)}{2f'(x_r)}\right]e_r^2 \tag{13.9}$$

Hence the error in x_{r+1} is proportional to the square of the error in x_r; should the process converge we say that it is **quadratically convergent**. In contrast, basic iteration formulae for which $g'(x) \neq 0$ are linearly convergent.

Example

Consider $x - \cos x = 0$.

Here $f(x) = x - \cos x$, $f'(x) = 1 + \sin x$ and $f''(x) = \cos x$ so that (13.5) becomes

$$x_{r+1} = x_r - \frac{(x_r - \cos x_r)}{(1 + \sin x_r)} \equiv x_r - c_r$$

Table 13.8 shows the first four iterations using an initial approximation of $x_0 = 0.8$.

Table13.8

r	x_r	$\cos x_r$	c_r	x_{r+1}	e_r
0	0.8	0.696706709	0.0601466939	0.739853306	0.0609148671
1	0.739853306	0.738567464	7.68043011E–4	0.739085263	7.68173253E–4
2	0.739085263	0.739085045	1.30075929E–7	0.739085133	1.301523E–7
3	0.739085133	0.739085133	–1.39118469E–10	0.739085133	–

Now $e_1 \cong 0.21e_0^2$ and $e_2 \cong 0.22e_1^2$. Notice that the number of significant figures which are correct is three for x_1 and six for x_2, illustrating the quadratic convergence: each step approximately doubles the number of correct significant figures.

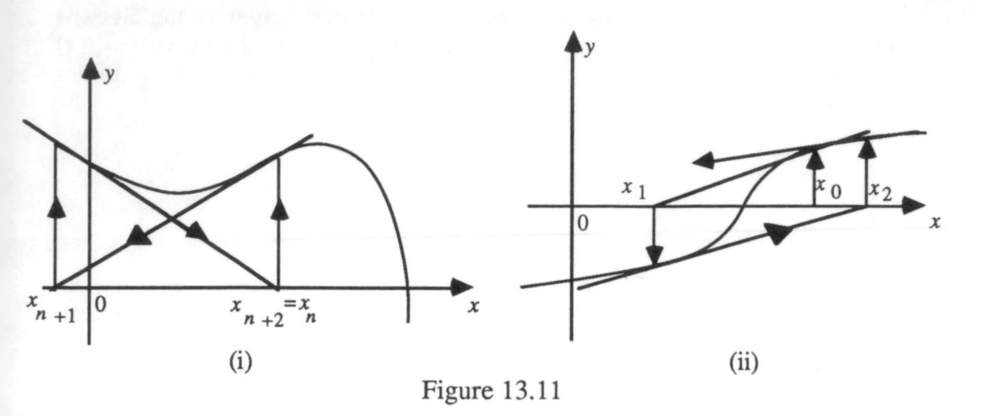

Figure 13.11

We must expect to pay a price for such an effective method; it is a method prone to failure under circumstances which are not always easy to predict in a specific example. Formula (13.9) indicates that if $f'(x)$ is small or even zero near the root then we can expect trouble. Figure 13.11 depicts some cases where failure to converge may result. In Figure 13.11(i) the iterations actually get into an oscillatory cycle because of the local minimum in the neighbourhood of the root; in Figure 13.11(ii) the point of inflection causes a close initial approximation to take us further away from the root.

It is wise to use the Newton-Raphson method as a fine tuning device: when we are sufficiently close to the root, as obtained by another method, so that we can be sure of no pitfalls, then applying the Newton-Raphson technique will rapidly provide high accuracy. Some computer routines combine the method with a safer technique, such as bisection, to keep the approximations from straying.

Secant Method

A problem with the Newton-Raphson approach, especially if a hand calculator is being used, is the effort involved in calculating the derivative at each stage. (Indeed, despite the fewer iterations needed, the total calculation effort to achieve a particular accuracy may be more than with a simple basic iteration).

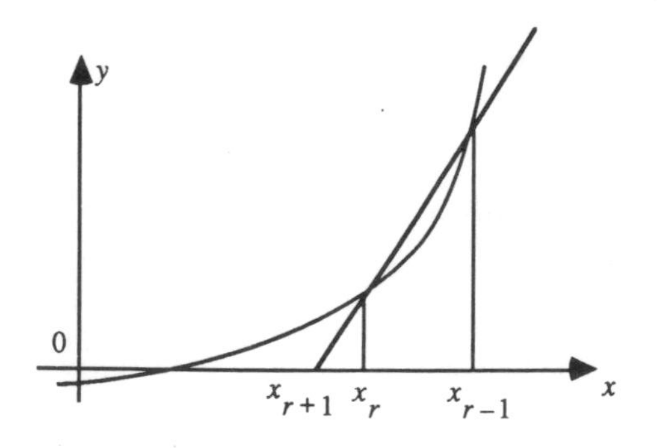

Figure 13.12

A variation on Newton-Raphson which avoids derivatives is the **Secant Method**, illustrated in Figure 13.12. It uses the secant to the curve $y = f(x)$ employing two previous approximations x_r and x_{r-1} to generate x_{r+1}, rather than using the tangent at x_r. If we substitute the approximation

$$f'(x_r) \cong \frac{f(x_{r-1}) - f(x_r)}{x_{r-1} - x_r}$$

into (13.5) we obtain

$$x_{r+1} = x_r - \frac{(x_{r-1} - x_r)f(x_r)}{f(x_{r-1}) - f(x_r)} \quad (13.10)$$

or, alternatively,

$$x_{r+1} = \frac{x_r f(x_{r-1}) - x_{r-1} f(x_r)}{f(x_{r-1}) - f(x_r)} \quad (13.11)$$

The form (13.10) is preferred for computations since $f(x_r)$ and $f(x_{r-1})$ will both be close to zero as we approach the root and it is the more stable of the two formulae. If any loss of significant figures is to occur, it is better that this is in the correction and not in the approximation. The form (13.11) does suggest a similarity with the method of false position: indeed, the secant method is sometimes called the method of linear extrapolation. It is a two-point method, whereas Newton-Raphson is a one-point method.

After using (13.10) to generate x_{r+1}, x_{r-1} takes on the value of x_r and x_r takes on the value of x_{r+1}; the old value of x_{r-1} is discarded.

The secant algorithm is

STEP 1 Put a counter, i, = 2 and input the maximum number of iterations, n

STEP 2 Input initial approximations x_0 and x_1

STEP 3 Calculate $y_0 = f(x_0)$ and $y_1 = f(x_1)$

STEP 4 While $i \le n$ carry out Steps 5 to 7

STEP 5 Calculate $x_2 = x_1 - y_1(x_1 - x_0)/(y_1 - y_0)$

STEP 6 If $|x_2 - x_1|$ is sufficiently small then output x_2 and stop; if not, go to Step 7

STEP 7 Increase i by 1 and put $x_0 = x_1$, $y_0 = y_1$, $x_1 = x_2$, $y_1 = y_2$ and go to Step 4

STEP 8 Output a message to indicate that the method has not converged after n iterations.

It is suggested that you write your own program to carry out the algorithm above.

Example
We take the equation $x - \cos x = 0$ and, for the purposes of comparison with the method of false position we take $x_0 = 0$, $x_1 = 1$. See Table 13.9; c_r is the correction to x_r in (13.10).

Table 13.9

r	x_{r-1}	x_r	x_{r+1}	c_r
1	0	1	0.685073357	0.3149226643
2	1	0.685073357	0.736298997	–5.12256402E–2
3	0.685073357	0.736298997	0.739119362	–2.8036441E–3
4	0.736298907	0.739119362	0.739085112	3.4249859E–5
5	0.739119362	0.739085112	0.739081133	–2.1006803E–8
6	0.739085112	0.739085133	0.739085133	–1.39698386E–10

Compare these results with Tables 13.3 and 13.8.

The convergence is slower than Newton-Raphson; it has been shown that e_{r+1} is proportional to e_r^{τ} where τ is the **golden ratio** whose value is approximately 1.618.

13.5 Polynomial equations

An equation of the form $p(x) = 0$ where $p(x)$ is a polynomial is the subject of this section. Such equations have certain special features which suggest specific methods of root-finding. In this section we look briefly at some of the problems involved and the methods in current use.

High degree polynomials, such as occur in the study of the stability of linear control systems, are often ill-conditioned in the sense shown below, making even their evaluation a process needing careful handling. J H Wilkinson studied the equation

$$(x-1)(x-2)(x-3) \;.........\; (x-20) = 0.$$

The roots of this equation are obvious. However, if the coefficient of x^{19}, which should be –210, is stored in the computer with a round-off error of 2^{-23} (not untypical of the storage error of many computers) then when the resulting equation is solved a totally different picture emerges. The first four roots are exact to 9 d.p., then accuracy deteriorates; the root $x = 9$ is obtained as 8.917250249, there is a root of 20.846908101 and the remaining roots appear as 5 pairs of complex conjugates with significant imaginary parts.

A second example of ill-conditioning occurs when considering the two equations

$$p_1(x) \equiv x^3 - 4x + 3.0 = 0$$

and

$$p_2(x) \equiv x^3 - 4x + 3.1 = 0$$

The first of these equations has three real roots at 1.3028, 1, –2.3028 whereas the second has only one real root at –2.3111 (If you study the functions and their derived functions, it is easy to see what the reason for the change is.).

The method of nested multiplication has already been introduced in Problem 8 (p. 28) as a means of reducing round-off error in the evaluation of a polynomial. Its use is clearly of importance here.

A second problem arises in the case of multiple roots. The equation $(x-1)(x-2)^2(x-3) = 0$ has a double root at $x = 2$, yet evaluating the polynomial at $x = 1.9$ and $x = 2.1$ gives in each case a negative result so that neither bisection nor false position could be employed to find it. Even when we used Newton-Raphson to locate the root its performance was very slow. (Whereas for the equation $(x-1)(x-2)(x-3) = 0$ an initial guess of 1.6 led to successive approximations of 2.4615385, 1.96354331, 2.0000973 and 2; for the given equation, $(x-1)(x-2)^2(x-3) = 0$, six iterations had produced 1.995372154 with the error being approximately halved at each step: Newton-Raphson was being reduced to a linearly convergent method.) Following a strategy of Ralston and Rabinowitz we used the iteration formula

$$x_{r+1} = x_r - m\,f(x_r)/f\,'(x_r) \tag{13.12}$$

where m is the multiplicity of the root, in this case 2. This gave successive approximations of 2.0941176, 1.9915124, 2.00000001, 2, indicating that quadratic convergence has been restored. Of course, we would have to know that the root near 2 was a double root in order to apply (13.12).

Now we consider a method of dividing a polynomial by a linear or a quadratic factor: the method of **synthetic division**.

Synthetic Division by a Linear Factor

The polynomial $p(x) = a_0x^3 + a_1x^2 + a_2x + a_3$ can be nested as

$$[(a_0x + a_1)x + a_2]x + a_3$$

and this latter form requires only 3 multiplications and 3 additions, as opposed to the $3 + 2 + 1 = 6$ multiplications and 3 additions of the first form. This is clearly more efficient and less round-off error is likely.

If we consider the division of the polynomial by a factor $(x - x_1)$ in its original form we have

$$p(x) = (x - x_1)(b_0x^2 + b_1x + b_2) + b_3$$

Multiplying out the R.H.S. and comparing coefficients of like powers of x,

$$\begin{aligned}
x^3 &: a_0 = b_0 && \Rightarrow b_0 = a_0 \\
x^2 &: a_1 = b_1 - b_0x_1 && \Rightarrow b_1 = a_1 + b_0x_1 \\
x &: a_2 = b_2 - b_1x_1 && \Rightarrow b_2 = a_2 + b_1x_1 \\
& \quad a_3 = b_3 - b_2x_1 && \Rightarrow b_3 = a_3 + b_2x_1
\end{aligned}$$

Apart from b_0, all the coefficients may be generated from the *recurrence formula*

$$b_r = a_r + b_{r-1}\,x_1$$

Putting $x = x_1$ in the nested form gives

$$[(a_0x_1 + a_1)x_1 + a_2]x_1 + a_3$$

or

$$[(b_0x_1 + a_1)x_1 + a_2]x_1 + a_3$$

or

$$[(b_1x_1 + a_2)x_1] + a_3$$

or

$$b_2x_1 + a_3$$

You should be able to see that the processes are the same.

Example 1

First we carry out, in traditional fashion, the division of $2x^3 + 3x^2 - 10x + 6$ by $x - 2$

$$\begin{array}{rl} & 2x^2 + 7x + 4 \\ \hline x - 2 \;) & \mathbf{2}x^3 + \mathbf{3}x^2 - \mathbf{10}x + \mathbf{6} \\ & 2x^3 - 4x^2 \\ \hline & \mathbf{7}x^2 - 10x \\ & 7x^2 - \mathbf{14}x \\ \hline & \mathbf{4}x + 6 \\ & 4x - \mathbf{8} \\ \hline & \mathbf{14} \end{array}$$

We see that the *quotient* is $2x^2 + 7x + 4$ and the *remainder* is 14.

In general terms $\quad p(x) = (x - 2)(2x^2 + 7x + 4) + 14$

therefore $\quad p(2) = 14$

and so if $\quad p(x) = (x - a)q(x) + r$

then $\quad p(a) = r$

Notice that the only essential quantities in the process are the coefficients in bold-face type. We now lay them out so that their roles are apparent.

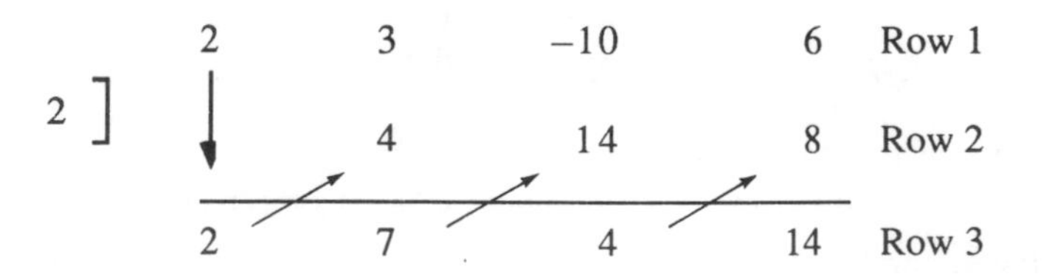

We proceed as follows. First the coefficients of the polynomial are written in Row 1, the number **–2** in the divisor is reversed in sign and placed to the left of the symbol] in Row 2 and the coefficient of x^3 is repeated in Row 3. Then this number is multiplied by the *reversed sign number* **2** from Row 2 to give the **4** in Row 2. This is added to the number above to give the entry **7** in Row 3 and the steps repeated until the *remainder* **14** is reached in Row 3. The first three numbers in that row are the coefficients in the *quotient.*

Example 2

We work through the example $(3x^3 + 2x - 6) \div (x + 2)$

	3	0	2	−6
−2]		−6	12	−28
	3	−6	14	−34

This gives a quotient of $3x^2 - 6x + 14$ and a remainder of −34.

It may be shown that, having divided $p(x)$ by $(x - a)$ to obtain a quotient $q(x)$ and a remainder $r = p(a)$, we may divide $q(x)$ by $(x - a)$ to obtain $p'(a)$.

The technique may be applied with the Newton-Raphson method as in the following example.

Example 3

To solve $p(x) = x^3 + 9x^2 + 23x + 14 = 0$, given that one root is near $x_0 = -0.7$, we divide by $x + 0.7$ as follows:

	1	9	23	14
−0.7]		−0.7	−5.81	−12.033
	1	8.3	17.19	[1.967 $= p(x_0)$]
−0.7]		−0.7	−5.32	
	1	7.6	[11.87 $= p'(x_0)$]	

Then we obtain

$$x_1 = x_0 - \frac{p(x_0)}{p'(x_0)} = -0.7 - \frac{1.967}{11.87} = -0.866 \qquad \text{(3 d.p.)}$$

Repetition of this process yields

$$x_2 = -0.88485 \text{ (5 d.p.)} \quad \text{and} \quad x_3 = -0.88509 \text{ (5 d.p.)}$$

Bearing in mind the small correction we would suspect this last approximation to be a very good one, a verification coming from the value of $p(x_3) = 0.000001$ (6 d.p.).

We may then divide $p(x)$ by the factor $(x + 0.88509)$ to give $(x^2 + 8.11491x + 15.81756)$ and this can be solved by the usual formula to yield the other roots as −4.86081 and −3.25410 (5 d.p.); their accuracy may be checked likewise, using nested multiplication (of course), and the values of $p(x)$ are −0.000009 and 0.000011 respectively, both to 6 d.p.

The above scheme is sometimes referred to as the **Birge-Vieta** method.

A program for the Birge-Vieta method follows:

```
>
  10   CLS:@%=10
  20   PRINTTAB(10,2)"Birge-Vieta"
  30   INPUTTAB(0,4)"Enter accuracy E ",E
  40   INPUT"Enter number of iterations N ",N
  50   INPUT"Enter initial guess X0 ",X0
  60   DIM A(N,4)
  70   FOR I=0 TO 3
  80     PRINT"Enter A";I;
  90     INPUTA(I,0)
 100     NEXT I
 110   REPEAT
 120     LET A(0,1)=X0
 130     LET A(0,2)=A(0,0)
 140     FOR I=1 TO N
 150       LET A(I,1)=A(I-1,2)*A(0,1)
 160       LET A(I,2)=A(I,1)+A(I,0)
 170       NEXT I
 180     LET PA=A(3,2)
 190     LET A(0,4)=A(0,0)
 200     FOR I=1 TO N-1
 210       A(I,3)=A(I-1,4)*A(0,1)
 220       A(I,4)=A(I,3)+A(I,2)
 230       NEXT I
 240     DPA=A(2,4)
 250     XN=X0-PA/DPA
 260     IF ABS(XN-X0)>E THEN X0=XN
 270     UNTIL ABS(XN-X0)<=E
 280   PRINTTAB(0,15)"The root is ";XN
 290   END
```

Evaluation of Quadratic Factors

We know that a polynomial with real coefficients has roots which are either real or occur in complex conjugate pairs so that if $(x - a - ib)$ and $(x - a + ib)$ are factors of $p(x)$, so is their product $(x^2 - 2ax + a^2 + b^2)$.

Suppose we divide $p(x)$ by $x^2 + cx + d$

where $p(x) = a_0x^n + a_1x^{n-1} + ... + a_n$ giving a remainder $R(x)$, i.e.

$$p(x) = (x^2 + cx + d)q(x) + R(x) \tag{13.13}$$

where

$$q(x) = b_0x^{n-2} + b_1x^{n-3} + ... + b_{n-2}$$

and

$$R(x) = b_{n-1}(x + c) + b_n$$

In general, $R(x)$ will not vanish unless $q(x)$ is an *exact* factor of $p(x)$.

Comparing coefficients in (13.13) we have

$$\begin{aligned}
b_0 &= a_0 \\
b_1 &= a_1 - cb_0 \\
b_2 &= a_2 - cb_1 - db_0 \\
&\vdots \\
b_r &= a_r - cb_{r-1} - db_{r-2} \\
&\vdots \\
b_{n-1} &= a_{n-1} - cb_{n-2} - db_{n-3} \\
b_n &= a_n - cb_{n-1} - db_{n-2}
\end{aligned}$$

We require that b_{n-1} and b_n should vanish *simultaneously*. The following technique is presented as a method of achieving this.

Bairstow's Method

We consider this method via an example.

$$x^4 + 5x^3 + 12x^2 + 14x + 8 = 0$$

First, we carry out a synthetic division using the first approximation $x^2 + 1.970x + 1.965$ to obtain a quotient of $q(x) = x^2 + 3.03x + 4.0659$ and a remainder of $0.0362x + 0.0105$ (i.e. $rx + s$).

We then take $xq(x)$ and $q(x)$ and synthetically divide these by $x^2 + 1.970x + 1.965$ as follows:

	$xq(x)$				$q(x)$		
	1	3.03	4.066	0	1	3.03	4.066
−1.970]		−1.970	−2.088			−1.970	
−1.965]			−1.965	−2.083			−1.965
	1	1.060	0.013	−2.083	1	1.060	2.101
			$\parallel$	$\parallel$		$\parallel$	$\parallel$
			α	β		γ	δ

We then solve the equations

$$\begin{aligned}
r &= \alpha\Delta c + \gamma\Delta d \\
s &= \beta\Delta c + \delta\Delta d
\end{aligned}$$

for the corrections Δc, Δd to our estimates c_0, d_0.

In our example we solve

$$\begin{aligned}
0.0362 &= 0.013\Delta c + 1.06\Delta d \\
0.0105 &= -2.083\Delta c + 2.101\Delta d
\end{aligned}$$

giving $\Delta c = 0.0291$, $\Delta d = 0.0337$.

Therefore $$c_1 = 1.970 + 0.0291 = 1.9991$$
$$d_1 = 1.965 + 0.337 = 1.9987$$

and a better approximation of $x^2 + 1.9991x + 1.9987$.

Note that $q(x)$ is obtained ready for further quadratic factors to be found.

The process of dividing a polynomial by a known factor is called **deflation**. Each time that a new root is located the corresponding factor is divided out; unfortunately, errors can build up and the later factors (and roots) can include large errors. There are ways in which this build-up of errors can be reduced.

Some methods which are in current use attempt to overcome the problem by finding all the roots simultaneously; a popular example of this is the **quotient-difference algorithm**, but this is beyond our scope.

Problems

Section 13.1

1 By sketching suitable functions, obtain information about the number, the nature and, where possible, the location of the roots of the following equations.

(i) $x^3 + 2x^2 + 1 = 0$ (ii) $x^3 - 2x^2 - 1 = 0$
(iii) $100 - x - (2/x) = 0$ (iv) $7 - x - (9/x) = 0$
(v) $\sin x - e^{-2x} = 0$ (vi) $x - 1 - \sin 2x = 0$

2 Find the number and nature of the roots of

(i) $3x^3 - 2x^2 + 4x - 1 = 0$
(ii) $3x^3 + 6x^2 - 2x - 4 = 0$
(iii) $3x^3 - 8x^2 - 7x - 1 = 0$
(iv) $x^4 - 8x^3 - 8x^2 + 7x + 6 = 0$

Section 13.2

3 For each of the equations below, verify that there is a root between the two values of x given in parentheses following the equation. Use three steps of the bisection method to improve the estimation of the root.

(i) $1.748x^2 - 3.5x - 5.254 = 0$ (2.8, 3.2)
(ii) $0.222x^3 - 1.3x^2 + 2.07x - 0.7 = 0$ (0.4, 0.6)
(iii) $1.316x^5 - 17.36x^4 + 83.2x^3 - 176.18x^2 + 158.7x - 46.67 = 0$ (4.5, 5)
(iv) $3.12 - 7.321x + 5.7655x^2 - 1.23459x^3 = 0$ (2.8, 3)
(v) $\tan x = 1.1x$ (0.4, 0.6)
(vi) $\ln x = 0.6$ (1, 2)

4 Repeat Problem 3 using the method of false position.

Section 13.3

5 Solve, using Basic Iteration, the equation $xe^x = 4$ correct to 2 d.p. Then solve $xe^x = 2$, $xe^x = 10$ to the same accuracy.

6 Find algebraically the roots of the equation

$$x^4 + x^2 = 20$$

Three possible rearrangements with their scale factors are

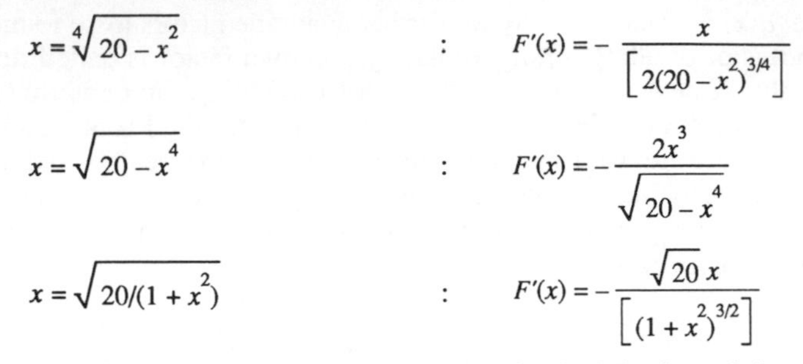

$$x = \sqrt[4]{20 - x^2} \quad : \quad F'(x) = -\frac{x}{\left[2(20 - x^2)^{3/4}\right]}$$

$$x = \sqrt{20 - x^4} \quad : \quad F'(x) = -\frac{2x^3}{\sqrt{20 - x^4}}$$

$$x = \sqrt{20/(1 + x^2)} \quad : \quad F'(x) = -\frac{\sqrt{20}\, x}{\left[(1 + x^2)^{3/2}\right]}$$

Verify that the scale factors are as quoted. Predict the behaviour of the corresponding iterative formulae and check by carrying out 3 iterations with each to find each root.

7 Find algebraically the roots of the equation $x^2 - 3x + 2 = 0$. Find four rearrangements of the equation and, where possible, the scale factor associated with each; hence predict the behaviour of the companion iterative formulae. Check your predictions by using the formulae you have obtained.

8 Show that $x = (x^2 + 2)/(2x - 1)$ is a rearrangement of $x^2 - x - 2 = 0$. Find algebraically the roots of this equation.
Find the scale factor for this rearrangement at the two roots.
Then use Basic Iteration with $x_0 = -2, 0, 1, 3$.

9 Find the positive root of the equation $\sinh x = 5x$ by iteration and the other roots by inspection.

10 It is suggested that the formula $x_{n+1} = 2x_n - Ax_n^2$ can be used to find the reciprocal of A without the division operation. Show that if the formula converges then it does so to $1/A$ and determine the limits on the initial guess, x_0, in order that convergence does take place. Check your results in the case $A = 9$ with $x_0 = 0.2$ and 1.0 respectively.

Section 13.4

11 Obtain the Newton-Raphson formula for the equations of Problems 7 and 8; comment.

12 Obtain the Newton-Raphson formula for finding the solution of $x^r = A$. What do you notice as r increases? Which method would you prefer for the hand computation of $\sqrt[3]{972}$: bisection, basic iteration or Newton-Raphson?

13 Find the first positive root of $\tan u = u$.

14 Write down the Newton-Raphson formula for the equation $x^4 - 12x^3 + 52.38x^2 - 98.28x + 67.0761 = 0$. Try four iterations from $x_0 = 2$,

3, 4 respectively. To understand what has happened repeat the exercise for the equation $x^2 - 4x + 4.1 = 0$, using $x_0 = 1, 3, 2$ respectively. What conclusions can you draw?

15 Kepler's equation, used for computing orbits of satellites, is

$$M = x - E \sin x$$

Given $E = 0.2$, solve this when

(i) $M = 0.5$ (ii) $M = 0.8$

correct to 4 decimal places.

16 Solve approximately the equation $\sin \omega t - e^{-at} = 0$ (arising from the motion of a planetary gear system used in automatic transmission). In particular, determine the smallest root correct to 4 d.p., taking $\omega = 0.573$ and $a = 0.01$.

17 Freudenstein's equation relating the output crank angle φ to the input crank angle θ of a 4-bar mechanism is

$$\frac{D}{C}\cos\theta - \frac{D}{A}\cos\varphi + \frac{(D^2 + A^2 - B^2 + C^2)}{2AC} - \cos(\theta - \varphi) = 0$$

where A, B, C and D are the lengths of the input crank, coupling link output crank and fixed link respectively. See Figure 13.13.

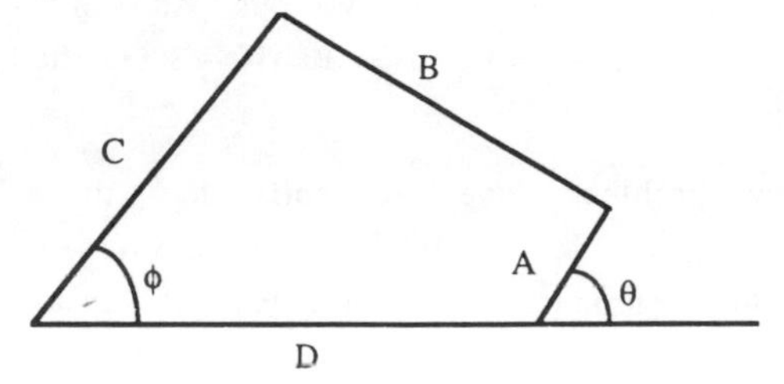

Figure 13.13

If $A = 5$ cm, $B = 10$ cm, $C = 10$ cm, $D = 10$ cm, program a computer to produce a table of values of φ against θ. [The derived function with respect to φ of $\cos(\theta - \varphi)$ is $+\sin(\theta - \varphi)$.]

18 A sphere of density ρ_1 and radius a floats on a fluid of density ρ_0. If it is submerged to a depth h ($< a$) it can be shown that the volume submerged is $(\pi/3)(3ah^2 - h^3)$. See Figure 13.14.

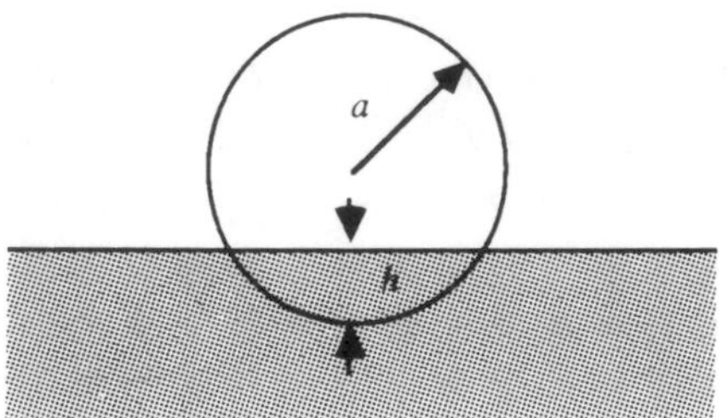

Figure 13.14

If $\rho_0 = 1$ and $\rho_1 = 0.4$ find a suitable approximation to the ratio $\alpha = h/a$ and use the Newton-Raphson method to find a to 2 d.p.

19 A technique for resolving two close roots is as follows:
having located their position approximately, solve $x = x + f'(x) - 1$ by Newton-Raphson for x^*; then evaluate

$$d = \sqrt{\frac{2[x^* - f(x^*)]}{f''(x^*)}}$$

and take $x^* - d$ and $x^* + d$ as starting approximations for the two original roots. For the equation $(x - 2)(x - 1.001)(x - 0.999) = 0$ try the usual Newton-Raphson method and then use the above technique.

Section 13.5

20 Carry out the Birge-Vieta method on the equations of Problem 2.

21 Using Bairstow's method, find the roots of
(i) $x^4 - 2x^2 - x + 3 = 0$
(ii) $x^4 + x^2 - x + 1 = 0$

22 Given that $x^2 - 3.9x + 4.8$ is an approximate factor of $x^4 - 4x^3 + 4x^2 + 4x - 5 = 0$, use Bairstow's method to improve the approximation.

23 In pipe-flow problems one frequently has to solve the equation $c_5D^5 + c_1D + c_0 = 0$. If $c_5 = 1000$, $c_1 = -3$, $c_0 = -9.04$, find a first root using the Newton-Raphson method and then apply the Bairstow's method to find one quadratic factor.

24 Given a polynomial equation

$$a_nx^n + a_{n-1}x^{n-1} + ... + a_1x + a_0 = 0$$

show that the root of maximum modulus, x^*_{max}, satisfies the inequality

$$\left|x^*_{max}\right| \le \sqrt{\left[\frac{a_{n-1}}{a_n}\right]^2 - 2\left[\frac{a_{n-2}}{a_n}\right]}$$

25 Consider the function $p_0(x) = x^3 + a_2x^2 + a_1x + a_0$. If x_1, x_2, x_3 are the real roots of $p_0(x) = 0$ show that $a_2 = -(x_1 + x_2 + x_3)$, $a_1 = x_1x_2 + x_2x_3 + x_3x_1$, $a_0 = -x_1x_2x_3$. If $|x_1| >> |x_2| >> |x_3|$, show that $x_1 \cong -a_2$, $x_2 \cong -a_1/a_2$ and $x_3 \cong -a_0/a_1$.

Then show that, when $y = x^2$,

$$p_1(y) \equiv -p_0(-x)\,p_0(x)$$

$$\equiv y^3 + (2a_1 - a_2^2)y^2 + (a_1^2 - 2a_0a_2)y - a_0^2$$

Show further that the roots of $p_1(y) = 0$ are $y_1 = x_1^2, y_2 = x_2^2, y_3 = x_3^2$. Finally, show that the roots of $p_2(z) = -p_1(-y).p_1(y)$ are $z_1 = x_1^4, z_2 = x_2^4, z_3 = x_3^4$. The roots z_1, z_2, z_3 are separated more widely than were the roots of x_1, x_2, x_3. Indicate how you could iterate on this process to find the roots x_1, x_2, x_3. This is **Graeffe's root-squaring** technique.

Computer-based Examples

26 Take the programs listed in Chapter 13 and make them more robust and user-friendly. In particular, ensure that the input is requested by the program in a fashion intelligible to the user, that the output is suitably labelled, that suitable checks are carried out on the input and that the values of the function are monitored to see whether they are close enough to zero.

27 Verify your programs by using them on the functions of Problem 3.

28 Consider the equation $\tan x = \alpha x$ where $\alpha > 0$. For the values $\alpha = 0.1(0.1)1.0$, find the first three positive roots of the equation.

29 Find the roots of the equation $xe^x = A$ for $A = 1(1)10$.

30 Using any suitable method, find the roots of the following equations

(i) $x^4 - 3x^3 + x^2 + 3x - 2 = 0$

(ii) $x^5 - 4x^4 + 4x^3 + 2x^2 - 5x + 2 = 0$

(iii) $x^5 - 2x^4 - 2x^3 + 4x^2 + x - 2 = 0$

(iv) $x^6 - 5x^5 + 8x^4 - 2x^3 - 7x^2 + 7x - 2 = 0$

(v) $x^6 - 4x^5 + 2x^4 + 8x^3 - 7x^2 - 4x + 4 = 0$

(vi) $x^6 - 6x^4 + 9x^2 - 4 = 0$

14

PARTIAL DIFFERENTIATION

So far we have dealt primarily with the behaviour of functions of one variable; in this chapter we start with a practical example where the need to determine the nature of such a functional relationship from experimental results leads to the study of functions of more than one variable. Most of the results we derive for functions of two variables can be easily extended to more than two variables but we concentrate on cases which involve two variables because (a) we have the possibility of geometrical interpretation of results, and (b) the significant step in our thinking is from one variable to two variables, the steps from two to three or more being relatively simple.

14.1 Functions of Two Variables – Surfaces

We turn our attention to the study of functions of two variables to see what useful results we can obtain. We have seen that just as a function of one variable can be represented by a line on a piece of paper, so a function of two variables $f(x, y)$ can be represented by a surface in three dimensions. We often write $z = f(x, y)$ and we see from Figure 14.1 how the three-dimensional representation of a point $z_0 = f(x_0, y_0)$ ties up with the idea of the point (x_0, y_0, z_0) on the surface of $f(x, y)$.

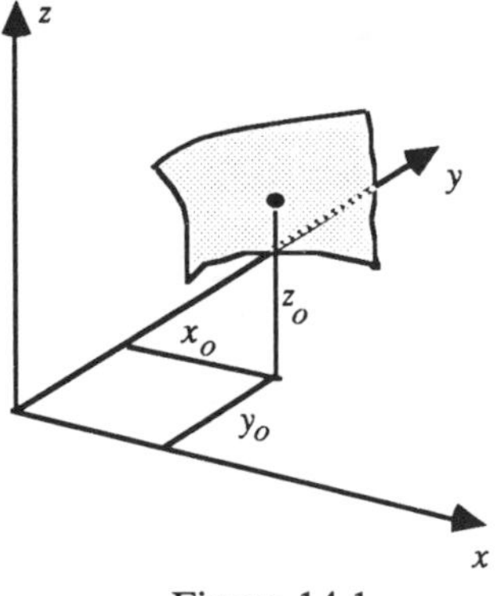

Figure 14.1

Formally we say that *if to each ordered pair of values of two independent variables x and y, each defined over some range, there corresponds* **one and only one value** *of a third (dependent) variable z, then z is a* **function** *of x and y in that range.*

For example

(i) $z = \sqrt{x^2 + y^2}$ defined for all values of x and y

(ii) $z = \sqrt{x^2 + y^2 - 1}$ defined for all values of x and y satisfying $x^2 + y^2 \geq 1$

Great care would be needed in interpreting the function $z = \tan^{-1}(y/x)$, for example.

Evaluation of the function at a point is straightforward. For example, if

$$f(x, y) = x^2 - 2y^2 + 3x + 1$$

then $$f(1, 2) = 1 - 8 + 3 + 1 = -3$$

and $$f(2, -1) = 4 - 2 + 6 + 1 = 9$$

The above are all examples of an **explicit** definition of the function. An example of an **implicit** definition is $x^2 + xz + y^2 + \sin z = 0$.

Notice that the specification of the plane $y = x$ must not be confused with the intersection line $y = x$ which, in three dimensions, should be $y = x$, $z = 0$.

A point free to roam in three dimensions has 3 **degrees of freedom** (d.o.f.). The equation confining it to a surface/plane means it is reduced to 2 degrees of freedom and to restrict it to a curve/line with 1 d.o.f. needs a *second* equation.

If we intersect a surface with a vertical plane then the intersection is a curve representing a function of one variable; see Figure 14.2

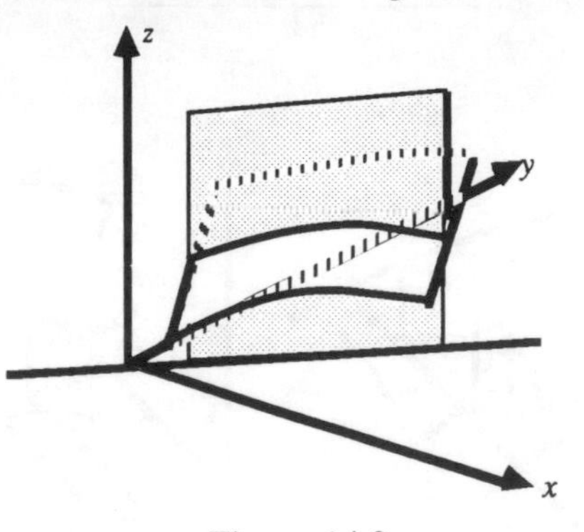

Figure 14.2

14.2 Techniques of Partial Differentiation

Recalling the problem of curve fitting by least squares, we seek the minimum of $S(a_0, a_1)$ and this would indicate some kind of differentiation, which in its turn would suggest the need for limit of a function. This calls for caution since we can approach the point (x_0, y_0, z_0) from a point (x, y, z) along the surface of $f(x, y)$ by a *whole host* of routes, not just by two as for a function of one variable. (See Problems 10 and 11 at the end of this chapter.)

We can say that the condition for continuity of a function at (a, b) is

$$\lim_{(x,\, y) \to (a,\, b)} f(x, y) = f(a, b)$$

and we take the symbol '$\to$' to imply *by any route whatsoever*; [we assume that $f(a, b)$ exists and is finite]. Geometrically, for a function which is continuous over some range of values of x and y, the surface which represents it has no holes in that range nor any vertical cliff faces.

Regarding differentiation as being bound up with some slope, we ask: what slope do we look for at a point on the surface? If we take a short needle and place its centre on the point in question then it can indicate an infinite number of slopes depending on its direction relative to the x and y axes. Since the function is defined in terms of x and y we take two principal slopes at a point: the first obtained by keeping x fixed and varying y, and the second by keeping y fixed and varying x.

In Figure 14.3(i), by proceeding along the surface keeping our value of x fixed we have effectively a function of y only and we can define the slope at (x_0, y_0) in the obvious way as follows: *assuming that the limit exists, the* **partial derivative** *at (x_0, y_0) of $f(x, y)$ with respect to y*,written $f_y(x_0, y_0)$

or

$$\left.\frac{\partial f}{\partial y}\right|_{(x_0,\, y_0)}$$

is equal to

$$\lim_{k \to 0} \frac{f(x_0, y_0 + k) - f(x_0, y_0)}{k} \tag{14.1a}$$

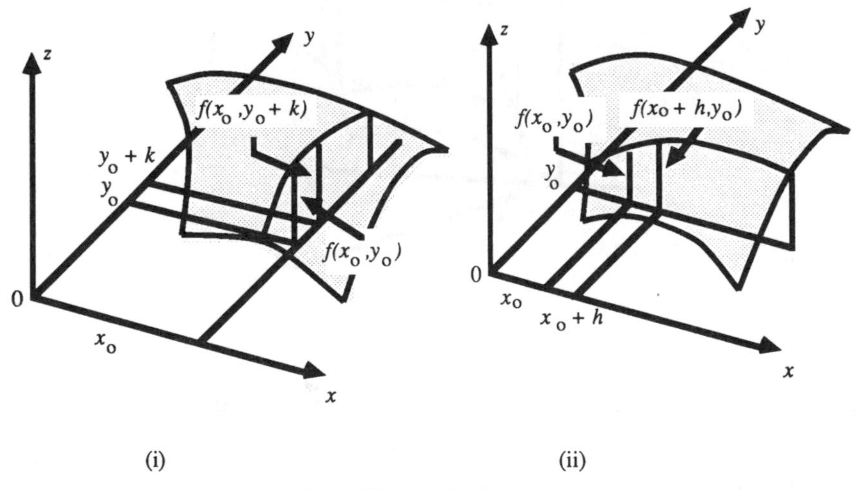

Figure 14.3

In a similar way, from Figure 14.3(ii) we may define $f_x(x_0, y_0)$ by the expression

$$\lim_{h \to 0} \frac{f(x_0 + h, y_0) - f(x_0, y_0)}{h} \tag{14.1b}$$

You should check out this definition on Figure 14.3(ii).

In practice, to differentiate partially with respect to x, say, we treat y, wherever it occurs, as a constant. Conversely, to differentiate partially with respect to y, we treat x everywhere as a constant.

Example

1 $\quad z = ax^2 + 2hxy + by^2 + 2gx + 2fy + c$

$$f_x \equiv z_x \equiv \frac{\partial z}{\partial x} = 2ax + 2hy + 2g$$

$$f_y \equiv z_y \equiv \frac{\partial z}{\partial y} = 2hx + 2by + 2f$$

2 $\quad z = x^2\sqrt{x^2 + y^2}$

$$\frac{\partial z}{\partial x} = 2x\sqrt{x^2 + y^2} + x^2 \cdot \frac{1}{2}\frac{2x}{\sqrt{x^2 + y^2}}$$

$$z_y \equiv \frac{\partial z}{\partial y} = x^2 \cdot \frac{1}{2}\frac{2y}{\sqrt{x^2 + y^2}} \equiv f_y(x, y)$$

$$\text{e.g. } f_y(1, 2) = 1 \cdot \frac{1}{2} \cdot \frac{2 \cdot 2}{\sqrt{5}} = \frac{2}{\sqrt{5}}$$

Continuity theorem

An application of the previous result is the **continuity theorem** of fluid flow. On the macro-scale the theorem says that in some region of the field of flow, the amount of fluid entering the volume of the region in a given time must balance that amount which flows out of that volume in the same time (unless build-up of fluid occurs). On the small scale, it gives a necessary relationship between components of velocity. Let the flow be two-dimensional, i.e. no flow in the z–direction and let the x and y components of velocity be $u(x, y)$ and $v(x, y)$ respectively. We take an imaginary cuboidal region (as shown in Figure 14.4) of the space in which the fluid is flowing and consider the flow through its faces.

We know there is no flow through the top and bottom faces and we first concentrate on the flow in the x-direction; refer to Figure 14.4(i).

The centre of the cuboid is at (x_0, y_0) and it is of height 1. We assume that the dimensions δx, δy are small. Then the flow into the region through the face

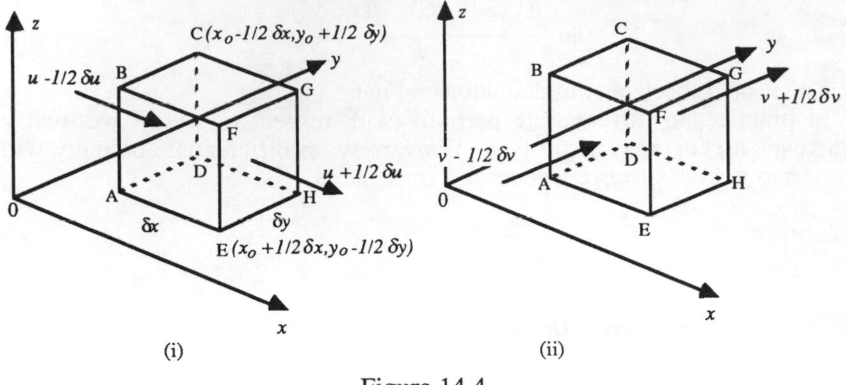

Figure 14.4

ABCD (of area δy . 1) depends on the x-component of velocity at that face; the x-coordinate of points on ABCD is $(x_0 - \frac{1}{2}\delta x)$ but the y-coordinates vary slightly. We assume that the average velocity in the x-direction is $u(x_0 - \frac{1}{2}\delta x, y_0)$, i.e. that at the centre of the face. Then the flow rate through ABCD is $u(x_0 - \frac{1}{2}\delta x, y_0).\delta y.1$; likewise the flow rate out through EFGH is $u(x_0 + \frac{1}{2}\delta x, y_0).\delta y.1$ which gives a net flow out of the cuboid in the x–direction of $[u(x_0 + \frac{1}{2}\delta x, y_0) - u(x_0 - \frac{1}{2}\delta x, y_0)].\delta y.1$.

A similar argument applies to the flow rate in the y–direction [Figure 14.4(ii)] and you should see that we can obtain the net flow rate out of the cuboid (which is, of course, zero) as

$$[u(x_0 + \tfrac{1}{2}\delta x, y_0) - u(x_0 - \tfrac{1}{2}\delta x, y_0)].\delta y.1 +$$

$$[v(x_0, y_0 + \tfrac{1}{2}\delta y) - v(x_0, y_0 - \tfrac{1}{2}\delta y)].\delta x.1 = 0$$

Dividing by $\delta x\, \delta y$. 1 (the volume of the cuboid) we obtain the equation

$$\frac{u(x_0 + \frac{1}{2}\delta x, y_0) - u(x_0 - \frac{1}{2}\delta x, y_0)}{\delta x} + \frac{v(x_0, y_0 + \frac{1}{2}\delta y) - v(x_0, y_0 - \frac{1}{2}\delta y)}{\delta y} = 0$$

If we now take the limiting case when δx and $\delta y \to 0$ simultaneously (what happens to the cuboid?) we obtain $u_x(x_0, y_0) + v_y(x_0, y_0) = 0$ or, more generally

$$\frac{\partial u}{\partial x} + \frac{\partial v}{\partial y} = 0 \qquad (14.2)$$

This equation means that the velocity components u and v are not independent and knowledge of one of them helps to determine the other.

Higher Derivatives

By differentiating z_x partially with respect to x, we obtain

$$z_{xx} \quad \text{or} \quad \frac{\partial^2 z}{\partial x^2} \equiv \frac{\partial}{\partial x}\left[\frac{\partial z}{\partial x}\right]$$

(provided the necessary limits exist). In a similar fashion we can obtain the other second derivatives z_{xy}, z_{yx}, z_{yy} where the first suffix refers to the first differentiation; the alternative notations are

$$\frac{\partial^2 f}{\partial y\,\partial x}, \quad \frac{\partial^2 f}{\partial x\,\partial y}, \quad \frac{\partial^2 f}{\partial y^2}$$

In **most** cases $z_{xy} \equiv z_{yx}$. Likewise, in most cases, there are *four* third partial derivatives,

$$z_{xxx}, z_{yxx}, z_{xyy}, z_{yyy} \quad \text{or} \quad \frac{\partial^3 z}{\partial x^3}, \quad \frac{\partial^3 z}{\partial x^2\,\partial y}, \quad \frac{\partial^3 z}{\partial y^2\,\partial x}, \quad \frac{\partial^3 z}{\partial y^3}$$

Notation

In thermodynamics, partial differentiation plays an important role. The **entropy** S of a gas is given by

$$S = C_v \ln p + C_p \ln v + A \tag{14.3a}$$

where C_p, C_v and A are constants. We can substitute for v from the gas law $pv = RT$ to obtain

$$S = (C_v - C_p)\ln p + C_p \ln T + B \tag{14.3b}$$

where B is a constant.

Suppose we calculate $\partial S/\partial p$ from equation (14.3a), then v is held constant and we obtain

$$\frac{\partial S}{\partial p} = \frac{C_v}{p}$$

To find $\partial S/\partial p$ from equation (14.3b) means that T is held constant and we obtain

$$\frac{\partial S}{\partial p} = \frac{C_v - C_p}{p}$$

The difficulty here is that the notation $\partial S/\partial p$ is not sufficiently precise. For equation (14.3a) we write

$$\left[\frac{\partial S}{\partial p}\right]_v$$

to indicate that v is held constant and for equation (14.3b) we write

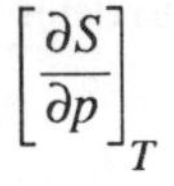

$$\left[\frac{\partial S}{\partial p}\right]_T$$

Examples

1 $z = ax^2 + 2hxy + by^2 + 2gx + 2fy + c$
From page 369 we find $z_{xx} = 2a,\ z_{xy} = 2h = z_{yx},\ z_{yy} = 2b$

2 $z = x^2\sqrt{x^2 + y^2}$
From page 369

$$z_x = (x^2 + y^2)^{-1/2}\,(3x^3 + 2y^2x)$$
$$z_y = x^2y(x^2 + y^2)^{-1/2}$$

Now

$$z_{xx} = (x^2 + y^2)^{-1/2}\,(9x^2 + 2y^2) - \tfrac{1}{2}(x^2 + y^2)^{-3/2}\,.\,2x(3x^3 + 2y^2x)$$
$$= (x^2 + y^2)^{-3/2}\,[(9x^2 + 2y^2)(x^2 + y^2) - 3x^4 - 2x^2y^2]$$
$$= (x^2 + y^2)^{-3/2}\,(6x^4 + 9x^2y^2 + 2y^4)$$

Likewise, as you can verify,

$$z_{xy} = xy(x^2 + 2y^2)(x^2 + y^2)^{-3/2} = z_{yx}$$
$$z_{yy} = x^4(x^2 + y^2)^{-3/2}$$

14.3 Stationary Points on the Surface Representing $f(x, y)$

Suppose, in theory, we take a small abstract coin (small in relation to surface changes) and move it on the surface representing $f(x, y)$. At the tops of rounded hills (*local maxima*) and the bottoms of rounded valleys (*local minima*), it will be horizontal; it will be above the surface at a local maximum and below it at a local minimum. We need to find a necessary condition for the coin to be horizontal at a point (x_0, y_0); this is that the slopes at (x_0, y_0), $\partial f/\partial x$ and $\partial f/\partial y$, are **both** zero. If we are concerned with small movements on the surface away from (x_0, y_0) we can approximate the function by a plane which is tangential to the surface at (x_0, y_0). In Figure 7.28 we may take Q to be (x_0, y_0, z_0) and recast equation (7.30) to give

$$f(x_0+h, y_0+k) \cong f(x_0, y_0) + h\left.\frac{\partial f}{\partial x}\right|_{(x_0, y_0)} + k\left.\frac{\partial f}{\partial y}\right|_{(x_0, y_0)} \qquad (14.4)$$

This is the **tangent plane approximation** to $f(x, y)$ at (x_0, y_0) and has obvious similarity to the one-variable case.

We shall be interested in the change in $f(x, y)$ as we move the small distance from (x_0, y_0) to (x_0+h, y_0+k). This change is

$$\delta f \equiv f(x_0+h, y_0+k) - f(x_0, y_0)$$

and we therefore have

$$\delta f \simeq hf_x(x_0, y_0) + kf_y(x_0, y_0) \tag{14.5}$$

We return to this idea in Section 14.4.

For the moment, note that if

$$\left.\frac{\partial f}{\partial x}\right|_{(x_0, y_0)} = 0 = \left.\frac{\partial f}{\partial y}\right|_{(x_0, y_0)}$$

then the tangent plane is horizontal at (x_0, y_0). (Try for yourself by experiment.)

A necessary condition for (x_0, y_0) to be a stationary point of $f(x, y)$ is that

$$f_x(x_0, y_0) = 0 = f_y(x_0, y_0) \tag{14.6}$$

Notice the connection between different notations.

A *local maximum* occurs when, holding x_0 fixed, the intersection of the plane $x = x_0$ with the surface shows a local maximum for the resulting function of y, **and** the intersection of the plane $y = y_0$ shows a local maximum for the resulting function of x. A local minimum is defined similarly.

There is a third case where we obtain a local maximum in one direction and a local minimum in the other; such a point is known as a **saddle point**, a name which you will see is clearly justified if you examine a saddle or, failing that, a camel's back. (There are other possible unusual features which satisfy $z_x = z_y = 0$ but we shall not deal with them.)

Example 1

Show that the function $f(x, y) = 2x^2 + 2xy - y^3$ has stationary points at (0, 0) and $(\frac{1}{6}, -\frac{1}{3})$ and determine their nature.

Now $$f_x = 4x + 2y; \quad f_y = 2x - 3y^2$$

Stationary points occur where $4x + 2y = 0$ and $2x - 3y^2 = 0$; then $y = -2x$, and, substituting in the second relationship we obtain $2x - 12x^2 = 0$, i.e. $2x(1 - 6x) = 0$ from which the stationary points follow (always be careful not to miss any).

Now keep $$x = \frac{1}{6}$$

so that $$f\left[\frac{1}{6}, y\right] = \frac{1}{18} + \frac{1}{3}y - y^3 = g(y), \text{ say}$$

Then $g'(y) = (1/3) - 3y^2$ and this is zero when $y = -1/3$; further, $g''(y) = -6y$ and hence $g''(-\frac{1}{3}) > 0$ indicating a local minimum. Now put $y = -1/3$ so that

$$f\left[x, -\frac{1}{3}\right] = 2x^2 - \frac{2}{3}x + \frac{1}{27} = h(x), \text{ say}$$

Then $h'(x) = 4x - (2/3)$, $h''(x) = 4 > 0$ indicating a local minimum. Hence $(\frac{1}{6}, -\frac{1}{3})$ is a local minimum; but *you* see what troubles arise with (0, 0).

A criterion which involves second derivatives is stated without proof in Problem 18; the criterion follows from the Taylor's series for $f(x, y)$.

Example 2

$f(x, y) = 3(x^2 + y^2)^{1/2}$

Now $\sqrt{x^2 + y^2}$ represents the distance of the point (x, y) on the x-y plane from the origin. This is clearly least when $x = y = 0$ but has **no** maximum value.

For reference

$$f_x = \frac{3x}{(x^2 + y^2)^{1/2}}, \quad f_y = \frac{3y}{(x^2 + y^2)^{1/2}}$$

14.4 Small Increments and Differentials

If we rewrite relationship (14.5), page 373, in a different notation we have the approximation

$$\delta f \cong \frac{\partial f}{\partial x}\,\delta x + \frac{\partial f}{\partial y}\,\delta y \tag{14.7}$$

We may imagine that the change δx causes a change in $f(x, y)$, the rate of response being $\partial f/\partial x$; likewise the change δy gives rise to a second change in $f(x, y)$ with response rate $\partial f/\partial y$, and the total change δf is the sum of these two.

We can derive this result in a somewhat different way. Let $u = f(x, y)$ and let x, y be changed by small amounts, $\delta x, \delta y$; we seek the resulting (small change) in u, i.e. δu.

$$\begin{aligned}\delta u &= f(x + \delta x, y + \delta y) - f(x, y)\\ &= f(x + \delta x, y + \delta y) - f(x, y + \delta y) + f(x, y + \delta y) - f(x, y)\\ &= f_x(x, y + \delta y).\delta x + f_y(x, y).\delta y + 2^{\text{nd}} \text{ order terms in } \delta x, \delta y\\ &= [f_x(x, y) + 0(\delta y)]\delta x + f_y(x, y).\delta y + 2^{\text{nd}} \text{ order terms}\end{aligned}$$

therefore

$$\delta u \cong f_x \delta x + f_y \delta y$$

We call the quantity

$$du = \frac{\partial f}{\partial x}\,\delta x + \frac{\partial f}{\partial y}\,\delta y \tag{14.8}$$

the **differential** for the increments δx, δy (sometimes called the **total differential**). We shall see its uses a little later on. For the moment we use the approximate formula (14.7) to estimate errors. Notice the difference between δu and du.

Example

The area S of triangle ABC is calculated from the formula $S = \frac{1}{2}bc \sin A$; the length b, c and the angle A are measured as 3.00 cm, 4.00 cm and 30° respectively. Find the error and percentage error in S as calculated from the measurements, given that b and c each has a maximum error of ± 0.05 cm and A one of ± 0.1°.

First, we have the result

$$S = \tfrac{1}{2}\,.\,3\,.\,4\,.\,\tfrac{1}{2} = 3 \text{ sq cm}$$

Now,

$$\delta S \cong \frac{\partial S}{\partial b}\delta b + \frac{\partial S}{\partial c}\delta c + \frac{\partial S}{\partial A}\delta A \tag{14.9}$$

and

$$\frac{\partial S}{\partial b} = \tfrac{1}{2}c \sin A, \quad \frac{\partial S}{\partial c} = \tfrac{1}{2}b \sin A, \quad \frac{\partial S}{\partial A} = \tfrac{1}{2}bc \cos A$$

We compute these rates at the measured values of b, c and A, therefore

$$\frac{\partial S}{\partial b} = \tfrac{1}{2}\,.\,4\,.\,\tfrac{1}{2} = 1; \quad \frac{\partial S}{\partial c} = \frac{3}{4}, \quad \frac{\partial S}{\partial A} = 3\sqrt{3}$$

Hence

$$|\delta S| \cong 1 \times 0.05 + \frac{3}{4}\,.\,0.05 + 3\sqrt{3}\;0.1(\pi/180)$$

$$= 0.05 \times \frac{7}{4} + 3\sqrt{3} \times \frac{\pi}{1800} \cong 0.097$$

and this is sufficient precision for our purposes.

Then $S = (3.000 \pm 0.097)$ sq cm

To find the percentage error, $\delta S/S \times 100$ we may take logarithms.

Let

$$u = \log S = \log \tfrac{1}{2} + \log b + \log c + \log \sin A$$

then

$$\delta u\left[\cong\frac{\mathrm{d}u}{\mathrm{d}S}\,\delta S=\frac{1}{S}\,\delta S\right]\cong\frac{\partial u}{\partial b}\delta b+\frac{\partial u}{\partial c}\delta c+\frac{\partial u}{\partial A}\,\delta A=\frac{1}{b}\delta b+\frac{1}{c}\delta c+\frac{\cos A}{\sin A}\delta A$$

Hence

$$\frac{\delta S}{S}\times 100\cong\frac{\delta b}{b}\times 100+\frac{\delta c}{c}\times 100+\cot A\ \delta A\times 100$$

$$\frac{\delta S}{S}\times 100\cong\frac{0.05}{3.00}\times 100+\frac{0.05}{4.00}\times 100+\sqrt{3}\times[0.1\ .\ (\pi/180]\times 100$$

$$\cong 1.667+1.250+0.302$$

$$\cong 3.219\%$$

Differentials and their applications

Given increments δx, δy the differential given by (14.8) is the change in u predicted by the tangent plane approximation. Before proceeding further, we note that $\partial x/\partial x = 1$, $\partial x/\partial y = 0$, if x, y are independent and hence, regarding x as a function of x and y, (equation 14.8) gives

$$\mathrm{d}x=\frac{\partial x}{\partial x}\delta x+\frac{\partial x}{\partial y}\delta y=\delta x$$

Similarly we may show $\mathrm{d}y = \delta y$, and we may rewrite (14.8) as

$$\mathrm{d}u=\frac{\partial u}{\partial x}\mathrm{d}x+\frac{\partial u}{\partial y}\mathrm{d}y \tag{14.10}$$

Example 1

The ideal gas law in thermodynamics is $pV = RT$ where R is constant. Hence $p = RT/V = p(T, V)$. This gives

$$\mathrm{d}p=\frac{\partial p}{\partial T}\mathrm{d}T+\frac{\partial p}{\partial V}\,\mathrm{d}V=\frac{R}{V}\mathrm{d}T-\frac{RT}{V^2}\mathrm{d}V$$

Along an isobar, p is constant and hence $\mathrm{d}p = 0$ irrespective of changes in V, T and so

$$0=\frac{R}{V}\mathrm{d}T-\frac{RT}{V^2}\mathrm{d}V$$

It follows that $\mathrm{d}V/\mathrm{d}T = V/T$ and this is the differential equation for the family of isobars. We have **not** treated $\mathrm{d}V/\mathrm{d}T$ as a *fraction*; the ideas go much deeper.)

Example 2

A **streamline** is an imaginary line in a fluid across which no flow takes place, i.e. the velocity vector at a point is tangential to the streamline passing through that point [Figure 14.5(i)].

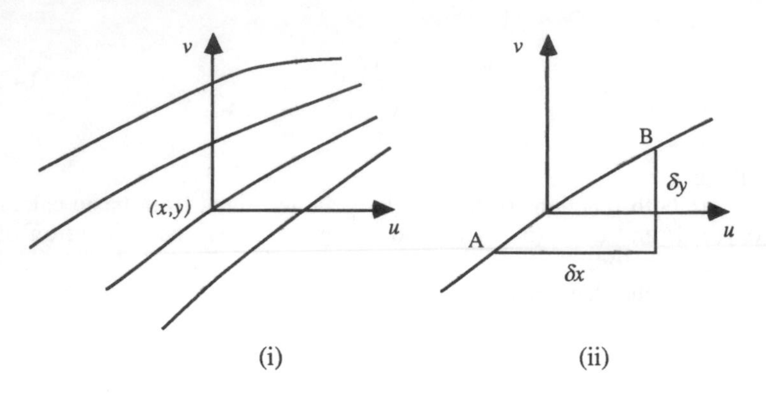

Figure 14.5

In Figure 14.5(ii) we consider a small portion AB of the streamline, small enough to assume it to be a straight line. The average velocity in the x-direction across AB can be taken as $u(x, y)$, by application of the *mean-value theorem*, and that in the y-direction as $v(x, y)$. In order that the *total* flow rate across AB should be zero, we require $u(x, y)\delta y - v(x, y)\delta x = 0$.

Now suppose the streamline to be a contour line specified by some equation $\psi(x, y) = C$, a constant; then $\psi_B - \psi_A = 0$ and $d\psi = (\partial\psi/\partial x)dx + (\partial\psi/\partial y)dy = 0$. By ψ_B we mean the value of $\psi(x, y)$ evaluated by B.

Now if we compare this with the relationship $udy - vdx = 0$ it is possible to choose ψ so that $\partial\psi/\partial x = -v$, $\partial\psi/\partial y = u$. (An alternative convention is $\partial\psi/\partial x = v$, $\partial\psi/\partial y = -u$.)

Then ψ is called a *stream function* for the flow.

For the problem of flow in a corner we may take $\psi(x, y) = 3xy$ and hence the fluid will flow along curves $3xy$ = constant; these are rectangular hyperbolae. It is found that this is in reasonable agreement with experiment, provided we do not go close to the origin. What are the formulae for u and v?

In general it is then possible under certain restrictions to find an equation satisfied by ψ. We have

$$\frac{\partial u}{\partial x} = \frac{\partial^2 \psi}{\partial x\,\partial y} \quad \text{and} \quad \frac{\partial v}{\partial y} = -\frac{\partial^2 \psi}{\partial y\,\partial x}$$

which from the equation of continuity (14.2) ties in with the equality of the mixed second derivatives. Furthermore,

$$\frac{\partial u}{\partial y} = \frac{\partial^2 \psi}{\partial y^2} \quad \text{and} \quad \frac{\partial v}{\partial x} = -\frac{\partial^2 \psi}{\partial x^2}$$

For flows which are *irrotational*, that is, particles of fluid do not spin about their axes, $\partial u/\partial y = \partial v/\partial x$ (consult a textbook on Fluid Dynamics to see why) and this leads to **Laplace's equation**, which is

$$\frac{\partial^2\psi}{\partial x^2}+\frac{\partial^2\psi}{\partial y^2}=0 \tag{14.11}$$

Chain Rule

If x and y are both functions of the single variable t then, from basic calculus, $dx = (dx/dt)dt$ and $dy = (dy/dt)dt$ and $z(x, y)$ will also be a function of t, therefore $dz = (dz/dt)dt$.

Then we have the chain rule

$$\frac{dz}{dt}=\frac{\partial z}{\partial x}\frac{dx}{dt}+\frac{\partial z}{\partial y}\frac{dy}{dt} \tag{14.12}$$

This yields

$$dz=\frac{\partial z}{\partial x}\frac{dx}{dt}dt+\frac{\partial z}{\partial y}\frac{dy}{dt}dt=\frac{\partial z}{\partial x}dx+\frac{\partial z}{\partial y}dy$$

showing that a similar relationship to (14.8) holds in this case.

We may regard (14.12) in the following light. A change in t causes a change in x with a response rate dx/dt; this change in x causes a change in z with a response rate $\partial z/\partial x$. Likewise the change in t causes a change in y with a response rate dy/dt and this causes a separate change in z with response rate $\partial z/\partial y$. The total response rate of z to the change in t is given by (14.12). We call dz/dt the **total derivative** of z with respect to t.

The chain rule may be extended in cases where x and y are both functions of two further variables s and t. For example, suppose $z = x^2 + 2y^2$ where $x = 2s + 3t$, $y = s - 4t^3$. Then

$$\begin{aligned}\frac{\partial z}{\partial s}&=\frac{\partial z}{\partial x}\cdot\frac{\partial x}{\partial s}+\frac{\partial z}{\partial y}\cdot\frac{\partial y}{\partial s}\\ &= 2x.2 + 4y.1\\ &= 12s + 12t - 16t^3\end{aligned}$$

Notice that since $z = z(x, y)$, $x = x(s, t)$ and $y = y(s, t)$, all the derivatives are partial derivatives.

Similarly,

$$\begin{aligned}\frac{\partial z}{\partial t}&=\frac{\partial z}{\partial x}\cdot\frac{\partial x}{\partial t}+\frac{\partial z}{\partial y}\cdot\frac{\partial y}{\partial t}=2x.3+4y(-12t^2)\\ &= 12s + 18t - 48st^2 + 192t^5\end{aligned}$$

Example

The radius, r, of a cylinder decreases at the rate of 3 mm/sec and the height increases at the rate of 4 mm/sec. Find the rate of change of the volume of the cylinder *at the instant* when $r = 60$ mm, $h = 100$ mm.

Now $V = \pi r^2 h$ and therefore we obtain

$$\frac{dV}{dt} = 2\pi rh\frac{dr}{dt} + \pi r^2\frac{dh}{dt}$$

$$= 2\pi \,.\, 60 \,.\, 100(-3) + \pi \,.\, 3600[4]$$
$$= -36000\pi + 14400\pi$$
$$= -21600\pi$$

Hence the instantaneous *decrease* of volume is 21600π mm^3/sec.

Implicit Differentiation

Suppose x and y are not independent, but connected by some equation $f(x, y) = 0$. If $\partial f/\partial x$ and $\partial f/\partial y$ exist, then since $df = 0$, it can be shown that

$$\frac{dy}{dx} = -\left[\frac{\partial f}{\partial x}\right]\Bigg/\left[\frac{\partial f}{\partial y}\right] \tag{14.13}$$

The proof follows from the approximation

$$df \cong \frac{\partial f}{\partial x}dx + \frac{\partial f}{\partial y}dy$$

Example 1

Consider $f(x, y) = x^3 + y^3 - 2xy = 0$
We use the chain rule in the form

$$\frac{df}{dx} = \frac{\partial f}{\partial x} + \frac{\partial f}{\partial y} \cdot \frac{dy}{dx}$$

to obtain

$$\frac{df}{dx} = 3x^2 - 2y + (3y^2 - 2x)\frac{dy}{dx}$$

But, since $f(x, y) = 0$, $\dfrac{\partial f}{\partial x} = 0$ and therefore $\dfrac{dy}{dx} = -\dfrac{3x^2 - 2y}{3y^2 - 2x}$

Example 2

Consider the ellipse $x^2 + 3y^2 = 4$
Let

$$f(x, y) = x^2 + 3y^2 - 4 = 0$$

then

$$\frac{\partial f}{\partial x} = 2x \quad \text{and} \quad \frac{\partial f}{\partial y} = 6y$$

Hence

$$\frac{dy}{dx} = -\frac{2x}{6y}$$

Compare this with the (so far unjustified) technique of implicit differentiation quoted in Section 12.1, page 312.

An application of this idea is the **envelope** to a family of curves.

Any family of curves (e.g. streamlines) will have an arbitrary constant, C, in their defining equation, $f(x, y, C) = 0$. The envelope is found by eliminating C from that equation and the equation

$$\frac{\partial}{\partial C} f(x, y, C) = 0$$

Example 3

A family of straight lines is given by

$$y = Cx - C^2 \qquad (14.14)$$

Then

$$f(x, y, C) = Cx - C^2 - y \quad \text{and} \quad \frac{\partial f}{\partial C} = x - 2C$$

We eliminate C between (14.14) and the equation $x - 2C = 0$, to obtain

$$y = \frac{x^2}{2} - \frac{x^2}{4} = \frac{x^2}{4}$$

The situation is depicted in Figure 14.6.

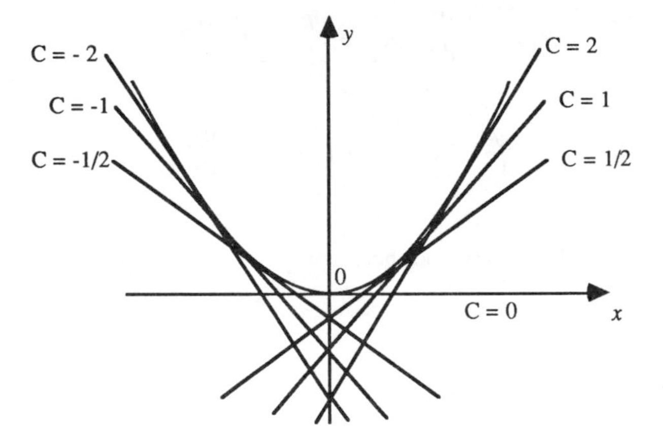

Figure 14.6

Each line of the family is a tangent to the parabola at some point and at every point of the parabola one of the family is a tangent there.

A second application is in **orthogonal trajectories** which arise in connection with stream functions. Let the streamlines for flow in a vortex be

$$x^2 + y^2 = a^2 \qquad (14.15)$$

Now $f(x, y, a) = x^2 + y^2 - a^2$ and

$$\frac{df}{dx} = \frac{\partial f}{\partial x} + \frac{\partial f}{\partial y} \cdot \frac{dy}{dx}$$

$$= 2x + 2y\frac{dy}{dx} = 0 \qquad (14.16)$$

This gives a slope of

$$\frac{dy}{dx} = -\frac{x}{y}$$

Curves which intersect the family (14.15) everywhere at right angles have slopes $+ y/x$ and thus have equation

$$x\frac{dy}{dx} - y = 0$$

It is seen that the family of straight lines $y = Cx$ satisfies this last equation. These *orthogonal curves* represent the velocity potential for the flow (Figure 14.7). Note: had (14.16) contained the parameter a then it would have been necessary to eliminate a between (14.15) and (14.16) before determining dy/dx.

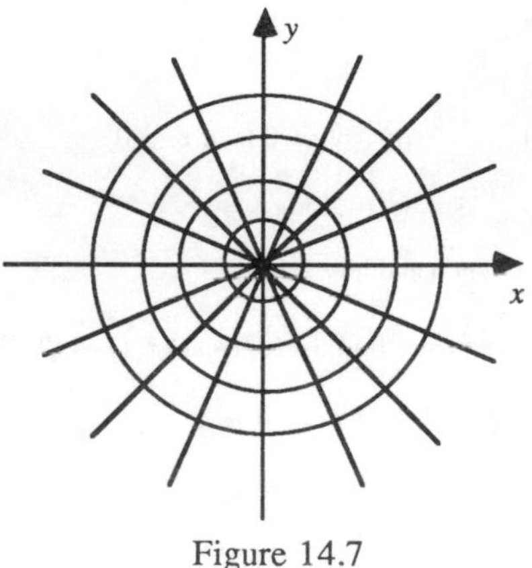

Figure 14.7

Problems

Section 14.1

1 Describe geometrically the functions

(i) $z = x^2 - y^2$ (ii) $z = \sqrt{x^2 + y^2 - 2x + 1}$

(iii) $z = \sqrt{x^2 + y^2 - 2x + 2}$ (iv) $z = 1 + \sqrt{4x - x^2 - y^2}$

On what range of values is each of these functions defined?

2 Evaluate $f(1, 2), f(-2, -1), f(0, 0)$ for

(i) $f(x, y) = x^2 + 3y^2 - x + 1$ (ii) $f(x, y) = xy/(x^2 + y^2)$

(iii) $f(x, y) = \log(1 + x^2 + 2y)$ (iv) $f(x, y) = 3x\,e^y + y^2\,e^{2x}$

3 Consider the intersection of $z = x^2 + y^2 + 2y + 1$ with the planes $x = 1, x = 2, x = a; y = 1, y = 2, y = b; x + y + z = 1$

4 The density of a gas at a point (x, y, z) at time t is given by $\rho(x, y, z, t) = (2x + 3y - z)\sin t$. Evaluate $\rho(1, -2, 1, 4)$.

Section 14.2

5 (i) Find $\displaystyle\lim_{(x, y)\to(1, 2)} \frac{x}{x^2 + y^2 + 2}$

(ii) Find $\displaystyle\lim_{(x, y)\to(1, 2)} \frac{(x-1)^3}{(x-1)^2 + (y-2)}$

(a) by first letting $x \to 1$ then letting $y \to 2$

(b) by reversing the order of taking limits.

6 Find $\displaystyle\lim_{(x, y)\to(0, 0)} \frac{x^2 - y^2}{x^2 + y^2}$ by

(a) $x \to 0$ then $y \to 0$ (b) $y \to 0$ then $x \to 0$ (c) along line $y = x$

(d) along $y = \frac{1}{2}x$ (e) along $y = x^2$

(f) put $x = r\cos\theta$, $y = r\sin\theta$ and let $r \to 0$.

7 By converting to polar coordinates find the limits as $(x, y) \to (0, 0)$ of

(i) $\dfrac{x^2y^2}{x^2 + y^2}$ (ii) $\dfrac{x + 2y}{xy}$ (iii) $\dfrac{1}{x^2 + y^2}$ (iv) $(x^2 + y^2)/xy$

8 Investigate the continuity of

(i) $\dfrac{x}{x^2 + y^2 + 1}$ (ii) $\dfrac{x^2y^2}{x^2 + y^2}$ (iii) $\dfrac{x - y}{x + y}$ (iv) $\dfrac{x + y}{x - y}$

(v) $f(x, y) = \begin{cases} \dfrac{x^3y^3}{x^2 + y^2} & (x, y) \neq (0, 0) \\ 10 & (x, y) = (0, 0) \end{cases}$

What value could be given to $f(0, 0)$ in part (ii) to make $f(x, y)$ continuous?

9 Find the first partial derivatives of

(i) $x^2 + y^3 + 2z^2$ (ii) x^2y^3/z (iii) $(1 + x^2y)e^{3z}$ (iv) $\sin(xz + y)$

(v) $e^{(xy+2y^2)}$ (vi) $\frac{1}{2}xy \sin z$ (vii) $\tan^{-1}(x/y)$ (viii) $(x/y^2) - (y/x^2)$

10 Find the first and second partial derivatives of the following functions and evaluate them at (0, 0), (1, 2), (–1, 0)

(i) x^3y^4 (ii) x^2/y^3 (iii) $2x \cos y$ (iv) $\sin(x + y)$

(v) $2y$ (vi) $e^x + \cos y$ (vii) $(1 + x^2y)e^{3y}$ (viii) $\tan^{-1}(y/3x)$

(ix) $\sqrt{x^2 - y^2}$

11 (i) Show that $\phi = Ae^{-(1/2)kt} \sin pt \cos qx$ satisfies the equation

$$\frac{\partial^2\phi}{\partial x^2} = \frac{1}{c^2}\left[\frac{\partial^2\phi}{\partial t^2} + k\frac{\partial\phi}{\partial t}\right]$$

provided that $p^2 = c^2 q^2 - \frac{1}{4}k^2$

(ii) Show that $V = (Ar^n + Br^{-n})\cos(n\theta - \alpha)$, where A, B, n, α are constants, satisfies the equation

$$\frac{\partial^2 V}{\partial r^2} + \frac{1}{r}\frac{\partial V}{\partial r} + \frac{1}{r^2}\frac{\partial^2 V}{\partial\theta^2} = 0$$

(iii) Find values of the parameter n so that

$$V = r^n(3\cos^2\theta - 1)$$

satisfies

$$\frac{\partial}{\partial r}\left[r^2\frac{\partial V}{\partial r}\right] + \frac{1}{\sin\theta}\frac{\partial}{\partial\theta}\left[\sin\theta\frac{\partial V}{\partial\theta}\right] = 0$$

12 (i) If $y = \phi(x - ct) + \psi(x + ct)$ where ϕ, ψ are arbitrary functions, show that

$$\frac{\partial^2 y}{\partial x^2} = \frac{1}{c^2}\frac{\partial^2 y}{\partial t^2}$$

(ii) If $z = xF(y/x) + f(y/x)$, F and f being arbitrary functions, show that

$$x\frac{\partial z}{\partial x} + y\frac{\partial z}{\partial y} = z - f$$

and deduce that

$$x^2 z_{xx} + 2xyz_{xy} + y^2 z_{yy} = 0$$

13 (i) If $V = \dfrac{x^3y^3}{x^3 + y^3}$, show that

(a) $x\dfrac{\partial V}{\partial x} + y\dfrac{\partial V}{\partial y} = 3V$ (b) $x^2\dfrac{\partial^2 V}{\partial x^2} + 2xy\dfrac{\partial^2 V}{\partial x\,\partial y} + y^2\dfrac{\partial^2 V}{\partial y^2} = 6V$

(ii) If $z = xf(y/x) + g(x/y)$, show that $x\dfrac{\partial z}{\partial x} + y\dfrac{\partial z}{\partial y} = xf(y/x)$ (LU)

14 (a) The relation $\left[p + \dfrac{a}{v^2}\right](v - b) = RT$, in which a, B and R are constants, is given.

Find an expression for $\left.\dfrac{\partial p}{\partial T}\right]_v$ and show that $T\left.\dfrac{\partial p}{\partial T}\right]_v - p = \dfrac{a}{v^2}$

What does this become for the ideal gas law $pv = RT$?

(b) If $x = r\cos\theta$ and $y = r\sin\theta$, find $\left.\dfrac{\partial x}{\partial r}\right]_\theta$ and show that $\left.\dfrac{\partial r}{\partial x}\right]_y = \cos\theta$ (EC)

15 At time t, the displacement y of a point at distance x from one end of a vibrating string is given by $y_{tt} = c^2 y_{xx}$ where c is a constant. Show that this equation is satisfied by

$$y = B\sin px\,.\,\sin(cpt + a)$$

where B, p and a are constants.

Section 14.3

16 Find the equation of the tangent plane to the surface $x^2yz + 3y^2 = 2xz^2 - 8z$ at the point $(1, 2, -1)$. Show that the equation of the normal to the surface at the same point can be expressed parametrically in the form $x = 1 - 6t$, $y = 2 + 11t$, $z = 14t - 1$. (EC)

17 Find the location and nature of the stationary points of

(i) $z = 34x^2 - 24xy + 41y^2$
(ii) $z = x^2 + y^2 + 6x - 4y + 25$
(iii) $z = x^3 + 4x^2 + 3y^2 + 5x - 6y$
(iv) $z = x^4 + y^4 + 4xy$
(v) $z = \ln(x^2 + y^2) - x - 2y$
(vi) $z = x + xy^2 - y - x^2y$
(vii) $z = 1/(x^2 + y^2 + 4)$
(viii) $z = xy(4x + 2y + 1)$
(ix) $z = x^4 + y^4 + 2x^2y^2$

18 The criterion for determining the nature of stationary points at (a, b) may be stated in terms of

$$D \equiv f_{xx}(a, b)f_{yy}(a, b) - [f_{xy}(a, b)]^2$$

$D > 0 \Rightarrow$ local maximum or local minimum

$f_{xx}(a, b) > 0 \Rightarrow$ local minimum, $f_{xx}(a, b) < 0 \Rightarrow$ local maximum

$D < 0 \Rightarrow$ saddle point

$D = 0$ needs further investigation.

Carry out these tests on (i), (ii), (iii), (iv), (vi), (viii), (ix) of Problem 22.

19 (i) Find the stationary values of $x^3 + ay^2 - 6axy$ where $a > 0$ and determine whether they are maxima, minima or saddle points.

(ii) Prove that the function $xy(3 - x - y)$ has a maximum when $x = y = 1$

(iii) Find the greatest value of the function $(x^2 - y^2)\exp(-x^2 - 2y^2)$ (LU)

20 Derive the expression
$$\frac{|ax_1 + by_1 + cz_1 + d|}{\sqrt{(a^2 + b^2 + c^2)}}$$
for the shortest distance, D, of the point (x_1, y_1, z_1) from the plane $ax + by + cz + d = 0$.

[Hint: Consider $D^2 = (x - x_1)^2 + (y - y_1)^2 + (z - z_1)^2$; use the equation of the plane to substitute for z and obtain an expression $g(x, y)$. Find the least value of $g(x, y)$.]

Find the point nearest to the origin which lies on the plane $x - 2y + 2z = 16$.

21 (i) The trough of uniform cross-section shown in Figure 14.8(i) is to have cross-sectional perimeter l. Find the dimensions for maximum cross-sectional area.

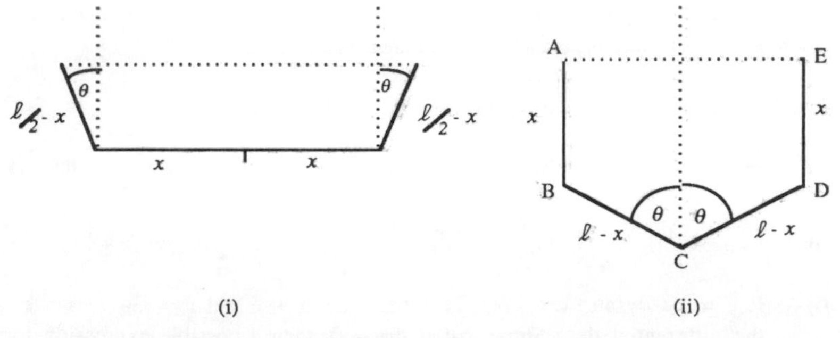

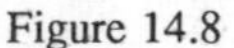
Figure 14.8

(ii) A rectangular sheet of metal of width $2l$ is bent to form a trough without ends. The cross-section is a polygon ABCDE as shown in Figure 14.8(ii). Prove that as x and θ vary, the maximum value of the cross-sectional area is $l^2/\sqrt{3}$.

22 On the pressure-volume diagram for a fluid one particular curve of constant temperature always has a point of inflexion, together with a horizontal tangent; this point is called the *critical* point.

A van der Waal's gas obeys the law
$$p = \frac{RT}{V - b} - \frac{a}{V^2}$$
where p is pressure, V is specific volume, T is temperature, and a and b are constants. Show that at the critical point for this gas
$$V = 3b \quad \text{and} \quad \frac{pV}{RT} = \frac{3}{8}$$
(EC)

Section 14.4

23 The relationship between the resonant frequency f of a series-timed circuit and the inductance L and capacitance C of the circuit is given by

$$f = \frac{1}{2\pi\sqrt{LC}}$$

(a) Find the approximate percentage error in f if the measurements of inductance and capacitance are liable to maximum errors of 1.25% and 0.75% respectively.

(b) Find the maximum percentage error in L if the frequency is in error by ±0.5% and the capacitance is liable to an error of ±1.5%. (EC)

24 A cylindrical hole of diameter 6 in and height 4 in is to be cut in a block of metal by a process in which the maximum error in diameter is 0.003 in and in height is 0.002 in. What is the largest possible error in the volume of the cavity? (EC)

25 The breaking weight W of a cantilever beam is given by the formula $Wl = kbd^2$, where b is the breadth, l the length, d the depth and k a constant depending on the material of the beam. If the length is increased by 1% and the breadth by 5%, how much should the depth be altered to keep the breaking weight unchanged?

26 Find the total differential du given that $u = f(x, y)$ where $f(x, y)$ is

(i) $\dfrac{1}{x^2y} + xy$ (ii) $e^{-(x+y)}\sin(x + y)$ (iii) $\sqrt{x^2 + y^2}\tan^{-1}(y/x)$

(iv) $x^3y + 3xy^4$ (v) $e^{(x^2+y^2)}$ (vi) $\cos(x + 2y)$

27 (i) If $u = \partial\psi/\partial y$ and $v = -\partial\psi/\partial x$ where $u = x + y$ and $v = x - y$, write down the differential dψ. Show that if d$\psi = 0$, then a possible expression for ψ is $y^2/2 + xy - x^2/2$. Sketch the streamlines for the flow represented by the velocities u and v.

(ii) For the stream function in part (i) show that

$$\frac{\partial^2\psi}{\partial x^2} + \frac{\partial^2\psi}{\partial y^2} = 0$$

28 The power consumed in an electrical resistor is given by $P = E^2/R$ watts. If $E = 100$ volts and $R = 10$ ohms, by approximately how much does P change if E is decreased by 2 volts and R is increased by 0.5 ohms? Compare your result with the exact answer.

29 The rate of flow of gas in a pipe is given by $v = Cd^{1/2}T^{-5/6}$, where C is a constant, d is the pipe diameter and T is the absolute gas temperature. The measurement of d is subject to a maximum error of ±1.6% and that of T to one of ±1.2%. Find approximately the maximum percentage error in the value of v.

30 Show that the total surface area of a cone of base radius r and height h is given by

$$S = \pi r^2 + \pi r\sqrt{r^2 + h^2}$$

Find the rate at which the surface area is increasing when $h = 5$ cm and $r = 4$ cm if h and r are both increasing at a rate of 0.5 cm/sec.

31 If x increases at a rate of 2 cm/sec at the instant when $x = 2$ cm and $y = 1$ cm, find the rate at which y must be changing in order that $u = (x^2 + y^2)/(x + y)$ shall be neither increasing nor decreasing when x and y have these values.

32 The equation $e_N = 2\sqrt{kTBR}$ arises in the study of thermal noise. If T, B and R are all functions of time t, find the formula for de_N/dt.

33 In a triode valve the anode current i is given by $i = C(V_a + \mu V_g)^{3/2}$ where V_g is the grid voltage and V_a the anode voltage. If V_g increases at the rate of 0.1 volt/s and V_a at 0.3 volt/s, find the rate of change of i when $V_g = 8$ volts and $V_a = 240$ volts.

34 A gas obeys the law $PV = RT$ where P is its pressure in N/m^2, V its volume in m^3 and T its temperature in °K. If the volume decreases by 0.4 m^3/s and the temperature increases by 4°K/s, find the rate of increase of the pressure; take $R = 8$ Nm/°K.

35 Find dy/dx from the following implicit relationships

(i) $x^3 + y^3 = 8$ (ii) $(x + y)\sin xy = 1$

(iii) $x \cos y + y \cos x = 2$ (iv) $e^{y/x} \ln(x + y) = 6$

36 Find the envelope of

(i) the straight lines $y = cx + c^3$ (ii) the curves $\dfrac{x^2}{\alpha} + \dfrac{y^2}{1-\alpha} = 1$

(iii) the lines $ux + (1 - u)y + u(u - 1) = 0$

Plot some of the lines and the envelope.

37 Show that the envelope of the family of curves with parameter α defined by $x \sin \alpha + y \cos \alpha = 4$ is a circle.

38 The path of a projectile is given by

$$y = x \tan \theta - \frac{gx^2}{2V^2}\sec^2 \theta$$

where θ is the angle of projection and V is the velocity of the projectile (see Figure 14.9)

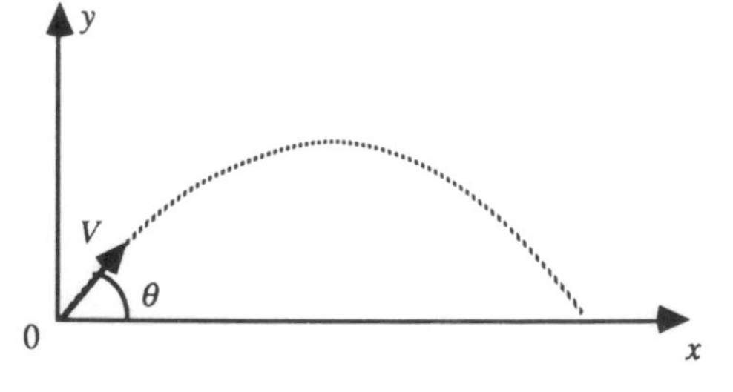

Figure 14.9

Regarding θ as a parameter, show that the enveloping parabola of projectiles projected from the origin with constant velocity V is $g^2x^2 = V^4 - 2gV^2y$. This is sometimes called the **parabola of safety** since no projectile can penetrate beyond it, no matter what value of θ is taken.

39 A long concrete wedge whose cross-section is a circular sector of angle α and radius R has one of its plane faces kept at temperature θ_1 and the other at θ_2 while the curved surface is insulated. See Figure 14.10. Except near the ends, it can be shown that the steady-state temperature is given by

$$\theta = \theta_1 + \frac{\theta_2 - \theta_1}{\alpha}\tan^{-1}(y/x)$$

Find the equation of the isotherms (constant temperature lines) and show that these are the orthogonal trajectories of the lines of flow $x^2 + y^2 = k^2$. Draw both families of curves.

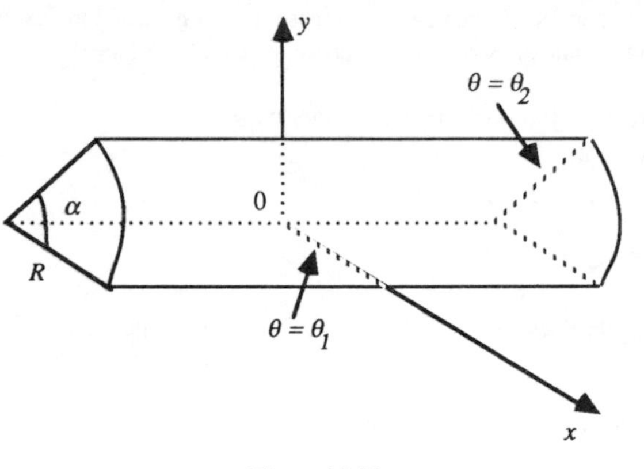

Figure 14.10

40 Show that in the following fluid flow problems the streamlines $\psi(x, y) = c$ and the equipotentials $\varphi(x, y) = k$ form orthogonal trajectories

(i) $\varphi = x^2 - y^2, \psi = 2xy$

(ii) φ and ψ are given by $\varphi + i\psi = -Uz$ where $z = x + iy$

(iii) φ and ψ are given by $\varphi + i\psi = U(z + 4/z)$

41 Find $\partial z/\partial x$ and $\partial z/\partial y$ given that z is defined implicitly in terms of x and y by the equation $f(x, y) = 2x^2 - 3xy^2z + z^3 = C$ (EC)

42 Find dv/dt when

(i) $v = xy + 2, \ x = t^2, y = 1/t$

(ii) $v = x^2 + y^2, \ x = t/(1 + t), \ y = t^2/(1 + t)$

43 Find $\partial z/\partial x$ and $\partial z/\partial y$ when

(i) $x^2 + 2y^2 - z^2 = 4$ (ii) $x + y + z = \ln z$

(iii) $z = x^2 + 8xy + 2$ (iv) $x^2 + 2yz + 2zx = 2$

Hint: write each equation in the form $F(x, y, z) = 0$ and note that $dF/dx = 0 = dF/dy$

44 A beam of light, parallel to the x-axis and coming from $x = -\infty$, hits the circular mirror $r = a$, $-\pi/3 \le \theta \le \pi/3$ and is reflected. Show that the envelope of the reflected ray (the **caustic curve**) is given by the equation

$$x = \tfrac{1}{2}a \cos\theta.(3 - 2\cos^2\theta), \qquad y = a \sin^3\theta$$

Sketch the caustic curve.

45 (a) The five variables p, v, H, S and T are connected by the relations

$$\mathrm{d}H = T\mathrm{d}S + v\mathrm{d}p \qquad \text{and} \left[\frac{\partial S}{\partial p}\right]_T = -\left[\frac{\partial v}{\partial T}\right]_p$$

(i) Explain the meaning of the expressions $\left[\dfrac{\partial}{\partial p}\right]_T$ and $\left[\dfrac{\partial}{\partial T}\right]_p$

(ii) Show that $T\left[\dfrac{\partial v}{\partial T}\right]_p = v - \left[\dfrac{\partial H}{\partial p}\right]_T$

(iii) In the case when $pv = RT$, where R is constant, find an expression for

$T\left[\dfrac{\partial v}{\partial T}\right]_p\left[\dfrac{\partial p}{\partial T}\right]_v$, which does not involve T.

(b) In a gauge the quantity E is determined from R, β and α by

$E = 2R[\sin(\tfrac{1}{2}\beta) - \sin(\tfrac{1}{2}\alpha)]$

If $R = [4 \pm 0.001]$cm; $\beta = [10° \pm 0.1']$; $\alpha = [9° \pm 0.1']$, using partial differentiation, find an approximate value for the greatest error in E.

(c) The x coordinate of a point on a plate is given in terms of r and θ by $x = r\cos\theta$. In a particular configuration r is known to lie in the range 3.98 to 4.01 and θ is known to lie in the range 59°30′ to 60°40′.
Using partial differentiation with $r = 4$ and $\theta = 60°$ as a basis, find the possible range of values of x correct to 2 decimal places. (EC)

46 (a) Determine the derivative $\mathrm{d}z/\mathrm{d}x$ given that

$$z = x^2 + 3xy - 2y^2$$
$$\text{and} \quad x^2 + 2y^2 + \sin(2xy) = 0$$

(b) If

$$g(x, y) = (1/\sqrt{y}) \exp[-(x - a)^2/4y]$$

where $y \ge 0$ and a is a constant, find $\partial^2 g/\partial x^2$ and $\partial g/\partial y$, and hence verify that $g(x, y)$ is a solution of the equation

$$\frac{\partial^2 g}{\partial x^2} = \frac{\partial g}{\partial y}$$ (EC)

47 (a) If $v = \exp(-\frac{1}{2}x^2)\sin^2(3t)$, show that the approximate percentage error in a value of v arising from errors δx in x and δt in t is given by $100\ [-\ x\delta x + 6\cot(3t)\delta t]$.

(b) A function S is defined in terms of p and v by

$$S = a \ln p + b \ln v + c \qquad \text{(A)}$$

in which a, b and c are constants.

In this definition p, v and another variable T are connected by

$$pv = RT \qquad \text{(B)}$$

where R is constant.

(i) Find expressions for $\left.\frac{\partial S}{\partial p}\right]_v$ and $\left.\frac{\partial S}{\partial p}\right]_T$

(ii) A new variable Q is introduced where

$$dQ = a dT + p dv \qquad \text{(C)}$$

and

$$dS = \frac{1}{T} dQ \qquad \text{(D)}$$

Eliminate Q, S and T between the four relations (A), (B), (C) and (D) and hence show that $b - a = R$. (EC)

48 (a) The reversible adiabatic expansion of a gas is given by

$$c_p \frac{dv}{v} + c_v \frac{dp}{p} = 0$$

in which c_p and c_v are constants.

(i) Solve the above differential equation and show that the solution can be written in the form pv^k = constant

(ii) What is k?

(iii) If particular values are p_0 and v_0, show that

$$\frac{p}{p_0} = \left[\frac{v_0}{v}\right]^k$$

(b) The side c of a triangular mechanism is calculated from

$$c = \frac{b \sin C}{\sin B}$$

It is known that B and C are 38° and 26° to the nearest degree and b can be in error by 0.5%. Find the greatest percentage error in the calculated value of c. (EC)

15

APPROXIMATION OF EXPERIMENTAL DATA

15.1 Least Squares Straight Line Fitting

Figure 15.1 shows a set of plotted results from an experiment. We wish to fit a particular relationship to the data; for example, a straight line $y = a_0 + a_1x$ or a parabola $y = a_0 + a_1x + a_2x^2$.

Once we have decided on the *kind* of relationship we wish to fit, then we have to decide on the coefficients that determine the particular curve of its kind which fits the data best. But what is *best* in this context? If we were to fit a straight line to the data of Figure 15.1, we could use a transparent straight-edge and move it on the paper until it seemed to split the data points roughly equally above and below the line. However, our idea of best fit and someone else's might not coincide and if the line of best fit were to be used to make predictions about y for non-observed values of x, the results could be significantly different.

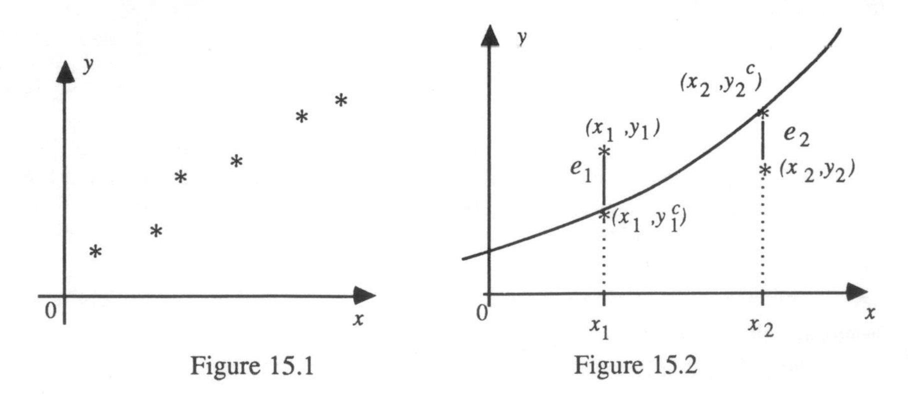

Figure 15.1 Figure 15.2

To overcome this subjective element, we need an algebraically defined criterion and we embark on this using Figure 15.2. We generalise the situation to a relationship $y(x)$ being fitted to n data points and we examine the behaviour of $y(x)$ near two of them. We assume that the experiment is one in which we can regard the measurement of x values to be *exact* (or nearly so) and the measurement of y values to be *subject to error* (e.g. x might be time, and

y voltage). Then for each observed x_i, $i = 1, 2, \ldots, n$, there is an observed y_i and, if we assume a particular relationship $y(x)$, a calculated value y_i^c where $y_i^c = y(x_i)$. We measure the discrepancy, e_i, between the observed and calculated quantities in each case; $e_i = y_i - y_i^c$. In Figure 15.2, $e_1 > 0$, $e_2 < 0$. In an ideal case, where all the data points lie *exactly* on the curve of a known relationship $y(x)$, $e_i = 0$ for all points. However, in general we must expect a discrepancy at each point and we seek that curve of a chosen kind for which the total discrepancy (i.e. the *sums* of discrepancies at each data point) is least.

The algebraic sum of discrepancies is a poor choice since it permits partial cancellation between positive and negative quantities and makes the situation seem better than it is. It would be wrong to say that the data of Figure 15.3 fitted the straight line well yet the sum of the discrepancies is nearly zero.

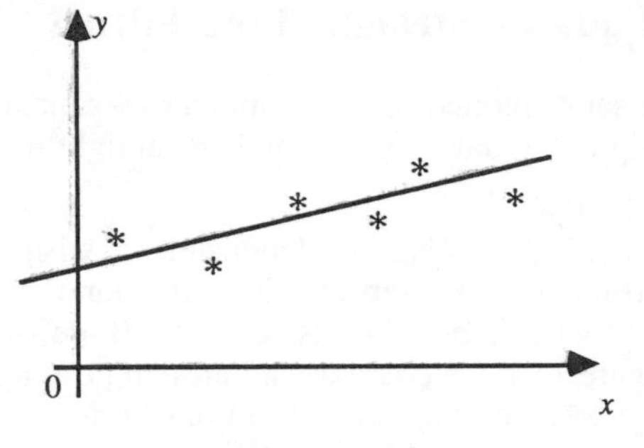

Figure 15.3

For simplicity with later calculations the **least squares criterion** is used: this calculates

$$S = \sum_{i=1}^{n} e_i^2 \tag{15.1}$$

for the n data points and seeks that curve of its class which *minimises* S. We note that this gives undue emphasis to points which are well separated from the majority but this is a risk which has to be weighed against the objectiveness of the criterion and its easy adaptation to programming for a computer (Problem 1). Problem 2 shows that the attempt to fit a straight line $y = a_0 + a_1x$ to given data points provides a value for S which depends on a_0 and a_1; we merely note in passing that under most circumstances we shall always be able to find a_0 and a_1 no matter how badly the data is fitted by a straight line. For the moment, we state that, in general, S will be a *function* of a_0 and a_1, viz $S = S(a_0, a_1)$ and we seek to find those values of a_0 and a_1 which minimise S. Problem 4 shows that for the straight line fit S is given by the formula

$$S = \sum_{i=1}^{n} (y_i - a_0 - a_1x_i)^2 \tag{15.2}$$

[There is no maximum value of S since we could merely keep the slope a_1 fixed and increase (or decrease) a_0, which would increase S without limit.]

We use the result that the derivative of a sum is the sum of the separate derivatives to find the values of a_0 and a_1 which minimise S.

Now
$$\frac{\partial S}{\partial a_0} = \sum_{i=1}^{n} 2(y_i - a_0 - a_1 x_i).(-1)$$

and
$$\frac{\partial S}{\partial a_1} = \sum_{i=1}^{n} 2(y_i - a_0 - a_1 x_i).(-x_i)$$

Consider the case $n = 2$, then
$$S = (y_1 - a_0 - a_1 x_1)^2 + (y_2 - a_0 - a_1 x_2)^2$$

and
$$\frac{\partial S}{\partial a_1} = 2(y_1 - a_0 - a_1 x_1)(-x_1) + 2(y_2 - a_0 - a_1 x_2)(-x_2)$$

You should be able to see how this result generalises; the same ideas apply for $\partial S/\partial a_0$.

For a minimum S (a maximum S does not exist) we require
$$\frac{\partial S}{\partial a_0} = 0 = \frac{\partial S}{\partial a_1}$$

i.e.
$$\sum_{1}^{n} (y_i - a_0 - a_1 x_i) = 0 = \sum_{1}^{n} x_i (y_i - a_0 - a_1 x_i)$$

Hence
$$\left.\begin{aligned} \sum y_i &= \sum a_0 + \sum a_1 x_i \\ \sum x_i y_i &= \sum x_i a_0 + \sum a_1 x_i^2 \end{aligned}\right\}$$

or
$$\left.\begin{aligned} \sum y_i &= n a_0 + a_1 \sum x_i \\ \sum x_i y_i &= a_0 \sum x_i + a_1 \sum x_i^2 \end{aligned}\right\} \qquad (15.3)$$

These are called the **normal equations**.

All we need to do is to evaluate each of the sums to produce numbers and then solve these two equations for a_0 and a_1.

Example

The readings in Table 15.1 were obtained for the specific heat of ethyl alcohol.

Table 15.1

x temperature °C	0	10	20	30	40	50
y specific heat	0.50680	0.54544	0.56617	0.58743	0.62984	0.66330

Now $$\sum y_i = 0.50680 + \ldots + 0.66330 = 3.49898$$

$$\sum x_i = 150 \qquad \sum x_i y_i = 92.7593 \qquad \sum x_i^2 = 5500$$

Hence equations (15.3) give
$$\begin{cases} 3.49898 = 6a_0 + 150a_1 \\ 92.7593 = 150a_0 + 5500a_1 \end{cases}$$

and solving we find
$$a_0 = 0.50766; \quad a_1 = 0.00311$$

Thus the straight line fit is
$$y = 0.50766 + 0.00311x$$

A flow chart is given in Figure 15.4 for the fitting of a linear relationship $y = a + bx$ to n pairs of observed x and y values.

We read the data into two one-dimensional arrays, calculate the values of a and b and write them out. Arrays are used in case the x and y values might be needed for calculations (these are omitted in this discussion).

The sum Σx will be represented by SX, Σy by SY, Σx^2 by SXX, Σxy by SXY.

A program to carry out Least Squares Straight Line Fit is given below.

```
>
   10   CLS
   20   PRINTTAB(10,2)"Least Squares Straight Line Fit"
   30   @%=&20209
   40   SX=0:SY=0:SXX=0:SXY=0
   50   INPUTTAB(0,5)"Enter number of points ",N
   60   FOR I = 1 TO N
   70     INPUT"Enter X value ",X
   80     INPUT"Enter Y value ",Y
   90     SX=SX+X
  100     SY=SY+Y
  110     SXX=SXX+(X*X)
  120     SXY=SXY+(X*Y)
  130     NEXT I
  140   B=(SX*SY-N*SXY)/(SX*SX-SXX*N)
  150   A=(SY-B*SX)/N
  160   PRINT"The equation of the straight line is
        Y = ";A;" + ";B;" X"
  170   END
```

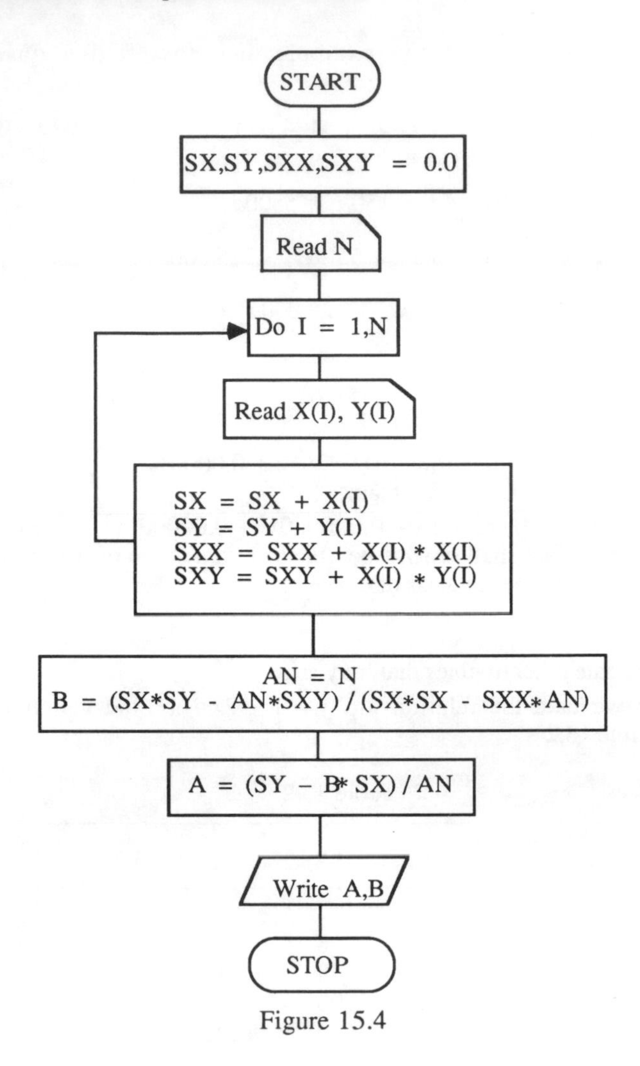

Figure 15.4

15.2 Fitting Other Curves

The normal equations for a quadratic fit are readily shown to be

$$\left.\begin{aligned}\sum y_i &= na_0 + a_1\sum x_i + a_2\sum x_i^2\\ \sum x_i y_i &= a_0\sum x_i + a_1\sum x_i^2 + a_2\sum x_i^3\\ \sum x_i^2 y_i &= a_0\sum x_i^2 + a_1\sum x_i^3 + a_2\sum x_i^4\end{aligned}\right\}\qquad(15.4)$$

We return to the data of the example and now find a quadratic fit $y = a_0 + a_1x + a_2x^2$. Now

$$\Sigma x_i^3 = 225\,000 \qquad \Sigma x_i^2 y_i = 3474.693 \qquad \Sigma x_i^4 = 979 \times 10^4$$

which gives from equations (15.4)

$$\begin{cases} 3.49898 = 6a_0 + 150a_1 + 5500a_2 \\ 92.7593 = 150a_0 + 5500a_1 + 225\,000a_2 \\ 3474.693 = 5500a_0 + 225\,000a_1 + 979 \times 10^4\, a_2 \end{cases}$$

Solving,

$$a_0 = 0.51039 \qquad a_1 = 0.00261 \qquad a_2 = 0.00001 \qquad \text{(5 d.p.)}$$

Hence

$$y = 0.51039 + 0.00261x + 0.00001x^2$$

We can similarly fit a cubic to obtain

$$y = 0.507662 + 0.003858x - 0.000060x^2 + 0.000001x^3 \qquad \text{(6 d.p.)}$$

Remember that we shall *always* be able to find a curve of best fit (no matter how poor a fit it is) for each class of curve.

Example

Let us investigate other troubles that may arise.

Suppose we take the relationship $y = 1 + 2x + x^2$ and generate the data points of Table 15.2

Table 15.2

	x	y	x^2	xy	x^2y	x^3	x^4
	1	4	1	4	4	1	1
	2	9	4	18	36	8	16
	3	16	9	48	144	27	81
	4	25	16	100	400	64	256
	5	36	25	180	900	125	625
Σ	15	90	55	350	1484	225	979

Fitting a straight line $y = a_0 + a_1x$ we obtain

$$\begin{cases} 90 = 5a_0 + 15a_1 \\ 350 = 15a_0 + 55a_1 \end{cases}$$

and hence

$$y = -6 + 8x$$

Fitting a quadratic $y = a_0 + a_1x + a_2x^2$ yields

$$\begin{cases} 90 = 5a_0 + 15a_1 + 55a_2 \\ 350 = 15a_0 + 55a_1 + 225a_2 \\ 1484 = 55a_0 + 225a_1 + 979a_2 \end{cases}$$

and hence

$$y = 1.00134 + 1.99145x + 1.01105x^2$$

[compare the true polynomial $y = 1 + 2x + x^2$]
Fitting a cubic provides

$$y = 1.00085 + 2.00125x + 1.00134x^2 + 0.00003x^3$$

and fitting a quartic

$$y = 1.00102 + 1.99913x + 0.99935x^2 + 0.00013x^3 + 0.00007x^4$$

Were we to try to fit a quintic we should expect that since we have only *five data points* to find *six coefficients* we would get infinitely many answers, but from the process we would, in fact, get *one* answer.

The trouble here is **round-off error**. If we had worked with *exact* arithmetic we should have got *exactly* $y = 1 + 2x + x^2$.

Often we need to fit curves other than polynomials. The normal equations are then more difficult to calculate. Suppose, for example, we wish to fit the curve $y = a + (b/x)$ to n data points (x_i, y_i). The quantity to be minimised is

$$S = \sum_{i=1}^{n} (y_i - a - b/x_i)^2$$

Now

$$\frac{\partial S}{\partial a} = -\sum_{i=1}^{n} 2(y_i - a - b/x_i)$$

and

$$\frac{\partial S}{\partial b} = -\sum_{i=1}^{n} \frac{2}{x_i}(y_i - a - b/x_i)$$

and hence the normal equations become

$$\sum_{i=1}^{n} y_i = na + b\sum_{i=1}^{n} \frac{1}{x_i}$$

$$\sum_{i=1}^{n} \frac{y_i}{x_i} = a\sum_{i=1}^{n} \frac{1}{x_i} + b\sum_{i=1}^{n} \frac{1}{x_i^2}$$

We must then calculate y_i/x_i, $1/x_i$, $1/x_i^2$ and form their sums.

It is sometimes more useful to use *logarithms* on the relationship to be fitted. For example, the relationship $y = ax^n$ becomes $\log y = \log a + n \log x$ and the relationship $y = ae^{bx}$ becomes $\log_e y = \log_e a + bx$.

We can transform the original data into a suitable form before fitting the line Y = A + BX. Note that in these cases the quantity being minimised is

$$\sum_{i=1}^{n} (\log y_i - \log a - n \log x_i)^2 \quad \text{and} \quad \sum_{i=1}^{n} (\log y_i - \log a - bx_i)^2$$

respectively and the least squares best fit may not agree with that obtained from fitting the curve directly.

15.3 Finite Differences and Interpolation

In Section 2.5 we introduced the idea of differences of a tabulated function, the notation for forward differences

$$\Delta f(x_0) = f(x_0 + h) - f(x_0)$$

and the first steps in interpolation. We now re-examine these differences and bear in mind their obvious similarity to differentiation.

Since engineers often have to estimate values of a quantity for which we have only a certain number of values tabulated, it is important to obtain estimates which are as accurate as the data permits. Table 15.3 is a typical difference table.

Table 15.3

x	$f(x)$	Δ	Δ^2	Δ^3
x_0	$f(x_0)$			
		$\Delta f(x_0)$		
x_0+h	$f(x_0+h)$		$\Delta^2 f(x_0)$	
		$\Delta f(x_0+h)$		$\Delta^3 f(x_0)$
x_0+2h	$f(x_0+2h)$		$\Delta^2 f(x_0+h)$	
		$\Delta f(x_0+2h)$		$\Delta^3 f(x_0+h)$
x_0+3h	$f(x_0+3h)$		$\Delta^2 f(x_0+2h)$	
		$\Delta f(x_0+3h)$		
x_0+4h	$f(x_0+4h)$			

Some books write $\Delta_h f(x_0)$ to emphasise the step size, h

An alternative notation is shown in Table 15.4

Since $\Delta f(x_0) = f(x_0 + h) - f(x_0)$ we have

$$f(x_0 + h) = f(x_0) + \Delta f(x_0) \tag{15.5a}$$

and

$$\begin{aligned} f(x_0 + 2h) &= f(x_0 + h) + \Delta f(x_0 + h) \\ &= f(x_0) + \Delta f(x_0) + \Delta f(x_0 + h) \end{aligned} \tag{15.5b}$$

Table 15.4

x	$f(x)$	Δ	Δ^2	Δ^3
x_0	f_0			
		Δf_0		
x_1	f_1		$\Delta^2 f_0$	
		Δf_1		$\Delta^3 f_0$
x_2	f_2		$\Delta^2 f_1$	
		Δf_2		$\Delta^3 f_1$
x_3	f_3		$\Delta^2 f_2$	
		Δf_3		
x_4	f_4			

[Now $f(x) = x^2$ is a *function* taking values of x and producing values x^2; differentiation is an **operation** taking a function and producing a new function and d/dx is called an **operator**.]

We now treat Δ as an *operator* on $f(x)$; this means it can be manipulated by the usual rules of algebra, provided it precedes the entity on which it operates. Therefore we can operate on the equation (15.5a) to obtain

$$\Delta f(x_0 + h) = \Delta f(x_0) + \Delta[\Delta f(x_0)] = \Delta f(x_0) + \Delta^2 f(x_0)$$

and so

$$f(x_0 + 2h) = f(x_0) + 2\Delta f(x_0) + \Delta^2 f(x_0)$$

In a similar way we can show that

$$f(x_0 + 3h) = f(x_0) + 3\Delta f(x_0) + 3\Delta^2 f(x_0) + \Delta^3 f(x_0)$$

and by induction that

$$f(x_0 + nh) = f(x_0) + n\Delta f(x_0) + \frac{n(n-1)}{2!}\Delta^2 f(x_0) + \ldots + \Delta^n f(x_0) \qquad (15.6)$$

where n is a positive integer. Note the strong similarity between (15.6) and the binomial theorem. Can we extend this result for non-integer values of n to give us a possible formula for interpolation?

Interpolation using forward differences

First we approach the problem of interpolation by a different route. We know that the linear interpolation formula $f(x_0 + ph) = f(x_0) + p\Delta f(x_0)$ is *exact* for a *linear* function $f(x) = bx + c$. If we wish to interpolate a quadratic function $f(x) = ax^2 + bx + c$ using this formula we would expect an error since we cannot approximate accurately a quadratic function by a straight line. To find this error we consider simply $f(x) = ax^2$ (why?) and find that

$$\Delta f(x_0) = a(x_0 + h)^2 - ax_0^2 \equiv 2ahx_0 + ah^2$$

and that

$$\Delta^2 f(x_0) = 2ah^2$$

Linear interpolation gives

$$f(x_0 + ph) = ax_0^2 + p(2ahx_0 + ah^2)$$

However, the true value is

$$a(x_0 + ph)^2 \equiv ax_0^2 + 2aphx_0 + ap^2h^2$$

and so the error is

$$ah^2(p - p^2)$$

(You should be able to show that the maximum error occurs when $p = \frac{1}{2}$ and is then $\frac{1}{4}ah^2$.) If we add on this correction term which can be written $\frac{1}{2}p(p-1)\Delta^2 f(x_0)$, we have a formula exact for a quadratic polynomial, viz

$$f(x_0 + ph) = f(x_0) + p\Delta f(x_0) + \tfrac{1}{2}p(p-1)\Delta^2 f(x_0)$$

A similar exercise that you could try is to establish the correction term to make the formula correct for a cubic polynomial: the correction is

$$\frac{1}{6}p(p-1)(p-2)\Delta^3 f(x_0)$$

Therefore, for a cubic polynomial

$$f(x_0 + ph) = f(x_0) + p\Delta f(x_0) + \frac{1}{2!}p(p-1)\Delta^2 f(x_0)$$
$$+ \frac{1}{3!}p(p-1)(p-2)\Delta^3 f(x_0)$$

Notice that the formula (15.6) seems to be true, at least as far as the term in $\Delta^3 f(x_0)$, if we substitute p instead of n and we therefore believe that (15.6) is also true for non-integral values of n.

A proof of the formula can be accomplished using the idea of operators. We introduce the *shift operator* E where $\mathrm{E}f(x_0) = f(x_0 + h)$ with the obvious extension

$$\mathrm{E}^2 f(x_0) = \mathrm{E}[\mathrm{E}f(x_0)] = \mathrm{E}[f(x_0 + h)] = f(x_0 + 2h)$$

and so on. Now

$$(1 + \Delta)f(x_0) = f(x_0) + \Delta f(x_0) = f(x_0) + f(x_0 + h) - f(x_0) = f(x_0 + h)$$

and we see that the operator $1 + \Delta$ has the same effect as the operator E. This allows us to identify them by an operator equation: $1 + \Delta = \mathrm{E}$. It can be shown directly (try it) that

$$(1 + \Delta)^2 = 1 + 2\Delta + \Delta^2 = \mathrm{E}^2$$

and so on.

It is fairly clear what interpretation to put on $\mathrm{E}^n f(x_0)$ if n is a positive integer but what about E^{-1} or $\mathrm{E}^{1/2}$? In fact we assume that

$$E^{1/2}[E^{1/2}f(x_0)] \equiv Ef(x_0) \quad \text{and} \quad E^{-1}[Ef(x_0)] \equiv E[E^{-1}f(x_0)] = f(x_0)$$

Interpretations for other values of n are similar.

Then for any *rational value* of p

$$f(x_0 + ph) = E^p f(x_0) = (1 + \Delta)^p f(x_0)$$

$$= \left\{ 1 + p\Delta + \frac{p(p-1)}{1.2}\Delta^2 + \ldots + \frac{p(p-1)\ldots(p-r+1)}{1.2\ \ldots\ldots\ r}\Delta^r + \ \ldots\ldots\ldots \right\} f(x_0)$$

(by the binomial expansion)

that is

$$f(x_0 + ph) = f(x_0) + p\Delta f(x_0) + \frac{p(p-1)}{2!}\Delta^2 f(x_0) + \frac{p(p-1)(p-2)}{3!}\Delta^3 f(x_0) + \ldots \quad (15.7)$$

Equation (15.7) is the **Newton-Gregory forward difference interpolation formula** which can be used in interpolating a function at values between tabulated points. Let us apply the formula to a specific example.

Example

The angle θ, turned through by a shaft, was measured; see Table 15.5.

Table 15.5

t (sec)	0	0.2	0.4	0.6	0.8	1.0
θ (rads)	–0.002	0.058	0.149	0.283	0.470	0.717

We first form a difference table which is shown as Table 15.6.

Table 15.6

t	θ	Δ	Δ^2	Δ^3	Δ^4
0	–0.002				
		0.060			
0.2	0.058		0.031		
		0.091		0.012	
0.4	0.149		0.043		–0.002
		0.134		0.010	
0.6	0.283		0.053		–0.003
		0.187		0.007	
0.8	0.470		0.060		
		0.247			
1.0	0.717				

It would seem unwise to carry the tabulation further. We estimate θ when
(i) $t = 0.3$ (ii) $t = 0.15$ (iii) $t = 0.7$ (iv) $t = 0.9$

(i) 0.3 lies half-way between 0.2 and 0.4, and using $x_0 = 0.2$, $h = 0.2$ then $0.3 = x_0 + ph$ gives $p = \frac{1}{2}$.

Using the Newton-Gregory forward difference formula we obtain

$$\theta(0.2 + \tfrac{1}{2} \times 0.2) \cong 0.058 + \tfrac{1}{2} \times 0.091 + \frac{\frac{1}{2}(-\frac{1}{2})}{2} \times 0.043$$

$$+ \frac{\frac{1}{2}(-\frac{1}{2})(-\frac{3}{2})}{6} \times (0.010)$$

$$= 0.058 + 0.0455 - 0.005375 + 0.000625$$

i.e. $\theta(0.3) = 0.099$ (3 d.p.)

We quote the result to 3 d.p. since we cannot hope to achieve an accuracy greater than the tabulated values.

(ii) Here we take $x_0 = 0$ and since $h = 0.2$, then $p = \frac{3}{4}$. Similarly

$$\theta(0 + \tfrac{3}{4} \times 0.2) \cong -0.002 + 0.0435 - 0.00290625 + 0.00046875$$

i.e. $\theta(0.15) = 0.039$ (3 d.p.)

(iii) This time take $x_0 = 0.6$ with $p = \frac{1}{2}$. Here, we can only pick up the first and second differences and

$$\theta(0.6 + \tfrac{1}{2} \times 0.2) \cong 0.283 + 0.0935 - 0.0075 = 0.369 \quad \text{(3 d.p.)}$$

i.e. $\theta(0.7) = 0.369$ (3 d.p.)

(iv) $h = 0.2$, $x_0 = 0.8$, $p = \frac{1}{2}$. We are restricted to linear interpolation:

$$\theta(0.9) \cong 0.470 + \tfrac{1}{2}(0.247) = 0.5593(5) \quad \text{(3 d.p.)}$$

Hence the formula ceases to be of use at the end of a table and we wonder if a similar formula exists which allows us to go back into a table to pick up values.

15.4 Other Interpolation Formulae

Referring to Table 15.3 we can label the entries in terms of so-called **backward differences**. The backward difference $f(x_0) - f(x_0 - h)$ can be written as

$$\nabla f(x_0) \quad \text{or} \quad \nabla f_0 \tag{15.8}$$

It is to be emphasised that the entry values are unchanged; it is the labels which differ. Table 15.7 shows the changes.

Table 15.7

x	$f(x)$	∇	∇^2	∇^3
x_0	f_0			
		∇f_1		
x_0+h	f_1		$\nabla^2 f_2$	
		∇f_2		$\nabla^3 f_3$
x_0+2h	f_2		$\nabla^2 f_3$	
		∇f_3		$\nabla^3 f_4$
x_0+3h	f_3		$\nabla^2 f_4$	
		∇f_4		
x_0+4h	f_4			

Now $E^{-1} = f(x_0 - h)$ and therefore

$$(1 - E^{-1})f(x_0) = f(x_0) - f(x_0 - h) \equiv \nabla f(x_0)$$

Hence $$\nabla = 1 - E^{-1}$$

and so $$E^{-1} = 1 - \nabla, \quad E = (1 - \nabla)^{-1}$$

and $$f(x_0 + ph) = (1 - \nabla)^{-p} f(x_0)$$

This gives the **Newton-Gregory backward difference interpolation formula**

$$f(x_0 + ph) = f(x_0) + p\nabla f(x_0) + \frac{p(p+1)}{2!}\nabla^2 f(x_0) + \frac{p(p+1)(p+2)}{3!}\nabla^3 f(x_0) + \ldots \tag{15.9}$$

(The differences used if we choose $x_0 + 3h$ or $x_0 + 4h$ as the base are shown in Table 15.7 by dashed lines.)

Example

We can therefore apply formula (15.9) to case (iv) of the example in the previous section where the forward interpolation formula reduced to linear interpolation; we have $h = 0.2$, $x_0 = 0.8$, $p = \frac{1}{2}$.

Then

$$\begin{aligned}\theta(0.9) &= 0.470 + \tfrac{1}{2} \times 0.187 + \frac{\frac{1}{2}\cdot\frac{3}{2}}{2} \times 0.053 + \frac{\frac{1}{2}\cdot\frac{3}{2}\cdot\frac{5}{2}}{6} \times 0.010 \\ &= 0.470 + 0.0935 + 0.019875 + 0.003125 \\ &= 0.586(5) \qquad \text{(3 d.p.)}\end{aligned}$$

Alternatively, we may take $x_0 = 1.0$ whence $p = -\frac{1}{2}$ to obtain

$$\theta(0.9) = 0.717 + (-\tfrac{1}{2}) \times 0.247 + \frac{(-\frac{1}{2})(\frac{1}{2})}{2} \times 0.060 + \frac{(-\frac{1}{2})(\frac{1}{2})(\frac{3}{2})}{6} \times 0.007$$

$$= 0.586(5) \qquad \text{(3 d.p.)} \quad \text{which is the same result as before.}$$

Central differences

Earlier we have seen that a central average gave more accurate results for numerical differentiation and we might expect a similar result of affairs for interpolation.

We define the **central difference**

$$\delta f\,(x_0 + \tfrac{1}{2}h) \equiv f\,(x_0 + h) - f\,(x_0)$$

and sometimes write it as $\delta f_{1/2}$. Then if we shift the centre of attention from x_0 to $x_r = x_0 + rh$,

$$\delta f[x_0 + (r + \tfrac{1}{2})h] \equiv f[x_0 + (r + 1)h] - f(x_0 + rh)$$

which is written $\delta f_{r+1/2}$. It follows that $\delta f_{r+1/2} \equiv \Delta f_r - \nabla f_{r+1}$. Table 15.8 shows the same entries as Tables 15.3 and 15.7 with their new labels.

Since $\delta^2 f_r \equiv \delta(\delta f_r)$, it follows that

$$\begin{aligned}\delta^2 f_1 &\equiv \delta(f_{r+1/2} - f_{r-1/2})\\ &\equiv (f_{r+1} - f_r) - (f_r - f_{r-1})\\ &\equiv f_{r+1} - 2f_r + f_{r-1}\end{aligned}$$

Table 15.8

x	$f(x)$	δ	δ^2	δ^3
x_0	f_0			
		$\delta f_{1/2}$		
x_0+h	f_1		$\delta^2 f_1$	
		$\delta f_{3/2}$		$\delta^3 f_{3/2}$
x_0+2h	f_2		$\delta^2 f_2$	
		$\delta f_{5/2}$		$\delta^3 f_{5/2}$
x_0+3h	f_3		$\delta^2 f_3$	
		$\delta f_{7/2}$		
x_0+4h	f_4			

Check this result against the values of Table 15.6. Observe how the suffices change as we move from column to column in Table 15.8.

We now state without proof two interpolation formulae which use central differences.

Bessel's Formula

$$f(x_0 + ph) = f_0 + p\delta f_{1/2} + \frac{1}{2.2!}p(p-1)(\delta^2 f_0 + \delta^2 f_1)$$

$$+\frac{1}{3!}p(p-1)(p-\tfrac{1}{2})\delta^3 f_{1/2} + \frac{1}{2.4!}(p+1)p(p-1)(p-2)(\delta^4 f_0 + \delta^4 f_1)$$

$$+ \ldots \qquad (15.10)$$

Everett's Formula

$$f(x_0 + ph) = q[f_0 + \frac{1}{3!}(q^2-1)\delta^2 f_0 + \frac{1}{5!}(q^2-1)(q^2-4)\delta^4 f_0 + \ldots]$$

$$+ p[f_1 + \frac{1}{3!}(p^2-1)\delta^2 f_1 + \frac{1}{5!}(p^2-1)(p^2-4)\delta^4 f_1 + \ldots] \qquad (15.11)$$

where $q = 1 - p$

Example

If we return to Table 15.6 we shall apply both formulae to the evaluation of $\theta(0.5)$. You should compare the results with estimates from the Newton-Gregory formulae.

(i) *Bessel's formula*

Here $x_0 = 0.4$ and $p = \frac{1}{2}$. Then

$$\theta(0.5) = 0.149 + \frac{1}{2}\times 0.134 + \frac{1}{4}\cdot\frac{1}{2}\cdot\left[-\frac{1}{2}\right](0.043 + 0.053) + \frac{1}{6}\cdot\frac{1}{2}\cdot$$

$$\left[-\frac{1}{2}\right](0).(0.010) + \frac{1}{48}\cdot\left[\frac{3}{2}\right]\cdot\left[\frac{1}{2}\right]\cdot\left[-\frac{1}{2}\right]\cdot\left[-\frac{3}{2}\right](-0.002 - 0.003)$$

$$= 0.149 + 0.067 - 0.006 + 0 - 0.0000585$$

$$= 0.210 \qquad \text{(3 d.p.)}$$

The way differences are picked up for use follows the pattern below.

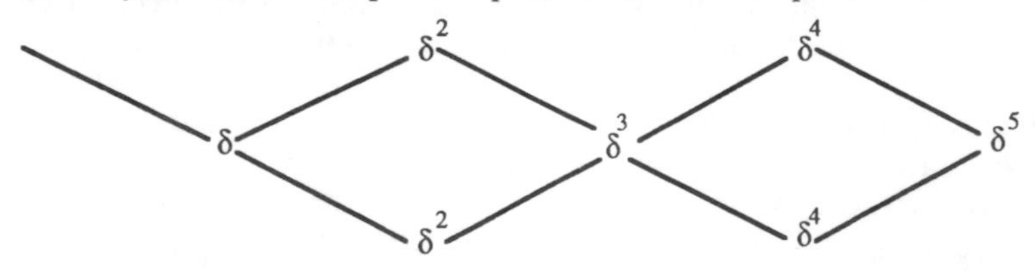

(ii) *Everett's Formula*

Here $x_0 = 0.4$ and $p = \frac{1}{2}$, $q = \frac{1}{2}$. Then we obtain

$$\theta(0.5) = \frac{1}{2}\left[0.149 + \frac{1}{6}\left[-\frac{3}{4}\right](0.043) + \frac{1}{120}\left[-\frac{3}{4}\right]\cdot\left[-\frac{15}{4}\right](-0.002)\right]$$

$$+\frac{1}{2}\left[0.283 + \frac{1}{6}\left[-\frac{3}{4}\right](0.053) + \frac{1}{120}\left[-\frac{3}{4}\right]\cdot\left[-\frac{15}{4}\right](-0.003)\right]$$

$$= \frac{1}{2}\left[0.149 - 0.005375 - 0.000046875\right] + \frac{1}{2}\left[0.283 - 0.006625 - 0.0000703125\right]$$

$$= 0.0717891 + 0.1381523 \qquad \text{(3 d.p.)}$$

$$= 0.210 \qquad \text{(3 d.p.)}$$

Choice of Formulae

It would seem advisable to give some guidance as to which formula to use. It is clear that the central difference based formulae produce as accurate an estimation as the Newton-Gregory formulae using fewer terms. In general, the central difference formulae should be used where possible because of the rapid decay of the successive terms; however, at the beginning of a table, the Newton-Gregory forward difference formula is used and its backward difference counterpart is used at the end of a table. To emphasise this, if we wanted to find $f(0.1)$, $f(0.3), f(0.55), f(0.93)$ from the table

x	0.0	0.2	0.4	0.6	0.8	1.0
$f(x)$	0.0	0.008	0.064	0.216	0.516	1.000

we would use Newton-Gregory forward difference, central difference, central difference and Newton-Gregory backward difference formulae respectively.

Lagrange Interpolation

Two questions should spring to mind: if in a formula we neglect differences higher than, say, the fourth, is this equivalent to assuming we can approximate the curve by a fourth order polynomial? What happens if the tabulated values are not equally spaced? The answer to the first question is *Yes* and this leads us to develop a different method of interpolation to overcome the difficulty raised by the second question. We make use of the result that there is a unique polynomial of Nth degree passing through $(N + 1)$ specified (different) points. We quote without proof the **Lagrange polynomial** through the N points

$$(x_1, f_1),\ (x_2, f_2),\ \ldots,\ (x_N, f_N)$$

This polynomial is

$$\sum_{r=1}^{N} f_r \prod_{\substack{k=1 \\ k \neq r}}^{N} \frac{(x - x_k)}{(x_r - x_k)} \tag{15.12}$$

where $\prod_{\substack{k=1 \\ k \neq r}}^{N}$ means the product of terms like the one quoted where k takes values 1, 2, ..., N but excluding $k = r$. For example, the *three-point formula* is

$$\frac{(x - x_2)(x - x_3)}{(x_1 - x_2)(x_1 - x_3)} f_1 + \frac{(x - x_1)(x - x_3)}{(x_2 - x_1)(x_2 - x_3)} f_2 + \frac{(x - x_1)(x - x_2)}{(x_3 - x_1)(x_3 - x_2)} f_3$$

Observe that this uses only function values and does not require a table of differences (which would require that these function values be equally spaced).

Example

Suppose we return to Table 15.6 and use the values $\theta(0.2)$, $\theta(0.4)$, $\theta(0.6)$ to estimate $\theta(0.5)$, then $x_1 = 0.2$, $x_2 = 0.4$, $x_3 = 0.6$, $f_1 = 0.058$, $f_2 = 0.149$, $f_3 = 0.283$ and $x = 0.5$. Then we have the approximation

$$\theta(0.5) \cong \frac{(0.5 - 0.4)(0.5 - 0.6)}{(-0.2)(-0.4)}(0.058) + \frac{(0.5 - 0.2)(0.5 - 0.6)}{(0.2)(-0.2)}(0.149)$$

$$+ \frac{(0.5 - 0.2)(0.5 - 0.4)}{(0.4)(0.2)}(0.283)$$

$$= -0.00725 + 0.11175 + 0.106125$$

$$= 0.211 \qquad \text{(3 d.p.)}$$

Bearing in mind the fact that the third differences are not quite constant, it might have been better to take the four-point formula: you should try to derive it in general and apply it to this example.

A program to carry out Lagrange interpolation is presented overleaf.

Disadvantages

The disadvantages of Lagrange interpolation are:

(i) that there is no easy check on its accuracy as can be the case when a table of differences is formed

(ii) that we have to decide in advance what degree of polynomial to fit

(iii) that to go to a higher degree polynomial we have to start from scratch whereas in the case of a difference-based formula we simply have to add on extra terms.

Before leaving this section we note the hazards involved in extrapolation.

```
>
 10   CLS
 20   PRINTTAB(10,2)"Lagrange Interpolation"
 30   N=6
 40   DIM X(N),F(N)
 50   PRINTTAB(1,4)"DATA  time      theta"
 60   FOR I=1 TO N
 70     READ X(I),F(I)
 80     PRINTX(I),F(I)
 90     NEXT I
100   INPUTTAB(0,N+7)"Enter time to estimate for ",X
110   LET term=0
120   FOR r=1 TO N
130     LET tot=1
140     FOR k=1 TO N
150       IF k<>r THEN t=(X-X(k))/(X(r)-X(k))
160       IF k<>r THEN LET tot=tot*t
170       NEXT k
180     LET term=term+(tot*F(r))
190     NEXT r
200   PRINTTAB(0,N+12)"The interpolated value for "
      ;X;" is ";term
210   END
220   DATA 0,-0.002,0.2,0.058,0.4,0.149,0.6,0.283
230   DATA 0.8,0.470,1,0.717
```

Example
The difference table (Table 15.9) is given below:

Table 15.9

x	$f(x)$	∇	∇^2	∇^3
0	0.0			
		0.4		
0.25	0.04		−0.8	
		−0.4		1.0
0.50	0.0		+0.2	
		−0.2		
0.75	−0.2			

We wish to estimate $f(0.9)$ and $f(1.0)$ using (i) all points and cubic interpolation, and (ii) the last three points and quadratic interpolation.

(i) For the cubic interpolation, in order to obtain $f(0.9)$ we use the Newton-Gregory backward formula with $x_0 = 0.75$, $p = 0.6$. Then

$$f(0.9) \cong -0.2 + (0.6)(-0.2) + \frac{(0.6)(1.6)}{2}(0.2) + \frac{(0.6)(1.6)(2.6)}{6}(1.0)$$

$$= -0.2 - 0.12 + 0.096 + 0.416 = 0.19 \qquad \text{(2 d.p.)}$$

Similarly $f(1.0) = -0.2 - 0.2 + 0.2 + 1.0 = 0.80$ (2 d.p.)
[We could have found $f(1.0)$ by continuing the difference table.]

(ii) For quadratic interpolation, we merely omit the last term in each case to obtain $f(0.9) = -0.22$ (2 d.p.) and $f(1.0) = -0.20$ (2 d.p.)

Now suppose we try Lagrange interpolation on (0.5, 0.0), (0.75, –0.2) and an additional point (1.25, 0.4). You should be able to show that revised estimates are $f(0.9) = -0.16$ (2 d.p.) and $f(1.0) = -0.067$ (3 d.p.).

Clearly something is amiss. In fact the tabular values came from a sine-type function and the attempts at extrapolation can be seen to fail by superimposing their predictions on the curve as shown in Figure 15.5.

This example emphasises two points:
(i) extrapolation is always risky
(ii) if the data is badly approximated by low degree polynomials then a higher-order extrapolation formula is not necessarily any better.

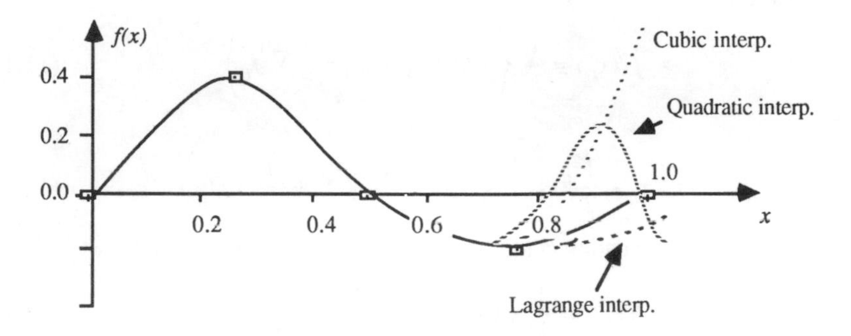

Figure 15.5

15.5 Numerical Differentiation

In Section 3.4 we developed ad hoc some formulae for differentiation from a table of values. We can, in fact, derive such formulae and more accurate ones by differentiating suitable interpolation formulae.

For example, if we differentiate successive terms in (15.7) with respect to p, we obtain the equation

$$hf'(x_0 + ph) = 0 + \Delta f(x_0) + \frac{2p-1}{2!}\Delta^2 f(x_0) + \frac{3p^2 - 6p + 2}{3!}\Delta^3 f(x_0) + \ldots$$

You should ask whether such differentiation is acceptable since the terms on the right-hand side are theoretically infinite in number; our only reply at present is that provided the terms still decay reasonably rapidly we shall be satisfied with the results. You might even ask whether we can even contemplate differentiating with respect to p; but, once the table of values is given, the differences are automatically determined and all that can cause $f(x_0 + ph)$ to vary is a change in p.

In the special case $p = 0$ we obtain

$$f'(x_0) \cong \frac{1}{h}\left\{\Delta f(x_0) - \frac{1}{2}\Delta^2 f(x_0) + \frac{1}{3}\Delta^3 f(x_0) - \frac{1}{4}\Delta^4 f(x_0) \ldots\right\} \quad (15.13)$$

which gives the derivative *at a tabulated point* x_0.

If we put $p = \frac{1}{2}$ we produce

$$f'(x_0 + \frac{1}{2}h) \cong \frac{1}{h}\left\{\Delta f(x_0) - \frac{1}{24}\Delta^3 f(x_0) + \frac{1}{24}\Delta^4 f(x_0) \ldots\right\} \quad (15.14)$$

which gives the derivative *mid-way between tabulated points*. You should notice that the term $\Delta^2 f(x_0)$ has vanished.

As might be expected, these formulae are best suited at the beginning of a table. At the end of a table we use the corresponding backward differences formula, viz

$$f'(x_0) \cong \frac{1}{h}\left\{\nabla f_0 + \frac{1}{2}\nabla^2 f_0 + \frac{1}{3}\nabla^3 f_0\right\} \quad (15.15)$$

or

$$f'(x_0 - \frac{1}{2}h) \cong \frac{1}{h}\left\{\nabla f_0 - \frac{1}{24}\nabla^3 f_0 - \frac{1}{24}\nabla^4 f_0\right\} \quad (15.16)$$

Similarly, by differentiating central difference formulae we find that

$$f'(x_0) \cong \frac{1}{h}\left\{\mu\delta f_0 - \frac{1}{6}\mu\delta^3 f_0 + \frac{1}{30}\mu\delta^5 f_0\right\} \quad (15.17)$$

$$f'(x_0 + \frac{1}{2}h) \cong \frac{1}{h}\left\{\delta f_{1/2} - \frac{1}{24}\delta^3 f_{1/2} + \frac{3}{640}\delta^5 f_{1/2}\right\} \quad (15.18)$$

[Here μ is the averaging operator defined by $\mu f_p = \frac{1}{2}(f_{p+1/2} + f_{p-1/2})$]

If we rewrite the first two terms on the right-hand side of (15.17) we obtain

$$f'(x_0) \cong \frac{1}{2h}\left\{f(x_0 + h) - f(x_0 - h)\right\} - \frac{1}{6h}\mu\delta^3 f_0$$

and we see that we have a correction term to our central formula of Section 3.4.

You should check that (15.13) and (15.15) produce similar results.

We say in Problem 35 on page 101 that repeated application of numerical differentiation yielded successively less accurate values.

We quote some approximate formulae for higher derivatives:

$$f''(x_0) \cong \frac{1}{h^2}\left\{\delta^2 f_0 - \frac{1}{12}\delta^4 f_0\right\} = \frac{f(x_0+h) - 2f(x_0) + f(x_0-h)}{h^2} - \frac{1}{12h^2}\delta^4 f_0$$

$$f'''(x_0) \cong \frac{1}{h^3}\mu\delta^3 f_0 \qquad (15.19)$$

$$f^{(\mathrm{iv})}(x_0) \cong \frac{1}{h^4}\delta^4 f_0$$

Example 1

We apply the above results to Table 15.10, which is part of a table of reciprocals so that we can check our results.

We endeavour to find the first four derivatives of $f(x)$ at $x = 1.0$; we expect values of $-1, 2, -6, 24$.

Using (15.13) we have, ignoring the terms in δ^4 (why?),

$$f'(1.0) \cong \frac{1}{0.1}\left\{-0.0909 - \frac{1}{2}(+0.0151) + \frac{1}{3}(-0.0034)\right\} = -0.986 \qquad \text{(3 d.p.)}$$

Table 15.10

x	$f(x)$	δ	δ^2	δ^3	δ^4
0.7	1.4290				
		–0.1790			
0.8	1.2500		+0.0401		
		–0.1389		–0.0123	
0.9	1.1111		+0.0278		+0.0047
		–0.1111		–0.0076	
1.0	1.0000		+0.0202		+0.0025
		–0.0909		–0.0051	
1.1	0.9091		+0.0151		+0.0017
		–0.0758		–0.0034	
1.2	0.8333		+0.0117		
		–0.0641			
1.3	0.7692				

From (15.15)

$$f'(1.0) \cong \frac{1}{0.1}\left\{-0.1111 + \frac{1}{2}(+0.0278) + \frac{1}{3}(-0.0123)\right\} = -1.013 \qquad \text{(3 d.p.)}$$

From (15.17)

$$f'(1.0) \cong \frac{1}{0.1}\left\{\frac{1}{2}\left[-0.0909 - 0.1111\right] - \frac{1}{6}\cdot\frac{1}{2}\left[-0.0051 - 0.0076\right]\right\} = -0.999$$

(3 d.p.)

To find the higher derivatives, we use (15.19)

$$f'(1.0) \cong \frac{1}{0.01}\left[+0.0202 - \frac{1}{12}(+0.0025)\right] = 2.00 \qquad \text{(2 d.p.)}$$

$$f'''(1.0) \cong \frac{1}{0.001}\left\{\frac{1}{2}(-0.0076 - 0.0051)\right\} = -6.35 \qquad \text{(2 d.p.)}$$

$$f^{(iv)}(1.0) \cong \frac{1}{0.0001}(+0.0025) = 25.0$$

It is evident that the central difference formulae are the most accurate and that the higher the derivative the fewer decimal places we can achieve and the less the accuracy.

We need some way of estimating the error in these interpolation and differentiation formulae and to do this we would have to resort to calculus as is so often the case.

Problems

Section 15.1

1 A straight line $y = 3x - 2$ is to be fitted to the data points which follow

x	0	1	2	3
y	−1.9	0.9	4.4	6.9

Plot the points and draw the straight line and notice the outlier – the point much further from the line than the others. Tabulate for each point e_i, $|e_i|$ and e_i^2 and find the sums of each quantity. Then remove the outlier from the calculations and repeat. Comment on your results.

2 Fit the straight line $y = a_0 + a_1x$ to the data points

x	−1	0	1
y	2	3	4

by calculating S from (15.1). Show that $S = 2(a_1 - 1)^2 + 3(a_0 - 3)^2$ and hence find the values a_0 and a_1 for which S is least. Verify your conclusions graphically.

3 Repeat the calculations of Problem 2 with y values 1.9, 3.1, 4.1

4 For a straight line $y = a_0 + a_1x$, write down an expression for e_i and deduce a form of S when the data points are $(x_1, y_1), \ldots, (x_n, y_n)$.

Section 15.2

5 Find the normal equations for fitting by the method of least squares the following curves to n observations (x_i, y_i):

(i) $y = a + bx^3$ (ii) $y = a + bx^2 + cx^3$ (iii) $y = 1 + ax$

(iv) $y = ae^{bx}$, a fixed (v) $y = \ln(a + bx)$ (vi) $y = a/x$

6 When unloaded reinforced concrete columns dry, compressive stresses occur due to shrinkage. The following table shows drying shrinkage in 10^{-6}cm/cm, y, measured against % reinforcement x.

x	0	2	4.5	8.5
y	475	380	190	75

Fit a least squares straight line and a quadratic.

7 The Highway Code gives the values of the distance D travelled by a car before coming to rest from a speed V

V (miles/hr)	20	30	40	50	60
D (ft)	45	75	120	175	240

Find a relationship of the form $D = a + bV + cV^2$ which represents the given data. Comment upon the validity of the expression for speeds below 20 miles/hr. Suggest a simple relation between D and V which could be applied for such speeds. (EC)

8 Fit by least squares a straight line to the data

x	0.5	0.6	0.7	0.8
y	3.8	2.1	–0.3	–1.7

Reject the point furthest from this line (called the *regression* line of y on x) and determine a new line to fit the remaining data.

9 Write down the appropriate normal equations and hence fit by least squares the following:

(i) $y = ax + \dfrac{b}{x}$ to the data

x	2	3	4	5
y	5.3	7.2	9.0	10.6

(ii) $y = ax^2 + b\sqrt{x}$ to the data

x	1	2	3	4
y	–5.2	–2.0	6.0	17.5

(iii) $y = a/x^2$ to

x	1	2	3	4	5
y	0.0224	0.0055	0.0030	0.0015	0.0010

(a) by direct application

(b) by transforming the equation into a linear relationship.

10 Linearise the relationship $y = ae^{b/x}$ and fit this to the data

x	1	2	3	4	5
y	30.3	7.41	4.12	3.18	2.73

11 Use finite differences to find the degree of the polynomial which would best fit the data given. Fit a polynomial of this degree by the method of least squares; find and estimate the closeness of fit.

x	1	2	3	4	5
y	2.00	1.18	1.20	2.05	3.75

12 The following pairs of values of S and T were obtained experimentally.

T	–2	–1	0	1	2
S	0	2	4	5	7

Assuming the values of S only to be in error, use the method of least squares to fit to the data

(i) $S = a + bT + cT^2$ (ii) $S = a + bT$ (iii) $S = a + bT + cT^2 + dT^3$

To determine which of these curves fits the data best, evaluate for each curve

$$\Omega = \frac{S_{\min}}{n - m}$$

where $S_{\min}$ is the minimised sum of squares, n is the number of observations and m is the number of coefficients in the relationship governing the curve. The least value of Ω implies the best fit.

13 It is believed that the variables s and t, whose corresponding values are given by the table below, are connected by a relation of the form $s = \lambda t^k$, where λ and k are constants. Use the method of least squares to determine the best values of λ and k, with as much accuracy as the data justifies.

t	1.0	1.2	1.4	1.6	1.8
s	84.45	68.64	55.76	45.33	36.80

(LU)

Section 15.3

14 Given the following table for a function $f(x)$ find $f(3.15)$

x	3.0	3.2	3.4	3.6	3.8	4.0
$f(x)$	3.0103	3.4242	3.8021	4.1497	4.4716	4.7712

Estimate the accuracy of your answer. (EC)

15 Values of $f(x) = \tan x$ are given for $x = 35°(2°)45°$. Find tan 36°.

$x°$	35	37	39	41	43	45
$\tan x$	0.70021	0.75355	0.80978	0.86929	0.93252	1.00000

Section 15.4

16 Use the following table of $f(x) = \tan x$ to interpolate tan 85°30′ and tan 87°30′.

$x°$	85	86	87	88	89
$f(x)$	11.430	14.301	19.081	28.636	57.290

Note how poorly the values compare with tabular values 12.706 and 22.904. In cases like this where $f(x)$ changes very rapidly, the reciprocal function $g(x) = 1/[f(x)]$ will change more slowly. Form a difference table for $g(x)$, obtain $g(85°30')$ and $g(87°30')$, and hence obtain more accurate values of tan 85°30′ and tan 87°30′.

17 The values of $f(x)$, a low degree polynomial, are given in the table

x	2	3	4	5	6	7	8	9	10
$f(x)$	15	40	85	165	259	400	585	820	1111

It is suspected that there is a transposition error in one of the values of $f(x)$.
By differencing, locate and correct this error.
Hence, by using an appropriate interpolation formula, evaluate $f(x)$ when $x = 2.5$. (LU)

18 From the table find sin 37°30′ using both Bessel and Everett interpolation formulae.

x	$\sin x$	x	$\sin x$
0	0.00000	40	0.64279
10	0.17365	50	0.76604
20	0.34202	60	0.86603
30	0.50000	70	0.93969

19 The table gives the values of a function $f(x)$ for values of x between 1.0 and 1.5

x	$f(x)$	x	$f(x)$
1.0	1.54308	1.3	1.97091
1.1	1.66852	1.4	2.15090
1.2	1.81066	1.5	2.35241

Calculate the values of $f(1.05)$ and $f(1.47)$. (LU)

20 A fourth degree polynomial is tabulated as follows

x	0	0.1	0.2	0.3	0.4
y	1.0000	0.9208	0.6928	0.3448	−0.0752

x	0.5	0.6	0.7	0.8	0.9
y	−0.5000	−0.8452	−0.9992	−0.8432	−0.2312

Show from a difference table that there is an error and use the corrected table with the Stirling interpolation formula

$$f_p = f_0 + \tfrac{1}{2}p(\delta f_{1/2} + \delta f_{-1/2}) + \tfrac{1}{2}p^2\delta^2 f_0 + \frac{p(p^2-1)}{2.3!}(\delta^3 f_{1/2} + \delta^3 f_{-1/2})$$

$$+ \frac{p^2(p^2-1)}{4!}\delta^4 f_0 + \ldots$$

to find the value of y when $x = 0.45$.

21 Define the shift operator E and the forward difference operator Δ when applied to the function $f(x) = f(x_0 + ph)$. Hence obtain Newton's forward interpolation formula

$$f_p = f_0 + p\Delta f_0 + \frac{p(p-1)}{2}\Delta^2 f_0 + \ldots + \binom{p}{r}\Delta^r f_0 + \ldots$$

A polynomial function is given by the following table.

x	0	1	2	3	4	5	6
f	0	3	14	39	84	155	258

Make a difference table and explain how the correctness of the arithmetic may be checked. Use this table to find f when $x = 1.5$ and when $x = 7$. (LU)

22 Given the following data

x	2	3	4	5	6
$f(x)$	9.00	4.00	2.25	1.44	1.00

evaluate $f(4.5)$ using

(a) 3-point formula with $x_1 = 3, x_2 = 4, x_3 = 5$

(b) 3-point formula with $x_1 = 4, x_2 = 5, x_3 = 6$

(c) 4-point formula with $x_1 = 3, x_2 = 4, x_3 = 5, x_4 = 6$

23 The values of a function $f(x)$ are given for $x = x_i$ $(i = 1, 2, \ldots, n)$. Assuming Lagrange's interpolation polynomial of degree $(n-1)$ is of the form

$$\sum_{i=1}^{n} l_i(x) f(x_i) \qquad \text{prove that} \qquad \sum_{i=1}^{n} l_i(x) \equiv 1$$

A function $f(x)$ is given by the following table of values

x	-2	0	3	4
$f(x)$	-14	8	-4	4

Determine $f(1)$ and $f(2)$. (LU)

Section 15.5

24 The following is a table of $f(x) = \sin x$ (x in radians).

x	0.7	0.8	0.9	1.0	1.1	1.2	1.3
$\sin x$	.644218	.717356	.783327	.841471	.891207	.932039	.963558

Using formulae (15.13), (15.15) and (15.17), show that at $x = 1.0$, $f'(x) = 0.54052$, 0.54011, 0.54030 respectively. Look up cos(1.0) and compare with the above.

Show at $x = 1.0$, $\quad f'' = -0.8415 \quad f''' = -0.538 \quad f^{(iv)} = 0.85$

What answers should we get? Note how accuracy is lost in numerical differentiation.

Find $(d/dx)(\sin x)$ at $x = 1.05$ and compare with $\cos(1.05) = 0.497571$.

25 Given the following table for $\cos x$

x	0.499	0.500	0.501
$\cos x$	0.878062	0.877563	0.877103

determine the value of $d/dx(\cos x)$ at $x = 0.5$. Estimate the possible error in your answer. (EC)

26 For the data of Problem 14 find d^2f/dx^2 at $x = 3.6$.
Estimate the accuracy of your answer. (EC)

16

APPROXIMATION OF FUNCTIONS

16.1 The Mean Value Theorem

This theorem has many applications in its own right and is also a basis for other theorems. Before we state it we need a basic result used in its proof.

Rolle's Theorem
If a function $g(x)$ is continuous in the closed interval $[a, b]$ and differentiable in the open interval (a, b) and if, further, $g(a) = 0 = g(b)$ then there is at least one point ξ in the interval $a < \xi < b$ such that $g'(\xi) = 0$.

Geometrically this means that if a continuous curve whose formula has a continuous first derivative intersects the x-axis at two points, it has a horizontal tangent at some intermediate point (see Figure 16.1).

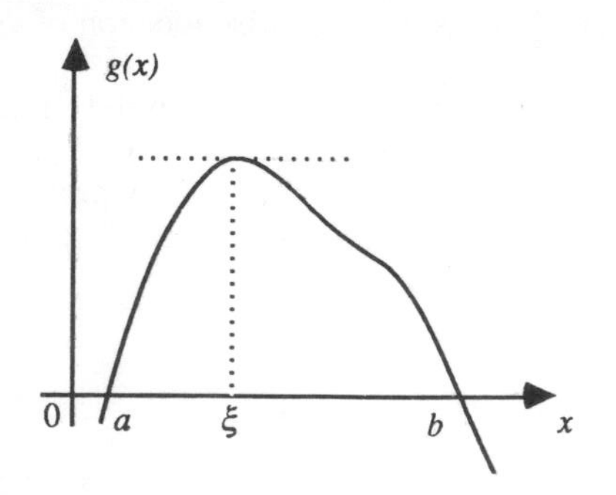

Figure 16.1

(This can be compared with the **Intermediate Value Theorem**, touched on in Section 3.3; *if $f(x)$ be continuous in $[a, b]$ then for any value c between $f(a)$ and $f(b)$ there is at least one point value ξ in the interval $a \le \xi \le b$ such that $f(\xi) = c$.*)

What Rolle's Theorem, however, does not tell us is where the point ξ is precisely. In individual examples, however, we may be able to establish this directly, by differentiation.

Proof of Rolle's Theorem
We assume the result that if M and m are the least upper bound and greatest lower bound respectively of a continuous function in some closed interval then the function achieves those values somewhere in the interval. Clearly if $m = M$ then the function $g(x)$ takes a constant value in the interval $[a, b]$ and $g'(x) \equiv 0$ there. Otherwise, we have at least one of m and M non-zero. We prove the result for the case $M \neq 0$; let $M > 0$ then there is a point ξ in (a, b) at which $g(\xi) = M$ (we have excluded $\xi = a$ and $\xi = b$; (why?). Now $g(x) \leq g(\xi)$ for all $x \in [a, b]$ and so for any small enough number $h > 0$, $g(\xi + h) - g(\xi) \leq 0$ and $g(\xi - h) - g(\xi) \leq 0$. Hence $[g(\xi + h) - g(\xi)]/h \leq 0$ and $[g(\xi) - g(\xi - h)]/h \geq 0$; if we take the limiting cases as $h \to 0$ then the first quotient stays negative or zero and the second stays positive or zero. But, in the limit both approach $g'(\xi)$ and so we conclude that $g'(\xi) = 0$.

Mean Value Theorem (MVT)
If $f(x)$ is a function continuous in $[a, b]$ and differentiable in (a, b) then there is at least one value ξ where $a < \xi < b$ for which $f'(\xi) = [f(b) - f(a)]/(b - a)$.

Geometrically, this means that there is a point ξ at which the tangent to the curve of $f(x)$ is parallel to the chord joining the end-points as shown in Figure 16.2. Again we are told nothing as to the precise location of the point ξ.

As an analogy imagine two cars setting off from the same place at the same time to arrive at a second place at the same later time. One car travels at a constant speed, the other is free to go at what speed it likes, stopping if it so desires. Then the MVT says that at some time the speed of the second car is equal to that of the first.

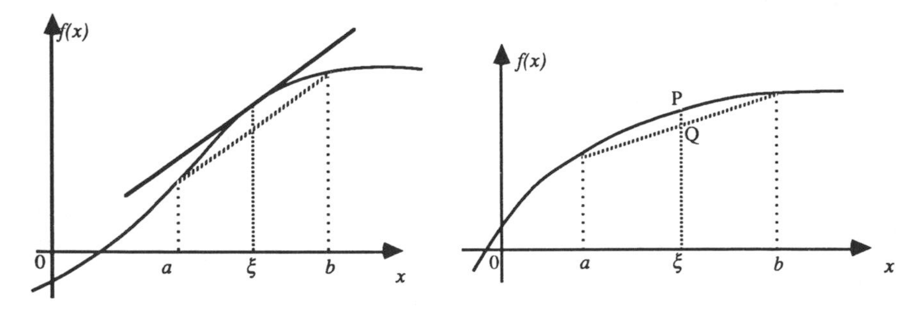

Figure 16.2 Figure 16.3

Proof of MVT

The proof of the theorem comes by applying Rolle's theorem to the function

$$g(x) = f(x) - \left\{ f(a) + \frac{f(b) - f(a)}{b - a}(x - a) \right\}$$

The function $g(x)$ represents the vertical distance PQ shown in Figure 16.3.

Now this function satisfies the first two conditions of Rolle's theorem because we are given that $f(x)$ does and so also does the other function in the braces { }. Further,

$$g(a) = f(a) - \{f(a) + 0\} = 0$$

and
$$g(b) = f(b) - \{f(a) + f(b) - f(a)\} = 0$$

Then, by Rolle's theorem, there is a value ξ where $a < \xi < b$ with $g'(\xi) = 0$.

But
$$g'(x) = f'(x) - \left\{ \frac{f(b) - f(a)}{b - a} \right\}$$

hence
$$0 = g'(\xi) = f'(\xi) - \left\{ \frac{f(b) - f(a)}{b - a} \right\}$$

Thus
$$f'(\xi) = \frac{f(b) - f(a)}{b - a}$$

and the existence of ξ is established.

Example 1

(i) Let $f(x) = x^3 + 3$ and $[a, b] = [1, 2]$. Then

$$\frac{f(b) - f(a)}{b - a} = \frac{11 - 4}{2 - 1} = 7$$

But $f'(x) = 3x^2$ and so ξ is such that $3\xi^2 = 7$, which gives $\xi = \pm 1.5275$ (4 d.p.) and only the value $+1.5275$ lies inside the interval [sketch $f(x)$ and check].

(ii) Let $f(x) = |x|$ and $[a, b] = [-1, 2]$. Then

$$\frac{f(b) - f(a)}{b - a} = \frac{2 - 1}{2 - (-1)} = \frac{1}{3}$$

yet nowhere in $[-1, 2]$ does $f'(x) = 1/3$. Why should this be? Check whether $f(x)$ obeys the conditions of MVT.

Example 2

One application of MVT is to estimate the maximum truncation error in linear interpolation. If we replace $f(x)$ in $[a, b]$ by

$$\phi(x) = f(a) + \left\{ \frac{f(b) - f(a)}{b - a} \right\} (x - a)$$

then $\phi(a) = f(a)$, and $\phi(b) = f(b)$ and ϕ is of the form $\alpha + \beta x$.

The error at any point in $[a, b]$ is

$$f(x)-\phi(x)=(x-a)\left[\frac{f(x)-f(a)}{x-a}-\left\{\frac{f(b)-f(a)}{b-a}\right\}\right]$$

Now, by MVT the right-hand side is equal to

$$(x-a)[f'(\xi_1)-f'(\xi_2)]$$

where both ξ_1 and $\xi_2 \in (a, b)$. Applying MVT to the expression

$$\frac{[f'(\xi_1)-f'(\xi_2)]}{\xi_1-\xi_2}$$

we find it to be equal to $f''(\psi)$ where ψ lies between ξ_1 and ξ_2, [and therefore in (a, b)].

Then $f(x) - \phi(x) = (x - a)f''(\psi)(\xi_1 - \xi_2)$. Now, the worst that $|\xi_1 - \xi_2|$ can be is $(b - a)$ as can $|x - a|$; therefore $|f(x) - \phi(x)| \leq |f''(\psi)|(b - a)^2$. If we can find a number M such that $|f''(x)| \leq M$ for all $x \in (a, b)$ then

$$|f(x)-\phi(x)| \leq M(b-a)^2 \tag{16.1}$$

Since (16.1) is a general result it is quite likely that the actual error is much less than $M(b - a)^2$ but at least we have a starting point. Unless we have a means of determining $f''(x)$ this estimation is of little use, anyway.

Example 3

We have seen that $(\sin\theta)/\theta \to 1$ as $\theta \to 0$, and so we may approximate $\sin\theta$ by θ. Suppose we regard this approximation as satisfactory if it is at most 0.01 in error. Let $a = 0$ and note that, since $f''(x) = -\sin x$, $M = 1$; then the maximum error in the approximation is b^2 and if we require this to be at most 0.01 we must restrict b to be ≤ 0.1. By symmetry we conclude that the acceptable interval is $[-0.1, 0.1]$. In fact, $\sin 0.38° = 0.3709$ (4 d.p.) and so the approximation is still acceptable here.

For interpolation given only tabulated values, errors may be estimated from tables designed for the purpose.

16.2 Polynomial Approximations

We turn our attention to the approximation of a function $f(x)$ by a straight line so that we may predict values of $f(x)$ without too much calculation. Certainly, from the point of view of using a computer the calculation of polynomial values is straightforward and therefore polynomials are the most commonly used approximating functions (a straight line is, of course, just a first-order polynomial).

The Tangent Approximation
The equation of the chord to the function $f(x)$ shown in Figure 16.4(i) is

$$y = f(x_0) + \left\{\frac{f(x_0 + h) - f(x_0)}{h}\right\}(x - x_0)$$

and we see that provided h is small and the slope $f'(x)$ does not vary too much in the interval $[x_0, x_0 + h]$, then the chord is a reasonable approximation to the curve. Note that to draw the chord we need to know two values of $f(x)$.

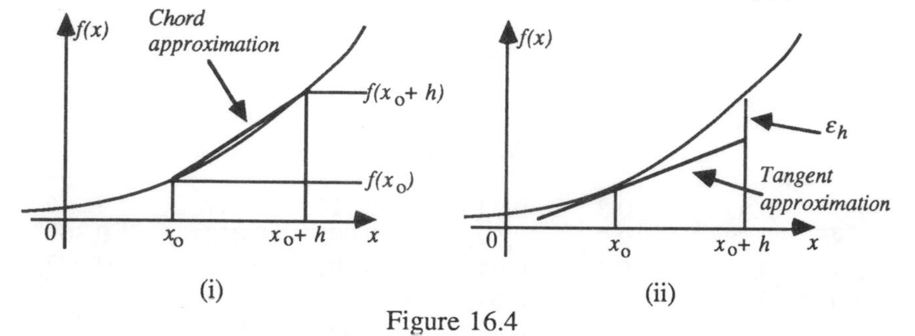

Figure 16.4

If we let $h \to 0$ then the equation of the straight line becomes

$$y = f(x_0) + f'(x_0)(x - x_0) \qquad (16.2)$$

which is called the **Tangent Approximation** to $f(x)$ at x_0 [see Figure 16.4(ii)] and requires knowledge of the function at the point x_0 only. Note the special case where $x_0 = 0$, viz

$$y = f(0) + f'(0)$$

We can use the MVT to put an upper bound on the error involved in this approximation. The discrepancy in the values at $(x_0 + h)$ [see Figure 16.4(ii)], is $\varepsilon_h \equiv f(x_0 + h) - [f(x_0) + f'(x_0)h]$.

Now

$$\frac{\varepsilon_h}{h} = \frac{f(x_0 + h) - f(x_0)}{h} - f'(x_0)$$

$$= f'(\xi) - f'(x_0)$$

[by MVT where ξ is in the interval $(x_0, x_0 + h)$].

A second application of MVT yields

$$\frac{\varepsilon_h}{h(\xi - x_0)} = \frac{f'(\xi) - f'(x_0)}{\xi - x_0} = f''(\psi)$$

where ψ lies between x_0 and ξ and hence $\psi \in (x_0, x_0 + h)$.

If we replace $f''(\psi)$ by M, the greatest value of $|f''(x)|$ in $[x_0, x_0 + h]$, then $|\varepsilon_h| \le Mh|\xi - x_0| < Mh^2$.

We shall now apply the tangent approximation to a specific function and see how realistic is this upper bound on $|\varepsilon_h|$.

Example
We wish to approximate $y = \ln x$ with $x_0 = 2$ as base; we quote the results in Table 16.1 to 3 d.p. The tangent approximation (16.2) becomes

$$y = \ln 2 + \frac{1}{2}(x - 2)$$

Table 16.1

x	$\ln x$	$x - 2$	Tangent Approxn.	Error	Predicted Max. Error	M
1.7	0.531	−0.3	0.543	0.012	0.031	0.346
1.8	0.588	−0.2	0.593	0.005	0.012	0.309
1.9	0.642	−0.1	0.643	0.001	0.003	0.277
2.0	0.693	0	0.693	0	0	–
2.1	0.742	0.1	0.743	0.001	0.003	0.25
2.2	0.789	0.2	0.793	0.004	0.010	0.25
2.3	0.833	0.3	0.843	0.010	0.023	0.25

In this example $f''(x) = -(1/x^2)$ and the maximum value of $|f''(x)|$ occurs at the left-hand end of each interval. A schematic view of the approximation is shown in Figure 16.5.

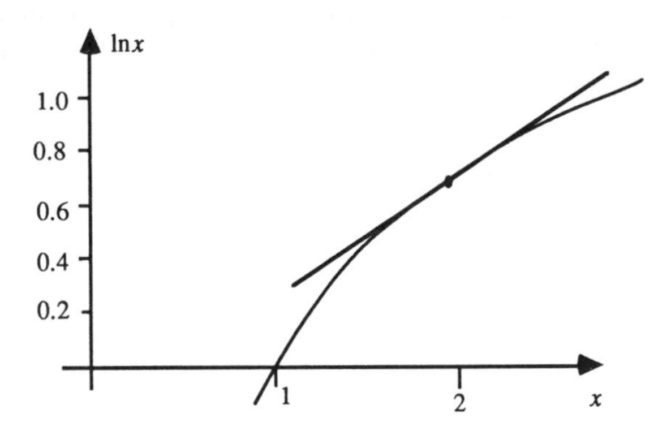

Figure 16.5

The error is in all cases $\leq \frac{1}{2}Mh^2$ rather than Mh^2 which was the value quoted. It is clear, as expected, that the larger h, the larger the error. The error seems to be roughly proportional to h^2; this means that the error becomes rapidly worse as we increase h. We have, in effect, matched the value $f(x_0)$ and slope $f'(x_0)$;

we would expect to obtain a better approximation by making $f''(x_0)$ also match.

Applications of the Tangent Approximation

(i) *Basic iteration*

We may apply the Tangent Approximation to verify the convergence criterion for Basic Iteration stated in Section 13.4. Suppose we have the equation $x = F(x)$, with root $x = a$ which we attempt to solve by the iterative formula $x_{n+1} = F(x_n)$. Using the tangent approximation we obtain $F(x_n) = F(a) + F'(a)(x_n - a)$; therefore, since $a = F(a)$,

$\varepsilon_{n+1} = x_{n+1} - a = F'(a)(x_n - a)$ and so

$$|\varepsilon_{n+1}| = |F'(a)| \,.\, |x_n - a| = |F'(a)| \,.\, |\varepsilon_n|$$

If the iterative method is to converge, $|\varepsilon_{n+1}|$ must be less than $|\varepsilon_n|$, and this requires $|F'(a)| < 1$ or, if the function is well-behaved, $|F'(x)| < 1$ near $x = a$.

(ii) *Newton-Raphson iterative method*

Let $f(a + h) = 0$ and let h be small, then

$$0 = f(a + h) \cong f(a) + hf'(a)$$

therefore, $h \cong -f(a)/f'(a)$ and hence $a - f(a)/f'(a)$ is, in general, a better approximation to $a + h$ than was a.

The Quadratic Approximation

Let $q(x) \equiv \alpha + \beta(x - x_0) + \gamma(x - x_0)^2$ where $\alpha, \beta, \gamma, x_0$ are constants, we choose this form for $q(x)$ by analogy with the tangent approximation: it is still a perfectly general quadratic. Given the function $f(x)$ we require that if $q(x)$ is to approximate $f(x)$ near $x = x_0$ then $q(x_0) = f(x_0)$, $q'(x_0) = f'(x_0)$ and $q''(x_0) = f''(x_0)$.

Applying these constraints we find that

$$\alpha = f(x_0) \qquad \beta = f'(x_0) \qquad 2\gamma = f''(x_0)$$

Hence

$$f(x) \cong f(x_0) + f'(x_0)(x - x_0) + \tfrac{1}{2}f''(x_0)(x - x_0)^2 \tag{16.3}$$

is the required quadratic approximation to $f(x)$ at $x = x_0$.

Note the special case when $x_0 = 0$:

$$f(x) \cong f(0) + xf'(0) + \frac{x^2}{2}f''(0) \tag{16.4}$$

Example

For $f(x) = \ln x$ with $x_0 = 2$ as base we have

$$f(x) \cong \ln 2 + \frac{1}{2}(x - 2) + \frac{1}{2}.(-\frac{1}{4})(x - 2)^2$$

In effect we have added on a correction term to the tangent approximation. Results are quoted in Table 16.2 to 4 d.p.

Table 16.2

x	$\ln x$	$x-2$	$(x-2)^2$	Quadratic Approxn.	Error
1.7	0.5306	−0.3	0.09	0.5320	0.0014
1.8	0.5878	−0.2	0.04	0.5881	0.0003
1.9	0.6419	−0.1	0.01	0.6420	−0.0001
2.0	0.6931	0	0	0.6931	0
2.1	0.7419	0.1	0.01	0.7418	−0.0001
2.2	0.7885	0.2	0.04	0.7881	−0.0004
2.3	0.8329	0.3	0.09	0.8320	−0.0009

We see clearly the improved accuracy as compared with Table 16.1, but it is not too easy to estimate the error in general; in fact, it turns out to be proportional to h^3.

Application

We can use the quadratic approximation to show the speed of convergence of the Newton-Raphson method.

Let the root of $f(x) = 0$ that we seek be $x = a$, i.e. $f(a) = 0$. Then the error at the $(n+1)^{\text{th}}$ step is given by

$$\varepsilon_{n+1} = a - x_{n+1} = a - \left\{x_n - \frac{f(x_n)}{f'(x_n)}\right\} = a - x_n - \left\{\frac{f(a) - f(x_n)}{f'(x_n)}\right\}$$

From the quadratic approximation, we have

$$f(a) - f(x_n) \cong f'(x_n)(a - x_n) + \frac{1}{2}f''(x_n)(a - x_n)^2$$

and hence $$\varepsilon_{n+1} \cong -\frac{1}{2}\frac{f''(x_n)}{f'(x_n)}(a - x_n)^2 = -\frac{1}{2}\frac{f''(x_n)}{f'(x_n)}\varepsilon_n^{\,2}$$

It follows that $|\varepsilon_{n+1}|$ is proportional to $|\varepsilon_n|^2$ which shows the quadratic nature of the convergence. What restrictions are there on our demonstration?

We might suppose that by matching higher derivatives of $f(x)$ at x_0, we could obtain a succession of better approximations. This is so for a suitable function $f(x)$; we would then extend the argument to an **infinite series**

$$a_0 + a_1x + a_2x^2 + a_3x^3 + \ldots$$

where the coefficients a_i have to be determined in each case. We must now turn our attention to such series to see what conditions are imposed on $f(x)$ if it is to be approximated by such a power series.

16.3 Infinite Series

The entity $1 + 2 + 3 + 4 + 5 + \ldots + 101$ is called a **finite series** and the expression $1 + 2x + 13x^2 + \ldots + 0.76x^{12}$ is called a **finite power series**: in essence, we are adding a collection of terms in a definite order. Such series are easy to handle and to evaluate. But what happens if there is an infinite number of terms to be added together?

Consider the expression

$$1 + \frac{1}{2} + \frac{1}{4} + \frac{1}{8} + \frac{1}{16} + \ldots$$

Ignoring for the moment the first term, we can see that the remaining terms can be geometrically interpreted as the area formed by taking half a square of unit area (Figure 16.6) then adding half of the remainder, then adding half of what is left, and so on. No matter how many times we perform these operations there will always be a small area left to divide into two parts. In this way we can see that

$$\frac{1}{2} + \frac{1}{4} + \frac{1}{8} + \frac{1}{16} + \ldots$$

can be interpreted as 1, in the sense that if we add enough of the terms together the result is as close to 1 as desired. We can construct a sequence of **partial sums** (i.e. the sum of the first term, of the first two terms, of the first three terms, etc) and study their behaviour. For the series

$$\frac{1}{2} + \frac{1}{4} + \frac{1}{8} + \frac{1}{16} + \ldots$$

the partial sums are

$$S_1 = \frac{1}{2} \qquad S_2 = \frac{3}{4} \qquad S_3 = \frac{7}{8} \qquad S_4 = \frac{15}{16} \qquad \text{etc}$$

and we can see that $S_n = (2^n - 1)/2^n$. Now, since $\{S_n\} \equiv \{(2^n - 1)/2^n\} \rightarrow 1$ as $n \rightarrow \infty$ we say that the **sum** of the series below is 1.

$$\frac{1}{2} + \frac{1}{4} + \frac{1}{8} + \frac{1}{16} + \ldots$$

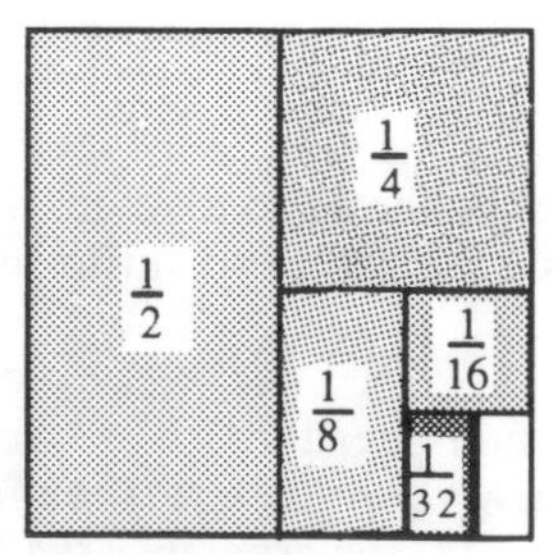

Figure 16.6

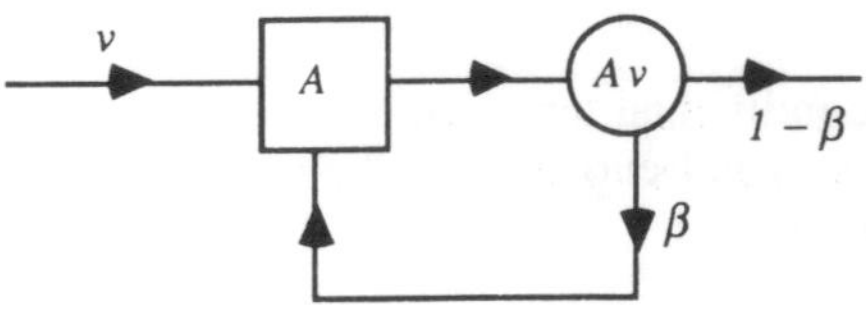

Figure 16.7

In general, *if the sequence $\{S_n\}$ of partial sums of an infinite series converges to a value, S, the series is said to have a sum S or to converge to the sum S.* It remains to establish suitable tests on the terms of a series in order to determine whether it converges.

For the moment we hope that you are familiar with arithmetic and geometric progressions. An application of a geometrical progression occurs in an electrical context.

Example (see Figure 16.7)

A feedback amplifier operates by taking a fraction β of its total amplified output and feeding it back. Let the amplifying factor be A then the input voltage v is amplified to Av; of this βAv is fed back into the amplifier and hence $Av - \beta Av = (1 - \beta)Av$ is allowed to be output. Continuing in this way, the total useful output voltage V is seen to be $(1 - \beta)Av + (1 - \beta)\beta A^2 v + (1 - \beta)\beta^2 A^3 v + \ldots$, that is,

$$V = (1 - \beta)Av(1 + \beta A + \beta^2 A^2 + \ldots)$$

$$= \frac{(1 - \beta)Av}{(1 - \beta A)} \qquad \text{since } 0 < \beta < 1$$

The effective amplification is $(1 - \beta)A/(1 - \beta A)$ which is called the *gain* of the amplifier. What happens if $\beta = 0$ or if $\beta = 1$?

Convergence Tests

If we consider $1 + 2 + 3 + 4 + \ldots$ it is clear that the partial sums increase without limit and the infinite series is said to **diverge**. The series $1 - 1 + 1 - 1 + 1 \ldots$ is said to **oscillate finitely** (its partial sums are alternatively 1 and 0); the series $1 - 2 + 3 - 4 + 5 \ldots$ **oscillates infinitely**. We need some tests to determine into which of these categories to place a particular series.

(i) Divergence criterion

*If $u_n \not\to 0$ as $n \to \infty$ the series does **not** converge.* For example, the series $1 + 2 + 3 + 4 + 5 + \ldots$. However, the converse does **not** follow: if we consider the **harmonic series**

$$1 + \frac{1}{2} + \frac{1}{3} + \frac{1}{4} + \frac{1}{5} + \ldots$$

the individual terms are decreasing but the series is known to diverge. It is clearly not enough that the terms get smaller: they must do so sufficiently rapidly.

(ii) Comparison tests (for series whose terms are non-negative)

(a) *If the series under test is such that each term is less than the corresponding term in a series which is known to converge, then the test series converges.* (The sum of the test series is unknown.)

Example
The series

$$1+\frac{1}{2^2}+\frac{1}{3^3}+\frac{1}{4^4}+\ldots$$

may be compared with the series

$$1+\frac{1}{2^2}+\frac{1}{2^3}+\frac{1}{2^4}+\ldots$$

which is known to converge; in fact, the sum of the second series is

$$\frac{1}{2}+\frac{1}{2^1}+\frac{1}{2^2}+\frac{1}{2^3}+\frac{1}{2^4}+\ldots$$

which, by the sum of a G.P. gives

$$\frac{1}{2}+\frac{\frac{1}{2}}{1-\frac{1}{2}}=1\tfrac{1}{2}$$

Term for term, the test series is smaller than the second series and hence it converges (to what, we cannot say at the moment).

(b) *If the series under test is such that each term is larger than the corresponding term in a series which is known to be divergent, then the test series diverges.*

Example

The series $$1.01+\frac{1.01}{2}+\frac{1.01}{3}+\frac{1.01}{4}+\ldots$$

when compared with the divergent series below is seen to be divergent.

$$1+\frac{1}{2}+\frac{1}{3}+\frac{1}{4}+\ldots$$

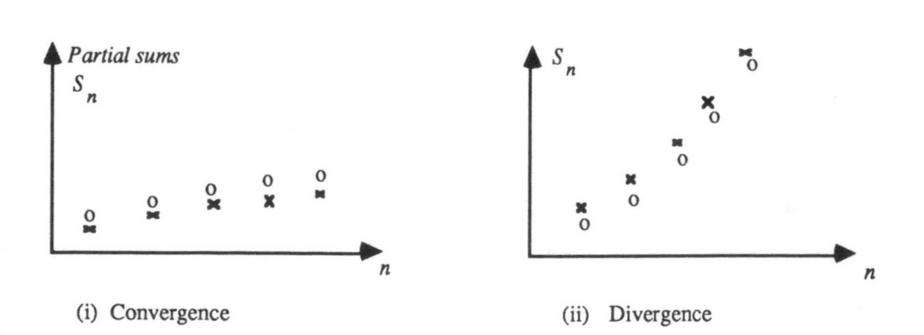

(i) Convergence (ii) Divergence

Figure 16.8

The tests are depicted schematically in Figure 16.8; the test series is shown by crosses, and the known series by circles.

The drawback to these tests is in choosing which series to use as a comparison; it is of no use, for example, knowing that our test series is term for term less than a known divergent series. There is a 'grey' area in which these tests are not sensitive enough to pick up the convergence or otherwise of a series. However a useful class of series for comparison purposes is

$$\sum_{r=1}^{\infty} \frac{1}{r^p} = 1 + \frac{1}{2^p} + \frac{1}{3^p} + \ldots$$

where there is convergence for $p > 1$ and divergence for $p \leq 1$.

It is not always a straightforward matter to effect a comparison and it is preferable often to compare successive terms of a series with themselves.

(iii) **D'Alembert's ratio test**

Let the series be written $u_1 + u_2 + u_3 + \ldots + u_n + u_{n+1} + \ldots$. If we can find an expression for u_n (and hence u_{n+1}) we form the ratio

$$\left| \frac{u_n}{u_{n+1}} \right|$$

and look at its long-term behaviour (as with sequences, it is only the *tail* which matters).

If
$$\left| \frac{u_n}{u_{n+1}} \right| \to l$$

as $n \to \infty$ (where l is a number) then $l > 1 \Rightarrow$ convergence of the series and $l < 1 \Rightarrow$ divergence. If $l = 1$, the test is not sufficiently sensitive to help.

Example 1

(a) $1 + \frac{3}{2} + \frac{5}{2^2} + \frac{7}{2^3} + \frac{9}{2^4} + \ldots$

Here

$$u_n = \frac{2n-1}{2^{n-1}} \quad \text{and} \quad u_{n+1} = \frac{2n+1}{2^n}$$

therefore

$$\left| \frac{u_n}{u_{n+1}} \right| = \frac{2(2n-1)}{2n+1} \to 2$$

as $n \to \infty$.

Hence $l = 2$ and the series converges.

(b) $1 + \frac{1}{2.1} + \frac{1}{2.2} + \frac{1}{2.3} + \ldots$

Now

$$u_n = \frac{1}{2(n-1)} \qquad u_{n+1} = \frac{1}{2n}$$

and so as $n \to \infty$

$$\left|\frac{u_n}{u_{n+1}}\right| = \frac{2n}{2n-1} \to 1$$

We can conclude nothing about the series.

Other more sensitive (and more complicated) tests exist but we shall not develop them here.

If each term u_n in a series is replaced by $|u_n|$ and the new series converges, the first series is called **absolutely convergent**. An absolutely convergent series is also convergent, but the converse is not always true, and when it is not the series is called **conditionally convergent**.

Example 2

(a) The series

$$1 - \frac{1}{2} + \frac{1}{3} - \frac{1}{4} + \frac{1}{5} + \ldots$$

can be shown to be convergent, but the series below is divergent.

$$1 + \frac{1}{2} + \frac{1}{3} + \frac{1}{4} + \frac{1}{5} + \ldots$$

(b) The series

$$1 + \frac{1}{2^2} + \frac{1}{3^2} + \frac{1}{4^2} + \ldots$$

is convergent and therefore the series below is absolutely convergent.

$$1 - \frac{1}{2^2} + \frac{1}{3^2} - \frac{1}{4^2} + \ldots$$

Power Series

Consider the three power series

(i) $1 + x + 2x^2 + 3x^3 + 4x^4 + \ldots$

(ii) $1 + x + \frac{1}{2}x^2 + \frac{1}{3}x^3 + \frac{1}{4}x^4 + \ldots$

(iii) $1 + x + \frac{1}{2}x^2 + \frac{1}{2^2}x^3 + \frac{1}{2^3}x^4 + \ldots$

Two factors affect the way the terms grow or decrease: the coefficients and the increasing powers of x. Different values of x affect whether a series converges or not. If $|x| > 1$, the values of the powers of x grow in magnitude and if $|x| < 1$ they diminish. We have already seen that if the terms of a series do not decrease then the series must diverge and hence for a power series with such terms as coefficients, $|x|$ must be < 1 if the power series is to stand a chance of converging. Suppose in the above three series that we put $x = 1$, then we know that, of the resulting series the first two diverge, and it is easy to show that the third converges. If $x = 1.5$, we cannot be sure that the third series still converges, but it is certain that the first two do not. If $x = \frac{1}{2}$ the third series will converge and we need to check the first two again. It should be clear that the bigger $|x|$ the less likely the series is to converge and in fact we define a **radius of convergence** for a power series $a_0 + a_1x + a_2x^2 + a_3x^3 + \ldots$ as that *largest* value ρ such that, for all x which satisfy $|x| < \rho$ the series converges and if $|x| > \rho$ the series diverges; notice that this range of x is symmetrical about $x = 0$. If a series converged for $-1 < x < 2$ and diverged elsewhere, the radius of convergence would be taken as 1. Refer to Figure 16.9.

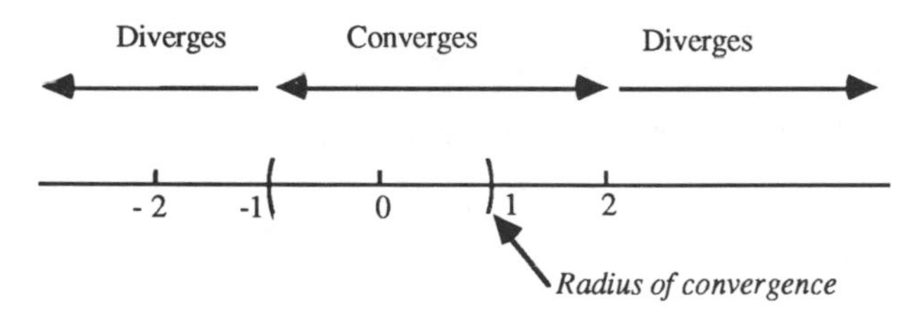

Figure 16.9

In effect, the coefficients determine how far we can allow x to stray from $x = 0$ before the terms fail to decay rapidly enough for convergence.

We test for convergence of a power series by the ratio test.

Consider the series

$$a_0 + a_1x + a_2x^2 + a_3x^3 + \ldots$$

Now

$$u_n = a_{n-1}\,x^{n-1}, \quad u_{n+1} = a_nx^n$$

and therefore

$$\left|\frac{u_n}{u_{n+1}}\right| = \left|\frac{a_{n-1}\,x^{n-1}}{a_nx^n}\right| = \left|\frac{a_{n-1}}{a_n}\right| \cdot \frac{1}{|x|}$$

If $\left|\frac{a_{n-1}}{a_n}\right| \to l$ then in order for $\left|\frac{u_n}{u_{n+1}}\right|$ to approach a limit $l > 1$ we require
as $n \to \infty$

$|x| < l$. In this case therefore we have convergence for $|x| < l$.

If $|x| > l$ we have divergence, but if $|x| = l$ we must make a special check.

Examples

(i) $1 + x + 2x^2 + 3x^3 + 4x^4 + \ldots$

$\left|\frac{a_{n-1}}{a_n}\right| = \frac{n-1}{n} \to 1$ and hence all we require is $|x| < 1$, a somewhat surprising result, since x = 1 clearly gives a divergent series

(ii) $1 + x + \frac{1}{2}x^2 + \frac{1}{3}x^3 + \frac{1}{4}x^4 + \ldots$

$\left|\frac{a_{n-1}}{a_n}\right| = \frac{n}{n-1} \to 1$ and again $\rho = 1$

(iii) $1 + x + \frac{1}{2}x^2 + \frac{1}{2^2}x^3 + \frac{1}{2^3}x^4 + \ldots$

$\left|\frac{a_{n-1}}{a_n}\right| = \frac{2^{n-1}}{2^{n-2}} = 2 \to 2$ and so $|x| < 2 \Rightarrow$ convergence

When $x = 2$, the series is $1 + 2 + 2 + 2 + 2 + \ldots$ which in fact diverges.

Two further examples will help consolidate ideas.

(iv) $1 - x + \frac{1}{2}x^2 - \frac{1}{3}x^3 + \ldots$

(The fact that the coefficients alternate in sign does not affect the test.)

$\left|\frac{a_{n-1}}{a_n}\right| = \frac{n}{n-1} \to 1$ and again $\rho = 1$

We know that if $x = -1$ we have a divergent series, but if $x = 1$ it can be shown that the resulting series converges.

(v) $1 + x + \frac{x^2}{2!} + \frac{x^3}{3!} + \ldots$

$\left|\frac{a_{n-1}}{a_n}\right| = \frac{n!}{(n-1)!} = n$ and so $\left|\frac{u_n}{u_{n+1}}\right| = \frac{n}{|x|}$

and as $n \to \infty$, this ratio tends to a value > 1 for any (fixed) value of x. This series converges for all x and has an *infinite* radius of convergence.

Practical Convergence of Series

It is all very well to say that a power series will converge for a particular value of x, but if we want to compute the sum of that series to within 1% of its true value we may need to calculate so many terms as to render the operation tedious to say the least. (There are sometimes algebraic methods of speeding up the convergence.)

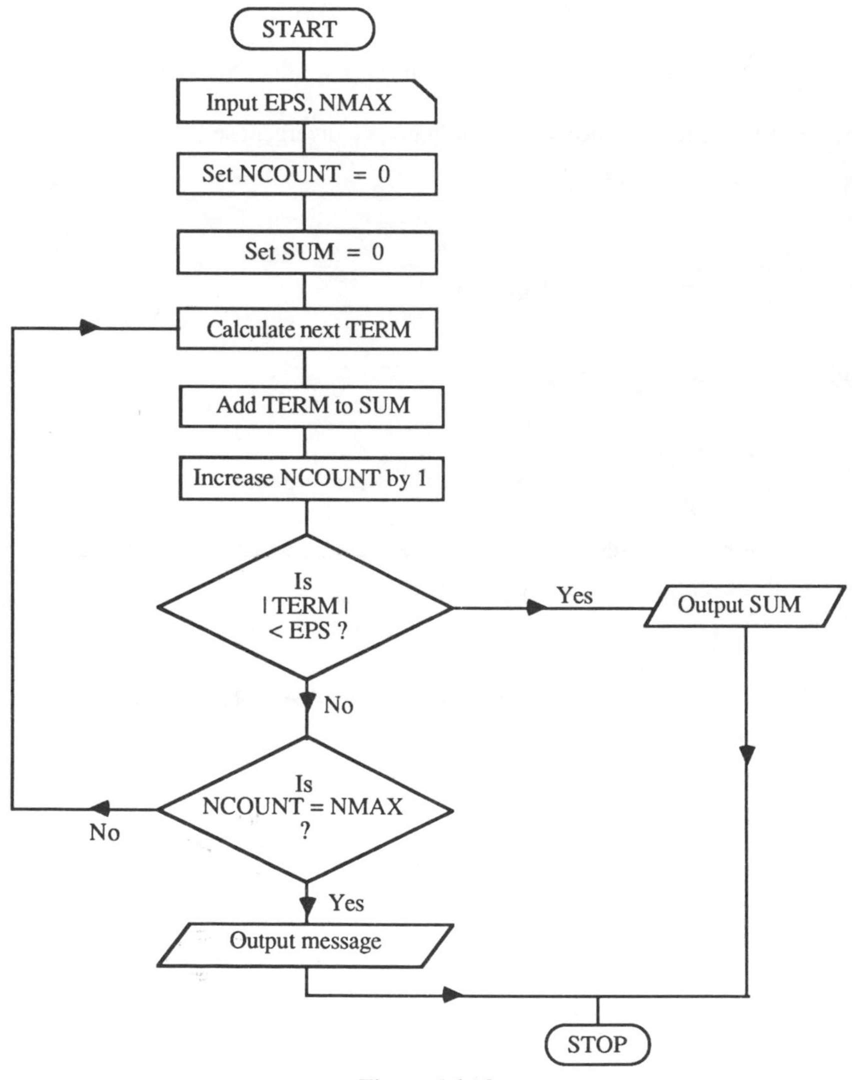

Figure 16.10

In Figure 16.10 we flow-chart a general method of calculating the sum of a series correct to within a specified amount, EPS. If this has not been achieved after NMAX terms have been added together, a suitable message is output. For a particular series it is sometimes possible to calculate each term from its predecessor by a general formula.

16.4 Representation of a Function by a Power Series: Maclaurin's Series

In the section on the tangent and quadratic approximations to a function we saw the special case where the function $f(x) \cong f(0) + xf'(0)$. We shall assume the general **Maclaurin expansion**

$$f(x) = f(0) + xf'(0) + \frac{x^2}{2!}f''(0) + \ldots + \frac{x^n}{n!}f^{(n)}(0) + \ldots \qquad (16.5)$$

Since $f(0), f'(0), f''(0)$ are constants we have a power series representation of $f(x)$. We can test that such power series converge and we assume that, if they do, they converge to the functions they represent.

Example 1: Sine series

Consider $f(x) = \sin x$. Note that $f(0) = 0$

$$\begin{aligned} f'(x) &= \cos x & \text{therefore} \quad f'(0) &= +1 \\ f''(x) &= -\sin x & f''(0) &= 0 \\ f'''(x) &= -\cos x & f'''(0) &= -1 \\ f^{iv}(x) &= \sin x & f^{iv}(0) &= 0 \\ f^{v}(x) &= \cos x & f^{v}(0) &= 1 \end{aligned}$$

and so on.

Hence

$$\sin x = x - \frac{x^3}{3!} + \frac{x^5}{5!} - \ldots + (-1)^{n-1}\frac{x^{2n-1}}{(2n-1)!}\ldots \qquad (16.6)$$

and thus $\sin x \cong x$ for small values of x (alternatively, $(\sin x)/x \to 1$ as $x \to 0$).

(Note the effect of replacing x by $-x$ on both sides.)

We now ask for what values of x the series converges.

The ratio

$$\left|\frac{u_n}{u_{n+1}}\right| = \frac{(2n+1)!}{(2n-1)!}\cdot\left|\frac{-x^{2n-1}}{x^{2n-1}}\right| = \frac{(2n+1)(2n)}{|x|^2} \to \infty \quad \text{as } n \to \infty$$

therefore the series converges for all values of x

Example 2: Binomial Expansion

Here

$$\begin{aligned} f(x) &= (1+x)^s \quad ; & f(0) &= 1 \\ f'(x) &= s(1+x)^{s-1} \quad ; & f'(0) &= s \end{aligned}$$

$$f''(x) = s(s-1)(1+x)^{s-2} \quad ; \quad f''(0) = s(s-1)$$

Similarly, $f'''(0) = s(s-1)(s-2)$; $f^{iv}(0) = s(s-1)(s-2)(s-3)$ etc

Then

$$(1+x)^s = 1 + sx + \frac{s(s-1)}{1.2}x^2 + \frac{s(s-1)(s-2)}{1.2.3}x^3 + \frac{s(s-1)(s-2)(s-3)}{1.2.3.4}x^4 + \ldots \tag{16.7}$$

This is the so-called **binomial expansion**.

The coefficient of x^r is

$$\frac{s(s-1)(s-2)\ldots(s-r+1)}{1.2.3\ldots r} \quad \text{or} \quad \frac{s!}{r!(s-r)!}$$

which is sometimes written

$${}^sC_r \quad \text{or} \quad \binom{s}{r}$$

To test for convergence we form

$$\left|\frac{u_n}{u_{n+1}}\right| = \left|\frac{s!}{n!(s-n)!}\frac{(n+1)!(s-n-1)!}{s!}\right|\frac{1}{|x|}$$

$$= \left|\frac{n+1}{s-n}\right| \cdot \frac{1}{|x|} \to \frac{1}{|x|} \quad \text{as } n \to \infty$$

The expansion is valid for $|x| < 1$.

If $x = -1$, $(1+x)^s = 0$, but the right-hand side of (16.7) is

$$1 - s + \frac{s(s-1)}{1.2} - \frac{s(s-1)(s-2)}{1.2.3} + \ldots$$

Now if $s = -2$, the series becomes

$$1 + 2 + 3 + 4 + \ldots$$

which clearly diverges, showing no representation for $x = -1$.

What do you think happens if $x = +1$?

Series representations for other functions can be derived directly or by some algebraic devices. Other series may be formed by simple manipulations.

Example 3: Cosine Series

The series for $\cos x$ can be obtained by differentiating (16.6) term by term, therefore

$$\cos x = 1 - \frac{x^2}{2!} + \frac{x^4}{4!} - \ldots \tag{16.8}$$

Differentiating tends to destroy good behaviour and it is necessary to check directly that this series converges; it can be shown by the ratio test that for all values of x it does converge.

Example 4: Exponential Series

$$e^{ix} = \cos x + i \sin x$$

$$= \left(1 - \frac{x^2}{2!} + \frac{x^4}{4!} - \ldots\right) + i\left(x - \frac{x^3}{3!} + \frac{x^5}{5!} - \ldots\right)$$

(assumes addition valid)

$$= \left(1 + ix - \frac{x^2}{2!} - \frac{ix^3}{3!} + \frac{x^4}{4!} + \frac{ix^5}{5!} - \ldots\right)$$

$$= 1 + ix + \frac{(ix)^2}{2!} + \frac{(ix)^3}{3!} + \frac{(ix)^4}{4!} + \frac{(ix)^5}{5!} + \ldots$$

If we replace ix by x we obtain the exponential series:

$$e^x = 1 + x + \frac{x^2}{2!} + \frac{x^3}{3!} + \frac{x^4}{4!} + \frac{x^5}{5!} + \ldots..$$

Similarly,

$$e^{-x} = 1 - x + \frac{x^2}{2!} - \frac{x^3}{3!} + \frac{x^4}{4!} - \frac{x^5}{5!} + \ldots.. \qquad (16.9)$$

Both series converge for all values of x.

The following is a program for calculating e^x via Maclaurin's series.

```
>
   10  CLS
   20  PRINTTAB(10,1)"Maclaurin's Series for EXP(X)"
   30  INPUTTAB(0,4)"Enter X ",X
   40  INPUT"Enter accuracy ",A
   50  INPUT"Enter number of terms ",N
   60  T=X:S=1+X:M=1
   70  @%=&20209
   80  REPEAT
   90    M=M+1
  100    T=T*X/M
  110    IF ABS(T)<A THEN 150
  120    IF M=N THEN 170
  130    S=S+T
  140    UNTIL FALSE
  150  PRINT "After ";M;" terms the sum is ";S
  160  END
  170  PRINT "After ";M;" terms the sum is ";S;
       " and the last term was ";T
  180  END
```

Example 5: Hyperbolic Series

$\cosh x = \frac{1}{2}(e^x + e^{-x})$

therefore

$$\cosh x = 1 + \frac{x^2}{2!} + \frac{x^4}{4!} \cdots \qquad (16.10)$$

Similarly, or by differentiation

$$\sinh x = x + \frac{x^3}{3!} + \frac{x^5}{5!} + \ldots \qquad (16.11)$$

Example 6: Logarithmic Series

Since ln(0) is not defined we cannot expand ln x by a Maclaurin's series. Instead, we expand $\ln(1 + x)$. You should show that

$$\ln(1 + x) = x - \frac{x^2}{2} + \frac{x^3}{3} - \frac{x^4}{4} \cdots \qquad (16.12)$$

and that the series converges for $|x| < 1$. We know it cannot converge for $x = -1$ but it does in fact converge for $x = 1$.

In each of these examples you should try to write down the nth term of the series.

Example 7

In addition to providing an approximate formula for a function Maclaurin's series is useful for calculating approximate values of that function for small values of x (provided that these values are within the range of convergence). The following three calculations illustrate thc ideas.

(i) $\sin 0.1^c$ $= 0.1 - \frac{(0.1)^3}{6} + \frac{(0.1)^5}{120} + \ldots$

$= 0.1 - 0.000167 + 0.0000001 - \ldots$

$= 0.0998$ (4 d.p.) (This is correct to 4 d.p.)

(ii) $\sin 0.5^c$ $= 0.5 - \frac{(0.5)^3}{6} + \frac{(0.5)^5}{120} - \ldots$

$= 0.5 - 0.02083 + 0.00006 - \ldots$

$= 0.4798$ (4 d.p.) (0.4794 to 4 d.p.)

(iii) $\sin 1^c$ $=1 - \frac{1}{6} + \frac{1}{120} - \ldots$

$= 1 - 0.1667 + 0.0083 - \ldots$

$= 0.8416$ (4 d.p.) (0.8415 by calculator)

We see that $\sin 0.5^c$ takes more terms than $\sin 0.1^c$ to achieve comparable accuracy yet $\sin 1^c$ seems to give a better result with only three terms. However, the first term neglected is $-(1/7!) = -0.0002$ and including this will not improve matters. What is happening is that this last series is converging more slowly than the others and it will take many more terms than the others to achieve 5 d.p. accuracy.

In Figure 16.11 we see the way in which successive Maclaurin approximations give a greater range of agreement.

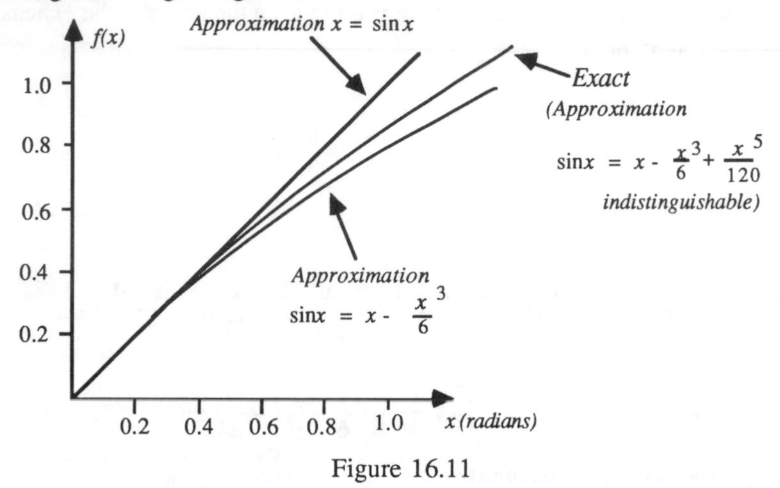

Figure 16.11

We conclude with an example to show the manipulations often necessary to obtain a Maclaurin approximation.

Example 8

Find the Maclaurin expansion for $f(x) = e^{\tan^{-1} x}$.

It would be unwise to attempt to differentiate the function as it stands. We first tackle therefore the expansion of $u = \tan^{-1}x$; again we would be rash to differentiate u many times. Instead note that since $du/dx = 1/(1 + x^2)$ we have $(1 + x^2)(du/dx) = 1$. Differentiating this equation we obtain

$$2x\frac{du}{dx} + (1 + x^2)\frac{d^2u}{dx^2} = 0$$

and, repeating the operation,

$$2\frac{du}{dx} + 4x\frac{d^2u}{dx^2} + (1 + x^2)\frac{d^3u}{dx^3} = 0$$

$$6\frac{d^2u}{dx^2} + 6x\frac{d^3u}{dx^3} + (1 + x^2)\frac{d^4u}{dx^4} = 0 \quad \text{etc}$$

Now, $u(0) = 0$, $u'(0) = 1$ and hence $2 \,.\, 0 \,.\, u'(0) + 1 \,.\, u''(0) = 0$ whence $u''(0) = 0$; also $2 \,.\, 1 + 0 + 1 \,.\, u'''(0) = 0$, therefore $u'''(0) = -2$.

Similarly $u^{(iv)}(0) = 0$.

Proceeding in this way we obtain

$$\tan^{-1}x = x - \frac{x^3}{3} + \frac{x^5}{5} \ldots \qquad (16.13)$$

Suppose we decide in advance that we want to keep the terms of the expansion of $f(x)$ as far as x^4 only.

We have

$$f(x) = e^u \cong 1 + u + \frac{u^2}{2!} + \frac{u^3}{3!} + \frac{u^4}{4!}$$

and we need no further terms (why?).

Then

$$e^{\tan^{-1}x} \cong 1 + (x - \frac{x^3}{3}) + \frac{1}{2}(x - \frac{x^3}{3})^2 + \frac{1}{6}(x - \frac{x^3}{3})^3 + \frac{1}{24}(x - \frac{x^3}{3})^4 + \ldots$$

$$\cong 1 + x - \frac{x^3}{3} + \frac{1}{2}x^2(1 - \frac{2x^2}{3} + \frac{x^4}{9}) + \frac{1}{6}x^3(1) + \frac{1}{24}x^4(1)$$

(neglecting terms of order x^5)

$$\cong 1 + x + \frac{1}{2}x^2 - \frac{1}{6}x^3 - \frac{7}{24}x^4$$

(neglecting terms of order x^5)

This expansion may be derived in a neater way which involves integration.

Applications of Maclaurin's Series

(i) The piston problem of page 307 can be simplified by using approximations if m is large. The expression for

$$\cos\phi = \left[1 - \frac{\sin^2\theta}{m^2}\right]^{1/2} \cong 1 - \frac{\sin^2\theta}{2m^2} - \frac{\sin^4\theta}{8m^4}$$

Now when $m = 4$ the last term is $(-\sin^4\theta)/2048$ which has magnitude no greater than 1/2048 and can thus be neglected. Hence, we have a simplified form for x using (12.1) and you should show that

$$x = r\left[(1 - \cos\theta) + \frac{1}{2m} - \frac{\cos^2\theta}{2m}\right]$$

On differentiating we have

the velocity $\quad v \cong \omega r(\sin\theta + \frac{\sin 2\theta}{2m})$

$$\text{the acceleration } f \cong \omega^2 r(\cos\theta + \frac{\cos 2\theta}{m})$$

This last form can be got by neglecting $\sin^4\theta$ and $\sin^2\theta$.

(ii) Webb's approximation for sec θ which arises in the theory of buckling of struts (where the equation θ = sec θ has to be solved) follows from the Maclaurin approximation.

$$\sec\theta = \frac{1}{\cos\theta} \cong \frac{1}{1-\frac{1}{2}\theta^2} \cong 1 + \frac{1}{2}\theta^2$$

Webb's approximation is

$$\sec\theta \cong \frac{1 + \frac{4\theta^2}{\pi^2} \times 0.26}{1 - \frac{4\theta^2}{\pi^2}}$$

You should check its accuracy for θ = 0(15°)90° by writing an appropriate computer program.

16.5 The Taylor Series

When using a Maclaurin series which is based on $x = 0$, to estimate the function at values of x far away from $x = 0$, we usually have to take more terms to achieve a particular accuracy than were we to make the estimation at a value of x nearer $x = 0$. Indeed, we cannot take values of x outside the range of convergence. It would seem reasonable to attempt a shift of base from $x = 0$ to another value x_0. This could suggest a generalisation of the tangent and quadratic approximations met earlier.

We can turn the tangent approximation into an exact equality thus:

$$f(x_0 + h) = f(x_0) + hf'(x_0) + \frac{h^2}{2!}f''(\xi)$$

where $x_0 < \xi < x_0 + h$.

The quadratic approximation can be converted to

$$f(x_0 + h) = f(x_0) + hf'(x_0) + \frac{h^2}{2!}f''(x_0) + \frac{h^3}{3!}f''(\xi)$$

where $x_0 < \xi < x_0 + h$.

Then by analogy (the rigorous proof involves a generalisation of the MVT) we have the result that if $f(x)$ is continuous in $[x_0, x_0 + h]$ and possesses derivatives up to the $(n+1)^{\text{th}}$ order in (x_0, x_0+h) then

$$f(x_0+h) = f(x_0) + hf'(x_0) + \frac{h^2}{2!}f''(x_0) + \ldots + \frac{h^n}{n!}f^{(n)}(x_0)$$

$$+ \frac{h^{n+1}}{(n+1)!}f^{(n+1)}(\xi) \qquad (16.14)$$

with $x_0 < \xi < x_0 + h$.
This is known as **Taylor's Theorem**. The corresponding Maclaurin's Theorem is deduced by putting $x_0 = 0$. We can estimate the maximum error in truncating a Taylor series when estimating a particular value as in the following.

Example

Consider $f(x) = 1/(1-x) = 1 + x + x^2 + x^3 + \ldots$ when expanded about $x_0 = 0$.

If we truncate the series at the term shown we commit an error of magnitude $(x^4/4!)f^{(iv)}(\xi)$. Now $f^{(iv)}(x) = 4!/(1-x)^5$ and if we are interested in estimating $f(0.3)$ then $0 < \xi < 0.3$. The largest value of $f^{(iv)}(x)$ occurs at $x = 0.3$ when it has the vlaue $4!/(0.7)^5$; then the magnitude of the error committed in estimating $f(0.3)$ as $1 + 0.3 + 0.09 + 0.027 = 1.417$ is at most

$$\frac{(0.3)^4}{4!} \cdot \frac{4!}{(0.7)^5} = 0.0481$$

In fact, $f(0.3) = 1/0.7 = 1.429$ (3 d.p.) and the error committed is actually 0.012 which is well within the maximum quoted. Were we unable to do the above analysis we should have had to quote our estimate as 1.417 ± 0.012.

Let us repeat the calculations where the expansion is based on $x = 0.1$.

The expansion is $f(x) = 1.1111 + 1.235(x-0.1) + 1.372(x-0.1)^2 + 1.524(x-0.1)^3$ where coefficients are kept to 3 d.p. This estimates $f(0.3)$ as 1.425 (3 d.p.) with a smaller error of 0.004. The maximum error from the formula is

$$\frac{(0.2)^4}{4!} \cdot \frac{4!}{(0.7)^5} = 0.0095$$

It should scarcely be necessary to remark that the error

$$R_{n+1} = \frac{h^{n+1}}{(n+1)!}f^{(n+1)}(\xi)$$

must decrease in magnitude as $n \to \infty$ if the series representation is to be exact. There are cases where R_{n+1} increases indefinitely as $n \to \infty$ and then the series is of little use. This really brings us back to the idea of convergence except that we must remember that the general Taylor expansion (16.14) can be written

$$f(x) = f(x_0) + (x - x_0)f'(x_0) + \frac{(x - x_0)^2}{2!}f''(x_0) + \ldots$$

$$+ \frac{(x - x_0)^{n+1}}{(n + 1)!}f^{(n+1)}(\xi) \qquad (16.15a)$$

We can also write the general Taylor expansion as an infinite series

$$f(x) = f(x_0) + (x - x_0)f'(x_0) + \frac{(x - x_0)^2}{2!}f''(x_0) + \ldots$$

$$+ \frac{(x - x_0)^n}{n!}f^{(n)}(x_0) + \ldots \qquad (16.15b)$$

16.6 Applications of Taylor's Series

(i) *Numerical Differentiation*

Suppose we wish to estimate $f'(a)$ and $f''(a)$. By Taylor's series we have

$$f(a + h) = f(a) + hf'(a) + \frac{h^2}{2!}f''(a) + \frac{h^3}{3!}f'''(a) + \ldots \qquad (16.16a)$$

$$f(a - h) = f(a) - hf'(a) + \frac{h^2}{2!}f''(a) - \frac{h^3}{3!}f'''(a) + \ldots \qquad (16.16b)$$

Subtracting (16.16b) from (16.16a), (is this justified?)

$$f(a + h) - f(a - h) = 2hf'(a) + \frac{2h^3}{3!}f'''(a) + \ldots$$

$$= 2hf'(a) + 0(h^3)$$

therefore

$$\frac{f(a + h) - f(a - h)}{2h} = f'(a) + 0(h^2)$$

Similarly, adding (16.16a) and (16.16b) gives

$$\frac{f(a + h) - 2f(a) + f(a - h)}{h^2} = f''(a) + 0(h^2)$$

Higher accuracy formulae can be found by expanding $f(a + 2h)$, $f(a - 2h)$ etc and taking suitable combinations of these expansions.

(ii) *Newton's formula for curvature at the origin*

Suppose the curve of $y = f(x)$ touches the x-axis at the origin so that

$f(0) = 0 = f'(0)$. Then the radius of curvature of the curve at the origin is $\{1 + [f'(0)]^2\}^{3/2}/f''(0) = 1/f''(0)$. If the curve is given in parametric form, the following manipulations help.

$$y = f(0) = xf'(0) + \tfrac{1}{2}x^2 f''(0) + 0(x^3) = \tfrac{1}{2}x^2 f''(0) + 0(x^3)$$

[since $f(0)$ and $f'(0)$ are both zero]

Hence

$$\frac{2y}{x^2} = f''(0) + 0(x)$$

and

$$\lim_{x \to 0} \frac{2y}{x^2} = f''(0)$$

and so the radius of curvature at the origin is $\lim\limits_{x \to 0} \dfrac{x^2}{2y}$, which is **Newton's formula for curvature at the origin.**

Example

The cycloid $x = 2(\theta + \sin\theta)$, $y = 2(1 - \cos\theta)$ touches the x-axis at the origin since, when $\theta = 0$, $x = 0$, $y = 0$, and

$$\frac{dy}{dx} = \left[\frac{dy}{d\theta}\right] \Big/ \left[\frac{dx}{d\theta}\right] = 0$$

Then the curvature at the origin

$$= \lim_{x \to 0} \frac{4(\theta + \sin\theta)^2}{4(1 - \cos\theta)}$$

$$= \lim_{x \to 0} \frac{\left[1 + \dfrac{\sin\theta}{\theta}\right]^2 . \theta^2}{2\sin^2\dfrac{\theta}{2}}$$

$$= \lim_{x \to 0} \left[2\left[1 + \frac{\sin\theta}{\theta}\right]^2 \Big/ \left[\frac{\sin\frac{1}{2}\theta}{\frac{1}{2}\theta}\right]^2 \right]$$

$= 2.2^2/1$
$= 8$

(iii) *Indeterminate forms: L'Hôpital's Rule*

In the above example we had a potential limiting form 0/0; equally awkward is the form ∞/∞. Although there are various methods of dealing with such forms we shall here develop a simple rule.

We suppose that for two differentiable functions, $f(x)$ and $g(x)$, $f(0) = g(0) = 0$. Then

$$\frac{f(x) - f(0)}{x} \cdot \frac{x}{g(x) - g(0)} = \frac{f(x)}{g(x)}$$

If we proceed to the limit as $x \to 0$ we obtain

$$\lim_{x \to 0} \frac{f(x)}{g(x)} = \frac{f'(0)}{g'(0)}$$

which is **L'Hôpital's Rule**. (What case do we have to beware?)

Examples

(a) $\lim_{x \to 0} \frac{1 - e^x}{x}$

Here $1 - e^0 = 0$ and so the conditions of the rule are obeyed. $f'(x) = -e^x$ and $g'(x) = 1$ so that the limit is $-e^0/1 = -1$. We could have deduced the same result by expanding the numerator as a Maclaurin's series.

(b) $\lim_{x \to 0} \frac{x - \sin x}{x^2}$

Here $f(0) = 0 = g(0)$, but $f'(x) = 1 - \cos x$, and so $f'(0) = g'(0) = 0$ and an application of L'Hôpital's rule gives 0/0. The basic principle is, *when in doubt apply the rule again* (**provided** the differentiations can be done). Hence $f''(x) = \sin x$, $g''(x) = 2$ and so the limit is

$$\frac{f''(0)}{g''(0)} = 0$$

The generalisation of L'Hôpital's rule follows from an extension of MVT.

(iv) *Small Errors*

Suppose $\delta f \equiv f(a + \delta x) - f(a)$, then by Taylor's expansion, if h is small,

$$\delta f \cong \delta x f'(a) \tag{16.17}$$

If any small error, δx, has been made in the measurement of x then (16.17) gives the approximate error in $f(x)$. We call $\delta f/f(a)$ the **relative error**; it is approximately

$$\delta x \,.\, \frac{f'(a)}{f(a)}$$

the **percentage error** is

$$\left[\frac{\delta f}{f(a)} \times 100\right]\% \cong \left[\delta x \frac{f'(a)}{f(a)} \times 100\right]\%$$

Example 1

A tower 50 metres high stands on one bank of a river, the elevation of the top from the point on the opposite bank being 30° (see Figure 16.12). Find the percentage error in the width of the river as measured via the angle which is subject to an error of 1′.

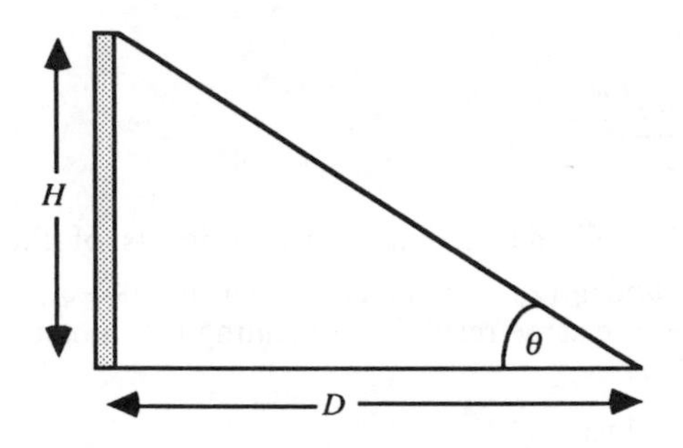

Figure 16.12

If D is the width of the river, H the height of the tower and θ the angle of elevation then

$$D(\theta) = H \cot \theta$$

Since

$$D'(\theta) = -H \operatorname{cosec}^2 \theta$$

the percentage error is

$$\frac{-H \operatorname{cosec}^2 \theta}{H \cot \theta} \,.\, \delta\theta \times 100 = -\frac{100}{\cos \theta \sin \theta} \delta\theta$$

and since $\theta = 30°$, this error

$$= \frac{-100}{\frac{\sqrt{3}}{2} \cdot \frac{1}{2}} \cdot \frac{1}{60} \cdot \frac{\pi}{180} = \frac{-400\pi}{10800\sqrt{3}}$$

$$= -0.067\% \quad \text{(3 d.p.)}$$

This result shows the scaling of the error by the function $D(\theta)$.

Example 2

The period of a simple pendulum if $T = 2\pi\sqrt{l/g}$. We use this to estimate g; suppose a small percentage error of +0.2% is made in measuring the length and −0.1% in measuring T and we seek the effect on g.

Now $g = (4\pi^2/T^2)l$ and it helps to take logarithms to obtain $\ln g = \ln 4\pi^2 + \ln l - 2\ln T$. Differentiating both sides w.r.t. g

$$\frac{1}{g} = 0 + \frac{1}{l}\frac{dl}{dg} - \frac{2}{T}\frac{dT}{dg}$$

and replacing derivatives by their approximations

$$\frac{\delta g}{g} \cong \frac{\delta l}{l} - \frac{2\delta T}{T}$$

and hence

$$\frac{dg}{g} \times 100 \cong 0.2 - 2(-0.1) = 0.4\%$$

This suggests that T must be measured twice as accurately as l to have as little effect on the value of g.

Optimal Values: A revisit

We saw in Section 12.3 the difficulty encountered with x^3 and x^4 in the determination of their turning points. Remember that the definition of a turning point of $f(x)$ is that $f'(x) = 0$. Then, if h is small, Taylor's expansion (16.14) becomes

$$\delta f = f(x_0 + h) - f(x_0) = hf'(x_0) + \frac{h^2}{2!}f''(x_0) + \ldots \cong \frac{h^2}{2!}f''(x_0)$$

Now, if $f''(x_0) > 0$, $\delta f > 0$ irrespective of whether h is positive or negative, and so, *locally*, $f(x_0)$ is the smallest value; conversely, if $f''(x_0) < 0$, $f(x_0)$ is a local maximum. If $f''(x_0) = 0$ then the approximation is too coarse to determine the details of the behaviour at x_0. We have to take the next term of the Taylor series to get

$$\delta f \cong \frac{h^3}{3!}f'''(x_0)$$

Here a different state of affairs obtains: if $f'''(x_0) > 0$ then δf takes opposite signs on either side of x_0 [likewise if $f'''(x_0) < 0$] which implies a point of inflection. If $f'''(x_0) = 0$ we have to proceed further along the Taylor series to obtain $\delta f \cong (h^4/4!)f^{(iv)}(x_0)$. It is clear that behaviour will continue to alternate. We have the rule:

if the first non-vanishing derivative of $f(x)$ at $x = x_0$ is of even order then there is a local maximum or minimum at x_0 (local minimum if the value of that derivative *is positive, local maximum if it is negative); if the first non-vanishing derivative is of odd order there is a point of infection at x_0.*

Note the flatter behaviour near $x = 0$ of x^4 than x^2 as shown in Figure 16.13.

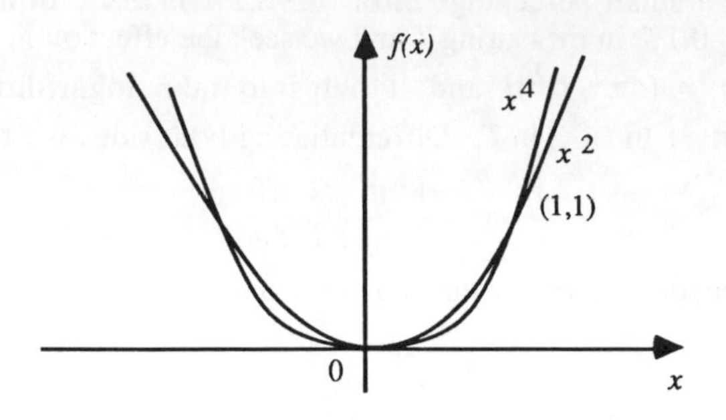

Figure 16.13

Problems

Section 16.1

1 For small values of x we can approximate $e^x \cong 1 + x$. What will be the maximum error in this truncation over the interval $0 < x < 0.1$?

2 Use the MVT to show that for $a > b \geq 0$

$$\frac{a-b}{1+b} > \log\left[\frac{1+a}{1+b}\right] > \frac{a-b}{1+a}$$

and hence find upper and lower bounds for the value of log(1.01). (EC)

Section 16.2

3 Find the tangent and quadratic approximations at the points $x = 0$ and $x = x_0$ to the functions

(a) e^x (b) $\ln(1+x)$ (c) $\cos x$ (d) $x^3 - 2x^2 + 7x - 5$

4 Using your expansions at $x = 0$ obtained in Problem 1, estimate $f(0.1)$, $f(0.25)$, $f(0.5)$, for each function and compare with the estimates obtained from your approximations based at the points $x = 0.2$, $x = 0.3$ and $x = 0.6$ respectively. Comment on your results.

Section 16.3

5 Use the ratio test to show the convergence or divergence of the following series

(i) $\frac{1}{1!} + \frac{1}{2!} + \ldots + \frac{1}{n!} + \ldots$ (ii) $\frac{3}{1} + \frac{3^2}{2} + \ldots + \frac{3^n}{n} + \ldots$

(iii) $\frac{1}{\sqrt{3}} + \frac{3}{3} + \frac{5}{(\sqrt{3})^3} + \ldots + \frac{(2n-1)}{(\sqrt{3})^n} + \ldots$

(iv) $1 + \frac{1.2}{1.3} + \frac{1.2.3}{1.3.5} + \ldots + \frac{1.2.3 \ldots n}{1.3.5 \ldots (2n-1)} + \ldots$

Use a computer program to evaluate the first 20 terms of each series and the corresponding partial sums.

6 Show that the ratio test is inconclusive for the following series and use the comparison test to show convergence or divergence.

(i) $1 + \frac{3}{2.4} + \frac{7}{4.9} + \ldots + \frac{2^r - 1}{2^{r-1}.r^2} + \ldots$ (compare with $\sum \frac{2}{r^2}$)

(ii) $\frac{1}{3^2} + \frac{1}{5^2} + \frac{1}{7^2} + \ldots + \frac{1}{(2n+1)^2} + \ldots$ (compare with $\sum \frac{1}{r^2}$)

(iii) $1 + \frac{2^2+1}{2^3+1} + \frac{3^2+1}{3^3+1} + \frac{4^2+1}{4^3+1} + \ldots + \frac{n^2+1}{n^3+1} + \ldots$

(iv) $\frac{\ln 2}{2} + \frac{\ln 3}{3} + \ldots + \frac{\ln n}{n} + \ldots$

7 Show that the series

$$\sum_{n=1}^{\infty} \frac{(-1)^{n+1}}{2n-1}$$

converges. How many terms of the series are needed in order to obtain in the sum an error which does not exceed 0.002 in magnitude? (EC)

8 Find the interval of convergence of

(i) $\frac{x-2}{1} + \frac{(x-2)^2}{2} + \frac{(x-2)^3}{3} + \ldots + \frac{(x-2)^n}{n} + \ldots$

(ii) $\frac{x+1}{\sqrt{1}} + \frac{(x+1)^2}{\sqrt{2}} + \frac{(x+1)^3}{\sqrt{3}} + \ldots + \frac{(x+1)^n}{\sqrt{n}} + \ldots$

(iii) $x - \frac{x^3}{3} + \frac{x^5}{5} - \frac{x^7}{7} + \ldots$ (iv) $x - \frac{x^2}{4} + \frac{x^3}{9} - \frac{x^4}{16} + \ldots$

9 Find the sum of any one of the above series for a selection of values of x using a computer program.

10 An infinite series $S(x)$ is defined by

$$S(x) = 1 + \frac{x^2}{2} + \frac{x^4}{2.4} + \frac{x^6}{2.4.6} + \ldots$$

For what values of x is the series convergent? (EC)

11 Compare the two series below as methods of calculating π. Write a computer program to help in your investigations.

(i) $\dfrac{\pi}{4} = 1 - \dfrac{1}{3} + \dfrac{1}{5} - \dfrac{1}{7} + \ldots$ (ii) $\dfrac{\pi^2}{6} = 1 + \dfrac{1}{2^2} + \dfrac{1}{3^2} + \dfrac{1}{4^2} + \ldots$

Section 16.4

12 Obtain the following Maclaurin expansions:

(i) $\cos^2 x = 1 - x^2 + \dfrac{1}{3}x^4 - \dfrac{2}{45}x^6 + \ldots$

(ii) $\sin^2 x = x^2 - \dfrac{1}{3}x^4 + \dfrac{2}{45}x^6 - \dfrac{1}{315}x^8 + \ldots$

(iii) $2^x = 1 + x \ln 2 + \dfrac{1}{2}(x \ln 2)^2 + \dfrac{1}{6}(x \ln 2)^3 + \ldots$

(iv) $a^{mx} = 1 + mx \ln a + (mx \ln a)^2/2! + (mx \ln a)^3/3! + \ldots$ (a and m constants)

(v) $\ln(1 + \cos x) = \ln 2 - \dfrac{1}{4}x^2 - \dfrac{1}{96}x^4 - \ldots$

13 Show that if c is large compared with x, $s = c \sinh(x/c)$ may be replaced approximately by $s = x + (x^3/6c^2)$, and $y = c \cosh(x/c)$ becomes approximately the equation of a parabola.

14 State the expansions, in ascending powers of x, of $\ln(1 + x)$, $\sin x$ and e^x. Deduce the following results:

(i) $\ln(1 - x^3) = -x^3 - \dfrac{x^6}{2} - \dfrac{x^9}{3} - \ldots$

(ii) $\ln(1 + x + x^2) = x + \dfrac{x^2}{2} - \dfrac{2}{3}x^3 + \dfrac{1}{4}x^4 + \ldots$

(iii) $\ln(1 + \sin x) = x - \dfrac{x^2}{2} + \dfrac{x^3}{6} - \ldots$

(iv) $e^x \ln(1 + x) = x + \dfrac{x^2}{2!} + \dfrac{2x^3}{3!} + \dfrac{9x^5}{5!} + \ldots$

15 If $y = \ln \cos x$ prove that

$$\frac{d^3y}{dx^3} + 2\frac{d^2y}{dx^2}\frac{dy}{dx} = 0$$

Hence, or otherwise, obtain the Maclaurin expansion of y as far as the term in x^4, and, by the substitution $x = \pi/4$, deduce the approximate relation

$$\ln 2 \cong \frac{\pi^2}{16}(1 + \frac{\pi^2}{96})$$

Show that

$$\lim_{x \to 0} \frac{\ln \cos x}{x^2} = -\frac{1}{2}$$

16 Use Maclaurin's Theorem to expand $\ln(1 + x)$, $\sin x$ and $\cos x$ in ascending powers of x. Hence find the series for $\ln[(1/x)\sin x]$ and $\ln \cos x$ as far as the terms in x^4, and show that if x is small, $\tan x \cong x\, e^{(x^2/3)}$

17 (i) Show that

$$1 - x^2 + x^4 + \ldots + (-1)^{n-1} x^{2n-2} + \frac{(-1)^n x^{2n}}{1 + x^2} = \frac{1}{1 + x^2}$$

(ii) Hence expand $\tan^{-1} y$ as an infinite series in powers of y.
For what range of values of y is the series convergent? (EC)

Hint: $\frac{d}{du}\tan^{-1} u = \frac{1}{1 + u^2}$

18 Expand

$$\sqrt{(1 + ax + bx^2 + cx^3)} - \sqrt{(1 + ax + bx^2)}$$

in ascending powers of x as far as the term in x^5. (LU)

19 State the Maclaurin expansion for a function $f(x)$ and use it to obtain the series for $\cos x$ in powers of x. Hence, or otherwise, show that for sufficiently small values of x

$$\sec x = 1 + \frac{x^2}{2!} + \frac{5x^4}{4!} + \ldots$$

Use this expansion to obtain the value of sec 2° correct to five decimal places. (LU)

20 Use Maclaurin's theorem to obtain the series expansion for $\ln(1 + x)$ in terms of x and state the range of values of x for which this expansion is valid. Deduce that

$$\ln\left[1 + \frac{1}{n}\right] = 2\left[\frac{1}{2n + 1} + \frac{1}{3(2n + 1)^3} + \frac{1}{5(2n + 1)^5} + \ldots\right]$$

and hence calculate ln 10 correct to five significant figures given that ln 3 = 1.09860. (LU)

21 If $y = \sin^{-1} x$ show that $(1 - x^2)y_{n+2} - (2n + 1)xy_{n+1} - n^2 y_n = 0$ where y_n denotes $d^n y/dx^n$, and obtain the Maclaurin series

$$y = x + \frac{1^2 x^3}{3!} + \frac{1^2 . 3^2 . x^5}{5!} + \ldots$$ (EC)

22 Show that $x = \ln 2 + a/8$ is an approximate solution of

$$2e^x \sinh x = 3 + a$$

where a is small. Find a better approximation by including terms in a^2. (EC)

23 Use any method to obtain the expression

$$\sinh^{-1} x = x - \frac{x^3}{6} \ldots$$

If a telegraph cable has series resistance R and shunt leakance G, its attenuation α_1 is given by $\alpha_1{}^2 = RG$. A model of the cable is constructed in the laboratory by connecting n networks; the attenuation α_2 of this model is given by

$$\sinh^2 \left[\frac{\alpha_2}{2n}\right] = \frac{RG}{4n^2}$$

Show that, if $(\sqrt{RG}/2n)$ is small, the difference between the attenuations α_2 and α_1 approximates to $\alpha_1{}^3/24n^2$. (EC)

24 The function $f(x)$ is defined by $f(x) = \exp(-kx).\sin(\omega x)$

(a) (i) Verify that $f(x)$ satisfies the differential equation

$$f''(x) + 2kf'(x) + (k^2 + \omega^2)f(x) = 0 \qquad \text{(A)}$$

Find the value of $f''(0)$ by using this differential equation, and by differentiating (A) twice find the values of $f'''(0)$ and $f^{(iv)}(0)$. Hence obtain the Maclaurin series for $f(x)$ as far as the term in x^4.

(ii) Use the standard series for $\exp(x)$ and $\sin(x)$ to obtain by an alternative method the Maclaurin series for $f(x)$ found in (i).

Which of these methods is preferable? Give reasons.

(b) In the case $k = 2$, $\omega = 4$ find the first positive values of x for which $f(x)$ has

(i) a zero (ii) a local maximum. (EC)

25 A simply supported uniform beam of length L is subjected to a uniformly distributed load W and a compressive force P at each end. The deflection D at the mid-point is given by

$$D = \frac{WEI}{P^2}[\sec(\tfrac{1}{2}mL) - 1] - \frac{WL^2}{8P}$$

in which $m^2 = P/EI$, where E and I are constants.
Using the standard Maclaurin series

$$\sec x = 1 + \frac{x^2}{2!} + \frac{5x^4}{4!} + \ldots$$

show that as $P \to 0$, $D \to \dfrac{5WL^4}{384EI}$ (EC)

26 (i) Write down the power series expansions for $\exp(x)$ and $\exp(ax) - 1$ as far as the terms in x^3.

(ii) In a problem on motion with quadratic resistance the value of x is given by

$$\left[1 + \frac{1}{2k}\right] x - \frac{1}{4k^2}(\exp(2kx) - 1) = 0$$

If k is sufficiently small that k^4 and higher powers of k may be neglected, show that this equation reduces to

$$x(2kx^2 + 3x - 6) = 0$$

Show that the positive solution of this is $x = 2 - (8k/3) + O(k^2)$ (EC)

27 The efficiency of a particular thermal-cycle is given by

$$\eta = 1 - \frac{1}{x(1 + x)} \ln(1 + x + x^2) \qquad (1)$$

in which x is a non-dimensional variable.
Expand the functions in (1) in ascending powers of x and deduce the series expansion of η up to and including the term in x^2. (EC)

Section 16.5

28 Expand each of the functions below about the point given

(a) e^x about $x = 1$ (b) $2/(1 + x)$ about $x = 2$

(c) $\sin x$ about $x = \pi/6$

State in each case for what range of x the expansion is permissible.

29 (i) Prove that

$$\sin x = \sin a + (x - a)\cos a - \frac{(x-a)^2 \sin a}{2!} - \frac{(x-a)^3 \cos \xi}{3!}$$

where $a \leq \xi \leq x$

(ii) Use the relationship in (i) to evaluate sin 51° given that sin 45° = $1/\sqrt{2}$. To how many decimal places is the answer accurate? (EC)

30 For small values of x the approximations $e^x \cong 1 + x$, $\sin x \cong x$ are sometimes employed. Use the error term from Taylor's expansion to estimate how large a value of x can be used such that the error in the approximation is less than 0.01.

31 The hydraulic radius χ used in open channel flow is defined by χ = cross-sectional area of water flowing/wetted perimeter. In a rectangular channel the depth d is small compared with the width w. Show that $\chi \cong d - 2d^2/w$ and estimate the error involved in using this formula when $d = 0.1$ m and $w = 5$ m.

32 Given that $\pi/90 = 0.03491$ use four terms of Taylor's series to estimate tan 43° to 4 d.p.

33 Show that

$$\sqrt{a+h} = \sqrt{a} + \frac{1}{2}\frac{h}{\sqrt{a}} - \frac{1}{8}\frac{h^2}{a\sqrt{a}} + \frac{1}{16}\frac{h^3}{a^2\sqrt{a}} - \frac{5}{128}\frac{h^4}{a^3\sqrt{a}}$$

and hence, with $a = 9$, $h = 1$, show that $\sqrt{10} = 3.16228$ correct to five decimal places.

34 Use Taylor's expansion to show that:

(i) $y = \dfrac{\ln(2-x)}{x^3 - 3x + 2}$ behaves like $-\dfrac{1}{3}\dfrac{1}{(x-1)}$ for x near 1.

(ii) $y = \dfrac{(1+\cos x)^2}{\sin x}$ behaves like $-\dfrac{1}{4}(x-\pi)^3$ for x near π.

Hence sketch a graph of y against x near the relevant value in each case.

Section 16.6

35 Use L'Hôpital's rule or other methods to establish:

(i) $\lim\limits_{x\to 0} \dfrac{\sin^2 x}{x}$ (ii) $\lim\limits_{x\to 0} \dfrac{\ln(1+x)}{x}$

(iii) $\lim\limits_{x\to \frac{1}{2}} \dfrac{\ln 2x}{2x-1}$ (iv) $\lim\limits_{x\to 0} \dfrac{\sec x - 1}{x \sin x}$

(v) $\lim\limits_{\lambda\to -1} \dfrac{a^{\lambda+1} - 1}{\lambda + 1} = \ln a \quad (a > 0)$ (vi) $\lim\limits_{x\to 1} \dfrac{\ln \cos 2\pi x}{1 + \cos \pi x} = -4$

(vii) $\lim\limits_{\theta\to 0} \dfrac{\tan 2\theta - 2\tan\theta}{\sin 2\theta - 2\sin\theta} = -2$

36 The measurement of the side of a cube is in error by –1%. What is the error in the estimated volume of the cube?

37 The density of a spherical ball-bearing is estimated by measuring the radius as 3.00 ± 0.02 cm and weighing the sphere as 30 ± 0.01 gm. What is a reasonable estimate of the density?

38 The stress on a shaft of diameter d under a torque T is given by $S = 16T/\pi d^3$. Find the approximate % error in the calculated stress if the measured value of d is liable to an error of 2%.

39 Find the curvature at the origin of the curves

(i) $y = \dfrac{x^2}{1+x^2}$ (ii) $y^2 = 4ax$ (iii) $\begin{cases} x = \cos\theta \\ y = 2(\sin\theta - 1) \end{cases}$

40 Obtain the expansion of $\tan\theta$ in powers of θ as far as the term in θ^5. A column of length l has a vertical load P and a horizontal load F at the top. The transverse deflection is

$$\delta = \frac{Fl}{P}\left[\frac{\tan ml}{ml} - 1\right]$$

where $m^2 = P/EI$. Show that as $P \to 0$, $\delta \to Fl^3/3EI$ and that a small value of P leads to an increase of this by about $(40Pl^2/EI)\%$ (LU)

41 The current i in an electrical circuit at time t is given by

$$i = \frac{E}{R}(1 - e^{-Rt/L})$$

Use L'Hôpital's rule to derive a suitable formula for i when R is negligibly small. (EC)

42 Show that, as far as the term in x^5

$$\ln\left[\frac{1 + \tan(\frac{1}{2}x)}{1 - \tan(\frac{1}{2}x)}\right] = x + \frac{x^3}{6} + \frac{x^5}{24} + \ldots$$

Hence, or otherwise, prove that

$$\lim_{x \to 0}\left\{\frac{\ln\dfrac{1 + \tan(\frac{1}{2}x)}{1 - \tan(\frac{1}{2}x)} - \sin x}{x^3}\right\} = \frac{1}{3}$$ (LU)

43 (a) Given that $y^3 + 2y = 3x \sin x$ find dy/dx in terms of x and y. Derive an expression for d^2y/dx^2, but do not simplify it.

(b) The radius of curvature of a cycloid at the origin is given by

$$\lim_{\theta \to 0} \frac{(\theta + \sin\theta)^2}{1 - \cos\theta}$$

Evaluate this radius using L'Hopital's rule. (EC)

17

TECHNIQUES OF INTEGRATION

A slender rod is fixed at its base A and is subjected to an increasing vertical load P at its free upper end 0. The rod remains upright until a critical load is reached; then the rod buckles into a deflected profile. Figure 17.1 depicts the situation for a load somewhat in excess of the critical value.

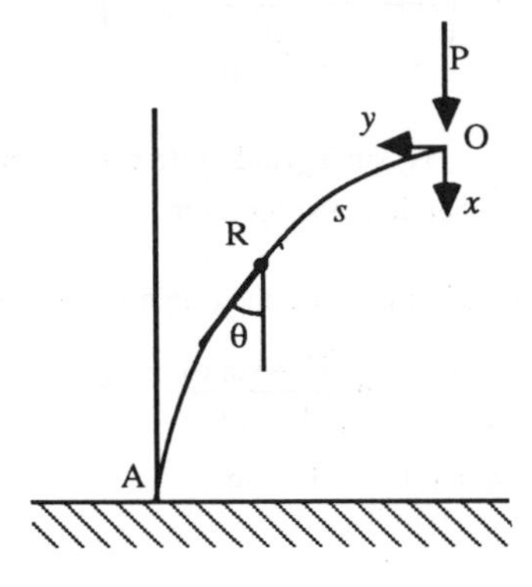

Figure 17.1

For most materials, the length, L, of the rod will not change appreciably during the bending. We take coordinate axes as shown and measure distance s along the rod from the upper end. The angle to the downwards vertical made by the tangent to the deflected profile at some point R is denoted by θ. The curvature of the rod is measured by $\dfrac{d\theta}{ds}$ and the equation of the deflected profile is

$$EI\frac{d\theta}{ds} = -Py \tag{17.1}$$

where EI is the flexural rigidity of the rod.

We differentiate (17.1) with respect to s and use the result that $dy/ds = \sin\theta$ to obtain

$$EI\frac{d^2\theta}{ds^2} = -P\sin\theta \tag{17.2}$$

Denoting $k^2 = P/EI$, we multiply (17.2) by $\frac{d\theta}{ds}$ and integrate the resulting equation with respect to s to give

$$\int \frac{d^2\theta}{ds^2} \cdot \frac{d\theta}{ds} ds = -k^2 \sin\theta \frac{d\theta}{ds} ds$$

Noting that $\frac{d}{ds}\left\{\left[\frac{d\theta}{ds}\right]^2\right\} = 2\frac{d^2\theta}{ds^2} \cdot \frac{d\theta}{ds}$, we obtain

$$\frac{1}{2}\left[\frac{d\theta}{ds}\right]^2 = k^2 \cos\theta + C, \quad C \text{ constant.}$$

Let $\theta = \alpha$ at the upper end of the rod, where $\frac{d\theta}{ds} = 0$

Then $C = -k^2 \cos\alpha$ so that

$$\frac{d\theta}{ds} = -k\,[2(\cos\theta - \cos\alpha)]^{1/2}$$

because $d\theta/ds$ is negative for this example.

Hence we can find an expression for the length of the rod:

$$L = \int_0^L ds = \frac{1}{k}\int_0^\alpha \frac{d\theta}{[2(\cos\theta - \cos\alpha)]^{1/2}} = \frac{1}{k} I_\alpha \qquad (17.3)$$

Provided that we can evaluate I_α, we can obtain a value for k and hence P; if we specify the angle α we shall then be able to find the ratio of P to the critical load. How do you evaluate I_α? The purpose of this chapter is to provide some answers.

Analytical methods of integration cannot always be relied upon. Some integrals defy obvious methods of attack as, for example, $I = \int_0^4 e^{-x^2}\, dx$. Some integrals are not posed in a form specified by a table of values, perhaps as the results of an experiment. Even when an indefinite integral can be evaluated after a long and tedious search, the resulting formula may not be easy to interpret.

Some of the methods studied in this chapter produce approximate values for the evaluation of $\int_a^b f(x)\,dx$. The estimation of the error in the approximate value is not an easy matter and is examined in Section 17.3.

In this chapter we introduce the ideas of numerical integration – or quadrature as it is often known. There are more powerful methods available to deal with specialised problems, for example, infinite integrals or integrands which contain a singularity: these are beyond our remit. Note that in all cases we approximate

$$I = \int_a^b f(x)\,dx \quad \text{by a sum} \quad S = \sum_{i=0}^{n} a_i f(x_i)$$

where $x_0, x_1, \ldots x_n$ are points in the interval $[a, b]$ and the a_i are coefficients.

17.1 Newton-Cotes Formulae

In this section we examine a class of simple integration formulae which are based on replacing the integrand $f(x)$ by a polynomial $p_n(x)$ of degree n that has the same value as $f(x)$ at each end of $(n+1)$ equally-spaced points x_i in the interval $[a, b]$. Most of the formulae are of the *closed* type, that is, they used $x_0 = a$, $x_n = b$ and $x_i = a + ih$ where $h = (b-a)/n$. The *open* formula set $x_{-1} = a$ and $x_{n+1} = b$ and the selected points are where $x_i = a + (i+1)h$, $h = (b-a)/(n+2)$.

The simplest open formula, known as the *mid-point rule* selects one value of x, viz.$\frac{1}{2}(a+b)$, and is illustrated in Figure 17.2(a). In geometric terms, the area under $f(x)$ is approximated by the area of the rectangle whose height is $f(\frac{1}{2}(a+b))$. It can be described via $h = \frac{1}{2}(b-a)$ as

$$\int_{x_{-1}}^{x_1} f(x)\,dx \cong 2h\,f(x_0) \qquad (17.4)$$

[This corresponds to $n = 0$ in the general formulation above.]

For simplicity and for comparison with the closed formulae which dominate this chapter, the algorithm has been described using $h = x_1 - x_0$ as

$$\int_{x_0}^{x_1} f(x)\,dx \cong h\,.\,f(x_0 + \tfrac{1}{2}h) \qquad (17.5)$$

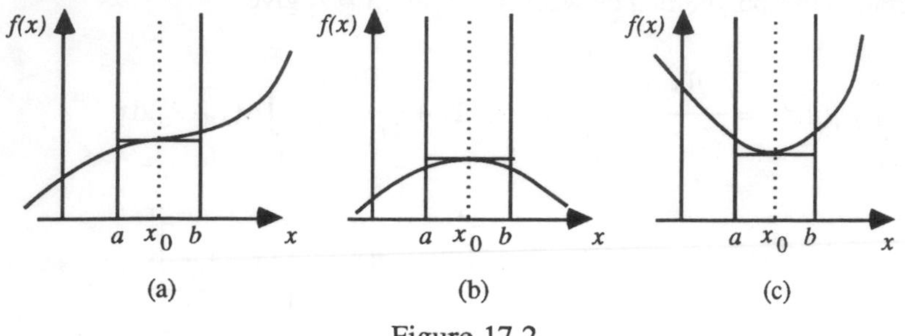

Figure 17.2

The situation depicted in Figure 17.2(a) shows the rule in its most favourable light, since the over-estimate of the area from a to x_0 is partly balanced by the under-estimate of the area from x_0 to b. Figures 17.2(b) and (c) show extreme examples where the rule fails to give a good approximation. We shall illustrate the rules on four simple examples which can be integrated analytically. These are

$$I_1 = \int_0^2 x^2\,dx = 8/3 \qquad\qquad I_2 = \int_0^1 (1 - x^3)dx = 0.75$$

$$I_3 = \int_{\pi/4}^{\pi/2} \cos x\,dx = 0.2929 \text{ (4 d.p.)} \qquad\qquad I_4 = \int_1^3 \frac{1}{x^2}dx = \frac{2}{3}$$

The mid-point rule (17.5) is applied to the first of these problems by first evaluating $f(1) = 1^2 = 1$ and then obtaining the estimate $S = 2.\ 1.\ 1 = 2$.

Then, for the second problem we evaluate $f(\frac{1}{2}) = 1 - (\frac{1}{2})^3 = 7/8$ and produce

$S = 2.\ \frac{1}{2}.\ 7/8 = 0.875.$

In the third example, we evaluate $f(3\pi/8) = \cos(3\pi/8)$ and obtain

$$S = 2.\ \pi/8.\ \cos 3\pi/8 = 0.3006 \quad \text{(4 d.p.)}$$

Finally, to estimate I_4 we evaluate $f(2) = 1/4$ and obtain

$$S = 2.\ 1.\ 1/4 = 0.5$$

The unsatisfactory nature of these approximations is mainly due to the simplicity of the formula.

Trapezoidal rule

We now turn to the closed formulae and consider the approximation of $f(x)$ by a first degree polynomial $p_1(x)$. The Lagrange formulation of that polynomial which agrees with $f(x)$ at $x = a = x_0$ and $x = b = x_1$ is

$$p_1(x) = \frac{(x - x_1)}{(x_0 - x_1)}f(x_0) + \frac{(x - x_0)}{(x_1 - x_0)}f(x_1) \tag{17.6}$$

Integrating this polynomial between $x = x_0$ and $x = x_1$ gives

$$\int_{x_0}^{x_1} p_1(x)\mathrm{d}x = \frac{f(x_0)}{(x_0 - x_1)} \int_{x_0}^{x_1} (x - x_1)\mathrm{d}x + \frac{f(x_1)}{(x_1 - x_0)} \int_{x_0}^{x_1} (x - x_0)\mathrm{d}x$$

$$= \frac{f(x_0)}{(x_0 - x_1)} \left[\frac{(x - x_1)^2}{2} \right]_{x_0}^{x_1} + \frac{f(x_1)}{(x_1 - x_0)} \left[\frac{(x - x_0)^2}{2} \right]_{x_0}^{x_1}$$

$$= \left[\frac{x_1 - x_0}{2} \right] [f(x_0) + f(x_1)]$$

Hence

$$\int_a^b f(x)\,\mathrm{d}x \cong \frac{1}{2} h\, [f(x_0) + f(x_1)] \tag{17.7}$$

which is the *trapezoidal rule*. It is shown geometrically in Figure 17.3.
For our four examples, the trapezoidal rule yields respectively

$I_1 \cong \frac{1}{2} \cdot 2[0 + 4] = 4$ $\qquad$ $I_2 \cong \frac{1}{2} \cdot 1[1 + 0] = 0.5$

$I_3 \cong \frac{1}{2} \cdot \frac{\pi}{4}[\frac{1}{\sqrt{2}} + 0] = 0.2777$ (4 d.p.) $\qquad$ $I_4 \cong \frac{1}{2} \cdot 2[\,1 + \frac{1}{9}] = 1.1111$ (4 d.p.)

These results do not look very impressive.

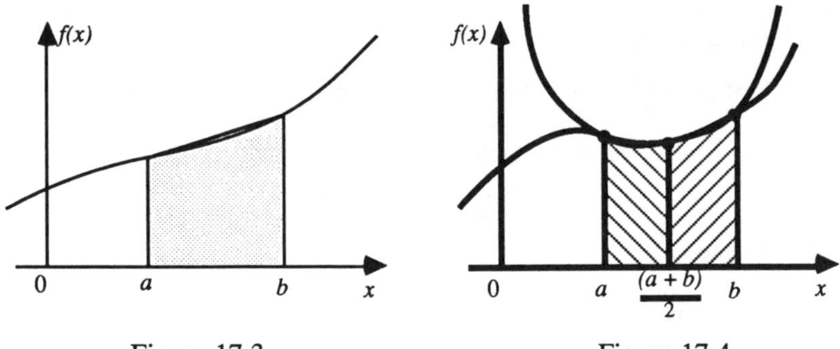

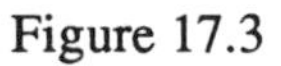

Figure 17.3 $\qquad$ Figure 17.4

Simpson's rule

The Lagrange form of the interpolating polynomial of second degree which agrees with $f(x)$ at $x_0 = a$, $x_1 = \frac{1}{2}(a + b)$, $x_2 = b$ is

$$p_2(x) = \frac{(x-x_1)(x-x_2)}{(x_0-x_1)(x_0-x_2)}f(x_0) + \frac{(x-x_0)(x-x_2)}{(x_1-x_0)(x_1-x_2)}f(x_1)$$
$$+ \frac{(x-x_0)(x-x_1)}{(x_2-x_0)(x_2-x_1)}f(x_2) \tag{17.8}$$

Note that $x_1 = x_0 + h$, where $h = \frac{1}{2}(x_2 - x_0)$.

It is easy to check that (17.8) can be written as

$$p_2(x) = w_0 f(x_0) + w_1 f(x_1) + w_2 f(x_2)$$

where $w_0 + w_1 + w_2 = 1$

and to verify that $p_2(x)$ is indeed a second degree polynomial agreeing with $f(x)$ at the required points.

Integrating $p_2(x)$ between x_0 and x_2 leads to the approximation

$$\int_a^b f(x)\mathrm{d}x \cong \frac{1}{3}h[f(x_0) + 4f(x_1) + f(x_2)] \tag{17.9}$$

This is the well-known *Simpson's rule*, illustrated in Figure 17.4. Note that h is half the value of that for the trapezoidal rule.

Simpson's rule applied to the four examples gives

$$I_1 \cong \frac{1}{3} \cdot 1[0^2 + 4 \times 1^2 + 2^2] = 8/3$$

$$I_2 \cong \frac{1}{3} \cdot \frac{1}{2}[1 + 4 \times \frac{7}{8} + 0) = 0.75$$

$$I_3 \cong \frac{1}{3} \cdot \frac{\pi}{8}\left[\frac{1}{\sqrt{2}} + 4 \times 0.38268 + 0\right] = 0.2929 \quad \text{(4 d.p.)}$$

$$I_4 \cong \frac{1}{3} \cdot 1\left[1 + 4 \times \frac{1}{4} + \frac{1}{9}\right] = 0.7037 \quad \text{(4 d.p.)}$$

We see at once how much more accurate the Simpson's rule is than the trapezoidal rule. We would have expected to get an exact answer for I_1 because the interpolating polynomial coincides with the given function $f(x)$, but it may come as a surprise to find that we also get the correct result for I_2, since we are in effect approximating a cubic function by a quadratic.

The error in any of our estimates so far can be attributed to two causes: (i) round-off error, which is an ever-present worry in any numerical calculation, and (ii) truncation error, due to the fact that the function $f(x)$ is being approximated by a polynomial of low degree. At the moment, round-off error is not significant and truncation error is responsible for virtually all the error in our estimates.

It can be shown that the truncation error in the trapezoidal rule is proportional to h^2, whereas that for Simpson's rule is proportion to h^4.

We define the degree of accuracy of an integration formula as the positive integer n such that all polynomials of degree $\leq n$ are integrated exactly via the formula, whereas at least one polynomial of degree $(n+1)$ is not. We see that the trapezoidal rule has degree of accuracy 1 and Simpson's rule has degree 3.

The following program carries out Simpson integration.

```
>
 10  CLS
 20  PRINTTAB(20,1)"Simpson Integration"
 30  DEF FNA(X)=1/(3*X^3+2)
 40  INPUTTAB(0,5)"Enter (even) number of strips ",N
 50  N=INT(N/2)*2
 60  PRINT N;" Strips"
 70  INPUT"Enter limit A ",A
 80  INPUT"Enter limit B ",B
 90  H=(B-A)/N
100  S=0:X=A
110  F1=FNA(X)
120  FOR I=2 TO N STEP 2
130    X=X+H
140    F2=FNA(X)
150    X=X+H
160    F3=FNA(X)
170    S=S+F1+4*F2+F3
180    F1=F3
190    NEXT I
200  S=S*H/3
210  PRINT "The solution is ";S
220  END
```

17.2 Errors in the Newton-Cotes Formulae

We consider first the trapezoidal rule. Suppose the integrand $f(x)$ has an indefinite integral $F(x)$, i.e. $F'(x) = f(x)$; then

$$I = \int_{x_0}^{x_1} f(x)\mathrm{d}x = F(x_1) - F(x_0)$$

The trapezoidal approximation is $I_T = h[f(x_0) + f(x_1)]$

where $h = \frac{1}{2}(x_1 - x_0)$ and the error in the approximation is $\varepsilon_T = I_T - I$.

Assuming that the function $f(x)$ can be expanded in a Taylor series about $x = x_0$ then

$$f(x_1) = f(x_0) + hf'(x_0) + \frac{h^2}{2!}f''(x_0) + \frac{h^3}{3!}f'''(x_0) + \dots$$

Similarly,

$$F(x_1) = F(x_0) + hF'(x_0) + \frac{h^2}{2!}F''(x_0) + \frac{h^3}{3!}F'''(x_0) + \dots$$

$$= F(x_0) + hf(x_0) + \frac{h^2}{2!}f'(x_0) + \frac{h^3}{3!}f''(x_0) + \dots$$

Therefore $$\varepsilon_T = \frac{h}{2}\left[2f(x_0) + hf'(x_0) + \frac{h^2}{2!}f''(x_0) + \frac{h^3}{3!}f'''(x_0) + \dots\right]$$

$$-\left[hf(x_0) + \frac{h^2}{2!}f'(x_0) + \frac{h^3}{3!}f''(x_0) + \dots\right]$$

i.e. $$\varepsilon_T = -\frac{h^3}{12}f''(x_0) + 0(h^4) \qquad (17.10)$$

We can employ the Mean Value Theorem to quote the following result

$$\varepsilon_T = -\frac{h^3}{12}f''(\xi) \quad \text{where } x_0 \le \xi \le x_1 \qquad (17.11)$$

The difficulty is that we are not told **where** ξ is in the interval $[x_0, x_1]$, we are merely told that such a ξ exists. We proceed a little further by noting that

$$|\varepsilon_T| < \frac{h^3}{12}M_T \qquad (17.12)$$

where M_T is the maximum value of $|f''(x)|$ as x moves through the interval $[x_0, x_1] = [a, b]$. If $f''(x)$ is positive throughout the interval $[a, b]$ then we can say that

$$\varepsilon_T < \frac{h^3}{12}M_T$$

where M_T is the maximum value of $f''(x)$ in $[a, b]$.

A similar consideration applies if $f''(x) < 0$ throughout $[a, b]$. For example,

$$I_3 = \int_{\pi/4}^{\pi/2} \cos x \, dx = 0.2929 \quad \text{(4 d.p.)}$$

We have found that $I_T = 0.2777$ (4 d.p.) so that

$$\varepsilon_T = I_T - I \cong -0.0152$$

(Since $f(x) = \cos x$, $f''(x) = -\cos x$ and $M_T = \cos \pi/4 = 1/\sqrt{2}$.)

Formula (17.12) predicts that

$$|\varepsilon_T| < \frac{1}{12} \cdot \left[\frac{\pi}{4}\right]^3 \cdot \frac{1}{\sqrt{2}} = 0.0285 \quad \text{(4 d.p.)}$$

which is clearly true, although somewhat pessimistic.

In a similar fashion, the error in the Simpson rule may be shown to be given by

$$\varepsilon_s = I_s - I = -\frac{h^5}{90} f^{(iv)}(\xi) \tag{17.13}$$

where $x_0 \le \xi \le x_1$.

Then

$$|\varepsilon_s| < \frac{h^5}{90} M_s \tag{17.14}$$

where M_s is the maximum value of $|f^{(iv)}(x)|$ in $[x_0, x_1]$. Bear in mind that, unless we can evaluate easily M_T and M_s, the formulae (17.12) and (17.14) are of little use.

Corrected trapezoidal rule

Provided that we can differentiate $f(x)$ once, we have a simple modification available known as the *corrected trapezoidal rule.*

$$\int_{x_0}^{x_1} f(x)\mathrm{d}x \cong \frac{1}{2} h[f(x_0) + f(x_1)] - \frac{1}{12} h^2[f'(x_1) + f'(x_0)] \tag{17.15}$$

For example, with $I_1 = \int_0^2 x^2 \, \mathrm{d}x = \frac{8}{3}$, $f'(x) = 2x$

and the correction is $\frac{1}{12} \cdot 2^2 . [4 - 0] = \frac{4}{3}$ so that the corrected trapezoidal rule

gives the value $4 - \frac{4}{3} = \frac{8}{3}$ which is exact.

For $I_2 = \int_0^1 (1 - x^3)\mathrm{d}x = 0.75$, $f'(x) = -3x^2$ and the correction of

$\frac{1}{12} \cdot 1^2[-3 + 0] = -\frac{1}{4}$ leads to the new approximation $0.5 + 0.25 = 0.75$, which is again exact.

The third example, $I_3 = \int_{\pi/4}^{\pi/2} \cos x \,\mathrm{d}x = 0.2929$ (4 d.p.) has $f'(x) =$

$-\sin x$ and the correction is $\frac{1}{12} \cdot \left[\frac{\pi}{4}\right]^2 \left[-1 + \frac{1}{\sqrt{2}}\right] = -0.0151$ (4 d.p.) and the modified value is $0.2777 + 0.0151 = 0.2928$ (4 d.p.).

For $I_4 = \int_1^3 \frac{1}{x^2}\,\mathrm{d}x = \frac{2}{3}$, $f'(x) = -\frac{2}{x^3}$; with the correction

$\frac{1}{12} \cdot 2^2 \left[-\frac{2}{27} + \frac{2}{1}\right] = 0.6420$ (4 d.p.)

we produce a revised approximation of $1.1111 - 0.6420 = 0.4691$ (4 d.p.).

The general gain in accuracy is apparent.

Comparison of (17.15) with (17.10) shows that we are, in effect, approximating $f''(x_0)$ by $\frac{1}{h}[f'(x_1) - f'(x_0)]$.

17.3 Composite Rules

Perhaps we were a little hasty in Section 17.1 when we compared the trapezoidal approximations unfavourably to the Simpson's estimates. The Simpson rule required three evaluations of the integrand $f(x)$ and the trapezoidal rule required but two. Suppose we divide the interval $[a, b]$ into two strips of equal width as shown in Figure 17.5(a).

Geometrically, it seems that the area under the two trapezoidal elements is closer to the required area under $f(x)$ than is the single trapezoidal. (But perhaps we have chosen a graph to suit our purposes – can you think of an example where our argument is false?)

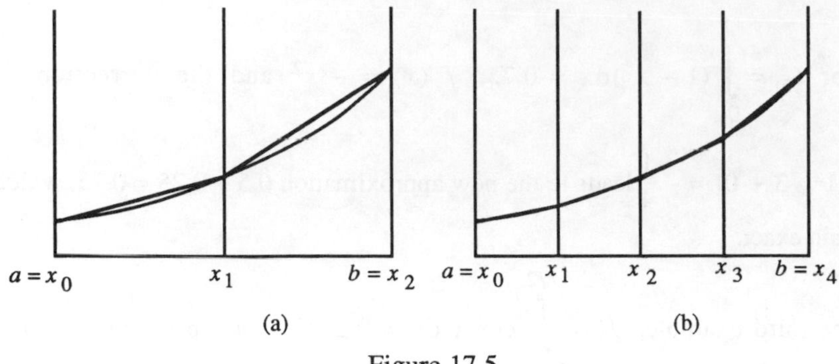

Figure 17.5

In effect we apply the trapezoidal rule once to the interval $[x_0, x_1]$ and once to the interval $[x_1, x_2]$, then we add the results. Therefore, the new estimate we obtain is

$$I_T = \frac{h}{2}[f(x_0) + 2f(x_1) + f(x_2)] \tag{17.16}$$

where $h = \frac{1}{2}(b - a)$.

Applying (17.16) to our four examples,

(i) $I_1 = \int_0^2 x^2 \, dx \cong \frac{1}{2}.1[0 + 2 \times 1 + 4] = 3$

(ii) $I_2 = \int_0^1 (1 - x^3)dx \cong \frac{1}{2} \cdot \frac{1}{2}[1 + 2 \times \frac{7}{8} + 0] = 0.6875$

(iii) $I_3 = \int_{\pi/4}^{\pi/2} \cos x \, dx \cong \frac{1}{2} \cdot \frac{\pi}{8}[0.7071 + 2 \times \cos\frac{3\pi}{8} + 0] = 0.2891$

(iv) $I_4 = \int_1^3 \frac{1}{x^2} dx \cong \frac{1}{2} \cdot 1[1 + 2 \times \frac{1}{4} + \frac{1}{9}] = 0.8056$

where results are quoted to 4 d.p.

We see at once that these estimates are much better than using one trapezoidal, but *for the same computational effort* they are not as accurate as Simpson's rule (17.9). However, we shall continue with the trapezoidal rule for the moment. Suppose we divide the interval $[a, b]$ into 4 strips of equal width as in Figure 17.5(b); then the new rule becomes

$$I_T = \frac{h}{2}\,[f(x_0) + 2f(x_1) + 2f(x_2) + 2f(x_3) + f(x_4)] \tag{17.17}$$

where $h = (b - a)/4$.

Applying (17.17) to three of the examples,

(i) $I_1 \cong \frac{1}{2} \cdot \frac{1}{2}[0 + 2 \times \frac{1}{4} + 2 \times 1 + 2 \times \frac{9}{4} + 4] = 2.75$

(ii) $I_2 \cong 0.7344$ (4 d.p.)

(iii) $I_3 \cong 0.2920$ (4 d.p.)

Note that an efficient means of calculation is to group the central terms, e.g. for the fourth example

$$I_4 \cong \frac{1}{2} \cdot \frac{1}{2}\left[1 + 2\left[\frac{4}{9} + \frac{1}{4} + \frac{4}{25}\right] + \frac{1}{9}\right] = 0.705$$

It is instructive to see what happens to the errors in the approximations as we progress from 1 through 2 to 4 strips. Table 17.1 indicates that doubling the number of strips divides the error by 4 (exactly or approximately). Results are quoted to 4 d.p.

Table 17.1

Error\Problem	(I_1)	(I_2)	(I_3)	(I_4)
1 strip	1.3333	–0.2500	–0.0152	0.4444
2 strips	0.3333	–0.0625	–0.0038	0.1389
4 strips	0.0833	–0.0156	–0.0009	0.0383

We quote a formula for the general composite trapezoidal rule based on N equal sub-intervals

$$\int_a^b f(x)\mathrm{d}x \cong \frac{1}{2}h\left[f_0 + 2\sum_{j=1}^{N-1} f_j + f_N\right] \tag{17.18}$$

where $h = (b - a)/N$, $x_0 = a$, $x_j = x_0 + jh$, $f_j = f(x_j)$.

The maximum error in the trapezoidal estimate is given by

$$|\varepsilon_T| \le \frac{(b-a)}{12}h^2 M_T \tag{17.19}$$

where M_T is the greatest value of $|f''(x)|$ in $[a, b]$ with the same proviso as was attached to (17.12).

The maximum error in approximating I_3 via (17.18) with $N = 4$ is

$$\frac{1}{12} \cdot \left[\frac{\pi}{4}\right] \cdot \left[\frac{\pi}{6}\right]^2 \cdot \frac{1}{\sqrt{2}} = 0.0018 \quad \text{(4 d.p.)}$$

and the actual error is smaller by a factor approaching 2. Generally, we can expect this global error formula to be more pessimistic as the number of strips increases; to apply (17.12) to each of the N strips is too time-consuming.

We can use (17.19) in a reverse sense to help us to decide the number of strips required to guarantee a specific accuracy. Suppose we want to estimate I_2 correct to 2 d.p., i.e. we require $|\varepsilon_T| < 0.005$ and, since $(b-a) = 2$ is fixed,

$$|\varepsilon_T| \leq \frac{(b-a)}{12} \,.\, h^2 \,.\, M_T \leq 0.005 \qquad \text{becomes} \qquad \frac{2}{12} \,.\, h^2 \,.\, 6 < \frac{1}{200}$$

and since $Nh = (b-a)$ it follows that

$$N^2 > 200 \quad \text{or} \quad N \geq 14.14 \quad (2 \text{ d.p.})$$

But N must be an integer so that we choose $N = 15$ (or perhaps 16 for convenience of calculating h.)

The estimate of I_2 obtained on a BBC B microcomputer was 0.7489 (4 d.p.), which is within the requirement on accuracy. In fact, this accuracy was achieved using only 8 strips; this is, of course, the price we pay for the relatively simple formula (17.19).

Suppose we now require the trapezoidal rule to estimate I_2 correct to 6 d.p.; this leads to the requirement that

$$N^2 > 2 \times 10^6 \quad \text{or} \quad N > 1414.2, \quad \text{say } N = 1415$$

The trouble is that so large a number of strips would involve 1416 function evaluations, at least two multiplications and some 1415 additions. The potential for round-off error is large and could well off-set the decreased truncation error. Figure 17.6 on page 467 shows the total error as a function of the number of strips. The composite Simpson's rule is stated as

$$\int_a^b f(x)\mathrm{d}x \cong \frac{1}{3}h\left[f_0 + 2\sum_{j=1}^{N-1} f_{2j} + 4\sum_{j=1}^{N} f_{2j-1} + f_{2N}\right] \qquad (17.20)$$

where $h = (b-a)/2N$, $x_0 = a$, $x_j = x_0 + jh$, $f_j = f(x_j)$.

Note that the number of strips, $2N$, is an *even* number.

For example, in the case $N = 4$,

$$\int_{x_0}^{x_8} f(x)\mathrm{d}x \cong \frac{1}{3}h[f_0 + 4f_1 + f_2] + \frac{1}{3}h[f_2 + 4f_3 + f_4]$$

$$+ \frac{1}{3}h[f_4 + 4f_5 + f_6] + \frac{1}{3}h[f_6 + 4f_7 + f_8]$$

$$= \frac{1}{3}h[f_0 + 2[f_2 + f_4 + f_6] + 4[f_1 + f_3 + f_5 + f_7] + f_8]$$

Problem 4 shows that doubling the number of strips approximately divides the error by 16. The maximum error in the Simpson's estimate is given by

$$|\varepsilon_s| < \frac{(b-a)}{180} h^4 M_s \qquad (17.21)$$

where M_s is the maximum value of $|f^{(iv)}(x)|$ in $[a, b]$.

To estimate I_4 correct to 5 d.p., (17.21) can be shown to lead to the requirement that the number of strips must exceed 12.78 (2 d.p.) and since it must be even we choose 14 strips so that $N = 7$. (By contrast we found that the number of strips for the trapezoidal rule to guarantee the same accuracy was 283). Figure 17.6 shows graphs of total error against the number of strips for both trapezoidal and Simpsons' rules.

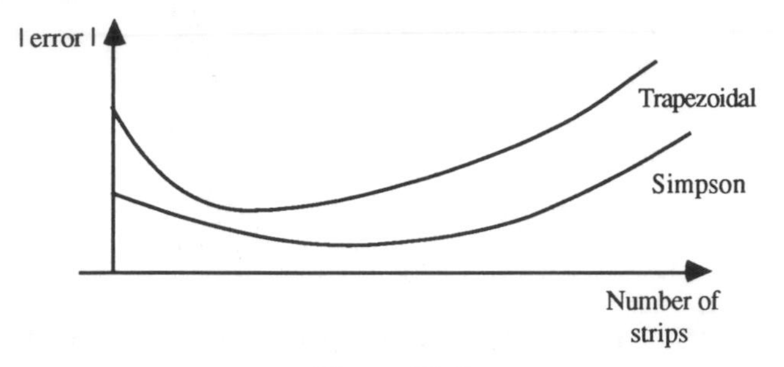

Figure 17.6

17.4 Analytical Methods of Integration

In this section we look briefly at some of the calculus-based methods of integration. It is assumed that you are familiar with most of the basic techniques and we have provided some problems, numbers 21 to 23, to give you some revision practice. We have included here some examples which may be new to you.

Example 1 $I = \int 3 \ln x \, dx$

The trick here is to use the method of integration by parts.

$$I = \int \ln x \, . \, 3dx = \ln x \, . \, 3x - \int \frac{1}{x} \, 3x \, dx$$

$$= 3x \ln x - 3x + C$$

Example 2 $I = \int \cos^{-1} x \, dx = \int \cos^{-1} x \, . \, 1 \, . \, dx$

$$= \cos^{-1} x \, . \, x - \int \frac{-1 \, . \, x}{\sqrt{1 - x^2}} dx$$

$$= x \cos^{-1} x - (1 - x^2)^{1/2} + C$$

Example 3 $$I = \int_0^{\pi/2} \sin^4 x \cos^3 x \, dx$$

There is an 'overlap' of $\cos x$ here. The clue is the pattern $\cos x\, \mathrm{d}x$. We substitute $s = \sin x$ so that $\dfrac{\mathrm{d}s}{\mathrm{d}x} = \cos x$ and $\cos^2 x = 1 - s^2$. When $x = 0$, $s = 0$ and when $x = \pi/2$, $s = 1$. Then

$$I = \int_0^1 s^4(1 - s^2)\mathrm{d}s = \left[\frac{s^5}{5} - \frac{s^7}{7}\right]_0^1 = \frac{2}{35}$$

Example 4

$$I = \int \frac{1}{a + b\cos^2\theta + c\sin^2\theta}\,\mathrm{d}\theta$$

For a function of $\cos^2\theta$ and/or $\sin^2\theta$ we use the substitution $t = \tan\theta$

Put $t = \tan\theta$, therefore

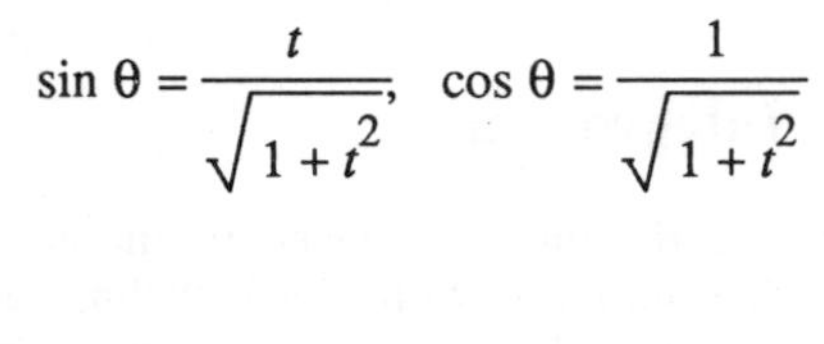

$$\sin\theta = \frac{t}{\sqrt{1 + t^2}}, \quad \cos\theta = \frac{1}{\sqrt{1 + t^2}}$$

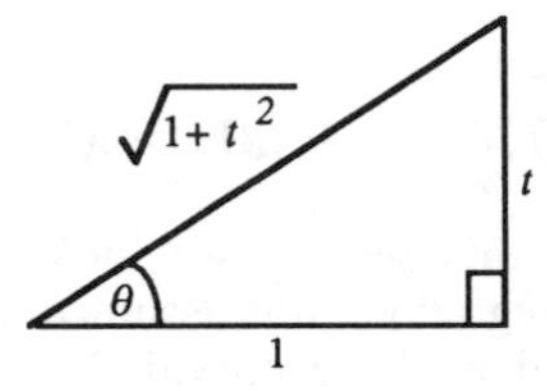

$$\frac{\mathrm{d}t}{\mathrm{d}\theta} = \sec^2\theta = 1 + t^2$$

therefore $$\frac{\mathrm{d}\theta}{\mathrm{d}t} = \frac{1}{1 + t^2}$$

therefore $$I = \int \frac{\mathrm{d}t}{a(1 + t^2) + b + ct^2}$$

which can be evaluated by standard techniques.

Example 5

$$I = \int \frac{1}{a + b\cos\theta + c\sin\theta}\,\mathrm{d}\theta$$

For a function of $\sin\theta$ and $\cos\theta$ we use the substitution $t = \tan\theta/2$

Put $t = \tan\frac{1}{2}\theta$, therefore

$$\sin\frac{\theta}{2} = \frac{t}{\sqrt{1+t^2}}, \quad \cos\frac{\theta}{2} = \frac{1}{\sqrt{1+t^2}}$$

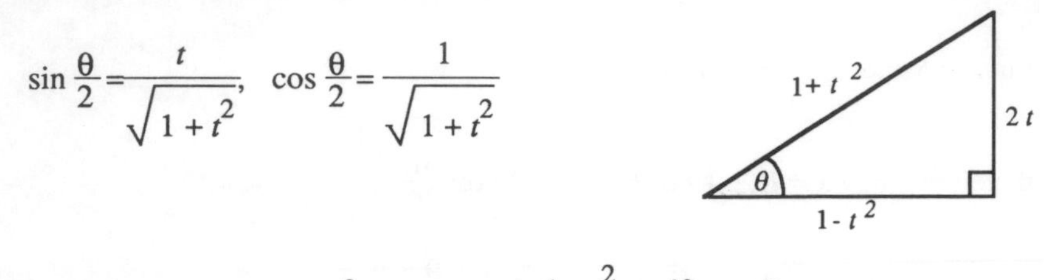

therefore $\quad \sin\theta = \dfrac{2t}{1+t^2} \quad \cos\theta = \dfrac{1-t^2}{1+t^2} \quad \dfrac{d\theta}{dt} = \dfrac{2}{1+t^2}$

therefore

$$I = \int \frac{2dt}{(1+t^2)\left[a + b\dfrac{(1-t^2)}{(1+t^2)} + c\,.\,\dfrac{2t}{1+t^2}\right]}$$

$$= \int \frac{2dt}{a(1+t^2) + b(1-t^2) + 2ct}$$

$$= \int \frac{2dt}{t^2(a-b) + 2ct + (a+b)}$$

which can be evaluated by standard techniques.

Example 6

Assume the earth is a sphere of radius R, producing a force on a particle outside it such that the acceleration of the particle varies inversely as the square of its distance from the earth's centre. The value of this acceleration at the earth's surface is denoted by g. The particle is released from rest at a height R above the earth: when it is at height x above the earth its speed is $\sqrt{[gR(R-x)/(R+x)]}$. Integrate by using $x = R\cos\theta$ to find the time taken to reach the earth.

Now $\dfrac{dx}{dt} = -\left[\dfrac{gR(R-x)}{x+R}\right]^{1/2}$. Let T be time to reach earth.

Therefore $T = -\displaystyle\int_R^0 \left[\frac{R+x}{gR(R-x)}\right]^{1/2} dx$

Put $x = R\cos\theta$; $\dfrac{dx}{d\theta} = -R\sin\theta$

therefore $R + x = R(1 + \cos\theta) = R\,.\,2\cos^2\dfrac{\theta}{2}$

and $R - x = R(1 - \cos\theta) = R\,.\,2\sin^2\dfrac{\theta}{2}$

When $x = 0$, $\cos\theta = 0 \Rightarrow \theta = \dfrac{\pi}{2}$

$x = R$, $\cos\theta = 1 \Rightarrow \theta = 0$
(continuity of substitution)

therefore

$$T = -\int_0^{\pi/2}\left[\frac{R\,.\,2\cos^2\theta/2}{gR\,.\,R\,.\,2\sin^2\theta/2}\right]^{1/2}(-R\sin\theta)d\theta$$

$$= -\sqrt{\frac{R}{g}}\int_0^{\pi/2} -\frac{\cos\theta/2\,.\,\sin\theta}{\sin\theta/2}d\theta$$

$$= \sqrt{\frac{R}{g}}\int_0^{\pi/2}\frac{\cos\theta/2\,.\,2\sin\theta/2\,.\,\cos\theta/2}{\sin\theta/2}d\theta$$

$$= \sqrt{\frac{R}{g}}\int_0^{\pi/2} 2\cos^2\frac{\theta}{2}d\theta$$

$$= \sqrt{\frac{R}{g}}\int_0^{\pi/2}(1 + \cos\theta)d\theta$$

$$= \sqrt{\frac{R}{g}}\Big[\theta + \sin\theta\Big]_0^{\pi/2}$$

$$= \sqrt{\frac{R}{g}}\left(\frac{\pi}{2} + 1\right)$$

If you noticed, the integrand became infinite at $x = R$.

Infinite Integrals

We now consider cases where there is a *singularity* in the range of integration or where the range of integration is *infinite.*

(a) *Infinite Limits*

$$\int_a^\infty f(x)\mathrm{d}x = \lim_{X\to\infty} \int_a^X f(x)\mathrm{d}x$$

Example 1

$$\int_1^\infty \frac{1}{x^3}\mathrm{d}x = \lim_{X\to\infty} \int_1^X \frac{1}{x^3}\mathrm{d}x = \lim_{X\to\infty} \left[-\frac{1}{2x^2}\right]_1^X$$

$$= \lim_{X\to\infty} \left[\frac{1}{2} - \frac{1}{2X^2}\right] = \frac{1}{2}$$

Example 2

$$\int_0^\infty x^{1/2}\,\mathrm{d}x = \lim_{X\to\infty} \int_0^X x^{1/2}\,\mathrm{d}x = \lim_{X\to\infty} \left[\frac{2}{3}x^{3/2}\right]_0^X$$

$$= \lim_{X\to\infty} \left[\frac{2}{3}X^{3/2}\right] \quad \text{which diverges}$$

Likewise

$$\int_{-\infty}^\infty = \lim_{X\to\infty} \int_a^X + \lim_{X\to\infty} \int_{-X}^a \quad \textit{if both exist.}$$

(b) *Suppose f(x) goes infinite in the range of integration*

Example 1

$$\int_0^2 \frac{\mathrm{d}x}{\sqrt{2-x}} = \lim_{\epsilon\to 0} \int_0^{2-\epsilon} \frac{\mathrm{d}x}{\sqrt{2-x}} = \lim_{\epsilon\to 0} \left[-2(2-x)^{1/2}\right]_0^{2-\epsilon}$$

(since when $x = 2$ the integrand is infinite)

$$= 2 \lim_{\epsilon\to 0} \left[2^{1/2} - \epsilon^{1/2}\right] = 2 \,.\, \sqrt{2}$$

Example 2

$$\int_{-1}^{1} x^{-1/3}\,dx = \lim_{\epsilon\to 0}\int_{-1}^{-\epsilon} x^{-1/3}\,dx + \lim_{\epsilon\to 0}\int_{\epsilon}^{1} x^{-1/3}\,dx$$

(since when $x = 0$ the integrand is infinite)

$$= \lim_{\epsilon\to 0}\left[\frac{3}{2}x^{2/3}\right]_{-1}^{-\epsilon} + \lim_{\epsilon\to 0}\left[\frac{3}{2}x^{2/3}\right]_{\epsilon}^{1}$$

$$= \lim_{\epsilon\to 0}\left[\frac{3}{2}\epsilon^{2/3} - \frac{3}{2}\right] + \lim_{\epsilon\to 0}\left[\frac{3}{2} - \frac{3}{2}\epsilon^{2/3}\right]$$

$$= -\frac{3}{2} + \frac{3}{2} = 0$$

Problems

Sections 17.1, 17.2

1 Use formulae (17.5), (17.7) and (17.9) to estimate the values of each of the following definite integrals. Where possible, calculate the value analytically and compare the actual error with the maximum error bound.

(a) $\int_1^2 e^x\,dx$ (b) $\int_1^4 e^x\,dx$ (c) $\int_1^{10} e^x\,dx$

(d) $\int_1^2 \frac{1}{x}dx$ (e) $\int_1^4 \frac{1}{x}dx$ (f) $\int_1^{10} \frac{1}{x}dx$

(g) $\int_0^2 \sqrt{x}\,dx$ (h) $\int_0^4 \sqrt{x}\,dx$ (i) $\int_0^{10} \sqrt{x}\,dx$

(j) $\int_0^2 e^{-x}\,dx$ (k) $\int_0^4 e^{-x}\,dx$ (l) $\int_0^{10} e^{-x}\,dx$

(m) $\int_0^{\pi/2} \sin x\,dx$ (n) $\int_0^{\pi/2} \cos x\,dx$ (o) $\int_0^{\pi} \sin x\,dx$

(p) $\int_0^{\pi} \cos x\,dx$ (q) $\int_0^{\pi/2} \cos^2 t\,dt$ (r) $\int_0^{\pi/2} \frac{1}{2}(1 + \cos 2t)\,dt$

(s) $\int_0^1 \sqrt{x} \sin x \, dx$ (t) $\int_{-1}^{1} \frac{x^8 \sqrt{1-x^2}}{(2-x)^{6.3}} dx$ (u) $\int_0^{\pi/2} \sin 4x \, dx$

(v) $\int_0^{\pi/2} \sin 10x \, dx$

2 Apply the corrected trapezoidal rule to the integrals in Problem 1. What improvements do you notice?

Section 17.3

3 For the integrals in Problem 1, estimate where possible the number of strips for the composite trapezoidal rule to achieve 1, 2, 3, 4, 5 and 6 decimal places of accuracy. Verify, if possible, that this accuracy is achieved and, by experimenting, find the number of strips actually needed for the accuracy required.

4 Repeat Problem 3 for the composite Simpson's rule.

5 Is it more efficient to estimate $I = \int_0^3 \cosh x \, dx$ directly or to employ the identity

$$I = \frac{1}{2}\left[\int_0^3 e^x \, dx + \int_0^3 e^{-x} \, dx\right] ?$$

Does the same conclusion hold for $\int_0^3 \sinh x \, dx$?

6 The function $f(x)$ is given by the table below. Obtain the best estimate you can for

$$\int_0^{0.8} f(x) \, dx$$

x	0	0.2	0.4	0.6	0.8
$f(x)$	1.0000	1.0811	1.3374	1.8107	2.5775

Knowing that $f(x)$ is cosh $2x$, comment on your results.

7 Estimate $\int_1^{\infty} \frac{1}{x^2} dx$ and compare with the analytical result; repeat for $\int_1^{\infty} \frac{1}{x^4} dx$

8 Obtain estimates of $I = \frac{2}{\sqrt{\pi}} \int_0^x e^{-t^2}\, dt$ for $x = 0, 0.1, 0.2, \ldots, 3$. What do your results suggest for $x \to \infty$?

9 Estimate $I_1 = \int_0^{2\pi} \ln(1 + x) \sin 8x \, dx$ and $I_2 = \int_0^{2\pi} e^{-2x} \sin 8x \, dx$

10 What error is involved when

$$\int_0^1 \frac{dx}{1 + x^2}$$

is evaluated to 4 d.p. by Simpson's rule dividing the interval into four equal parts? (EC)

11 Determine numerically

$$\int_0^{\pi/4} \sin x \, dx$$

by using Simpson's rule with a step length so chosen that the truncation error is less than 0.001. Find a bound for the round-off error in your answer. (EC)

12 Evaluate

$$\int_0^1 \sqrt{x} \, dx$$

by using Simpsons' rule with 4 strips.
Put $x = \sin^2 \theta$ and evaluate the new integral, again with 4 strips. Why might you expect the second result to be better than the first?

13 The flow across part of a boundary layer in a viscous fluid is given by the definite integral

$$\int_0^{0.8} 1.4(1 - \exp(-4x^2))dx$$

Evaluate this integral using Simpson's rule with 4 intervals. Perform all calculations to five decimal places. (EC)

14 A particle moves along a straight line so that at time t its distance s from a fixed point of the line is given by

$$\frac{ds}{dt} = t\sqrt{8 - t^3}$$

Use Simpson's rule with 8 strips to calculate the approximate distance travelled by the particle from $t = 0$ to $t = 2$. (LU)

15 Apply Simpson's rule with 8 strips to evaluate the integral $\int_0^{0.4} \sqrt{1+x^3}\,dx$

using four figure tables in your calculations.
Obtain an approximate value of this integral by using the binomial expansion of $(1+x^3)^{1/2}$ as far as the term in x^6 and integrating this expansion. (LU)

16 The quarter-perimeter of the ellipse $\frac{x^2}{a^2}+\frac{y^2}{b^2}=1$ is $\int_0^a \sqrt{\frac{a^2-m^2x^2}{a^2-x^2}}\,dx$

where $m^2 = 1 - b^2/a^2$.
Use infinite series to evaluate the integral in the case when $b = 0.8a$.

17 The rate at which radiant energy leaves unit area of the surface of a black body depends upon the wavelength. For wavelengths in the range λ m to $(\lambda + d\lambda)$ m the rate is given by

$$E_\lambda \, d\lambda = \frac{2\pi hc^2}{\lambda^5} \frac{d\lambda}{e^{hc/k\lambda T} - 1}$$

where $h = 6.6256 \times 10^{-34}$ Js (Planck's constant)
$c = 2.9979 \times 10^8$ ms^{-1} (speed of light)
$k = 1.3805 \times 10^{-23}$ JK^{-1} (Boltzmann's constant)
T = absolute temperature in °K

The Stefan-Boltzmann law states that the total rate of energy radiated at an absolute temperature T °K is

$$E = \int_0^\infty E_\lambda \, d\lambda = \sigma T^4$$

where $\sigma = 5.6697 \times 10^{-8}$ Wm^{-2} K^{-4} (Stefan's constant).
Is the Stefan-Boltzmann law valid?
The luminous efficiency of a surface is the ratio of the energy radiated in the visible spectrum $(4 \times 10^{-7} \le \lambda \le 7 \times 10^{-7})$ to the total energy radiated.
Find the luminous efficiency for T = 100°K (50°K) 400°K.

18 Find the value of the integral I_α in the introductory example to this chapter, for $\alpha = 0\ (\pi/12)2\pi/3$. For each value of α calculate the displacements x_α and y_α of the upper end relative to the fixed end by noting that $\frac{dx}{ds} = \cos\theta$.

The critical load P_{cr} at which buckling occurs is given by $P_{cr} = \frac{\pi^2 EI}{4\ell^2}$

Hence for each value of α calculate the ratio P/P_{cr}.

19 The Debye function, which occurs in thermodynamics, can be expressed as

$$D(x) = \frac{3}{x^3}\int_0^x \frac{y^3}{e^y - 1}\,dy$$

Evaluate the function when $x = 0.5, 1, 10, 20, 50, 80, 100$.

20 A simple pendulum comprises a heavy bob of mass m at the end of a rigid bar of negligible weight which is attached to a fixed point. If θ is the angle made by the bar with the downwards vertical, produce a table of θ against time for different initial angular displacements θ_0. Do not assume that $\sin\theta$ can be approximated by θ.

Section 17.4

21 Use suitable trigonometric substitutions to evaluate:

(i) $\displaystyle\int_0^1 \frac{dx}{(x^2+1)^{3/2}}$ (ii) $\displaystyle\int_0^1 \sqrt{4 - x^2}\,dx$

22 Find $\int xe^{(1+i)x}\,dx$ and hence obtain $\int xe^x \cos x\,dx$ and $\int xe^x \sin x\,dx$

23 Find $\int (x^2 - 4)^{1/2}\,dx$ and $\int x^3(x^2 - 4)^{1/2}\,dx$

24 Evaluate (i) $\int x \tan^{-1} x\,dx$ (ii) $\displaystyle\int_0^{\pi/2} \frac{dx}{3\cos x + 4\sin x + 5}$

25 (i) Evaluate the integrals

(a) $\displaystyle\int_0^1 \frac{(x+1)dx}{x^2 - 2x + 4}$ (b) $\displaystyle\int_0^{\pi} e^x \sin(x/2)dx$ (c) $\displaystyle\int_0^1 \frac{e^x\,dx}{2 + e^x}$

(ii) In determining the fugacity of a gas which obeys Van der Waal's equation we must find

$$\int \left[\frac{2a}{RTV^2} - \frac{V}{(V-b)^2}\right] dV$$

where R, a, b and T are constants. Evaluate this, using the substitution $z = (V - b)$ in the second part of the integrand.

26 A spherical surface of radius a metres, with centre O, carries an electric charge of surface density σ coulombs/square metre, symmetrically distributed with respect to an axis OX. The potential at a point P on this axis, such that OP = x metres, is given by

$$\int_0^{\pi} \frac{a^2 \sigma \sin\theta}{2\varepsilon_0(a^2 + x^2 - 2ax\cos\theta)^{1/2}} \mathrm{d}\theta \text{ volts}$$

Evaluate the integral, with ε_0 and σ constant, in the case when $x > a$.

27 Evaluate $\int \sqrt{x} \ln x \, \mathrm{d}x$

28 (i) Evaluate (a) $\int_1^2 \frac{\mathrm{d}x}{\sqrt{1 + 2x - x^2}}$ (b) $\int_0^{\pi/2} \cos^2 x \sin^5 x \, \mathrm{d}x$

(ii) Find $C = \int x\mathrm{e}^{-x} \cos x \, \mathrm{d}x$ and $S = \int x\mathrm{e}^{-x} \sin x \, \mathrm{d}x$

(Hint: Consider $C + \mathrm{i}S$)

29 Find the indefinite integrals of $\frac{x}{(x^2 + x + 1)^{1/2}}$ and $\frac{x}{(x^2 + x + 1)}$

30 Evaluate (i) $\int_2^3 \frac{x}{(x-1)(x+2)} \mathrm{d}x$ (ii) $\int_{-2}^0 \frac{1}{\sqrt{x^2 + 4x + 5}} \mathrm{d}x$

31 Integrate with respect to x: (i) $x^2 \mathrm{e}^{-x}$ (ii) $\frac{\sqrt{1 - x^2}}{1 + x}$

32 Evaluate (i) $\int_0^1 \frac{x + 2}{x^2 - 2x + 4} \mathrm{d}x$ (ii) $\int \sec^3 x \, \mathrm{d}x$

33 Write down the expressions for $\cos\theta$ and $\sin\theta$ in terms of $\mathrm{e}^{\mathrm{i}\theta}$ and $\mathrm{e}^{-\mathrm{i}\theta}$ and use these forms to expand $\sin^4 x$ in cosines of multiples of x. Hence evaluate $\int_0^{\pi/4} \sin^4 x \, \mathrm{d}x$

34 The force on a body distant x from a fixed point is $k(x^2 + a^2)^{-3/2}$. Find the work done in moving the body from $x = a$ to $x = 2a$.

35 Evaluate the integrals

(i) $\int_0^{\pi} e^{-x} \cos 3x \, dx$ (ii) $\int \frac{1}{\sqrt{2x^2 + 3x - 1}} dx$ (iii) $\int \sinh^3 x \, dx$

36 Show that

(i) $\int_0^a f(x)dx = \int_0^a f(a-x)dx$

(ii) $\int_0^{\pi/2} \ln \sin x \, dx = \int_0^{\pi/2} \ln \cos x \, dx = \frac{1}{2}\int_0^{\pi/2} \ln \sin 2x \, dx - \frac{\pi}{4} \ln 2$

Deduce that $\int_0^{\pi/2} \ln \sin x \, dx = -\frac{\pi}{2} \ln 2$ (LU)

37 (a) Evaluate the integral $\int_3^4 \frac{3x-4}{(x-2)(x^2+4)} dx$ correct to 3 significant figures.

(b) Evaluate

$$I = \int_0^{0.1} x \sin x \, dx$$

to 5 decimal places using integration by parts. Then employ a Maclaurin series for $\sin x$ to obtain an infinite power series for the integrand and use term-by-term integration showing intermediate steps to verify your earlier result. (EC)

38 (i) Use integration by parts to show that

$$\int \exp(ax) \cos bx \, dx = \frac{1}{a} \exp(ax) \cos bx + \frac{b}{a} \int \exp(ax) \sin bx \, dx$$

and derive a similar result for $\int \exp(ax) \sin bx \, dx$

(ii) Use these two results to find the 'standard integral' result for $\int \exp(ax) \cos bx \, dx$

(iii) Find the value of $\int_0^{\infty} \exp(-3x) \cos 2x \, dx$ correct to 2 decimal places. (EC)

39 (a) Show that

$$\int_0^{\infty} x \exp(-x^2) dx = \frac{1}{2}$$

It can be shown that, if $I_n = \int_0^{\infty} x^n \exp(-x^2)\mathrm{d}x$

then $(n+1)I_n = 2I_{n+2} \qquad n \geq 1$

Use this result to find $\int_0^{\infty} x^5 \exp(-x^2)\mathrm{d}x$

(b) Use the substitution $x = \sinh t$ to determine the indefinite integral

$$\int \frac{x^2}{\sqrt{x^2+1}}\mathrm{d}x$$

(EC)

40 Show that

$$\int_0^{\pi/2} x^n \cos x \,\mathrm{d}x = \left[\frac{\pi}{2}\right]^n - n\int_0^{\pi/2} x^{n-1} \sin x \,\mathrm{d}x \qquad (n \geq 1)$$

Hence show further that, if $I_n = \int_0^{\pi/2} x^n \cos x \,\mathrm{d}x$ then

$$I_n = \left[\frac{\pi}{2}\right]^n - n(n-1)I_{n-2} \qquad (n \geq 2)$$

Find $\int_0^{\pi/2} x^6 \cos x \,\mathrm{d}x$ correct to 2 decimal places. (EC)

41 (a) Use the substitution $x = a \tan \theta$ to show that

$$\int \frac{\mathrm{d}x}{(x^2+a^2)^{3/2}} = \frac{1}{a^2}\sin\left[\tan^{-1}\frac{x}{a}\right]$$

The work done W in moving a certain body from $x = a$ to $x = 2a$ is given by

$$W = \int_a^{2a} \frac{k\,\mathrm{d}x}{(x^2+a^2)^{3/2}}$$

Show that $W = \frac{k}{a^2}\left[\frac{2}{\sqrt{5}} - \frac{1}{\sqrt{2}}\right]$

(b) (i) Show that

$$\int_0^{2\pi} x \sin nx \,\mathrm{d}x = -\frac{2\pi}{n}$$

where n is a positive integer.

(ii) Find, in terms of n, the integral given by

$$a_n = \int_0^{2\pi} x^2 \cos nx \, dx$$

where n is a positive integer and show that $a_{2n} = a_n/4$ (EC)

42 Given that $$I_n = \int_0^{\pi/2} \cos^n x \, dx$$

derive the reduction formula $$I_n = \left[\frac{n-1}{n}\right] I_{n-2} \qquad n \geq 2$$

Hence, after a suitable trigonometrical substitution, evaluate

$$\int_0^1 (1 - y^2)^4 dy$$

correct to 3 decimal places. (EC)

43 Evaluate (a) $\int \sin^{-1} x \, dx$ (b) $\int \tan^{-1} x \, dx$

44 Evaluate

(i) $\int \frac{dx}{3 + 5\cos x}$ (ii) $\int \frac{dx}{1 - \cos x + \sin x}$ (iii) $\int \frac{dt}{1 + \sin^2 t}$

(iv) $\int \frac{dt}{25 - 24\sin^2 t}$ (v) $\int \frac{2 + \sin x}{5\sin x - 4} dx$

45 Consider $\int_\epsilon^1 x \ln x \, dx$ and hence evaluate $\int_0^1 x \ln x \, dx$

46 Evaluate $$\int_0^\infty x \, e^{-ax} \, dx \qquad (a > 0)$$

47 Find (i) $\int_0^1 \frac{dx}{\sqrt{1 - x^2}}$ (ii) $\int_0^1 \frac{dx}{\sqrt{x}}$ (iii) $\int_0^\infty e^{-2\theta} \sin\theta \, d\theta$

48 Given that $\int_0^\infty e^{-x^2} dx = \frac{1}{2}\sqrt{\pi}$, show that $\int_0^\infty x^{-1/2} e^{-x} dx = \sqrt{\pi}$ (EC)

18

APPLICATIONS OF DEFINITE INTEGRATION

In this chapter we study some of the applications which involve the formulation and evaluation of definite integrals.

18.1 Plane Areas and Volumes of Revolution

(a) Area under a curve

The value of

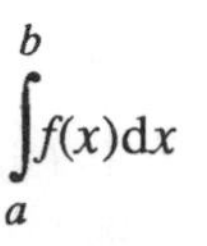

$$\int_a^b f(x)\mathrm{d}x$$

can represent the area under the curve $y = f(x)$ between the ordinates $x = a$ and $x = b$ (Figure 18.1).

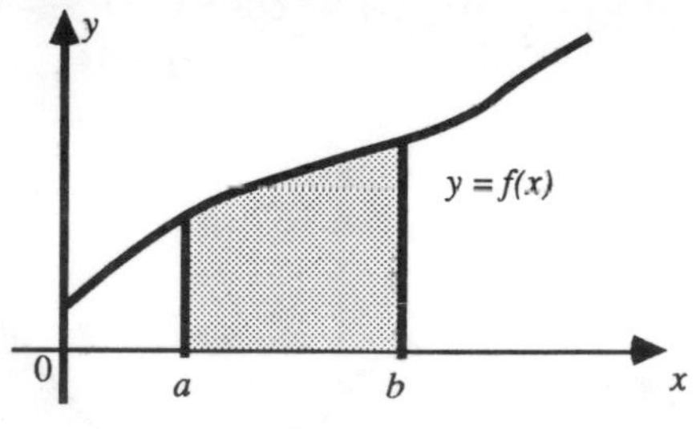

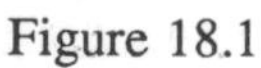

Figure 18.1

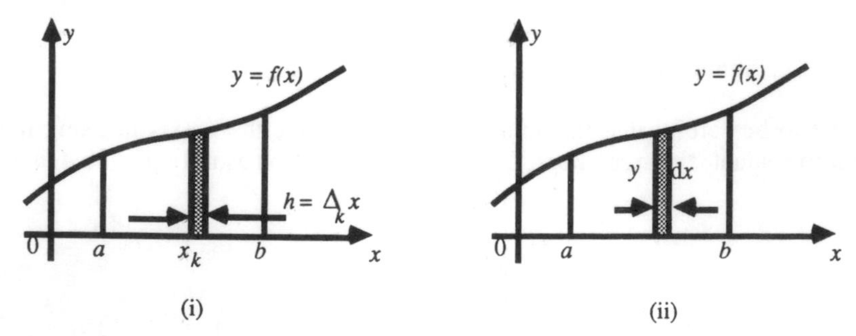

Figure 18.2

The integral is really the limit of the sum of the areas of the strips of width $\Delta_k x$ which can be drawn under the curve – a typical element of area being shown in Figure 18.2(i).

$$\int_a^b f(x)\mathrm{d}x = \lim_{n\to\infty} \sum_{k=1}^{n} f(x_k).\Delta_k x$$

We can therefore say that $\int_a^b y\ \mathrm{d}x$ is the area under the curve $y = f(x)$ between $x = a$ and $x = b$ and can think of it as being made up of the limit of the sum of the elements of area $y\ \mathrm{d}x$ between a and b as shown in Figure 18.2(ii).

Example 1

A propeller blade has profile $y = \sin 2x - \sqrt{3}\sin x$ between $x = 0$ and $x = \pi/6$. Find the area of the cross-section.

The graph of this function is sketched in Figure 18.3.

The area under the graph $= \displaystyle\int_0^{\pi/6} y\ \mathrm{d}x = \int_0^{\pi/6} (\sin 2x - \sqrt{3}\sin x)\mathrm{d}x$

$$= \left[-\frac{1}{2}\cos 2x + \sqrt{3}\cos x\right]_0^{\pi/6} = \frac{7}{4} - \sqrt{3}$$

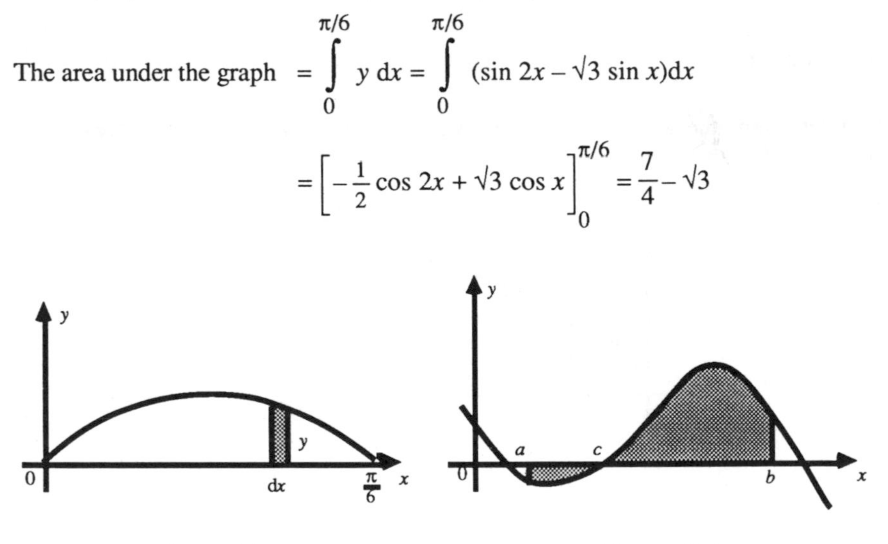

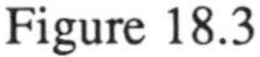

Figure 18.3

Figure 18.4

We have to be careful that the area lies entirely above the x-axis in using the integral to evaluate the area. For a function having a graph as in Figure 18.4 then

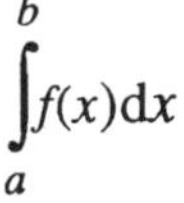

$$\int_a^b f(x)\mathrm{d}x$$

would give (the area between c and b) – (area between a and c).

(b) **Volume of revolution**
The idea of a definite integral being the limit of a sum of elements is also used here. As shown in Figure 18.5(i) we think of the area under the curve $y = f(x)$ as being revolved through 360° about the x-axis and the resulting volume is made up of elemental discs of volume $(\pi y^2)\mathrm{d}x$.

$$\text{Hence the volume, } V = \int_a^b \pi y^2 \mathrm{d}x$$

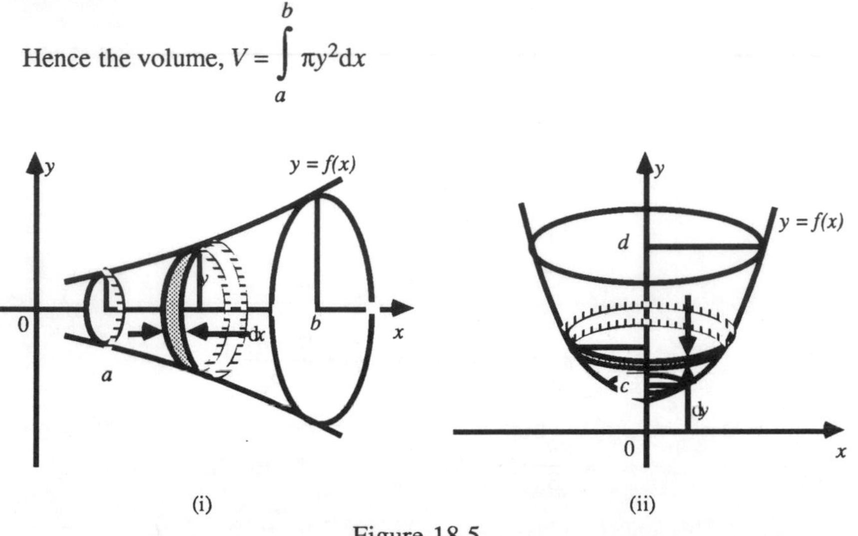

Figure 18.5

Similarly we have for revolution about the y-axis [Figure 18.5(ii)]

$$\text{Volume} = \int_c^d \pi x^2 \mathrm{d}y$$

Example 2
Find the volume of the cap of a sphere as depicted in Figure 18.6.

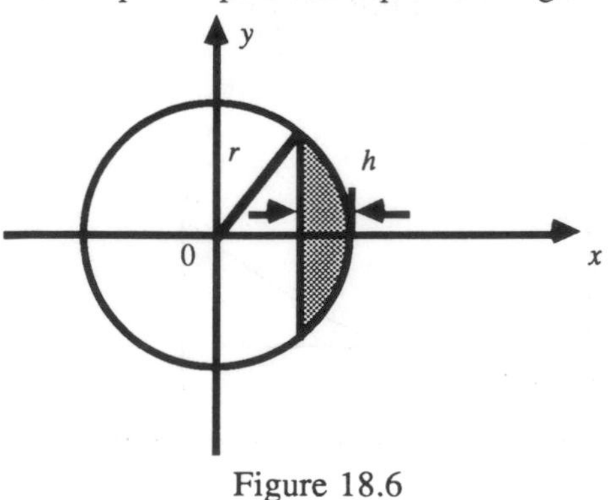

Figure 18.6

The circle is $x^2 + y^2 = r^2$, that is, $y^2 = r^2 - x^2$ is the equation of the circumference.

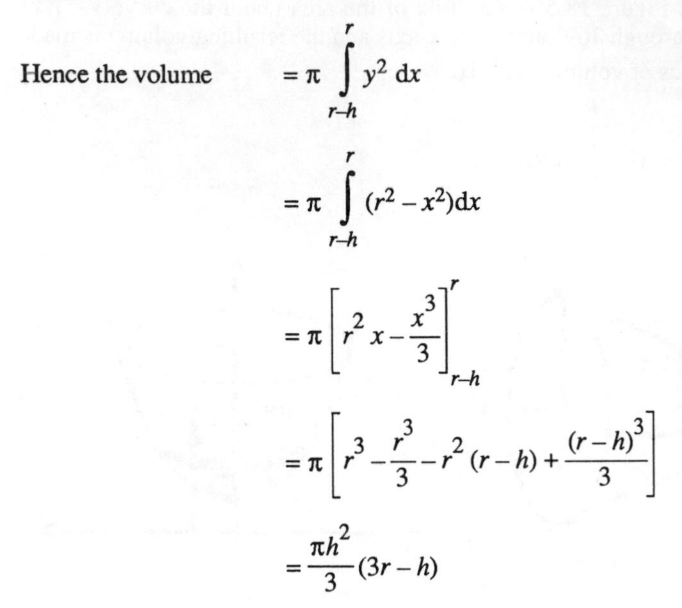

Hence the volume $= \pi \int_{r-h}^{r} y^2 \, dx$

$$= \pi \int_{r-h}^{r} (r^2 - x^2) dx$$

$$= \pi \left[r^2 x - \frac{x^3}{3} \right]_{r-h}^{r}$$

$$= \pi \left[r^3 - \frac{r^3}{3} - r^2 (r - h) + \frac{(r-h)^3}{3} \right]$$

$$= \frac{\pi h^2}{3} (3r - h)$$

18.2 Length of Arc of a Plane Curve

The perimeter or length of arc of a curve is easily obtained from first principles. Looking at Figure 18.7 we see that the length of arc between A and B is the limit of the sum of small arcs δs.

That is

$$\text{Perimeter} = \int_{A}^{B} ds$$

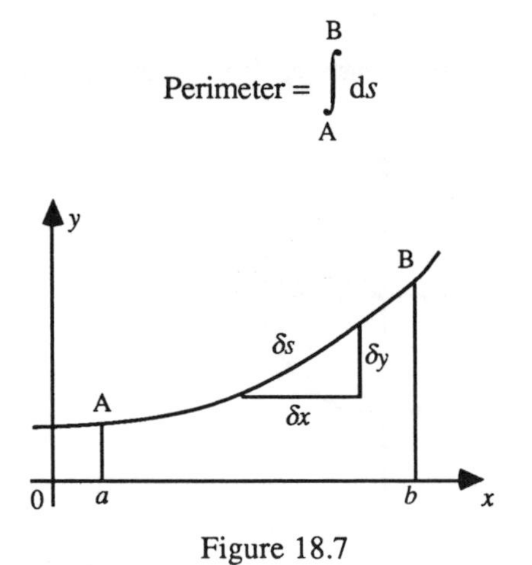

Figure 18.7

But $(\delta s)^2 = (\delta x)^2 + (\delta y)^2$ (for small enough arcs)

or
$$\left[\frac{\delta s}{\delta x}\right]^2 = 1 + \left[\frac{\delta y}{\delta x}\right]^2$$

Hence
$$\frac{\delta s}{\delta x} = \sqrt{1 + \left[\frac{\delta y}{\delta x}\right]^2}$$

Similarly
$$\frac{\delta s}{\delta y} = \sqrt{1 + \left[\frac{\delta x}{\delta y}\right]^2}$$

Hence
$$\int_A^B ds = \int_a^b \frac{ds}{dx} dx = \int_a^b \sqrt{1 + \left[\frac{dy}{dx}\right]^2}\, dx$$

(We can obtain a similar expression for the perimeter as an integral over y.)

Example

Find the length of the curve $y = \ln x$ from $x = 1$ to $x = 2\sqrt{2}$.

The length,
$$l = \int_{x=1}^{x=2\sqrt{2}} \sqrt{1 + \left[\frac{1}{x}\right]^2}\, dx$$
$$= \int_1^{2\sqrt{2}} \frac{\sqrt{x^2 + 1}}{x} dx$$

We must now use the techniques we have learned to evaluate this integral.

Put
$$u^2 = x^2 + 1, \quad \frac{du}{dx} = \frac{x}{u}$$

Hence
$$l = \int_{\sqrt{2}}^{3} \frac{u}{x} \cdot \frac{u}{x}\, du = \int_{\sqrt{2}}^{3} \frac{u^2}{u^2 - 1} du$$
$$= \int_{\sqrt{2}}^{3} \left[1 + \frac{1}{u^2 - 1}\right] du$$

$$= \frac{1}{2} \int_{\sqrt{2}}^{3} \left[2 + \frac{1}{u-1} - \frac{1}{u+1} \right] du$$

$$= \frac{1}{2} \left[2u + \ln \frac{u-1}{u+1} \right]_{\sqrt{2}}^{3}$$

$$= 3 - \sqrt{2} - \frac{1}{2} \ln 2 - \frac{1}{2} \ln \left[\frac{\sqrt{2}-1}{\sqrt{2}+1} \right]$$

18.3 Area of Surface of Revolution

Consider Figure 18.8(i). The solid is divided into elemental discs.
Area of curved surface of elemental disc = $(2\pi y)\delta s$

$$\text{Surface area} = \int_{x=a}^{b} 2\pi y \, ds$$

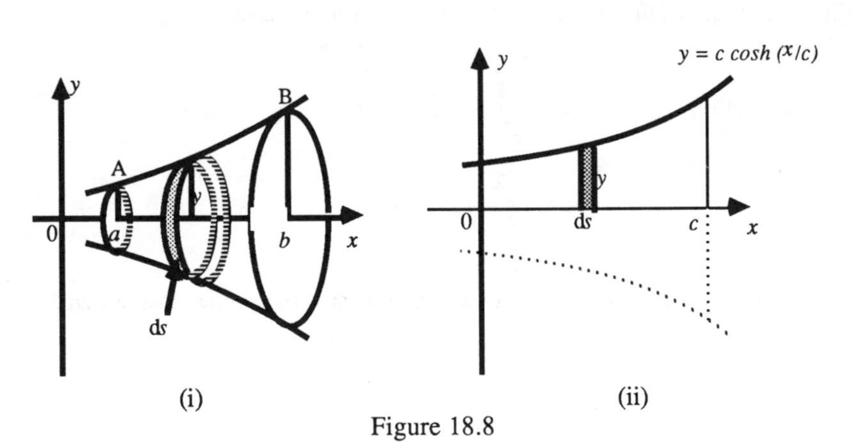

Figure 18.8

Example
The arc of the catenary $y = c \cosh (x/c)$ from $x = 0$ to $x = c$ rotates about the axis of x. Refer to Figure 18.8(ii) Find the area of surface formed.

$$\text{Area} = \int_{x=0}^{c} 2\pi y \, ds$$

Now $$\frac{ds}{dx} = \sqrt{1 + \left[\frac{dy}{dx}\right]^2} = \sqrt{1 + \sinh^2\frac{x}{c}} = \cosh\frac{x}{c}$$

therefore $$\text{Area} = \int_0^c 2\pi y \frac{ds}{dx}\,dx = 2\pi c \int_0^c \cosh^2\frac{x}{c}\,dx$$

$$= \pi c \int_0^c \left[1 + \cosh\frac{2x}{c}\right] dx$$

$$= \pi c \left[x + \frac{c}{2}\sinh\frac{2x}{c}\right]_0^c$$

$$= \pi c[(c + \frac{c}{2}\sinh 2) - 0]$$

$$= \pi c^2[1 + \frac{1}{4}(e^2 - \frac{1}{e^2})]$$

18.4 Centroids

In this section we consider the centroids of (i) a plane area (ii) a solid volume of revolution (iii) an arc and a surface of revolution. In each case the strategy is the same. We take an element of the object, calculate its first moment about a given axis and sum to get a total moment. This total moment will be equal to the product of the distance of the centroid from the given axis and the area, volume, length of arc or surface area as appropriate.

(a) **Centroid of a plane area**
We wish to determine the centroid of the area in the first quadrant bounded by the parabola $y = 8 - x^2$. Refer to Figure 18.9.

The element of area shaded is almost a rectangle with centroid at $(x, \frac{1}{2}y)$. The moment of this elementary area about the y-axis is $x.y\delta x$ and the total moment of the area bout the y-axis is

$$M_y = \int_0^{2\sqrt{2}} xy\,dx = \int_0^{2\sqrt{2}} x(8 - x^2)dx = \left[4x^2 - \frac{x^4}{4}\right]_0^{2\sqrt{2}} = 16$$

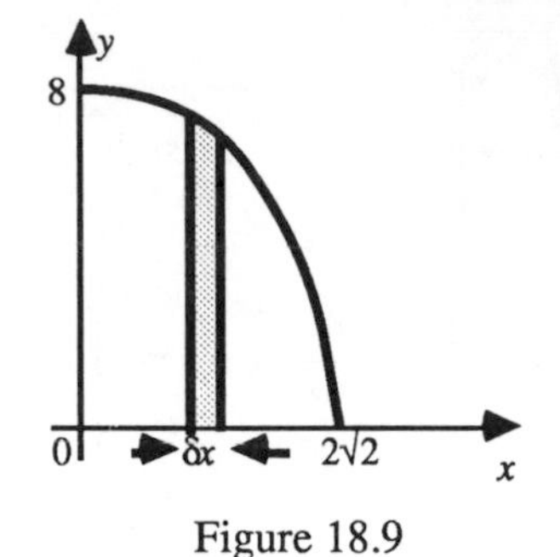

Figure 18.9

Similarly the first moment of the elementary area about the x-axis is $y.\frac{1}{2}y\,\delta x$ and the total moment is

$$M_x = \int_0^{2\sqrt{2}} y\,.\,\tfrac{1}{2}y\;\mathrm{d}x = \int_0^{2\sqrt{2}} \tfrac{1}{2}(8-x^2)^2 = \frac{512\sqrt{2}}{15}$$

The value of the area in question is

$$A = \int_0^{2\sqrt{2}} y\mathrm{d}x = \int_0^{2\sqrt{2}} (8-x^2)\mathrm{d}x = \frac{32\sqrt{2}}{3}$$

Hence $\quad \bar{x} = M_y/A = 16\sqrt{2}/15 \qquad \bar{y} = M_x/A = 16/5$

A simple check will show that the point $(\bar{x}, \bar{y})$ does indeed lie in the area under consideration.

(b) **First theorem of Pappus**

The volume swept out by revolving a plane area about an axis in its plane, which does not cut the area, is equal to the product of the area and the length of the path traced out by the centroid of the area.

If we consider the area of the semi-circle of radius a centred at the origin and lying in the upper half-plane then its area is obviously πa^2. The volume that is generated as the semi-circle revolves about the x-axis is that of a sphere of radius a, i.e. $\frac{4}{3}\pi a^3$. Now the centroid traces out a circular path of radius $\bar{y}$, i.e. a distance of $2\pi\bar{y}$. The theorem indicates that $\frac{4}{3}\pi a^3 = \frac{1}{2}\pi a^2.2\pi\bar{y}$ so that $\bar{y} = 4a/3\pi$. By symmetry, $\bar{x} = 0$ so that the centroid is at the point $(0, 4a/3\pi)$.

(c) **Centroid of a volume of revolution**

Consider the solid obtained by revolving the area given in subsection (a) about the x-axis; then we can imagine an elemental disc as shaded in Figure 18.10

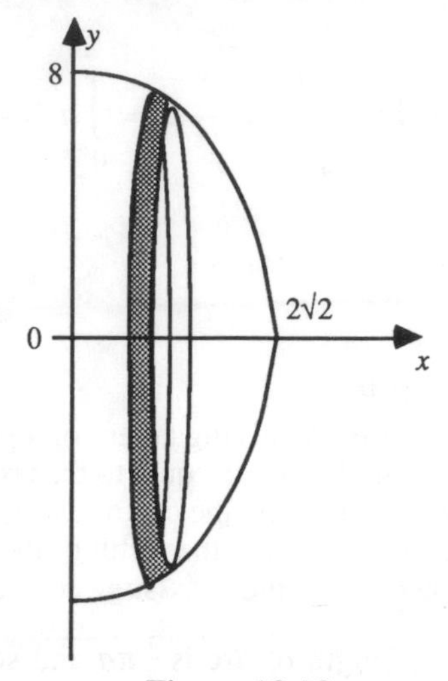

Figure 18.10

The disc has volume $\pi y^2 \delta x$ and its first moment about the y-axis is $x.\pi y^2 \delta x$. The total moment of the solid about the y-axis is

$$M_y = \int_0^{2\sqrt{2}} x.\pi y^2 \mathrm{d}x = \int_0^{2\sqrt{2}} \pi x(8 - x^2)^2 \mathrm{d}x$$
$$= 256\pi/3$$

The volume of the solid is $V = \int_0^{2\sqrt{2}} \pi y^2 \, \mathrm{d}x = \dfrac{1024\sqrt{2}}{15}\pi$

Hence $\bar{x} = 4\sqrt{2}/5$

By symmetry, it is clear that $\bar{y} = 0$.

(d) **Centroid of an arc**

The arc of the circle $x^2 + y^2 = a^2$ in the first quadrant has length $\frac{1}{2}\pi a$. The moment of the element of arc of length δs about the x-axis is $y\delta s$ and therefore the total moment of the arc about the x-axis is

$$M_x = \int_0^a y \, \mathrm{d}s = \int_0^a y \sqrt{1 + \left[\frac{\mathrm{d}y}{\mathrm{d}x}\right]^2} \, \mathrm{d}x$$

$$= \int_0^a y\sqrt{1 + \frac{x^2}{y^2}}\,dx = \int_0^a a\,dx = a^2$$

hence $\bar{x} = a^2/(\frac{1}{2}\pi a) = 2a/\pi$.

By symmetry, $\bar{y} = 2a/\pi$

(e) **Second Theorem of Pappus**
The area of surface swept out by revolving an arc of a plane curve about an axis in its plane, which does not cut the arc, is equal to the product of the length of arc and the length of the patch traced out by the centroid of the arc.

We shall apply this theorem to obtain the result of the last subsection a second time. The arc, when rotated about the x-axis, sweeps out a hemisphere whose surface area is $2\pi a^2$. The length of arc is $\frac{1}{2}\pi a$ and so the length of the path traced by the centroid is $2\pi a^2/(\frac{1}{2}\pi a) = 4a$. But the centroid tracks round a circle of radius $\bar{y}$, i.e. a path of length $2\pi\bar{y}$. Hence $\bar{y} = 4a/2\pi = 2a/\pi$. In a similar fashion, you should be able to find $\bar{x}$.

18.5 Moments of Inertia

For the plane area under a plane curve $y = f(x)$ between $x = a$ and $x = b$ the *second moments of area* about Ox and Oy are given respectively by

$$S_x = \int_a^b \frac{1}{3}y^3\,dx, \qquad S_y = \int_a^b x^2\,y\,dx$$

If we regard the area as a laminar having a mass distribution then the moment of inertia of this lamina about Ox and Oy are given respectively by

$$I_x = \int_a^b \frac{1}{3}y^3\,m\,dx, \qquad I_y = \int_a^b x^2\,y\,m\,dx$$

where m is the mass/unit area.

The *radius of gyration*, k, is defined by

$$k^2 = \frac{\text{second moment of area}}{\text{total area}} = \frac{\text{moment of inertia}}{\text{total mass}}$$

Example
Find the moment of inertia with respect to the y-axis of the plane area between the parabola $y = 8 - x^2$ and the x-axis. Refer to Figure 18.11(i)

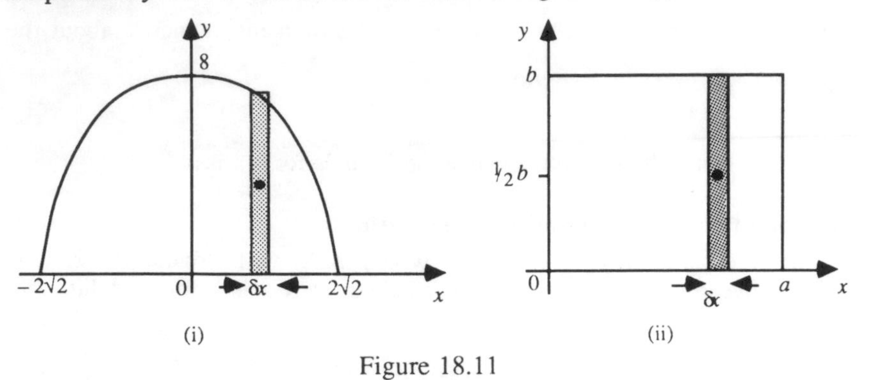

Figure 18.11

The second moment of the elementary strip which is shaded is $x^2.y\delta x$ and hence the total second moment and the total moment of inertia are respectively

$$S_y = \int_{-2\sqrt{2}}^{2\sqrt{2}} x^2.y\mathrm{d}x \qquad I_y = \int_{-2\sqrt{2}}^{2\sqrt{2}} m\, x^2.y\mathrm{d}x$$

where m is the mass per unit area.

Now
$$I_y = m \int_{-2\sqrt{2}}^{2\sqrt{2}} x^2(8 - x^2)\mathrm{d}x$$

$$= m\,.\,2 \int_{0}^{2\sqrt{2}} x^2(8 - x^2)\mathrm{d}x, \text{ by symmetry}$$

$$= \frac{128\sqrt{2}}{5} m$$

Noting that the total mass $M = 2 \int_{0}^{2\sqrt{2}} m\, y\, \mathrm{d}x = \dfrac{64\sqrt{2}}{3} m$

we can write
$$I_y = M\,.\,\frac{6}{5}.$$

In Figure 18.11(ii) we consider the moment of inertia of the rectangular strip, shown shaded, about the y-axis. The strip has area $b\delta x$ with centroid at height $\frac{1}{2}b$ above the x-axis. The relevant moment of inertia is therefore $mx^2\, b\delta x$ and for the rectangle $0 \leq x \leq a, 0 \leq y \leq b$ the total moment of inertia about the y–axis is $\int_0^a mx^2 b\,dx = \frac{1}{3}mba^3$. Similarly, the moment of inertia about the x–axis is $\frac{1}{3}m\,ab^3$.

You should now be able to see where the formula for I_x arises.

Moment of inertia of a solid of revolution

Consider the parabola $y^2 = 4x$, part of which is shown in Figure 18.12. The portion between $x = 0$ and $x = 2$ is revolved about the x-axis to form a solid.

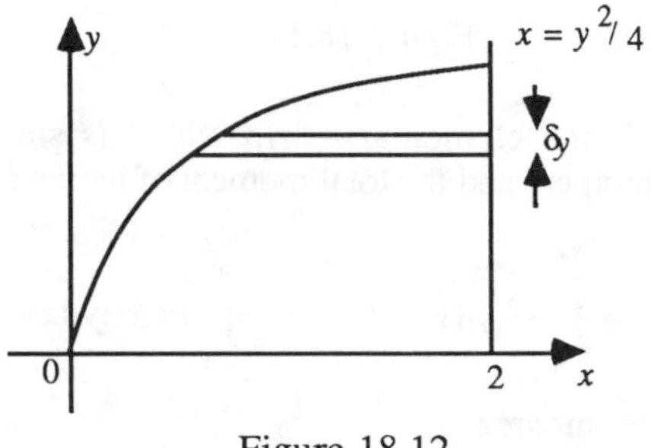

Figure 18.12

The strip shown has area $(2 - y^2/4)\delta y$ and produces an annulus of volume $2\pi y(2 - y^2/4)\delta y$ and its moment of inertia about the x-axis is $m2\pi y^3(2 - y^2/4)\delta y$. The total moment of inertia of the solid is

$$2\pi m \int_0^2 y^3(2 - y^2/4)\mathrm{d}y = \frac{32\pi m}{3}$$

Theorems on moments of inertia

I Perpendicular axes theorem for areas

Let Ox and Oy be two perpendicular axes in the plane of an area and Oz a line through O perpendicular to the plane. If k_x, k_y and k_z are the radii of gyration of the area about Ox, Oy and Oz respectively then

$$k_z^2 = k_x^2 + k_y^2$$

II Parallel axes theorem

The moment of inertia of an area or volume with respect to any axis is equal to the moment of inertia with respect to a parallel axis through the centroid plus the product of the area or volume and the square of the distance between the parallel axes.

Example
The moment of inertia of a circular disc of radius a about an axis through its centre and perpendicular to its plane is $\frac{1}{2}m\pi a^4$. It should be clear that, by symmetry, the moment of inertia of the disc about any diameter is independent of its direction. Hence, if we choose two perpendicular diameters and apply the perpendicular axes theorem we see that the moment of inertia about any diameter is $\frac{1}{4}m\pi a^4$. If we consider an axis parallel to a diameter but tangential to the circle then the distance between the axes is a. Hence by the parallel axes theorem the moment of inertia is $\frac{1}{4}m\pi a^4 + m\pi a^2.a^2 = \frac{5}{4}m\pi a^4$.

18.6 Further Applications

(a) Area and centroid of area in polar coordinates

Refer to Figure 18.13(i).

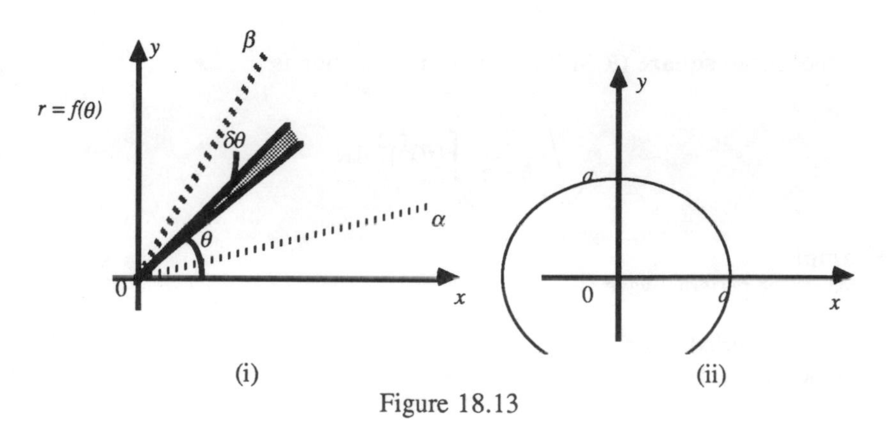

Figure 18.13

It can be shown that the area of the sector OAB is

$$A = \int_{\theta=\alpha}^{\theta=\beta} \frac{1}{2}r^2 d\theta$$

and that the position of the centroid $(\bar{x}, \bar{y})$ is given by

$$A\bar{x} = \int_{\alpha}^{\beta} \frac{1}{3}r^3 \cos\theta \, d\theta, \qquad A\bar{y} = \int_{\alpha}^{\beta} \frac{1}{3}r^3 \sin\theta \, d\theta$$

Example
Find the area and the centroid of a quadrant of a circle. In Figure 18.13(ii) we see that the circle boundary is $r = a$ so that

$$A = \int_0^{\pi/2} \frac{1}{2}a^2 \, d\theta = \frac{\pi a^2}{4}$$

Then $$\frac{\pi a^2}{4}\bar{x} = \int_0^{\pi/2} \frac{1}{3}a^3 \cos\theta \, d\theta = \frac{1}{3}a^3 \Big[\sin\theta\Big]_0^{\pi/2} = \frac{1}{3}a^3$$

so that $\bar{x} = \dfrac{4a}{3\pi}$. Similarly, $\bar{y} = \dfrac{4a}{3\pi}$.

(b) Mean values

The **mean value** of a function $f(x)$ over an interval $[a, b]$ is

$$\frac{1}{b-a}\int_a^b f(x)\mathrm{d}x$$

The **root mean square** (R.M.S.) value of the function is

$$\sqrt{\frac{1}{b-a}\int_a^b [f(x)]^2 \, \mathrm{d}x}$$

Example
Consider the periodic wave

$$y = 4 + 3\sin x$$

which has a period of 2π.
Over the interval $[0, 2\pi]$ its mean value is

$$\frac{1}{2\pi}\int_0^{2\pi} (4 + 3\sin x)\mathrm{d}x = \frac{1}{2\pi}\Big[4x - 3\cos x\Big]_0^{2\pi}$$

$$= \frac{1}{2\pi} \cdot 8\pi = 4$$

Its R.M.S. value is given by R where

$$R^2 = \frac{1}{2\pi}\int_0^{2\pi} (4 + 3\sin x)^2 \, \mathrm{d}x$$

$$= \frac{1}{2\pi}\int_0^{2\pi} (16 + 24 \sin x + 9 \sin^2 x)dx$$

$$= \frac{1}{2\pi}\int_0^{2\pi} (16 + 24 \sin x + \frac{9}{2}(1 - \cos 2x))dx$$

$$= \frac{1}{2\pi}\int_0^{2\pi} (20\frac{1}{2} + 24 \sin x - \frac{9}{2}\cos 2x)dx$$

$$= \frac{1}{2\pi} . 2\pi . 20\frac{1}{2} = \frac{41}{2}$$

$$R = 4.53 \quad \text{(3 d.p.)}$$

Problems

Section 18.1

1 (i) Sketch the curve $y^2 = x^2(4 - x)$ and find the finite area which is completely enclosed by part of this curve.

(ii) The portion of the curve $y^2 = 9/(2x + 5)$ between $x = 1$ and $x = 2$ is rotated about the x-axis to sweep out a solid of revolution. Find is volume.

2 The profile of an impeller blade is bounded by the lines $x = 0.1$, $y = 2x$, $y = e^{-x}$, $x = 1$ and the x-axis. The blade thickness t varies linearly with x, thus $t = (1.1 - x)\tau$ there τ is a constant. Find to 2 d.p. from tables, *or otherwise*, the point given by the equation $2x = e^{-x}$. Hence determine the volume of the blade, showing the result to be approximately $\tau/4$. (EC)

3 Find the area bounded by the curves $y^2 = 4x$ and $x^2 = 4y$.

4 Find the area under the curve given parametrically by $x = a(\theta - \sin \theta)$, $y = a(1 - \cos \theta)$ between the points $\theta = 0$ and $\theta = \pi$.

5 Show by integration that the volume of a hemisphere of radius a is $(2/3)\pi a^3$.

6 Find the volume obtained by revolving the following area about the axes Ox, Oy:

The area under $y = \sin x$ from $x = 0$ to $x = \pi$

7 (i) The area bounded by $y^2x = 4a^2(2a - x)$ and the ordinates $x = a$, $x = 2a$ revolves around Ox. Show that the volume generated is $4\pi a^3(2 \ln 2 - 1)$.

(ii) Prove that the volume of a segment of a sphere of height h and base radius c is $\pi h(3c^2 + h^2)/6$.

8 Sketch the curve whose equation is $y^2 = x^2(1 - x^2)$ and find the area enclosed by one loop of the curve. If this loop is rotated about the x-axis to generate a surface of revolution, show that this surface encloses a volume $2\pi/15$. (LU)

9 A circular flywheel of outer radius r_2 is designed to fit on a shaft of radius r_1. The thickness t of the flywheel varies linearly with the radial distance r from its centre, i.e. $t = ar + b$. The thickness is t_1 at the shaft ($r = r_1$) and is t_2 at the outer edge ($r = r_2$). It is required to find the volume of the flywheel by integration.
Choose a typical volume element and use this to represent the total volume in the form of a summation. Show clearly how this summation is translated into an integral.
Evaluate the integral to find an expression for the required volume. (EC)

Section 18.2

10 Show that for a catenary $y = c\cosh(x/c)$, the area under the curve between any two ordinates is c times the arc length of the curve between the corresponding points.

11 Sketch the curve $3ay^2 = x^2(a - x)$, a being a positive constant, and find the area and length of its loop.

12 Find the area of the loop of the curve whose equation is $ay^2 = (x - a)(x - 5a)^2$

13 Sketch the cycloid $x = a(\theta - \sin\theta)$, $y = a(1 - \cos\theta)$.
Prove that one arch of the cycloid is of length $8a$.

14 Find the length of the curve $x = \ln\sec y$ from $y = 0$ to $y = \pi/3$.

15 Sketch the curve $3y^2 = x^2(x + 1)$. Find the length and area of its loop.

Section 18.3

16 Find the surface area of the paraboloid of revolution formed by rotating about the y-axis that portion of the parabola $y = x^2/a$ between $x = 0$ and $x = a$.

17 A segment of height h is taken from a sphere of radius r as shown.

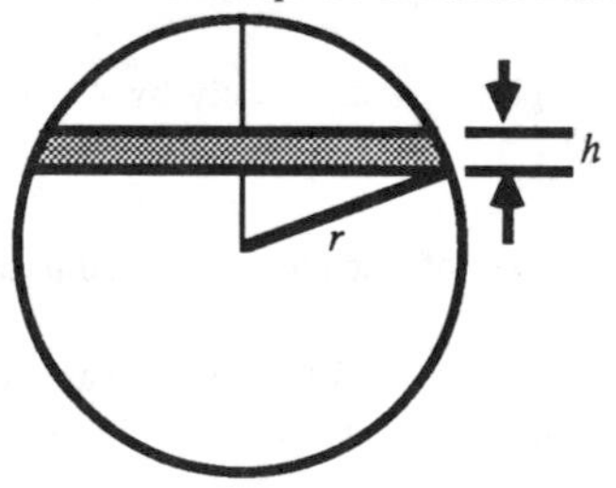

Find the curved surface area of this segment.

18 A parabolic mirror is formed by revolving the part of the parabola $y^2 = 4x$ from $x = 0$ to $x = 2$ about Ox. Find the surface area of the mirror so formed.

19 (i) Find the length of the curve $y = \ln \cos x$ from $x = 0$ to $x = \pi/3$.

(ii) The area enclosed by the parabola $y^2 = 6ax + 16a^2$ and the ordinates $x = 0$ and $x = 4a$ is rotated about the x-axis to form a solid of revolution. Show that the area of the curved surface of the solid is $436\pi a^2/9$. (LU)

20 Find the curved surface area when the following curves are rotated about the x-axis.

(i) $y = \dfrac{x^3}{6} + \dfrac{1}{2x}$ from $x = 1$ to $x = 2$ (ii) $y = 4 \cosh \dfrac{x}{4}$ from $x = -4$ to $x = 4$

(iii) $x = a(\theta - \sin\theta)$, $y = a(1 - \cos\theta)$ from $\theta = 0$ to $\theta = 2\pi$

Section 18.4

21 (i) Find the coordinates of the centroid of the area enclosed by the curve $y = 8/(x^2 + 4)$, the x-axis and the ordinates $x = -2\sqrt{3}$ and $x = +2\sqrt{3}$.

(ii) Sketch the curve $ay^2 = x^2(a - x)$ $(a > 0)$ and find the area of its loop.

22 Find the first two points (starting at $x = 0$) at which the curve $y = e^x \sin x$ crosses the axis of x. Find the height above the axis of x, of the centroid of the area bounded by the curve and the axis of x between these points.

23 The density at any point P of a cone of height h is proportional to the square of the perpendicular distance of that point from the base of the cone. Show that the centre of gravity of the cone is at a distance $\frac{1}{2}h$ from the vertex.

24 Find the coordinates of the centroid of the area between the parabola $y = x^2 - 7x + 12$ and the axes of x and y.

25 A rope is suspended between two vertical poles of equal height and takes the shape of the curve $y = 5 \cosh x/5$, where the origin O is mid-way between the feet of the poles and y is the height above the ground. If the rope is 450 metres long show that the distance between poles is 45 metres to the nearest metre. Find the height of the centroid of the rope above the ground to the nearest metre. (LU)

26 Sketch the graph of $r = a(3 - 2\cos\theta)$ where r, θ are polar coordinates and a is a constant. Find the area enclosed by this curve and the position of the centroid of the area. (LU)

27 Sketch the curve $ay^2 = x^2(a - x)$, where $a > 0$, and show that the tangents at the origin to this curve are perpendicular. Calculate the area A enclosed by the x-axis and that part of the curve which lies in the first quadrant. Calculate also the volume swept out when the area A is rotated through 2π about (i) the x-axis, (ii) the y-axis. Deduce the coordinates of the centroid of the area A. (LU)

28 The first moment of area about the x-axis of the region R contained between a curve and the x-axis is given by

$$\int_a^b \tfrac{1}{2}y^2 \, dx$$

(i) Explain carefully with a diagram, showing typical elements with dimensions y and δx, and using a Σ sign, how this integral is obtained.
(ii) Derive a formula for the first moment of area of R about the y-axis using the same elements.
(iii) Find the first moment of area about the x-axis of the first quadrant of the circle $x^2 + y^2 = a^2$. (EC)

29 The area of the region R between a curve and the x-axis is given by

$$\int_a^b y \, dx$$

(i) Explain using a diagram how this integral results from the summation of elements with dimensions y and δx.
(ii) Show that the first moment of area about the x-axis of this complete region R is given by

$$\int_a^b \tfrac{1}{2}y^2 \, dx$$

(iii) Locate the position of the centroid of the region bounded by the x-axis and that portion of the parabola $y = 3x - x^2$ which is above the x-axis. (EC)

Section 18.5

30 Find the second moment (a) about the x-axis, (b) about the y-axis of the area enclosed between the curves $y = x^{1/2}$ and $y = x$. Find also the volume of the solid of revolution obtained by rotating this area through 2π radians about the x-axis. (LU)

31 Find the moment of inertia of the plane area given about each coordinate axis
(i) $y = 16 - x^2, x = 0, y = 0$ (ii) $y^2 = 4ax, x = a$ (iii) $x^2 + 4y^2 = 4$

32 Using the parallel axes theorem, find the moment of inertia of the plane areas about the axes given below
(i) $y = 16 - x^2, x = 0, y = 0$ about $x = 16$ (ii) $y^2 = 4ax, x = a$ about $x = a$
(iii) $x^2 + 4y^2 = 4$ about a horizontal tangent.

33 A solid is generated by revolving the ellipse $x^2 + 4y^2 = 4$ about the x-axis. Find its moment of inertia about the x-axis.

34 The area bounded by the curve $y^2 = 4(2 - x)$ and the y-axis is rotated through two right angles about the x-axis. Show that the volume of the solid of revolution is 8π and find the radius of gyration of this solid about the y-axis.

35 A uniform lamina of mass M is in the shape of the area bounded by the curve $y = c\cosh(x/c)$, the coordinate axes and the line $x = c$. Find the distance of the centre of gravity of the lamina from the y-axis, and show that the moment of inertia of the lamina about the y-axis is $Mc^2(3 - 2\coth 1)$. (LU)

Section 18.6

36 If $e = 24\cos pt + 4\cos 3pt$ and $i = 4\sin pt + \sin 3pt$, prove that $ei = 52 \sin 2pt + 20\sin 4pt + 2\sin 6pt$. Hence show that the mean value of ei over the range $t = 0$ to $t = \pi/2p$ is 33.5

37 (i) By expressing $\sin^2 100\pi t$ in terms of $\cos 200\pi t$, find the mean value of y^2 between $t = 0$ and $t = 1/100$ where $y = 300\sin 100\pi t$
(ii) Find, by integration, the form factor (i.e. R.M..S Value/Mean Value) of the wave $e = E_1\sin\omega t + E_3\sin 3\omega t$ over the range $t = 0$ to $t = \pi/\omega$.

38 Find the area bounded by the curves
(i) $r^2 = a^2(1 + \cos 2\theta)$ (ii) $r = \cos\theta$ (iii) $r = 4a\sin^2\theta$

39 Find the centroid of the areas inside
(i) $r = a(1 - \sin\theta)$, $-\pi/2 \le \theta \le \pi/2$ (ii) $r = 2 + \cos\theta$, $0 \le \theta \le \pi$

40 Show that the length of arc of the curve $r = f(\theta)$ for $\alpha \le \theta \le \beta$ is

$$s = \int_\alpha^\beta \sqrt{r^2 + \left[\frac{dr}{d\theta}\right]^2}\, d\theta$$

Find the length of the curve $r = a(1 - \cos\theta)$.

41 A viscous fluid flows between two fixed horizontal plates a distance Y_0 apart. The velocity v at a distance y from the lower plate is given by

$$v = \frac{K}{2\mu}(Y_0\, y - y^2)$$

where K and μ are constants.
By choosing elements of 1 m across the flow and of height δy, explain carefully, using a sketch and a summation process, why the volume flow rate per metre width through the whole channel is given by

$$\int_0^{Y_0} v\, dy$$

Find the volume flow rate when $K = 12 \times 10^6$ N/m^3, $Y_0 = 0.01$ m, $\mu = 1 \times 10^4$ N s/m^2, paying particular attention to the units. (EC)

42 The diagram below shows a thin uniform straight rod AB of mass α per unit length and a body P of mass m. It is required to find the component of the force of gravitational attraction on P due to the rod in the direction parallel to the rod.
The rod is subdivided into elements δx of which QQ′ is typical. It may be assumed that,

using appropriate units, the force of attraction between two bodies of mass m_1 and m_2 separated a distance r is $(Gm_1m_2)/r^2$, where G is a constant.
Explain why the components of the force on P due to QQ′ are

$$\frac{Gm\,\alpha\delta x}{L^2}\cdot\frac{x}{L} \quad \text{and} \quad \frac{Gm\,\alpha\delta x}{L^2}\cdot\frac{h}{L}$$

Hence derive the result
$$F_{AB} = Gm\int_a^b \frac{\alpha x}{(x^2+h^2)^{3/2}}\,dx$$

Find this integral by means of the substitution $x = h\tan\phi$.

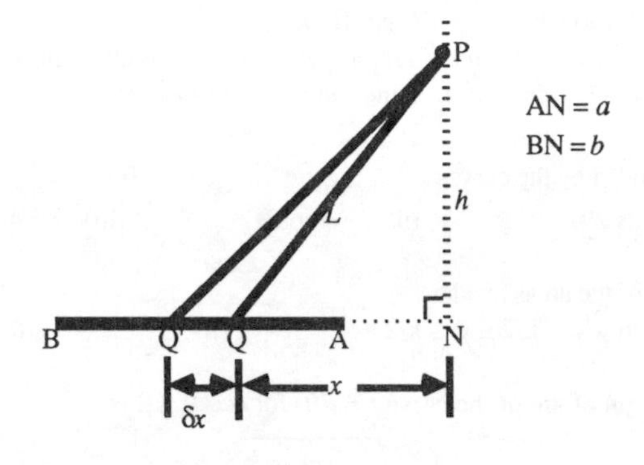

(EC)

43 A point body, P, of mass m is placed at a distance L along the normal from the centre of a very thin circular disc of radius R. This disc, which has a uniform thickness T, has a density ρ per unit volume. It is required to find the gravitational attraction of the disc on the body. Consider the disc to be divided into elements which take the form of rings as in the figure below.

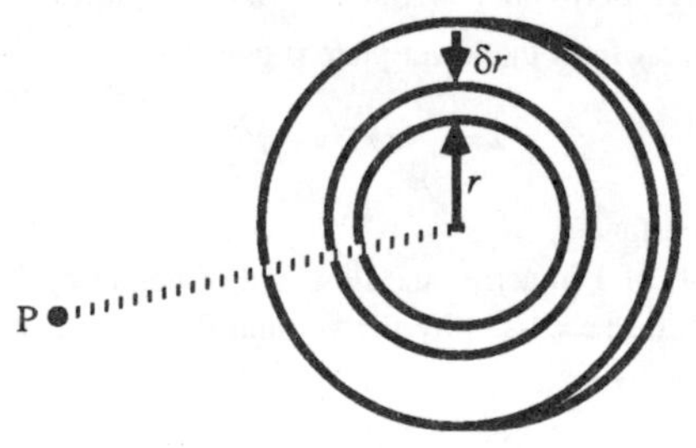

A typical element has a radius r and width δr.
It may be assumed that the force of attraction between two bodies of mass m_1 and m_2 a distance d apart is given by Gm_1m_2/d^2, where G is a constant.

(i) Establish the force of attraction on P in a direction normal to the disc due to this element.

(ii) Show how, by summing the forces due to all such elements, the total force on P is obtained. Indicate, clearly, how this summation is translated into a definite integral.

(iii) Evaluate this integral to obtain the total force. (EC)

19

DISCRETE PROBABILITY MODELS

19.1 Probability Distributions

We shall restrict our attention to discrete random variables. If an experiment has a given discrete sample space $S = \{x_1, x_2, x_3, ..., x_n\}$ and we assign probabilities $p_1, p_2, p_3, ..., p_n$ to these outcomes then we have defined for the experiment a **discrete probability distribution**. We call the function $p(X)$ which takes the values p_1, p_2, etc. when X takes the values x_1, x_2, etc. the **probability function** of X.

Example 1

For the tossing of a coin $S = \{H,T\}$ and $p(T) = \frac{1}{2}$. It should be clear that

$$0 \leq p(x_i) \leq 1 \qquad \text{(i)}$$

and, since $S = x_1 \cup x_2 \cup ... \cup x_n$, that

$$\sum_{i=1}^{n} p(x_i) = 1 \qquad \text{(ii)}$$

If A is any event then

$$p(A) = \sum_{x_i \in A} p(x_i)$$

where we sum the probabilities of all outcomes in A.

Provided conditions (i) and (ii) are satisfied then a function will be a probability function; it is clear that we try to choose a probability function which closely matches observed frequency (see later sections on Binomial and Poisson distributions). We can represent a discrete probability distribution by a **line diagram**.

Example 2

For the tossing of three coins we can specify outcomes as *HHH, HTH, HTT*, etc. each of which has equal probability of occurring of 1/8 if the coins are fair. Alternatively, we may specify the outcomes as no heads, 1 head, 2 heads, 3 heads, which have probabilities respectively of 1/8, 3/8, 3/8, 1/8 and we represent these on a line diagram shown in Figure 19.1.

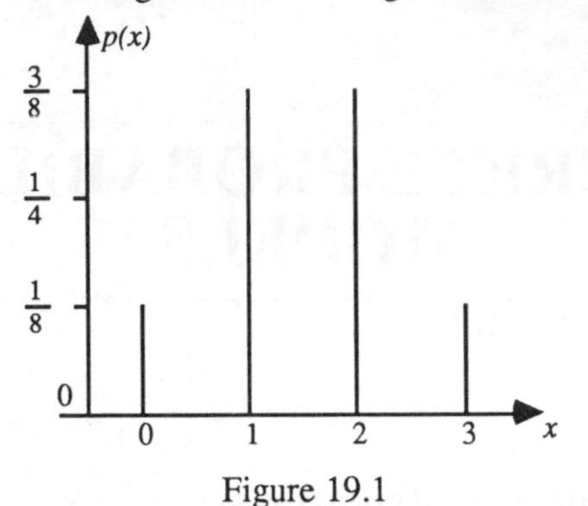

Figure 19.1

Example 3
The random variable X can take values 0, 1, 2, 3, 4, 5, 6 with probabilities 0.3, 0.14, 0.07, 0.08, 0.13, 0.18, 0.1. Verify that this is a description of a probability distribution, calculate $p(X \geq 3)$ and $p(X < 2)$ and display the distribution on a line diagram shown in Figure 19.2.

All probabilities are in the range [0, 1] and their sum is 1.

$$p(X \geq 3) = 0.08 + 0.13 + 0.18 + 0.1 = 0.49$$
$$p(X < 2) = 0.3 + 0.14 = 0.44$$

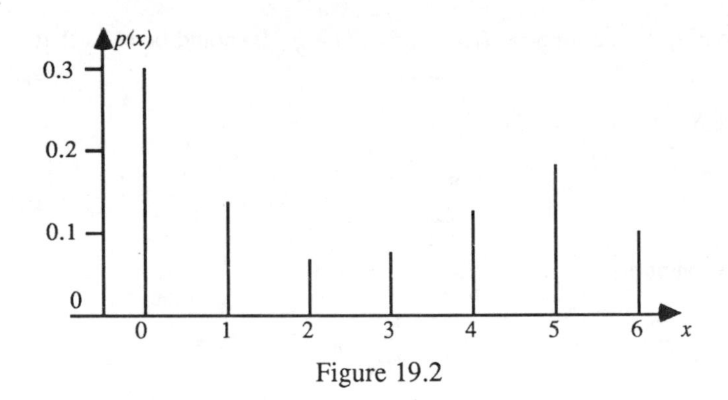

Figure 19.2

19.2 Mathematical Expectation

As usual in this chapter, we concentrate on discrete random variables. We have already seen that for a sample of n observations falling into s distinct values the arithmetic mean

$$\bar{x} = \sum_{r=1}^{s} \frac{f_r}{n} x_r \qquad (19.1)$$

Now we have already interpreted probability as long-range relative frequency and we might wonder what interpretation to put on the quantity

$$\sum_{r=1}^{s} p_r x_r$$

where the random variable X takes the values $x_1, x_2, \ldots, x_s$ with probabilities $p_1, p_2, \ldots, p_s$. We define the **expectation** or **expected value** of the discrete random variable X, written μ or $E(X)$, as

$$\sum_{r=1}^{s} p_s x_s \tag{19.2}$$

For example, in the throwing of a die the probability of each outcome is1/6 and so

$$E(X) = \frac{1}{6} \times 1 + \frac{1}{6} \times 2 + \frac{1}{6} \times 3 + \frac{1}{6} \times 4 + \frac{1}{6} \times 5 + \frac{1}{6} \times 6 = \frac{21}{6} = 3.5$$

What significance has this value ? Before we answer this question let us consider a more simple experiment: the tossing of an unbiased coin; here we can code the outcomes Head and Tail as 0 and 1 respectively. Each outcome has probability $\frac{1}{2}$ and the expected value is $(\frac{1}{2} \times 0 + \frac{1}{2} \times 1) = \frac{1}{2}$. But $\frac{1}{2}$ is not a value assumed by either outcome; rather it represents the average score if the experiment were to be repeated many times. We will not get a score of $\frac{1}{2}$ on one tossing, but the relative frequency of tails will approximate to $\frac{1}{2}$ after several tossings and the approximation should get closer the more times the coin is tossed. In effect, the expectation $\frac{1}{2}$ is the model prediction of the average score: for this reason, we use the alternative notation μ and speak of the **population mean**. The sample mean, $\bar{x}$, of n observations is then compared to the population mean since we are testing the sample observations against the model. (You might find it helpful to look back to Chapter 4.)

Returning to the throwing of a die, we can say that if we repeat the experiment many times we would expect the average score/throw to be 3.5 and we may compare the observed average to 3.5 ; if the discrepancy is large we may suspect that the die is not fair, since the model is built on the assumption of equiprobable outcomes, which leads to a population mean of 3.5.

Example

A game consists of you drawing a ball from a bag containing 6 white balls and 4 blue balls. If the ball is white you win 40p, if it is blue you lose 80p. The ball is replaced, the bag shaken and another ball drawn. What are your expected winnings from this game?

Let $x_1 = +40$p, $x_2 = -80$p. Then $p(x_1) = 6/10$, $p(x_2) = 4/10$

$$E(X) = p(x_1) \cdot x_1 + p(x_2) \cdot x_2 = (6/10) \cdot 40 + (4/10) \cdot -80 = -8\text{p}$$

On average you lose 8p per game, though you can either win 40p or lose 80p on each occasion. In n games you would expect to lose $8n$p and hence you would be unwise to play.

Properties of expectation

We state these properties without proof (they follow from the definition of expectation). X is a *discrete random variable* (d.r.v.)

(i) $E(aX + b) = aE(X) + b$ for any numerical constants a and b

(ii) $E(m) = m$ where m is a constant

(iii) If $E(X) = \mu$ then $E(X - \mu) = 0$

(iv) If Y is also a discrete random variable, $E(X + Y) = E(X) + E(Y)$

(v) If X and Y are independent, $E(XY) = E(X).\ E(Y)$

(vi) If $f(X)$ is any function of X then $p[f(X)] = \sum_{i=1}^{n} p_i f(x_i)$

for example, $$E(X^2) = \sum_{i=1}^{n} p_i x_i^2$$

As a result of this last property we may define the variance of the d.r.v. X,

$$\text{var}(X) = E(X - \mu)^2 \qquad (19.3)$$

Now
$$E[(X - \mu)^2] = \sum_{i=1}^{n} p_i (x_i - \mu)^2$$

$$= \sum_{i=1}^{n} p_i (x_i^2 - 2\mu x_i + \mu^2)$$

$$= \sum_{i=1}^{n} p_i x_i^2 - \sum_{i=1}^{n} 2\mu p_i x_i + \sum_{i=1}^{n} p_i \mu^2$$

$$= \sum_{i=1}^{n} p_i x_i^2 - 2\mu \sum_{i=1}^{n} p_i x_i + \mu^2 \sum_{i=1}^{n} p_i$$

$$= \text{E}(X^2) - 2\mu E(X) + \mu^2$$

$$= \text{E}(X^2) - 2[E(X)]^2 + [E(X)]^2$$

$$= \text{E}(X^2) - [E(X)]^2$$

Hence
$$\text{var}(X) = E(X^2) - [E(X)]^2 \qquad (19.4)$$

We usually denote population variance by σ^2.

It follows that $\text{var}(aX + b) = a^2 \text{ var}(X)$ (19.5)

where a and b are constants.

Further, if X and Y are independent, $\text{var}(X + Y) = \text{var}(X) + \text{var}(Y)$.

19.3 The Binomial Distribution

Suppose a trial has two outcomes; we can label these success and failure in some sense. We denote the probability of success by p and that of failure by $q(= 1 - p)$. We are interested in experiments which consist of several trials repeated. Some examples of such trials (sometimes called **Bernoulli trials**) are:

(i) a marksman firing a shot at a target; a success is a bull's-eye

(ii) a die is thrown: success is a six ; $p = 1/6$, $q = 5/6$

(iii) a large box of components contains 1% defectives – a component is selected: success is choosing a defective (if we are trying to test a theory which says there are defectives in the box); $p = 1/100$, $q = 99/100$.

If we repeat a **binomial** trial a **fixed** number of times, n, we can define a new discrete random variable which is the number of successes in the n trials.

Provided the probability of success is the same for each trial and the trials are independent of each other we define the experiment as a **binomial** experiment.

Note that four conditions must be satisfied:

(i) the trials must have only two outcomes

(ii) the number of trials must be fixed

(iii) the probability of success is the same for all trials

(iv) the trials are independent.

These form the assumptions of a **binomial model**. We have to decide on the applicability of the model in any given situation. For example, if the marksman fires five rounds at the target we might reasonably assume his chance of a bulls-eye remains constant unless some sudden change in the light occurs, but if he fires 100 rounds this may not be the case. Again, if two components are selected from the box the probability of the second one being defective is not quite the same as that for the first, but if the box contains a large number of components, conditions (iii) and (iv) are nearly satisfied.

Let us find the probability of 0, 1, 2, 3, 4 successes in an experiment consisting of up to 4 repeated independent Bernoulli trials each with probability of success p. We enumerate the probabilities as follows:

Number of trials Number of successes	1	2	3	4
0	q	q^2	q^3	q^4
1	p	$2pq$	$3pq^2$	$4pq^3$
2	0	p^2	$3p^2q$	$6p^2q^2$
3	0	0	p^3	$4p^3q$
4	0	0	0	p^4

For example, for two successes in three trials the possible outcomes are *SSF*, *SFS*, *FSS*. Since the trials are independent the respective probabilities of these outcomes are *ppq*, *pqp*, *qpp*, that is, each is p^2q; since these outcomes are mutually exclusive, the probability of two successes in three trials is $3p^2q$. You can check any of the others by a tree diagram. We prove one other case: 2 successes in 4 trials. Each of the possible outcomes comprises 2 successes and two failures; each has probability p^2q^2. Now we can imagine a row of 4 pigeon-holes into which we must put a letter from a set consisting of 2 *S*'s and 2 *F*'s. Once we decide where the *S*'s go, the *F*'s must go into the empty holes. These *S*'s can go in any 2 of the 4 holes and this can be accomplished in ${}^4C_2 = 6$ ways ; hence the probability we seek is $6p^2q^2$.

It should be clear that we can summarise (and generalise) the above to n trials and say that the probabilities of the various outcomes are just the terms in the expansion of $(p + q)^n$. It follows from this that the sum of these probabilities is 1.

Example
A die is thrown with a success being a "six". We now calculate the probabilities of the various outcomes if the trial is repeated 5 times. The probabilities are the terms in the expansion of $(q + p)^5$, i.e. q^5, $5pq^4$, $10p^2q^3$, $10p^3q^2$, $5p^4q$, p^5 where $p = 1/6$, $q = 5/6$, and these become

$$\frac{3125}{7776}, \frac{3125}{7776}, \frac{1250}{7776}, \frac{250}{7776}, \frac{25}{7776}, \frac{1}{7776} \text{ (which, you can check, sum to 1)}$$

Binomial Formula
An alternative generalisation is: if an experiment consists of n independent Bernoulli trials each with probability of success p and of failure q $(= 1 - p)$ the probability of the experiment yielding r successes is

$${}^nC_r\, p^r\, q^{n-r}, \quad r = 0, 1, 2, \ldots, n \tag{19.6}$$

[which is sometimes written $b(r;\ n, p)$] – this is the **binomial formula**.

This expression is just a typical term in the expansion of $(q + p)^n$. We argue that r successes have probability p^r of occurring and the $(n - r)$ failures have probability q^{n-r}. The outcome r successes followed by $(n - r)$ failures has the probability $p^r\, q^{n-r}$; this will be the probability of any arrangement of r successes and $(n - r)$ failures and as there are ${}^nC_r$ ways of choosing the positions of the r successes we obtain the probability given by (19.6).

Example 1
A die is thrown 56 times; the probability of obtaining at least 3 sixes

$$= 1 - (\text{probability of } 0, 1, 2 \text{ sixes})$$
$$= 1 - [(5/6)^{56} + 56(5/6)^{55} \,.\, (1/6) + {}^{56}C_2(5/6)^{54}(1/6)^2]$$

Example 2
In a manufacturing process it is found that on average 0.5% of the articles produced are defectives. If the articles are packed in cartons of 100, what is

(i) p(the carton is free from defectives)
(ii) p(there are at least 2 defectives in a carton)?

We may take a binomial model of 100 trials with $p = 1/200$.

(i) $p(0) = \left[\frac{99.5}{100}\right]^{100}$. This is easily evaluated as 0.606 (3 d.p.)

(ii) $p(1) = 100\left[\frac{99.5}{100}\right]^{99}\left[\frac{0.5}{100}\right] = 0.304$ (3 d.p.)

Hence $p(\geq 2) = 1 - p(0) - p(1) = 0.090$ (3 d.p.)

Example 3

A hurdler has a probability of 15/16 of clearing each hurdle; what is the probability that he knocks down less than 2 hurdles in a flight of 10?

Here the binomial model is not quite accurate since if a hurdler knocks one hurdle down he may be thrown out of his stride and the next hurdle may fall; the trials are not *strictly* independent. However, applying the model

$$p(0) = (15/16)^{10} = 0.524 \quad (3 \text{ d.p.})$$

and $$p(1) = 10(15/16)^9(1/16) = 0.350 \quad (3 \text{ d.p.})$$

Then $$p(\text{less than } 2) = p(0) + p(1) = 0.874 \quad (3 \text{ d.p.})$$

Expectation and variance of the binomial distribution

Now $$\mu = E(X) = \sum_{r=0}^{n} {}^nC_r\, p^r (1-p)^{n-r} \,.\, r$$

Consider an example. Let us repeat the trial 3 times; then

$$\mu = q^3 \,.\, 0 + 3pq^2 \,.\, 1 + 3p^2q \,.\, 2 + p^3 \,.\, 3 = 3pq^2 + 6p^2q + 3p^3$$
$$= 3p(q^2 + 2pq + p^2) = 3p(q+p)^2 = 3p$$

In general, since we would have in n trials with constant probability of success p an expected number of successes np we have

$$\mu = np$$

The number of successes is the sum of n random variables, each of which has two outcomes (0 for failure, 1 for success); the mean of one of these variables is $1 \times p + 0 \times q = p$ and by an extension of rule (iv) for expectations,

$$E(X) = np$$

For the variance of X, we have

$$\text{var}(X) = E[(X-\mu)^2] = \text{E}(X^2) - [E(X)]^2$$

It can be shown that $$\text{var}(X) = npq$$

Use of probability generating function

We can obtain these last two results by means of a **probability generating function** (p.g.f.) for the binomial distribution. We consider the expansion

$$(q + pt)^n \quad \text{where } t \text{ is a dummy variable.}$$

Note that the coefficient of t^r gives the probability of r successes in n trials

(P_r). If we write $(q + pt)^n = P_0 + P_1 t + P_2 t^2 + \ldots + P_n t^n$ then differentiation with respect to t yields

$$np(q + pt)^{n-1} = P_1 + 2P_2 t + \ldots + nP_n t^{n-1} \tag{19.7a}$$

If we now multiply throughout by t we obtain

$$npt(q + pt)^{n-1} = P_1 t + 2P_2 t^2 + \ldots + nP_n t^n$$

and a further differentiation yields

$$np(q + pt)^{n-1} + npt(n - 1)p(q + pt)^{n-2} = P_1 + 2^2 P_2 t + \ldots + n^2 P_n t^{n-1} \tag{19.7b}$$

If we put $t = 1$ in (19.7a) we have

$$np(q + p)^{n-1} = 1 \,.\, P_1 + 2 \,.\, P_2 + \ldots + nP_n$$

or $np \,.\, 1 = 0 \,.\, P_0 + 1 \,.\, P_1 + 2 \,.\, P_2 + \ldots + nP_n = \sum_{r=0}^{n} rP_r = E(X)$

If we put $t = 1$ in (19.7b) we obtain

$$np(q + p)^{n-1} + np(n - 1)p(q + p)^{n-2} = 1^2 P_1 + 2^2 P_2 + \ldots + n^2 P_n$$

that is,

$$np + np^2(n - 1) = 0^2 P_0 + 1^2 P_1 + 2^2 P_2 + \ldots + n^2 P_n = \sum_{r=0}^{n} r^2 P_r = E(X^2)$$

that is

$$E(X^2) = np + n^2p^2 - np^2 = n^2p^2 + np(1 - p) = n^2p^2 + npq$$

so that

$$\text{var}(X) = E(X^2) - [E(X)]^2 = n^2p^2 + npq - (np)^2 = npq$$

Example
A fair coin ($p = 1/2$) is tossed 64 times. The Expected Number of Heads = np = 64.(1/2) = 32. The variance is npq = 32.(1/2) = 16, giving a standard deviation of 4 heads.

General Examples

Example 1
Of a large number of mass-produced articles, one-tenth are defective; find the probability that a random sample of 20 will contain
(a) exactly 3 defective articles (b) at least 3 defective articles.
The model used is Binomial with $n = 20$, $p = 0.1$, $q = 0.9$

(a) $p(3) = {}^{20}C_3(1/10)^3(9/10)^{17} = 0.190$ (3 d.p.)

(b) $p(\text{at least } 3) = 1 - p(0, 1 \text{ or } 2) = 1 - p(0) - p(1) - p(2)$
$= 1 - (9/10)^{20} - 20(9/10)^{19} \,.\, (1/10) - [(20.19)/2][(9/10)^{18} \,.\, (1/10)^2]$
$= 0.323$ (3 d.p.)

Example 2 *(Activity sampling)*
A machine shop contains 4 operatives who use a particular machine. Random

checks showed that each needed the machine for 22% of the time. How many machines would be needed for 8 operatives if the men were not to spend more than 4% of their working time waiting for a spare machine?

Let N be the number of machines. We can use a Binomial model, counting a success as wanting a machine with $p = 0.22$; clearly, at least one man will be waiting for a machine at a given time if more than N men want to use a machine at that time. The probability of this is

$$\sum_{r=N+1}^{8} {}^8C_r\, p^r q^{8-r} = 1 - \sum_{r=0}^{N} {}^8C_r\, p^r q^{8-r}$$

which is $\not> 0.04$.

We require that $$\sum_{r=0}^{N} {}^8C_r\, p^r q^{8-r}$$

just exceeds 0.96 and look for the value of N which just tips the total over 0.96. We quote the results to 3 d.p.

$$p(0) = (0.78)^8 = 0.137$$
$$p(1) = 8\,.\,(0.78)^7(0.22) = 0.309$$

therefore $$p(0) + p(1) = 0.446$$

$$p(2) = \frac{8.7}{2}\,.\,(0.78)^6(0.22)^2 = 0.305$$

therefore $$p(0) + p(1) + p(2) = 0.751$$

Similarly

$$\sum_{r=0}^{3} p(r) = 0.923, \qquad \sum_{r=0}^{4} p(r) = 0.984$$

We see that with 3 machines the cumulative probability is less than 0.96 and with 4 machines the cumulative figure clears 0.96. Therefore we need 4 machines.

Example 3

Certain manufactured components are made containing 1% defective. They are made into assemblies which are classified as sub-standard if they contain one or more defective components.

(a) If not more than 10% of the assemblies are to be sub-standard what is the maximum number of components they can contain?

(b) If the assemblies consist of 40 components, what is the maximum percentage of defectives in the bulk of components if only 5% of the assemblies are to be sub-standard?

(c) A different kind of assembly is to be produced with the components in 40 pairs, a pair being defective only if both components which comprise it are defective. If the percentage of sub-standard assemblies is to be at most 5%, what maximum percentage of defective components is permissible?

(a) Let N be the number of components per assembly. At least 90% of assemblies must contain no defective components. The probability of no defectives in a random sample of N is

$$q^N = (1 - 0.01)^N = (0.99)^N$$

We require that $(0.99)^N \geq 0.90$

that is, $\quad N \ln(0.99) \geq \ln 0.90$

therefore $\quad N \leq 10.5$

Hence at most 10 components are allowed.

(b) Let α be the proportion of defectives allowed; then $(1 - \alpha)^{40} \geq 0.95$

that is, $\quad 40 \ln(1 - \alpha) \geq \ln 0.95$

therefore $\quad 1 - \alpha \geq 0.9987$

i.e. $\quad \alpha \leq 0.0013 = 0.13\% \quad$ (2 d.p.)

(c) If the components are paired, the probability of a defective pair is α^2. Then

$$(1 - \alpha^2)^{40} \geq 0.95$$

This leads to $\quad \alpha^2 \leq 0.0013$

so that $\quad \alpha \leq \sqrt{0.0013} \cong 0.036 \quad$ (3 d.p.)

that is $\quad \alpha \leq 3.6\% \quad$ (2 s.f.)

19.4 Poisson Distribution

Consider the following situations:
(i) the number of accidents/year in a particular factory
(ii) the number of faults in a length of cable
(iii) the number of cars crossing a bridge per hour
(iv) the number of bacteria in a square metre of material.

(i) In any year there are almost endless opportunities for accidents to occur but several factors help reduce the actual number to a minimum. We can calculate the average number of accidents/year based on a long period of experience.
(ii) Here we must talk in terms of, say, an average number of faults/kilometre.
(iii) In the design of bridges we want to avoid congestion: our only sensible measure is an average number of cars/hour.
(iv) This is a two-dimensional extension of (ii).

We say that for isolated events in space and time, that is, in those situations where the average number of events in a specified interval is λ, the probability of r such events occurring in that interval is given by

$$p(r) = \frac{\lambda^r e^{-\lambda}}{r!} \tag{19.8}$$

The resulting distribution is known as the **Poisson distribution**.

It can be shown that the variance of the distribution is λ so that **the mean is equal to the variance**.

Example 1

The number of goals scored in 500 league games were distributed as follows:

Goals/match	0	1	2	3	4	5	6	7	8
Frequency	52	121	129	90	42	45	18	1	2

Compare the frequency distribution to a Poisson distribution.

We first compute the average number of goals/match, $\lambda = 1173/500 = 2.346$. Then we calculate the Poisson frequencies: $500p(0)$, $500p(1)$, ..., $500p(8)$ and compare.

$p(0) = e^{-2.346}$ therefore $500p(0) = 500 \times 0.09575 = 48$

$p(1) = e^{-2.346} \times 2.346$ therefore $500p(1) = 500 \times 0.2246 = 112$

$p(2) = e^{-2.346} \times 2.346^2/2$ therefore $500p(2) = 500 \times 0.2635 = 132$

etc

and so we obtain the following theoretical frequencies:

48	112	132	103	60	28	11	4	1

Example 2

The average rate of telephone calls received at an exchange of 8 lines is 6 per minute. Find the probability that a caller is unable to make a connection if this is defined to occur when all lines are engaged within a minute of the time of the call.

Provided the overall rate of calls is constant, we can use the Poisson formulation. Our time unit is 1 minute and so $\lambda = 6$; formula (19.8) becomes

$$p(r) = \frac{6^r \,.\, e^{-6}}{r!}$$

Now the probability of not being able to make a call is the probability of there being at least 9 calls in any one interval of a minute. This is

$$\sum_{r=9}^{\infty} \frac{6^r \,.\, e^{-6}}{r!} = 1 - \sum_{r=0}^{8} \frac{6^r \,.\, e^{-6}}{r!}$$

We speed the calculations by noting that

$$p(r+1) = \frac{\lambda}{r+1} p(r)$$

$p(0) = e^{-6} = 0.0025$; all results in Table 19.1 are quoted to 4 d.p.

Hence the probability we seek is $1 - 0.8473 = 0.1527$ (4 d.p.).

Is this accuracy justifiable?

Table 19.1

r	$\lambda/(r+1)$	$p(r)$
0	6	0.0025
1	3	0.0149
2	2	0.0446
3	1.5	0.0892
4	1.2	0.1339
5	1.0	0.1606
6	6/7	0.1606
7	0.75	0.1377
8		0.1033
	Total	0.8473

Example 3
On average, 240 vehicles per hour pass through a check-point and a queue forms if more than 3 vehicles attempt to pass in a minute; what is the probability that a queue will form in any given minute? If the check-point is closed for five minutes, what is the probability that a queue of at least two cars will form in the five minutes?

The unit we work with is a minute, and the average number of cars per minute is 4. Hence we use the Poisson formula with $\lambda = 4$. The probability of a queue forming is the probability of at least 4 vehicles arriving in one minute which is

$$\sum_{r=4}^{\infty} p(r) = 1 - p(0) - p(1) - p(2) - p(3)$$

Now $p(0) = e^{-4} = 0.0183$

Table 19.2

r	$\lambda/(r+1)$	$p(r)$
0	4	0.0183
1	2	0.0733
2	4/3	0.1465
3	1	0.1954
	Total	0.4335

From Table 19.2

$$p(\text{queue}) = 1 - 0.4335 = 0.5665 \quad \text{(4 d.p.)}$$

In a period of five minutes the expected number of cars arriving is 20 and the probability of 2 or more arriving is

$$\sum_{r=2}^{\infty} \frac{20^r e^{-20}}{r!} = 1 - e^{-20} - 20e^{-20} = 1.0000 \quad (4 \text{ d.p.})$$

Example 4

A shop sells on average 16 articles of product A per month; if the shop re-orders each week, to what number must the shop-keeper make up his stock if the risk of his running out of stock is not to exceed 5%, 10%?

The model is Poisson (why?) with $\lambda = 4$ articles/week. If we stock x articles then the stock will run out if the demand exceeds x, so for the risk of running out not to exceed 5%, we require

$$\sum_{r=x+1}^{\infty} \frac{4^r e^{-4}}{r!} \leq 0.05$$

that is,

$$1 - \sum_{r=0}^{x} \frac{4^r e^{-4}}{r!} \leq 0.05$$

or

$$\sum_{r=0}^{x} \frac{4^r e^{-4}}{r!} \geq 0.95$$

We can proceed as before to accumulate probabilities until the total just exceeds 0.9. You can check for yourselves that for 5% risk the number of articles required is 8, and repeating the calculations for 10% risk gives 7.

The real crux of applying the Poisson distribution is the decision as to the value of λ by selecting the appropriate interval.

Poisson approximation to the Binomial distribution

It is clear that the Binomial formula (19.6) becomes tedious to operate in situations where n is large and p (or q) is very small; in addition, accuracy may be lost in the course of the calculations. We would like a somewhat simpler formula which is a good approximation to (19.6). Suppose then, we take the situation to the extreme. We seek a distribution which will approximate the Binomial distribution when n is very large and p (or q) is very small.

We achieve this by taking the limit of the Binomial distribution as $n \to \infty$, $p \to 0$ and $np = \lambda$ stays fixed. (np is the mean of the Binomial distribution). The case q very small is achieved by reversing *success* and *failure*.

We give a derivation of the formula for the probability of r successes in our new distribution; this derivation may be omitted if desired.

For the Binomial distribution, the probability formula (19.6)

$${}^nC_r\, p^r\, q^{n-r}$$

may be written

$$\frac{n(n-1)(n-2)\ldots(n-r+1)}{r!}\,\frac{(1-p)^n p^r}{(1-p)^r}$$

$$= \frac{1\,.\left[1-\frac{1}{n}\right].\left[1-\frac{2}{n}\right]\ldots\left[1-\frac{r-1}{n}\right]}{r!}\,.\left[1-\frac{\lambda}{n}\right]^n\frac{\lambda^r}{\left[1-\frac{\lambda}{n}\right]^r}$$

Now, as $n \to \infty$ and $\lambda = np$ stays fixed and finite, then p must $\to 0$,

$${}^nC_r\, p^r\, q^{n-r} \to \frac{\lambda^r}{r!}e^{-\lambda}$$

since

$$\left[1-\frac{1}{n}\right] \text{etc} \to 1, \quad \left[1-\frac{\lambda}{n}\right]^r \to 1 \quad \text{and} \quad \left[1-\frac{\lambda}{n}\right]^n \to e^{-\lambda} \qquad (19.8)$$

This is, of course, the Poisson formula (19.8).

Example 1
Certain mass-produced articles, of which 0.5% are defective, are packed in cartons each containing 100. What proportion of cartons are free from defective articles and what proportion contain two or more defectives?

Using Binomial $n = 100, p = 0.005$

$$p(0) = {}^{100}C_0\,(0.995)^{100} = 0.6058$$
$$p(1) = {}^{100}C_1\,(0.995)^{99}\,(0.005) = 0.3044$$
$$p(2) = {}^{100}C_2\,(0.995)^{98}\,(0.005)^2 = 0.0757$$

Using Poisson $\lambda = np = 0.5$

$$p(0) = e^{-0.5} = 0.6065$$
$$p(1) = (0.5)e^{-0.5} = 0.3033$$
$$p(2) = (1/8)e^{-0.5} = 0.0758$$

Using the Poisson approximation, the proportion free from defectives is 60.65% and the proportion containing 2 or more defectives is 9.02%.

If the number of items involved is not large enough, the approximation may be poor.

Example 2
The average number of defective articles produced by a machine is 1 in 20; a batch of 10 is tested: what is the probability that there will be at least three defectives in the batch?

For the Binomial model $p = 0.05$ with $\mu = np = 10\ .\ (1/20) = 0.5$ and variance $= npq = 0.5 \times 0.95 = 0.495$ (for Poisson, mean and variance are equal).

We use Poisson, since p is small, via the formula

$$p(r) = \frac{(0.5)^r\, e^{-0.5}}{r!}$$

$$p(0 \text{ defectives}) = e^{-0.5} = 0.6065$$

$$p(1) = 0.5\, e^{-0.5} = 0.3033 \qquad p(2) = [(0.5)^2/2]\, e^{-0.5} = 0.0758$$

therefore $p(> 2) = 1 - p(0) - p(1) - p(2) = 0.0144$ (4 d.p.)

For comparison, the Binomial model gives

$$p(0) = (0.95)^{10} = 0.5987$$

$$p(1) = 10(0.95)^9.0.05 = 0.3151$$

$$p(2) = \frac{10.9}{2}(0.95)^8(0.05)^2 = 0.0746$$

therefore $p(> 2) = 0.0116$ (4 d.p.)

The approximation is about 19% in error.

Problems

Section 19.1

1 The sample space $X = \{0, 1, 2, 3\}$ is given. Which of the following can be probability functions? (Determine the value of C where necessary.) For those which are, find the probabilities of the events.

(a) x is odd (b) x is at least 1 (c) x is 1 or 2

For those which are not, state why not.

(i) $\frac{1}{8}(1 + x)$ (ii) $\frac{1}{10}(1 + x)$ (iii) $\frac{2}{x}$ (iv) $\frac{C}{x^2 + 1}$ (v) $\frac{x^2 - 2}{4}$ (vi) $\frac{C}{2 + x}$

2 Find the probabilities associated with all the possible scores obtained when two dice (assumed fair) are thrown and draw a line diagram to represent the probability distribution.

3 A customer buys 20 transistors and tests a sample of 5 of them. If 4 of the transistors are faulty, how many of the possible samples of 5 will contain at least 1 faulty transistor? Find the probabilities of a sample containing 0, 1, 2, 3, or 4 faulty transistors. Draw a line diagram to represent the probability distribution.

4 If the sample space is $X = \{0, 1, 2, 3, 4\}$, verify that with a suitable choice of k, $p(X) = k/(X + 2)$ can be used as a probability function. Sketch the distribution.

5 The following is a simplified model representing the spread of measles within a family of three susceptible children A, B and C. If A catches measles, there is a period during which B may get the disease from him and the chance that B does in fact succumb is θ; similarly for C. The probabilities for B and C are independent. Thus in the first stage none, one or two of the children may catch the disease. If the first or last event occurs this completes the

matter, but if B, say, gets the disease and not C, there is a further period during which C may, with probability θ, catch the disease from B.
Show that, given that one child has measles, the chances that 0, 1, 2 of the other children get it are $(1-\theta)^2$, $2\theta(1-\theta)^2$, $\theta^2(3-2\theta)$. Obtain corresponding probabilities for families of four children.

Section 19.2

6 A prize of £200 is given for naming the three winning dogs, in correct order, in a race for which 10 dogs were entered. What is the expectation of a person having no inside knowledge of the ability of the dogs?

7 If a single die is thrown (assumed fair) write down the probability distribution of the scores X and obtain $E(X)$ for the distribution. Also obtain var(X).

8 A man buys a sweepstake ticket. The first prize is £10 000, the second prize £1000 and the third prize if £100. His probabilities of winning these prizes are 0.0001, 0.0008 and 0.005 respectively. What is a fair price to pay for the ticket if he can only win one prize?

9 Show that $E(X-\mu)^3 = E(X^3) - 3\mu E(X^2) + 2\mu^3$

10 The daily demand for a particular commodity at a grocery store has the following distribution

X	0	1	2	3	≥ 4
$p(X)$	0.36	0.35	0.19	0.09	0.01

Find the expected or average daily demand over a long period of time.

11 An event has a fixed probability p of success. Show that the expected number of failures before the first success occurs is

$$E = \sum_{r=0}^{\infty} rp(1-p)^r$$

Show that (i) $E = (1-p)/p$ and that (ii) the expected number of trials necessary for a success is $1/p$. [Hint for (i): find $E-(1-p)E$].

Section 19.3

12 A box contains ten components, all apparently sound, but four of them are, in fact, sub-standard. If three components are removed from the box altogether, what is the probability that two are sub-standard?

13 A sampling inspection plan operates as follows. Take a random sample of size ten from a large batch. If none of the sample is defective, then accept the batch. If more than one are defective, then reject the batch. If exactly one is defective, take another sample of size ten, and accept the batch only if this second sample contains no defectives. If a batch which is 5% defective is tested by this plan, what is the probability that
(i) it is accepted after the first sample (ii) it is accepted?

14 In a series of trials, an anti-aircraft battery had, on average 3 out of 5 successes in shooting down missiles, the trials taking place over a long period. What is the probability that if 8 came within range not more than 2 would get through?

15 Given a Binomial distribution, what is the value of p for the probability of 4 successes in 10 trials to be twice the probability of 2 successes in 10 trials?

16 On average, 3% of the articles produced by a certain manufacturer are defective. What is the probability that in a sample of 10 articles
(i) 2 are defective (ii) at least 3 are defective?

17 A successful attack by an interceptor depends upon
(a) the reliable operation of a computing system
(b) the transmission of correct directions
(c) the proper function of the striking mechanism.
When the probability of (a) is 0.7 and (b) is assured, the overall probability of success is 0.6. If the computing system is improved to 95% reliability, while (b) has only 0.8 probability and the probability of (c) is unchanged, what is the new overall probability of success?

18 Certain components are manufactured in quantity; the probability that any one taken at random will be faulty is q. Show that if we are to be 99% certain that any set of four components will contain at least two sound ones, then q must be given by the equation

$$300q^4 - 400q^3 + 1 = 0$$

19 In a survey of a day's production from 400 machines making similar components, 4 items selected at random from the output of each machine were inspected in detail. The number m of machines producing f faulty items was found to be

f	0	1	2	3	4
m	16	89	145	118	32

Represent this data by a binomial distribution and calculate the theoretical distribution of faulty components. (EC)

20 Of a large number of mass-produced articles, one in ten is defective. Find the probability that a random sample of 20 will contain
(i) exactly 2 defective articles (ii) at least 2 defective articles.

21 A radio transistor has a probability of functioning for at least 800 hours of 0.25. Find the probability that in a test sample of 20 transistors
(i) none function more than 800 hours (ii) only one functions more than 800 hours.

22 If two out of every ten vehicles turn right at a T-junction, what is the probability that
(i) three consecutive vehicles turn right
(ii) three out of five consecutive vehicles turn left?

23 For a binomial distribution with $n = 5$ and $p = 1/4$, tabulate the separate probabilities. Find the mean and variance of the distribution
(i) directly (ii) from the general formula $\mu = np, \sigma^2 = npq$

24 In an experiment the probability of success is 5/7; the experiment is repeated 12 times. Show that the most probable number of successes is 9 with probability of occurrence of 0.248. Draw a line diagram for the distribution.

25 In an experiment the probability of success is 1/4. Find the most probable number of successes and the associated probability in 15 trials. Draw a line diagram.

26 In the manufacture of screws by a certain process it was found that 5% of the screws were rejected because they failed to satisfy tolerance requirements. What was the probability that a sample of 12 screws contained
(a) exactly 2 (b) not more than 2 rejects? (LU)

27 A mass-produced article is packed in cartons each containing 40 articles. Seven hundred cartons were examined for defectives with the results given in the table.

Defective articles per carton	0	1	2	3	4	5	6	More than 6
Frequency	390	179	59	41	18	10	3	0

Obtain the proportion of defective articles p.
Assuming that the total number of articles examined is so large that p can be taken to represent the proportion of defectives in the population, use the Binomial distribution to calculate the probability that a random sample of 5 of these articles will contain 2 defectives. (LU)

Section 19.4

28 (i) In a factory, 2000 employees will be asked to use a barrier cream for which the probability of an initial reaction sufficient to cause absence from work is 0.001. Find the probability that more than two employees will be absent when the barrier cream is introduced.
(ii) The number of days n in a 100-day period when x accidents occurred in a factory is shown.

No. of accidents (x)	0	1	2	3	4	5
No. of days (n)	42	35	14	6	2	1

(a) Fit a Poisson distribution to this data to predict the probability of x accidents occurring.
(b) Compare the variance of the given and calculated distributions. (EC)

29 Telephone calls are received at a switchboard at an average rate of 1.5 calls per minute. The switchboard is operated by two telephonists, and calls have an equal chance of being dealt with by either. What is the probability that a given telephonist does not have to deal with any calls in a particular two-minute interval? Do not attempt to evaluate the resulting expression.

30 A machine-shop storekeeper finds that, over a long period, the average demand per week for a certain machine tool is 3. His stocking policy is to make up stock to 4 at the beginning of each week. Estimate the probability that he will fail to satisfy demand in a given week, and determine his stocking policy if the chance of running out is not to exceed 5%.

31 On average, telephone calls are received at an exchange at the rate of 3 per minute. The exchange has 4 lines. What is the probability of being unable to make a connection if this arises when all lines are engaged within a minute of the time of the call?

32 A factory uses a particular spare part at the average rate of 3 a week. If stocks are replenished weekly, use the Poisson distribution to determine the number which should be in stock at the beginning of each week so that not more than once every year on average (i.e. probability 1/52) will the stock be insufficient.

33 If the chance that any one of 10 telephone lines is busy at an instant is 0.2, what is the chance that 5 of the lines are busy? What is the most probable number of busy lines?

34 Refer to Problem 17, Chapter 4.
On the assumption that the number of vehicles passing in a two-minute interval follows a Poisson distribution with the same mean as this data, calculate the probability of less than 3 vehicles passing in a two-minute interval.

35 A process for manufacturing a device comprising s connected components can be taken to consist of s independent operations. The probability of r flaws in any component is Poisson with $\lambda = 1$. Show that the probability of a defective device (contains more than one flaw) is $1 - e^{-s}(1 + s)$.

36 Stoppages on a building site occur according to the Poisson law on the average once in three weeks. Obtain the probabilities of 0, 1, 2, 3, 4 stoppages occurring in five weeks and determine if the occurrence of 5 stoppages in five weeks is significant.

37 The probability of a fault in a component is 1/1000.
(i) Find approximately the probability that there will be 4 faulty components in a daily output of 2000.
(ii) Find the minimum daily output to ensure a probability of 95% of at least one faulty component.

38 If, on average, one in every thousand entries in a table contains an error, what is the probability that a set of 100 readings taken from the table are all correct?

39 Oranges are packed in crates of one gross (144). It is estimated that, on average, 1 in 96 oranges is bad. In a consignment of 1000 crates, how many will contain 2 or more bad oranges? (Use the Poisson approximation to the Binomial distribution.)

40 Describe the Binomial and Poisson distributions stating the circumstances in which each might be used.
Random samples of fifty are examined from each batch of a large consignment of manufactured articles and in such samples the average number of defective articles was found to be 3.1. A batch is rejected if the sample taken from it contains three or more defective articles. Assuming a Poisson distribution, show that the probability of a batch being rejected is approximately 0.6. Calculate the probability that of six batches sampled three or more will be rejected. Find also the probability that of 100 batches sampled at least half will be accepted. (LU)

41 The probability that any one machine will become defective in the small interval between time t and $t + \delta t$ after maintenance is $\alpha e^{-\alpha t}\delta t$, where α is a constant.

A firm possessing 10 similar machines institutes a weekly maintenance of the machines, each of which operates for 44 hours per week, and $\alpha = 1/400$ when the unit of time is one hour. Find:

(i) the probability that all ten machines will continue to function throughout the week

(ii) the probability that more than two machines will become defective before the weekly maintenance is due. (LU)

42 (a) The probability function for the Poisson distribution is $P(r) = \dfrac{\lambda^r \exp(-\lambda)}{r!}$

(i) Use this formula to show that $\sum_{r=0}^{\infty} P(r) = 1$.

(ii) A store-keeper finds that over a long period of time the average number of articles needed per week is 2. His policy is to make up his stock to 3 at the beginning of each week. What is the probability that he will fail to satisfy the demand in any given week?

(b) A machine makes a large number of components of which on average 4% are defective, the defective components occurring at random during production. If the components are packaged 60 per box, use the Poisson approximation to find the probability that a given box will contain

(i) no defective components (ii) 1 defective component

(iii) more than 2 defective components. (EC)

20

THE NORMAL DISTRIBUTION AND SIGNIFICANCE TESTS

20.1 Continuous Probability Distributions

Consider the experiment of spinning a mounted pointer and measuring the final angle of deflection made with some fixed direction (easiest to visualise as an unmagnetised compass needle); again consider the time taken to perform a certain calculation. In both cases the distribution may seem to be discrete because of the limitations of measuring devices, but, in theory, the angle can be any of an infinite number of possible angles in the interval [0, 360°] and the time taken can be any value in the range $(0, \infty)$. With the spun pointer, if each outcome has an equal probability of occurrence this is infinitesimally small and cannot be distinguished from zero. In general, for continuous distributions it makes little sense to talk about the probability of a single outcome. Rather, we confine our attention to probabilities of an interval of outcomes: for example, we may say that the probability of the angle at which the pointer settles being in the range 0 to 180° is 1/2, in the range 90° to 180° is 1/4, in the range 150° to 180° is 1/12, and in the general range $[\theta_1, \theta_2]$ is $(\theta_2 - \theta_1)/360$. These ideas lead us to the idea of a **probability density function** (p.d.f.) which, in this example, is defined by

$$\lim_{\delta\theta \to 0} \left\{ \frac{1}{\delta\theta} p(\theta_0 \le \theta \le \theta_0 + \delta\theta) \right\} = \rho(\theta_0)$$

There is a clear analogy between the discrete distribution and a light rod bearing isolated point loads and between the continuous distribution and a heavy rod where the weight is distributed continuously along its length: hence the term 'density function'. The properties that the density function $\rho(x)$ must obey are

(i) $\rho(x) \ge 0$ for all x in the sample space

(ii) $\int_{-\infty}^{\infty} \rho(x) \mathrm{d}x = 1$, or, if the interval of outcomes is $[a, b]$, $\int_a^b \rho(x) \mathrm{d}x = 1$

Notice how Σ for discrete distributions is replaced by $\int$ for continuous distributions.

The spun pointer experiment is an example of a **rectangular distribution** and the graph is shown in Figure 20.1. Note that the area under the graph is 1.

We can concoct any continuous distribution by defining a suitable p.d.f.

Example

Are any of the following functions p.d.f.'s?

(a) $\rho(x) = \frac{1}{2}x$, defined on $[1, 4]$ (b) $\rho(x) = \frac{1}{4} - x$, defined on $[-2, 2]$

(c) $\rho(x) = \lambda(1 - x)$, defined on $[\frac{1}{2}, \frac{3}{4}]$

Sketch the graphs of those which are.

(a) $\rho(x) \geq 0$ but $\int_1^4 \frac{1}{2}x\,dx = [\frac{1}{4}x^2]_1^4 = 4 - \frac{1}{4} \neq 1$, therefore not a p.d.f.

(b) $\int_{-2}^{2} (\frac{1}{4} - x)dx = \left[\frac{1}{4}x - \frac{x^2}{2}\right]_{-2}^{2} = (\frac{1}{2} - 2) - (-\frac{1}{2} - 2) = 1$, but $\rho(x) < 0$ for x in $[\frac{1}{4}, 2]$, therefore not a p.d.f.

(c) $\rho(x) \geq 0$ if $\lambda \geq 0$; $\int_{1/2}^{3/4} \lambda(1 - x)dx = \lambda\left[x - \frac{x^2}{2}\right]_{1/2}^{3/4}$

$$= \lambda\left\{\left[\frac{3}{4} - \frac{9}{32}\right] - \left[\frac{1}{2} - \frac{1}{8}\right]\right\}$$

$$= \lambda\left\{\frac{15}{32} - \frac{3}{8}\right\} = \lambda \cdot \frac{3}{32}$$

and thus $\lambda = 32/3$, if property (ii) is to be satisfied. Assuming this value for λ, the graph is shown in Figure 20.2. Again the area under the graph is 1.

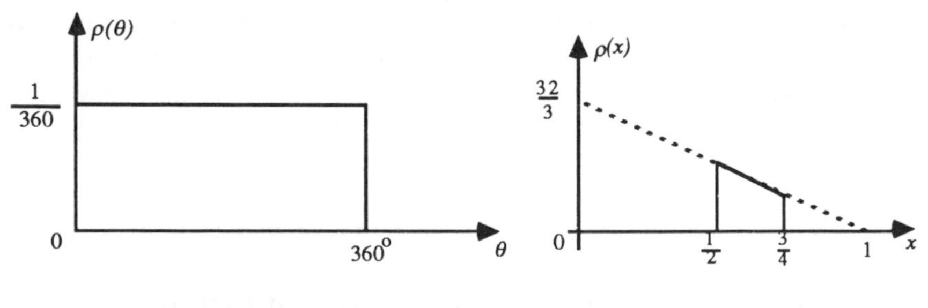

Figure 20.1 Figure 20.2

If, for example, we want $p(0.6 \leq x \leq 0.7)$, this is found as

$$\int_{0.6}^{0.7} \frac{32}{3}(1-x)\mathrm{d}x$$

i.e. the area under the graph between the ordinates 0.6 and 0.7.

Cumulative distribution function

We find it useful to talk in terms of a **cumulative distribution function** (c.d.f.) $F(x)$ which gives the probability of all events up to and including a particular value, that is

$$F(x_0) = p(X \leq x_0) = \int_{-\infty}^{x_0} \rho(x)\mathrm{d}x \tag{20.1}$$

$$\text{for interval } [a, b], F(x_0) = \int_{a}^{x_0} \rho(x)\mathrm{d}x$$

For a discrete distribution we write

$$F(x_r) = \sum_{i=0}^{r} p(x_i) \tag{20.1a}$$

For the rectangular distribution of Figure 20.1,

$$F(\theta) = \int_{0}^{\theta} \frac{1}{360}\mathrm{d}\theta = \frac{\theta}{360}$$

and for the p.d.f. of Figure 20.2,

$$F(x) = \int_{1/2}^{x} \frac{32}{3}(1-x)\mathrm{d}x = \frac{32}{3}\left[x - \frac{x^2}{2}\right] - 4$$

Note that

$$\int_{-\infty}^{1/2} \rho(x)\mathrm{d}x$$

is not relevant since x is defined only on $[\frac{1}{2}, \frac{3}{4}]$.

The graphs are shown in Figures 20.3(i) and (ii).

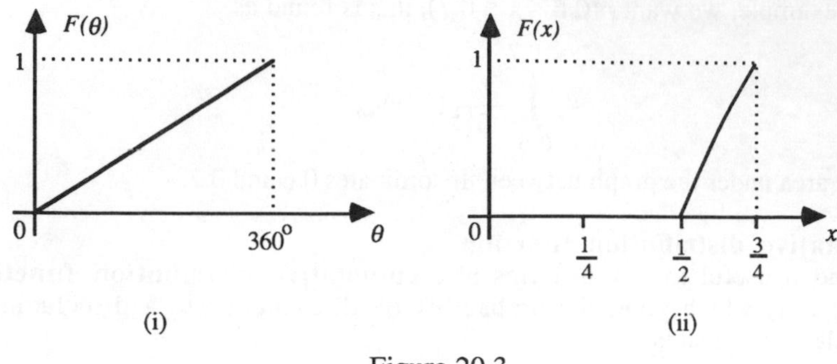

Figure 20.3

Measures of central tendency and spread

Other features which are analogous to those for discrete distributions are listed. X is assumed to be a continuous random variable defined on $(-\infty, \infty)$.

(i) $\mu = E(x) = \int_{-\infty}^{\infty} x\rho(x)\mathrm{d}x$

Properties (i) to (v) on page 504 hold here also.

(ii) For a unimodal distribution, the mode is found where $\dfrac{\mathrm{d}\rho(x)}{\mathrm{d}x} = 0$

(iii) The median M is such that $\int_{-\infty}^{M} \rho(x)\mathrm{d}x = \dfrac{1}{2} = \int_{M}^{\infty} \rho(x)\mathrm{d}x$

(iv) The variance $= \int_{-\infty}^{\infty} (x-\mu)^2\rho(x)\mathrm{d}x$

$$= \int_{-\infty}^{\infty} x^2\rho(x)\mathrm{d}x - 2\mu \int_{-\infty}^{\infty} x\rho(x)\mathrm{d}x + \mu^2 \int_{-\infty}^{\infty} \rho(x)\mathrm{d}x$$

$$= \int_{-\infty}^{\infty} x^2\rho(x)\mathrm{d}x - 2\mu \,.\, \mu + \mu^2 \,.\, 1$$

$$= \int_{-\infty}^{\infty} x^2\rho(x)\mathrm{d}x - \mu^2$$

$$= E(X^2) - [E(X)]^2$$

20.2 The Normal Distribution

A measurement which is subject to error has observational errors following a **normal** or **Gaussian distribution**. Many continuous distributions in practice, such as the heights or weights of a large group of people or the life of a class of electrical components follow approximately a normal distribution, a sketch of which is shown in Figure 20.4

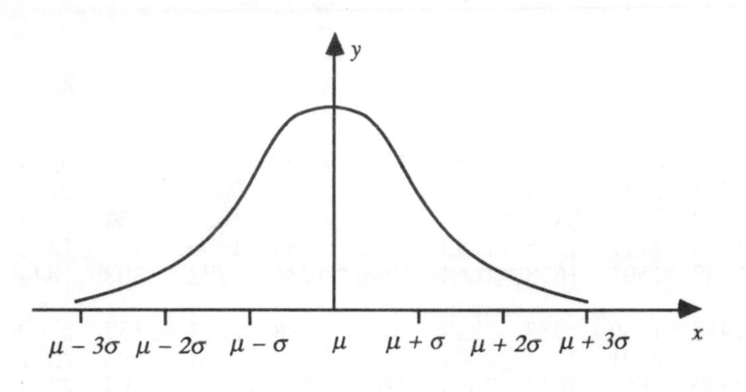

Figure 20.4

This distribution is defined entirely in terms of its mean μ and standard deviation σ. The defining formula is

$$y = \frac{1}{\sigma\sqrt{2\pi}} e^{-1/2[(x-\mu)/\sigma]^2} \tag{20.2}$$

or we sometimes say *x is normally distributed with mean μ and standard deviation* σ, written

$$x \sim N(\mu, \sigma^2) \tag{20.2a}$$

Note the symmetry of the distribution about μ and the rapid decay on either side.

Although a whole host of such distributions exists, we make use of the fact that, if $x \sim N(\mu, \sigma^2)$ then the variable

$$z = \frac{x-\mu}{\sigma} \sim N(0, 1)$$

$N(0, 1)$ is called the **standard normal distribution**; it is given by

$$y = \frac{1}{\sqrt{2\pi}} e^{-z^2/2}$$

and it has been tabulated in Table 20.1. It is of great significance, since we can easily convert statements about $N(\mu, \sigma^2)$ to ones about $N(0, 1)$, which can then be assessed.

The values quoted (to 4 d.p.) lie between 0 and 0.5 (why?).

Table 20.1

$$\frac{1}{\sigma\sqrt{2\pi}}\int_{\mu}^{x}\exp\left\{-\frac{(x-\mu)^2}{2\sigma^2}\right\}dx = \frac{1}{\sqrt{2\pi}}\int_{0}^{(x-\mu)/\sigma} e^{-z^2/2}\,dz$$

The Normal Probability Integral

$\frac{x-\mu}{\sigma}$	0	1	2	3	4	5	6	7	8	9
0	0000	0040	0080	0120	0160	0199	0239	0279	0319	0359
0.1	0398	0438	0478	0517	0557	0596	0636	0675	0714	0753
0.2	0793	0832	0871	0909	0948	0987	1026	1064	1103	1141
0.3	1179	1217	1255	1293	1331	1368	1406	1443	1480	1517
0.4	1555	1591	1628	1664	1700	1736	1772	1808	1844	1879
0.5	1915	1950	1985	2019	2054	2088	2123	2157	2190	2224
0.6	2257	2291	2324	2357	2389	2422	2454	2486	2517	2549
0.7	2580	2611	2642	2673	2703	2734	2764	2794	2822	2852
0.8	2881	2910	2939	2967	2995	3023	3051	3078	3106	3133
0.9	3159	3186	3212	3238	3264	3289	3315	3340	3365	3389
1.0	3413	3438	3461	3485	3508	3531	3554	3577	3599	3621
1.1	3643	3665	3686	3708	3729	3749	3770	3790	3810	3830
1.2	3849	3869	3888	3907	3925	3944	3962	3980	3997	4015
1.3	4032	4049	4066	4082	4099	4115	4131	4147	4162	4177
1.4	4192	4207	4222	4236	4251	4265	4279	4292	4306	4319
1.5	4332	4345	4357	4370	4382	4394	4406	4418	4429	4441
1.6	4452	4463	4474	4484	4495	4505	4515	4525	4535	4545
1.7	4554	4564	4573	4582	4591	4599	4608	4616	4625	4633
1.8	4641	4649	4656	4664	4671	4678	4686	4693	4699	4706
1.9	4713	4719	4726	4732	4738	4744	4750	4756	4761	4767
2.0	4772	4778	4783	4788	4793	4798	4803	4808	4812	4817
2.1	4821	4826	4830	4834	4838	4842	4846	4850	4854	4857
2.2	4861	4865	4868	4871	4875	4878	4881	4884	4887	4890
2.3	4893	4896	4898	4901	4904	4906	4909	4911	4913	4916
2.4	4918	4920	4922	4925	4927	4929	4931	4932	4934	4936
2.5	4938	4940	4941	4943	4946	4947	4948	4949	4951	4952
2.6	4953	4955	4956	4957	4959	4960	4961	4962	4963	4964
2.7	4965	4966	4967	4968	4969	4970	4971	4972	4973	4974
2.8	4974	4975	4976	4977	4977	4978	4979	4979	4980	4981
2.9	4981	4982	4982	4983	4984	4984	4985	4985	4986	4986
3.0	4987	4990	4993	4995	4997	4998	4998	4999	4999	4999

The probability of x lying in the range $[\mu-\sigma, \mu+\sigma]$ is approximately 68%; in $[\mu-2\sigma, \mu+2\sigma]$ is approximately 95% and in $[\mu-3\sigma, \mu+3\sigma]$ is approximately 99.7%. Put another way, the probability of x lying outside $[\mu-\sigma, \mu+\sigma] \cong 32\%$, outside $[\mu-2\sigma, \mu+2\sigma] \cong 5\%$ and outside $[\mu-3\sigma, \mu+3\sigma]$ is about $\frac{1}{4}\%$.

Table 20.1 gives the probability of x lying in the shaded area of Figure 20.5.

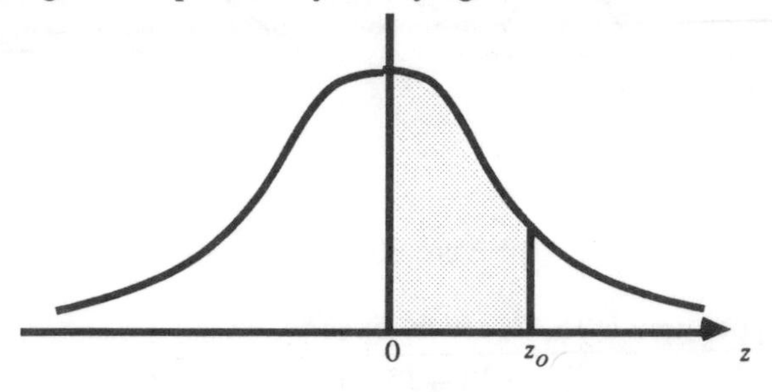

Figure 20.5

For example, $p(0 \leq z \leq 1.5) = 0.4332$ (= 43.32%)
$p(0 \leq z \leq 2.67) = 0.4962$
$p(0 \leq z \leq 3.6) = 0.4998$

You should confirm these values from Table 20.1.

We now apply these ideas to an example which illustrates how to compute probabilities for various kinds of interval relative to the mean.

Example

A machine produces components of mean diameter 15.35 mm with standard deviation 0.05 mm. The diameters are assumed normally distributed.

(a) Find the probabilities
  (i) that a component has diameter between 15.35 and 15.43 mm
  (ii) that a component has diameter between 15.23 and 15.35 mm
  (iii) that a component has diameter between 15.29 and 15.42 mm
  (iv) that a component has diameter between 15.37 and 15.44 mm
  (v) that a component has diameter between 15.27 and 15.33 mm

(b) If all components with diameters outside the range 15.28 to 15.40 mm are rejected, what proportion of components are rejected? What is the probability that of 4 components selected at least 3 will be rejected?

(c) If 28% of components have diameters less than a given value, what is this value?

The distribution of the diameters is shown in Figure 20.6(i).

(a) The first task in each problem is to code the variable x by using

$$z = \frac{x-\mu}{\sigma} = \frac{x-15.35}{0.05} \quad \text{so that, e.g. } 15.35 \equiv 0$$

The distribution of z is sketched in Figure 20.6(ii)

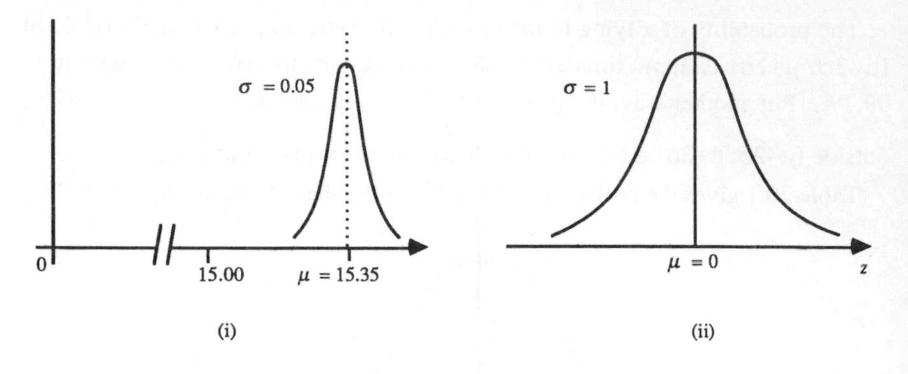

Figure 20.6

(i) The upper limit is 15.43, which gives

$$z = \frac{15.43 - 15.35}{0.05} = 1.6$$

We want the value of the shaded area in Figure 20.7(i). From the table we have that the area corresponding to $z = 1.6$ is 0.4452 and hence the probability we seek is 44.52% (2 d.p.). Note we are saying that 15.43 is 1.6 standard deviations above the mean.

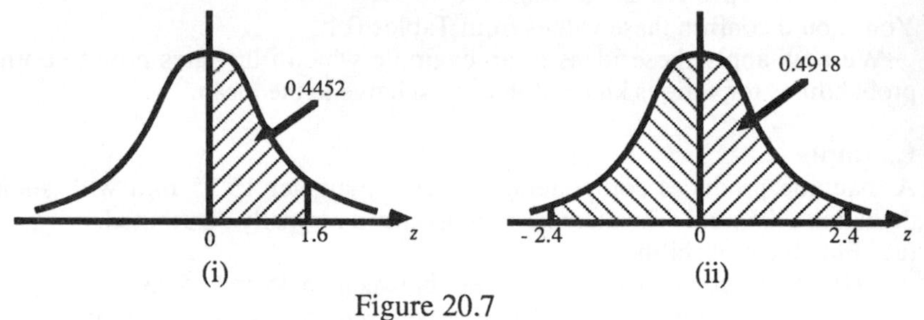

Figure 20.7

(ii) Here the lower limit is

$$z = \frac{15.23 - 15.35}{0.05} = -2.4$$

By the symmetry of the normal distribution, the shaded area we want is equal to the shaded area between 0 and +2.4 as shown in Figure 20.7(ii). From Table 20.1 we pick up 0.4918 or 49.18% and this is the probability required.

(iii) The limits are

$$z_1 = \frac{15.29 - 15.35}{0.05} = -1.2 \quad \text{and} \quad z_2 = \frac{15.42 - 15.35}{0.05} = 1.4$$

By reference to Figure 20.8(i) we can see that the area we want is the sum of two areas. The one between 0 and z_2 has the associated probability 0.4192. Now, the shaded area between –1.2 and 0 is equal to one between 0 and +1.2 which is 0.3849; the sum of these areas is 0.8041 and hence we obtain a probability of 80.41% (2 d.p.)

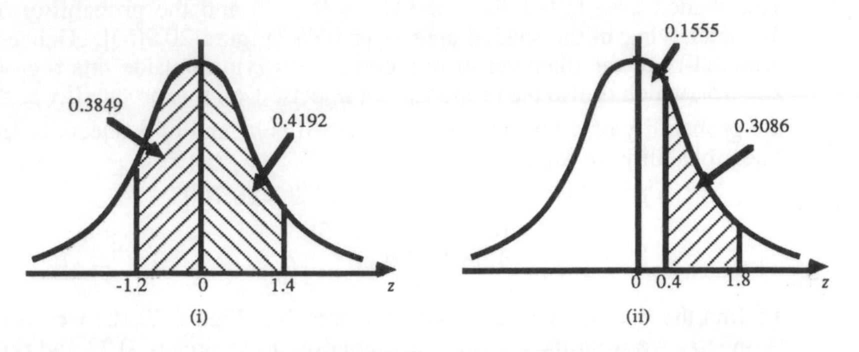

Figure 20.8

(iv) Here the limits are

$$z_1 = \frac{15.37 - 15.35}{0.05} = 0.4 \quad \text{and} \quad z_2 = \frac{15.44 - 15.35}{0.05} = 1.8$$

The shaded area we want is the difference between the area from 0 to 1.8 and the area from 0 to 0.4 [see Figure 20.8(ii)]. The first of these areas is 0.4641 and the second is 0.1555 and, therefore, we have the area difference of 0.3086 giving a probability of 30.86% (2 d.p.).

(v) In a like manner, these limits are

$$z_1 = \frac{15.27 - 15.35}{0.05} = -1.6 \quad \text{and} \quad z_2 = \frac{15.33 - 15.35}{0.05} = -0.4$$

By symmetry, we want the area between 0.4 and 1.6; this is 0.4452 – 0.1555 = 0.2897. Refer to Figure 20.9(i). The required probability is 28.97%.

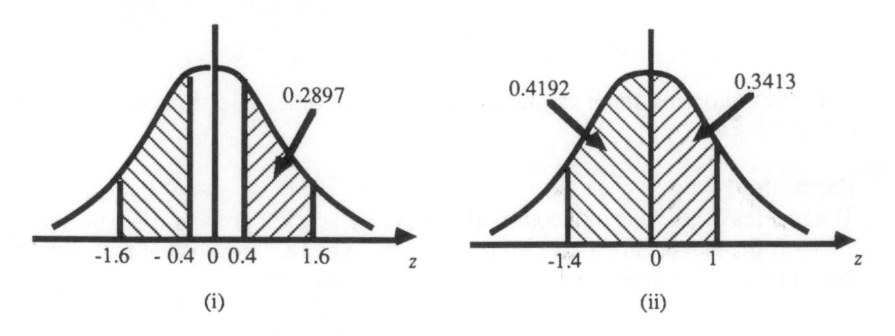

Figure 20.9

(b) We have limits of

$$z_1 = \frac{15.28 - 15.35}{0.05} = -1.4 \quad \text{and} \quad z_2 = \frac{15.40 - 15.35}{0.05} = 1.0$$

The shaded area is 0.4192 + 0.3413 = 0.7605 and the probability of a diameter lying in the shaded area is 76.05% [Figure 20.9(ii)]. Hence the probability of the diameter of one component lying outside this region is 23.95%, which is also the proportion of rejects. Call this probability p; then the probability of 4 rejects out of 4 selected is p^4 and of 3 rejects is $4p^3q$. The probability we want is

$$\begin{aligned} p^4 + 4p^3q &= (0.2395)^4 + 4(0.2395)^3(0.7605) \\ &= (0.2395)^3(3.2815) \\ &= \underline{4.58\%} \text{ (2 d.p.)} \end{aligned}$$

(c) To find the line cutting off 28% of the area (see Figure 20.10) we note by symmetry we want the z-score corresponding to an area of 0.22 and this is between 0.58 and 0.59, from the table. We choose the nearer value 0.58 and hence have $z_1 = -0.58$; decoding, we obtain

$$\begin{aligned} x &= \mu - 0.58\sigma \\ &= 15.35 - 0.58 \times 0.05 \\ &= 15.35 - 0.029 \\ &= \underline{15.32} \text{ (3 d.p.)} \end{aligned}$$

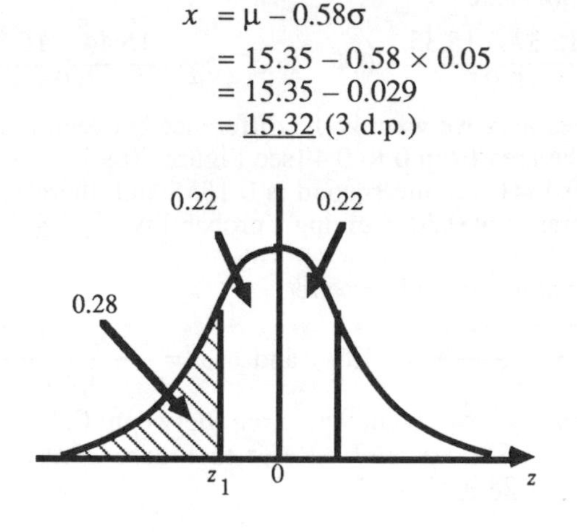

Figure 20.10

In the following three examples we illustrate other questions which might be asked.

Example 1

The masses of packets of a commodity have mean 18.2 gm and it is observed that 10% of the packets are above an acceptable level 18.7 gm. What is the standard deviation of the distribution, assuming it to be normal?

In this case we have $z_1 = \dfrac{x_1 - \mu}{\sigma} = \dfrac{18.7 - 18.2}{\sigma} = \dfrac{0.5}{\sigma}$

But, from Table 20.1, $z_1 = 1.28$ (2 d.p.), therefore

$$\sigma = \frac{0.5}{1.28} = 0.39 \text{ gm} \quad (2 \text{ d.p.}) \qquad \text{[see Figure 20.11(i)]}$$

It is debatable whether the result should be quoted as 0.4 gm if the mean of 18.2 is accurate only to the last figure quoted; were it 18.20 then we should be justified in retaining 2 d.p. for σ.

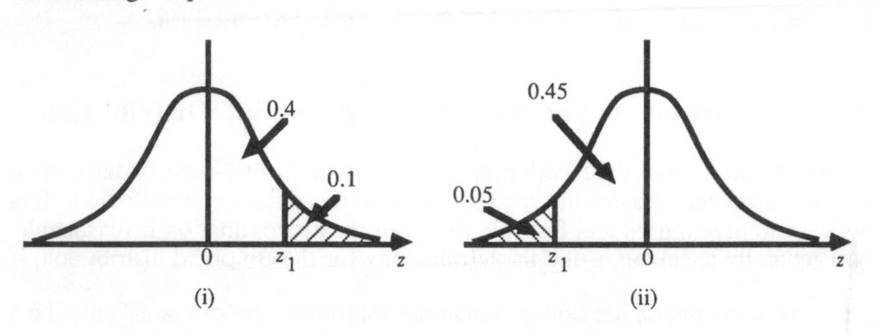

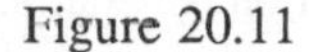

Figure 20.11

Example 2
The lengths of life of light bulbs produced in a particular factory have standard deviation 50 hours. If the lives are assumed normally distributed and 5% last less than 940 hours what is the mean life of the bulbs produced?

From the table we may deduce

$$z = -1.64 \qquad \text{[see Figure 20.11(ii)]}$$

But

$$z_1 = \frac{940 - \mu}{50}$$

therefore

$$\begin{aligned} \mu &= 940 + 1.64 \times 50 \\ &= 940 + 82 \\ &= 1022 \text{ hours} \end{aligned}$$

Example 3

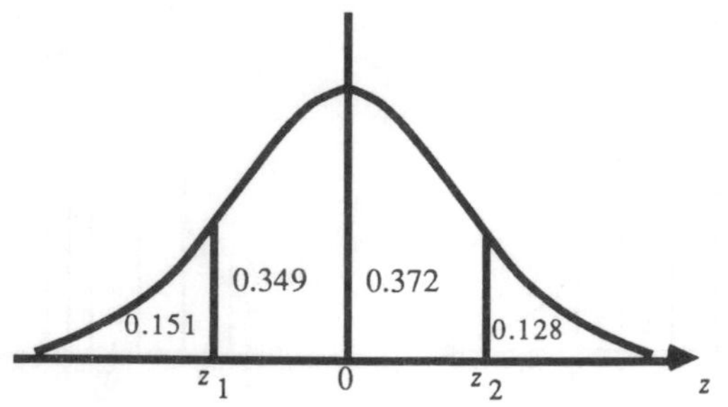

Figure 20.12

Given that the lengths of bolts produced by a machine are normally distributed, that 12.8% have lengths greater than 126 mm and 15.1% less than 124.7 mm, what are the mean and standard deviation of the distribution? [See Figure 20.12]

We obtain $z_2 = 1.14$ and $z_1 = -1.03$ (2 d.p.)

Hence $\dfrac{126 - \mu}{\sigma} = 1.14$ and $124.7 = -1.03\sigma + \mu$

Hence we solve to obtain $\mu = 125.3$ $\sigma = 0.6$ (2 d.p.)

20.3 Normal Approximation to Binomial Distribution

Although the normal distribution is continuous, it can be used under certain circumstances as a reasonable approximation to the Binomial distribution. It is not easy to give a hard and fast rule for when the approximation is reasonable, but, generally speaking, using the terminology for the Binomial distribution, if $p \cong \frac{1}{2}$, or if np and nq are both > 5 then the approximation can be taken to be a fair one. We show the approximation by examples.

Example 1

A die is thrown 18 times: what is the probability of 8 successful outcomes, if a successful outcome is defined to be a 'five' or 'six'?

The binomial model has $p = 1/3$, $q = 2/3$, $n = 18$ (note $np = 6$, $nq = 12$) which has mean $\mu = np = 6$ and standard deviation $\sigma = \sqrt{npq} = \sqrt{(1/3).(2/3).18}$ $= 2$. We now approximate this model by a normal distribution with $\mu = 6$, $\sigma = 2$.

We approximate the probability of 8 successes by the area under the curve between 7.5 and 8.5. Now we cannot score 7.5 successes, but we are effectively splitting the difference between the last case we want to exclude – 7 successes – and the 8 we want to include; a similar consideration obtains at the 8.5 mark [Figure 20.13(i)]. We are effectively approximating a histogram by the normal curve and approximating the rectangle based on 8 by the area mentioned above. The normal curve and the histogram are superimposed in Figure 20.13(ii).

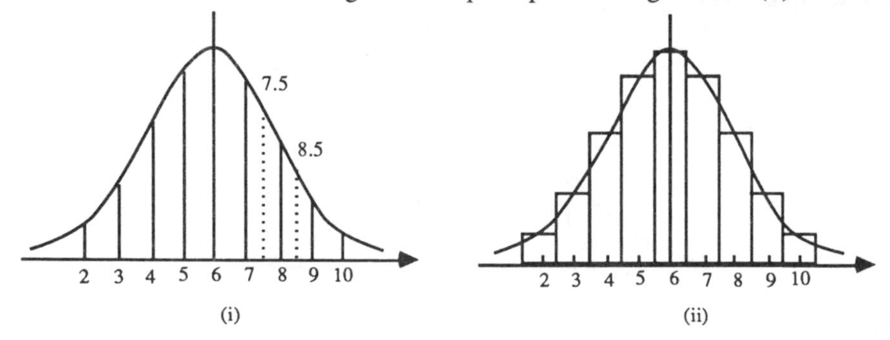

Figure 20.13

Now we find this area under the normal curve:
The limits are

$$z_1 = \frac{7.5 - 6}{2} = 0.75 \quad \text{and} \quad z_2 = \frac{8.5 - 6}{2} = 1.25$$

The required area is 0.3944 – 0.2734 = 0.1210 and the probability we obtain is therefore 12.10% (2 d.p.); the exact binomial result is

$$^{18}C_8\, p^8\, q^{10} = {}^{18}C_8(1/3)^8(2/3)^{10} = 11.57\% \quad (2 \text{ d.p.})$$

A comparison of Binomial (B) and Normal (N) probabilities is shown in Table 20.2, where r is the number of successes.

Table 20.2

r	0	1	2	3	4	5	6	7
B	0.0007	0.0051	0.0259	0.0690	0.1294	0.1818	0.1963	0.1682
N	0.0024	0.0092	0.0279	0.0655	0.1210	0.1747	0.1974	0.1747

r	8	9	10	11	12	13	≥ 14
B	0.1157	0.0643	0.0289	0.0105	0.0031	0.0007	0.0001
N	0.1210	0.0655	0.0279	0.0092	0.0024	0.0005	Negligible

Binomial probabilities shown sum to 1.000
Normal probabilities shown sum to 0.9993

Example 2
For the experiment above obtain the probabilities of
(i) at least 12 successes (ii) between 8 and 10 successes (i.e. 8, 9 or 10)
These results can be obtained by adding entries from Table 20.2 or directly, as follows:

(i) $p(\text{at least 12 successes}) \cong p(11.5 \leq \text{successes})$

For 11.5 successes, $z = \dfrac{11.5 - 6}{2} = 2.75$

The probability that z lies between 0 and 2.75 is 0.4970 and so the probability of at least 12 successes is at most 0.0030 or 0.30% (2 d.p.). The corresponding Binomial result is 0.39% (2 d.p.)

(ii) Corresponding to 7.5 successes is $z_1 = \dfrac{7.5 - 6}{2} = 0.75$

Corresponding to 10.5 successes is $z_2 = \dfrac{10.5 - 6}{2} = 2.25$

The probability that z lies between 0.75 and 2.25 is

$$0.4878 - 0.2734 = 0.2144 = 21.44\% \quad (2 \text{ d.p.})$$

The corresponding Binomial result is 20.89% (2 d.p.)

20.4 Sampling Distributions and Statistical Inference

We mentioned some of the reasons behind sampling in Chapter 4. Our aim is to use the sample as a basis for inferring certain things about the parent population. For the population we have certain **parameters**, for example, the mean, mode, standard deviation, etc. If similar quantities are computed from a sample, each is called a **sample statistic**. Whereas the value of a parameter is constant for a population, the corresponding statistic is variable from sample to sample. We can see this from an example: the sample {4, 3, 2, 6, 3} is taken from a population with mean $\mu = 3.2$. The mean of the sample is

$$\bar{x} = (1/5)(4 + 3 + 2 + 6 + 3) = 3.6$$

yet, if we took only the first three items as our sample, the sample mean $\bar{x} = 3$.

The sample mean $\bar{x}$ is calculated in the same way as the population mean μ. However, the sample standard deviation, s, of a sample of size n is calculated by the formula

$$s = \sqrt{\sum_{r=1}^{n} (x_r - \bar{x})^2 \Big/ (n-1)} \tag{20.3}$$

Now, for large values of n, this gives almost the same result as the formula

$$\sqrt{\frac{1}{n} \sum_{r=1}^{n} (x_r - \bar{x})^2}$$

and this latter formula is sometimes used in that case. But for small samples there is a marked discrepancy; the reason for choosing (20.3) is discussed in the next section. Again, s will vary, in general, from sample to sample.

There are three main inferences to draw from a sample and these are dealt with in the next three sections.

In Section 20.5 we study the estimation of a parameter from a sample statistic. We shall mainly be concerned with the estimation of μ, the population mean, from the observed sample mean, $\bar{x}$. Of course, we cannot do more than quote a range of possible values for μ, around $\bar{x}$, outside which the chances of μ taking such a value are small. Since statistical inference is *not exact* we must attach a *risk* to our interval: the risk of a wrong statement.

Another aspect of sampling discussed in Section 20.6 is the testing of hypotheses about, for example, the population mean. If the sample mean is markedly different from the expected, or claimed, population mean does this indicate that μ is not what it should be? Of course, we cannot be certain, but we can state a conclusion and attach the risk of it being wrong. This is a branch of **statistical decision-making**.

In Section 20.7 we compare sample means and other statistics. We turn for the rest of this section to the study of the behaviour of certain sampling statistics. On page 539 we deal with sums and differences.

Behaviour of sample means
A statistical population consists of the following six numbers, each printed on a ball: 1, 4, 4, 6, 6, 9. The balls are placed in a bag. The mean of this population $\mu = 5$ and the standard deviation $\sigma = \sqrt{6}$. Now consider the possible samples of size two we can draw from this population. There are ${}^6C_2$, that is, 15 of them.

For each sample we calculate the sample mean, $\bar{x}$. The possible samples and their means are shown below in Table 20.3. Notice we have to allow for the fact that there are two fours and two sixes, and although there are only six distinct pairings, some occur twice as often as others.

Table 20.3

Sample	Mean	Sample	Mean	Sample	Mean
(1, 4)	2.5	(4, 4)	4	(4, 6)	5
(1, 4)	2.5	(4, 6)	5	(4, 9)	6.5
(1, 6)	3.5	(4, 6)	5	(6, 6)	6
(1, 6)	3.5	(4, 9)	6.5	(6, 9)	7.5
(1, 9)	5	(4, 6)	5	(6, 9)	7.5

We have several distinct sample means; how well does any one of them represent the true population mean of 5?

We see that the samples (1, 9) and (4, 6) give exactly the population mean, yet the samples (1, 4) and (6, 9) give means differing by 2.5 from that of the population. The other samples give means 1 or 1.5 above or below the true mean. Had we presented the bag of balls to a third person and asked him to select a sample of two and find the sample mean, he would have no clue as to how accurate a prediction of the population mean this would be. The sample means constitute a random variable, the behaviour of which we now study.

First we compute its arithmetic mean, that is, the mean of the sample means. Since we have deliberately written down duplicates, we can simply average the entries in the above table, therefore

$$E(\bar{x}) = \frac{1}{15}(2.5 + 2.5 + 3.5 + 3.5 + 5 + 4 + 5 + 5 + 6.5 + 5 + 5 + 6.5 + 6 + 7.5 + 7.5) = 5$$

We see that the expectation of the sample means is equal to the population mean.

This is quite generally true: however the individual sample means may be above or below the population mean, their average is always equal to the population mean (assuming all sample means are used in the computation of the average). Now you can argue that in this example we could easily enumerate the possible samples, but suppose we took a sample of ten components produced by a machine and found the sample mean diameter. How could we be sure that the above result applies since we would have to measure all the components! You will have to accept this result, because although there is a large number of possible samples, we would get on average the population mean.

Let us now consider samples of four taken from the population of six numbered balls; it should be easy to see that there are again fifteen of them (why?) and they are shown in Table 20.4, together with their means.

Table 20.4

Sample	Mean	Sample	Mean	Sample	Mean
(4, 6, 6, 9)	6.25	(1, 6, 6, 9)	5.5	(1, 4, 6, 9)	5
(4, 6, 6, 9)	6.25	(1, 4, 6, 9)	5	(1, 4, 6, 6)	4.25
(4, 4, 6, 9)	5.75	(1, 4, 6, 9)	5	(1, 4, 4, 9)	4.5
(4, 4, 6, 9)	5.75	(1, 4, 6, 6)	4.25	(1, 4, 4, 6)	3.75
(4, 4, 6, 6)	5	(1, 4, 6, 9)	5	(1, 4, 4, 6)	3.75

Note this time there are again five sample means agreeing with the population mean, but the sample means are bunched more closely about that mean – the furthest being 1.25 away.

If we compute $E(\bar{x})$ we find it is

$$\frac{1}{15}(2 \times 6.25 + 2 \times 5.75 + 5.5 + 5 \times 5 + 2 \times 4.25 + 4.5 + 2 \times 3.75) = 5$$

We now compute the standard deviation of the sample means for samples of size two. This is denoted by $\sigma_{\bar{x}}$ and can be calculated to be

$$\sqrt{\frac{36}{15}} = \frac{6}{\sqrt{15}}$$

Here we divide by 15 and not 14 since we are treating the sample as a set of numbers at the moment.

If we compute $\sigma_{\bar{x}}$ for samples of size 4 we obtain

$$\sqrt{\frac{9}{15}} = \frac{3}{\sqrt{15}}$$

and we see that doubling the sample size halves $\sigma_{\bar{x}}$.

It is clear that the larger sample size gives us less chance of being too far from the population mean, but to choose, for example, a sample size two-thirds that of the population size would in practice be ridiculous.

We have so far considered sampling *without* replacement. The case where the population is very large compared with the sample size approximates sampling *with* replacement and this is the case we now pursue in the rest of the chapter.

From the general results for variances on page 118 we have

$$\text{var}(\bar{x}) = \text{var}\left[\frac{x_1 + x_2 + \ldots + x_n}{n}\right]$$

$$= \frac{1}{n^2}[\text{var}(x_1) + \text{var}(x_2) + \ldots + \text{var}(x_n)]$$

$$= \frac{1}{n^2}.n\sigma^2 \quad = \frac{\sigma^2}{n}$$

Then the **standard error of the mean** $\sigma_{\bar{x}} = \frac{\sigma}{\sqrt{n}}$ and generalises σ for a sample of size 1.

The relationship between the distributions of x and $\bar{x}$ is shown schematically in Figure 20.14.

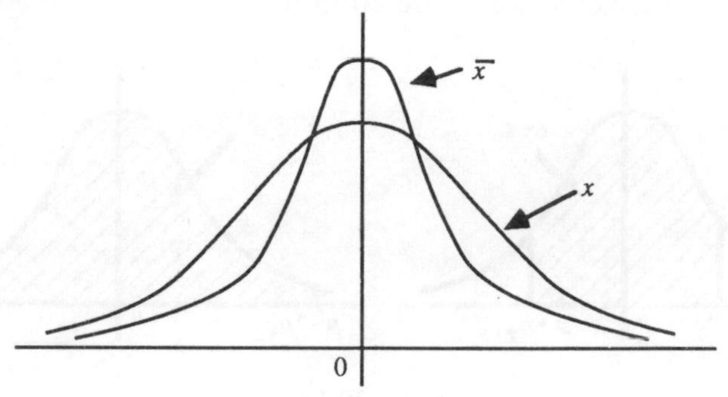

Figure 20.14

We state for such cases, with a sample of size n from which sample means $\bar{x}$ are calculated,

$$E(\bar{x}) = \mu, \quad \sigma_{\bar{x}} = \frac{\sigma}{\sqrt{n}} \tag{20.4}$$

As the sample size increases, so the sample means become more closely bunched about the mean and hence any one of them is a better guide to the population mean than would be the case for a smaller sample size.

We now make use of the important **central limit theorem**. This states that the sampling distribution of $\bar{x}$ is approximately normal if the sample size n is large, the approximation improving with increasing n. This applies irrespective of the parent population, provided σ is finite.

Now this is a quite remarkable result and one which has allowed sampling theory to proceed. It says that *no matter what* the shape of the original population sample means are *approximately normally distributed*. How large n

should be before the result holds is not clear-cut. A safe rule is that if $n \geq 30$ the theorem can be applied with confidence; for $n < 30$ we should need some assurance that the parent population is approximately normal.

Example 1
From the student registration forms for a university an experiment is carried out. 1000 samples of 64 forms are taken (with replacement) and the ages of the students noted for each sample: the average age of the population is known to be 19.8 with $\sigma = 1.6$ years. There are 1000 sample means; the first mean may be 21, the second 19.5 and so on. But the 1000 means will be approximately normally distributed ($n \geq 30$) and the average of these will be 19.8 years with a standard error of $1.6/\sqrt{64} = 0.2$ years.

By our results for the normal distribution, we know that approximately 68% of these sample means will lie in the range $\mu \pm \sigma_{\bar{x}} = 19.8 \pm 0.2$, that is, between 19.6 and 20.0 years [See Figure 20.15(i)].

Likewise approximately 95% of sample means will lie in the range $19.8 \pm (1.96 \times 0.2)$ years [see Figure 20.15(ii)].

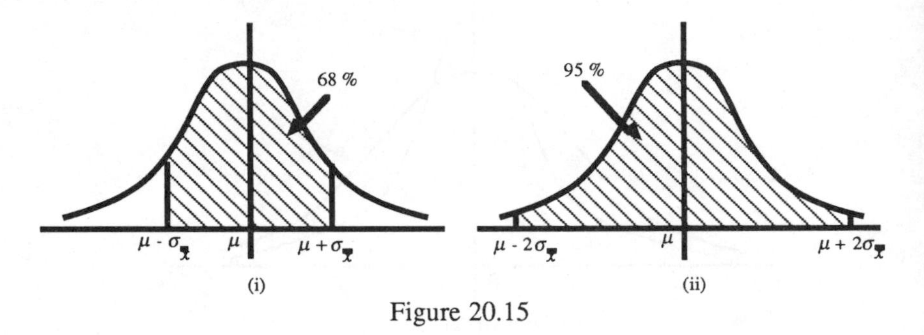

Figure 20.15

(Note there is a *double approximation*: the figures 68% and 95% are not exact and they would only be achieved approximately due to sampling variability.)

Example 2
The lives of certain machine components are normally distributed with mean $\mu = 5000$ hours and standard deviation $\sigma = 50$ hours. What is the probability
(i) that a single component will last longer than 5030 hours
(ii) that the mean life from a sample of 16 is greater than 5030 hours
(iii) that the mean life from a sample of 64 is greater than 5030 hours?

(i) We form $$z = \frac{5030 - 5000}{50} = 0.6$$

and from the tables, the probability of a z-score greater than 0.6 is $0.5 - 0.2257 = 0.2743$, giving a probability of 27.43% (2 d.p.)

(ii) In this case, the standard deviation σ is replaced by the standard error

$$\frac{\sigma}{\sqrt{n}} = \frac{50}{\sqrt{16}} = 12.5$$

The z-score now formed is $\dfrac{5030 - 5000}{12.5} = 2.4$

and the probability of a larger z-score is

$$0.5 - 0.4918 = 0.0082 \quad \text{or} \quad 0.82\% \ \text{(2 d.p.)}$$

(iii) the standard error is $50/\sqrt{64} = 6.25$ giving a z-score of

$$\frac{5030 - 5000}{6.25} = 4.8$$

and so the probability of the mean life from a sample of size 64 exceeding 5030 is negligibly small. Note that all we have done is to replace the standard deviation σ by the more general standard error $\sigma/\sqrt{n}$.

Distributions of sums and differences

Given two independent random variables X and Y with means μ_1 and μ_2 and variances $\sigma_1{}^2$ and $\sigma_2{}^2$ respectively, then the random variable $X + Y$ has mean $\mu_1 + \mu_2$ and variance $\sigma_1{}^2 + \sigma_2{}^2$ and the random variable $X - Y$ has mean $\mu_1 - \mu_2$ and variance $\sigma_1{}^2 + \sigma_2{}^2$. *These results apply only for independent variables*. Notice that we add variances in both cases.

Example 1

Two dice are thrown several times; the scores on each are recorded as X and Y. Find the mean and standard deviation of the differences in the scores.

For each die the mean score is 3.5 and the variance is $2\frac{11}{12}$. For the difference in the scores, the mean is $3.5 - 3.5 = 0$, and the variance is $2\frac{11}{12} + 2\frac{11}{12} = 5\frac{5}{6}$, which gives a standard deviation of 2.416 (3 d.p.).

Example 2

Batteries of type I have voltages which are distributed $N[6.0, (0.15)^2]$ whilst those of type II are distributed $N(12.0, (0.2)^2]$. A battery from each type is taken and the two connected in series, what is the probability that the combined voltage exceeds 17.4?

Let the voltages of the batteries of types I and II be x and y respectively. Then $(x + y)$ is distributed $N[18.0, (0.0225 + 0.04)]$, i.e. $N(18.0, 0.0625)$ which is $N[18.0, (0.25)^2]$. Then the critical z-score is

$$z = \frac{17.4 - 18.0}{0.25} = -2.4$$

The probability that $z > -2.4 = 1 - (\text{probability that } z < -2.4)$
$= 1 - 0.0082 = 0.9918$ (4 d.p.)

Differences between sample means

A most important distribution is the difference between the means of two independent samples from two different populations. Given a sample of size n_A from population A with mean μ_A and standard deviation σ_A and a second sample of size n_B from population B with parameters μ_B and σ_B, the difference in means is distributed with mean $\mu_A - \mu_B$ (provided the difference is taken the corresponding way) and variance $(\sigma^2{}_A/n_A) + (\sigma^2{}_B/n_B)$.

Example

A bolt of diameter y mm must fit a nut of internal diameter x mm, both selected at random from cartons. If x and y are distributed $N[1.0, (0.0009)^2]$ and $N[0.9965, (0.0012)^2]$ respectively, what is the probability that the nut and bolt would fail to fit, given that the clearance for fitting must be at least 0.0008 mm?

Let $w = x - y$, then we require $w > 0.0008$ for a fit. The distribution of w is $N(0.0035, 0.00000081 + 0.00000144)$, that is, $N[0.0035, (0.0015)^2]$. We require for failure that

$$w < \frac{0.0008 - 0.0035}{0.0015} = -1.8$$

The required probability is $0.5 - 0.4641 = 0.0359$

20.5 Estimation from a Sample

Point estimation

We first examine point estimation when a sample statistic is used to give an estimate of a population parameter; for example, $\bar{x}$ may be used to estimate μ and s^2 may be used to estimate σ^2. Often in statistical work we denote an estimated value by placing a *hat* on it: thus $\hat{\alpha}$ is an *estimated* value of α.

There are three properties of estimators worthy of attention: to define these we assume λ is a population parameter and l is a sample statistic used to estimate λ; l has a distribution with mean $E(l)$ and standard error σ.

(i) *Unbiasedness*

l is an **unbiased estimator** of λ, if $E(l) = \lambda$. If $E(l) \neq \lambda$, l is said to be a **biased estimator** of λ. Clearly we would prefer an estimator to be unbiased, for if an estimator is biased it will, on average, give too high or too low a value and the chances of obtaining a close estimate are decreased.

$\bar{x}$ is an unbiased estimator of μ since $E(\bar{x}) = \mu$.

For example, $$\frac{1}{n}\sum_{r=1}^{n}(x_r-\bar{x})^2$$

is a biased estimator of σ^2 since $$E\left[\frac{1}{n}\sum_{r=1}^{n}(x_r-\bar{x})^2\right]=\frac{\sigma^2(n-1)}{n}$$

however, $$\frac{1}{n-1}\sum_{r=1}^{n}(x_r-\bar{x})^2$$ is an unbiased estimator of σ^2.

(ii) *Efficiency*
Given two unbiased estimators which is to be preferred? Since any one value of the estimator may be too high or too low, we would prefer that value not to be too far from the true value of the population parameter. We therefore choose the estimator whose distribution has the less variance. In general, in a set of unbiased estimators, the one with least variance is called the *most efficient* estimator.

It may be shown that of all possible unbiased estimators of μ, $\bar{x}$ is the one with least variance and is therefore the most efficient estimator of μ.

(iii) *Consistency*
Since the ideal of all values of the estimator coinciding with the parameter to be estimated cannot be realised in practice, we look for an estimator which improves as the sample size n increases. If l is an unbiased estimator of λ and $\text{var}(l) \to 0$ as $n \to \infty$, then l is a **consistent estimator** of λ. For example,

$$\text{var}(\bar{x}) = \sigma^2/n \to 0 \text{ as } n \to \infty$$

so $\bar{x}$ is a consistent estimator of μ.

Sometimes we wish to pool information from two independent samples. Let the suffix 1 refer to the one sample and 2 refer to the other, then if both samples come from a population of mean μ and variance σ^2, the most useful unbiased estimators of mean and variance are

$$\hat{\mu}=\frac{n_1\bar{x}_1+n_2\bar{x}_2}{n_1+n_2} \quad \text{and} \quad \hat{\sigma}^2=\frac{(n_1-1)s_1^2+(n_2-1)s_2^2}{n_1+n_2-2}$$

You should show that these estimators are unbiased and note the obvious extension to more than two samples.

Interval estimation

Although we know that $\bar{x}$ is an unbiased, efficient and consistent estimator of μ, we need to do more than quote $\bar{x}$ as an approximate value of μ. Since we know that the sample means are normally distributed we may say that 95% of these sample means are expected to lie in the interval $[\mu - 1.96\sigma_{\bar{x}}, \mu + 1.96\sigma_{\bar{x}}]$. Alternatively, we may say that a single sample mean has a 95% probability of lying in that range or that it has a 5% chance of falling outside this interval. Yet another way of looking at the situation is that if for each sample we find $[\bar{x} - 1.96\sigma_{\bar{x}}, \bar{x} + 1.96\sigma_{\bar{x}}]$ then we would expect 95% of these intervals to contain μ.

We call the interval $[\bar{x} - 1.96\sigma_{\bar{x}}, \bar{x} + 1.96\sigma_{\bar{x}}]$ the 95% **confidence interval** for μ.

Sometimes we do not know σ^2 for the population and have to estimate it by the sample statistic s^2.

Example 1

A random sample of 100 bolts produced by a certain machine had a mean diameter of 12.5 mm with a standard deviation of 0.1 mm. Find

(i) the 95% confidence interval for the mean diameter of bolts produced by the machine

(ii) the 99% confidence interval for the mean diameter of bolts produced by the machine.

(i) The 95% (or 0.95) confidence interval for μ is $\bar{x} \pm 1.96\sigma_{\bar{x}}$

Now $\sigma_{\bar{x}} = \sigma/\sqrt{100}$ but we must approximate this by $s/\sqrt{100} = 0.01$ mm.

Then our confidence interval is

$$(12.50 \pm 0.0196) \text{ mm} \quad \text{or } (12.50 \pm 0.02) \text{ mm (2 d.p.)}$$

(ii) The 99% confidence interval for μ is $\bar{x} \pm 2.58\sigma_{\bar{x}}$. That is,

$$(12.50 \pm 0.0258) \text{ mm} \quad \text{or } (12.50 \pm 0.03)\text{mm (2 d.p.)}$$

Choice of sample size

The question which must always be asked in these kinds of problem is how large the sample should be? Now if we are given a maximum error which can be accepted, we can choose a large enough sample to try to cut down the error in our estimate, but we can never be quite sure that we shall exceed that maximum error.

For example, suppose we are asked to estimate the mean weight of bags of cement produced by a machine by estimation from a sample and the estimate is to be in error by no more than 50 gm. Clearly the larger the sample the more reliable the estimate, but whatever the size of the sample, we can never guarantee that the estimate will be within 50 gm of the correct value. All we can do is to try to ensure that the probability of the error exceeding 50 gm is less than, say, 5%.

We choose the sample size then so that $1.96\sigma_{\bar{x}}$ is just equal to 50 gm, that is,

$1.96\sigma/\sqrt{n} = 50$, or $n = (1.96\sigma/50)^2$.

Note that if we wished to ensure that the error in our estimate did not exceed the stated maximum on more than 1% of occasions, we should have to take a larger value of n, viz $(2.58\sigma/50)^2$.

Of course, if we have two populations with standard deviations σ_1, σ_2 where $\sigma_1 > \sigma_2$ then the sample sizes (n_1 and n_2) to achieve the same accuracy with the same confidence will be such that $n_1 > n_2$; this is intuitively obvious.

Example 2
Past records indicate that the lengths of rods produced by a machine are such that $\mu = 500$ cm and $\sigma = 5$ cm. Find the sample size needed if there should be a 99% confidence of the error in the sample estimate not exceeding 0.5 cm.

We have
$$\frac{2.58\sigma}{\sqrt{n}} = 0.5$$

that is,
$$\frac{2.58 \times 5}{\sqrt{n}} = 0.5$$

or
$$n = (25.8)^2 = 666, \text{ at least.}$$

20.6 Testing Statistical Hypotheses about the Population Mean

Having drawn a random sample and found its mean $\bar{x}$, we can state with a given confidence, limits within which we believe the population mean μ to lie. But what if we have been told what μ is supposed to be and this supposed value lies outside our confidence interval; do we believe the supposed value of μ or not? For example, if we throw a die 6000 times and we obtain 1250 6's we might suspect that the die were 'loaded'. If previous tests on a drug had shown it to be 80% effective and after modification of the drug sampling indicates it is 88% effective are we safe in concluding that the modified drug is more effective? (Notice we say 'safe' and not 'certain'.)

We now examine a specific example in detail.

A manufacturer claims that electrical components of a particular type last on average 1000 hours. A simple random sample of 100 components is taken and found to give an average life of 988 hours; what can we conclude about the manufacturer's claim? In statistical terms the claim is that if all the components made, and to be made, (forming an infinite population) were tested and the mean found, it would be 1000 hours.

Setting up of a statistical hypothesis

The first step is to set up a **statistical hypothesis**. This is usually of the form that the proportion of successes is unchanged, that the die is fair; that the components have the claimed average life (and, consequently, that any variability is due solely to sampling variations); this is called the **null hypothesis**, H_0. In this example, we state H_0: $\mu = 1000$.

We must state an **alternative hypothesis**, H_1, which in this case is H_1: $\mu < 1000$. Now you may ask why we did not have H_1: $\mu \neq 1000$. The reason is that the customer is unlikely to object if the component lasts more than 1000 hours: he is concerned only with getting at least 1000 hours from the component: his money's worth.

Now if the sample findings indicate a marked departure from what we had expected we would say that the departure is significant and we would tend to reject the null hypothesis, H_0; if the departure is not too large, we would tend either to accept H_0 or require further testing.

Risk of wrong judgement

In either eventuality we make a **statistical decision**. This decision must carry a **risk** of being wrong and a statement of such a risk should always be given alongside the statement of the decision.

In this example, we set a risk of wrong judgement of 5%. Assuming that a suitable sample size has been chosen we decide what statistic to use. In this section we are concerned with decisions concerning the population mean, μ, and we use as test statistic the sample mean $\bar{x}$. Having assumed that the null hypothesis H_0 is true, we then set limits on the values that $\bar{x}$ can take if we are to accept H_0. With a 5% risk, these limits are shown in Figure 20.16(i).

If $\bar{x}$ lies inside the **acceptance region** then we find no evidence to reject H_0; on the other hand, if $\bar{x}$ lies outside the acceptance region we reject H_0.

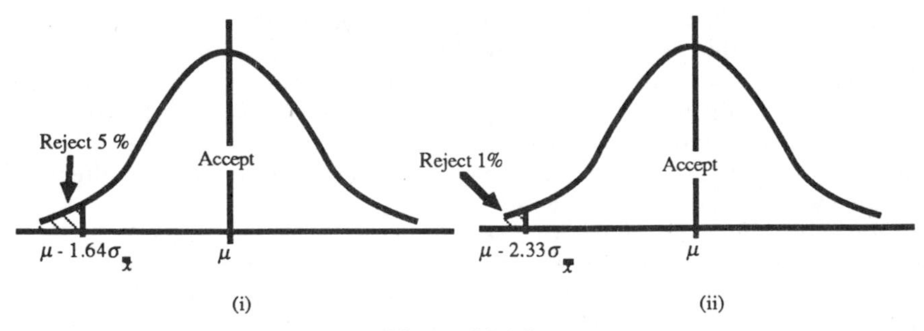

Figure 20.16

Notice here that we have a departure from the picture associated with confidence intervals in that the rejection region is on one side only; for this reason we are carrying out a **one-tailed test**.

It cannot be emphasised too strongly that it is possible that we *reject* a claim which should be *accepted* (the sample gives misleading evidence), and to offset this we may reduce the risk of a wrong decision to 1%. However, this has the effect of increasing the chances of not rejecting a claim when it should be rejected since the region of acceptance is now widened: see Figure 20.16 (ii).

The risk of rejecting H_0 when it should be accepted is called an α-risk and the error of judgement is called a **Type I error**. The error in accepting H_0 when it should be rejected is called a **Type II error**; the risk of acceptance is called a β-risk.

To decrease the chances of a Type I error we widen the acceptance region and cut down the rejection region; this increases the chances of a Type II error. However, this may be what is required, if we want to give a manufacturer the benefit of the doubt. On the other hand, it may be that we want to be stringent and cut down the chances of passing a wrong claim; in this eventuality we need to reduce Type II error by narrowing the acceptance region and increase the risk of rejecting H_0. The safest way of reducing Type II errors is to increase the sample size. Let us work through these steps with our example of the manufacturer's claim.

Example

We are concerned with the distribution of sample means which we know has mean μ and standard deviation $\sigma/\sqrt{n}$. Let us complete the data by saying that the population standard deviation is 50 hours. Then we have

$$\sigma_{\bar{x}} = \frac{50}{\sqrt{100}} = 5 \text{ hours}$$

We calculate $$z = \frac{\bar{x} - \mu}{\sigma_{\bar{x}}} = \frac{988 - 1000}{5} = \frac{-12}{5} = -2.4$$

From the table we find that the probability of a z-score of –2.4 or less, that is, of a sample mean of 988 or less is (0.5 – 0.4918) = 0.82%.

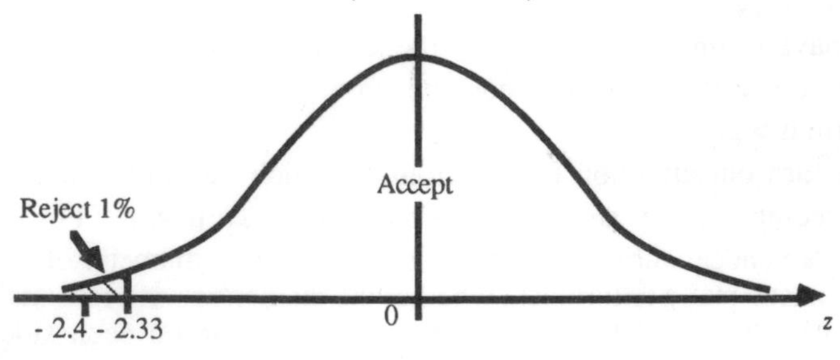

Figure 20.17

If we work to a 1% risk, we see that the claim must be rejected. Alternatively, we may note that 1% corresponds to $z = -2.33$ and our smaller z-score means rejection; refer to Figure 20.17.

Suppose that the manufacturer's claim is now modified to stating that the mean life of his components is 997 hours; how does this affect our decision?

The z-score is now $\quad z = \dfrac{988 - 997}{5} = \dfrac{-9}{5} = -1.8$

The corresponding probability is $(0.5 - 0.4641) = 3.59\%$ (2 d.p.).

If we work to a 5% risk, we shall still reject the null hypothesis, H_0; if we work to a 1% risk, we cannot reject H_0 on the evidence of the sample.

Example

Suppose a manufacturer claims his components last on average 1000 hours and that a sample of 100 had a mean life of 992 hours with a standard deviation of 40 hours. What can we conclude?

Here we have no information about the population standard deviation σ and we must use the sample standard deviation, s, in order to estimate $\sigma_{\bar{x}}$; we then have

$$\hat{\sigma}_{\bar{x}} = \frac{s}{\sqrt{n}} = \frac{40}{10} = 4$$

We calculate the probability that the sample mean of 100 components will be 992 hours or less.

We form $\quad z = \dfrac{\bar{x} - \mu}{\hat{\sigma}_{\bar{x}}} = \dfrac{992 - 1000}{4} = -2.0$

The probability that $\bar{x} \leq 992$ is $(0.5 - 0.4772) = 2.28\%$ (2 d.p.).

If we reject the claim, there is a risk of 2.28% that we are making a mistake and rejecting a valid claim. Perhaps a better term than *accepting the null hypothesis* is *reserving judgement.* Also, another way of stating that we have a risk of 5% and the null hypothesis is rejected is to say that the null hypothesis is rejected at the 5% **level of significance**.

Two-tailed tests

So far we have examined cases where the alternative hypothesis H_1 is of the form $\mu < \mu_0$; it is clear that similar considerations obtain for the cases where H_1 is of the form $\mu > \mu_0$.

We now turn our attention to situations in which the null hypothesis H_0: $\mu = \mu_0$ is accompanied by an alternative hypothesis H_1: $\mu \neq \mu_0$. For example, suppose we are endeavouring to manufacture nuts with a diameter of 40 mm, it would be almost equally unfortunate whether the diameters were too small or too large. Clearly there will be two rejection regions, one on each side of the acceptance region.

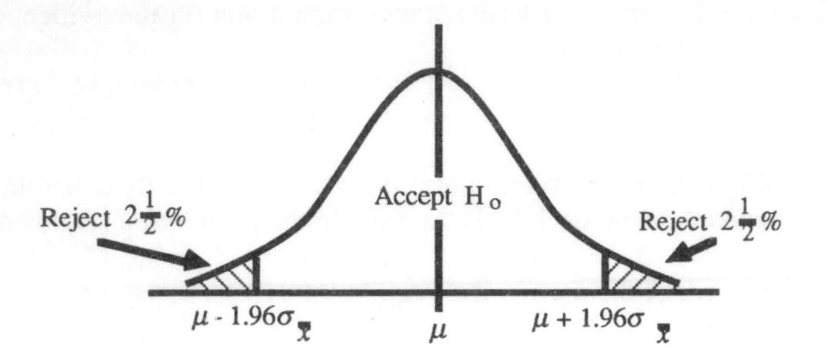

Figure 20.18

But if we work to a 5% level of significance, we are now saying that the total probability of falling in the rejection region is 5%, and since the acceptance region is symmetrical about the mean μ this will yield a $2\frac{1}{2}$% probability of landing on one side of the acceptance region; see Figure 20.18. We say we are carrying out a **two-tailed test**.

Example 1

A machine is supposed to pour on average 500 gm of a chemical into a jar. A sample of 64 jars is taken from a day's output of 5000 and the amount of chemical in a jar is found to have a mean of 504 gm with a standard deviation of 12.8 gm. Working to a 1% risk, is the machine correctly set?

Here the sample of 64 is sufficiently small with respect to the 5000 output that we can estimate the standard error of the mean as

$$\hat{\sigma}_{\bar{x}} = \frac{s}{\sqrt{n}} = \frac{12.8}{\sqrt{64}} = 1.6$$

The null hypothesis is H_0: $\mu = 500$, and the alternative hypothesis H_1: $\mu \neq 500$.

We form
$$z = \frac{\bar{x} - \mu}{\hat{\sigma}_{\bar{x}}} = \frac{504 - 500}{1.6} = 2.5$$

Now the acceptance region is shown in Figures 20.19(i) and (ii).

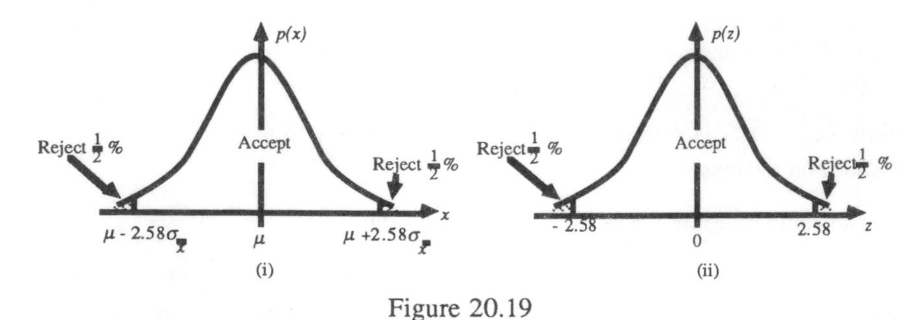

Figure 20.19

Clearly we are just inside the acceptance region and on the evidence of the sample, we reserve judgement.

It is important to notice that we have considered large-size samples here.

Example 2

A sample of 400 is taken from a population which is normally distributed with mean 6 and standard deviation 1.6. Is the sample truly random if its mean is 5.6?

The standard error of the mean $\hat{\sigma}_{\bar{x}} = \dfrac{1.6}{20} = 0.08$

The discrepancy $\bar{x} - \mu$ is 5 standard errors from the mean since

$$z = \frac{\bar{x} - \mu}{\hat{\sigma}_{\bar{x}}} = \frac{5.6 - 6}{0.08} = -5$$

Therefore we must conclude that the sample is not random.

20.7 Other Tests of Significance

The difference between sample means

Suppose we wish to compare mean scores from two samples. We have seen at the end of Section 20.4 that the sampling distribution for the difference in means $\bar{x}_A$ and $\bar{x}_B$ from two samples A and B containing n_A and n_B items respectively has expectation $\mu_A - \mu_B$, written μ_{A-B} and variance $(\sigma_A^2/n_A + \sigma_B^2/n_B)$, where σ_A^2 and σ_B^2 are the variances of the two populations (which may be equal if the samples are drawn from the same population or they may be estimated by sample variances if necessary). Let us consider an example.

Example

Two brands of component were sampled for the following results:

	Brand A	*Brand B*
Number in sample	50	40
Sample mean	10.7 mm	10.2 mm
Sample s.d.	1.5 mm	1.7 mm

At the 5% level of significance what can we say about the difference in sample means? We shall have to estimate population variances by sample variances.

We set $H_0\colon \mu_A = \mu_B$ and $H_1\colon \mu_A \neq \mu_B$

Now $(\bar{x}_A - \bar{x}_B)$ is observed to be $(10.7 - 10.2) = 0.5$ mm, but under H_0 should, on average, be 0.

The standard error, s.e. $= \sigma_{\bar{x}_A - \bar{x}_B}$ is such that

$$(\text{s.e.})^2 = \sigma_A^2/n_A + \sigma_B^2/n_B$$

(the samples are large enough for a normal distribution approximation).

Then

$$(\text{s.e.})^2 = \frac{(1.5)^2}{50} + \frac{(1.7)^2}{40} = \frac{2.25}{50} + \frac{2.89}{40} = 0.045 + 0.07225$$

$$= 0.1173 \quad (4 \text{ d.p.})$$

therefore $\text{s.e.} = 0.3425 \quad (4 \text{ d.p.})$

Now

$$z = \frac{(\bar{x}_A - \bar{x}_B) - (\mu_A - \mu_B)}{\text{s.e.}} = \frac{0.5 - 0}{0.3425} = 1.46 \quad (2 \text{ d.p.})$$

The probability associated with this z-score is $(0.5 - 0.4279) = 7.21\%$

We require to show that the means are significantly different and we therefore double this figure to 14.42%. At the 5% level of significance, we must accept H_0 and say that there is no significant difference in the sample means.

A sample proportion

We again assume a normal distribution applies.

The tests for a sample proportion follow the same path as those for a sample mean. All that is changed is the standard error, σ_p, which is now

$$\sigma_p = \sqrt{\frac{\pi(1-\pi)}{n}}$$

where π is the 'favourable' proportion and n the number in the sample.

The ideas of one-sided (or one-tailed) and two-sided testing hold here too and we provide one example to enable a comparison to be made.

Example

A medicine is claimed to be at least 85% effective in treatment. If, from a sample of 200 patients treated with the medicine, 160 showed improvement, do you consider the claim to be reasonable?

We set H_0: $\pi = 0.85$ and H_1: $\pi < 0.85$ (a higher effectiveness than 85% is acceptable), and work to a risk of 1%. The test is one-tailed.

The standard error is

$$\sigma_p = \sqrt{\frac{\pi(1-\pi)}{n}} = \sqrt{\frac{0.85 \times 0.15}{200}} = 0.0252 \quad (4 \text{ d.p.})$$

Then $$z = \frac{0.80 - 0.85}{0.02525} = \frac{-0.05}{0.02525} = -1.98 \quad (3 \text{ s.f.})$$

Now the probability under H_0 of the sample proportion being 0.80 or less is

$$(0.5 - 0.4761) = 0.0239 = 2.39\% \quad (2 \text{ d.p.})$$

At the 1% level of significance we must therefore reserve judgement (the critical figure being $z = 2.33$).

Differences between sample proportions

In this case the standard error is given by the formula

$$\sqrt{\frac{\pi_A(1-\pi_A)}{n_A}+\frac{\pi_B(1-\pi_B)}{n_B}}$$

Example

In the previous example suppose the drug is tested after some modifications and a further sample of 200 patients treated with the new version showed 180 were cured. With a risk of 1%, can we conclude anything about the effectiveness of the new version?

We set $H_0\colon \pi_A = \pi_B$ and $H_1\colon \pi_A \neq \pi_B$

The standard error is

$$\sqrt{\frac{0.90(0.10)}{200}+\frac{0.80(0.20)}{200}} = \sqrt{4.5\times 10^{-4} + 8.00\times 10^{-4}}$$

$$= \sqrt{12.5\times 10^{-4}}$$
$$= 0.0354 \quad \text{(4 d.p.)}$$

$$z = \frac{(0.90-0.80)-0}{0.0354}$$

$$= \frac{0.10}{0.0354}$$
$$= 2.82$$

Now for a two-tailed test the acceptance range for z is $[-2.58, 2.58]$ and we are outside this region so we must conclude that there is a difference in effectiveness of the two versions.

Problems

Section 20.1

1 Find the p.d.f. for the continuous distribution which is rectangular on the interval [1, 4] and zero elsewhere. Sketch the distribution and find the probabilities of the events:

(i) $x < 0$ (ii) $x < 2$ (iii) $1 < x < 2$ (iv) $x > 2$

(v) $x = 1\frac{1}{2}$ (vi) $x = 1\frac{1}{2}$ or $x = 2$ (vii) $x < 2$ and $x > 1$

2 Sketch the distribution whose p.d.f. is

$$\rho(x) = \begin{cases} k(2x+1), & 0 < x < 2 \\ 0, & \text{elsewhere} \end{cases}$$

Find the value of k. Find the probability of the events

$A =$ '$x < 1$', $B =$ '$x > \frac{1}{2}$' $C =$ '$x \in (\frac{1}{2}, 1)$'

3 Sketch the distribution whose p.d.f. is

$$\rho(x) = \begin{cases} ke^{-x}, & 0 < x < 1 \\ 0, & \text{elsewhere} \end{cases}$$

Find the value of k and find the probabilities that $x < \frac{1}{2}$, $x > \frac{3}{4}$ and $x \in (\frac{1}{2}, \frac{3}{4})$.

4 Obtain and plot the cumulative distribution function for the variable which has probability density function

$$\rho(x) = \frac{1}{4}(4 - 2x) \qquad 0 < x < 2$$

Find the probabilities of '$x < 1$', '$x > 1\frac{1}{2}$', '$x > \frac{3}{4}$'

Also find x such that $\rho(X \leq x) = \frac{1}{2}$.

5 Assuming that the number of vehicles passing a checkpoint is a Poisson variable with a given average of vehicles per hour, then the length of time interval between two cars passing (headway) is a variable which follows the **negative exponential distribution**. If λ is the number of events/unit time then the probability density function for the N intervals is $\rho(t) = \lambda e^{-\lambda t}$ where t is the number of unit time intervals (measured in seconds). Show that

$$\int_0^\infty \rho(t)dt = 1$$

and graph both the distribution and the cumulative distribution function. Find the probabilities that the headway is ≥ 8 seconds for vehicle flows of 100(100)1000 vehicles/hour; find the mean time headway and show that the mean headway for all intervals $\geq t$ is $t + (1/\lambda)$.

Section 20.2

6 Assuming that the variable z is distributed as $N(0, 1)$ use tables to find:

(a) $P(z < 1.82)$ (b) $P(z > 0.58)$ (c) $P(z > -2.03)$
(d) $P(z < -2.64)$ (e) $P(1.20 < z < 1.35)$ (f) $P(-2.35 < z < 0)$
(g) $P(-3 < z < -2)$ (h) $P(-0.05 < z < 0.25)$ (i) $P(|z| > 1.82)$
(j) $P(|z| < 0.38)$ (k) Find z' such that $P(z < z') = 0.75$
(l) Find z' such that $P(0 < z < z') = 0.1$
(m) Find z' such that $P(z < z') = 0.30$
(n) Find z' such that $P(1 < z < z') = 0.08$
(o) Find z' such that $P(-3 < z < z') = 4\%$
(p) If x is $N(60, 16)$, find $P(50 < x < 70)$
(q) If x is $N(0.36, 0.04)$, find $P(-0.1 < x < 0.5)$
(r) If T is $N(2.5 \times 10^3, 10^4)$, find T' such that $T > T'$ in only 1% of all cases
(s) If T is $N(230, 120)$, find T' such that $T > T'$ in 99.9% of all cases.

7 Components are manufactured to be 18 mm in length, but they are acceptable inside the limits $17\frac{15}{16}$ mm and $18\frac{1}{16}$ mm. Observation indicates that about $2\frac{1}{2}$% are rejected as too long and about $2\frac{1}{2}$% as too short. Assuming that the lengths are normally distributed about the mean of 18 mm, find the standard deviation of the distribution. Hence calculate the proportion of rejects if the tolerance limits are narrowed to $17\frac{61}{64}$ mm and $18\frac{3}{64}$ mm.

8 The average life of a 250 watt electric motor is 8 years, with a standard deviation of 2 years. The manufacturer replaces, free of charge, all motors that fail whilst under guarantee. Assuming that the motor lives are normally distributed, how long a guarantee should he provide if he is willing to replace no more than 2% of all the motors he sells?
What proportion of motors will still be serviceable after 11 years?

9 Cylinders of diameter 10 mm are to fit into holes whose diameters are normally distributed with mean 10.02 mm and variance 3×10^{-4} (mm)^2? How many holes out of 500 would you expect to be too small?

10 The following table gives the electrical resistance of 138 carbon rods tested at the same temperature.

Resistance in ohms (mid-class value)	310	311	312	313	314	315	316	317	318	319	320
Frequency	1	2	6	21	25	32	24	18	5	3	1

Can the data reasonably be regarded as a sample from a normal distribution?

11 A machine is producing screws whose lengths are normally distributed with mean 10.02 mm and standard deviation 0.03 mm. Limit gauges are used to reject all screws greater than 10.1 mm in length and all screws less than 9.9 mm in length. What proportion of the screws will be rejected?

12 The breaking strength in N/m^2 of 100 mild steel specimens are arranged in the following frequency distribution with a class interval of 1000 N/m^2, the intervals being denoted by their central points.

Breaking strength	Frequency
65 500	1
66 500	0
67 500	4
68 500	10
69 500	14
70 500	22
71 500	18
72 500	14
73 500	8
74 500	5
75 500	3
76 500	1

Find the equation of the normal curve which has the same mean, standard deviation as this distribution and show that the frequencies deduced from its curve are: 0, 2, 4, 9, 14, 19, 19, 15, 10, 5, 2, 1.

13 In a normal distribution the mean is 80, the standard deviation 15 and the total number of items 1000.
(a) Estimate the number of items between 65 and 95
(b) Find the probability that an item chosen at random will be greater than 100
(c) What is the least value that an item in the highest 30% can have?
(d) Between what values of the variable will the middle 60% of the frequencies lie?
(e) The probability that a certain value of the variable will be exceeded is 5%. What is the value?
(f) What are the quartiles of the distribution?

14 The ground manoeuvring times of 84 aircraft landing at an airport on one day were

Time (min)	2	3	4	5	6	7
Frequency	8	13	27	23	10	3

Calculate the mean and standard deviation of this sample and derive the corresponding normal frequency distribution. (EC)

15 Analysis of past data shows that the hub thickness of a particular type of gear is normally distributed about a mean thickness of 20.0 mm with a standard deviation of 0.4 mm.
(i) What is the probability that a gear chosen at random will have a thickness greater than 20.6 mm?
(ii) How many gears in a production run of 600 such gears will have a thickness between 18.9 and 19.5 mm? (EC)

16 (a) The lives of certain electrical components are known to follow a normal distribution with mean 1500 hours and standard deviation 300 hours.
(i) What is the probability that a component chosen at random lasts between 1500 and 1950 hours?
(ii) What is the percentage of components which last more than 2200 hours?
(iii) The components are 'guaranteed' to last 1000 hours. What proportion fail to last this guaranteed life?
(iv) What should be the guaranteed life-time if 99% are to satisfy it?

(b) The probability that a variable, from a population with normal distribution and having mean 0 and standard deviation 1, has a value in the interval (0, 1) is given by

$$I = \int_0^1 \frac{1}{\sqrt{2\pi}} \exp(-\tfrac{1}{2}x^2)\mathrm{d}x$$

Use Simpson's rule with 4 strips to find the value of I correct to three decimal places. Compare the value obtained with that found from the table of normal distribution probabilities. (EC)

17 A certain machine produces components to any required length specification but always with a standard deviation of 1.40 mm. At a particular setting it produces a mean length of 102.30 mm. Assume that the lengths follow a normal distribution.

(i) What proportion will need to be reworked because their lengths are greater than 105.00 mm?
(ii) What percentage will be rejected because their lengths are less than a stated minimum of 100.00 mm?
(iii) Between what limits will the lengths of 99% of the production lie?
(iv) To what value should the mean be increased in order to reduce the rejections in (ii) to 1%? (EC)

18 A machine is producing washers whose diameters have a standard deviation of 0.5 mm. If the tolerance allowed on diameter is ±0.8 mm, and diameter can be assumed to have a normal distribution, approximately what percentage of washers will be rejected?

19 Rods are made to a nominal length of 40 mm but in fact they form a normal distribution with mean 40.1 mm and standard deviation 0.3 mm. Each rod costs 6p to make and may be used immediately if its length lies between 39.8 and 40.2 mm. If its length is less than 39.8 mm, the rod cannot be used but has a scrap value of 1p. If its length exceeds 40.2 mm, it may be shortened and used at a further cost of 2p. Find the average cost per usable rod.

Section 20.3

20 Assume a gunner has a 50/50 chance of hitting a target each time he fires at one. Find the probability that after shooting at 12 targets he will have hit just 4 of them. Use the Binomial distribution and the Normal approximation and compare.

21 Use the Normal approximation to the Binomial to find the probabilities that in 100 spins of an unbiased coin, there will be
(i) exactly 40 heads (ii) fewer than 42 tails (iii) more than 40 heads
(iv) between 25 and 50 tails (v) exactly 20 heads (vi) fewer than 30 tails.

Section 20.4

22 A population consists of 6 numbers, 1, 2, 3, 6, 7, 8. Samples of size 2 are taken from this population (without replacement).
(a) Calculate the population mean and standard deviation
(b) Calculate the mean of each possible sample
(c) Calculate the mean of these means of samples
(d) Find the standard error of the means.
Comment on the results.

23 A random sample of size 50 is drawn from a Normal population; the population mean is 52 and standard deviation is 24. Find the probability that the sample mean will:
(a) lie between 50 and 55 (b) lie between 40 and 60.

24 It is known that of a large group of students, 25% live in lodgings. If a random sample of 100 is chosen from the group, what is the probability that
(a) more than 30 will live in lodgings (b) less than 20 will live in lodgings?

25 Large numbers of plates of two types are produced. One type has thicknesses x mm following $N(2.4,\ 1.3 \times 10^{-3})$; the other has thicknesses y mm following

$N(1.3, 0.8 \times 10^{-3})$. One plate of each type is taken at random and these are bolted together to give a plate of overall thickness $(x + y)$ mm. Between what limits can we expect 90% of all pairs to lie?

26 In a manufacturing process, a piston of circular cross-section has to fit into a similarly shaped cylinder. The distributions of diameters of pistons and cylinders are normal with parameters: pistons: mean diameter 104.2 mm, standard deviation 0.3 mm
cylinders: mean diameter 105.2 mm, standard deviation 0.4 mm
If the pistons and cylinders are selected at random for assembly
(a) what proportion of the pistons will not fit into cylinders?
(b) what is the chance that, in 100 pairs selected at random, all the pistons will fit? (EC)

27 A type of sensor has a mean life of 15 000 hours with standard deviation of 1000 hours. Three of these sensors are connected in series in a fire-warning device so that when one fails another takes over. Assuming that the life times are normally distributed, what is the probability that the device
(a) will function for 40 000 hours (b) will fail before 39 000 hours. (EC)

28 A manufactured product consists of a major part and three minor parts. The mass of the major part is normally distributed with mean mass 15 grams and standard deviation 1 gram. The mass of each minor part is normally distributed with mean mass 2 grams and standard deviation 0.2 gram. The product is rejected as unsuitable if its total mass is less than 19.5 grams or more than 22.5 grams. Find the proportion of the product which is unsuitable.

29 The weights of bags of sand have a distribution with mean 1.14 kN and standard deviation 0.08. If a delivery consists of 16 such bags, what is the probability that the average weight per bag is less than 1.12 kN?

30 Wire cables are formed from 10 separate wires, the strength of each wire being normally distributed with standard deviation of 0.245 kN about a mean of 6.45 kN. Assuming that the strength of each cable is the combined strength of the separate wires, what proportion of the cables have a breaking strain of less than 6.35 kN? Assuming the variability to remain unchanged, to what must the mean strength of the individual wires be increased if only 1 in 1000 cables is to have a breaking strain of less than 6.35 kN?

Section 20.5

31 Strength tests on steel parts gave a distribution with mean stress 27.56 and standard deviation 1.10 N/m^2. Find the standard error of the mean of 10 tests and find the probability that the mean of 10 tests will be less than 27.24 N/m^2.

32 A process produces items with a mean mass of 5.7 kg and standard deviation 2 kg. What are the 95% confidence limits for the mean of a sample of 50 items?

33 Strength tests conducted over a period of time have established that the breaking loads (in units of kN) of certain manufactured spars are distributed in a manner which is approximately normal, having a mean of 32.6 and a variance of 3.24.

(i) What is the probability that a single spar selected at random will have a strength in excess of 34×10^3 N?

(ii) Find the probability that the mean strength of a sample of 9 spars selected at random will be less than 34×10^3 N.

34 Measurement of the diameters of a random sample of ball bearings made by a certain machine during one week showed a mean of 0.824 cm, and a standard deviation of 0.042 cm. Find the 99% confidence limits for the mean diameter of all ball bearings made by this machine, assuming the normal distribution.

35 The standard deviation of the breaking load of certain cables is taken to be 150 N. A random sample of 5 cables is tested and has a mean breaking load of 2436 N. What are the 98% confidence limits for the mean breaking load of all such cables?

Section 20.6

36 The distribution of hourly output of a certain type of machine has mean 700 units and standard deviation 50 units. With a new method of operation a sample of 10 single hour test runs gave a mean of 750 units. Show that the new method has made a statistically significant change in the hourly rate.

37 The lengths in mm of 50 components were recorded and the following summary values noted:

$$\sum_{i=1}^{50} x_i = 75.6, \qquad \sum_{i=1}^{50} x_i^2 = 120.80$$

where x_i = length − 100.

(a) Estimate the mean of the normal distribution which is assumed to describe the variation in length, and give an approximate 95% confidence interval for this mean.

(b) If 5 of the components are laid in line, their ends just touching, provide an estimate of the mean of the overall length; give a 95% confidence interval.

Now assume the lengths are distributed as a $N[101.5, (0.4)^2]$ random variable. If the components are discarded when they are under 101 mm in length, what proportion will be discarded? With this variance, to what value should the mean length be increased if the proportion of discarded components is to be at most 0.01?

38 Routine strength tests have established that certain spars have a breaking stress normally distributed about 27.56 N/m^2 with a standard deviation of 1.01 N/m^2.

(a) What is the probability that the average breaking stress from a sample of 10 spars will be less than 27.24 N/m^2?

(b) Certain modifications are made in the construction of the spars and in order to test whether these modifications have changed the mean strength, a sample of 10 modified spars is tested. They gave a sample mean of 28.43 N/m^2. Does this suggest that the modifications have had a real effect?

Section 20.7

39 Two different processes are used to produce what should be identical items. Samples are taken from each to give the following measures (x) of some characteristic.

Process	A	B
Number in sample	144	96
Σx	2246	1354
$\bar{x}$	15.6	14.1
$\Sigma(x-\bar{x})^2$	548.2	426.4

Is there a significant difference in the means?

40 A machine produces items with a mean of 105 and variance 4. A sample of 400 items is found to have a mean of 104.45. Can this be regarded as a truly random sample? Justify your answer.

41 Population A has mean length 137 mm and variance 25 mm^2, and population B has mean 125 mm and variance 37 mm^2. Find the probability that the average of 6 values taken from A is less than the average of 5 from B.

42 The lengths of life of two types of electric light bulb were recorded giving the following results:

	Type 1	*Type 2*
Number in sample	$N_x = 60$	$N_y = 80$
Mean of sample	$\bar{x} = 1070$	$\bar{y} = 1060$
	$\Sigma(x-\bar{x})^2 = 5480$	$\Sigma(y-\bar{y})^2 = 5800$

Are the mean lives of the two types of bulb significantly different at the 1% level of significance?

43 Tests on the moisture content of a product produced by two processes gave the following results (expressed as percentages).

A 100 samples had mean 8.1 and standard deviation 3.0

B 60 samples had mean 6.6 and standard deviation 2.6

Find 95% confidence limits for the difference of the means of the two populations.

44 The television tubes of manufacturer A have a mean lifetime of 6.5 years and a standard deviation of 0.6 years, whilst those of manufacturer B have a mean lifetime of 6.0 years and a standard deviation of 0.8 years. What is the probability that a random sample of 20 tubes from manufacturer A will have a mean lifetime that is at least 1 year more than the mean lifetime of a sample of 16 tubes from manufacturer B?

45 The following table shows the number of employees and of serious accidents in one year in each of two factories.

Factory	*No. of employees*	*No. of accidents*
1	15 000	7
2	8 000	16

How strong is the evidence that the chance of an individual being involved in a serious accident is greater in the second factory than in the first?

46 In a random sample poll of 200 individuals in constituency A the number favouring the party X was 96. Give 95% confidence limits for the proportion favouring this party in the

constituency as a whole. In a neighbouring constituency B a random poll of 300 was taken and the number favouring party X was found to be 168. Does this indicate that a higher proportion favours party X in the second constituency than in the first? (LU)

47 X is a random variable representing the number of successes in n independent trials, such that the probability of success is a constant p. Show that the distribution of the proportion of successes $\hat{p} = x/n$ has a mean p and variance $p(1-p)/n$.
The results of a one hundred per cent inspection on a batch of 1000 components show that 190 were faulty. Do you consider these results are consistent with an expected proportion p of faulty components equal to 16%? (LU)

21

ORDINARY DIFFERENTIAL EQUATIONS – GENERAL IDEAS

Many mathematical models are based on differential equations. In this chapter we first carry out a case study which will illustrate the ideas inherent in the solution of problems involving ordinary differential equations; we then set down some general definitions and ideas.

21.1 Case Study – Newton's Law of Cooling

We have already discussed the problem of the cooling of a hot liquid in Chapter 1 where we showed the nature of the graph of the experimental results. We now proceed to obtain a solution to this problem via a mathematical model. Referring back to Section 1.1, we stated that the mathematical model equivalent to Newton's Law of Cooling was

$$\frac{\mathrm{d}\theta}{\mathrm{d}t} = -k(\theta - \theta_s), \quad \text{with } \theta = \theta_0 \text{ at time } t = 0$$

In this model equation $\theta(t)$ is the temperature of the liquid at time t, θ_s the temperature of the surrounding air, θ_0 the initial temperature and k a constant of proportionality whose value depends on the particular liquid involved. The equation is our mathematical model. It is one example of a **differential equation** which, in simplest terms, can be defined as an equation involving derivatives or rates of change. We shall examine three main methods of obtaining information about the nature of the relationship between θ and t.

(a) **The analogue computer approach**
The analogue computer is a device which can be made to produce an output voltage which varies with time in the same way as θ in our problem. This voltage can be displayed on a pen-recording trace or on a cathode-ray screen so that we can study the *qualitative* nature of the relationship between θ and t.

Furthermore, using the analogue computer, we can vary the parameters θ_s, θ_0

and k independently, so that we can study the role which each of these plays in the solution.

The circuit of an analogue computer contains high-gain direct current amplifiers (used as summers and integrators) and potentiometers operating on voltages varying with time. The diagrammatic representation and function of these is as follows.

(i) *Summers*

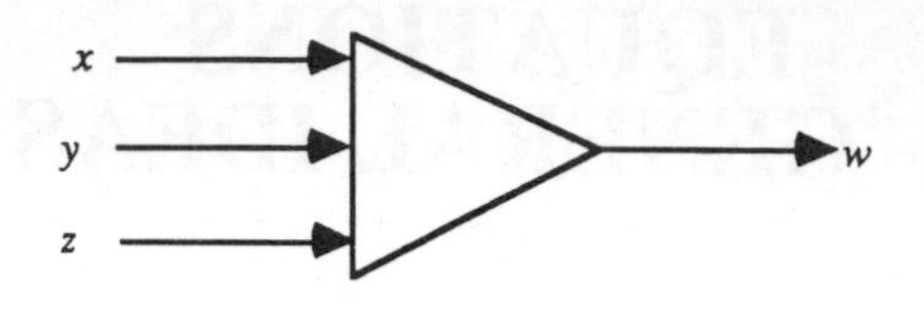

The input voltages to the amplifier are x, y and z. The output voltage produced is w and is such that

$$w = -(x + y + z)$$

Notice the reversal of sign.
In fact, if the only input is x, then the output is given by $w = -x$ so that the summer can be used to multiply an input by -1.

(ii) *Integrators*

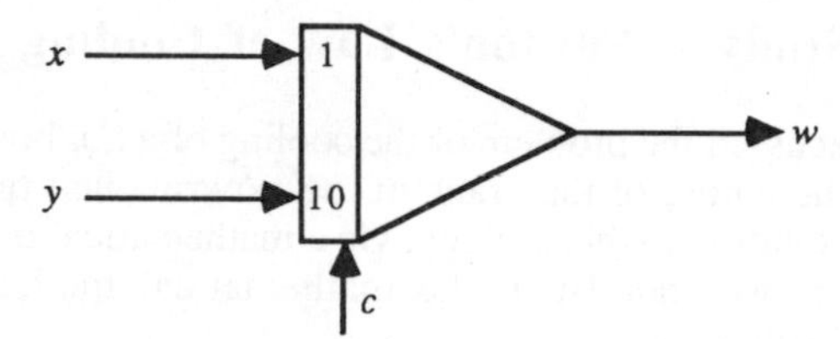

Here the input voltages are x and y. The integrator has the facility to multiply these voltages by the numbers in the box (here they are shown as 1 and 10), add them together, and then integrate with respect to time. It produces an output voltage w, where

$$w = -\int_0^t (x + 10y)\mathrm{d}t - c$$

t is time and c is the *initial condition* so referred to because when $t = 0$, $w = -c$. Notice again the sign reversal.

(iii) *Coefficient Multipliers*

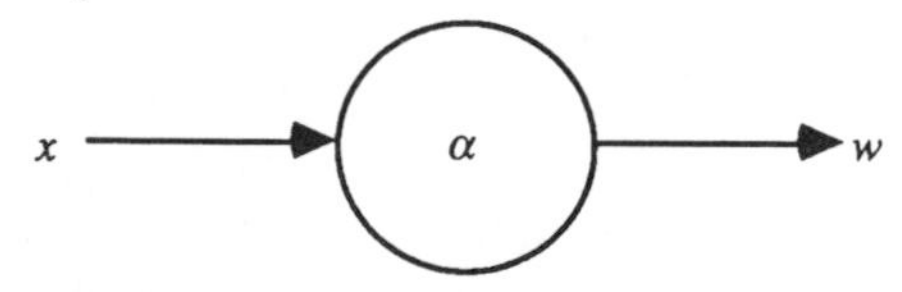

The coefficient multipliers are potentiometers which can be used to reduce a voltage. Here we have shown an input voltage x and output voltage w.

The relation between these is of the form

$$w = \alpha x \quad (0 \le \alpha \le 1)$$

The above discussion, then, summarises the operations which can be performed by the analogue computer and which can be used to simulate the solution of a differential equation. We return to the differential equation we were considering, that is,

$$\frac{d\theta}{dt} = -k(\theta - \theta_s), \quad \text{with } \theta = \theta_0 \text{ at time } t = 0$$

It can be written in integrated form as

$$\theta = -\int_0^t k(\theta - \theta_s)dt + \theta_0$$

The R.H.S. can be broken down as follows: $\theta - \theta_s$ is multiplied by k, the result is then integrated with an initial condition θ_0. So we have a scheme as shown in Figure 21.1.

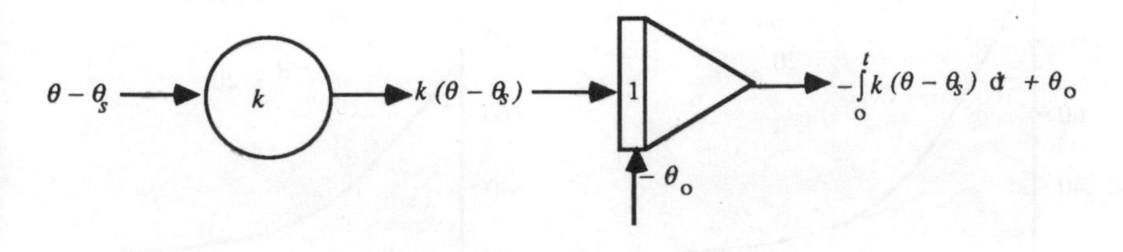

Figure 21.1

To produce $\theta - \theta_s$, which is $-(\theta_s - \theta)$, (remembering the sign reversal) we need a summer

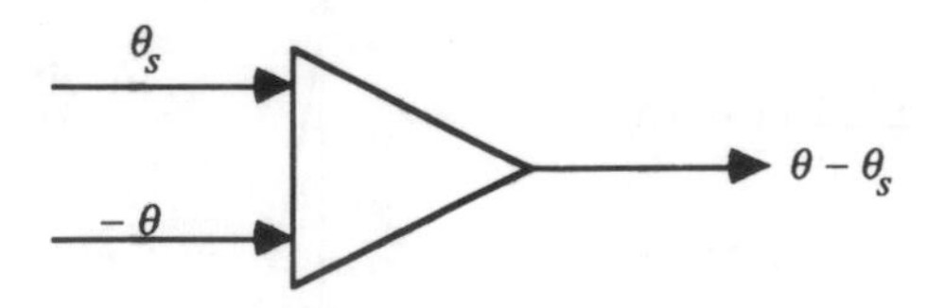

To produce $-\theta$ we can use another summer

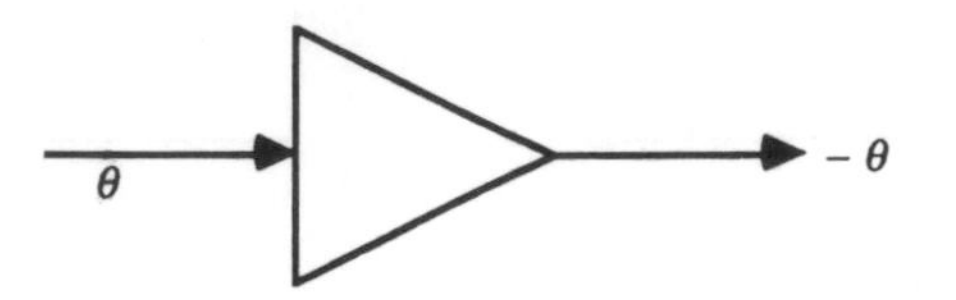

The full circuit is shown in Figure 21.2

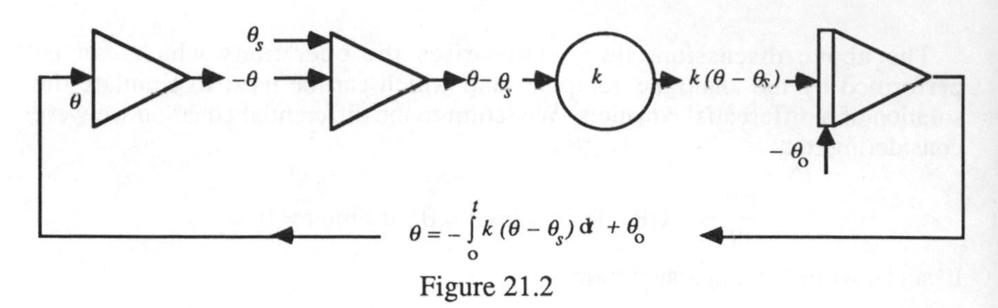

Figure 21.2

Note that we can vary the three parameters θ_s, k and θ_0 at different points of the circuit. A typical set of traces from the analogue computer set up in this way is sketched in Figure 21.3. In these diagrams, the axes have been inserted for clarity although we would not see these on the oscilloscope.

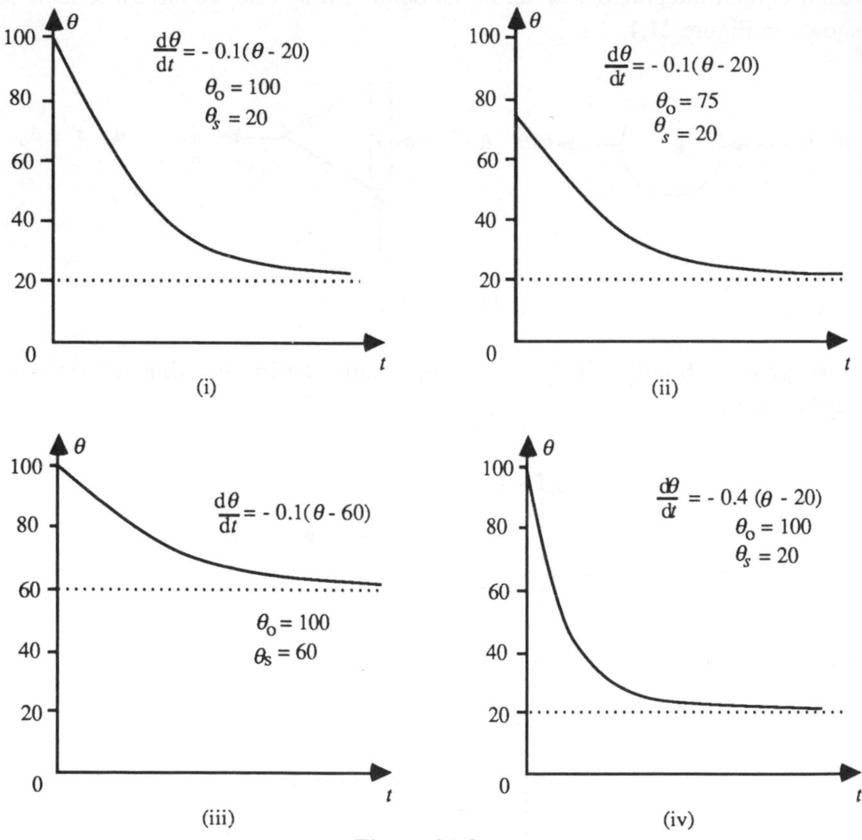

Figure 21.3

Notice that, in each case, the general shape of the curves is the same and this gives rise to the idea of a **general solution** of the differential equation from which a **particular solution** is found by selecting particular values of θ_0, θ_s and k. It is seen from the traces that

(i) decreasing the initial temperature θ_0 lessens the time taken to reach a specified temperature

(ii) increasing θ_s decreases the time taken to cool to a specified temperature

(iii) increasing k steepens the rate of decrease of temperature.

But, whatever the set of values taken for the three parameters θ_0, θ_s and k, the general form of the solution curve is an initial rapid decay of temperature followed by a more gradual approach to the temperature of the surroundings and this agrees with the experimental curve in Figure 1.4.

The same shape of curve is seen in many other physical systems. For example, when a beam of light passes through an absorbing medium, the model equation is $dI/dx = -\mu I$ where x is the distance from the light source, I is the intensity of the beam and μ is the absorption coefficient for the medium. Similarly, the rate of decay of a radioactive substance is governed by $dm/dt = -\lambda m$ where m is the amount of the substance remaining at time t. It is therefore clearly important to be able to get some more quantitative information about this kind of equation and to this end we now describe a numerical approach to the problem.

(b) Step-by-step method

Graph plot from the digital computer of the numerical solution of $d\theta/dt = -0.1(\theta - 20)$

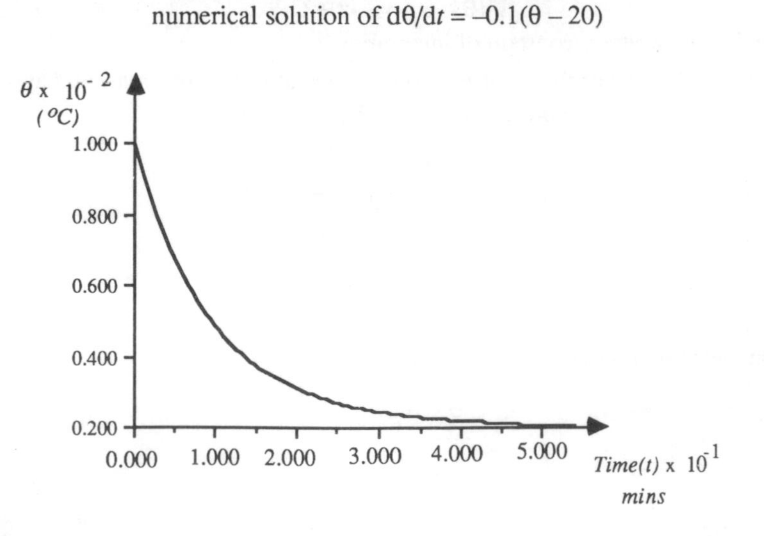

Figure 21.4

We seek here to produce a table of the values of θ at definite values of t. There are several methods available **but** the *general principle* is that, given a starting value of θ_0, the temperature, θ, of the liquid after a small interval of time is estimated to be θ_1; this estimated value is then used to estimate θ_2, the temperature after the next time interval, and so on. The values of θ_0, θ_s and k would be read as data accompanying a program for a digital computer and the results could be output as a table of values or via a graph plotter in the form shown in Figure 21.4. To obtain the solution under different conditions we can alter the input data.

It is seen that the graph thus produced has the same form as before; since we are using approximate methods, we must expect an error in the results. We shall examine later in this chapter a number of such methods and their accuracy.

(c) Analytical method

We can also use calculus to obtain a formula for θ in terms of t. This is known, of course, as an analytical method of solving a differential equation and our solution is achieved as follows. If we rearrange the equation

$$\frac{d\theta}{dt} = -k(\theta - \theta_s) \tag{21.1}$$

as

$$\frac{1}{\theta - \theta_s}\frac{d\theta}{dt} = -k$$

we can integrate both sides with respect to t to give

$$\ln(\theta - \theta_s) = -kt + C$$

where C is an arbitrary constant of integration.

To find the value of this arbitrary constant we have the information that $\theta = \theta_0$ when $t = 0$. Substituting these values into the equation gives

$$\ln(\theta_0 - \theta_s) = C$$

Therefore

$$\ln(\theta - \theta_s) = -kt + \ln(\theta_0 - \theta_s)$$

or

$$\ln\left[\frac{\theta - \theta_s}{\theta_0 - \theta_s}\right] = -kt$$

and on taking antilogarithms of both sides we obtain

$$\frac{\theta - \theta_s}{\theta_0 - \theta_s} = e^{-kt}$$

Rearranging,

$$\theta = \theta_s + (\theta_0 - \theta_s)e^{-kt} \tag{21.2}$$

Then the value of θ at any time t can be found by substitution in (21.2); sketching this form of solution we get Figure 21.5. The solution comprises a

transient term $(\theta_0 - \theta_s)e^{-kt}$ which decays to zero leaving a **steady state solution** $\theta = \theta_s$. Note that this steady state solution can be obtained from (21.1) by putting $d\theta/dt = 0$.

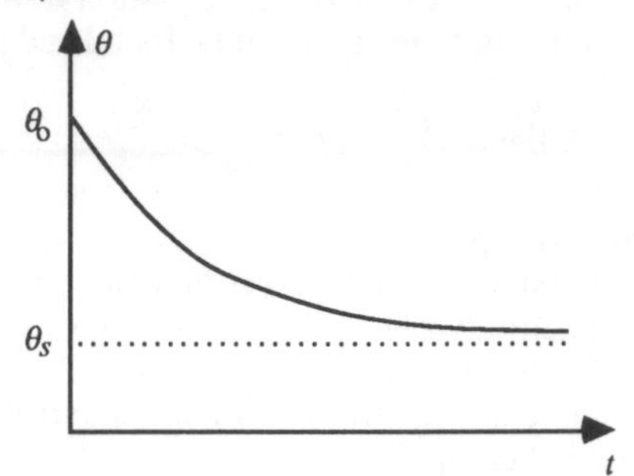

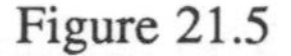
Figure 21.5

Figure 21.5 is similar to the traces obtained from both the digital and analogue computers. We can, of course, obtain numerical values of θ by substituting particular values for k, θ_0 and θ_s into the formula (21.2).

Comparison of methods

Of the above methods of solving differential equations, it would be fair to say that the analytical solution is the most desirable since it gives an exact form of solution for any starting temperature θ_0 and any surrounding temperature θ_s. The analogue solution gives an idea of the solution enabling us to see how θ varies with time, at least in a qualitative way, for different values of θ_0 and θ_s. However, it is more difficult to obtain accurate quantitative answers. On the other hand, the digital computer will give us a table of values but requires to be told values for both the starting temperature θ_0 and the surrounding temperature θ_s. It gives us a *particular* solution rather than a *general* solution and the values for θ are not necessarily exact to the **last figure** quoted.

Hence if we were confronted with a differential equation we would aim as far as possible to get analytical expressions as our solution but in many practical cases we might not be able to find an analytical solution; furthermore, in some cases where we can find an analytical solution, this solution is of such a form that it cannot be used for finding quantitative and qualitative results. Let us consider an example of this latter case.

Example

A particle of unit mass is moving in a straight line and is being attracted towards a fixed point O by a force μx, where x is its distance from O. There is a frictional force opposing the motion equal to kv where v is the velocity of the particle. We can show that the equation governing this motion when the particle is moving away from O is

$$v\frac{dv}{dx} = -kv - \mu x$$

Although it is outside our scope to solve this equation at the present time, the analytical expression giving v in terms of x can be found and is

$$\ln(v^2 + kvx + \mu x^2) - \left(\frac{k}{\omega}\right)\tan^{-1}\left(\frac{v}{\omega x} + \frac{k}{2\omega}\right) = \text{constant}$$

where $\omega^2 = \mu - k^2/4$, assumed positive.

The expression given above is much too complicated to be used for finding either qualitatively or quantitatively the explicit relationship between v and x without resorting to difficult numerical computation. Therefore, in this case it would be preferable to solve the equation in a non-analytical way from the start using analogue and/or digital computers.

21.2 Classification and General Features

Since we have said that the ideal situation is one where we can obtain an analytical solution to a differential equation, we shall spend some time on the different types of differential equations which are amenable to the analytical approach. These types are very important since they give us general types of solution and are of great use when developing theories in engineering subjects. However we recognise that we do not live in an ideal world and so we shall have to examine numerical and other techniques of solution which can be used as an alternative to, or as a substitute for, the analytical process.

First of all we must divide the differential equations we shall meet into different classes. We need some definitions for this task.

Definitions

1 An **ordinary differential equation** is an equation which contains only total derivatives and no partial derivatives. That is, it may contain dy/dx, d^2y/dx^2, etc but not $\partial y/\partial x$, $\partial^2 y/\partial z^2$, etc. In other words, there is only one independent variable in the problem of which the differential equation is the mathematical model.

2 $y = y(x)$ is a **solution** of the equation if it satisfies this equation.

3 A **first order equation** is one which contains dy/dx, y and x only (as distinct from a second order equation which contains d^2y/dx^2 and could contain dy/dx, y and x as well). In general, the **order** of a differential equation is the order of the highest derivative in the equation.

4 The **degree** of a differential equation is the power to which the highest derivative in that equation is raised, for example the following equation is *third* order and of *second* degree

$$\left[\frac{d^3y}{dx^3}\right]^2 + 2\frac{d^2y}{dx^2} + y\left[\frac{dy}{dx}\right]^4 = y + x$$

Arbitrary constants, general solutions and particular solutions
Consider the equation $y = x + A$, where A is arbitrary. We know that this equation can be represented by a straight line graph for any given value of A as follows (Figure 21.6).

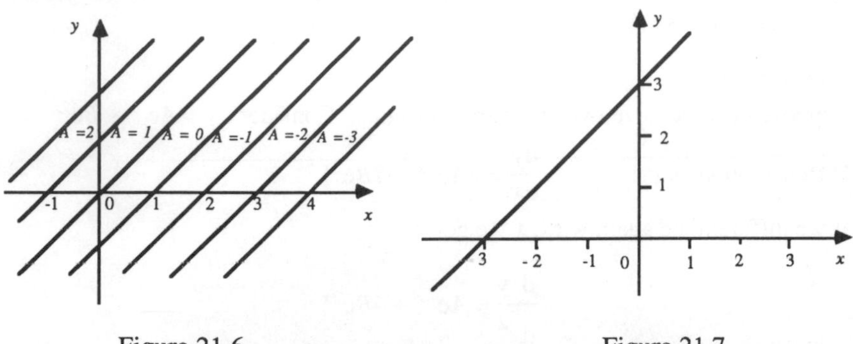

Figure 21.6 Figure 21.7

The straight lines obtained for different values of A are all parallel and, indeed, if we differentiate the given equation with respect to x, the arbitrary constant disappears and we get $dy/dx = 1$, showing that the slope always has value 1. Hence $dy/dx = 1$ is the differential equation of all these straight lines. Reversing the process, we integrate $dy/dx = 1$ w.r.t. x to obtain

$$y = x + \text{constant}$$

We say that this is the **solution** of the differential equation and we say it is a **general** solution because it represents any of the straight-line graphs which are such that $dy/dx = 1$. A **particular** solution will be obtained by taking a particular value for the constant of integration. This could be determined by specifying some point through which the straight line must pass. For example, we might require y to be 4 when $x = 1$.

There is only one of the lines that passes through the point (1, 4) and making this point satisfy our solution we get that the value of the constant = 4 – 1 = 3. The particular solution is then $y = x + 3$ (See Figure 21.7).

You will notice that to eliminate one arbitrary constant we needed to differentiate once. This is always the case.

Example 1
We start with the **family** of curves,

$$y = Ae^{-x}, \quad A \text{ arbitrary}$$

Differentiating with respect to x, we obtain $dy/dx = -Ae^{-x}$. We eliminate A by substituting from the equation of the family to obtain

$$\frac{dy}{dx} = -\left[\frac{y}{e^{-x}}\right] e^{-x}$$

or
$$\frac{dy}{dx} = -y$$

The whole family of curves $y = Ae^{-x}$ has an associated differential equation $dy/dx = -y$ and we say the **general solution** of the differential equation $dy/dx = -y$ is $y = Ae^{-x}$ where A is arbitrary. Again, a **particular solution** will be obtained by specifying some information which yields a particular value for A.

Example 2

Suppose now we have two arbitrary constants. Consider $y = Ae^{-x} + Be^{2x}$

Differentiation yields
$$\frac{dy}{dx} = -Ae^{-x} + 2Be^{2x}$$

If we differentiate again w.r.t. x we get

$$\frac{d^2y}{dx^2} = Ae^{-x} + 4Be^{2x}$$

We can now eliminate A and B from these two differential equations and you can verify that we obtain

$$\frac{d^2y}{dx^2} - \frac{dy}{dx} - 2y = 0$$

This is the second order differential equation of the whole family of curves $y = Ae^{-x} + Be^{2x}$. We could not eliminate both arbitrary constants without differentiating w.r.t. x twice, so producing an equation involving d^2y/dx^2 and therefore an equation of second order. In reverse we would expect a second order differential equation to give a general solution having *two* arbitrary constants since it must be integrated twice to produce y, each integration yielding one arbitrary constant.

In the same way, the general solution of a third order differential equation should contain 3 arbitrary constants and so on. We therefore postulate the following theorem which will not be proved here:

There are n arbitrary constants in the general solution of an nth order linear ordinary differential equation.

(A **linear** differential equation is of the form

$$a_0\frac{d^ny}{dx^n} + a_1\frac{d^{n-1}y}{dx^{n-1}} + a_2\frac{d^{n-2}y}{dx^{n-2}} + \dots + a_{n-1}\frac{dy}{dx} + a_ny = f(x)$$

that is, one where all derivatives and y are raised to the first power and where such terms as

$$y\frac{dy}{dx}, \quad \left[\frac{dy}{dx}\right]\left[\frac{d^3y}{dx^3}\right]$$

do not occur.)

This theorem is very important, and can be shown to be true even if the equation is not linear. In particular, it enables us to recognise that the solution we have obtained is the most general solution by seeing that it contains the requisite number of arbitrary constants.

In books on more advanced mathematics, theorems are proved known as **existence theorems** which give the precise conditions under which the equation has a solution and **uniqueness theorems** are proved which explore the validity of considering any general solution of a differential equation as being the only solution of the equation.

21.3 Graphical Solutions – Isoclines

Sometimes we can obtain a rough idea of the shape of solution curves by indicating their slopes locally. We illustrate the general principles by means of examples.

Example 1

$$\frac{dy}{dx} = x$$

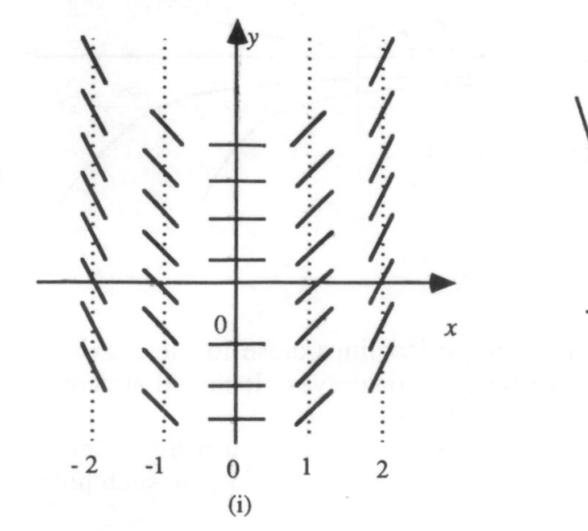

(i)

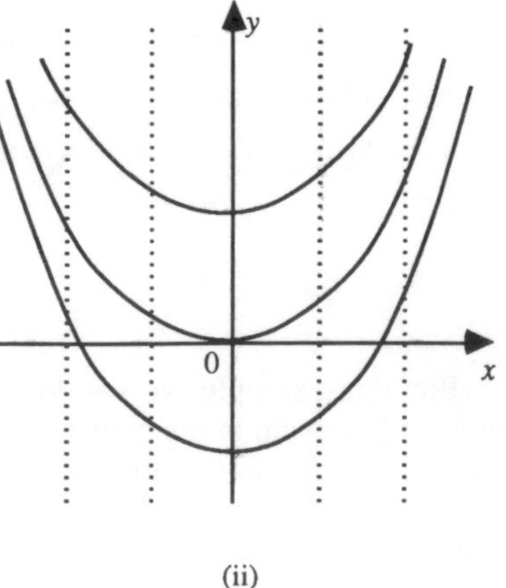

(ii)

Figure 21.8

First we sketch the curves on which dy/dx is constant. Now $dy/dx = c \Rightarrow x = c$ for this particular equation so the **isoclines** (lines of equal slope) are the straight lines x = constant, that is, parallel to the y-axis. In Figure 21.8(i) some of the lines are shown, together with the appropriate values of c. The short thick lines indicate the local direction of a solution curve which crosses that isocline at the relevant point. Note that the value of the slope at the point is the value of c associated with the isocline and so, for example, on the isocline $x = 2$, the slope at all points is 2.

The next stage is to trace the paths of some solution curves and we usually start at a point on $x = 0$ and move off in either direction. In Figure 21.8(ii) some solution curves are sketched. The analytical solution of the differential equation is

$$y = \tfrac{1}{2}x^2 + A, \quad A \text{ constant}$$

You should be able to see the relationship between isoclines, local slopes and solution curves.

Example 2

$$\frac{dy}{dx} = y$$

dy/dx = constant $\Rightarrow y$ = constant and the isoclines are the lines $y = c$. These iscolines are drawn for some values of c in Figure 21.9(i). Again we start at a point on $x = 0$ from which we branch out in each direction. Several solution curves are shown in Figure 21.9(ii). The analytical solution is $y = Ae^x$ and it can be seen from the figure that the solution curves are of exponential form.

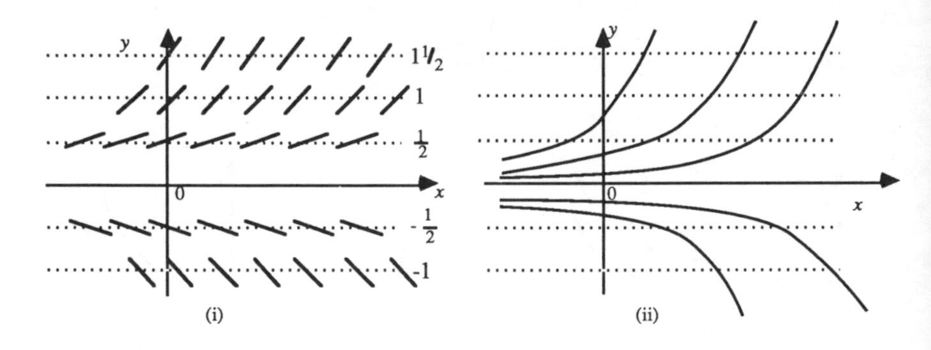

Figure 21.9

For this example we would require more isoclines close to the x-axis for greater precision in the construction of the solution curves. It should already be clear that this graphical method relies on skill and experience.

A point to remember is that $c = 0$ gives the locus of the stationary points on the solution curves, or as in this second example, where there are no such points in reality, it gives the *asymptote* to these curves.

There are many more complicated curves but we shall not consider them here, and, in any event, the method requires a certain amount of draughtsmanship.

Example 3

$$\frac{dy}{dx} = x + y$$

(Refer to Figure 21.10). The isoclines here are $x + y$ = constant.

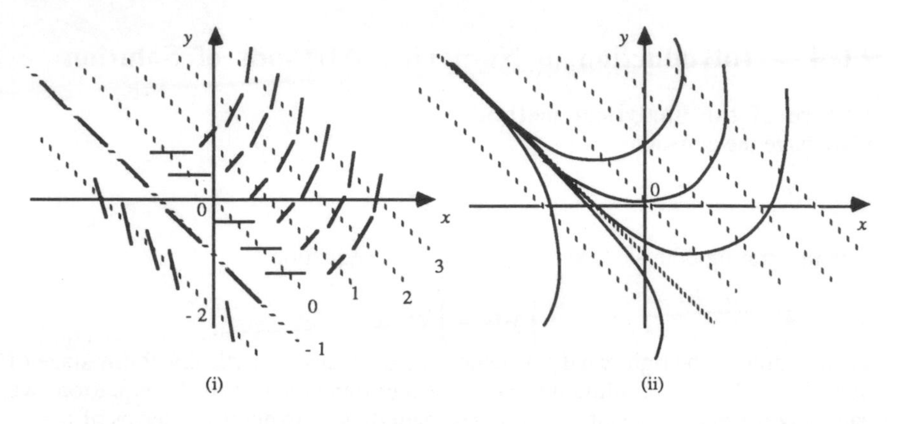

Figure 21.10

$y = -x - 1$ is an asymptote for large negative x, as can be deduced from the analytical solution $y = -x - 1 + Ce^x$.

Example 4

$$\frac{dy}{dx} = x^2 + y^2$$

(See Figure 21.11). The isoclines this time are $x^2 + y^2$ = constant.

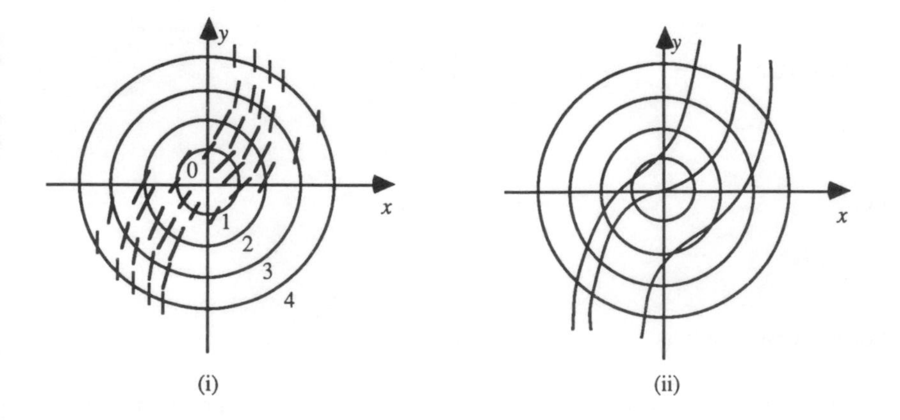

Figure 21.11

Here there is no simple analytical solution and the only place where $dy/dx = 0$ is at the origin. Where the local slope is tangential to its iscoline it can be shown that the solution curve which passes through the point has a point of inflection. These points of inflection move further away from the y-axis as $x^2 + y^2$ becomes larger.

21.4 Introduction to Numerical Methods of Solution

Failure of the analytical methods

If we have the equation

$$y\frac{dy}{dx} = e^{x^2}$$

then integrating both sides w.r.t. x produces the equation

$$\int y\,dy = \int e^{x^2}\,dx$$

The integral on the right-hand side cannot be evaluated analytically at this stage of our work. Therefore, although we have theoretically solved the equation, we cannot perform the integration and hence cannot get y explicitly in terms of x.

Given an initial pair of values of y and x, we could use an approximate method of integration to find an approximate value of y for some given value of x. For example, suppose we know that $y = 0$ when $x = 0$, then the last equation can be written

$$\int_0^y y\,dy = \int_0^x e^{x^2}\,dx$$

By this notation we mean that we integrate both sides, putting $y = 0$ when $x = 0$ (bottom limits correspond) and obtain a particular solution for y in terms of x.

We obtain
$$\left[\frac{y^2}{2}\right]_0^y = \int_0^x e^{x^2}\,dx$$

i.e.
$$\frac{y^2}{2} = \int_0^x e^{x^2}\,dx$$

For $x = 1$ we would get
$$\frac{y^2}{2} = \int_0^1 e^{x^2}\,dx$$

and using an approximate method of integration we find the approximate value of y when $x = 1$.

Similarly for $x = 2$, we obtain

$$\frac{y^2}{2} = \int_0^2 e^{x^2}\,dx$$

which produces an approximate value of y when $x = 2$.

Proceeding in the same way we can produce a whole table of values of y for corresponding values of x and we have found the solution. However the process is wasteful in time and it would be better to use a wholly numerical method in such a situation.

We now discuss a simple type of the step-by-step method mentioned when dealing with Newton's Law of Cooling in Section 21.1.

Euler's method

In the cooling problem we had to solve the equation

$$\frac{d\theta}{dt} = -0.1(\theta - 20) \text{ with } \theta = 100 \text{ when } t = 0 \qquad (21.3)$$

We want to estimate θ at certain specified times and what we shall do is to start at $\theta = 100$, $t = 0$ and step forward a small interval of time to find an estimate of θ at this new time. From this point we step forward again and estimate θ after another short interval of time and so on. The estimate of θ at the new time is found by evaluating $d\theta/dt$ at the old time using the differential equation (21.3) and estimating the new value of θ by the formula

$$\theta_{new} = \theta_{old} + h\left[\frac{d\theta}{dt}\right]_{old} \qquad (21.4)$$

where h is the step length. This is called **Euler's method.**

Essentially we are using a straight line approximation to the function in each interval as shown in Figure 21.12(i). This means that we shall obtain a graph for the estimated behaviour of θ consisting of straight-line segments as shown in Figure 21.12(ii).

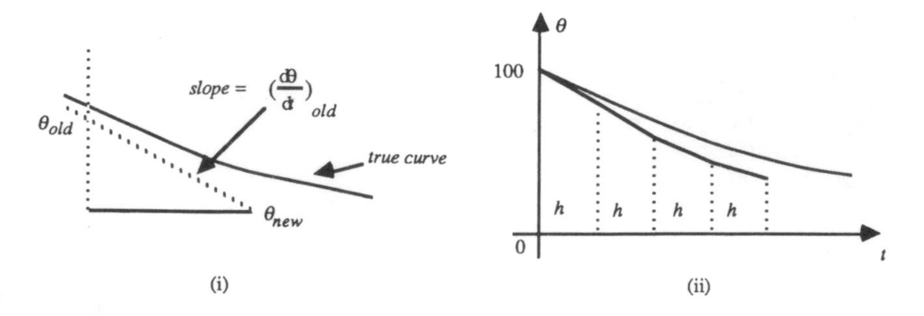

Figure 21.12

Let us do a few steps of the process mathematically. We shall assume that the constant k is such that we may use times in minutes.

We have $d\theta/dt = -0.1(\theta - 20)$ given that $\theta = 100$ when $t = 0$, i.e. $\theta(0) = 100$. Let us take $h = 1$. Then $d\theta/dt$ at $t = 0$ is $-0.1(100 - 20) = -8$.

Applying the Euler formula (21.4)

$$\theta(1) = \theta_1 = 100 + (1)\left[\frac{d\theta}{dt}\right]_{t=0} = 100 - 8 = 92$$

$$\theta(2) = \theta_2 = 92 + (1)\left[\frac{d\theta}{dt}\right]_{t=1} = 92 - 7.2 = 84.8$$

$$\theta(3) = \theta_3 = 84.8 + (1)\left[\frac{d\theta}{dt}\right]_{t=2} = 84.8 - 6.48 = 78.32$$

$$\theta(4) = \theta_4 = 78.32 - 5.832 = 72.488$$

This process could easily be used on a digital computer, the flow chart being as shown in Figure 21.13. A program based on the flow chart is shown below.

```
>
   10   CLS
   20   PRINTTAB(5,2)"Euler's Method applied to
        Newton's Law of Cooling"
   30   INPUTTAB(0,4)"Enter air temperature ",T2
   40   INPUT"Enter start temperature ",T1
   50   INPUT"Enter K ",K
   60   INPUT"Enter start time ", T0
   70   INPUT"Enter number of steps ",N
   80   INPUT"Enter the step size ",H
   90   PRINT"     TEMP","      TIME"
  100   @%=&20209
  110   T=T1
  120   T3=T0
  130   I=0
  140   REPEAT
  150     D=-K*(T-T2)
  160     T3=T3+H
  170     I=I+1
  180     T4=T+D*H
  190     PRINTT4,T3
  200     IF I=N THEN END
  210     T=T4
  220     UNTIL FALSE
```

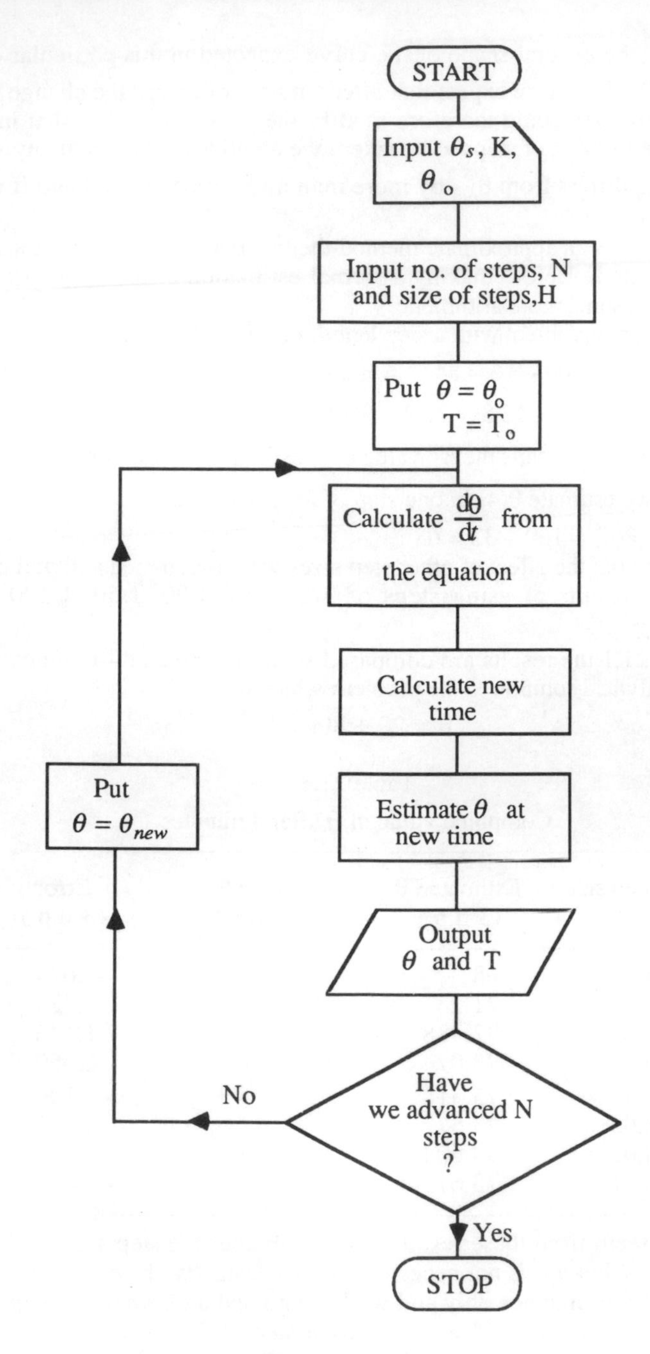
START
Input θ_s, K, θ_o
Input no. of steps, N and size of steps, H
Put $\theta = \theta_o$ $T = T_o$
Calculate $\frac{d\theta}{dt}$ from the equation
Calculate new time
Estimate θ at new time
Output θ and T
Have we advanced N steps ?
No
Yes
Put $\theta = \theta_{new}$
STOP

Figure 21.13

We know the general shape of the curve expected in this particular example (see Figure 21.3) and we expect that after a number of steps the change in θ will be very small. We could therefore modify the flow diagram so that instead of giving N, the number of steps to be taken, we could test whether at any stage the value of θ_{new} differs from θ_{old} by more than a specified amount and if not, stop the process.

Because this is an approximate method there is bound to be an error attached to the result quoted. Before making a formal estimation of the error expected we shall carry out some more arithmetic.

Take the same problem with a step length of 2 minutes. Then

$$\theta_1 = 100 - 16 = 84 \quad = \theta(2)$$

$$\theta_2 = 84 - 12.8 = 71.2 = \theta(4)$$

(The notation θ_k means the kth value of θ estimated by Euler's method.)

Suppose we estimate $\theta(4)$ in one step of length 4, then

$$\theta(4) = 100 - 32 = 68$$

In order to try out the effect of other step sizes we programmed a digital computer to show the results of using steps of 1/2, 1/10, 1/20, 1/50, 1/100 minutes respectively.

In Table 21.1 the results are compared with the value at 4 minutes obtained from the analytical solution to the problem which is

$$\theta = 20 + 80e^{-0.1t} \tag{21.5}$$

Table 21.1

Computed value of θ after 4 minutes

Step size h	Estimated θ (3 d.p.)	True θ (3 d.p.)	Error (3 d.p.)
4	68		–5.624
2	71.2		–2.424
1	72.488		–1.136
0.5	73.074		–0.550
0.1	73.518	73.624	–0.106
0.05	73.572		–0.052
0.02	73.604		–0.020
0.01	73.615		–0.009

It would seem from these results that the smaller the step size the better our approximation, but this is not necessarily true because we have done calculations over the first four minutes only and we have no real idea what will happen later. In fact there is a **round-off error** at each step of the calculations so that the more steps we take then the more there is a danger of cumulative round-off error. Clearly what is needed is a balance between poor approximation and the number of steps.

Discussion of round-off error

We cannot get much idea from Table 21.1 of the accumulation of error, since over the first four minutes the graph of θ against t does not differ much from a straight line. We need to look at the longer-term situation. If we ask at what time the temperature of the liquid will reach 22°, we can produce Table 21.2.

Table 21.2

t	Euler Estimates of θ with time steps 5.0 mins	1.0 mins	0.5 mins (3 d.p. where relevant)	Analytical value 3 d.p.
0	100	100	100	100
5	60	67.239	67.899	68.522
10	40	47.894	48.679	49.430
15	30	36.471	37.171	37.850
20	25	29.726	30.281	30.827
25	22.5	25.743	26.156	26.568
30	21.25	23.391	23.501	23.983
35	20.625	22.003	22.096	22.416
40	20.313	21.182	21.321	21.465

We see that for a step-size of 5 minutes, the temperature is estimated to reach 22° after a time between 25 and 30 minutes. For a step size of 1 minute, it is estimated to take just over 35 minutes; for a step of 0.5 minutes it takes a little longer but still just over 35 minutes. We can find an exact value from our analytical formula (21.5), as follows. We had

$$\theta = 20 + 80e^{-0.1t}$$

Now when $\theta = 22$, $22 = 20 + 80e^{-0.1t}$

so that $$2 = 80e^{-0.1t}$$

and $$e^{0.1t} = 40$$

Taking logarithms, $t = 10 \ln 40 = 36.889$ minutes (3 d.p.).

Now, although it takes approximately 3700 evaluations with a step size of 0.01 minutes, we did program the digital computer to work with this step size and it gave a value of 22° at a time between 37.07 and 37.08 minutes. We can, therefore, assume that this discrepancy from the analytical value is almost entirely due to round-off error.

It can be seen that decreasing the step size provides better accuracy at each estimate examined. Although one might expect errors to accumulate, we find that for a fixed step size, the errors decrease with increasing time. There are clearly other sources of error besides round-off error and we examine them in turn.

(i) *Formula error*

The approximation during each step that the slope of the function stays constant will lead to a wrong estimate of θ_{new}. This could be alleviated by reducing the step size if the formula gave a consistently better approximation.

(ii) *Inherited error*

It has to be recognised that we are handicapped before we step forward each time because θ_{old}, the starting value of θ, is itself an estimation (except for the initial value). Under certain conditions this may cause an accumulation of errors but under other conditions this may lead to a better approximation.

Problems

Section 21.2

1 What are the order and degree of the following differential equations?

(a) $\dfrac{dy}{dx} = \dfrac{\sqrt{1+x}}{\sqrt{1+y}}$ (b) $y\dfrac{d^2y}{dx^2} + \left[\dfrac{dy}{dx}\right]^2 = 0$ (c) $EI\dfrac{d^2y}{dx^2} = \frac{1}{2}wx^2 - \frac{1}{2}wlx$

(d) $\dfrac{d^2y}{dx^2} - \left\{1 + \left[\dfrac{dy}{dx}\right]^2\right\}^{3/2} = 0$ (e) $2\dfrac{dy}{dx}\dfrac{d^3y}{dx^3} - 3\left[\dfrac{d^2y}{dx^2}\right]^2 = 0$ (f) $\left[\dfrac{dy}{dx}\right]^3 - y = x$

2 Eliminate the arbitrary constants from the following equations to obtain the associated differential equations

(i) $y = Ax + B$ (ii) $y^2 = Ax$ (iii) $y = A\cos 2x + B\sin 2x$

(iv) $y = (A + Bx)e^{2x}$ (v) $x = \frac{1}{2}\ln(A + Bt)$

3 Consider the equation $dy/dx = 2x$. Is there a solution curve which passes through (0, 1) and (2, 5)? Is there a solution curve which passes through (1, 3) and (4, 17)?

4 The general solution of

$$\frac{d^2y}{dx^2} + 3\frac{dy}{dx} + 2y = 0$$

is

$$y = Ae^{-2x} + Be^{-x}$$

A and B constants. Explain why $y(0) = 1$ is not sufficient information to determine a unique solution. [Hint: consider the Maclaurin expansion of y.]

Would the sole condition $y(1) = 2$ change your argument?

5 Examine

$$y = x\frac{dy}{dx} + \left[\frac{dy}{dx}\right]^2$$

Show that a *general* solution to the equation is

$$y = Ax + A^2, \quad A \text{ constant}$$

To find the envelope of this family of curves we find the partial derivatives of both sides with respect to the parameter A. This gives $0 = x + 2A$. Hence find the envelope as $y = -\frac{1}{4}x^2$. Show that this is a **singular** solution of the equation, that is, it cannot be obtained from the general solution. Sketch some of the family of curves and their envelope.

If the boundary condition $y(3) = -2\frac{1}{4}$ were added would this give an unique solution?

6 In many branches of science the concept of an **orthogonal trajectory** is important. Remember from page 380 that this is concerned with finding a family of curves which is orthogonal to a given family of curves. That is to say, at each point where a member of the given family is intersected by a member of the new family, the intersection is at *right angles*. If the equation of the given family is $dy/dx = f(x, y)$ then the equation of the new family is $dy/dx = -1/[f(x, y)]$.
Show that for the family of circles $x^2 + y^2 = a^2$ the governing differential equation is $dy/dx = -x/y$. Show that the orthogonal family's differential equation is satisfied by $y = Kx$, K constant. Interpret geometrically, and give two examples from fluid flow of the applicability of the result.

7 (i) Show that the trajectory orthogonal to the family $xy = K$ is the family $x^2 - y^2 = B$ (K and B constants).
(ii) Show that the trajectory orthogonal to the family $x^2 + 9y^2 = K$ is the family $y = Ax^9$ (K and A constants).

Section 21.3

8 Sketch the isoclines and the solution curve for

(i) $\dfrac{dy}{dx} = 2y - 1$ (ii) $\dfrac{dy}{dx} = y - x$ (iii) $\dfrac{dy}{dx} = x^2 + y$ (iv) $\dfrac{dy}{dx} = y - e^{-x}$

9 Show that the locus of points of inflection on the solution curves of Problem 8 is as stated in the text.

22

FIRST ORDER DIFFERENTIAL EQUATIONS

22.1 Variables Separable

In this chapter we shall be considering only those differential equations which are of first order and first degree. That is, we consider differential equations of the form

$$\frac{dy}{dx} = f(x, y) \tag{22.1}$$

We can recognise those equations of the form (22.1) for which the variables can be separated by the fact that $f(x, y)$ factorises into $F(x).G(y)$, that is, $f(x, y)$ is the product of two terms, one containing x only, the other containing y only. For example,

$$\frac{dy}{dx} = x^2y + y$$

can be written
$$\frac{dy}{dx} = (x^2 + 1)y$$

and the equation
$$xy^2(1 + x^2)\frac{dy}{dx} - y^3 = 1$$

can be written
$$\frac{dy}{dx} = \frac{1 + y^3}{xy^2(1 + x^2)} = \frac{1}{x(1 + x^2)} \cdot \frac{(1 + y^3)}{y^2}$$

To solve differential equations of this form we note that if

$$\frac{dy}{dx} = F(x).G(y)$$

then
$$\frac{1}{G(y)}\frac{dy}{dx} = F(x)$$

We then integrate both sides with respect to x, that is

$$\int \frac{1}{G(y)} dy = \int F(x) dx$$

Each integration could produce an arbitrary constant but, remembering that a first order differential equation needs only one arbitrary constant for its general solution, we only need to put an arbitrary constant on one side of the equation as shown in the following example.

Example 1

$$\frac{\mathrm{d}y}{\mathrm{d}x} = x^2 y + y$$

That is,
$$\frac{\mathrm{d}y}{\mathrm{d}x} = (x^2 + 1)y$$

or
$$\frac{1}{y}\frac{\mathrm{d}y}{\mathrm{d}x} = x^2 + 1$$

Integrating both sides w.r.t. x

$$\int \frac{1}{y}\mathrm{d}y = \int (x^2 + 1)\mathrm{d}x$$

This yields $\ln y = \frac{x^3}{3} + x + A,$ where A is arbitrary

Taking antilogarithms of both sides gives

$$y = \mathrm{e}^{(x^3/3)+x+A} = \mathrm{e}^A \,.\, \mathrm{e}^{(x^3/3)+x}$$

Since A is arbitrary, e^A is also an arbitrary constant, and we can write $\mathrm{e}^A = B$, giving finally the general solution

$$y = B\mathrm{e}^{(x^3/3)+x}$$

Example 2

We now solve
$$xy^2(1 + x^2)\frac{\mathrm{d}y}{\mathrm{d}x} - y^3 = 1$$

which can be written
$$\frac{\mathrm{d}y}{\mathrm{d}x} = \frac{1}{x(1 + x^2)} \cdot \frac{(1 + y^3)}{y^2}$$

Separating the variables we get

$$\frac{y^2}{(1 + y^3)}\frac{\mathrm{d}y}{\mathrm{d}x} = \frac{1}{x(1 + x^2)}$$

Integrating both sides w.r.t. x produces

$$\int \frac{y^2}{(1 + y^3)}\mathrm{d}y = \int \frac{1}{x(1 + x^2)}\mathrm{d}x$$

This gives $$\frac{1}{3}\ln(1 + y^3) = \int\left[\frac{1}{x} - \frac{x}{1+x^2}\right] dx$$

$$= \ln x - \frac{1}{2}\ln(1 + x^2) + A$$

In this case, where all the other terms are in logarithmic form it is best to write the arbitrary constant similarly, so we put $A = \ln \alpha$ where α is arbitrary. Then

$$\frac{1}{3}\ln(1 + y^3) = \ln x - \frac{1}{2}\ln(1 + x^2) + \ln \alpha$$

Taking antilogs, we obtain the general solution in the form

$$(1 + y^3)^{1/3} = \frac{\alpha x}{\sqrt{1+x^2}}$$

Example 3

We now consider a practical example which can be modelled by an ordinary differential equation whose solution is obtained by the separation of variables method.

Liquid is contained in a cylindrical tank of cross-sectional area 1400 cm^2 and initially the depth of water is 50 cm. The tank is being emptied through an orifice of effective cross-sectional area 10 cm^2 (Figure 22.1). Find the time taken to empty the tank, *assuming* (i) that the effective area of the orifice remains constant, and (ii) that the velocity of the emerging liquid, v, is $\sqrt{2gh}$ cm/sec, when h cm is the depth of liquid in the tank. These are the assumptions for our physical model.

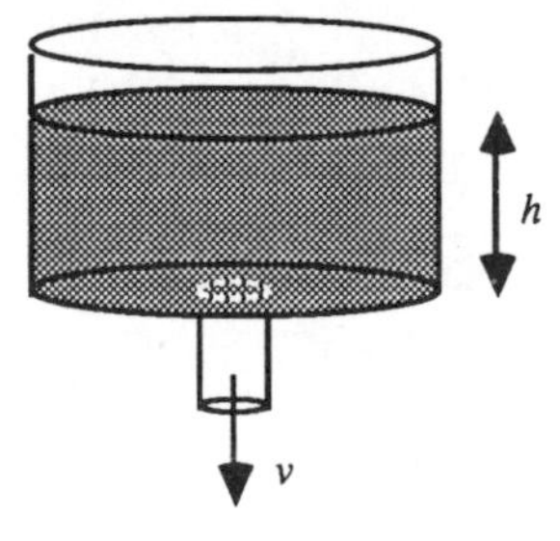

Figure 22.1

If h is the depth of liquid at time t sec, then the rate at which water is emerging through the orifice = $\sqrt{2gh}$. 10 cm^3/sec. At this time the surface of the liquid must be descending at a rate of $\sqrt{2gh}$ 10/1400 cm/sec. [This is the rate of loss of liquid in the tank divided by the cross-sectional area.]

Hence $$\frac{dh}{dt} = -\sqrt{2gh}\ .\ 10/1400$$

that is, $$\frac{dh}{dt} = -\frac{1}{\sqrt{10}}\sqrt{h} \quad (g \text{ is taken as } 980 \text{ cm/sec}^2)$$

Separating the variables

$$\frac{1}{\sqrt{h}}\frac{dh}{dt} = -\frac{1}{\sqrt{10}} \tag{22.2}$$

and integrating both sides with respect to t, we get

$$\int \frac{1}{\sqrt{h}}dh = \int -\frac{1}{\sqrt{10}}dt$$

or $$2\sqrt{h} = -\frac{1}{\sqrt{10}}t + A, \quad A \text{ any arbitrary constant.}$$

This is the general solution of equation (22.2) but we require a *particular* solution, since we were told that at time $t = 0$, the height $h = 50$ cm. Substituting these values into the general solution we find

$$2\sqrt{h} = -\frac{1}{\sqrt{10}}t + 10\sqrt{2}$$

When the tank is empty, then $h = 0$ and we find

$$0 = -\frac{1}{\sqrt{10}}t + 10\sqrt{2}$$

or $$t = 10\sqrt{20} \text{ secs}$$

Hence the tank empties in $10\sqrt{20}$ seconds $\cong$ 45 seconds.

22.2 Integrating Factor Method

Exact equations

If we write a first order differential equation in the form

$$P(x, y)dx + Q(x, y)dy = 0 \tag{22.3}$$

where $P(x, y)$ and $Q(x, y)$ are functions of x and y, then the equation is **exact** if we can find a function $\phi(x, y)$ such that the differential $d\phi$ is given by

$$d\phi = Pdx + Qdy$$

In such a case we have $$d\phi = 0$$

which integrates immediately to $\phi = \text{constant}$

For example $$(2x + 3x^2y)dx + (x^3)dy = 0$$

can be written $$2xdx + (3x^2ydx + x^3dy) = 0$$

i.e. $$\mathrm{d}(x^2) + \mathrm{d}(x^3y) = 0$$

The second term arises from two terms; it is the derivative of a product.

Therefore we have $$\mathrm{d}(x^2 + x^3y) = 0$$

and, integrating, we get $$x^2 + x^3y = \text{constant}$$

This is all very well in a simple example, but can we recognise when it is possible to group the terms in such a form? For example, the equation

$$y^3\,\mathrm{d}x + \frac{x}{y}\mathrm{d}y = 0$$

is not exact, but we might spend a long time discovering this if we first play around trying to find a suitable function $\phi(x, y)$. The clue is to start in reverse: if we assume we can find $\phi(x, y) = \text{constant}$, then we must have by differentiation

$$\mathrm{d}\phi(x, y) = 0$$

or $$\frac{\partial\phi}{\partial x}\mathrm{d}x + \frac{\partial\phi}{\partial y}\,\mathrm{d}y = 0$$

Comparing this with $P\mathrm{d}x + Q\mathrm{d}y$, we see that

$$P = \frac{\partial\phi}{\partial x} \quad \text{and} \quad Q = \frac{\partial\phi}{\partial y}$$

Since, for most cases $$\frac{\partial^2\phi}{\partial x\,\partial y} = \frac{\partial^2\phi}{\partial y\,\partial x}$$

then we must have $$\frac{\partial P}{\partial y} = \frac{\partial^2\phi}{\partial y\,\partial x} = \frac{\partial^2\phi}{\partial x\,\partial y} = \frac{\partial Q}{\partial x}$$

So, if the equation (22.3) is exact, then

$$\frac{\partial P}{\partial y} = \frac{\partial Q}{\partial x} \tag{22.4}$$

Taking our first example,

$$\left.\begin{aligned} P &= 2x + 3x^2y \qquad & \frac{\partial P}{\partial y} &= 3x^2 \\ Q &= x^3 & \frac{\partial Q}{\partial x} &= 3x^2 \end{aligned}\right\}\quad \text{equal}$$

Hence the equation is **exact**. The second example was such that

$$\left.\begin{aligned} P &= y^3 \qquad & \frac{\partial P}{\partial y} &= 3y^2 \\ Q &= \frac{x}{y} & \frac{\partial Q}{\partial x} &= \frac{1}{y} \end{aligned}\right\}\quad \text{not equal}$$

The equation is **not exact**.

So we can recognise straight away whether or not an equation is exact.

Integration of Exact Equations

We can now recognise when an equation is exact or not. Can we integrate it? In the case where (Pdx + Qdy) is exact, Pdx + Qdy = 0 can be written

$$\frac{\partial\phi}{\partial x}\mathrm{d}x + \frac{\partial\phi}{\partial y}\mathrm{d}y = 0$$

Thus if we keep y constant in the function P and integrate with respect to x we have the opposite of partial differentiation and this will give us ϕ except for an unknown function of y (which would disappear on partially differentiating w.r.t. x). If we now differentiate this expression for ϕ partially with respect to y we should get Q.

Example

$$(2x + 3x^2y)\mathrm{d}x + x^3\mathrm{d}y = 0$$

$$P = \frac{\partial\phi}{\partial x} = 2x + 3x^2y$$

therefore $\phi = x^2 + x^3y + f(y)$ where $f(y)$ is an unknown function of y.
We now differentiate with respect to y, therefore

$$\frac{\partial\phi}{\partial y} = x^3 + f'(y)$$

But $$\frac{\partial\phi}{\partial y} = Q = x^3$$

therefore $$f'(y) = 0$$

therefore $$f(y) = A, \text{ an arbitrary constant}$$

Hence $$\phi = x^2 + x^3y + A = \text{constant}$$

or $$x^2 + x^3y = \text{constant}$$

Notice we do not need 2 arbitrary constants, so we could miss out A if we liked.

Sometimes an equation is not exact in the form it is given, but can be made exact by multiplying by a suitable factor called an **integrating factor**. It is beyond our scope to be able to find these factors in general for such equations but an example will explain what is meant. Consider the equation

$$(4x^2y^2 + 2\ln y)\mathrm{d}x + \left[2x^3y + \frac{x}{y}\right]\mathrm{d}y = 0$$

$P = 4x^2y^2 + 2\ln y$ $\qquad$ $\dfrac{\partial P}{\partial y} = 8x^2y + \dfrac{2}{y}$

Not Exact

$Q = 2x^3y + \dfrac{x}{y}$ $\qquad$ $\dfrac{\partial Q}{\partial x} = 6x^2y + \dfrac{1}{y}$

But to the experienced eye, it would be seen that if we multiply the given equation by x we do get an exact equation. Thus

$$(4x^3y^2 + 2x \ln y)\mathrm{d}x + \left[2x^4y + \frac{x^2}{y}\right]\mathrm{d}y = 0$$

P now is $4x^3y^2 + 2x \ln y$ $\qquad \dfrac{\partial P}{\partial y} = 8x^3y + \dfrac{2x}{y}$

Exact

Q now is $2x^4y + \dfrac{x^2}{y}$ $\qquad \dfrac{\partial Q}{\partial x} = 8x^3y + \dfrac{2x}{y}$

Thus by multiplying by the **integrating factor** x we obtain an exact equation and we can proceed along the usual lines – you should complete the example yourself. Note that $2x$, $4x$, $-79.2x$ are also possible integrating factors (why?).

Linear Equations

Another type of equation which is theoretically possible to solve analytically is the *linear, first order differential equation.* The general form of such an equation is

$$\frac{\mathrm{d}y}{\mathrm{d}x} + py = q$$

where p and q are functions of x only. For example

$$\frac{\mathrm{d}y}{\mathrm{d}x} + x^2y = (x - 3) \qquad \frac{\mathrm{d}y}{\mathrm{d}x} - \frac{y}{(x+1)} = x^2e^x \qquad \frac{\mathrm{d}y}{\mathrm{d}x} + \tan x \,.\, y = \sec x$$

We have already met the idea of an integrating factor in this section on exact equations. This is a factor by which we multiply the differential equation in order to make it into a suitable form for integration. This is just the technique we use here.

We take the l.h.s. of the general linear equation and try to make it into the derivative of a product by multiplying it by some function of x. That is, we take $(\mathrm{d}y/\mathrm{d}x) + py$ and multiply by a function of x, say $I(x) = I$ for short.

This gives $\qquad I\dfrac{\mathrm{d}y}{\mathrm{d}x} + pIy$

The first term is the first part of

$$\frac{\mathrm{d}}{\mathrm{d}x}(Iy) = I\frac{\mathrm{d}y}{\mathrm{d}x} + \frac{\mathrm{d}I}{\mathrm{d}x}y$$

Comparing the second terms, we want

$$pIy = \frac{\mathrm{d}I}{\mathrm{d}x}y \quad \text{or} \quad \frac{\mathrm{d}I}{\mathrm{d}x} = pI$$

Separating the variables

$$\int \frac{\mathrm{d}I}{I} = \int p\mathrm{d}x$$

That is, $\ln I = \int p\mathrm{d}x$ or $I = \mathrm{e}^{\int p\mathrm{d}x}$

Thus our integrating factor (I.F.) can, theoretically, be found.

Take as an example $$\frac{\mathrm{d}y}{\mathrm{d}x} + x^2y = (x - 3)$$

The function p is x^2, therefore

$$\text{I.F. is } \mathrm{e}^{\int x^2\mathrm{d}x} = \mathrm{e}^{x^3/3}$$

Similarly, for the example

$$\frac{\mathrm{d}y}{\mathrm{d}x} - \frac{y}{(x+1)} = x^2\mathrm{e}^x$$

$$p \text{ is } -\frac{1}{(1+x)}$$

therefore $$\text{I.F. is } \mathrm{e}^{\int(-1/[1+x])\mathrm{d}x} = \mathrm{e}^{-\ln(1+x)}$$

Thus, in the general equation $(\mathrm{d}y/\mathrm{d}x) + py = q$, if we can integrate p with respect to x, we can find the integrating factor and multiplying the equation by this factor reduces the l.h.s. to the form

$$\frac{\mathrm{d}}{\mathrm{d}x}(\text{I.F.} \times y)$$

The r.h.s. will have become $q \times$ I.F. Then the differential equation becomes

$$\frac{\mathrm{d}}{\mathrm{d}x}(\text{I.F.} \times y) = q \times \text{I.F.}$$

Integrating both sides w.r.t. x, we get

$$y \times \text{I.F.} = \int (q \times \text{I.F.})\mathrm{d}x$$

Thus $$y = \frac{1}{\text{I.F.}}\int (q \times \text{I.F.})\mathrm{d}x$$ is the solution.

This solution will be general if it contains one arbitrary constant. This arbitrary constant will arise from the integral on the right-hand side.

Example 1

$$\frac{\mathrm{d}y}{\mathrm{d}x} + y = x \quad \text{for which} \quad \text{I.F.} = \mathrm{e}^{\int 1.\mathrm{d}x} = \mathrm{e}^x$$

Multiplying throughout by this we get

$$\mathrm{e}^x\frac{\mathrm{d}y}{\mathrm{d}x} + y\mathrm{e}^x = x\mathrm{e}^x \quad \text{or} \quad \frac{\mathrm{d}}{\mathrm{d}x}(y\mathrm{e}^x) = x\mathrm{e}^x$$

Therefore $$y\mathrm{e}^x = \int x\mathrm{e}^x\mathrm{d}x$$

that is, $$ye^x = xe^x - e^x + A$$

Hence $$y = x - 1 + Ae^{-x}$$

(Notice that Ae^{-x} satisfies the equation $(dy/dx) + y = 0$.)

Note that, once you are used to the method, there is no need to put in the step where you multiply the equation by the integrating factor. You can go straight to the following line

$$\frac{d}{dx}(y \times \text{I.F.}) = q \times \text{I.F.}$$

Example 2

$$x(x-1)\frac{dy}{dx} + y = (x+1)$$

We must first write this in the standard form $(dy/dx) + py = q$. In this case we have

$$\frac{dy}{dx} + \frac{1}{x(x-1)}y = \frac{x+1}{x(x-1)}$$

$$\text{I.F.} = e^{\int 1/(x(x-1))dx} = e^{\int (-1/x)+(1/(x-1))dx} = e^{\ln((x-1)/x)} = \frac{x-1}{x}$$

The equation therefore reduces to

$$\frac{d}{dx}\left[y \cdot \left[\frac{x-1}{x}\right]\right] = \frac{x+1}{x(x-1)} \cdot \left[\frac{x-1}{x}\right] = \frac{x+1}{x^2}$$

Integrating both sides

$$y\frac{(x-1)}{x} = \int \frac{x+1}{x^2}dx$$

$$= \int \left[\frac{1}{x} + \frac{1}{x^2}\right] dx$$

Hence $$y\frac{(x-1)}{x} = \ln x - \frac{1}{x} + A$$

or $$y = \frac{x}{x-1}\ln x - \frac{1}{x-1} + \frac{Ax}{x-1}$$

Which equation do you think $Ax/(x-1)$ satisfies?

Example 3 (See Figure 22.2)

A circuit consisting of a resistance R ohms and an inductance L henrys is connected to an e.m.f. voltage of $E \cos \omega t$. Find the current i amps at a time t after the circuit is closed. We assume connecting wires are resistance-free.

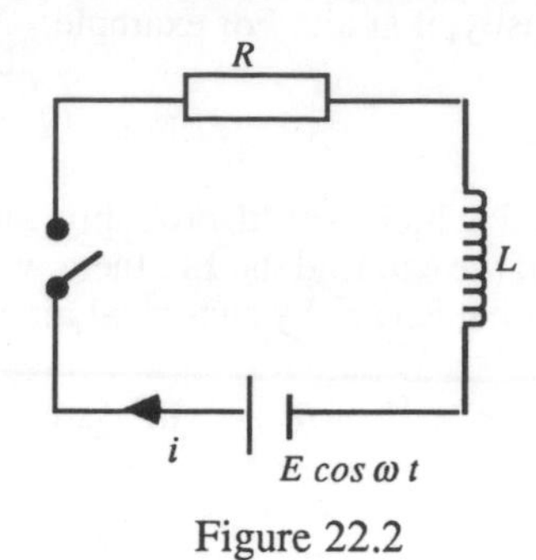

Figure 22.2

The voltage $E \cos \omega t$ must be equal to the voltage drops Ri across the resistance and $L(\mathrm{d}i/\mathrm{d}t)$ due to the inductance so that

$$L\frac{\mathrm{d}i}{\mathrm{d}t} + Ri = E \cos \omega t$$

The integrating factor $= \mathrm{e}^{\int (R/L)\mathrm{d}t} = \mathrm{e}^{(R/L)t}$

Hence
$$i\mathrm{e}^{(R/L)t} = \int \frac{E}{L} \cos \omega t \, \mathrm{e}^{(R/L)t} \, \mathrm{d}t$$

then
$$i\mathrm{e}^{(R/L)t} = \frac{\mathrm{e}^{(Rt/L)} E}{(R^2 + L^2 \omega^2)}(R \cos \omega t + L\omega \sin \omega t) + A$$

Since $i = 0$ when $t = 0$, then

$$0 = \frac{ER}{(R^2 + L^2 \omega^2)} + A$$

giving
$$A = \frac{-ER}{R^2 + L^2 \omega^2}$$

so that

$$i\mathrm{e}^{(R/L)t} = \frac{\mathrm{e}^{(Rt/L)} E}{(R^2 + L^2 \omega^2)}(R \cos \omega t + L\omega \sin \omega t) - \frac{ER}{(R^2 + L^2 \omega^2)}$$

therefore

$$i = \frac{E}{R^2 + L^2 \omega^2}(R \cos \omega t + L\omega \sin \omega t - R\mathrm{e}^{-(R/L)t})$$

Note that there is an oscillatory part to the solution and a decaying exponential or *transient* part. Where has this last part come from?

Failure of the integrating factor method

Now this all appears very simple but many things can go wrong. Firstly, we may not be able to find the I.F. easily, if at all. For example,

$$\frac{dy}{dx} + \sqrt{1 + x^3}\, y = 1 \qquad \text{I.F.} = e^{\int\sqrt{1+x^3}\,dx} = ?$$

This is a difficult integral which would probably make a solution by this method unattainable. Even if we can find the I.F. there will be many cases where we cannot integrate the resulting R.H.S. by analytical methods. For example,

$$\frac{dy}{dx} + 2xy = (x - 2) \qquad \text{I.F. is } e^{\int 2x dx} = e^{x^2}$$

The equation therefore reduces to

$$\frac{dy}{dx}(y \,.\, e^{x^2}) = (x - 2)e^{x^2} \qquad \text{or} \qquad ye^{x^2} = \int (x - 2)e^{x^2}\, dx$$

We can find $\int xe^{x^2}\, dx$

but $\int -2e^{x^2}\, dx$

is the stumbling block. We could evaluate the integral on the R.H.S. by numerical integration, but we must ask *should we*? Would it be better in these cases of failure to start from the beginning with a numerical technique? The answer is probably *Yes!*

22.3 Rocket Re-Entry: A Case Study

Before considering further numerical methods we should like to consider the following practical example which illustrates a failure of the analytical method.

A rocket re-entering the earth's atmosphere experiences a drag force proportional to the square of its velocity and proportional to the density of the air. This density, ρ, is approximated by the law $\rho = \rho_0\, e^{-\lambda z}$ where z is the height above the earth; ρ_0 is the density at sea-level and λ is a positive constant.

Let A be the cross-sectional area of the rocket, C_D a constant called the drag coefficient and let us assume that re-entry is vertical and without lift. From Newton's second law, the equation of motion is

$$mv\,\frac{dv}{dz} = -\,mg + \tfrac{1}{2}C_D\, A\rho v^2$$

Note that we are measuring acceleration upwards (See Figure 22.3). It turns out to be more convenient this way. Imagine that we throw a stone upwards and then consider only its falling to the ground. In this example, the rocket should slow down as it approaches the earth and so the net force must be upwards.

Now $$\frac{d}{dz}(v^2) = 2v\,\frac{dv}{dz}$$

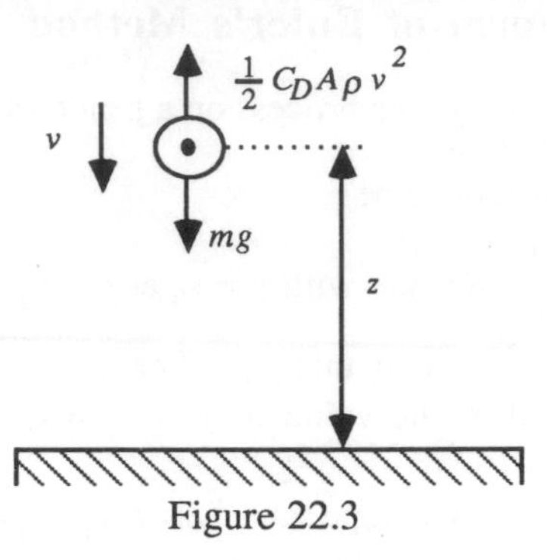

Figure 22.3

therefore, writing $V = v^2$ and $K = C_D A/m$ we obtain

$$\frac{\mathrm{d}}{\mathrm{d}z}(v^2) = -2g + \frac{C_D A}{m}\rho v^2$$

that is

$$\frac{\mathrm{d}V}{\mathrm{d}z} = -2g + K\rho_0 \,\mathrm{e}^{-\lambda z}\, V$$

or

$$\frac{\mathrm{d}V}{\mathrm{d}z} - K\rho_0 \,\mathrm{e}^{-\lambda z}\, V = -2g$$

The Integrating Factor is $\quad \mathrm{e}^{\int -K\rho_0 \mathrm{e}^{-\lambda z}\, \mathrm{d}z}$

$$= \mathrm{e}^{(K/\lambda)\rho_0 \mathrm{e}^{-\lambda z}}$$

Multiplying the equation of motion by this factor produces

$$\frac{\mathrm{d}}{\mathrm{d}z}\left(V\mathrm{e}^{(K\rho_0/\lambda)\mathrm{e}^{-\lambda z}}\right) = -2g\mathrm{e}^{(K/\lambda)\rho_0 \mathrm{e}^{-\lambda z}}$$

which, on integration, becomes

$$V\mathrm{e}^{(K\rho_0/\lambda)\mathrm{e}^{-\lambda z}} = \int -2g\mathrm{e}^{(K/\lambda)\rho_0 \mathrm{e}^{-\lambda z}}\,\mathrm{d}z$$

If we expand $\mathrm{e}^{(K/\lambda)\rho_0 \mathrm{e}^{-\lambda z}}$ as a Maclaurin's series and integrate term by term, we obtain

$$V = \mathrm{e}^{(-K\rho_0/\lambda)\mathrm{e}^{-\lambda z}}\left[-\frac{2g}{\lambda}\sum_{n=1}^{\infty}\frac{\left[\dfrac{\rho_0 K}{\lambda}\mathrm{e}^{-\lambda z}\right]^n}{n.n\,!} - 2gz + \text{constant}\right] \tag{22.5}$$

This formula would be very time-consuming to use as the basis for calculating V at several points on the rocket's descent. It is virtually impossible to use for qualitative information.

22.4 General Form of Euler's Method

Let us now put Euler's step-by-step process on a general footing for use with any first order differential equation.

Take the first order equation to be

$$\frac{dy}{dx} = f(x, y) \quad \text{with } y = y_0 \text{ at } x = x_0$$

The slope at x_0 is $f(x_0, y_0)$; call this f_0. Then, assuming the slope to be constant over a range of x, the value of y at $x_1 = x_0 + h$ is approximately $y_1 = y_0 + f_0.h$.

We can now estimate the slope at x_1 as $f(x_1, y_1) = f_1$, say. Then the estimated value of y at $x_2 = x_1 + h = x_0 + 2h$ is

$$y_2 = y_1 + f_1 h$$

This procedure is repeated giving the general expression

$$y_{n+1} = y_n + f_n h$$

or

$$y_{n+1} = y_n + h\left[\frac{dy}{dx}\right]_{x=x_n} \tag{22.6}$$

Example 1

$$\frac{dy}{dx} = 1 \text{ and } y = 1 \text{ when } x = 0$$

We can easily obtain the analytical solution of this equation as $y = x + 1$, the graph of which is shown in Figure 22.4(i). Note we advance from $x = 0$.

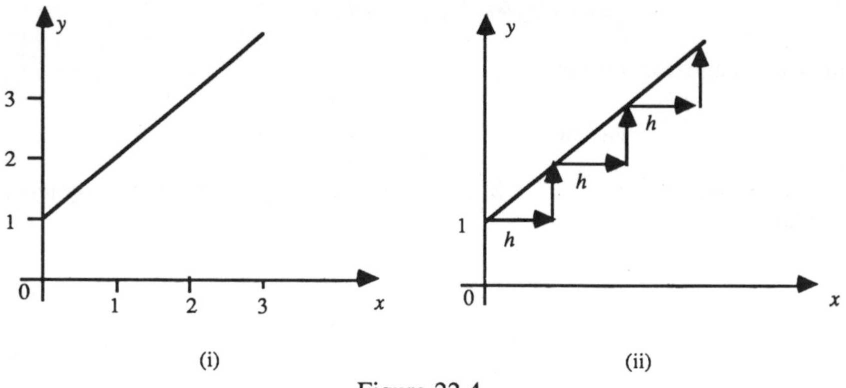

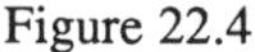
Figure 22.4

The step-by-step method is

$$y_{n+1} = y_n + h\left[\frac{dy}{dx}\right]_{x=x_n}$$

But dy/dx = 1 at all points, therefore

$$y_{n+1} = y_n + h$$

Starting at $y = 1, x = 0$, we shall get

$$y_1 = 1 + h$$
$$y_2 = (1 + h) + h = 1 + 2h$$
$$y_3 = (1 + 2h) + h = 1 + 3h$$
$$y_n = 1 + nh$$
$$y_{n+1} = 1 + (n + 1)h$$

The approximation is shown in Figure 22.4(ii). The approximation is, in this case, exact.

Example 2

$$\frac{dy}{dx} = x \text{ with } y = 1 \text{ when } x = 0$$

The analytical solution is $$y = \frac{x^2}{2} + 1 \qquad (22.7)$$

The step-by-step method gives

$$y_1 = 1 + h.(0) = 1$$
$$y_2 = 1 + h.h = 1 + h^2$$
$$y_3 = (1 + h^2) + h.2h = 1 + 3h^2$$
$$y_4 = (1 + 3h^2) + h.3h = 1 + 6h^2$$
$$y_5 = (1 + 6h^2) + h.4h = 1 + 10h^2$$

You should show that $$y_n = 1 + h^2.\tfrac{1}{2}n(n-1)$$

and hence that $$y_{n+1} = y_n + hx_n$$

The graphs of the analytical and step-by-step solutions with $h = 1$ and $h = 0.5$ are shown in Figure 22.5.

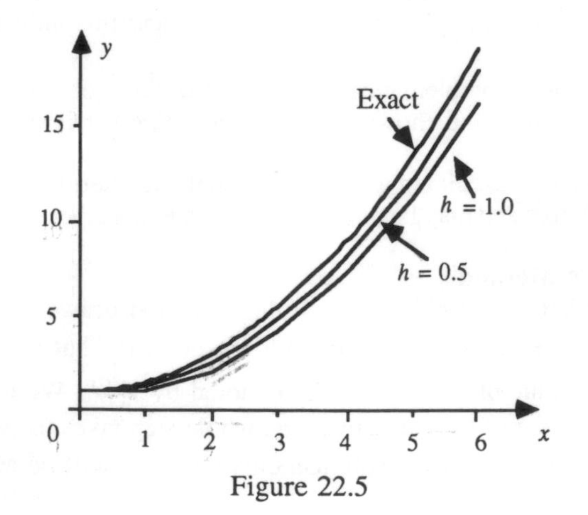

Figure 22.5

Note that in both these examples we started at $x = 0$; this was merely to simplify the arithmetic.

Error estimates for Euler method

In Example 2 the true value of $y_n = y(nh)$ is $1 + \frac{1}{2}n^2h^2$ (see equation (22.7)) and so the error accumulated by the nth stage is

$$\varepsilon_n = 1 + \frac{h^2 n}{2}(n-1) - (1 + \tfrac{1}{2}n^2h^2) = -\frac{nh^2}{2}$$

Suppose we seek the formula error over one step. If we start from a correct value y_n, the Euler method gives

$$y_{n+1} = y_n + hx_n = y_n + nh^2$$

However, algebraically,

$$y_n = \frac{x_n^2}{2} + 1 = \frac{(nh)^2}{2} + 1 \quad \text{from (22.7)}$$

and

$$y_{n+1} = \frac{x_{n+1}^2}{2} + 1 = \frac{(n+1)^2h^2}{2} + 1$$

therefore

$$y_{n+1} = y_n + nh^2 + \tfrac{1}{2}h^2$$

In proceeding from y_n to y_{n+1} the Euler method gives a result which is too small by an amount $\frac{1}{2}h^2$; hence, the total contribution from this source of error over n stages will be $-nh^2/2$. In other words, the formula error accounts for all the accumulated error (round-off not being considered).

We approximate $\Delta y_{n+1}/\Delta y_n$ by dy_{n+1}/dy_n in order to estimate the scale factor. In our present example, $dy_{n+1}/dy_n = 1 + h$, and this indicates a growth in effect ($h > 0$).

For the cooling liquid problem it can be shown that the successive estimates get closer to the true values and shows that the inherited error is stronger than the formula error.

The overall error associated with an estimate is therefore seen to be a combination of factors and may be difficult to evaluate in a practical problem.

Improved Euler Method

Graphically, we know that effectively Euler's method draws a tangent at the starting point P_0 (see Figure 22.6) and finds the point $P_1^{(0)}$ after a step h in the x-direction. The value of this ordinate is denoted by $y_1^{(0)}$. We have therefore assumed that the slope of the curve over the whole step takes the same value as that at P_0. Now the slope of the solution curve at $P_1^{(0)}$ will be approximately

$f(x_1, y_1^{(0)})$ and this slope is indicated by the line $P_1^{(0)}T_1$.

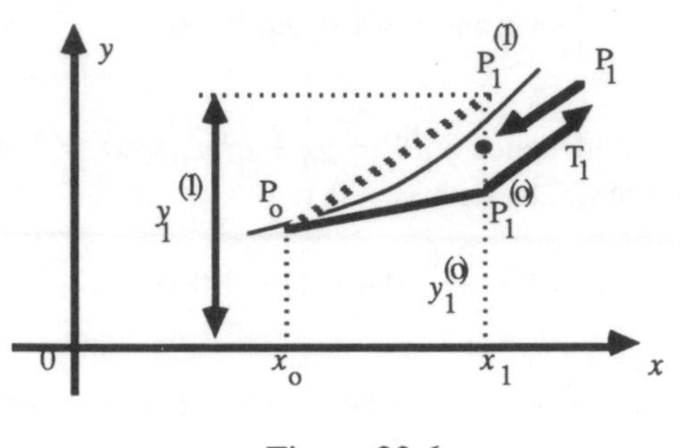

Figure 22.6

If we draw a parallel line through P_0 we get the line $P_0P_1^{(1)}$ and the ordinate of the point $P_1^{(1)}$ is given by

$$y_1^{(1)} = y_0 + hf(x_1, y_1^{(0)})$$

Here, $y_1^{(1)}$ denotes the first improved approximation to y_1.

We see that in this case $P_1^{(0)}$ and $P_1^{(1)}$ lie on either side of the exact solution so that half-way between these points should be a better approximation to the true value. This gives the point P_1 with ordinate

$$y_1 = \tfrac{1}{2}(y_1^{(0)} + y_1^{(1)})$$

which is our improved approximation to the value of y at $x = x_1$.

Now since, $\quad y_1^{(0)} = y_0 + hf(x_0, y_0)$

$$y_1 = \tfrac{1}{2}[y_0 + hf(x_0, y_0) + y_0 + hf(x_1, y_1^{(0)})]$$

that is

$$y_1 = y_0 + \frac{h}{2}[f(x_0, y_0) + f(x_1, y_1^{(0)})] \tag{22.8}$$

Formula (22.8) forms the basis for the **Improved Euler method**.

The next step will give

$$y_2 = y_1 + \frac{h}{2}[f(x_1, y_1) + f(x_2, y_2^{(0)})]$$

and, in general, $\quad y_{n+1} = y_n + \frac{h}{2}[f(x_n, y_n) + f(x_{n+1}, y_{n+1}^{(0)})]$

the value $y_{n+1}^{(0)}$ being obtained by Euler's method as

$$y_{n+1}^{(0)} = y_n + hf(x_n, y_n)$$

It can be shown that for $h < 1$ the error in using Euler's method is of the order of h^2 whereas that in the Improved Euler is of order h^3; thus the latter is a more accurate method.

Example

$$\frac{dy}{dx} = x \text{ and } y = 1 \text{ when } x = 0$$

Let us use $h = 0.1$.
Here $x_0 = 0$, $y_0 = 1$ and hence $y_1^{(0)} = y_0 + hf(x_0, y_0) = 1 + (0.1).(0) = 1$. Thus, using the formula (22.8) with $x_1 = 0.1$,

$$y_1 = 1 + \frac{0.1}{2}[0 + 0.1] = 1.005$$

The next step now gives

$$y_2^{(0)} = y_1 + hf(x_1, y_1) = 1.005 + (0.1)[0.1] = 1.015$$

and

$$y_2 = y_1 + \frac{h}{2}[f(x_1, y_1) + f(x_2, y_2^{(0)})] = 1.005 + \frac{0.1}{2}[0.1 + 0.2] = 1.020$$

Proceeding step-by-step we can construct Table 22.1

Table 22.1

x_n	y_n	$f(x_n, y_n)$	$y_{n+1}^{(0)}$	$f(x_{n+1}, y_{n+1}^{(0)})$	y_{n+1}
0.0	1.000	0.0	1.000	0.1	1.005
0.1	1.005	0.1	1.015	0.2	1.020
0.2	1.020	0.2	1.040	0.3	1.045
0.3	1.045	0.3	1.075	0.4	1.080
0.4	1.080	0.4	1.120	0.5	1.125
0.5	1.125	0.5	1.175	0.6	1.180
0.6	1.180	0.6	1.240	0.7	1.245
0.7	1.245				

In Table 22.2 we compare these results with the exact values obtained from the analytical solution $y = (x^2/2) + 1$ and with the values obtained using Euler's method.

Table 22.2

x	0.0	0.1	0.2	0.3	0.4	0.5	0.6	0.7
y_{exact}	1.0	1.005	1.02	1.045	1.080	1.125	1.180	1.245
y_{Euler}	1.0	1.0	1.01	1.03	1.06	1.10	1.15	1.21
$y_{\text{Improved Euler}}$	1.0	1.005	1.02	1.045	1.080	1.125	1.180	1.245

We can see that there is an increase in accuracy using the Improved Euler method as compared with straightforward Euler method but it must be remembered that to obtain this increase in accuracy we have to make twice as many function evaluations. Of course this is no great problem if we use a digital computer. A flow diagram is given in Figure 22.7.

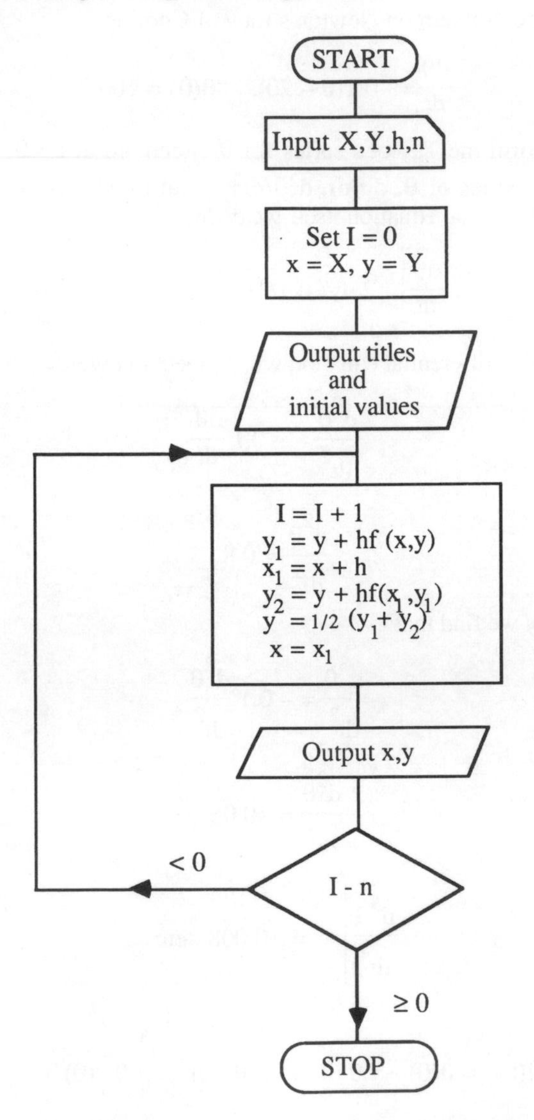

Here n is the number of estimations required and I is the counter which tells how many stages have been carried out to date.

Figure 22.7

22.5 Taylor Series Method

We start with the problem on Newton's Law of Cooling.

$$\frac{d\theta}{dt} = -0.1(\theta - 20); \qquad \theta(0) = 100$$

We aim to form the Taylor's series for $\theta(t)$ centred at $t = 0$. To do this, we shall need the values of θ, $d\theta/dt$, $d^2\theta/dt^2$ ... at $t = 0$. We know $\theta(0) = 100$ and from the differential equation itself we deduce

$$\left.\frac{d\theta}{dt}\right|_{t=0} \quad \text{is } -0.1(100 - 20) = -8$$

Differentiating the differential equation with respect to t yields

$$\frac{d^2\theta}{dt^2} = -0.1\frac{d\theta}{dt}$$

and so at $t = 0$, $$\frac{d^2\theta}{dt^2} = 0.8$$

In the same way we find that

$$\frac{d^3\theta}{dt^3} = -0.1\frac{d^2\theta}{dt^2}$$

and at $t = 0$, $$\frac{d^3\theta}{dt^3} = -0.08$$

Similarly $$\left.\frac{d^4\theta}{dt^4}\right|_{t=0} = +0.008 \quad \text{etc}$$

Hence $$\theta(t) = \theta(0) + t.\theta'(0) + \frac{t^2}{2!}\theta''(0) + \frac{t^3}{3!}\theta'''(0) + \frac{t^4}{4!}\theta^{iv}(0) + \ldots$$

$$= 100 - 8t + \frac{0.8t^2}{2} - 0.08\frac{t^3}{3!} + 0.008\frac{t^4}{4!} - \ldots \qquad (22.9)$$

In this example we can rearrange the right-hand side of (22.9) as

$$20 + 80(1 - 0.1t + 0.01\frac{t^2}{2!} - 0.001\frac{t^3}{3!} + 0.0001\frac{t^4}{4!} - \ldots)$$

that is $$20 + 80e^{-0.1t}$$
to recover the analytical solution. We can now see that the series on the right-hand side will converge for all values of t.

The next example shows the difficulty associated with finding the radius of convergence of the resulting series when the analytical solution is not known.

Example

$$\frac{dy}{dx} = x - y^2 \text{ with } y(0) = 0$$

or $$y' = x - y^2 \Rightarrow y'(0) = 0$$

Now
$$y'' = 1 - 2yy' \Rightarrow y''(0) = 1$$
$$y''' = 0 - 2(y')^2 - 2yy'' \Rightarrow y'''(0) = 0$$
$$y^{iv} = -4y'y'' - 2y'y'' - 2yy''' \Rightarrow y^{iv}(0) = 0$$
$$y^{v} = -6(y'')^2 - 6y'y''' - 2y'y''' - 2yy^{iv} \Rightarrow y^{v}(0) = -6$$
$$y^{vi} = -20y''y''' - 10y'y^{iv} - 2yy^{v} \Rightarrow y^{vi}(0) = 0$$

Therefore Taylor's series is

$$y = \frac{x^2}{2!} - \frac{6x^5}{5!} + \ldots = \frac{x^2}{2} - \frac{x^5}{20} + \ldots$$

Suppose we wish to estimate $y(0.1)$; the terms we have yield

$$y \cong \frac{(0.1)^2}{2} - \frac{(0.1)^5}{20} = 0.005 - 0.0000005$$
$$= 0.0050 \quad \text{(4 d.p.)}$$

Now the first term to be ignored will at best contain x^7 and may be of the form $a_7(x^7/7!)$ where a_7 is a constant [in this case a_7 is zero and the first missing term is $a_8(x^8/8!)$]. This gives us a clue as to how big x has to be before the loss of the term is significant in so far as it affects the accuracy to which we are working. But it is *only a clue* and unless we calculate that next term we cannot be sure. There are two factors to consider in using this series method:

(i) for what range of x will the series converge?
(ii) for what range of x will the series converge fast enough for practical computation?

22.6 Runge-Kutta Methods

We have just seen that, provided the series converges, the Taylor expansion provides an accurate estimation of the dependent variable, y. However it does require the evaluation of derivatives, which can be a tedious process. Methods have been developed which approximate the Taylor expansion by means of a combination of function evaluations. In this section we shall study one such method: the **Runge-Kutta fourth order method**. This requires four evaluations of $f(x,y)$ and the particular combination of these chosen then provides an estimate of y which is equivalent to the Taylor expansion as far as the term in $(x - x_0)^4$.

We may proceed via a geometrical interpretation. We assume $y_0 = y(x_0)$ is known and we seek $y_1 = y(x_1) = y(x_0 + h)$ [see Figure 22.8(i)]. At x_0, the slope of the curve is $y'(x_0) = f(x_0, y_0)$; this provides an estimate of y_1 as given by the Euler method. We calculate $k_1 = hf(x_0, y_0)$.

We now calculate the slope half-way along this line, that is, at the point A; this is $f(x_0 + h/2, y_0 + {}^1/2k_1)$.† Through (x_0, y_0) we draw a line with this slope to produce a new estimate of y_1 and calculate $k_2 = hf(x_0 + h/2, y_0 + {}^1/2k_1)$[see Figure 22.8(ii)].

We calculate the slope half-way along this line, i.e. at B. It is $f(x_0 + h/2, y_0 + {}^1/2\, k_2)$. Through (x_0, y_0) we draw a line with this slope to produce a third estimate of y, and calculate $k_3 = hf(x_0 + h/2, y_0 + {}^1/2k_2)$[see Figure 22.8(iii)].

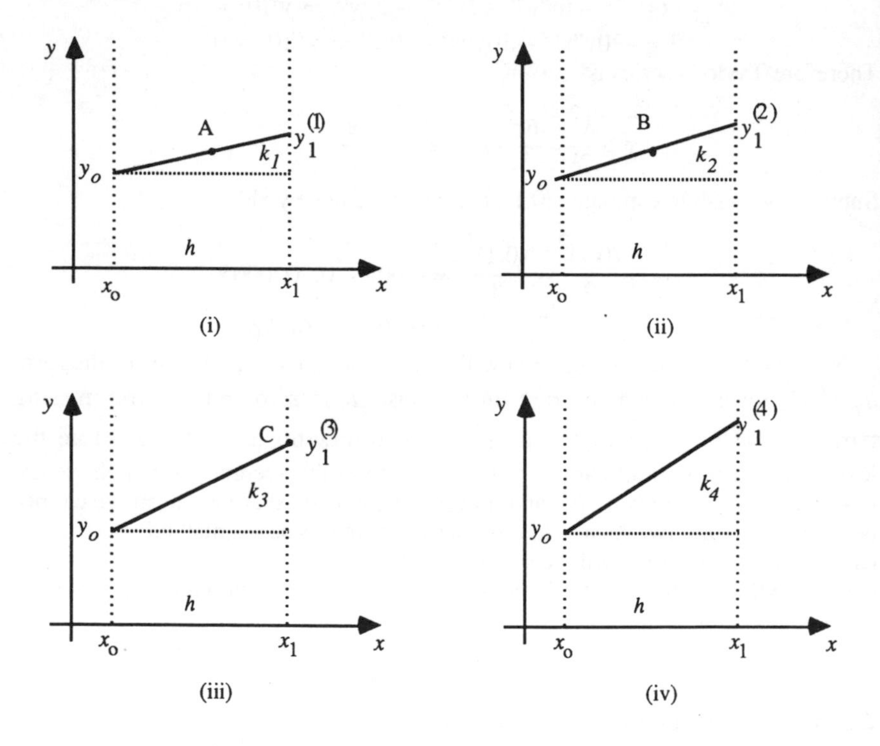

Figure 22.8

At this new estimated point, C, we calculate the slope $f(x_0 + h, y_0 + k_3)$ and draw through (x_0, y_0) a line with this slope to produce a fourth estimate of y, and calculate $k_4 = hf(x_0 + h, y_0 + k_3)$[see Figure 22.8(iv)].

† Note that this is not the slope of the straight line.

The composite diagram for the estimates is shown in Figure 22.9. In general, the true value of y_1 will lie between the two extreme estimates. We take the *weighted average*

$$k = \frac{1}{6}(k_1 + 2k_2 + 2k_3 + k_4)$$

and estimate y_1 as $y_0 + k$.

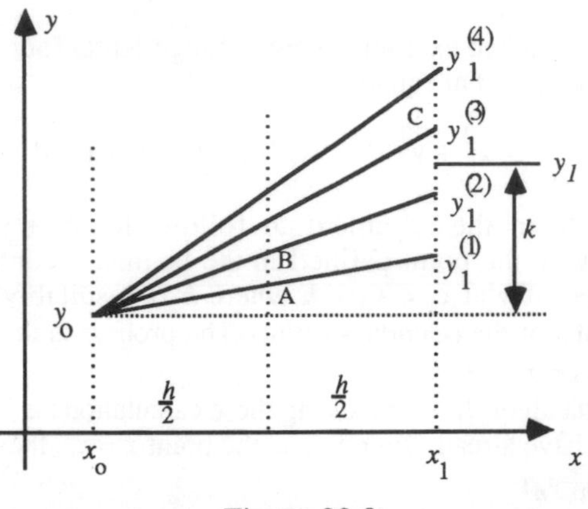

Figure 22.9

This seems a curious ritual, but our aim was to produce estimates of y_1 as accurate as the Taylor expansion to the terms in $(x_1 - x_0)^4$ and analysis of the Runge-Kutta formula has shown this to be the case. [If $f(x,y)$ is independent of y then you should be able to show that the Runge-Kutta formula becomes Simpson's Rule].

Example 1

$\frac{d\theta}{dt} = -0.1(\theta - 20)$; $\theta_0 = 100$ and we seek to estimate $\theta(4)$. In this example,

$f(\theta,t) = f(\theta)$ only.

1 $\dot{\theta}_0 = -8$, therefore $k_1 = 4(-8) = -32$

2 At A, $\dot{\theta} = -0.1(100 - 1/2 \,.\, 32 - 20) = -6.4$
therefore $k_2 = 4(-6.4) = -25.6$

3 At B, $\dot{\theta} = -0.1(100 - 1/2 \,.\, 25.6 - 20) = -6.72$
therefore $k_3 = 4(-6.72) = -26.88$

4 At C, $\dot{\theta} = -0.1(100 - 26.88 - 20) = -5.312$
therefore $k_4 = 4 \times (-5.312) = -21.248$

Taking the weighted average, we obtain

$$k = -\frac{1}{6}(32 + 2 \times 25.6 + 2 \times 26.88 + 21.248) = -\frac{1}{6}(158.208) = -26.368$$

therefore $\theta(4) = 100 - 26.368 = 73.63$ (2 d.p.).

This compares with the analytical value of 73.624 and an Euler estimate of 73.615 with a step size of 0.01 obtained using a digital computer.

Example 2

We next consider a worked example using a Runge-Kutta fourth order method to solve the ordinary differential equation

$$\frac{dy}{dx} = x^2 + y^2 \qquad \text{with } y = 0 \text{ when } x = 0$$

For convenience in the argument to follow, let $x^2 + y^2 = f(x, y)$. The object is to start with the point defined in the boundary condition and then to calculate the value of y at $x_1 = x_0 + h$ where h is a suitably chosen step size and x_0 the value of x at the boundary point. The process is then repeated to find the value of y at $x_2 = x_1 + h$.

The Runge-Kutta algorithm for making these calculations is:

(i) Suppose we have already found y_n at the point $x = x_n$, then calculate
$k_1 = h \times f(x_n, y_n)$

(ii) Use this value of k_1 to calculate k_2 from:

$$k_2 = h \times f(x_n + \tfrac{1}{2}h, y_n + \tfrac{1}{2}k_1)$$

(iii) Use this value of k_2 to calculate k_3 from:

$$k_3 = h \times f(x_n + \tfrac{1}{2}h, y_n + \tfrac{1}{2}k_2)$$

(iv) Use this value of k_3 to calculate k_4 from:
$k_4 = h \times f(x_n + h, y_n + k_3)$

(v) Calculate y_{n+1} from:
$y_{n+1} = y_n + (k_1 + 2k_2 + 2k_3 + k_4)/6$

(vi) Finally we have $x_{n+1} = x_n + h$

Hence starting with the boundary values x_0, y_0, we can apply the above rules to find x_1, y_1, and then repeat to find x_2, y_2, and so on.

In the example, start at Step (i). We have $x_0 = 0$, $y_0 = 0$ and let $h = 0.1$.
Then $k_1 = 0.1 \times (x_0^2 + y_0^2) = 0.0$
Step (ii) gives
$k_2 = 0.1 \times [(x_0 + 0.05)^2 + (y_0 + 1/2 \times 0.0)^2] = 0.00025$
Step (iii) gives
$k_3 = 0.1 \times [(x_0 + 0.05)^2 + (y_0 + 1/2 \times 0.00025)^2] = 0.25000156 \times 10^{-3}$
and Step (iv) gives
$k_4 = 0.1 \times [(x_0 + 0.1)^2 + (y_0 + 0.00025000156)^2] = 0.0010000063$
From Step (v) we get
$y_1 = y_0 + (0.0 + 2 \times 0.00025 + 2 \times 0.00025 + 0.001)/6 = 0.0003333349$

Now we repeat the process using $x_1 = 0.1$ and $y_1 = 0.0003333349$.

$k_1 = 0.1 \times [x_1^2 + y_1^2] = 0.0010000111$

$k_2 = 0.1 \times [(x_1 + 0.05)^2 + (y_1 + 1/2 \times 0.0010000111)^2] = 0.0022500694$

$k_3 = 0.1 \times [(x_1 + 0.05)^2 + (y_1 + 1/2 \times 0.0022500694)^2] = 0.0022502127$

$k_4 = 0.1 \times [(x_1 + 0.1)^2 + (y_1 + 0.0022502127)^2] = 0.0040006675$

$$y_2 = y_1 + \frac{1}{6}(k_1 + 2k_2 + 2k_3 + k_4) = 0.0026668754$$

$x_2 = x_1 + h = 0.2$

In Table 22.3 the two sets of steps above are concisely written and the process is continued until $x = x_{10} = 1.0$

Table 22.3

x	y	k_1	k_2	k_3	k_4
0.0	0.000000	0.000000	0.000250	0.0002500	0.001000
0.1	0.000333	0.001000	0.002250	0.0022502	0.004000
0.2	0.002667	0.0040007	0.006252	0.0062534	0.009008
0.3	0.009003	0.0090081	0.012268	0.0122729	0.016045
0.4	0.021359	0.0160456	0.020336	0.0203494	0.025174
0.5	0.041791	0.0251747	0.030545	0.0305756	0.036524
0.6	0.072448	0.0365249	0.043073	0.0431333	0.050336
0.7	0.115660	0.0503377	0.058233	0.0583460	0.067278
0.8	0.174081	0.0670304	0.076559	0.0767597	0.087292
0.9	0.250908	0.0872955	0.098926	0.0992723	0.112263
1.0	0.350233				

Case Study with Runge-Kutta

We return to the rocket re-entry problem of Section 22.3. This was programmed for a digital computer and the results are shown in Table 22.4. Two step sizes are shown.

The dashes indicate that the numerical solution misbehaved by increasing the velocity of the rocket because no check was built-in to prevent such occurrence. Note that Runge-Kutta is almost independent of the two step sizes.

To see the effect of gravity the computations were repeated without it and the results are also shown in Table 22.4. A sketch of this is Figure 22.10. Since the Runge-Kutta is expected to be the most accurate method of the three, we consider the effects of including gravity on its calculation. Above 30 000 m the discrepancy is less than 0.5% but this quickly builds up as the altitude decreases. It has been shown that the exclusion of gravity is less important for higher re-entry velocities. If we return to the analytical solution then omission of gravity produces (see Equation (22.5)):

$$V = \text{constant} \times e^{(-K\rho_0/\lambda)e^{-\lambda z}}$$

so that

$$v = \text{constant} \times e^{(-K\rho_0/2\lambda)e^{-\lambda z}}$$

Table 22.4

z (m)	Euler 250	Euler 10^3	Improved Euler 250	Improved Euler 10^3	Runge-Kutta 250	Runge-Kutta 10^3
50 000	10 000	10 000	10 000	10 000	10 000	10 000
45 000	8992	9013	8985	8985	8985	8985
40 000	7542	7575	7531	7531	7531	7531
35 000	5633	5646	5629	5631	5629	5629
30 000	3467	3409	3486	3497	3485	3485
25 000	1546	1398	1591	1621	1590	1589
20 000	425	305	466	511	464	465
15 000	145	141	150	168	149	149
10 000	107	107	107	111	107	107
5000	83	82	83	–	83	83
0	64	64	64	–	64	–
With g						
50 000	10 000	10 000	10 000	10 000	10 000	10 000
45 000	8987	9008	8980	8980	8980	8980
40 000	7532	7565	7521	7521	7522	7522
35 000	5619	5632	5615	5617	5615	5615
30 000	3450	3391	3469	3480	3468	3468
25 000	1523	1373	1569	1600	1567	1567
20 000	380	329	425	473	423	423
15 000	34	80	50	90	49	50
10 000	0	–	2	25	1	2
5000	0	–	0	–	0	–
0	–	–	0	–	0	–
Without g						

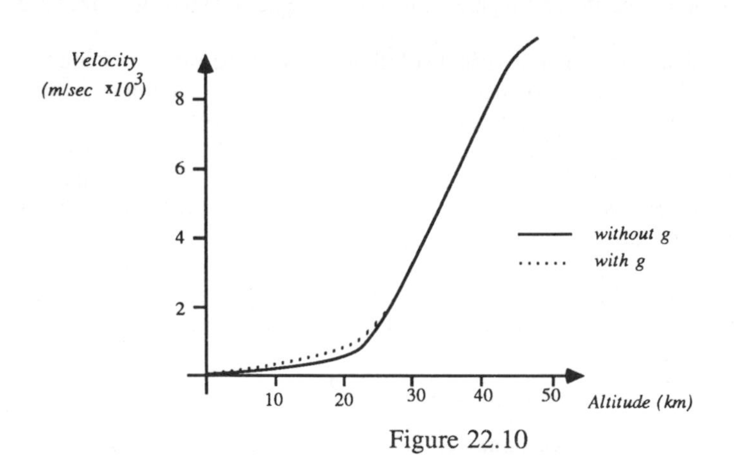

Figure 22.10

At altitudes of the order of 100 kilometres the exponential term as a whole ~1 if mg/C_DA is sufficiently large, and if v_e is the re-entry velocity at that speed

$$v = v_e\, e^{(-K\rho_0/2\lambda)e^{-\lambda z}}$$

This is one in a chain of approximations, but it illustrates the use of a simple method as a basis for the analysis of more complicated physical situations.

One question with which we leave this section is how reliable are the Runge-Kutta estimates. Write a computer program to check the tabulated estimates with gravity using Equation (22.5).

The following program will carry out the Runge-Kutta method.

```
>
   10  CLS
   20  REM Applied here to dy/dx=(X*X)+(Y*Y)
   30  PRINTTAB(10,2)"Runge-Kutta Method"
   40  INPUTTAB(0,5)"Enter initial X0 ",X0
   50  INPUT"Enter initial Y0 ",Y0
   60  INPUT"Enter step size ",H
   70  INPUT"Enter number of steps ",N
   80  FOR I = 1 TO N
   90    @%=&20609
  100    X=X0:Y=Y0
  110    GOSUB 290
  120    K1 = Z
  130    X=X0+0.5*H
  140    Y=Y0+0.5*K1
  150    GOSUB 290
  160    K2=Z
  170    Y=Y0+0.5*K2
  180    GOSUB 290
  190    K3=Z
  200    X=X0+H
  210    Y=Y0+K3
  220    GOSUB 290
  230    K4=Z
  240    X0=X0+H
  250    Y0=Y0+(K1+2*K2+2*K3+K4)/6
  260    PRINT TAB(0)X0;TAB(15)Y0
  270    NEXT I
  280  END
  290  Z=H*((X*X)+(Y*Y))
  300  RETURN
```

22.7 Comparison of Methods

Now we compare methods for the cooling liquid experiment. The step size for the numerical methods was 5 minutes. Values are quoted to 3 d.p. (Table 22.5).

Table 22.5

	Analytical	Euler	Improved Euler	Runge-Kutta	
0	100	100	100	100	
5	68.52	60	70	68.511	
10	49.432	40	51.25	49.430	
15	37.848	30	39.532	37.846	
20	30.824	25	32.208	30.827	Newton's Law
25	26.568	22.5	27.630	26.567	of Cooling
30	23.984	21.25	24.768	23.983	
35	22.416	20.625	22.881	22.417	
40	21.464	20.313	21.801	21.465	

You see that the Runge-Kutta method is the most accurate when compared with the "true" analytical solution. Try to produce a flow chart and write a computer program to obtain the above results.

Finally, we now give a summary comparison of the main methods used with a simple differential equation.

We seek the solution of $dy/dx = x + y$ when $y = 1$ at $x = 0$ for $x = 0(0.1)0.2$ (i.e. $h = 0.1$).

Analytical Solution

$y = 2e^x - x - 1$

x	e^x	y
0.1	1.10517	1.1103
0.2	1.22140	1.2428
0.3	1.34986	1.3997

Try to arrive at the analytical formula via the Taylor series approach.

Euler Method

$$y_1 = y_0 + h(x_0 + y_0)$$
$$y_1 = 1 + 0.1(0 + 1) = 1.1$$
$$y_2 = 1.1 + 0.1(0.1 + 1.1) = 1.22$$
$$y_3 = 1.22 + 0.1(0.2 + 1.22) = 1.362$$

Improved Euler Method

Let $f_0 = x_0 + y_0$ and $f_1 = x_1 + y_1$ where y_1 = previous estimate of y at $x_1 = x_0 + h$. Then a better estimate is obtained using the following formula

$$y_1 = y_0 + h(f_0 + f_1)/2 \qquad (22.10)$$

First, $y_1 = y_0 + hf_0$

x_0	y_0	x_1	y_1	f_0	f_1	y_1
0	1.0			1.0		1.1
		0.1	1.1		1.2	1.11
			1.11		1.21	1.1105
			1.1105		1.2105	1.1105
0.1	1.1105			1.2105		1.2316
		0.2	1.2316		1.4316	1.2426
			1.2426		1.4426	1.2432
			1.2432		1.4432	1.2432
0.2	1.2432			1.4432		1.3875
		0.3	1.3875		1.6875	1.3997
			1.3997		1.6997	1.4003
			1.4003		1.7003	1.4004
			1.4004		1.7004	1.4004

Note we have iterated by applying (22.9) successively until convergence of y_1 is obtained.

Runge-Kutta

$$k_1 = h(x_0 + y_0)$$
$$k_2 = h(x_0 + {}^1/2h + y_0 + {}^1/2k_1)$$
$$k_3 = h(x_0 + {}^1/2h + y_0 + {}^1/2k_2)$$
$$k_4 = h(x_0 + h + y_0 + k_3)$$
$$y_1 = y_0 + (k_1 + 2k_2 + 2k_3 + k_4)/6$$

x_0	y_0	k_1	k_2	k_3	k_4	y_1
0	1.0	0.1	0.11	0.1105	0.12105	1.1103
0.1	1.1103	0.12103	0.13208	0.13263	0.14429	1.2428
0.2	1.2428	0.14428	0.15649	0.15710	0.16999	1.3997

Problems

Section 22.1

1 (a) Solve the following differential equations

(i) $\sqrt{1 + x^2}\dfrac{dy}{dx} = xe^{-y}$ given $y = 0$ when $x = 0$

(ii) $\dfrac{dy}{dx} + \dfrac{1 + y^{3/2}}{xy^{1/2}(1 + x^2)} = 0$

(iii) $\dfrac{dy}{dx}\cos x = (\sin x + x \sec x)\cot y$

(iv) $x\dfrac{dy}{dx} = y \ln x$

(v) $y\dfrac{dy}{dx} + x = xy^4$ (vi) $x\dfrac{dy}{dx} = y(1 - y)$ given that $y = 2$ when $x = -4$

(vii) $\dfrac{dy}{dx} = (1 + 2x)\cos^2 y$ (viii) $xy\dfrac{dy}{dx} = 1 + x^2$

(ix) $xy(1 - x^2)\dfrac{dy}{dx} = 1 - y^2$

Show that the solution curves are symmetrical about each of the coordinate axes and that they all pass through four fixed points. (LU)

(b) Show by deriving it that the general solution of the differential equation

$$(1 + 2\cos\theta)\frac{dr}{d\theta} + 3r\sin\theta = 0$$

is $r = c(1 + 2\cos\theta)^{3/2}$ in which c is an arbitrary constant. (EC)

2 Solve the following differential equations

(i) $\dfrac{dy}{dx} = e^y \sin x$, given $y = 0$ when $x = 0$

(ii) $\dfrac{dy}{dx} = \dfrac{x e^y}{\sqrt{1 + x^2}}$, given $y = 0$ when $x = 3/4$

(iii) $t\theta(1 + t^2)\dfrac{d\theta}{dt} - \theta^2 = 1$, given $\theta = 0$ when $t = 1$

3 The rate at which a radio-active substance decays is proportional to n, the number of atoms present at time t. If the constant of proportionality is λ and initially there are N atoms present, express n as a function of t. Find the time taken for 1/4 of the initial amount to decay; find the time for the initial amount to be halved.

4 The relationship between the radial compressive stress p and the radius r in a thick circular cylinder is given by

$$2(p + A) = -r\frac{dp}{dr}, \quad A \text{ being a constant}$$

If $p = 0$ when $r = R$, solve this separable differential equation to give p in terms of r, A and R. (EC)

5 Liquid fills a spherical vessel of radius a to a depth h_0. At time $t = 0$ the fluid is allowed to drain out of an orifice at the lowest point of the vessel at a volume rate $k\sqrt{h}$ where k is a constant. Derive the differential equation for the variation of depth h with time and find an expression for the time taken to empty the vessel. (EC)

6 A hemispherical water tank of radius R is full at $t = 0$. Water escapes through a hole of radius a at the bottom for time $t > 0$. Find the depth of water as a function of time. (Assume that the velocity of emerging liquid is $\sqrt{2gh}$ where h is the depth of water.)

7 A particle moves along a straight line so that at time t sec its distance from a fixed point O is x cm and its velocity v cm/sec. If its acceleration is $(v^3 - v)$ cm/sec^2 directed away from O and $v = 3$ at $x = 0$, show that $x = \frac{1}{2}\ln 2 - \coth^{-1} v$, proving any formula which you use for an inverse hyperbolic function. (LU)

8 The mass M of a crystal growing in a solution is given at time t, by the differential equation

$$\frac{dM}{dt} = kM(M_1 - M)$$

where $M < M_1$. In this k is a constant and M_1 is the maximum attainable mass of the crystal.
Solve this equation to find an expression for M as a function of t given that $M = M_0$ when $t = 0$. (EC)

9 The charge q in a certain electrical circuit is given by the differential equation

$$\frac{dq}{dt} = \frac{E}{R^2 + L^2\omega^2}(R\cos\omega t + L\omega\sin\omega t - R\exp(-Rt/L))$$

in which R, L, ω and E are constants.
Derive a formula for q in terms of t given that $q = q_0$ at $t = 0$. (EC)

10 In a certain vertical motion the air resistance is assumed to be proportional to the square of the velocity. The equation of motion can be shown to be

(i) $\dfrac{dv}{dt} = -g - kv^2$ for motion vertically upwards

(ii) $\dfrac{dv}{dt} = g - kv^2$ for motion vertically downwards

where g and k are positive constants.
Solve the differential equation in both cases and find expressions for v in terms of t for both upward and downward motion. If an object is projected vertically upwards in the air with an initial velocity v_0, find the time which elapses before it reaches its maximum height. (EC)

11 The quantity z of a chemical diffusing through a membrane is governed by the differential equation

$$\frac{dz}{dt} = 2(3 - z)(2 - z) \quad \text{where } z < 2$$

If $z = 1$ when $t = 0$, solve the equation to find z as a function of t. Sketch a graph of z against t for $t \geq 0$. (EC)

12 A bacterial population of size P is known to have a rate of growth proportional to P itself. If between 9 am and 10 am the population doubles, at what time will P become 100 times what it was at 9 am?

Section 22.2

13 Test each of the following for exactness and solve the equation if it is exact.

(i) $(3x^2y - x^3)\mathrm{d}x + (x^3 + y)\mathrm{d}y = 0$ (ii) $(x + 2y)\mathrm{d}x + 2(y + x)\mathrm{d}y = 0$

(iii) $(r + \sin\theta + \cos\theta)\mathrm{d}r + r(\cos\theta - \sin\theta)\mathrm{d}\theta = 0$ (iv) $\theta\mathrm{d}r = (\mathrm{e}^{\theta} + 2r\theta - 2r)\mathrm{d}\theta$

14 Show that the following equation is not exact but that $1/y^4$ is an integrating factor and hence solve the equation

$$(y^3 - 2xy)\mathrm{d}x + (3x^2 - xy^2)\mathrm{d}y = 0$$

15 Prove that x is an integrating factor of the equation $(x^2 + y^2 + x)\mathrm{d}x + xy\mathrm{d}y = 0$ and solve.

16 Prove the equation $(2x^3 + 3y)\mathrm{d}x + (3x + y - 1)\mathrm{d}y = 0$ is exact and solve.

17 Prove that x^n is an integrating factor of $x\mathrm{d}y = (y + \ln x)\mathrm{d}x$ only for $n = -2$ and solve.

18 Solve the differential equation $(3x^2y - y^3)\mathrm{d}x + (x^3 - 3xy^2)\mathrm{d}y = 0$. (LU)

19 For each of the following differential equations, indicate which can be solved by integrating factor and which by separation of variables. Write down the integrating factor where applicable.

(i) $\dfrac{\mathrm{d}y}{\mathrm{d}x} = y\cos x$ (ii) $y\dfrac{\mathrm{d}y}{\mathrm{d}x} = y + 3$ (iii) $\dfrac{\mathrm{d}u}{\mathrm{d}t} = u\mathrm{e}^{2t} - \mathrm{e}^{t}$

(iv) $\dfrac{\mathrm{d}z}{\mathrm{d}t} = \sqrt{z(1+t)^2}$ (v) $\dfrac{\mathrm{d}\theta}{\mathrm{d}t} = \theta(\theta(t)) \times \mathrm{e}^{t}$ (vi) $v\dfrac{\mathrm{d}v}{\mathrm{d}t} = v^2\,\mathrm{e}^{-2t}$

(vii) $\dfrac{\mathrm{d}y}{\mathrm{d}x} + \dfrac{y}{x} = 4$

20 Solve the differential equations

(i) $x\dfrac{\mathrm{d}y}{\mathrm{d}x} + y = x^2$ (ii) $(x + 1)\dfrac{\mathrm{d}y}{\mathrm{d}x} + 3y = x + 1$

(iii) $\dfrac{\mathrm{d}y}{\mathrm{d}x} + y\cot x = \operatorname{cosec} x$

(iv) $(1 + t)\dfrac{\mathrm{d}\theta}{\mathrm{d}t} + (1 + 2t)\theta = (1 + t)^2$, given $\theta = 0$ when $t = 1$

(v) $t\cos t\dfrac{\mathrm{d}\theta}{\mathrm{d}t} + (t\sin t - \cos t)\theta - t^2 = 0$, given $\theta = 0$ when $t = \dfrac{\pi}{4}$

21 Solve the differential equations

(i) $\dfrac{\mathrm{d}y}{\mathrm{d}x} + y\cot x = \sin x$, where $y = 1$ at $x = \dfrac{\pi}{2}$

(ii) $x(x^2 - 1)\dfrac{dy}{dx} + y = x^3$ (iii) $\dfrac{dy}{dx} + 2y = e^x$

(iv) $\sin x \dfrac{dy}{dx} + 2y \cos x = \cos x$, given that $y = -\frac{1}{2}$ when $x = \dfrac{\pi}{2}$

(v) $x \dfrac{dy}{dx} + 2y = x^2$, given that $y = \frac{1}{2}$ when $x = 1$

(vi) $\dfrac{d^2z}{dt^2} + \dfrac{dz}{dt} = e^{2t}$ (vii) $(x + 4)\dfrac{dy}{dx} + 3y = 3x$

(viii) $\dfrac{d^2y}{dx^2} - \dfrac{dy}{dx} = x$, with the conditions $y = 2$, $\dfrac{dy}{dx} = 0$ at $x = 0$

22 If $\dfrac{dy}{dx} + 2y \tan x = \sin x$ and $y = 0$ when $x = \dfrac{\pi}{3}$ show that the maximum value of y is $\frac{1}{8}$. (EC)

23 Solve the differential equation $\dfrac{dy}{dx} + xy = \exp(-x^2/2)$. (EC)

24 A rocket, when full of fuel, is of mass M lbs and it burns fuel at a constant rate, m lbs/sec. The relative backward velocity of the gases is constant and equal to u. The rocket is ignited and moves vertically upwards from rest against an air resistance $km\ V$, where V is the velocity of the rocket. Using the equation of motion

$$(M - mt)\frac{dV}{dt} - mu = -(M - mt)g - km\ V$$

show that, for $Mg < mu$ and $k \neq 0$, $k \neq 1$,

$$V = \frac{u}{k} + \frac{g(M - mt)}{m(1 - k)} - \left[\frac{Mg}{m(1 - k)} + \frac{u}{k}\right]\left[1 - \frac{mt}{M}\right]^k$$

25 A circular coil of n turns of area A whose inductance is L henrys and resistance R ohms is rotated with angular velocity ω about a diameter perpendicular to a field of strength $HA\ m^{-1}$. The current i induced in the coil is given by

$$L\frac{di}{dt} + Ri = n\omega\ HA \cos \omega t$$

Find the current at time t, assuming that initially it is zero.

26 A body of mass m is projected with speed V on a rough table whose coefficient of friction is μ. If the air resistance is proportional to the (speed)^2, find the distance travelled before the body comes to rest.

27 (a) A body which is immersed in a slowly cooling fluid has a temperature difference θ, measured relative to a fixed temperature level, at a time t. The variable θ is given by the first order differential equation

$$\frac{d\theta}{dt} + 0.1\theta = 0.2(5 - t)$$

Initially, the temperature difference θ was 70.

(i) Use an integrating factor to show that the general solution is
$\theta = 30 - 2t + c\exp(-0.1)t$ where c is a constant.

(ii) Find the special solution incorporating the initial condition.

(b) Use Euler's method with two steps to estimate the value of θ when $t = 0.2$ for the differential equation in part (a). (EC)

28 A non-standard differential equation is given as

$$x^2\frac{dy}{dx} - xy = \frac{1}{y} \qquad (1)$$

It is required to change the dependent variable from y to z by substituting $y = \sqrt{z}$. Find $\frac{dy}{dx}$ in terms of z and $\frac{dz}{dx}$, and hence show that on substitution equation (1) becomes

$$\frac{dz}{dx} - \frac{2z}{x} = \frac{2}{x^2}$$

Find the general solution of this differential equation and hence write down the solution of (1). (EC)

29 (a) Solve the differential equation $(2x + 1)\frac{dy}{dx} + 4y = 3x + 2$ given that $y(0) = 2$.

(b) Find the general solution of the differential equation $x\frac{dy}{dx} - y = (x + 1)^2$

(c) Find the solution of the differential equation $t\frac{dy}{dt} + 2y = 2t$

given that $y(1) = 5$. Hence find $y(2)$. (EC)

30 A particle of mass m is attracted to a fixed point O by a force $m\mu x$ when it is at a distance x from O in a medium which offers a resistance mkv^2 to the motion (v being the velocity). Show that the equation of motion is

$$\frac{d(v^2)}{dx} + 2kv^2 + 2\mu x = 0$$

If the particle is initially at rest at a distance a from O show that it reaches O with velocity

$$\sqrt{\frac{\mu}{2k^2}\left[1 + e^{2ka}\,[2ka - 1]\right]}$$

31 A body is immersed in an atmosphere whose temperature in °C varies as $25(2 - t)°$. The temperature of the body is initially 60°C. Show that the governing equation for the temperature of the hot body, θ, is

$$\frac{d\theta}{dt} + k\theta = 25k(2 - t)$$

where k is some constant. If $k = 0.1$, obtain estimates of θ and compare with the analytical solution which you should show is

$$\theta = 300 - 25t - 240e^{-0.1t}$$

Try to sketch the solution and see whether it agrees with your formulation.

32 The differential equation for current i at time t in an $L - R$ series circuit is

$$L\frac{di}{dt} + Ri = E$$

If the applied voltage is $E = E_0(1 - e^{-\alpha t})$ and the initial current is i_0, obtain an expression for $i(t)$; assume that $\alpha \neq R/L$. (EC)

33 A reservoir has been contaminated by effluent from a factory. The capacity of the reservoir is 10^6 litres. The degree of contamination is 0.02% by weight. The (constant) average daily rate of consumption of water for non-drinking purposes is 2×10^4 litres and this is continuously replaced by pure water. How long will it be before the concentration of contaminant drops to the safe level of 10^{-5}%?

34 A room of volume V m^3 is supplied with fresh air containing some CO_2. Let $c(t)$ be the concentration of CO_2 in the room at time $t > 0$, c_0 the initial concentration and c_f the concentration in the fresh air supply; all three are measured in parts per 10 000. Let Q m^3/sec be the rate of fresh air supplied and Q_p m^3/sec the volume of CO_2 produced by people inside the room.
Derive a differential equation for the concentration $c(t)$ and solve it.
Sketch on the same axes the solution in the case

$$\left[\frac{Qc_f + 10\,000\,Q_p}{Q}\right] < c_0$$

and the solutions for the special cases:

(i) no people in the room and the fresh air free from CO_2

(ii) the initial concentration of CO_2 in the room is zero.

A room of volume 170 m^3 receives a total change of air every 30 minutes and has a CO_2 content of 0.03% without people present. The concentration of CO_2 in the outside air is also 0.03%. If the production of CO_2 per person is 4.7×10^{-6} m^3/sec, what is the maximum number of people allowed in the room at any one time if the concentration of CO_2 in the room is not to exceed 0.1%? Sketch the solution curve in this case for $c(t)$.
[Hint: the long-term solution only is important.]

Section 22.4

35 In the following six problems estimate the dependent variable at suitable equally spaced

values of the independent variable, using Euler's method. Compare your results with the analytical solution. (You should not advance the solution by more than 2 steps unless you program a digital computer.)

(i) $\frac{dy}{dx} = x^2 + 1;$ $y(0) = 1$ Estimate $y(1)$

(ii) $\frac{dy}{dx} = 2y - 1;$ $y(0) = 2$ Estimate $y(1)$

(iii) $y' + y = 0;$ $y(0) = 2, h = 0.2$ Estimate $y(1)$

(iv) $\frac{dy}{dx} = x^2 - y;$ $y(0) = 3$ Estimate $y(2)$

(v) $y^2 \frac{dy}{dx} = \frac{x + 3}{y + 1};$ $y(0) = 1$ Estimate $y(1)$

(vi) The differential equation for the current i at time t in an $L - R$ series circuit is

$$L \frac{di}{dt} + Ri = E$$

If $E = 240(1 - e^{-0.5t})$ and initially $i = 0$, find i at $t = 1$, given that $R = 100$ and $L = 100$.

36 A projectile is launched from the earth's surface with velocity V. Assuming no drag the equation of motion is

$$v \frac{dv}{dr} = -g \frac{R^2}{r^2}$$

where v is the velocity at a distance r from the centre of the earth of radius R. Take $g = 9.81$ m/sec^2, $R = 6.37 \times 10^6$ m and $V = 15\,000$ m/sec. [If you are *not* using a computer program, take $g = 10.0$ and $R = 6 \times 10^6$.] Find the velocity when $r = 2R$.

37 A liquid solution flows steadily along a straight tube in the x direction. Some of the solute contained in the solution diffuses through the tube wall thus reducing the concentration z within the tube. The concentration z is given by

$$\frac{dz}{dx} = -z(0.2 + z^{1/2}) \exp(-0.03x)$$

Given that $z = 1.5$ at $x = 2$, use Euler's method with a step length of 0.2 to estimate the value of z at $x = 2.4$. (EC)

38 Try the Problems 35 (i) – (vi) using the Improved Euler technique.

39 We can regard the Improved Euler method as giving one improvement to the Euler approximation. We may write it as

$$y_{n+1}^{(1)} = y_n + \frac{h}{2}[f(x_n, y_n) + f(x_{n+1}, y_{n+1}^{(0)})]$$

If we repeat the process we can obtain

$$y_{n+1}^{(2)} = y_n + \frac{h}{2}[f(x_n, y_n) + f(x_{n+1}, y_{n+1}^{(1)})]$$

and so on, thus forming an iterative procedure.
Flow chart this method, working to a suitable accuracy and carry out two steps on the equation $dy/dx = x^2 + 1$, $y(0) = 1$ with $h = 0.1$.
This is an example of a **predictor-corrector** method, the first formula being the predictor and the second the corrector.

40 The **Modified Euler** method is, using customary notation

$$y_{n+1} = y_n + hf[x_n + \tfrac{1}{2}h, y_n + \tfrac{1}{2}f(x_n, y_n)]$$

Estimate values of $y(0.1)$, $y(0.2)$ using this method on the equation $dy/dx = x^2 + 4$; $y(0) = 2$.

Section 22.5

41 For the equation $(dy/dx) + xy = 0$ with $y = 1$ at $x = 0$, show that

$$y = 1 - \frac{x^2}{2} + \frac{x^4}{8} - \frac{x^6}{48} + \frac{x^8}{384} - .....$$

and hence compute the solution at $x = 0.2$, 0.4, 0.8 correct to 4 d.p. Compare with the analytical solution.

42 Show, using Taylor's series, that the solution of $(dy/dx) = 2x - y$, $y(1) = 3$, near $x = 1$ is given numerically by

x	1.1	1.2	1.3
y	2.9145	2.8562	2.8225

Compare with the analytical solution.

43 How accurate is the approximation $y \cong (1/3)x^3$ to the solution of the differential equation

$$\frac{dy}{dx} = x^2 + y^2, \quad y(0) = 0 \ ?$$

44 A Taylor expansion can be used to solve a differential equation numerically by a step-by-step process. Explain this statement with reference to the equation

$$y'(x) = [x + y(x)]^3, \quad y(0) = 1$$

with step length 0.1. (EC)

45 The differential equation

$$\frac{dx}{dt} + 2x + x^3 = 4$$

has the initial condition $x = 1$ when $t = 0$. By using the first five non-zero terms of a Taylor series, obtain values of x at $t = 0.10$ and $t = 0.15$, correct to 3 d.p. (LU)

Section 22.6

46 Solve any of the problems given earlier on Euler and Improved Euler using Runge-Kutta and compare the results.

47 Compute the solution of $dv/dt = 21.5 - 1.2v^2$ with initial condition $v = 0$ at $t = 0$, using the Runge-Kutta fourth order method at $t = 0.001, 0.002$.

48 Solve the following

(i) $\dfrac{dy}{dx} = x - \sin x + y$ Initial condition: $x = 0, y = 1.0$

(ii) $\dfrac{dy}{dx} = \dfrac{2yx}{y^2 + x^2}$ Initial condition: $x = 0, y = 3$

(iii) $\dfrac{dv}{dx} = \dfrac{2x^2 + 3x}{4v}$ Initial condition: $x = 2, v = 1$

49 Consider $dy/dx = x + y$ with $y(0) = 2$; find $y(1)$.

50 The response y (> 0) of a certain hydraulic valve subject to sinusoidal input variation is given by

$$\frac{dy}{dt} = \sqrt{2\left[1 - \frac{y^2}{\sin^2 t}\right]} \quad \text{with } y_0 = 0,\ t_0 = 0$$

Why is the Taylor series method unsuitable as a solution? Show that

$$\left[\frac{dy}{dt}\right]_0 = \sqrt{\frac{2}{3}}$$

and hence use Runge-Kutta (fourth order) to obtain a solution at $t = 0.2$. (EC)

51 Why (apart from round-off error) should the Runge-Kutta method give exact results for the equation $dy/dx = 2(x + 1)$; $y(0) = 1$? Show that an analytical solution of this equation and of $dy/dx = 2y/(x + 1)$ is $y = (x + 1)^2$. Compute $y(1)$, $y(2)$ by Runge-Kutta for each equation and compare the results. What can you conclude?

52 The current i in an LR circuit at any time t after a switch is thrown at $t = 0$ can be expressed by the equation

$$\frac{di}{dt} = (E \sin \omega t - Ri)/L$$

where $E = 50$ volts, $L = 1$ henry, $\omega = 300$, $R = 50$ ohms and the initial condition is that at $t = 0$, $i = 0$. Solve the differential equation numerically using the Runge-Kutta method and compare your answers with the analytical solution

$$i = \frac{E}{Z^2}(R \sin \omega t - \omega L \cos \omega t + \omega L e^{-Rt/L})$$

where

$$Z = \sqrt{R^2 + \omega^2 L^2}$$

53 A quantity of 10kg of material is dumped into a vessel containing 60 kg of water. The concentration of the solution, c, in percentage at any time, t, is expressed as

$$(60 - 1.212c)\, dc/dt = (k/3)(200 - 14c)(100 - 4c)$$

where k, the *mass transfer coefficient*, is equal to 0.0589. The initial condition is that at $t = 0$, $c = 0$. Find the c-t relation by the Runge-Kutta and Euler methods and compare the results with the analytical solution.

23

SECOND AND HIGHER ORDER DIFFERENTIAL EQUATIONS

Vibrations are induced in many situations, for example when lorries cross a bridge, when a ship sets sail, when machines are set in motion. As in many other situations it is important to be able to predict the nature of these vibrations; failure to do so accurately may result in the snapping of a shaft, the breaking away of an engine from its mountings or, on a more spectacular scale, the collapse of a suspension bridge (as happened at Tacoma Narrows in 1940). Many of these vibrations can be modelled reasonably accurately by linear, second order differential equations with constant coefficients (especially if the amplitude of the vibrations is small). In any event, such equations provide a starting point for the study of more complicated vibrations.

23.1 Case Study

The purpose of automobile suspension springing is two-fold: it must maintain the wheels in contact with the road to provide the degree of adhesion necessary to allow safe acceleration, cornering and braking; also it must provide a large measure of isolation for the vehicle and passengers from the road irregularities and from other possible causes of discomfort. With the advent of the pneumatic tyre even a simplified physical model becomes complicated. In Figure 23.1 the body of the car is the sprung mass M and its connections to the tyres of masses m_1 and m_2 are represented by a spring and a dashpot to signify damping of oscillations (this damping is called shock absorption). Each tyre is assumed to have an elasticity which can be represented as a spring connection with the road. We have taken only a simple two-dimensional model but even so it comprises three separate bodies; each of these needs six coordinates to have its position in space specified completely (three to fix the centre of gravity and three to specify its orientation about the c.g.) and so our model would require 18 coordinates – we say that the system possesses 18 degrees of freedom. As well as translational motion in three directions, each rigid body is capable of three rotational motions: *pitch, roll and yaw*. Most of the motions are unimportant, but even if we confine our attention to a purely vertical translation with the axes parallel we still have 2 degrees of freedom.

Consider just one wheel in vertical motion where the tyre is replaced by a solid wheel so that any irregularities in the road are transmitted without loss of form to the vehicle (with railways the track is relatively smooth).

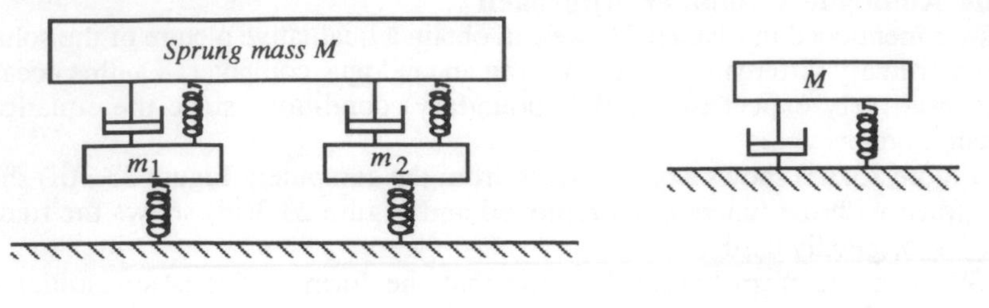

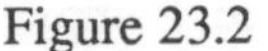

Figure 23.1 Figure 23.2

In effect, we have the situation depicted in Figure 23.2. Assume that the vehicle moves with a constant velocity, v. Let the profile of the road be described by $y(x)$; then the vertical velocity of the vehicle is

$$\frac{dy}{dt} = \frac{dy}{dx} \cdot \frac{dx}{dt} = \frac{dy}{dx} v$$

and its vertical acceleration is

$$\frac{d^2y}{dt^2} = \frac{d}{dt}\left[\frac{dy}{dt}\right] = \frac{d}{dx}\left[\frac{dy}{dx} \cdot v\right] v = v^2 \frac{d^2y}{dx^2}$$

Therefore we can treat the vertical acceleration of the vehicle, and hence the vertical force on the vehicle due to the road, as a function of t or of x.

If we assume that the elastic effect of the suspension can be represented as a linear spring (i.e. it exerts a restoring force mn^2y *directly proportional* to the displacement), **and if** we assume the damping effect of the suspension is viscous [i.e. the force it exerts $2mk\dot{y}$ is *proportional* to the vertical velocity], then the equation obtained by equating force to (mass × acceleration) is

$$m\ddot{y} = -mn^2y - 2km\dot{y} + g(t)$$

where n^2 is the spring stiffness of the suspension and $g(t)$ is the force due to the uneven road. This equation can be rearranged to give

$$\ddot{y} + \frac{2km}{m}\dot{y} + \frac{mn^2}{m}y = \frac{1}{m}g(t)$$

or, relabelling coefficients,

$$\ddot{y} + b\dot{y} + cy = f(t) \qquad (23.1)$$

It should be emphasised that this *linear, second order, ordinary differential equation with constant coefficients* can represent fairly accurately many forms of electrical or mechanical vibrations. Our task is to study the kinds of solution it can possess. Later in this chapter we shall look at techniques to handle vibrations which are governed by other equations, including those which are non-linear, as is sometimes the case when the amplitudes are large.

The Analogue Computer Approach
As we mentioned in Chapter 21, we can obtain a qualitative picture of the solution to an ordinary differential equation using an analogue computer. On this occasion we intuitively expect two initial (boundary) conditions since the equation is second order.

Figure 23.3(i) shows a typical trace from the computer; Figure 23.3(ii) shows the trace with the function $f(t)$ removed and Figure 23.3(iii) shows the trace of the $f(t)$ originally used.

We can see from Figure 23.3(iii) that the form of $f(t)$ is sinusoidal - an idealised bumpy road. Figure 23.3(i) seems to follow this shape, but superimposed on it is an oscillation which decays with time. Physically, this suggests that the *shock* of the bumpy road causes the car to oscillate rapidly and then the damping effect of the suspension causes these oscillations to die out rapidly and the car to follow the profile of the road in its motion. The *linearity* of the equation suggests a superimposing of solutions and since we know that part of the total trace is due directly to $f(t)$ we wonder from what the remainder arises. In fact, it is the so-called **natural vibration** or **free vibration** of the system: the motion which results when there is *no* imposed force and the system is slightly disturbed from an equilibrium position. When there is an external force acting, the motion is said to be **forced.**

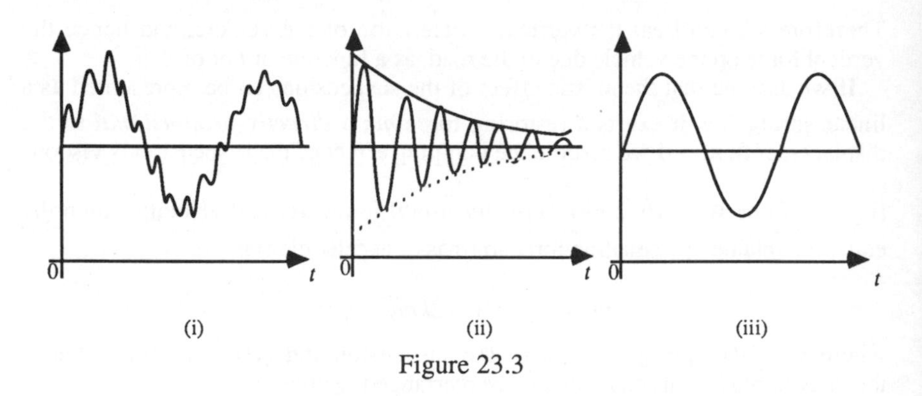

Figure 23.3

If we recall from Chapter 1 the example of the oscillating spring, we remember that the amplitude of the oscillations was damped out with time: the motion of the spring corresponded to Figure 23.3(ii).

We shall begin our study by examining the case of **free** oscillations, but first we shall rewrite Equation (23.1) as

$$\ddot{y} + 2k\dot{y} + n^2y = f(t) \qquad (23.2)$$

This is merely for algebraic convenience, as we shall see.

Case (i): no damping (**simple harmonic motion**)

Here $$\ddot{y} + n^2y = 0 \qquad (23.3a)$$

The analogue trace Figure 23.4(i) shows two waveforms: the upper one represents the displacement, y, and the lower one represents the velocity, $\dot{y}$.

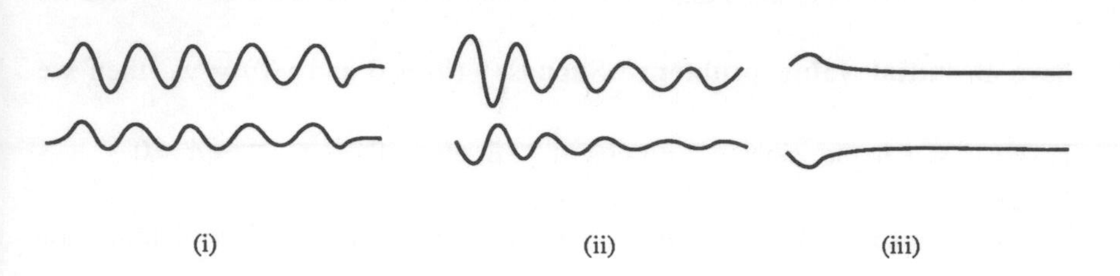

(i) (ii) (iii)

Figure 23.4

This suggests a sine or cosine solution and it is not too hard to see that $y = A \sin nt$ and $y = B \cos nt$, where, A and B are constants, both satisfy Equation (23.3a). As we increase the damping coefficient k from zero, we obtain

Case (ii): light damping **(underdamping)**
[see Figure 23.4(ii)]

Here
$$\ddot{y} + 2k\dot{y} + n^2y = 0 \tag{23.3b}$$

The amplitudes of both the displacement and the velocity decrease with time in an exponential form.

Case (iii): heavy damping **(overdamping)**
For large values of k the waveforms cease to show any oscillatory behaviour [Figure 23.4(iii)] and we have overdamping. The disturbance fails to cause the system to pass through equilibrium, merely approaching it asymptotically as $t \to \infty$. This condition is met with a ballistic galvanometer . (Why might it not be desirable for automobile suspensions?) The larger the value of k, the longer the system takes to reach any particular position near equilibrium (why have we excluded the equilibrium itself?).
There are therefore three distinct cases to consider and we take each in turn.

23.2 Free Oscillations: Analytical Approach

The linearity of Equation (23.3a) means that the sum of any two solutions is also a solution. Hence we have a solution
$$y = A \sin nt + B \cos nt \tag{23.4}$$

But from Chapter 21 we know that the general solution of the equation contains two arbitrary constants. In fact, $A \sin nt$ and $B \cos nt$ are linearly independent solutions of (23.3a). We can, therefore, take the general solution to be (23.4) and the particular solution we want will be found by prescribing two conditions.

If these are both values of y for two (different) values of t, they are called **boundary conditions** and we have a **boundary value problem**. For the moment, we consider the situation when y and $\dot{y}$ are prescribed at $t = 0$ and we have an **initial value problem**. Suppose $\dot{y}(0) = 0$ and $y(0) = y_0$, then we have $y_0 = 0 + B$ and, since $\dot{y} = nA \cos nt - nB \sin nt$, $0 = nA - 0 \Rightarrow A = 0$.

Hence
$$y = y_0 \cos nt$$
and we have simple harmonic motion about $y = 0$ (equilibrium) with amplitude y_0.

Now we turn to the case of damping which is described by (23.3b). We note that the functions which reproduce themselves (save for a multiplicative constant) on differentiation are exponential (sine and cosine can each be defined as a linear combination of exponentials). We therefore substitute $y = e^{mt}$ into (23.3b) and obtain

$$m^2 e^{mt} + 2kme^{mt} + n^2 e^{mt} = 0$$

Since e^{mt} is never zero we may divide the equation by this factor to obtain the **auxiliary equation**

$$m^2 + 2km + n^2 = 0 \qquad (23.5)$$

The general solution of (23.5) is

$$m = -k \pm \sqrt{k^2 - n^2}$$

so there are three cases to consider.

Case (a): $k < n$

Let $k^2 - n^2 = -p^2$, then $m = -k \pm ip$ and we have two possibilities: $y = e^{(-k+ip)t}$ and $y = e^{(-k-ip)t}$ to give a general solution

$$y = A' e^{(-k+ip)t} + B' e^{(-k-ip)t}$$

where A' and B' are constants. The values of A' and B' will be determined by the initial conditions. The solution is more readily interpreted if we note that the general solution can be written as

$$y = e^{-kt} (A' e^{ipt} + B' e^{-ipt})$$

and by using the identity

$$e^{i\theta} \equiv \cos\theta + i \sin\theta$$

we have

$$y = e^{-kt}[(A' + B') \cos pt + i(A' - B') \sin pt]$$

which we can write finally as

$$y = e^{-kt}(A \sin pt + B \cos pt) \qquad (23.6a)$$

Example $\ddot{y} + 8\dot{y} + 25y = 0$, with $y(0) = y_0 \quad \dot{y}(0) = 0$

The auxiliary equation is

$$m^2 + 8m + 25 = 0$$

Solving this we get

$$m = -4 \pm \sqrt{4^2 - 5^2} = -4 \pm 3\mathrm{i}$$

Thus, in the above notation, we have $k = 4$, $p = 3$ and the solution is

$$y = \mathrm{e}^{-4t}(A \sin 3t + B \cos 3t)$$

Now $\quad \dot{y} = \mathrm{e}^{-4t}[(-4A - 3B) \sin 3t + (-4B + 3A) \cos 3t]$

Applying the initial conditions:

$$y_0 = 1(0 + B) \Rightarrow B = y_0$$

so that

$$A = \frac{4y_0}{3}$$

to give the solution $\quad y = y_0 \mathrm{e}^{-4t} (\frac{4}{3} \sin 3t + \cos 3t)$

whose graph resembles Figure 23.4(ii) – a case of light damping.

Case (b): $k > n$

Let $k^2 - n^2 = q^2$ so then $m = -k \pm q$ to give a general solution

$$y = C\mathrm{e}^{(-k+q)t} + D\mathrm{e}^{(-k-q)t} \qquad (23.6b)$$

where C and D are constants.

Example $\ddot{y} + 4\dot{y} + 3y = 0$ with $y(0) = y_0$, $\dot{y}(0) = u$

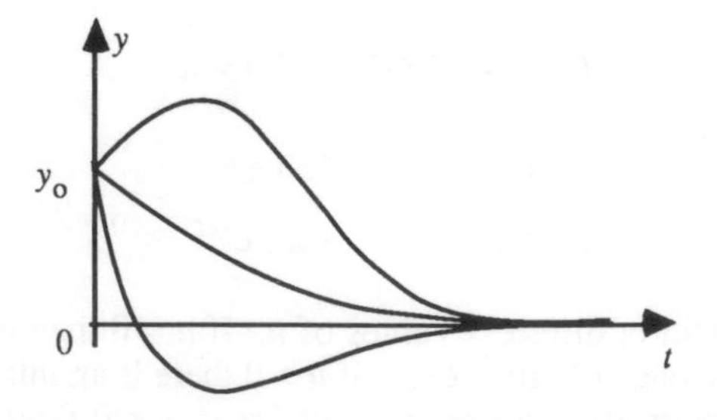

Figure 23.5

Here $\qquad q^2 = 4 - 3 = 1$

The auxiliary equation $\qquad m^2 + 4m + 3 = 0$

leads to $\qquad y = C\mathrm{e}^{-t} + D\mathrm{e}^{-3t}$

Now $$\dot{y} = -Ce^{-t} - 3De^{-3t}$$
and you can check that the initial conditions lead to the solution

$$y = \frac{1}{2}\left\{(3y_0 + u)e^{-t} - (u + y_0)e^{-3t}\right\}$$

The possible motions are shown in Figure 23.5; the particular motion will depend on the sign and size of y_0 and u.

Case (c): $k = n$
$m = -k$, and we obtain $y = E\,e^{-kt}$; this contains only one arbitrary constant, E. To attempt to find another linearly independent solution we use the technique of **Variation of Parameters.**

In this method we replace the constant E by a general function $f(t)$ so that

$$y = f(t)e^{-kt}$$

which yields $$\dot{y} = [-kf(t) + f'(t)]\,e^{-kt}$$

and $$\ddot{y} = [k^2f(t) - 2kf'(t) + f''(t)]\,e^{-kt}$$

Substitution in (23.3b) provides

$$e^{-kt}\,[k^2f(t) - 2kf'(t) + f''(t) - 2k^2f(t) + 2kf'(t) + k^2f(t)] \equiv 0$$

that is, $$e^{-kt}[f''(t)] \equiv 0$$

which implies that $$f''(t) \equiv 0$$

Hence $$f(t) = Et + F$$

where E and F are constants so that

$$y = (Et + F)\,e^{-kt} \qquad (23.6c)$$

Example
Suppose we have the conditions

$$y(0) = y_0\ (> 0), \quad \dot{y}(0) = u$$

Then $$\dot{y} = (E - kEt - kF)\,e^{-kt}$$

to give $$y = [(u + ky_0)t + y_0]\,e^{-kt}$$

We examine the effect of different values of u. If $u = 0$ then we obtain a graph similar to the middle one of Figure 23.5; if $u > 0$ there is an initial motion further away from equilibrium (the top graph), whilst if $u < 0$ (and $u + ky_o < 0$) the graph at the bottom depicts a slight overshooting.

This case is called **critical damping** and the motion is said to be **dead beat.** There is a similarity between critical damping and heavy damping and the distinction is that the *critical* value of k is the *least* value to prevent one complete oscillation.

We now summarise the results obtained. The auxiliary equation of the differential equation

$$\frac{d^2y}{dt^2} + 2k\frac{dy}{dt} + n^2y = 0$$

is

$$m^2 + 2km + n^2 = 0$$

which can be solved to give $m = -k \pm \sqrt{k^2 - n^2}$

with the three cases

(i) $k < n \Rightarrow m = -k \pm ip$ $\Rightarrow$ solution $y = e^{-kt}[A \sin pt + B \cos pt]$

(ii) $k > n \Rightarrow m = -k \pm q$ $\Rightarrow$ solution $y = Ce^{(-k+q)t} + D\,e^{(-k-q)t}$

(iii) $k = n \Rightarrow m = -k$ $\Rightarrow$ solution $y = (Et + F)\,e^{-kt}$

23.3 Forced Vibrations: Complementary Function and Particular Integral

We return to equation (23.2): $\ddot{y} + 2k\dot{y} + n^2y = f(t)$

We shall utilise the result that, *if $y_1(t)$ is a particular solution (called a* **particular integral**) *of* (23.2) *and $y_2(t)$ (called the* **complementary function**) *is the general solution of the associated homogeneous equation*

$$\ddot{y} + 2k\dot{y} + n^2y = 0$$

then the general solution of (23.2) *is*

$$y^*(t) \equiv y_1(t) + y_2(t)$$

The proof of this result is in two parts: first we show that $y^*(t)$ is a solution of (23.2) and then we show that any other solution differs only in the arbitrary constants of the complementary function. We know that

$$\ddot{y}_1 + 2k\dot{y}_1 + n^2y_1 = f(t)$$

and

$$\ddot{y}_2 + 2k\dot{y}_2 + n^2y_2 = 0$$

Since $\dot{y}^* = \dot{y}_1 + \dot{y}_2$ and $\ddot{y}^* = \ddot{y}_1 + \ddot{y}_2$

it follows that

$$\begin{aligned} \ddot{y}^* + 2k\dot{y}^* + n^2y^* &\equiv (\ddot{y}_1 + 2k\dot{y}_1 + n^2y_1) + (\ddot{y}_2 + 2k\dot{y}_2 + n^2y_2) \\ &= f(t) + 0 \\ &= f(t) \end{aligned}$$

Further, let $y_s(t)$ be a second solution of (23.2) and consider

$$y_d(t) \equiv y^*(t) - y_s(t); \quad \ddot{y}_d + 2k\dot{y}_d + n^2y_d \equiv f(t) - f(t) = 0$$

and so y_d is a solution of the associated homogeneous equation and simply means a relabelling of the coefficients of $y_2(t)$. If

$$y_2(t) = A\,e^{m_1 t} + B\,e^{m_2 t}$$

then any solution of (23.2) can be written as

$$y(t) = y_1(t) + A\,e^{m_1 t} + B\,e^{m_2 t}$$

The particular solution we want (i.e. the choice of the values of A and B) is determined by the initial (or boundary) conditions.

The rules may be summarised:

(i) find the complementary function (C.F.)
(ii) spot *any* particular integral (P.I.)
(iii) add (i) and (ii) to form the general solution (G.S.)
(iv) fit the initial (or boundary) conditions to obtain the particular solution.

It is important to note that the initial (or boundary) conditions are applied only at the last stage.

Example $\ddot{y} + 8\dot{y} + 25y = 4; \quad y(0) = 2,\ \dot{y}(0) = 1$

(i) C.F. This has been found in an earlier example to be

$$y_{CF} = e^{-4t}(A \sin 3t + B \cos 3t)$$

(ii) P.I. We *spot* that y = constant is a possible particular integral. Substitution yields

$$y_{PI} = 4/25$$

(iii) G.S. Addition gives

$$y = e^{-4t}(A \sin 3t + B \cos 3t) + 4/25$$

(iv) $y(0) = 2 \Rightarrow 2 = 1(0 + B) + 4/25 \Rightarrow B = 46/25$

Since $\dot{y} = e^{-4t}[(-4A - 3B) \sin 3t + (-4B + 3A) \cos 3t]$

$$\dot{y}(0) = 1 \Rightarrow 1 = 1[0 + (-4B + 3A)] \Rightarrow A = \frac{209}{75}$$

The solution we require is therefore

$$y = e^{-4t}\left(\frac{209}{75} \sin 3t + \frac{46}{25} \cos 3t\right) + \frac{4}{25}$$

– do not be surprised by such unwieldy coefficients.

As long as we remember the scheme (i) to (iv) we can turn our attention now to (ii), that is, finding the particular integral which clearly depends (unlike the complementary function) on the form of $f(t)$. We shall develop techniques for finding the particular integral in the next section and in Chapter 24.

23.4 Particular Integral: Trial Solutions

This technique is applicable when each term of $f(t)$ has a finite number of linearly independent derivatives; the functions we should consider are: a constant, a polynomial, sine and cosine, exponential. We cannot cope, for instance, with $1/x^2$ since this has an infinite number of linearly independent derivatives (in the sense that none of them can be expressed as a linear combination of the others).

We have a general result that *if $f(t)$ can be expressed as $f_1(t) + f_2(t)$ then a particular integral (P.I.) of $f(t)$ is the sum of a P.I. of $f_1(t)$ and a P.I. of $f_2(t)$.*

We consider again the differential equation

$$\ddot{y} + 2k\dot{y} + n^2y = f(t) \tag{23.2}$$

(i) $f(t) = c$, constant. We try $y = A$, and since $\dot{y} = 0 = \ddot{y}$ it should be clear that $y = c/n^2$ is the particular integral we seek.

(ii) $f(t) = a_nt^n + a_{n-1}t^{n-1} + ... + a_1t + a_o$. We try a polynomial for y of the same form, but all powers of t lower than t^n must be present.

Example $\quad \ddot{y} + 2\dot{y} + y = 3t^2 + 4$

We try $y = At^2 + Bt + C$ (note the presence of the term in t)

Then $\quad \dot{y} = 2At + B$ and $\ddot{y} = 2A$

to give, on substitution in the differential equation,

$$2A + 4At + 2B + At^2 + Bt + C = 3t^2 + 4$$

Equating coefficients of like powers of t we have

$$t^2:\ A = 3$$

$$t:\ 4A + B = 0 \Rightarrow B = -12$$

$$\text{constant}:\ 2A + 2B + C = 4 \Rightarrow C = 22$$

Hence the P.I. is $\quad y = 3t^2 - 12t + 22$

(iii) $f(t) = Ae^{\lambda t}$. We try $y = Ce^{\lambda t}$, C constant.

Example 1 $\quad \ddot{y} + 4\dot{y} + 4y = 2e^{3t}$

We try $y = Ce^{3t}$ so that $\dot{y} = 3Ce^{3t}$ and $\ddot{y} = 9Ce^{3t}$

Substitution yields $\quad (9C + 12C + 4C)e^{3t} = 2e^{3t}$

Hence $C = 2/25$ to give a P.I. of $\quad y = \frac{2}{25}e^{3t}$

Example 2 $\ddot{y} + 4\dot{y} + 3y = e^{-t}$

We try $y = Ce^{-t}$ so that $\dot{y} = -Ce^{-t}$ and $\ddot{y} = Ce^{-t}$, therefore

$$(C - 4C + 3C)e^{-t} = e^{-t}$$

The bracket on the left is zero, so the method has failed.

The trouble here is that e^{-t} is part of the complementary function. The method of variation of parameters will show that $y = (Ct)e^{-t}$ is the form to try. This gives

$$\dot{y} = (C - Ct)e^{-t} \text{ and } \ddot{y} = (-2C + Ct)e^{-t}$$

which on substitution leads to

$$(-2C + Ct + 4C - 4Ct + 3Ct)e^{-t} = e^{-t}$$

that is, $C = \frac{1}{2}$ and $y = \frac{1}{2}te^{-t}$ is the P.I. required.

Example 3 $\ddot{y} + 4\dot{y} + 4y = e^{-2t}$

It would seem that since e^{-2t} is part of the C.F., that we should try $y = Cte^{-2t}$, but fate is not often kind to the foolhardy. You can check that such a substitution yields $0 = 1$. The trouble here is that -2 is a double root of the auxiliary equation.

Variation of parameters or intuition suggests the trial of $y = Ct^2e^{-2t}$ and you should check the wisdom of this choice.

(iv) $f(t) = A \sin pt$ or $B \cos pt$

This is an important case practically and we consider first the undamped forced oscillations, i.e. we seek to solve $\ddot{y} + n^2y = F \cos pt$.

We know that the only linearly independent derivatives of sin pt are of the form $\alpha \cos pt$ and $\beta \sin pt$, where α and β are constants; consequently we try for a P.I. in the form

$$y = \alpha \cos pt + \beta \sin pt$$

Then $$\dot{y} = -p\alpha \sin pt + p\beta \cos pt$$

and $$\ddot{y} = -p^2\alpha \cos pt - p^2\beta \cos pt$$

Substitution into the equation of motion produces

$$(n^2 - p^2)(\alpha \cos pt + \beta \sin pt) = F \cos pt$$

hence $\beta = 0$ and $\alpha = F/(n^2 - p^2)$ so that the particular integral is

$$y = [F/(n^2 - p^2)] \cos pt$$

Now the complementary function is

$$y = B \cos nt + C \sin nt$$

and so the general solution becomes

$$y = \underbrace{B \cos nt + C \sin nt}_{\text{free oscillations}} + \underbrace{[F/(n^2 - p^2)] \cos pt}_{\text{forced oscillations}}$$

Let us suppose that the body undergoing these forced oscillations is released from the origin at rest, i.e. at $t = 0$, $y = 0 = \dot{y}$: then you can check that the particular solution is

$$y = \frac{F}{n^2 - p^2}\left[\cos pt - \cos nt\right]$$

We see that F, the amplitude of the forcing term, has been magnified by a factor $1/(n^2 - p^2)$ in the P.I. which represents the forced oscillation. It is clear that our solution is invalid when $p = n$, i.e. when the frequency of the forcing motion is equal to the frequency of the natural oscillations of the body. (The graph of magnification factor against p is shown in Figure 23.6.) Before dealing with that case, we examine the behaviour for other values of p. For p near zero, the forcing frequency is very slow and the body movement will approximate to free oscillations subject to a *static* force F; on the other hand, for very high frequencies, the body is simply unable to follow the forcing motion and the amplitude is very small. For values of p approaching n, the amplitude of the vibration becomes very large since the forcing motion is almost in sympathy with the natural oscillation and can push the body almost at the right time in the right direction.

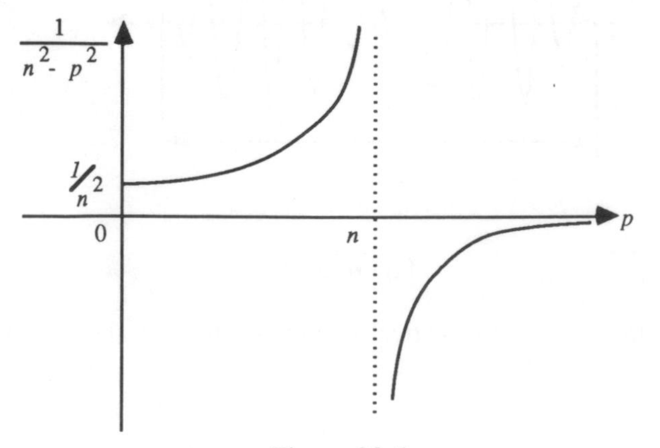

Figure 23.6

If the forcing term is $F \cos pt$ or $F \sin pt$, and if p is almost equal to n, then we have the phenomenon of **beats.** Let us take the case with a forcing term of $F \sin pt$. The solution is now

$$y = \frac{F}{n^2 - p^2}\left[\sin pt - \frac{p}{n}\sin nt\right]$$

Suppose that p is almost equal to n, then we have approximately

$$y \cong \frac{F}{n^2 - p^2}\left[\sin pt - \sin nt\right] = \frac{2F}{(n-p)(n+p)}\cos\left(\frac{p+n}{2}\right)t\,\sin\left(\frac{p-n}{2}\right)t$$

Denoting the small quantity $p - n$ by 2ε, so that

$$\frac{p+n}{2} = n + \varepsilon \quad\text{and}\quad \frac{p-n}{2} = \varepsilon$$

we can write
$$y \cong \frac{-F}{2n\varepsilon}\cos nt \sin \varepsilon t$$

Since ε is a small quantity, the period $2\pi/\varepsilon$ of the term $\sin \varepsilon t$ is large and we can regard y as a periodic function $\cos nt$ with slowly varying amplitude as shown in Figure 23.7 where the dashed curves are the curves to which the actual wave periodically rises. A similar analysis can, of course, be carried out for a forcing term of $F \cos pt$ and you should examine this for yourself.

Beats can be heard when an electric generator starts up or in a power house when the hum of the generator and that of the others on the line are almost of the same frequency.

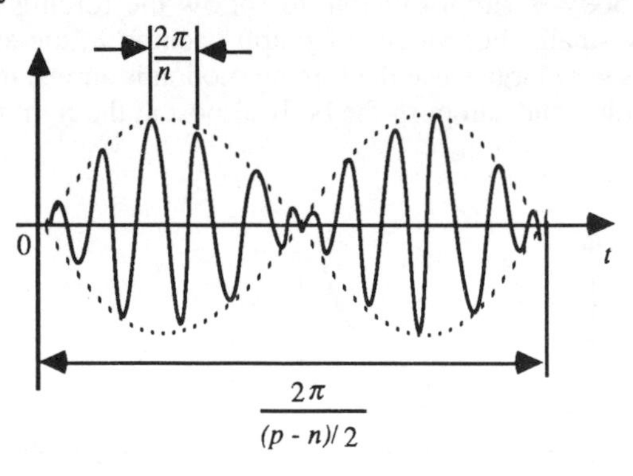

Figure 23.7

Now we return to the mathematical expression for y in the case of a forcing function $F \cos pt$. We had

$$y = \frac{F}{n^2 - p^2}\left[\cos pt - \sin nt\right]$$

which can be rearranged to

$$y = \frac{F}{n^2 - p^2} \, 2 \sin\left(\frac{p+n}{2}\right)t \,.\, \sin\left(\frac{n-p}{2}\right)t$$

Let $p = n + \Delta n$ so that

$$y \cong \frac{F}{n\,\Delta n} \sin\left(n + \tfrac{1}{2}\Delta n\right)t \,.\, \sin\frac{\Delta n}{2}t \tag{23.7}$$

Let t be fixed in (23.7) and $\Delta n \rightarrow 0$;

Now, $$\left(\sin\frac{\Delta n}{2}t\right) \Big/ \left(\frac{\Delta n}{2}\right) \rightarrow t$$

and $$\sin\left(n + \tfrac{1}{2}\Delta n\right)t \rightarrow \sin nt$$

as $\Delta n \rightarrow 0$, therefore $$y \rightarrow \frac{F}{2n} t \sin nt$$

(Why should we have presumed this from a trial solution for the case $p = n$?)

We see that the amplitude will grow indefinitely with time. This is the case of **resonance** and a well-known application is the requirement for a group of marching soldiers to break step as they cross a bridge in case their 'p' should equal the 'n' of the bridge. Other examples of resonance will be found in many problems.

If we now introduce a little damping the danger is avoided; of course, the larger the damping, the less the maximum amplitude at $p = n$. The graph of magnification factor against p is shown in Figure 23.8.

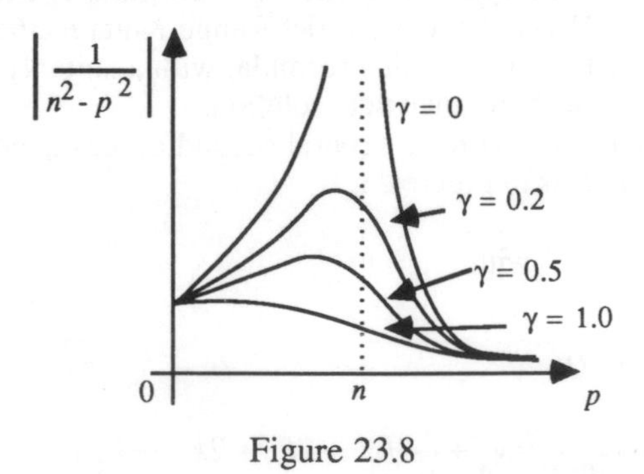

Figure 23.8

The ratio $\gamma = k/k_c$ measures the fraction of critical damping. We have shown the magnification factor to be positive in this case, so that a comparison is more readily seen.

23.5 Numerical Solutions of Initial Value Problems

The techniques we have studied so far cope with linear equations with constant coefficients. The numerical method we shall describe can cope with most second order differential equations; there are many numerical methods but we restrict ourselves to one.

First we consider the equation

$$\ddot{y} + 2k\dot{y} + n^2y = f(t)$$

we write $\dot{y} = v$ and then cast the equation in the form

$$\dot{v} = -2kv - n^2y + f(t)$$

We have, therefore, two first order differential equations to be solved simultaneously. In essence, we start at $t = 0$ knowing y and v; the equation $\dot{y} = v$ allows us to predict y at Δt by

$$y_{new} \cong y_{old} + \Delta t\left(\frac{dy}{dt}\right)_{t=0}$$

(Euler's method) or by a more accurate formula. Likewise, the equation $\dot{v} = -2kv - n^2y + f(t)$ allows us to predict v at Δt by

$$v_{new} \cong v_{old} + \Delta t\left(\frac{dv}{dt}\right)_{t=0}$$

or some more sophisticated formula. Thus at $t = \Delta t$, we know y and v and we are ready to step forward again. The choice of formula is subject to the same criteria as in Chapter 21 and the fourth order Runge-Kutta method is popular.

As you can verify for yourself, this formula, with a suitably chosen step size, gives good agreement with the analytical solution.

First we quote the formula for a general second order equation and show the flow chart for the process in Figure 23.9.

Consider $\ddot{y} = f(t, y, v)$ with $\dot{y} = v$

so that $\dot{v} = f(t, y, v)$

Then $$v_{n+1} = v_n + \frac{1}{6}(k_1 + 2k_2 + 2k_3 + k_4)$$

and $$y_{n+1} = y_n + \frac{1}{6}(l_1 + 2l_2 + 2l_3 + l_4)$$

where k_1, l_1, etc are given by the following formulae

$$
\begin{aligned}
k_1 &= \Delta t \,.\, f(t_n, y_n, v_n) & l_1 &= \Delta t \,.\, v_n \\
k_2 &= \Delta t \,.\, f(t_n + \frac{\Delta t}{2}, y_n + \frac{l_1}{2}, v_n + \frac{k_1}{2}) & l_2 &= \Delta t\,(v_n + \frac{k_1}{2}) \\
k_3 &= \Delta t \,.\, f(t_n + \frac{\Delta t}{2}, y_n + \frac{l_2}{2}, v_n + \frac{k_2}{2}) & l_3 &= \Delta t\,(v_n + \frac{k_2}{2}) \\
k_4 &= \Delta t \,.\, f(t_n + \Delta t, y_n + l_3, v_n + k_3) & l_4 &= \Delta t\,(v_n + k_3)
\end{aligned}
\tag{23.8}
$$

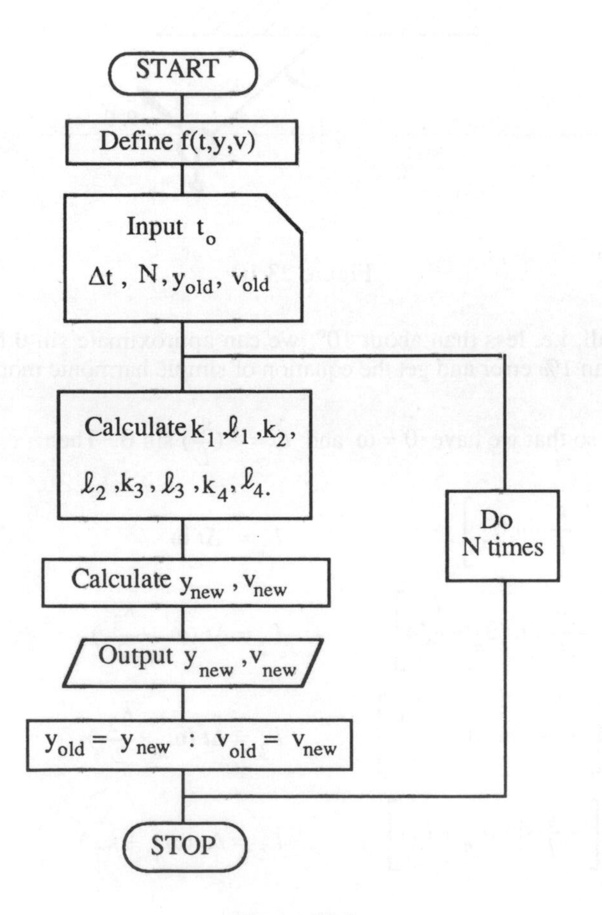

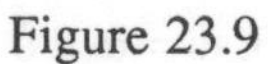
Figure 23.9

Example
We consider the application of the Runge-Kutta formula to the oscillations of a simple pendulum when the oscillations are sufficiently large in amplitude to

prevent the usual assumption of a linear governing equation.

We now consider a simple pendulum consisting of a heavy bob at the end of a slender rod of negligible mass as shown in Figure 23.10.

The equation of motion neglecting air resistance etc is

$$ml\ddot{\theta} = -mg \sin \theta$$

or

$$\ddot{\theta} = -\frac{g}{l} \sin \theta$$

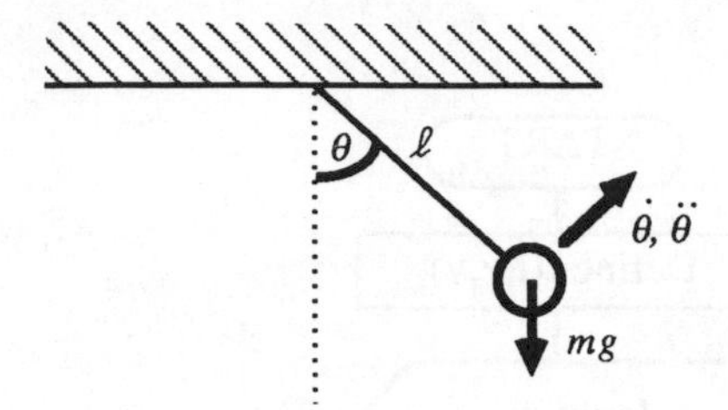

Figure 23.10

[If θ is small, i.e. less than about 10°, we can approximate sin θ by θ(radians) with less than 1% error and get the equation of simple harmonic motion.]

Let $\omega = \dot{\theta}$ so that we have $\dot{\theta} = \omega$ and $\dot{\omega} = -(\frac{g}{l}) \sin \theta$. Then

$$k_1 = \Delta t\left[-\frac{g}{l}\sin \theta_n\right] \qquad l_1 = \Delta t\, \omega_n$$

$$k_2 = \Delta t\left[-\frac{g}{l}\sin (\theta_n + \frac{l_1}{2})\right] \qquad l_2 = \Delta t\, (\omega_n + \frac{k_1}{2})$$

$$k_3 = \Delta t\left[-\frac{g}{l}\sin (\theta_n + \frac{l_2}{2})\right] \qquad l_3 = \Delta t\, (\omega_n + \frac{k_2}{2})$$

$$k_4 = \Delta t\left[-\frac{g}{l}\sin (\theta_n + l_3)\right] \qquad l_4 = \Delta t\, (\omega_n + k_3)$$

If we assume that, at $t = 0$, $\theta = \pi/3$, $\omega = \dot{\theta} = 0$ and that $l = 0.3$ metre, where the equation of motion is

$$\ddot{\theta} = -(\frac{g}{l})\,\theta$$

for small amplitudes, the period of oscillation would be

$$T = 2\pi\sqrt{\frac{l}{g}} \cong 1 \text{ sec}$$

so we take a step size of 0.05 sec as a first attempt. Then

$k_1 = -0.05(32.7 \,.\, \sin\frac{\pi}{3}) = -1.416$ $\qquad l_1 = 0.05(0) = 0$

$k_2 = -0.05[32.7 \,.\, \sin(\frac{\pi}{3} + 0)] = -1.416$ $\qquad l_2 = 0.05(0 - 0.708) = -0.035$

$k_3 = -0.05[32.7 \,.\, \sin(\frac{\pi}{3} - 0.0175)] = -1.401$ $\qquad l_3 = 0.05(0 - 0.708) = -0.035$

$k_4 = -0.05[32.7 \,.\, \sin(\frac{\pi}{3} - 0.035)] = -1.386$ $\qquad l_4 = 0.05(0 - 1.401) = -0.070$

Hence $\qquad \theta_{new} = 1.012; \; \omega_{new} = -1.406$

The analytical approach would be to write $\ddot{\theta}$ as $\omega(\mathrm{d}\omega/\mathrm{d}\theta)$ and obtain

$$\omega\frac{\mathrm{d}\omega}{\mathrm{d}\theta} = -\frac{g}{l}\sin\theta$$

that is, $$\int \omega\,\mathrm{d}\omega = -\int \frac{g}{l}\sin\theta\,\mathrm{d}\theta$$

which gives $$\tfrac{1}{2}\omega^2 = \frac{g}{l}\cos\theta + \text{constant}$$

When $t = 0$, $\omega = 0$, $\theta = \theta_0$, therefore

$$0 = +\frac{g}{l}\cos\theta_0 + \text{constant}$$

and so $$\tfrac{1}{2}\omega^2 = \frac{g}{l}\cos\theta - \frac{g}{l}\cos\theta_0$$

that is, $$\frac{\mathrm{d}\theta}{\mathrm{d}t} \equiv \omega = \sqrt{\frac{2g}{l}(\cos\theta - \cos\theta_0)}$$

and hence $$t = \int_{\theta_0}^{\theta} \frac{\mathrm{d}\theta}{\sqrt{\frac{2g}{l}(\cos\theta - \cos\theta_0)}}$$

The integration then poses problems which we cannot tackle here; notice, however, that the integrand is infinite at $\theta = \theta_0$.

Comparison Between Analytical and Numerical Approaches
Similar remarks to those made in Chapter 21 apply here. Wherever possible, the analytical approach should be used, but for some problems linearisation of the model equation may give a false picture of the behaviour of the independent variable; alternatively, the solution of the full equation may be difficult to obtain. In such cases, the Runge-Kutta method is useful. To check the validity of its results we could halve the step size and repeat until sufficiently close agreement is reached. As before, the analogue solution of linear problems is a helpful guide to the behaviour to be analysed.

23.6 Boundary Value Problems

We intend to introduce this class of problem by an example. In Figure 23.11 a straight uniform column of length l is hinged at both ends and an axial load P is applied.

The problem is to find the deflected profile $y(x)$ which is satisfied approximately by the d.e.

$$\frac{d^2y}{dx^2} + \frac{Py}{EI} = 0 \tag{23.9}$$

where EI is a constant for a given column.

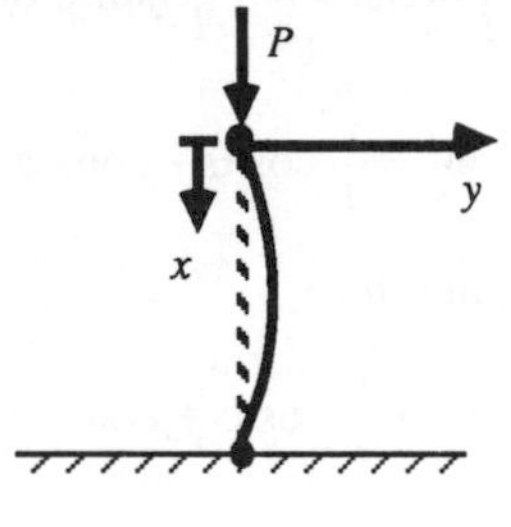

Figure 23.11

The solution must satisfy the boundary conditions: $y = 0$ at $x = 0$ and at $x = l$.
The general solution of Equation (23.9) is

$$y = A \sin\sqrt{\frac{P}{EI}}\,x + B \cos\sqrt{\frac{P}{EI}}\,x$$

Applying $y = 0$ at $x = 0$ we find $B = 0$, and from the other boundary condition we find that

$$A \sin\sqrt{\frac{P}{EI}}\,l = 0$$

Since $A = 0$ gives $y(x) \equiv 0$ we seek other solutions; we know $\sin n\pi = 0$ and so we conclude

$$\sqrt{\frac{P}{EI}}\, l = n\pi$$

and so we may find for a fixed beam several possible loads P_n which satisfy the d.e. and boundary conditions. The solutions are given by

$$P_n = \frac{n^2\pi^2}{l^2} EI, \qquad n = 1,2,3,....$$

Such loads are called **critical loads** and in practice the first such load is the one which causes the column to buckle (according to the theory which produced Equation (23.11).

Such boundary value problems do not easily allow a solution via Laplace transforms. Furthermore the numerical solution of such problems is usually much more complicated than the solution for the corresponding initial value problem. We shall not consider numerical solutions at this stage, but give two boundary value problems for you to try by analytical methods (see Problems 31 and 32).

Problems

Section 23.2

1 Write down and solve the auxiliary equations for each of the following differential equations

(i) $\ddot{y} + 5\dot{y} - 6y = 0$ (ii) $\ddot{y} - 16\dot{y} + 64y = 0$

(iii) $\ddot{y} + 6\dot{y} + 10y = 0$ (iv) $4\ddot{y} + 4\dot{y} + y = 0$

(v) $\dddot{y} - \ddot{y} - \dot{y} - 2y = 0$ (Hint: you should get a cubic equation)

(vi) $\ddot{y} + \dot{y} + 7y = 0$ (vii) $5\ddot{y} + 4\dot{y} - \frac{13}{4}y = 0$

Hence write down the general solutions in the form of (23.6a), (23.6b) or (23.6c). In the case of (v), try to guess what form the solution will take.

2 A uniform flexible cable hangs under its own weight w/unit length. Show that the profile $y(x)$ of the cable satisfies the differential equation

$$\frac{d^2y}{dx^2} = \frac{w}{H}\sqrt{1 + \left(\frac{dy}{dx}\right)^2}$$

where H is the horizontal tension in the cable. Write $p = dy/dx$ and by choosing suitable boundary conditions obtain a formula for the profile.

3 An object moves in a straight line (which we can take as the x-axis) with constant speed V. A missile, initially a distance D away, at right angles to the x-axis homes in on its target with constant speed nV, $n > 1$, in a direction always toward the target. Show that

the equation of the path of the missile can be written

$$\frac{d^2x}{dy^2} = \frac{1}{ny}\sqrt{1 + \left(\frac{dx}{dy}\right)^2}$$

and show that the target is hit after a time $\frac{n}{n^2 - 1}\frac{D}{V}$.

4 Solve the equation $\frac{d^3y}{dx^3} + 4\frac{d^2y}{dx^2} + 4\frac{dy}{dx} + y = 0$

5 In a mechanical vibration problem we have the following parameters: mass of spring, spring stiffness, damping constant. What are the analogues of these quantities for
(i) a torsional vibration problem (ii) an *LCR* circuit?

6 A mass *m*, constrained to move along the *x*-axis, is attracted towards the origin with a force proportional to its distance from the origin. Find the motion
(i) if it starts from rest at $x = x_0$
(ii) if it starts from the origin with velocity v_0. (EC)

7 Solve the differential equation $\frac{d^4y}{dx^4} = 16y$ (EC)

Section 23.4

8 Find the Particular Integrals of the following differential equations

(i) $\frac{d^2y}{dx^2} + 4\frac{dy}{dx} + 5y = 2e^{-3x}$ (Try $y = k\,e^{-3x}$)

(ii) $4\frac{d^2y}{dx^2} - 4\frac{dy}{dx} + y = x^2 - x + 2$ (Try $y = \alpha x^2 + \beta x + \gamma$)

(iii) $\frac{d^2y}{dx^2} - 3\frac{dy}{dx} + 2y = \sin 2x$ (Try $y = \alpha \sin 2x + \beta \cos 2x$)

(iv) $\frac{d^2y}{dx^2} + \frac{dy}{dx} = 2e^{-2x} + 3x$ (Try $y = k\,e^{-2x} + \alpha x^2 + \beta x + \gamma$)

(v) $\frac{d^2y}{dx^2} + 4\frac{dy}{dx} + 4y = 2e^{-2x}$ (This time $y = k\,e^{-2x}$ will not work)

(vi) $\frac{d^2y}{dx^2} + 16y = 2\sin 4x$ (We shall have to try something different from $y = \alpha \sin 4x + \beta \cos 4x$)

9 Solve the differential equations

(i) $\dfrac{d^2y}{dx^2} + 4\dfrac{dy}{dx} + 5y = 8\cos x$ given that $y = 0$ when $x = 0$ and $dy/dx = 3$ when $x = 0$

(ii) $\dfrac{d^2y}{dx^2} - \dfrac{dy}{dx} - 2y = 2e^{2x}$ given that $y = 0$ and $dy/dx = 0$ when $x = 0$

(iii) $\dfrac{d^2x}{dt^2} + 4x = 2e^t$ given $x = 0$ when $t = 0$ and $dx/dt = 3$ when $t = 0$

(iv) $9\dfrac{d^2x}{dt^2} + x = \sin\dfrac{t}{3}$

(v) $\dfrac{d^2x}{dt^2} + 4\dfrac{dx}{dt} + x = t^3 - 2$

(vi) $\dfrac{d^2y}{dx^2} + 3\dfrac{dy}{dx} - 4y = e^{-x}\cos x$ with $y = 1$ when $x = 0$ and $y \to 0$ when $x \to \infty$. (EC)

10 Consider the equation $$\frac{d^2y}{dx^2} + 4y = \sin 2x$$

Show that the complementary function is
$$y = C\cos 2x + D\sin 2x$$
Assume now that C and D are not constants and show that
$$\frac{dy}{dx} = -2C\sin 2x + 2D\cos 2x + C'\cos 2x + D'\sin 2x$$

where $$C' \equiv \frac{dC}{dx}, \quad D' \equiv \frac{dD}{dx}$$

Since we have two arbitrary functions C and D to satisfy one equation, we can make the additional condition that
$$C'\cos 2x + D'\sin 2x = 0$$
Now find d^2y/dx^2 and substitute in the differential equation and you should obtain
$$-2C'\sin 2x + 2D'\cos 2x = \sin 2x$$
Solve the last two equations simultaneously and show that
$$C = -\frac{1}{4}x + \frac{1}{16}\sin 4x + A$$
$$D = -\frac{1}{16}\cos 4x + B$$
Hence show that the general solution is
$$y = A\cos 2x + F\sin 2x - \frac{1}{4}x\cos 2x$$

where $$F = B + \frac{1}{16}$$

(Notice that we could have tried $y = kx \cos 2x$ for the P.I.)

This is another method of Variation of Parameters. Repeat this approach for the equations

(a) $\dfrac{d^2y}{dx^2} + 2\dfrac{dy}{dx} + y = 2e^{-x}$ (b) $\dfrac{d^2y}{dx^2} - 4\dfrac{dy}{dx} = \dfrac{1}{2}e^{4x}$

11 A uniform beam of infinite length rests on an elastic foundation so that a principal axis of the beam is vertical. The reaction from the foundation at any point on the beam can be assumed proportional to the deformation there. If there is a uniformly distributed load of w/unit length acting over a length $2a$, it may be shown that the equation for the deflected profile is

$$EI\frac{d^4y}{dx^4} + ky = f(x)$$

where EI is the (constant) flexural rigidity of the beam, k is a constant of proportionality, and $f(x)$ represents the load on the beam.

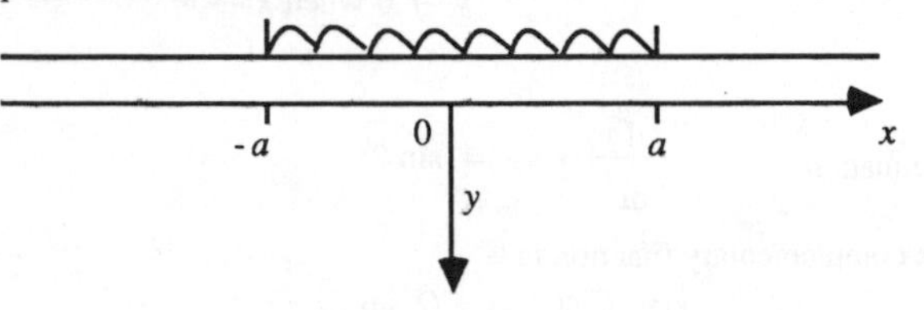

Using symmetry and splitting the profile into two parts, find the eight boundary conditions for the two resulting equations. Hence determine the deflected profile.
(Hint: it helps to put $\beta^4 = k/EI$)

12 Solve the equation $$\frac{d^2y}{dx^2} + 4\frac{dy}{dx} - 5y = 2e^{-2x}$$
given that, at $x = 0$, $y = 0$ and $dy/dx = 1$. (EC)

13 Solve completely the differential equation

$$\frac{d^2s}{dt^2} + 2k\frac{ds}{dt} + 4s = \cos t$$

in each of the cases (a) $k = 2^1/2$, (b) $k = 2$, (c) $k = 1$. (LU)

14 Obtain the general solutions of the differential equations

(i) $\dfrac{d^2y}{dx^2} + 5\dfrac{dy}{dx} + 4y = e^{-2x} + \sin x$ (ii) $\dfrac{d^2y}{dx^2} - 6\dfrac{dy}{dx} + 10y = x + 1$ (LU)

15 The current, I, in a circuit is given by the differential equation

$$\frac{d^2I}{dt^2} + 4\frac{dI}{dt} + 2504I = 250.4$$

If $I = 0$ and $dI/dt = 0$ when $t = 0$, find I in terms of t. Find also the maximum current in the circuit. (LU)

16 A body is attached by a spring to a point which is oscillating with sinusoidal motion. The displacement s of the body from a fixed reference point at any time t is given by the differential equation

$$\frac{d^2s}{dt^2} + 4s = \lambda \sin t$$

where λ is a constant. Given that $s = 0$ when $t = 0$, and $ds/dt = 2\lambda/3$ when $t = 0$, solve the equation for s in terms of t. Show that the ratio of the maximum and minimum displacements is numerically equal to $5\sqrt{5} : 4$. (LU)

17 An electric cable has resistance R per unit length and the resistance per unit length of the insulation is n^2R, where n and R are constants. The voltage v and current i at a distance x from one end satisfy the equations

$$\frac{dv}{dx} + Ri = 0 \quad \text{and} \quad \frac{di}{dx} + \frac{v}{n^2R} = 0$$

If current is supplied at one end A of a cable, of length l, at a voltage V and the cable is insulated at the other end, show that the current entering at A is

$$\frac{V}{nR} \tanh \frac{l}{n}$$ (LU)

18 A body is restricted to move along a straight line, its motion being governed by the equation $d^2x/dt^2 = 2 \sin 3t$, where x is the displacement from a fixed origin and t is the elapsed time. If the initial displacement is 2 and the initial velocity 3, solve the equation to find x.
Establish the time taken for the body to reach its maximum velocity on the first occasion after $t = 0$, and find this maximum velocity. (EC)

19 In an oscillatory system with damping the following differential equation occurs

$$\frac{d^2x}{dt^2} + 4\frac{dx}{dt} + 8x = 20 \sin 2t$$

(i) Find the general solution of this differential equation.
(ii) Find the particular solution given that $x = 0$ and $dx/dt = 0$ when $t = 0$.
(iii) Identify the transient part and the steady-state part of the solution.
(iv) In the steady-state show that the maximum displacements from the centre line are $\sqrt{5}$, and find at what values of t these occur. (EC)

20 The expression for the electric charge q in a particular circuit is given by the differential equation

$$\frac{d^2q}{dt^2} + 2k\frac{dq}{dt} + 4q = \cos t$$

where k is a constant.

(i) Find the complementary function in each of the three cases $k = 5/2$, $k = 1$, $k = 2$. State, giving reasons, which are 'transients'.

(ii) In the case $k = 1$, solve the differential equation completely given that $q = 0$ and $\dot{q} = 1$ when $t = 0$.

(iii) Show that the 'steady-state' can be written in the form $R(\sin t + \alpha)$, giving the numerical values of R and α. Suggest a physical interpretation of R. (EC)

21 In an electric circuit an inductor L, a resistor R and a capacitor C are connected in series to a generator supplying an e.m.f., $E(t)$. The current i flowing in the circuit and the charge q on the capacitor plates are related by $i = dq/dt$. It can be shown that

$$L\frac{d^2q}{dt^2} + R\frac{dq}{dt} + \frac{q}{C} = E(t)$$

In a particular situation $L = 0.005$, $R = 10$, $C = 10^{-4}$ and $E(t) = 50 \sin 1000\, t$. Initially $t = 0$ when $i = 0$ and $q = 0$.

(i) Find a formula for the charge q on the capacitor at time t.

(ii) Find the maximum value of the charge when the steady-state is reached. (EC)

22 (i) The differential equation satisfied by the displacement x of a particle performing free damped oscillations is given by

$$\frac{d^2x}{dt^2} + 2\lambda\frac{dx}{dt} + (\lambda^2 + n^2)x = 0$$

If λ and n are positive constants solve this differential equation given that $x = b$ and $dx/dt = 0$ when $t = 0$.

(ii) A forcing function $2b \sin t$ is applied to a particular system for which the differential equation is

$$\ddot{x} + 4\dot{x} + 5x = 2b \sin t$$

($\dot{x}$ and $\ddot{x}$ denote dx/dt and d^2x/dt^2 respectively)

If the initial conditions are $x = b$ and $\dot{x} = 0$ when $t = 0$, show that the solution of this is

$$x = \frac{1}{4}b\exp(-2t)(9\sin t + 5\cos t) + \frac{1}{4}b(\sin t - \cos t)$$

(iii) Show that when t is very large the particle will describe a motion with a displacement given by $x = R\sin(t - \alpha)$. Find R and α. (EC)

23 The equation of motion of a vibrating galvanometer is given by

$$I\frac{d^2\theta}{dt^2} + \alpha\frac{d\theta}{dt} + c\,\theta = Gi_m\cos(\omega t)$$

where I, α, c and G are positive constant factors of the galvanometer and i_m is the constant amplitude of the alternating current flowing.

(i) Establish the complementary function for this differential equation and give one sketch to indicate the behaviour of this function when $\alpha^2 - 4Ic > 0$ and another when $\alpha^2 - 4Ic < 0$.

(ii) Find the particular integral. Express this in the form $R\cos(\omega t - \phi)$ and show that

$$R = \frac{Gi_m}{\sqrt{(c - I\omega^2)^2 + \omega^2\alpha^2}}$$

(iii) For given particular values of c, I, and α what value of ω^2 will ensure that R is a maximum? (EC)

24 Find the general solution of the differential equation

$$\frac{d^2y}{dx^2} + 5\frac{dy}{dx} + 6y = 6x + 13\sin 2x$$

Find the particular solution which satisfies $y = 0$ and $dy/dx = 0$ when $x = 0$.
Describe very briefly the behaviour of the solution as x increases without bound. (EC)

25 A beam of length l is clamped horizontally at each end, so that $y = 0$ and $dy/dx = 0$ at $x = 0$ and also at $x = l$. The loading on the beam is given by

$$w(x) = 2 + 2\sin\left(\frac{\pi x}{l}\right)$$

and the differential equation satisfied at any point of the beam is

$$EI\frac{d^4y}{dx^4} = w(x)$$

in which E and I are constants. Find the particular solution satisfying the given end conditions. (EC)

Section 23.5 (Unless you use a computer, advance two steps only)

26 If, in the example on page 633, we assume air damping to be proportional to the square of the angular velocity of the pendulum, then the governing d.e. is

$$\frac{d^2\theta}{dt^2} + 2k\dot{\theta}^2 + \frac{g}{l}\sin\theta = 0$$

Evaluate numerically for $t = 0\ (0.05)\ 2.5$ minutes, θ and $\dot{\theta}$.

27 In the circuit shown on the next page a coil is wound around an iron core; the core is in series with a resistance and an e.m.f. source. The self-inductance of the coil is L and the non-linearity of the magnetisation of the iron leads to the d.e.

$$\frac{d\phi}{dt} = E - 5\phi - 0.03\phi^3$$

where ϕ is the flux in the core and t is in milliseconds. Take $E = 30$ and solve.

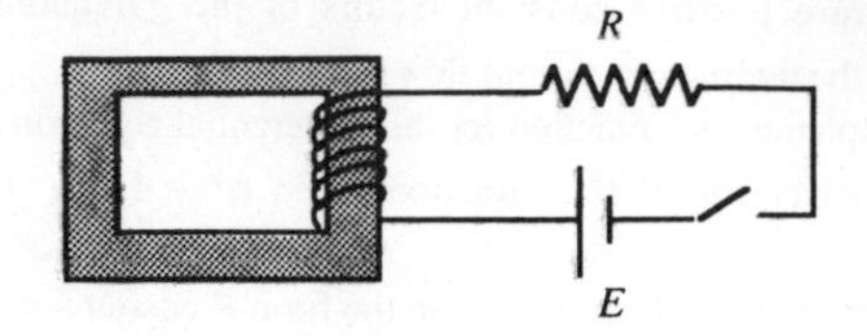

28 Van der Pol's equation, which arises in electronics, is

$$\ddot{x} + (1 - x^2)\dot{x} + x = 0$$

With initial conditions $x = 0.5$, $\dot{x} = 0$ at $t = 0$, solve for $x, \dot{x}$ and $\ddot{x}$ at $t = 0\ (0.1)\ 4$.

29 A simple free vibrating system has a mass subjected to *Coulomb Friction* so that the equation of motion is

$$m\ddot{x} + n^2 x = \begin{cases} -A, & \dot{x} > 0 \\ +A, & \dot{x} < 0 \end{cases} \quad \text{where } A \text{ is a constant}$$

At $t = 0, \dot{x} = 0, x = 3$ m. Take $m = 1$, $n = 0.8$ and $A = 2$ and solve for $t = 0(0.1)5s$.

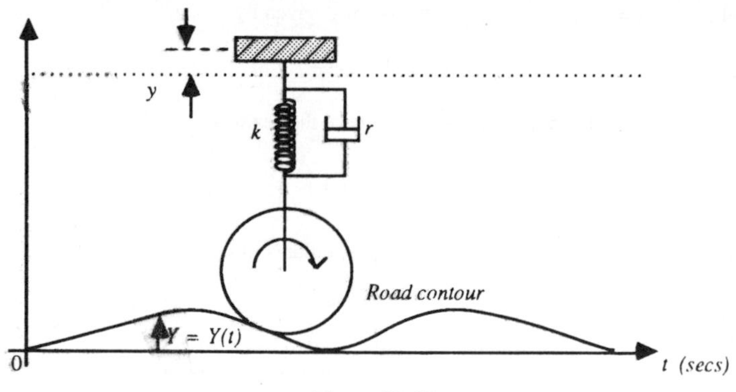

Figure 23.12

30 Figure 23.12 shows a wheel of a vehicle suspension which is connected to the body by a spring of stiffness coefficient k with a damping system (coefficient r) and it can be shown that the displacement of the body, when the wheel travels along a road of contour $Y = Y(t)$ is governed by

$$M\frac{d^2y}{dt^2} = -k(y - Y) - r\left(\frac{dy}{dt} - \frac{dY}{dt}\right)$$

where $y = 0$ and $dy/dt = 0$ when $t = 0$.

Take suitable values for M, k and r with $Y = {}^1/2\ (1 - \cos 3t)$ and solve for $t = 0\ (0.1)10$ seconds.

Section 23.6

31 A beam of length l is freely supported at its ends and weighs w kg/unit length. The differential equation governing the deflection of the beam is

$$EI\frac{d^2y}{dx^2} = w\left[\frac{x^2}{2} - \frac{lx}{2}\right]$$

Solve this equation by integration given that $y = 0$ when $x = 0$ and $y = 0$ when $x = l$.

32 Figure 23.13 shows a very long cooling fin of thickness t and length l attached to a hot wall which is maintained at a temperature of 100°C.

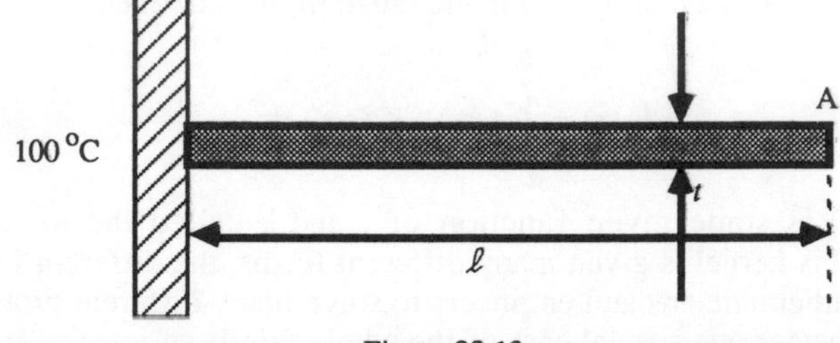

Figure 23.13

Heat is conducted steadily along the fin and is lost from the sides by convection to the surrounding air (which is at a tempertature of 70°C). The fin temperature θ, assumed to depend only on the distance x along the fin, obeys the differential equation

$$kt\frac{d^2\theta}{dx^2} = 2h\,(\theta - \theta_{air})$$

where k and h are constants.

We assume that the fin is long enough that the end A is at the same temperature as the surrounding air, so that we have the boundary conditions $\theta = 100$ when $x = 0$; $\theta = 70$ when $x \to \infty$.

Solve the equation, given that $h = 2\ \mathrm{W/m^2K}$, $k = 20$ W/mK, $t = 0.75$ cm (Beware of the boundary condition $\theta = 70$ when $x \to \infty$).

24

LAPLACE TRANSFORMS

Integral Transform methods have been used increasingly in recent times in the solution of differential equations, both ordinary and partial. The integral transform $F(s)$ of a function $f(t)$ of t in the range $[a, b]$ is defined as

$$F(s) = \int_a^b f(t) \,.\, K(s,t)\, \mathrm{d}t$$

where $K(s,t)$ is some given function of t and s called the **kernel** of the transform. This kernel is given many different forms, the different forms being chosen by mathematicians and engineers to solve many different problems. We shall only consider one special case of the whole family of integral transforms - we choose $K(s,t)$ to be e^{-st} and the limits of integration as 0 and ∞. $F(s)$ is then called the **Laplace Transform** of $f(t)$. It is particularly useful in the solution of ordinary differential equations where the initial conditions (at $t = 0$) are given. We shall first of all have to study the properties of the Laplace transform before we can begin to apply it in the solution of differential equations.

24.1 The Laplace Transform

As indicated above we define the Laplace Transform $F(s)$ of a function $f(t)$ by the relation

$$F(s) = \int_0^\infty \mathrm{e}^{-st} f(t)\mathrm{d}t \qquad (24.1)$$

it being assumed that the integral exists. (This places restrictions on s which we do not deal with here.)

The idea behind Laplace transforms is that we replace the process of integration by one of algebraic manipulation of linear equations. Just as the logarithmic function has an inverse which is a function, so Laplace transforms have a unique inverse. Further, there are tables of transforms just as there are tables of logarithms. But note that "logarithm" is a *function* operating on *numbers* whereas "Laplace Transform" is an *operator* applied to *functions*.

We shall now build up a table of transforms of given functions of t.

Example 1 $f(t) = 1$

$$F(s) = \int_0^\infty e^{-st} \,.\, 1 \,.\, dt = \left[-\frac{e^{-st}}{s} \right]_0^\infty = \frac{1}{s}$$

it being assumed that $s > 0$ in order that the integral exists.

Example 2 $f(t) = t$

$$F(s) = \int_0^\infty e^{-st}\, t \; dt = \left[-\frac{t\, e^{-st}}{s} \right]_0^\infty + \frac{1}{s} \int_0^\infty e^{-st}\, dt \qquad \text{(by parts)}$$

$$= 0 + \frac{1}{s} \,.\, \frac{1}{s} = \frac{1}{s^2} \qquad \text{(again } s > 0\text{)}$$

Example 3 $f(t) = t^2$

$$F(s) = \int_0^\infty e^{-st}\, t^2\, dt$$

We could complete this integral by integration by parts, but consider the following technique. We know that

$$\frac{1}{s^2} = \int_0^\infty e^{-st}\, t\; dt$$

If we differentiate both sides partially with respect to s we have

$$-\frac{2}{s^3} = \frac{\partial}{\partial s} \int_0^\infty e^{-st}\, t\; dt$$

We shall assume that

$$\frac{\partial}{\partial s} \int_0^\infty e^{-st}\, t\; dt = \int_0^\infty \frac{\partial}{\partial s} (e^{-st}\, t)\, dt$$

This gives

$$\int_0^\infty -t\, e^{-st}\, t\; dt = -\int_0^\infty e^{-st}\, t^2\; dt$$

Hence
$$-\frac{2}{s^3} = -\int_0^\infty e^{-st} t^2 \, dt$$

giving
$$F(s) = \int_0^\infty e^{-st} t^2 \, dt = \frac{2}{s^3} \qquad (s > 0)$$

Example 4 $f(t) = t^n$
We differentiate the result of Example 1 n times with respect to the parameter s to obtain

$$\frac{\partial^n}{\partial s^n} \int_0^\infty e^{-st} \, dt = \frac{\partial^n}{\partial s^n} \left(\frac{1}{s}\right)$$

so that
$$\int_0^\infty (-t)^n e^{-st} \, dt = (-1)^n \frac{n!}{s^{n+1}}$$

Cancelling the factor $(-1)^n$, we obtain

$$F(s) = \int_0^\infty e^{-st} t^n \, dt = \frac{n!}{s^{n+1}} \qquad (n \geq 0, s > 0)$$

Example 5 $f(t) = e^{at}$

$$F(t) = \int_0^\infty e^{-st} e^{at} \, dt = \int_0^\infty e^{-(s-a)t} \, dt = \left[-\frac{e^{-(s-a)t}}{(s-a)}\right]_0^\infty$$

$$= \frac{1}{s-a} \qquad (\text{provided } s > a)$$

The Linear Property
If c_1 and c_2 are any constants and $f_1(t)$ and $f_2(t)$ are functions with Laplace transforms $F_1(s)$ and $F_2(s)$ respectively, then the Laplace transform of the function $c_1 f_1(t) + c_2 f_2(t)$ is
$$c_1 F_1(s) + c_2 F_2(s)$$

Example 6 $f(t) = \cos kt$

Now $$\cos kt = \tfrac{1}{2}(e^{ikt} + e^{-ikt})$$

Hence $$F(s) = \mathfrak{L}\left\{\frac{1}{2}(e^{ikt} + e^{-ikt})\right\}$$

$$= \mathfrak{L}(\tfrac{1}{2}e^{ikt}) + \mathfrak{L}(\tfrac{1}{2}e^{-ikt})$$

$$= \frac{1}{2}\frac{1}{(s-ik)} + \frac{1}{2}\frac{1}{(s+ik)} \qquad \text{(using Example 5)}$$

$$= \frac{1}{2}\frac{2s}{s^2+k^2} = \frac{s}{s^2+k^2}$$

($\mathfrak{L}$ stands for Laplace transform.)

Example 7 $f(t) = \cosh kt$

Since $$\cosh kt = \tfrac{1}{2}(e^{kt} + e^{-kt})$$

then $$F(s) = \frac{1}{2}\left(\frac{1}{s-k} + \frac{1}{s+k}\right) = \frac{s}{s^2-k^2}$$

Notice that we could have said $\cosh kt = \cos ikt$ Hence using Example 6

$$F(s) = \frac{s}{s^2+(ik)^2} = \frac{s}{s^2-k^2}$$

We can now write a table of Laplace transforms. (You should prove any not previously obtained.)

Table 24.1

$f(t)$	$F(s)$	$f(t)$	$F(s)$
1	$\frac{1}{s}$	$\sin kt$	$\frac{k}{s^2+k^2}$
t^n	$\frac{n!}{s^{n+1}}$	$\cosh kt$	$\frac{s}{s^2-k^2}$
e^{at}	$\frac{1}{s-a}$	$\sinh kt$	$\frac{k}{s^2-k^2}$
$\cos kt$	$\frac{s}{s^2+k^2}$		

24.2 Further Transforms

We can extend the table of transforms by employing two further theorems.

Translation or Shift Theorem

This states that *if $F(s)$ is the Laplace transform of a function $f(t)$, then the Laplace transform of* $e^{+at}f(t)$ *is $F(s-a)$. In other words, s is replaced by $(s-a)$ in the Laplace transform of $f(t)$.*

Example 1

We have seen that the L.T. of 1 is $1/s$ and that of e^{at} is $1/(s-a)$.

Example 2 Find the L.T. of $e^{-t}\cos 3t$

If
$$f(t) = \cos 3t$$

then
$$F(s) = \frac{s}{s^2+9}$$

Thus for $e^{-t}\cos 3t$ (with $a = -1$) the Laplace transform is

$$F(s+1) = \frac{s+1}{(s+1)^2+9}$$

Example 3 Find the L.T. of $t^3 e^{2t}$

If
$$f(t) = t^3$$

then
$$F(s) = \frac{3!}{s^4}$$

Thus, using the theorem, the L.T. of $t^3 e^{2t} = F(s-2) = \dfrac{3!}{(s-2)^4}$

Example 4 Find the L.T. of $e^{-2t}(3\cosh 4t - 2\sinh 4t)$

Taking $f(t) = 3\cosh 4t - 2\sinh 4t$, then

$$F(s) = \frac{3s}{s^2-16} - 2.\frac{4}{s^2-16} = \frac{3s-8}{s^2-16}$$

Thus, the L.T. of $e^{-2t}(3\cosh 4t - 2\sinh 4t)$ is

$$\frac{3(s+2)-8}{(s+2)^2-16} = \frac{3s-2}{s^2+4s-12}$$

Proof of the Theorem

By definition $$F(s) = \int_0^\infty f(t)\,e^{-st}\,dt$$

The Laplace Transform of $f(t)\,e^{at}$ is

$$I = \int_0^\infty f(t)\,e^{at}\,e^{-st}\,dt = \int_0^\infty f(t)\,e^{-(s-a)t}\,dt$$

Now put $S = s - a$, then

$$I = \int_0^\infty f(t)\,e^{-St}\,dt = F(S) = F(s-a)$$

"Multiplication by t^n" Theorem

This states that *if $F(s)$ is the Laplace transform of $f(t)$, then the L.T. of $t^n f(t)$ is*

$$(-1)^n \frac{d^n F(s)}{ds^n}$$

To make the theorem clear, we take a few examples.

Example 1 Find the L.T. of $t^3 e^{2t}$

Taking $f(t) = e^{2t}$, then $F(s) = 1/(s-2)$. Thus, using the theorem

$$\mathcal{L}\{t^3 e^{2t}\} = (-1)^3 \frac{d^3}{ds^3}\left[\frac{1}{s-2}\right] = -\frac{d^2}{ds^2}\left[\frac{-1}{(s-2)^2}\right] = -\frac{d}{ds}\left[\frac{2}{(s-2)^3}\right] = \frac{6}{(s-2)^4}$$

the same result as obtained earlier by another method.

Example 2 Find the L.T. of $t \sin 2t$

Taking $f(t) = \sin 2t$, then $$F(s) = \frac{2}{s^2+4}$$

thus the L.T. of $t \sin 2t$ is

$$(-1)\frac{d}{ds}\left(\frac{2}{s^2+4}\right) = \frac{4s}{(s^2+4)^2}$$

Example 3 Find the L.T. of $te^{-2t}\cos 3t$

We need to use both theorems here.

Taking, first of all, $f(t) = \cos 3t$, then $F(s) = \dfrac{s}{s^2 + 9}$

Thus the L.T. of $e^{-2t}\cos 3t$ is $\dfrac{s+2}{(s+2)^2+9} = \dfrac{s+2}{s^2+4s+13}$

Now use the second theorem, and the L.T. of $te^{-2t}\cos 3t$

$$= (-1)\frac{d}{ds}\left(\frac{s+2}{s^2+4s+13}\right)$$

$$= -\frac{(s^2+4s+13)-(s+2)(2s+4)}{(s^2+4s+13)^2}$$

$$= \frac{s^2+4s-5}{(s^2+4s+13)^2}$$

The table of Laplace transforms is extended as follows. Those that we have not proved you should obtain by any convenient method.

Table 24.2

$f(t)$	$F(s)$	$f(t)$	$F(s)$
$e^{at}\cos kt$	$\dfrac{(s-a)}{(s-a)^2+k^2}$	te^{-at}	$\dfrac{1}{(s+a)^2}$
$e^{at}\sin kt$	$\dfrac{k}{(s-a)^2+k^2}$	$e^{at}f(t)$	$F(s-a)$
$t\cos kt$	$\dfrac{(s^2-k^2)}{(s^2+k^2)^2}$	$t^n f(t)$	$(-1)^n\dfrac{d^nF(s)}{ds^n}$
$t\sin kt$	$\dfrac{2ks}{(s^2+k^2)^2}$		

24.3 Solution of Ordinary Differential Equations

Laplace Transform of Derivatives

If we use Laplace transforms to solve differential equations we shall need to know the Laplace transforms of derivatives of $f(t)$. These are obtained as follows. The L.T. of df/dt is

$$\int_0^\infty e^{-st}\frac{df}{dt}\,dt = \left[e^{-st}f\right]_0^\infty + s\int_0^\infty e^{-st}f(t)\,dt = -f(0) + sF(s) \quad (24.2a)$$

where $f(0)$ as usual denotes the value of $f(t)$ when $t = 0$ and we have assumed

that $$\lim_{t\to\infty}(e^{-st}f) = 0$$

This is usually the case for physical problems for suitable values of s.

Similarly the L.T. of d^2f/dt^2 is

$$\int_0^\infty e^{-st}\frac{d^2f}{dt^2}\,dt = \left[e^{-st}\frac{df}{dt}\right]_0^\infty + s\int_0^\infty e^{-st}\frac{df}{dt}\,dt$$

$$= -f'(0) + s[-f(0) + sF(s)]$$

$$= -f'(0) - sf(0) + s^2F(s) \quad (24.2b)$$

$f'(0)$ denoting the value of df/dt at $t = 0$. We have assumed that

$$\lim_{t\to\infty}\left(e^{-st}\frac{df}{dt}\right) = 0$$

In general, assuming that

$$\lim_{t\to\infty}\left(e^{-st}\frac{d^{r-1}f}{dt^{r-1}}\right) = 0 \quad \text{for } r = 1,2,\ldots, n,$$

the L.T. of d^nf/dt^n is

$$-f^{n-1}(0) - sf^{n-2}(0) - \ldots. - s^{n-1}f(0) + s^nF(s) \quad (24.2c)$$

Example

We study the circuit of Figure 24.1 with inductance L, resistance R and a constant electromotive force E, with zero current at $t = 0$.

The equation governing the current i as a function of time t is

$$L\frac{di}{dt} + Ri = E, \quad \text{given } i = 0 \text{ when } t = 0$$

Writing down the Laplace transform of each term we get the following equation

$$L[-i(0) + sI(s)] + RI(s) = \frac{E}{s}$$

where $I(s)$ is the L.T. of $i(t)$.

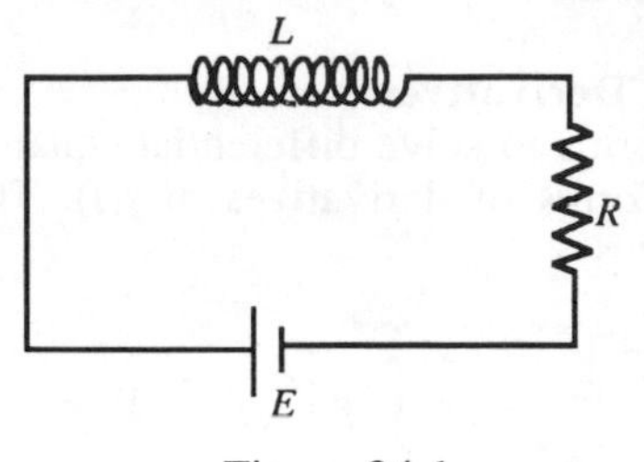

Figure 24.1

We are given $i(0) = 0$ and so

$$I(s)[Ls + R] = \frac{E}{s}$$

Hence
$$I(s) = \frac{E}{s(Ls + R)}$$

This is the L.T. of $i(t)$ and we should like to find its inverse to find the current at any time. We look at our table and ask - what function gives this L.T.? There is none there just like this but suppose we put the R.H.S. into partial fractions.

$$I(s) = \frac{\frac{E}{R}}{s} - \frac{\frac{LE}{R}}{Ls + R}$$

We can see in the table that these are L.T.'s of the form $1/s$ and $1/(s - a)$ so we write

$$I(s) = \frac{E}{R} \cdot \frac{1}{s} - \frac{E}{R} \cdot \frac{1}{s + \frac{R}{L}}$$

$1/s$ is the transform of unity and $1/[s + (R/L)]$ is the transform of $e^{-(R/L)t}$. Hence the inverse is

$$i(t) = \frac{E}{R} - \frac{E}{R}e^{-(R/L)t} = \frac{E}{R}\left[1 - e^{-(R/L)t}\right]$$

Thus we have solved the differential equation. The method has not involved integration (this was taken care of by writing down the Laplace transforms since in effect we multiplied by e^{-st} and integrated from 0 to ∞) and the technique reduced to one of algebraic manipulation.

Example 1

Solve
$$\frac{d^2y}{dt^2} + 4y = 12t$$

given that $y = 0$ and $dy/dt = 9$ when $t = 0$.

Writing the Laplace transform of each term we get

$$-y'(0) - sy(0) + s^2 Y(s) + 4Y(s) = 12.\frac{1}{s^2}$$

where $Y(s)$ is the L.T. of $y(t)$. Now $y'(0) = 9$ and $y(0) = 0$, so we obtain

$$(s^2 + 4)\,Y(s) = \frac{12}{s^2} + 9$$

or

$$Y(s) = \frac{9s^2 + 12}{s^2(s^2 + 4)} = \frac{3}{s^2} + \frac{6}{s^2 + 4}$$

$$= 3.\frac{1}{s^2} + 3.\frac{2}{s^2 + 2^2}$$

Hence

$$y(t) = 3t + 3\sin 2t$$

Example 2

Solve

$$\frac{d^2y}{dt^2} - 4\frac{dy}{dt} + 5y = e^{2t}$$

given that $y = 0$ and $dy/dt = 0$ when $t = 0$.

Writing down the Laplace transforms we get

$$-y'(0) - sy(0) + s^2 Y(s) - 4[-y(0) + sY(s)] + 5Y(s) = \frac{1}{s-2}$$

so that

$$(s^2 - 4s + 5)\,Y(s) = \frac{1}{s-2}$$

Hence

$$Y(s) = \frac{1}{(s-2)(s^2 - 4s + 5)} = \frac{1}{s-2} - \frac{s-2}{s^2 - 4s + 5}$$

The term

$$\frac{s-2}{s^2 - 4s + 5} = \frac{s-2}{(s-2)^2 + 1}$$

which is of the form $S/(S^2 + 1^2)$ with $S = s - 2$.

Remember that the shift theorem replaces s by $(s - 2)$ when the function is multiplied by e^{2t}. The inverse function of $S/(S^2 + 1^2)$ is $\cos t$, so that the inverse of $(s - 2)/[(s - 2)^2 + 1^2]$ is $e^{2t}\cos t$. Thus we obtain

$$y(t) = e^{2t} - e^{2t}\cos t$$

Example 3

Solve $$\frac{d^3x}{dt^3} - \frac{d^2x}{dt^2} + 2x = e^{-t}$$

given that $x = 1$, $dx/dt = 0$ and $d^2x/dt^2 = -2$ when $t = 0$.
Taking the Laplace transform of each term gives

$$-x''(0) - sx'(0) - s^2x(0) + s^3X(s) - [-x'(0) - sx(0) + s^2X(s)] + 2X(s) = \frac{1}{s+1}$$

which reduces to

$$(s^3 - s^2 + 2)X(s) = \frac{1}{s+1} + s^2 - s - 2 = \frac{s^3 - 3s - 1}{s+1}$$

Hence

$$X(s) = \frac{s^3 - 3s - 1}{(s+1)(s^3 - s^2 + 2)} = \frac{s^3 - 3s - 1}{(s+1)^2(s^2 - 2s + 2)}$$

$$= \frac{4}{25}\frac{1}{s+1} + \frac{1}{5}\frac{1}{(s+1)^2} + \frac{21}{25}\frac{(s-1)}{(s-1)^2 + 1^2} - \frac{22}{25}\frac{1}{(s-1)^2 + 1^2}$$

The inverse of each term then gives

$$x(t) = \frac{4}{25}e^{-t} + \frac{1}{5}te^{-t} + \frac{21}{25}e^{t}\cos t - \frac{22}{25}e^{t}\sin t$$

You will have noticed that in finding the inverses we are looking in the table of transforms and trying to find something similar to that which we have got. Then by completing the square in the denominator and other manipulation we suitably modify our terms. In the last example, after putting $X(s)$ into partial fractions, we have a term

$$\frac{1}{25}\frac{(21s - 43)}{(s^2 - 2s + 2)} = \frac{1}{25}\frac{(21s - 43)}{(s-1)^2 + 1}$$ (on completing the square in the denominator)

The nearest we have to this are the forms

$$\frac{(s-a)}{(s-a)^2 + k^2} \quad \text{and} \quad \frac{k}{(s-a)^2 + k^2}$$

which appear in our table as transforms of $e^{at}\cos kt$ and $e^{at}\sin kt$ respectively. Hence we split up the numerator to get such forms as

$$\frac{1}{25}\frac{21(s-1) - 22}{(s-1)^2 + 1^2} = \frac{21}{25}\frac{s-1}{(s-1)^2 + 1} - \frac{22}{25}\frac{1}{(s-1)^2 + 1^2}$$

One further point when using Laplace transforms: always check that your solution satisfies the boundary conditions on the problem.

24.4 Comparison of Methods of Solution

Consider the following problem which we shall solve by each method in turn: a mass m rests on a horizontal table and is attached to one end of a light spring which, when stretched, exerts a tension $m\omega^2$ times the extension (ω is constant) – refer to Figure 24.2. The other end of the spring is now moved with constant velocity u along the table in a direction away from the mass. The table offers a frictional resistance on the mass of amount mk times its speed (k is constant). Show that if $k = 2\omega$ then the extension of the spring after time t is

$$x = \frac{u}{\omega}[2 - (2 + \omega t)\,\mathrm{e}^{-\omega t}]$$

The position of the mass after time t is $(l + x) - ut$, the speed after time t is $\dot{x} - u$., and the acceleration after time t is $\ddot{x}$

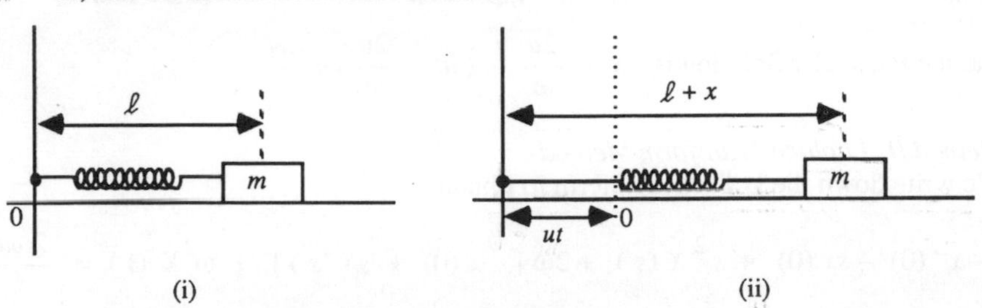

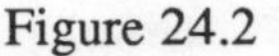
Figure 24.2

Hence $\quad m\ddot{x} = -$ (Resistance + Tension) $= -[mk(\dot{x} - u) + m\omega^2 x]$

Therefore $$\ddot{x} + k\dot{x} + \omega^2 x = ku$$

Since $k = 2\omega$, then $$\ddot{x} + 2\omega\dot{x} + \omega^2 x = 2\omega u$$

and $x = 0$ when $t = 0$, $\dot{x} = u$ when $t = 0$ since speed $= 0$ when $t = 0$.

Method I: Trial Method

$$\ddot{x} + 2\omega\dot{x} + \omega^2 x = 2\omega u$$

For the C.F. $(m^2 + 2\omega m + \omega^2) = 0$, i.e. $(m + \omega)^2 = 0$

Then the C.F. is $\quad x = (At + B)\mathrm{e}^{-\omega t}$

For the P.I. we try $x = C$ giving $\quad 0 + 0 + \omega^2 C = 2\omega u$

Hence $$C = \frac{2u}{\omega}$$

Therefore $\quad \text{P.I.} = \dfrac{2u}{\omega}$

Thus the solution is $\quad x = (At + B)\,e^{-\omega t} + \dfrac{2u}{\omega}$

We know that $x = 0$ when $t = 0$ which implies that $0 = B + 2u/\omega$

hence $\quad B = -\dfrac{2u}{\omega}$

Also $\dot{x} = u$ when $t = 0$.

Now $\quad \dot{x} = -\omega(At + B)\,e^{-\omega t} + A\,e^{-\omega t} \Rightarrow u = -\omega B + A$

Thus $\quad A = u + \omega B = -u$

and the Particular Solution is $\quad x = \dfrac{2u}{\omega} - \left(ut + \dfrac{2u}{\omega}\right)e^{-\omega t}$

Method II: Laplace Transform Method
We write down the L.T. of each term to obtain

$$[-x'(0) - sx(0) + s^2X(s)] + 2\omega[-x(0) + sX(s)] + \omega^2X(s) = \frac{2\omega u}{s}$$

Hence $\quad (s^2 + 2\omega s + \omega^2)X(s) = \dfrac{2\omega u}{s} + u = \dfrac{u(2\omega + s)}{s}$

Thus $\quad X(s) = \dfrac{u(2\omega + s)}{s(s+\omega)^2} = \dfrac{2u/\omega}{s} - \dfrac{u}{(s+\omega)^2} - \dfrac{2u/\omega}{(s+\omega)}$

The inverse is, as before, $\quad x = \dfrac{2u}{\omega} - ut\,e^{-\omega t} - \dfrac{2u}{\omega}e^{-\omega t}$

Discussion
There is little difference in the length and difficulty of the solutions but we observe the following:

> The Laplace Transform Method eliminates the need to find the constants A and B. However it does involve us in finding the partial fractions which is not such an easy task. Nevertheless it is probably the best method to try if we are given the *initial conditions* as was the case here.

Since this was an initial value problem we could have solved the equation by the method given in Section 23.5. As stated earlier, the analytical approach should be used wherever possible unless we have to so simplify the model of the particular physical situation that the results produced give a false picture.

Example

Let us consider one further example. A body is dropped from a great height above the Earth and falls freely. It is a fair approximation to say that the air resistance is proportional to the square of the velocity. Hence the equation of motion is

mass × acceleration = mass × acceleration due to gravity – mass × k × (velocity)^2

↑ net force ↑ gravitational force ↑ air resistance (k constant)

Hence measuring y downwards from the starting position we get

$$\frac{d^2y}{dt^2} = g - k\left(\frac{dy}{dt}\right)^2$$

That is,

$$\frac{d^2y}{dt^2} + k\left(\frac{dy}{dt}\right)^2 = g$$

where $y = 0$ when $t = 0$, and $\dfrac{dy}{dt} = 0$ when $t = 0$.

This is a non-linear second order differential equation because of the presence of the term $(dy/dt)^2$. Hence we cannot use the Laplace transform method of solution.

Furthermore the trial method does not easily yield a solution. We would be led therefore to use the Runge-Kutta method or some other suitable numerical method. However an analytical solution is possible – we can write the equation in the form

$$\frac{d}{dt}\left(\frac{dy}{dt}\right) + k\left(\frac{dy}{dt}\right)^2 = g$$

which is

$$\frac{dv}{dt} + kv^2 = g \qquad \left[v = \frac{dy}{dt}\right]$$

and it is possible to integrate this equation by separation of variables giving

$$\ln\left[\frac{1 + \sqrt{\frac{k}{g}}\,v}{1 - \sqrt{\frac{k}{g}}\,v}\right] = 2\sqrt{gk}\,t$$

We can rearrange this to get

$$v = \sqrt{\frac{g}{k}}\left\{\frac{e^{2\sqrt{gk}\,t} - 1}{e^{2\sqrt{gk}\,t} + 1}\right\}$$

and we can integrate both sides with respect to t to find y in terms of t.

However this is by no means easy.

An alternative is to write d^2y/dt^2 as

$$\frac{d}{dy}\left(\frac{dy}{dt}\right) \cdot \frac{dy}{dt} = v\,\frac{dv}{dy}$$

The equation then becomes $\quad v\dfrac{dv}{dy} + kv^2 = g$

or

$$\frac{d}{dy}\left(\frac{v^2}{2}\right) + kv^2 = g$$

giving, by use of the Integrating Factor method

$$v^2 = \frac{g}{k}\,(1 - e^{-2ky})$$

Again it is possible but by no means easy to find y in terms of t from this equation.

It would seem, therefore, that for any second order equation which is non-linear, the most straightforward method is to use a numerical approach.

24.5 Simultaneous Ordinary Differential Equations

In various problems in science and engineering more than one differential equation is required to describe the system. For example, consider the coupled electrical circuits shown in Figure 24.3 where E represents the e.m.f. in a circuit, R the resistance and L the inductance. M is the mutual inductance of the coils and i_1, i_2 represent the currents. We assume that $M^2 < L_1L_2$.

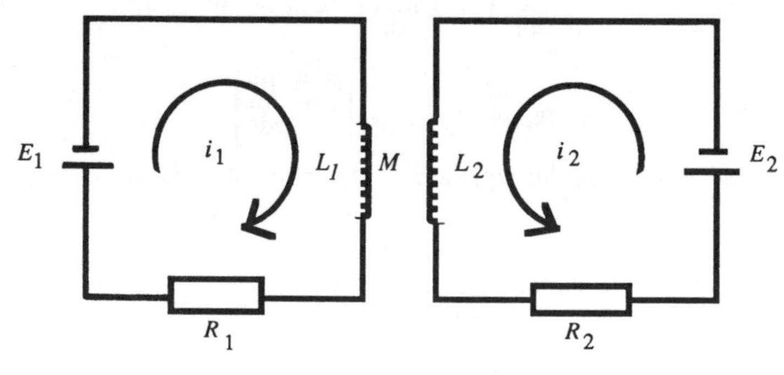

Figure 24.3

We know from Kirchhoff's Laws that

$$R_1 i_1 = E_1 - L_1\frac{di_1}{dt} - M\frac{di_2}{dt}$$

and
$$R_2 i_2 = E_2 - L_2 \frac{di_2}{dt} - M \frac{di_1}{dt}$$

Thus we have a pair of simultaneous differential equations in the two variables i_1 and i_2 each dependent on time.

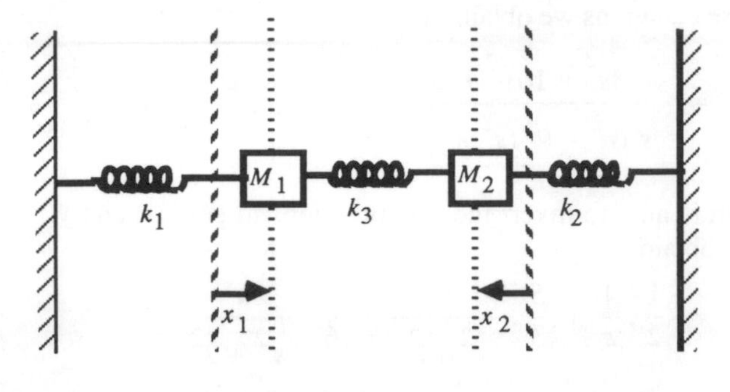

Figure 24.4

Similarly in a mechanical vibration system such as the one shown in Figure 24.4, where k_1,k_2,k_3 are the elastic constants of the springs, we obtain the two simultaneous differential equations

$$M_1 \frac{d^2 x_1}{dt^2} = -k_1 x_1 - k_3 (x_1 + x_2)$$

$$M_2 \frac{d^2 x_2}{dt^2} = -k_2 x_2 + k_3 (x_1 + x_2)$$

to describe the vibrational motion of the masses M_1 and M_2. (The springs at the outside are stretched, the central one is compressed.) The Laplace transform method is particularly useful for solving such problems, if the initial values of the variables are given. The extension of the method to cover such problems will be clear if we take an example.

Example
Solve

$$\frac{d^2 x}{dt^2} - \frac{dy}{dt} + 2x = 1 \qquad \frac{dx}{dt} + \frac{d^2 y}{dt^2} + 2y = 0$$

given that $x = 3$, $y = 0$, $dx/dt = 0$, $dy/dt = 0$ when $t = 0$.
Taking the L.T. of each term we obtain

$$[-x'(0) - sx(0) + s^2X(s)] - [-y(0) + sY(s)] + 2X(s) = 1/s$$

and
$$[-x(0) + sX(s)] + [-y'(0) - sy(0) + s^2Y(s)] + 2Y(s) = 0$$

giving

$$(s^2 + 2)X(s) - sY(s) = \frac{1}{s} + 3s$$

$$sX(s) + (s^2 + 2)Y(s) = 3$$

Solving these equations we obtain

$$X(s) = \frac{3s^4 + 10s^2 + 2}{s(s^2+1)(s^2+4)}, \quad Y(s) = \frac{5}{(s^2+1)(s^2+4)}$$

The problem is now to invert these expressions to get $x(t)$ and $y(t)$. Via partial fractions we obtain

$$X(s) = \frac{1}{2}\cdot\frac{1}{s} + \frac{5}{3}\cdot\frac{s}{s^2+1^2} + \frac{5}{6}\cdot\frac{s}{s^2+2^2}$$

$$Y(s) = \frac{5}{3}\cdot\frac{1}{s^2+1^2} - \frac{5}{6}\cdot\frac{2}{s^2+2^2}$$

we find the inverses are

$$x(t) = \frac{1}{2} + \frac{5}{3}\cos t + \frac{5}{6}\cos 2t \qquad y(t) = \frac{5}{3}\sin t - \frac{5}{6}\sin 2t$$

24.6 Further Forcing Terms

In this section we consider some more forcing terms and examine how to build up our stock of techniques.

(a) Unit Step Function

Figure 24.5(a) depicts the unit step function defined by

$$u(t) = \begin{cases} 0, & t < 0 \\ 1, & t \geq 0 \end{cases} \quad \text{whilst} \quad u(t-a) = \begin{cases} 0, & t < a \\ 1, & t \geq a \end{cases}$$

the generalised step function, is shown in Figure 24.5(b).
The Laplace transform of $u(t-a)$ is found directly to be

$$\int_0^\infty e^{-st} u(t-a)\, dt = \int_a^\infty e^{-st}\, dt = \frac{1}{s} e^{-as} \qquad (24.3)$$

It follows immediately that the transform of $u(t)$ is $1/s$.

Other functions can be expressed in terms of step functions. For example, the square pulse shown in Figure 24.6(a) is defined by

$$f(t) = K[u(t-a) - u(t-b)]$$

You can see this more easily if you sketch $-u(t-b)$.

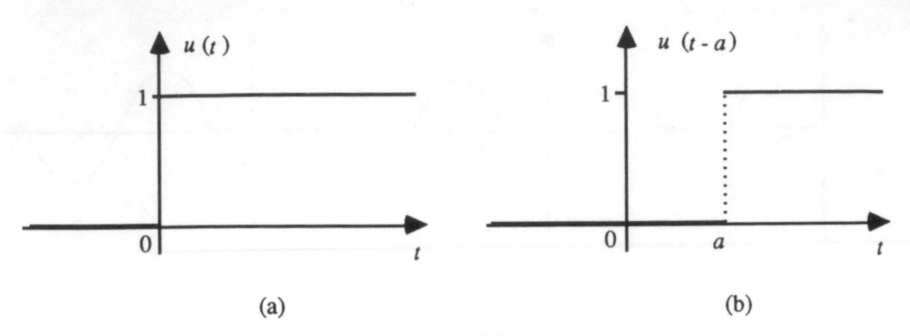

Figure 24.5

The square wave shown in Figure 24.6(b) is defined by

$$g(t) = u(t) - 2u(t-a) + 2u(t-2a) - 2u(t-3a) + \dots$$

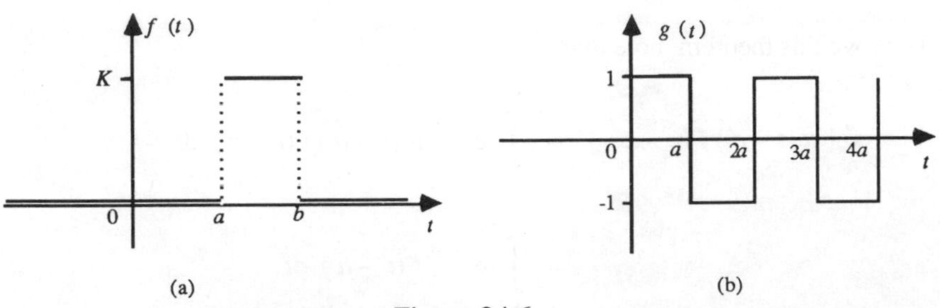

Figure 24.6

Verify that the transform of $f(t)$ is $F(s) = \frac{K}{s}(e^{-as} - e^{-bs})$

and that the transform of $g(t)$ is

$$G(s) = \frac{1}{s} - \frac{2}{s}e^{-as} + \frac{2}{s}e^{-2as} - \frac{2}{s}e^{-3as} + \dots$$

$$= \frac{1}{s} - \frac{2}{s}e^{-as} \cdot \frac{1}{1+e^{-as}} = \frac{1}{s}\frac{(1-e^{-as})}{(1+e^{-as})}$$

(b) **Step functions combined with other functions**

The function $\sin t$ is combined with $u(t - \pi/2)$ to form $f(t) = u(t - \pi/2)\sin t$, which is graphed in Figure 24.7(a); the effect has been to *suppress* $\sin t$ for $t < \pi/2$. However, if we consider $g(t) = u(t - \pi/2)\,\sin(t - \pi/2)$ a different effect emerges. Firstly, the function $\sin(t - \pi/2)$ is shown in Figure 24.7(b): it is merely the sine function shifted through $\pi/2$. The graph of $g(t)$ is displayed in Figure 26.7(c) and you can see that we have essentially shifted the graph of $\sin t$, $t > 0$, through $\pi/2$.

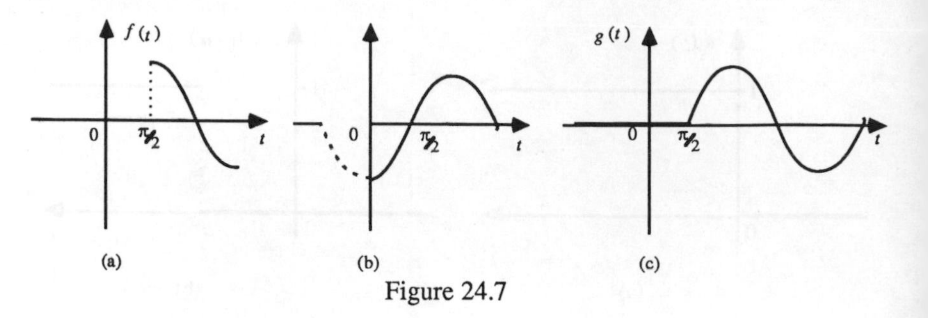

Figure 24.7

To find the transform of $g(t)$ we make use of the following theorem.

Second Shift Theorem

If $\mathcal{L}\{f(t)\} = F(s)$ then $\mathcal{L}\{u(t-a)f(t-a)\} = e^{-as}F(s)$ (24.4)

To prove this theorem, note that

$$\mathcal{L}\{u(t-a)f(t-a)\} = \int_0^\infty e^{-st}u(t-a)f(t-a)\,dt$$

$$= \int_a^\infty e^{-st}f(t-a)\,dt$$

$$= \int_0^\infty e^{-s(v+a)}f(v)\,dv \qquad (v = t-a)$$

$$= e^{-as}\int_0^\infty e^{-sv}f(v)\,dv = e^{-as}F(s)$$

Example

A series RC circuit has an applied voltage of the form shown in Figure 24.8(a). The circuit equation is

$$Ri(t) + \frac{1}{C}\int_0^\tau i(\tau)\,d\tau = v(t)$$

where $i(t)$ is the current in the circuit and $v(t)$ is the applied voltage which can be specified as $v(t) = v_0\,[u(t-a) - u(t-b)]$

Applying Laplace transform and assuming that the initial condition is $i = 0$ at $t = 0$, then we obtain the following result.

$$R\,I(s) + \frac{1}{C}\cdot\frac{1}{s}\,I(s) = v_0\cdot\frac{1}{s}\,[\mathrm{e}^{-as} - \mathrm{e}^{-bs}]$$

so that $$I(s) = \frac{v_0}{R}\cdot\frac{(\mathrm{e}^{-as} - \mathrm{e}^{-bs})}{s + 1/CR}$$

Now, if $$G(s) = \frac{v_0}{R}\cdot\frac{1}{s + 1/CR}$$

we know that $$g(t) = \frac{v_0}{R}\,\mathrm{e}^{-t/RC}$$

Applying the second shift theorem we obtain

$$i(t) = \mathcal{L}^{-1}[\mathrm{e}^{-as}G(s)] - \mathcal{L}^{-1}[\mathrm{e}^{-bs}G(s)]$$

$$= \frac{v_0}{R}\,[\mathrm{e}^{-(t-a)/CR}\,u(t-a) - \mathrm{e}^{-(t-b)/CR}\,u(t-b)]$$

Hence $$i(t) = \begin{cases} 0 & t < a \\ D\,\mathrm{e}^{-t/RC} & a \le t < b \\ (D - E)\,\mathrm{e}^{-t/RC} & t \ge b \end{cases}$$

where $D = \frac{v_0}{R}\,\mathrm{e}^{a/CR}$ and $E = \frac{v_0}{R}\,\mathrm{e}^{b/CR}$

The graph of $i(t)$ is shown in Figure 24.8(b).

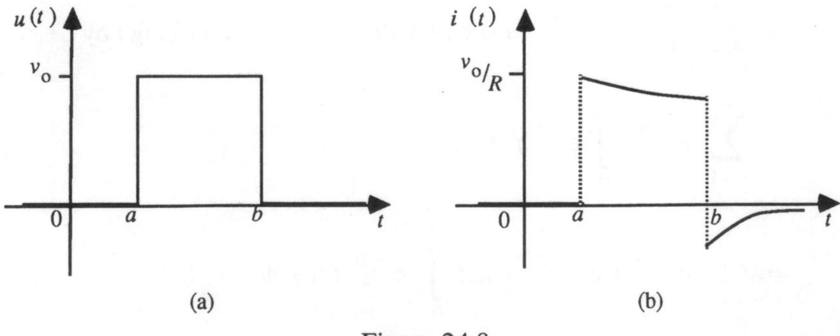

Figure 24.8

(c) **Periodic Functions**

Consider the half-wave rectifier output for an input of of sin ωt; this is shown in Figure 24.9 on the next page.

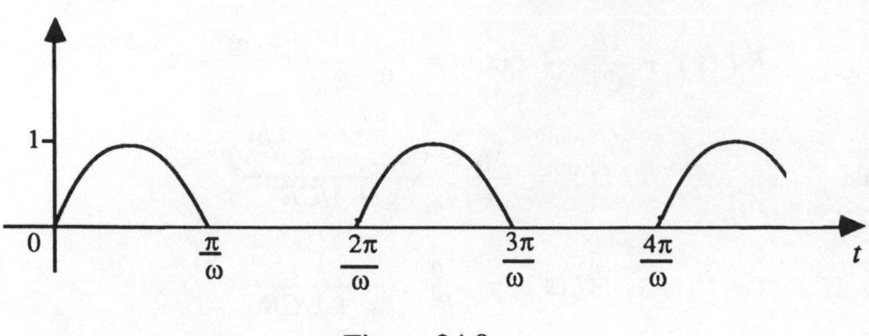

Figure 24.9

The function is periodic, with period $2\pi/\omega$, and is defined by

$$f(t) = \begin{cases} \sin \omega t & 0 \leq t < \pi/\omega \\ 0 & \pi/\omega \leq t < 2\pi/\omega \end{cases}$$

$$f(t + 2\pi/\omega) = f(t)$$

In general, a function of period T defined on $[0,T]$ by $f(t)$ can be specified fully via $f(t + T) = f(t)$. Then

$$F(s) = \int_0^\infty e^{-st} f(t)\, dt = \int_0^T e^{-st} f(t)\, dt + \int_T^{2T} e^{-st} f(t)\, dt + \ldots$$

$$= \sum_{r=0}^{\infty} \int_{rT}^{(r+1)T} e^{-st} f(t)\, dt$$

$$= \sum_{r=0}^{\infty} \int_0^T e^{-s(t+rT)} f(t+rT)\, dt \qquad \text{(replacing } t \text{ by } t + rT\text{)}$$

$$= \sum_{r=0}^{\infty} e^{-srT} \int_0^T e^{-st} f(t)\, dt$$

$$= (1 + e^{-sT} + e^{-2sT} + \ldots) \int_0^T e^{-st} f(t)\, dt$$

i.e.

$$F(s) = \frac{1}{1 - e^{-sT}} \int_0^T e^{-st} f(t)\, dt \qquad (24.5)$$

Hence, for the half-rectified sine wave we obtain the following equation

$$F(s) = \frac{1}{1 - e^{-2\pi s/\omega}} \left[\int_0^{\pi/\omega} e^{-st} \sin \omega t \, dt + \int_{\pi/\omega}^{2\pi/\omega} e^{-st} . 0 \, dt \right]$$

It can be shown readily that

$$I = \int_0^{\pi/\omega} e^{-st} \sin \omega t \, dt = \frac{\omega}{s^2 + \omega^2} [1 + e^{-s\pi/\omega}]$$

so that

$$F(s) = \frac{\omega}{(s^2 + \omega^2)} \frac{[1 + e^{-s\pi/\omega}]}{[1 - e^{-2\pi s/\omega}]} = \frac{\omega}{(s^2 + \omega^2)[1 - e^{-s\pi/\omega}]}$$

It would be instructive for you to derive the transform of $g(t)$ shown in Figure 24.6(b) using result (24.5).

Example
The variable $\theta(t)$ satisfies the differential equation

$$\frac{d^2\theta}{dt^2} + \omega^2\theta = g(t)$$

where $g(t)$ is a periodic function defined by

$$g(t) = \begin{cases} a, & 0 < t < T/2 \\ 0, & T/2 < t < T \end{cases}$$

$$g(t + T) = g(t)$$

Now $$G(s) = \frac{1}{(1 - e^{-sT})} . \frac{a(1 - e^{-sT/2})}{s} = \frac{a}{s[1 + e^{-sT/2}]}$$, applying (24.5)

Suppose that at $t = 0$, $\theta = 0$, $d\theta/dt = 0$
Then if the Laplace transform of $\theta(t)$ is $\Theta(s)$ we find that the transforms of $d\theta/dt$ and $d^2\theta/dt^2$ are $s\,\Theta(s)$ and $s^2\,\Theta(s)$ respectively. Transforming the differential equation we obtain

$$(s^2 + \omega^2)\,\Theta(s) = \frac{a}{s[1 + e^{-sT/2}]}$$

so that

$$\Theta(s) = \frac{a}{s(s^2 + \omega^2)} [1 - e^{-sT/2} + e^{-2sT/2} - e^{-3sT/2} + \dots]$$

A comprehensive Table of Laplace transforms is found in the Appendix.

Problems

Section 24.1

1 Find the transforms of the following functions

(i) $e^{-bt} \sinh at$ (ii) $e^{-bt} \cosh at$

(iii) $\sin(at - b)$ (iv) $(at - \sin at)/a^3$

(v) $(\cos at - \cos bt)/(b^2 - a^2)$ (vi) $t \cos at$

(vii) te^{-2t} (viii) $t^2 e^{-2t}$

Section 24.2

2 Find the transforms of the following functions

(i) $(\sin at - at \cos at)/2a^3$ (ii) $e^{-at} t^{n-1} / (n-1)!$

(ii) $t \cos kt$ iv) $t \sinh kt$

(v) $e^{-2t} \cos kt$ (vi) $(t \sin t)\, e^{-t}$

Section 24.3

3 Solve the following equations with the conditions given at $t = 0$ in each case

(i) $\dfrac{di}{dt} + 2i = 100 \cos t$ given that $i = 0$

(ii) $\dfrac{d^2r}{dt^2} + 4\dfrac{dr}{dt} + 3r = 0$ given that $r = 2$, $dr/dt = 3$

(iii) $\dfrac{d^2x}{dt^2} + 6\dfrac{dx}{dt} + 9x = 0$ given that $x = 2$, $\dot{x} = 4$

(iv) $\dfrac{d^2x}{dt^2} + 9x = t$ given that $x = 1$, $\dot{x} = 2$

(v) $\dfrac{d^2x}{dt^2} + n^2x = \cos nt$ given that $x = x_0$, $\dot{x} = x_1$

(vi) $\dfrac{d^2x}{dt^2} + 4x = 5 \cos 2t$ given that $x = 1$, $\dot{x} = 3$

4 Obtain, in each case, the Laplace transform of the dependent variable and so find the solution

(i) $2\dfrac{d^2y}{dt^2} - 5\dfrac{dy}{dt} - 3y = e^{3t}$ when $y = 1$ and $dy/dt = 0$ at $t = 0$

(ii) $\dfrac{d^2y}{dt^2} + 4\dfrac{dy}{dt} + 8y = 1$ if $y = 0$ and $dy/dt = 0$ at $t = 0$

(iii) $\dfrac{d^2\theta}{dt^2} + 2\dfrac{d\theta}{dt} + \theta = \sin 2t$ if $\theta = d\theta/dt = 0$ at $t = 0$

(iv) $\dfrac{d^2x}{dt^2} + 9x = \cos 3t$ if $x = 2$ and $dx/dt = -5$ when $t = 0$

(v) $\dfrac{dy}{dt} + y = t^2 e^{-t}$ if $y = 10$ when $t = 0$

(vi) $\dfrac{d^2u}{dt^2} + 2\dfrac{du}{dt} + 5u = e^{-t}$ if $u = du/dt = 0$ at $t = 0$

(vii) $\dfrac{d^4y}{dt^4} - k^4 y = 0$ if $y = dy/dt = d^2y/dt^2 = d^3y/dt^3 = 0$ at $t = 0$

Section 24.4

5 Try completing the second example of Section 24.4 by the analytical methods described in the section and compare with the Runge-Kutta solution for the first 10 seconds of the motion.

Section 24.5

Find x and y, each as a function of t, in the following cases:

6
$$\left.\begin{aligned} \frac{dx}{dt} + 2x + y &= 0 \\ \frac{dy}{dt} + x + 2y &= 0 \end{aligned}\right\} \text{if } x = 1 \text{ and } y = 0 \text{ at } t = 0$$

7
$$\left.\begin{aligned} \frac{dx}{dt} + 5x - 2y &= t \\ \frac{dy}{dt} + 2x + y &= 0 \end{aligned}\right\} \text{if } x = y = 0 \text{ at } t = 0$$

8
$$\left.\begin{aligned} \frac{d^2x}{dt^2} - \frac{dy}{dt} + 2x &= 1 \\ \frac{dx}{dt} + \frac{d^2y}{dt^2} + 2y &= 0 \end{aligned}\right\} \text{if } x = 3,\ y = dx/dt = dy/dt = 0 \text{ when } t = 0.$$

9 In the coupled electrical circuit of Figure 24.3, with $E_1 = E$, $E_2 = 0$, $R_1 = R_2 = R$, $L_1 = L_2 = L$ and $M^2 < L^2$, let $i_1 = i_2 = 0$ at $t = 0$; show that

$$i_1 = \frac{E}{R}(1 - e^{-kt} \cosh mt)$$

$$i_2 = -\frac{E}{R}\, e^{-kt} \sinh mt$$

where $m = RM/(L^2 - M^2)$, $k = RL/(L^2 - M^2)$.

10 In the mechanical vibrational system of Figure 24.4 suppose $\dot{x}_1(0) = \dot{x}_2(0) = 0$, $x_1(0) = x_1^{\,o}$ and $x_2(0) = x_2^{\,o}$, where $x_1^{\,o}, x_2^{\,o}$ are unknown fixed values. Take $k_1 = k_2 = k_3 = k$ and $M_1 = M_2 = 1$ and solve for x_1 and x_2.

11 In Problem 6 substitute for y in the second equation and obtain a second order differential equation in x only. Solve this and then obtain y. (Be careful which equation you use to obtain y.)
Repeat for Problem 7.

12 Solve the system

$$3\frac{dx}{dt} - 2x + 4y = 7e^{t} + 9e^{-t}$$

$$2x + 3\frac{dy}{dt} - 4y = 5e^{t} - 9e^{-t}$$

with the condition $x = 0$ and $y = 1$ at $t = 0$. (LU)

Section 24.6

13 Represent the following functions using unit step functions; then find their Laplace transforms.

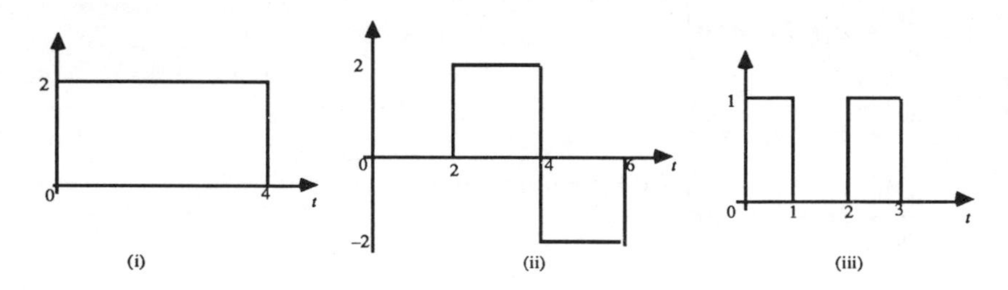

(i) (ii) (iii)

14 Prove that, under suitable conditions on s, $\mathcal{L}\{\int_0^t f(\tau)d\tau\} = \frac{1}{s}\mathcal{L}\{f(t)\}$.

Hence find the Laplace transforms of the following:

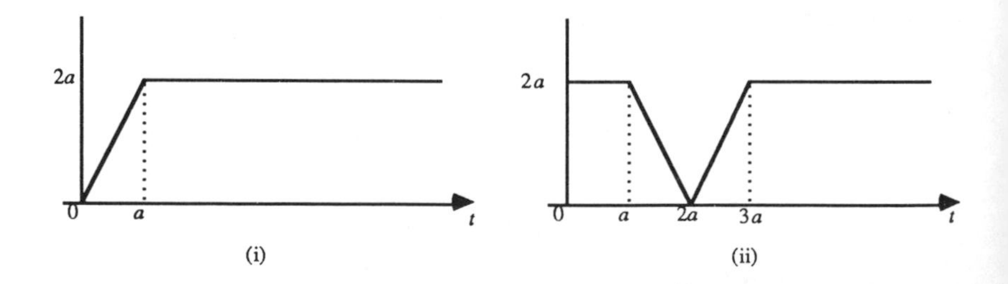

(i) (ii)

15 Find $f(t)$ if $F(s)$ is
(i) $(e^{-s} - e^{-2s})/s$ (ii) $2e^{-s}/s^2$ (iii) e^{-as}/s^3
(iv) $(e^{-s} - e^{-2s} + e^{-3s} - e^{-4s})/s^2$

16 Find the Laplace transforms of
(i) $(t-2)\,u(t-2)$ (ii) $t\,u(t-2)$ (iii) $(t-2)^2\,u(t-2)$
(iv) $u(t-\pi)\sin t$ (v) $e^{-at}\,u(t-1)$

17 Find the Laplace transforms of the following if each function is zero outside the given interval.
(i) $4t,\ 0 < t < 4$ (ii) $2t^2,\ 0 < t < 5$
(iii) $A\sin\alpha t,\ 0 < t < \pi/\alpha$ (iv) $B\cos\alpha t,\ 0 < t < \pi/\alpha$
(v) $B\cos\alpha t,\ 0 < t < 2\pi/\alpha$

18 Find the inverse Laplace transforms of
(i) $e^{-2\pi s}/(s^2 + \omega^2)$ (ii) $e^{-as}/(s^2 + 2s + 2)$

19 Show that, under suitable conditions, $\mathcal{L}\left\{\frac{f(t)}{t}\right\} = \int_s^\infty F(\sigma)\,d\sigma$

Hence find $f(t)$ if $F(s)$ is
(i) $\dfrac{1}{(s^2+\omega^2)^2}$ (ii) $\dfrac{s}{(s^2-9)^2}$ (iii) $\ln\left[\dfrac{s+2}{s+1}\right]$

20 The following functions have period 2π. Find their Laplace transforms
(i) $t,\ 0 < t \le 2\pi$ (ii) $\pi - t,\ 0 < t \le 2\pi$ (iii) $e^t,\ 0 < t \le 2\pi$

(iv) $\sin\omega t,\ 0 < t \le 2\pi$ (v) $\begin{cases} 0, & 0 < t \le \pi \\ t-\pi, & \pi < t \le 2\pi \end{cases}$ (vi) $\begin{cases} 1, & 0 < t < \pi \\ -1, & \pi < t \le 2\pi \end{cases}$

(vii) $\begin{cases} t, & 0 < t \le \pi \\ \pi - t, & \pi < t \le 2\pi \end{cases}$

25

EPILOGUE

Now that we have reached the end of the book, it is time to take stock of what has been achieved. Chapter 1 was entitled *Why Mathematics?* and it was intended to set the scene for the approach of this book by stressing both the concept of mathematical modelling and the dangers inherent in working with real data. These themes should have remained uppermost in your minds as you progressed through the text material and the problems.

Subsequent chapters presented fundamental mathematical concepts and techniques which form part of the basis of an undergraduate engineer's training. In today's world an engineer without this mathematical foundation will be unable to cope with many of the new ideas and techniques that are already becoming an essential part of their field of study. Tomorrow's world will be even more demanding.

It has to be emphasised that we have only laid the foundations of your mathematical knowledge; much more remains to be learned. However, if the foundations are laid properly you have a good chance of building a solid construction to cope with the future requirements of an engineering career.

We emphasise the importance of an integrated approach to the application of mathematics to engineering. Computing and numerical methods have grown in importance alongside their analytical cousins; however, without a working understanding of the analytical background of the former, you could be reduced to a 'black box' operator and that is not a satisfactory position from which to acquire new skills and knowledge.

If you have worked through the problems in this book you will have gained expertise in the practical application of mathematics to engineering. This is important in the same way that a designer of reinforced concrete beams should have made at least one such beam during his undergraduate career. Engineering mathematics should not be a spectator sport, and even if for much of your career you rely on software packages you have at least the knowledge to look critically at the results from using such packages.

Finally, we hope that you have enjoyed working through this book and we wish you well in your further study.

APPENDIX

GRAPHS OF TRIGONOMETRIC FUNCTIONS

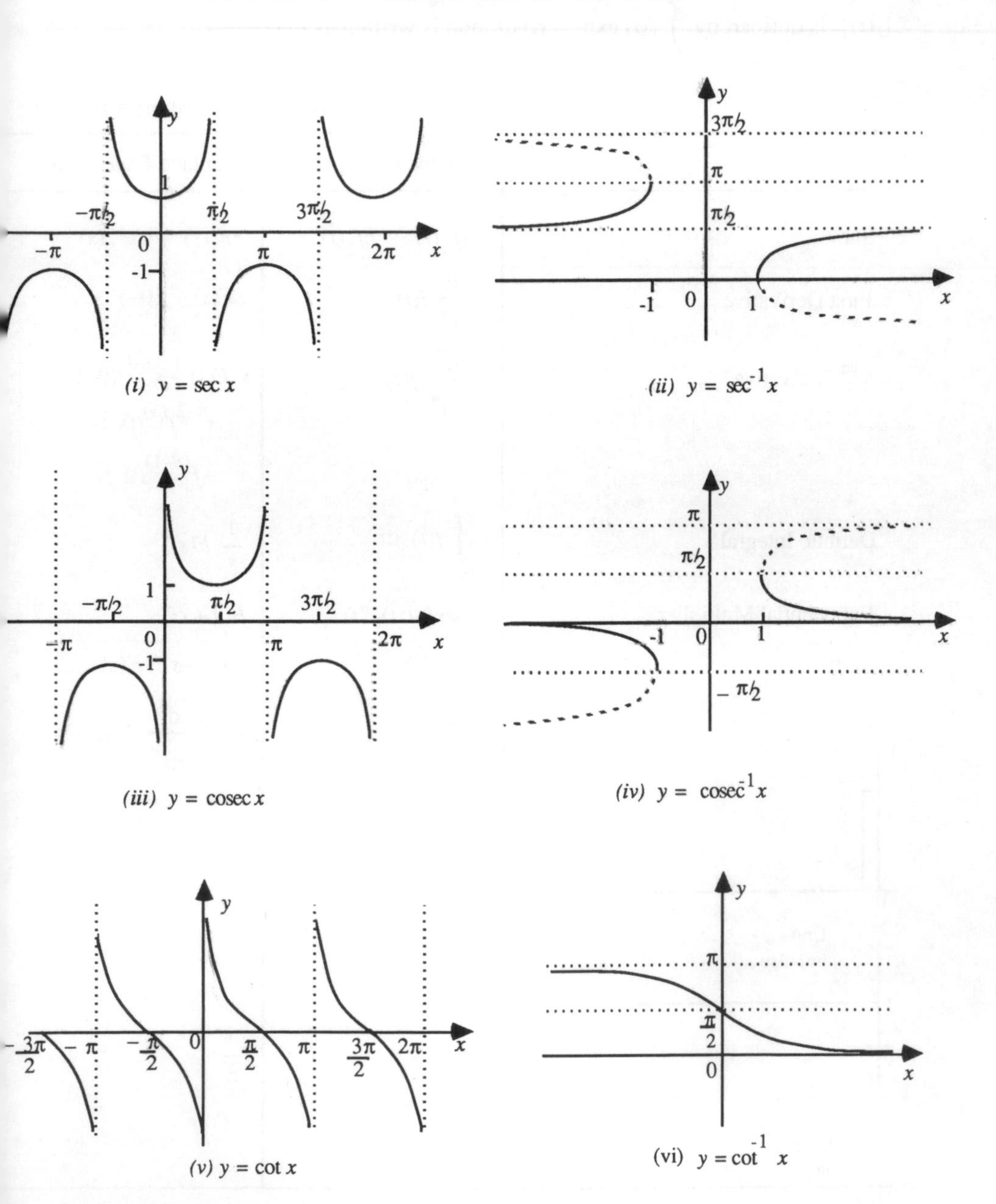

(i) $y = \sec x$

(ii) $y = \sec^{-1} x$

(iii) $y = \operatorname{cosec} x$

(iv) $y = \operatorname{cosec}^{-1} x$

(v) $y = \cot x$

(vi) $y = \cot^{-1} x$

TABLE OF LAPLACE TRANSFORMS

This table is adapted from that used in the Part 2 Examination of the Engineering Council.

$\mathfrak{L}\{f(t)\}$ is defined by $\int_{0-}^{\infty} f(t)\exp(-st)\,\mathrm{d}t$ and is written as $F(s)$.

Initial conditions are those just prior to $t = 0$.

	Time Function	Laplace Transform
Sum	$af_1(t) + bf_2(t)$	$aF_1(s) + bF_2(s)$
First Derivative	$\frac{\mathrm{d}}{\mathrm{d}t} f(t)$	$sF(s) - f(0-)$
n^{th} Derivative	$\frac{\mathrm{d}^n}{\mathrm{d}t^n} f(t)$	$s^nF(s) - s^{n-1}f(0-) - s^{n-2}f^{(1)}(0-) \ldots - f^{(n-1)}(0-)$
Definite Integral	$\int_{0-}^{t} f(t)\,\mathrm{d}t$	$\frac{1}{s}F(s)$
Exponential Multiplier	$\exp(-\alpha t)f(t)$	$F(s+\alpha)$
Shift	$f(t-T)$	$\exp(-sT)F(s)$
	$t^nf(t)$	$(-1)^n \frac{\mathrm{d}^n}{\mathrm{d}s^n} F(s)$
1. 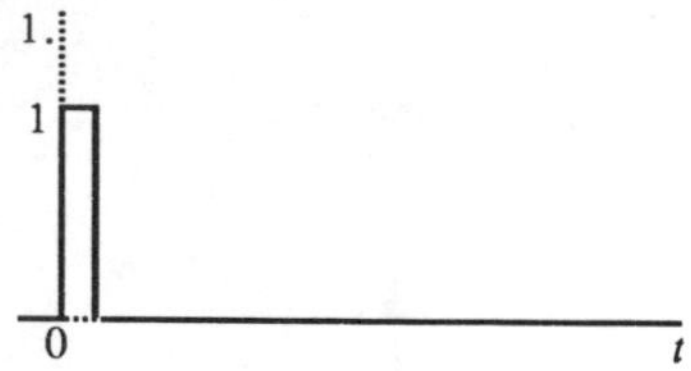 Unit impulse	$\delta(t)$	1
2. 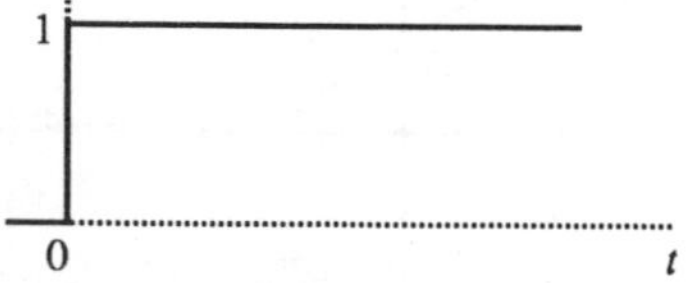 Unit step	$u(t)$	$\frac{1}{s}$

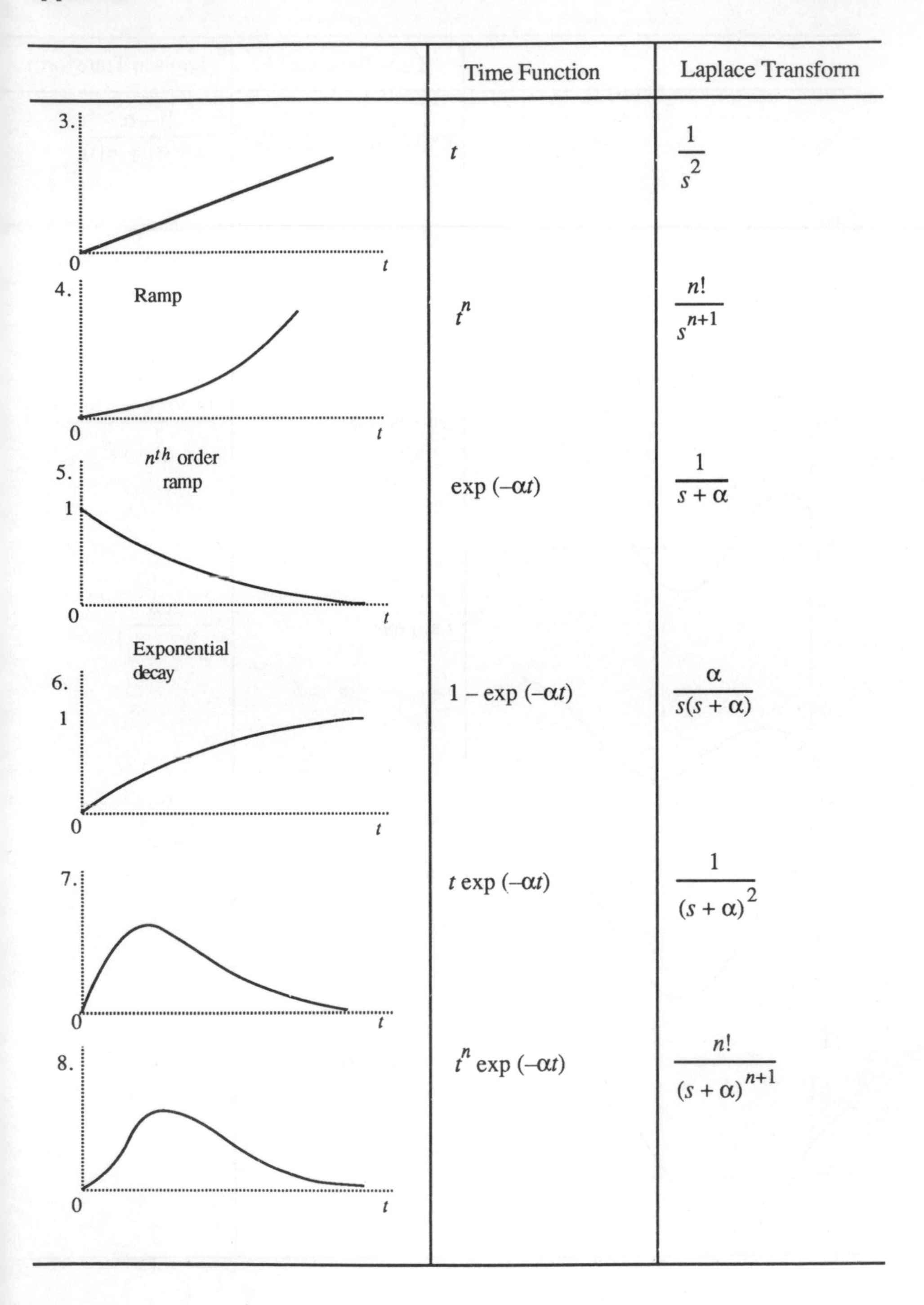

	Time Function	Laplace Transform
3. Ramp	t	$\frac{1}{s^2}$
4. n^{th} order ramp	t^n	$\frac{n!}{s^{n+1}}$
5. Exponential decay	$\exp(-\alpha t)$	$\frac{1}{s+\alpha}$
6.	$1 - \exp(-\alpha t)$	$\frac{\alpha}{s(s+\alpha)}$
7.	$t \exp(-\alpha t)$	$\frac{1}{(s+\alpha)^2}$
8.	$t^n \exp(-\alpha t)$	$\frac{n!}{(s+\alpha)^{n+1}}$

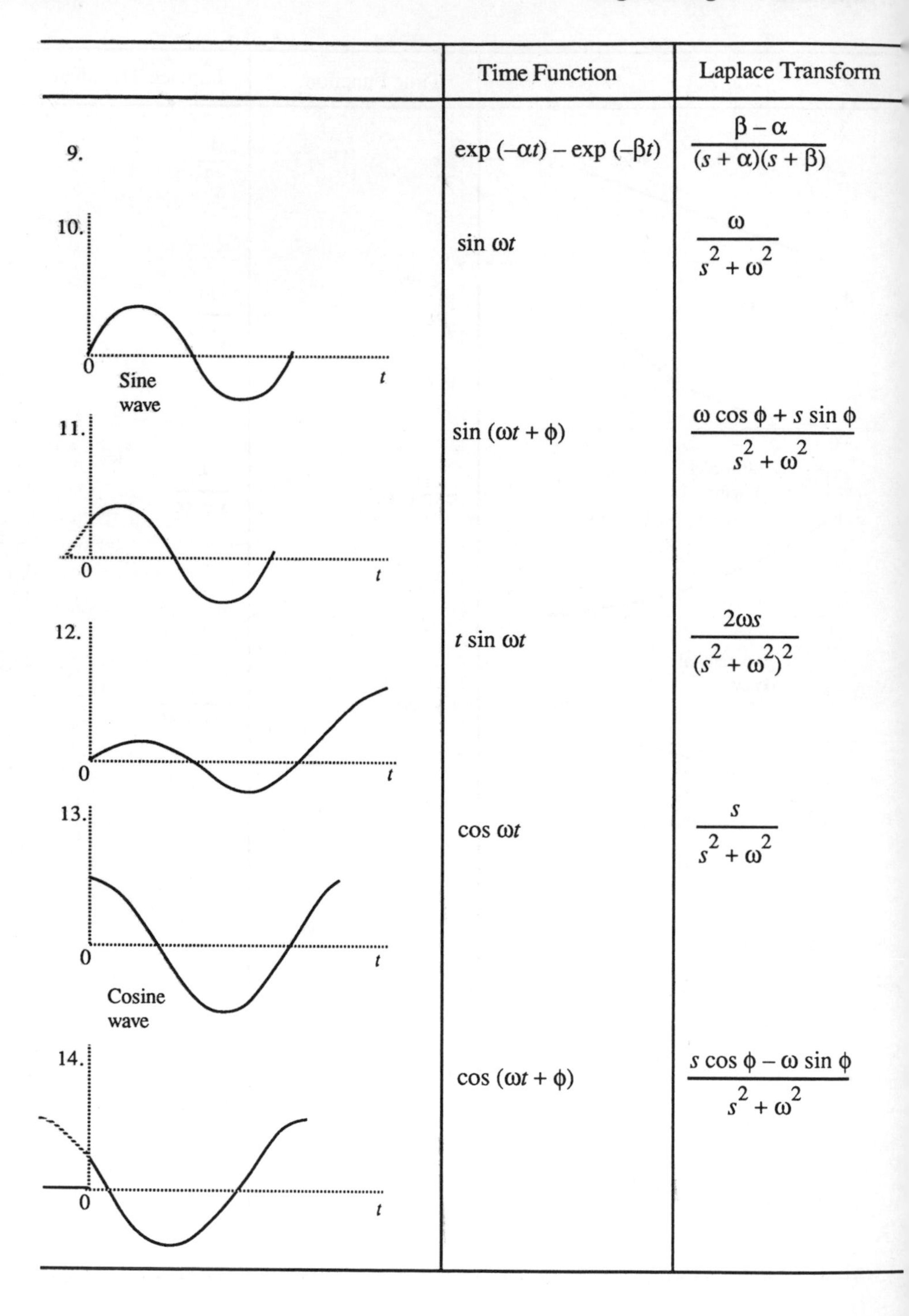

	Time Function	Laplace Transform
9.	$\exp(-\alpha t) - \exp(-\beta t)$	$\dfrac{\beta - \alpha}{(s + \alpha)(s + \beta)}$
10. Sine wave	$\sin \omega t$	$\dfrac{\omega}{s^2 + \omega^2}$
11.	$\sin(\omega t + \phi)$	$\dfrac{\omega \cos \phi + s \sin \phi}{s^2 + \omega^2}$
12.	$t \sin \omega t$	$\dfrac{2\omega s}{(s^2 + \omega^2)^2}$
13. Cosine wave	$\cos \omega t$	$\dfrac{s}{s^2 + \omega^2}$
14.	$\cos(\omega t + \phi)$	$\dfrac{s \cos \phi - \omega \sin \phi}{s^2 + \omega^2}$

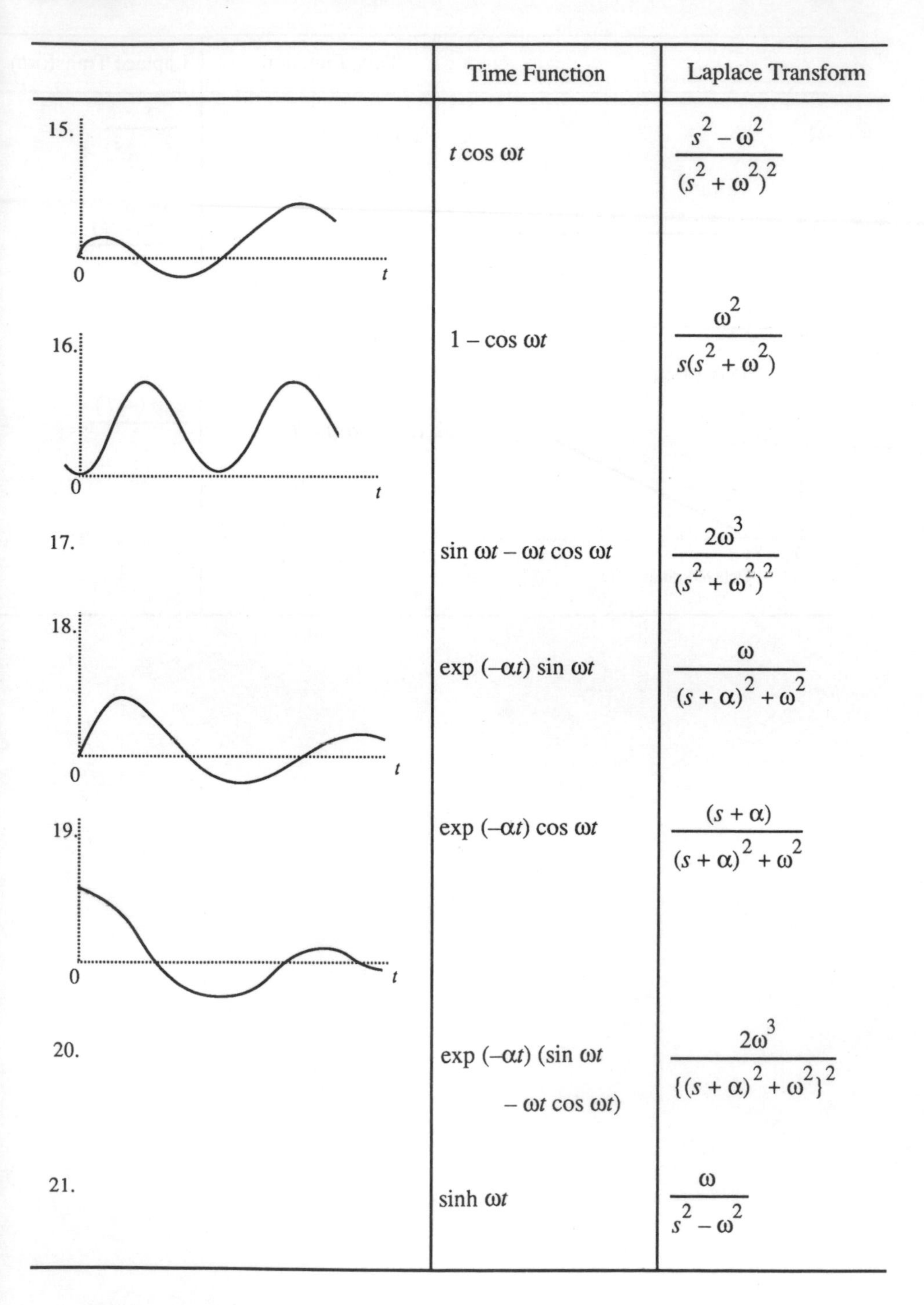

	Time Function	Laplace Transform
15.	$t \cos \omega t$	$\frac{s^2 - \omega^2}{(s^2 + \omega^2)^2}$
16.	$1 - \cos \omega t$	$\frac{\omega^2}{s(s^2 + \omega^2)}$
17.	$\sin \omega t - \omega t \cos \omega t$	$\frac{2\omega^3}{(s^2 + \omega^2)^2}$
18.	$\exp(-\alpha t) \sin \omega t$	$\frac{\omega}{(s + \alpha)^2 + \omega^2}$
19.	$\exp(-\alpha t) \cos \omega t$	$\frac{(s + \alpha)}{(s + \alpha)^2 + \omega^2}$
20.	$\exp(-\alpha t)(\sin \omega t - \omega t \cos \omega t)$	$\frac{2\omega^3}{\{(s + \alpha)^2 + \omega^2\}^2}$
21.	$\sinh \omega t$	$\frac{\omega}{s^2 - \omega^2}$

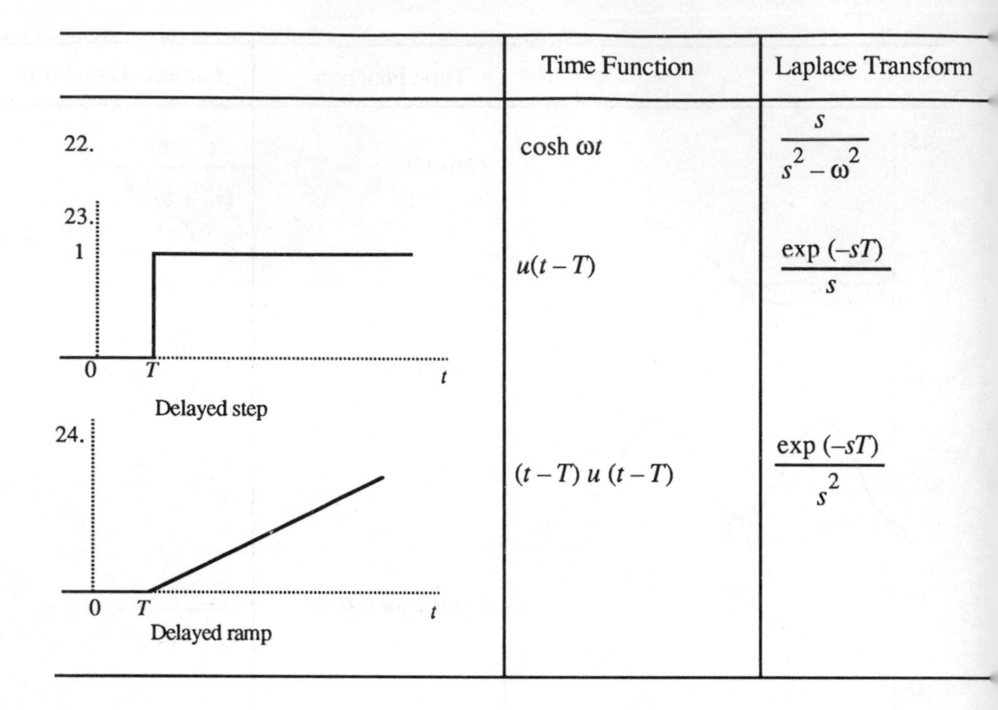

	Time Function	Laplace Transform
22.	$\cosh \omega t$	$\dfrac{s}{s^2 - \omega^2}$
23. 1, 0, T, t Delayed step	$u(t - T)$	$\dfrac{\exp(-sT)}{s}$
24. 0, T, t Delayed ramp	$(t - T)\, u\, (t - T)$	$\dfrac{\exp(-sT)}{s^2}$

ANSWERS

Chapter 1

Section 1.2

1 1.83, largest = 1.89, smallest = 1.77, relative error ≤ 0.0336

2 Addition 6.5, 6.6, 6.4, absolute error ≤ 0.1; subtraction, 1.9, 2.0, 1.8, absolute error ≤ 0.1; multiplication, 9.66, 9.99, 9.34, relative error ≤ 0.034

3 4.996, 0.042, 0.84%

Section 1.3

8 Direct form has 5 multiplications and 3 additions. Nested form has 3 multiplications and 3 additions. Nested form is quicker and has less round-off.

11 The angle between two lines.

Section 1.4

14 Identical; $x = 0, y = -2$; $x = 0, y = -2$; no solution; $x = 2, y = 0$

15 $x = -1999, y = 1002$; no solution (199.8 changed to 200)
$x = -1995, y = 1000$ (299.6 changed to 300).

Section 1.5

16 34 (31.71); 0.7 (0.756)

17 –1.5 (–1.381); 15 (14.99); –0.013 (–0.018); 1.5 (1.4772)

19 Sum = 0.91, product = 32.7

Chapter 2

Section 2.1

2 (a) $a < 1$ (b) $b \ge -3/4$ (c) $p > 21/10$ (d) $x < \frac{1}{2}$ or $x > 4$
(e) $-2 < y < 5$ (f) $x = 0$

3 (a) $-4 < x < 4$, $(-4, 4)$ (b) $1 < x < 5$, $(1, 5)$ (c) $-3\frac{2}{3} < x < 2\frac{1}{3}$, $(-3\frac{2}{3}, 2\frac{1}{3})$

4 (a) $|x - 2| < 5$ (b) $|x - 6| < 3$

5 999

7 (a) $0.281 < x < 0.686$ and $-1.781 > x > -2.186$ (b) $-0.5275 < x < 2.5275$
(c) $x \ge 1, x \le -1$ (d) $x > 4, x < 0$ (e) $1/100 > x > 1/200$

Section 2.2

9 (a) $\{1,2,3,4,5\}$, $\{1,2,5,6,20\}$ (b) $\{1,2,3\}$, $\{-1,8,6,-2\}$
(c) $\{1.5, 1.6, 1.7, 1.8, 1.9\}$, $\{2,3,5,7,9\}$ (d) $\{Z\}$, $\{Z\}$ (e) $\{Z\}$, $\{Z\}$

10 (a) one-many (b) many-many (c) one-one (d) many-many (e) many-many

11 (a) does (b) does not (c) does not (d) does not

12 (a) $-\infty \le x \le \infty$ (b) $x \ge 1$ (c) $-1 \le x \le 1$ (d) $x \ne 0$ (e) $x \ge 1, x \le -1$

13 (a) one-one (b) one-one (c) many-one (d) one-one (e) many-one

15 (a), (b) and (d)

16 one-one, $f^{-1} = x$; one-one, $f^{-1} = (1/x) - 2$; one-one, $f^{-1} = \sin^{-1} x$; many-one

17 (b) and (c) **18** (b) $\sqrt[3]{x+3}$ (c) x

19 $f+g = x^2 + x - 4$, all x; $f-g = x - x^2 - 2$, all x; $f.g = (x^2-1)(x-3)$, all x; $f \div g = (x-3)/(x^2-1)$, $x \neq \pm 1$; $g \div f = (x^2-1)/(x-3)$, $x \neq 3$

20 $f+g = x^2 - 3x$, all x; $f-g = x^2 - 3x + 2$, all x; $f.g = (x-1)^3$, all x; $f \div g = x - 1$, $x \neq 1$; $g \div f = 1/(x-1)$, $x \neq 1$

21 $f(g) = x^2 - 4$; $g(f) = x^2 - 6x + 8$; $f(g) = (x-2)^2$; $g(f) = x^2 - 2x$

22 $f(x)$ must lie in the domain; all x; $x = 0,1$; $x = 1$; $x = 1,3$

23 $f(x) = x$, all x; $g(x) = x$, all x

Section 2.3

24 0.7078, –0.1001, 0.7493, 2.5607, 1.2764, –1.5040

Section 2.4

29 3.7622, 0.5211, –0.2913, 1.1948, 1.7627, 0.2554, –0.2554, 0.8671, –1.4436

30 41/9, –40/41

Section 2.5

33 $\cos 11\pi/12 = 0$ by linear interpolation, should be –0.9659
$\cos 160° = -0.9107$, $\cos 170° = -0.9553$, not accurate, interval too large.

34 0.5505 by mean differences, 0.5504 by linear interpolation.

35 Useful as far as third differences.

38 Third differences. **39** 74.75 (Better, 75.00 due to inaccuracy of tables)

40 $\rho(5500) = 0.6982$, $\rho(20\ 000)$ obtained by assuming third differences are constant and extending table.

Chapter 3

Introduction

1 Equality when $a = b$ 2 $-\sqrt{2}$ is also a root

3 $x_{n+1} = \frac{1}{3}(2x_n + \frac{A}{x_n^2})$, $x_{n+1} = \frac{1}{3}\left[x_n + 2\sqrt{\frac{A}{x_n}}\right]$, $^3\sqrt{10} = 2.154435$ (6 d.p.),

formula could give $^3\sqrt{8}$ but it diverges.

Section 3.1

4 (i) $\{2,12,30,56,\ldots\}$ (ii) $\{0,2,6,16,42,110,\ldots\}$

5 2, 10, 26, 82

6 (i) convergent, limit = 3 (ii) convergent, limit = 2 (iii) divergent
(iv) convergent, limit = 2

7 (i) divergent (ii) convergent, limit = 0 (iii) convergent, limit = 1
(iv) convergent, limit = 10^{-4} (v) convergent, limit = 2 (vi) divergent (vii) divergent

8 $(-\infty, 1 - 2^9/9) \cup (1 + 2^{10}/10, \infty)$, $(-\infty, 1 - 2^{999}/999) \cup (1 + 2^{1000}/1000, \infty)$

9 $\{2\frac{1}{2}, 2\frac{1}{3}, 2\frac{1}{4}, \ldots\}$; limits are 1, 1, 2; limits are 1,1; $\{2\frac{1}{2}, 4\frac{1}{3}, 6\frac{1}{4}, \ldots\}$,

$\lim\{u_n\} = 1$, others do not exist; $(4\frac{1}{2}, 3\frac{2}{3}, 3\frac{9}{20}, \ldots\}$, 1, 2, 3, 2, $\frac{1}{2}$

Section 3.3

13 6, 2, ∞ **14** $x = 0$; $x = 1$; none; none; $x = 1$

16 Split into $(x^2)(\sin x) + (2\cos x) = f.g + h$ and use rules.

19 $\to 0$, $\to 2\cos x$ **20** $\to \infty$, $\to 0$

Section 3.4

21 33.4×10^{-6}, 33.4×10^{-6}, 33.5×10^{-6} **22** 0.09, 0.49, 2.26

23 -0.1135×10^{-3}, -0.0865×10^{-3}, -0.0537×10^{-3} **24** 0.3, 0.5625, 1.06, 1.18

25 [1.6, 2.4], 0.507; [1.8, 2.2], 0.502; [1.9, 2.1], 0.500; [2.6, 3.4], 0.335; [2.8, 3.2], 0.334; [2.9, 3.1], 0.3335; [4.6, 5.4], 0.200; [4.8, 5.2], 0.200; [4.9, 5.1], 0.200; derived function = $1/x$

26 Derived function = $-1/x^2$ **27** $f'(0) = 0$, no advantage in straddling, odd function.

28 2.3524, 2.1293 **29** $\sqrt[3]{9} = 2.0801$ **30** $\sqrt{17} = 4.12$ (2 d.p.)

31 $f'(x) = 3ax^2 + 2bx + c$ **32** $-\sin x$, $2\cos 2x$, $-3\sin 3x$

35 Estimates are 0.5650, 0.5575, 0.5400, 0.5225, 0.5150; then −0.625, −0.875, −0.625; then −12.5, 0, 12.5; finally, 625; accuracy rapidly gets worse.

36 $g'(x_2) = [f(x_3) - f(x_1)]/2h$, more accurate since the curve is approximated by a parabola over three points.

Section 3.5

37 4 strips give Over-Estimate 284 and Under-Estimate 220; 8 strips give 271 and 239 respectively.

38 For 4 strips we obtain 1502 and 1246 as the estimates; for 8 strips we obtain $1453\frac{7}{8}$ and $1325\frac{7}{8}$. The error is $512/n$ and this suggests 5120 strips for 1 d.p. accuracy and 51200 strips for 2 d.p. accuracy

39 Value −2 predicted. Formula does not allow for discontinuity at $x = 2$

41 $0 \le I \le \pi/2$; $\frac{3}{8} \le I \le \frac{1}{2}$; $3/10 \le I \le 3$ **42** $\frac{1}{2}$; $1/\sqrt{3}$; $4/\sqrt[3]{4}$ or $\sqrt[3]{16}$

Chapter 4

Section 4.3

5

Group	35.0–35.4	35.5–35.9	36.0–36.4	36.5–36.9	37.0–37.4
Frequency	5	1	11	13	15
Group	37.5–37.9	38.0–38.4	38.5–38.9	39.0–39.4	39.5–39.9
Frequency	13	13	6	1	2
Group	35.0–35.9	36.0–36.9	37.0–37.9	38.0–38.9	39.0–39.9
Frequency	6	24	28	19	3

Information is lost in making the groups too wide

6 (a) (i) 3.474, 2.516 (ii) 10 is a possible number

(iii)

Group	2.500–2.599	2.600–2.699	2.700–2.799	2.800–2.899	2.900–2.999
Frequency	1	4	10	10	10
Group	3.000–3.099	3.100–3.199	3.200–3.299	3.300–3.399	3.400–3.499
Frequency	12	14	11	5	3

(b) (i) 1.77, 1.21 (ii) 12 is a possible number

(iii)

Group	1.20–1.24	1.25–1.29	1.30–1.34	1.35–1.39	1.40–1.44	1.45–1.49
Frequency	1	1	5	3	5	5
Group	1.50–1.54	1.55–1.59	1.60–1.64	1.65–1.69	1.70–1.74	1.75–1.79
Frequency	11	4	7	6	0	2

Section 4.4

7 (i) 50 (ii) $58\frac{5}{6}\%$

8

Rejects	0	1	2	3	4	5
Frequency	5	9	11	7	3	1
Cumulative frequency	5	14	25	32	35	36

(i) 2 (ii) 2 (iii) 1.92

9 37.3, 37.0–37.4, 37.3; 3.018, 3.100–3.199, 3.055; 1.51, 1.50–1.54, 1.51

10 116.1

Group	61–70	71–80	81–90	91–100	101–110	111–120
Cumulative frequency	2	10	27	55	91	131
Group	121–130	131–140	141–150	151–160	161–170	
Cumulative frequency	165	194	213	223	225	

11 $8\frac{4}{9}$

Section 4.5

12 (b) 14.2% (c) 0.785, 0.63–1.06 **13** 10.64, 0.709, 0.892

14 1000.2, 7.09, modal group = 997.5, median group = 997.5, mean deviation = 5.60, inter-quartile range = 995.8–1004.2

15 32.9, 0.369, median = 32.91 **16** mean = 6.22378, s.d. = 0.2592

17 3.41, 1.98 **18** 0.913, 0.150

19 Modal group = 12, median group = 12, mean = 12.24, s.d. = 2.74, grouping has condensed the information.

20

Group	1.00-1.49	1.50-1.99	2.00-2.49	2.50-2.99	3.00-3.49	3.50-3.99	4.00-4.49
Frequency	11	3	12	1	6	1	1

mean = 2.17, s.d. = 0.84, median group = 2.00–2.49 = modal group, mean deviation = 0.66, inter-quartile range = 1.44–2.48

21 0.084; 0.0071; 0.011; 0.580

22 Variance (grouped) = 0.0167, Variance (raw) = 0.01691, Sheppard's correction does not improve on the calculated variance in this case.

Chapter 5

Section 5.1

1 (a) {3,4,5,6} (b) {2,3,4}

(c) {(1,4), (2,3), (2,4), (3,2), (3,3), (3,4), (4,3), (4,2), (4,1), and negative combinations}
(d) {1,3}

2 (a) $A \cup B = \{1,2,3,4,5,6,7,8,9,11\}$, $A \cap B = \{1,3,5,7\}$
(b) $A \cup B = \{0,\pm1,\pm2,\pm3,\pm4,\pm5,\pm6,7,8,9\}$, $A \cap B = \{4,5\}$
(c) $A \cup B = \{0, \pm1, \pm2, \pm3, \pm4, 5, \ldots, 14, \pi, \sqrt{2}\}$, $A \cap B = \{0,5,6,8,9,10\}$

4 (i) and (ii) are true

5 (i) $\{a,d\}$ (ii) $\{a,c,d,e,f,g\}$ (iii) $\{a,b,d,f\}$ (iv) $\emptyset$ (v) $\{b,c\}$
(vi) $\{e,f\}$ (vii) $\{c,g\}$ (viii) $\{a,b,d\}$ (ix) $\{g\}$
(x) $\{(a,c), (a,e), (a,g), (b,c), (b,e), (b,g), (c,c), (c,e), (c,g), (d,c), (d,e), (d,g)\}$
(xi) $\{b,c,g\} = C'$

6 (i) $B \subseteq A$ (ii) $A \subseteq B$ (iii) $A = \emptyset$ (iv) always true
(v) $A \subseteq B$ (vi) $A = B$

7 (iii) $\emptyset$ (iv) {7}, {2,4}

8 (i) $\{(1,c), (1,d), (1,4), (1,5), (2,c), (2,d), (2,4), (2,5)\}$
(ii) as (i) (iii) $\emptyset$ (iv) as (iii)

Section 5.4

12 1/216, 5/18

13 (i) (a) 0.6 (b) 0.9 (c) 0.7
(ii) (a) 33/95 (b) 1/190 (c) 81/95 (d) 62/95 (e) 48/95
(f) 153/190 (g) 14/95
(iii) (a) 68/95 (b) 27/95

14 25/69, 28/69, 16/69

15 (i) 1/169 (ii) 1/221

16 $\frac{1}{2}$

17 (i) 5/9 (ii) (a) 2/105 (b) 4/10

18 0.098

19 7/40

20 (a) 1/32 (b) 7/40 (c) 7/10

21 (a) 0.813 (b) 0.013

22 4/13

23 (a) 1/36 (b) 1/24 (c) 19/36

24 347/2048

25 (i) (a) 125/1296 (b) $5^9/6^{10}$ (c) $5^{r-1}/6^r$ (ii) ${}^{28}C_3/{}^{51}C_3$

26 (ii) 4/9

27 (i) 21/25 (ii) 1.68

28 (i) (a) 8/27 (b) 16/81 (ii) (a) 0.896 (b) 0.544

29 $$\frac{{}^rC_n \; {}^{N-r}C_{s-n}}{{}^NC_s} \cong {}^sC_n \left[\frac{r}{N}\right]^n \left[\frac{N-r}{N}\right]^{s-n}; \quad \hat{N} = \frac{rs}{n}; \; 250$$

30 (i) 0.05 (ii) 0.60 (iii) 0.35

31 (a) (i) 0.2852 (ii) 0.7361 (b) (i) $\frac{1}{48}$ (ii) $\frac{35}{48}$ (iii) $\frac{13}{48}$

32 (i) 0.8438 (ii) 0.6278

Chapter 6

(**Note:** where truth tables are required the last column only is given and is in the form of a row; reading this left to right is equivalent to reading the last column top to bottom)

Section 6.1

1 (i) and (ii)

2 (i) It is not hot (ii) It is hot and I am tired (iii) It is hot or I am tired
(iv) I am tired or it is not hot. (v) It is not hot and I am not tired.

3 (i) $p \wedge q$; (ii) $p \wedge \sim q$; (iii) $\sim p \wedge \sim q$; (iv) $\sim (\sim p \vee \sim q)$.

4 (i) FFTF; (ii) FTTT; (iii) FFTF; (iv)FTTT;
(v) TTTTTFFF; (vi) TFTFTFFF; (vii) TTTTTFFF;
(viii) FFFFFTTT; (ix) FFFFFTTT.

5 (i) Neither (ii) Tautology (iii) Neither (iv) Contradiction

10 (i) TFFF
(ii) TFTTFTFF
(iii) TFFTFTTF

12 (i) If a geometric figure is a square, it has four right angles.
If a geometric figure does not have four right angles, it is not a square.
If a geometric figure is not a square, it does not have four right angles.
(ii) If an enginer is good at mathematics, he has not got practical skills.
If an engineer has practical skills, he is not good at mathematics.
If an engineer is not good at mathematics, he has not got practical skills.

Section 6.2

15 (ii), (iii) and (v) are valid

Section 6.3

18 (i) True; $x = 0$. (ii) False (iii) True; $y = -x$. (iv) False
(v) True; eg $x = 2, y = 4$. (vi) False; $x = 0$.

20 (i) $(\forall x)\ (\forall y)\ [M(x) \wedge F(x,y) \rightarrow R(y)]$
(ii) $(\forall x)\ (\exists y)\ [J(x) \wedge F(x,y) \rightarrow R(y)]$
(iii) $[(\exists x)\ (\exists y)\ [R(x) \wedge F(x,y) \rightarrow J(y)] \rightarrow [(\forall x)\ (\forall y)\ [R(x) \wedge F(x,y) \rightarrow J(y)]$

Section 6.4

24 (i)

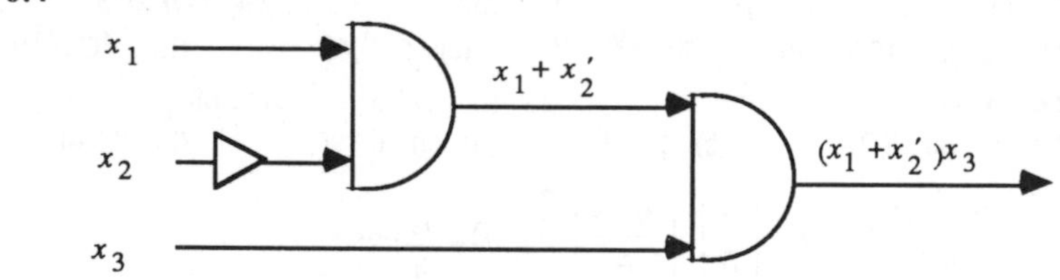

(The above is given as an illustrative example.)

25 $x'y + z$

26 Expressions are equivalent

Section 6.5

30 (i) x_1x_2' (ii) $x_1x_2 + x_1x_3$
(iii) $x_1x_2 + x_1'x_3'$ (iv) $x_1x_3 + x_1x_2' + x_1'x_3$
(v) $x_3'x_4' + x_1x_3x_4 + x_1'x_2x_4'$

Chapter 7

Section 7.2

2 $x + 2y < 6$ has no effect; $x + 2y < 3$ will cut down the region.

3 (0,0), (2,1), (3,0), (0,2); no solution if new condition added.
4 (i) $x \geq 0, y \leq 0$ (ii) $x + 3y > 3$ (iii) $0 \leq y \leq 2, 2x + 3y \geq 6, 4x + 3y \leq 12$
5 $c = 6, c = 2$ 6 $c = 9, c = 6$ 7 $26X, 16Y$; $20X, 20Y$

Section 7.4

8 (a) Parabola (b) Hyperbola (c) Ellipse (d) Touches axis of x at origin
(e) Touches x–axis at $x = 3$ (f) Asymptotes are $x = 2, x = 3, y = 0$
(g) Asymptotes are $x = 1, x = 2, y = 0$ (h) $y = -1$ is asymptote
(i) $y = 0$ is asymptote (j) $x = 0$ and $y = 1 - x$ are asymptotes
(k) $x = 0$ and $y = 1 - x$ are asymptotes

Section 7.5

9 (i) $r^2 = \sec^2 \theta(1 - \tan^2 \theta)$ (ii) $r^2 = \tan^2 \theta$ (iii) $r^4 = 4/(1 - \sin 2\theta)$
10 (i) $x^2 + y^2 = 3x + 4y$ (ii) $x^4 - 6x^2y^2 + y^4 = 2(x^2 + y^2)$
(iii) $x^6 + x^4y^2 = (x^2 - y^2)^2$
11 Figure of 8; circle; circle; figure of 8; four leaf clover; four leaf clover; three leaf clover
14 $r = (9 \pm 3\sqrt{5})/2, \cos \theta = (7 \pm 3\sqrt{5})/2$; curves do not cross at the origin
16 $x' = (x + y\sqrt{3})/2, y' = (-x\sqrt{3} + y)/2$; $x' = y, y' = -x$
17 $2x'y' = 1$

Section 7.6

19 $x = 2ct_1t_2/(t_1 + t_2), y = 2c/(t_1 + t_2), (t_1 + t_2 \neq 0)$
20 $(x - a)^2 + y^2 = a^2$, circle 21 $x = 50(\cos 7t + \cos 6t), y = 50(\sin 7t + \sin 6t)$
22 $x^{2/3} + y^{2/3} = 1$ 23 $x = x_1 + \mu(x_2 - x_1), y = y_1 + \mu(y_2 - y_1)$

Section 7.7

24 (i) and (ii) represent planes (iii) represents a line (iv) represents a cylinder
(v) represents a circle
25 (i) $(2,3,z)$ (ii) $(0,y,0)$ (iii) $(1 - z, 1 + z, z)$
26 Intersect if not all three of a/a', b/b', c/c' are equal, planes must not be parallel
27 Ellipsoid 28 $(d/a, d/b, d/c)$

Chapter 8

Section 8.2

3 (a) $(-7, -2, 2)$ (b) $\sqrt{158}/2$ 4 $(3/7, 6/7, -2/7)$ 5 $(\sqrt{3} - 1)/4$
6 $(2, -6, 3)$, 7 7 $\sqrt{6}, 5\sqrt{6}/2, \sqrt{114}/2$ 8 $m = -19/11, n = -53/22$

Section 8.3

10 $(\mathbf{i} + \mathbf{j} + \mathbf{k})/\sqrt{3}$ 11 $(5\mathbf{i} + 3\mathbf{j}).(2\mathbf{i} + 5\mathbf{j}) = 25$ Nm 13 −11

15 (a) $11\frac{5}{6}$ (c) $\mathbf{x} = \frac{1}{\alpha}\mathbf{w} - \left[\frac{\mathbf{w}\,.\,\mathbf{u}}{\alpha(\alpha + \mathbf{v}\,.\,\mathbf{u})}\right]\mathbf{v}$

(d) (i) $\hat{\mathbf{n}} = (1/\sqrt{14})(3, -2, 1)$ (ii) $\mathbf{r} = (1,2,3) + \lambda\hat{\mathbf{n}}$ (iii) $\lambda = 2/\sqrt{14}$
16 (b) (i) 106° 16′ (ii) 5 (iii) 2.492 (c) $x + 2y - 3z = 1, y = 1/2$

Section 8.4

17 −4**i** + 5**j** + 13**k** **18** $\sqrt{107/2}$ **21** **b** = (41/157, 95/157, 21/157)

22 2**i** − **j** + **k** **23** (a) Area of rectangle (b) 45° 35′

24 (b) $\frac{1}{157}(41, 95, 21)$ (c) 4.53

25 (a) (i) 11 √3/2 (ii) 35° 13', 93°, 57° 47' (iii) 5.099, 3.742, 6.481
(c) (3/2, 1/2)

26 (a) −7**i** − 5**j**, −**i** + 4**j** + 3**k**, −8**i** − **j** + 3**k**

(b) (i) 75.64° (ii) 67.41° (iii) 7.81 (c) $\mathbf{x} = \frac{12}{17}\mathbf{i} - \frac{8}{17}\mathbf{j} + \frac{8}{17}\mathbf{k}$

Section 8.5

27 18 **31** each side = 7**j** − 7**k**

33 24**i** + 7**j** − 5**k**, 15**i** + 15**j** − 15**k** **34** (μ/56π)(−6**i** − 9**j** + 11**k**)

Chapter 9

Section 9.2

1 (a) 1/2, 3/2, 1 (b) 0, 1/3, 0
(c) equation (iv) = (i) − (ii) − (iii), 18/16, 30/16, 21/16, −3/16, 27/16
(d) no solution (e) 81/11, −4/5, −73/55 (f) 1/2, 1/3, 1/4

2 (a) i, 1 − i (b) (289 + 119i)/221, −2 − i

3 Determinants are all zero

5 (i) $p = 4, q \neq 1$ (ii) $p = 4, q = 1$ (iii) $p \neq 4, q \neq 1$

6 (a) 1/6 (b) (i) $k \neq 4$ or −1 (iii) $k = 4$ or −1
$k = -1$ (−λ/7, −3λ/7, λ) $k = 4$ (μ, 6μ, −14μ/21)

7 4, 8, 10; 4.28, 8.08, 9.88 **8** 1,1,1,1

9 Both sets of values almost satisfy the equations. Equations ill-conditioned

Section 9.3

10 (a) 3.36, 1.10, 1.24 (b) 5.00, −5.07, 0.99 (c) 1.59, −0.31, −0.20
(d) 2.28, 0.21, −0.44

11 $(x_1, x_2, x_3) = (3/2, 5/2, -2)$

12 (a) (1/16, 5/8, 41/16) (b) (ii) No solutions

13 (a) 3.3610, 1.1026, 1.2384 (b) 5.01, −5.08, 0.99

Section 9.4

15 (i) True solution is 1,2,3 (ii) True solution is 3,2,1

Chapter 10

Section 10.1

1 (i) $[\mathbf{A} + \mathbf{B}] = \begin{bmatrix} 3 & -5 \\ 2 & 7 \\ -4 & -2 \end{bmatrix}$, $[\mathbf{A} - \mathbf{B}] = \begin{bmatrix} 5 & -5 \\ 2 & -7 \\ -8 & 8 \end{bmatrix}$ (ii) $[\mathbf{AB}] = \begin{bmatrix} 20 \\ 29 \end{bmatrix}$

(iii) $[\mathbf{BA}] = \begin{bmatrix} 3 & 8 \\ 8 & -12 \end{bmatrix}$, $[\mathbf{AB}] = \begin{bmatrix} -13 & 6 & -5 \\ 21 & 2 & 11 \\ 6 & -4 & 2 \end{bmatrix}$ (iv) $[\mathbf{BA}] = \begin{bmatrix} -22 & 4 & 42 \\ -8 & 16 & 48 \\ -1 & -2 & -3 \\ 0 & 0 & 0 \end{bmatrix}$

(v) $[\mathbf{A} + \mathbf{B}] = \begin{bmatrix} 2 & 2 & -4 \\ 2 & 1 & 8 \\ 2 & -1 & 5 \end{bmatrix}$, $[\mathbf{A} - \mathbf{B}] = \begin{bmatrix} 2 & -4 & 6 \\ -2 & 1 & -4 \\ 0 & 1 & -3 \end{bmatrix}$

$[\mathbf{AB}] = \begin{bmatrix} -1 & 5 & -12 \\ 4 & -2 & 14 \\ 1 & 2 & -1 \end{bmatrix}$, $[\mathbf{BA}] = \begin{bmatrix} -5 & 3 & 1 \\ 10 & -2 & 8 \\ 6 & -2 & 3 \end{bmatrix}$

$[\mathbf{A} + \mathbf{B}][\mathbf{A} - \mathbf{B}] = \begin{bmatrix} 0 & -10 & 16 \\ 2 & 1 & -16 \\ 6 & -4 & 1 \end{bmatrix}$, $\mathbf{A}^2 - \mathbf{B}^2 = \begin{bmatrix} 4 & -8 & 3 \\ -4 & 1 & -10 \\ 1 & 0 & -3 \end{bmatrix}$

Section 10.2

8 (a) pre-multiply by $\begin{bmatrix} 1 & 0 & 3 \\ 0 & 1 & 0 \\ 0 & 0 & 1 \end{bmatrix}$ (b) post-multiply by $\begin{bmatrix} 1 & 0 & 1 \\ 0 & -1/2 & 0 \\ 0 & 0 & 1 \end{bmatrix}$

(c) pre-multiply by $\begin{bmatrix} 1 & 0 & 0 \\ 0 & 1 & -1 \\ 0 & 0 & 1 \end{bmatrix}$

Section 10.3

10 (i) –35 (ii) 0.765 (iii) –30 (iv) –232

Section 10.4

14 $|\mathbf{P}| = 1$, $|\mathbf{Q}| = 48$

16 (i) $\begin{bmatrix} -4/11 & 3/11 \\ 1/11 & 2/11 \end{bmatrix}$ (iii) $\dfrac{1}{14}\begin{bmatrix} 5 & 3 & 1 \\ 3 & 6 & -5 \\ 1 & -5 & 3 \end{bmatrix}$ (iv) $\begin{bmatrix} 7 & -3 & -3 \\ -1 & 1 & 0 \\ -1 & 0 & 1 \end{bmatrix}$

(vi) $\dfrac{1}{29}\begin{bmatrix} 1 & -2 & 9 & 4 \\ -21 & 13 & 14 & 3 \\ 13 & 3 & 1 & -6 \\ -19 & 9 & 3 & 11 \end{bmatrix}$ (ii) and (v) not possible

17 $\mathbf{A} = (2\mathbf{B}^2 - \mathbf{I})^{-1}$

19 $\mathbf{A}^{-1} = \dfrac{1}{\alpha^2 + \beta^2}\begin{bmatrix} \alpha & -\beta \\ \beta & \alpha \end{bmatrix}$; $\alpha^2 + \beta^2 = 1$; rotation of axes, θ anti-clockwise

21 (i) $\mathbf{A}^{-1} = \dfrac{1}{22}\begin{bmatrix} 1 & -5 \\ 4 & 2 \end{bmatrix}$, $\mathbf{B}^{-1} = \dfrac{1}{12}\begin{bmatrix} 6 & -3 \\ 2 & 1 \end{bmatrix}$ (ii) $\mathbf{X} = \dfrac{1}{264}\begin{bmatrix} 114 & -45 \\ 16 & -4 \end{bmatrix}$

22 $\begin{bmatrix} 1/a & 0 & 0 \\ -b/a^2 & 1/a & -b/a^2 \\ 0 & 0 & 1/a \end{bmatrix}$; $a = \pm 1, b = 0$ **23** $\begin{bmatrix} 107 & -26 \\ 91 & 10 \end{bmatrix}$

Section 10.5

24 $\frac{1}{78}\begin{bmatrix} 15 & -3 & 9 \\ -7 & 17 & 1 \\ -2 & 16 & -22 \end{bmatrix}$; $\frac{1}{2}$, 3/2, 1 **25** $\begin{bmatrix} 7 & -3 & -3 \\ -1 & 1 & 0 \\ -1 & 0 & 1 \end{bmatrix}$

26 5/11, 7/11; 1, –2, 5 **27** $\begin{bmatrix} \cos\alpha & \sin\alpha & 0 \\ -\sin\alpha & \cos\alpha & 0 \\ 0 & 0 & 1 \end{bmatrix}$

Chapter 11

Section 11.1

2 $3\angle 0°$, $2\angle 90°$, $1\angle \pi$, $2\angle(-\pi/2)$, $13\angle \tan^{-1}(12/5)$, $2\angle(5\pi/6)$, $10\angle -126°52$, $2\angle(-\pi/4)$, $4\angle(\pi/3)$, $\sqrt{13}\angle 146°19$, $\sqrt{2}\angle(-3\pi/4)$, $2.236\angle(-63°26)$

3 $1 + i$, $-1.532 + 1.2856i$, $-\sqrt{3} - i$, $2.868 - 4.096i$

Section 11.2

4 $5 - 5i$ **5** (i) $-2 \pm 3i$, $3.606\angle \pm 123°41$

6 (i) $u = 2$ or -3 (ii) $z = (1 + i)/2$ (iii) $x = -5, y = -2$

10 10, $26 + 2i$, $(5 - 6i)/7$, $(24 + 7i)/25$ **11** $2\angle(2\pi/3)$, $5\angle 53°8$, $2\sqrt{5}\angle -26°34$

12 $\arg \bar{z} = -\arg z$ **13** $z = (3 \pm \sqrt{23})(1 + i)/2$

14 (i) $z = -1 \pm \sqrt{3i - 4}$

(ii) (a) arguments differ by 0 or π (b) real parts equal, imaginary parts equal and opposite

17 $22i$, 6/25

20 (i) $2 + 2\sqrt{3}i$ (ii) $-108\sqrt{2}(1 + i)$

21 (i) $3\angle 90°$ (ii) $2\sqrt{2}\angle 105°$ (iii) $4\angle 60°$

Section 11.3

22 $\cos 5\theta = 16\cos^5\theta - 20\cos^3\theta + 5\cos\theta$, $\sin 5\theta = 16\sin^5\theta - 20\sin^3\theta + 5\sin\theta$; $\cos 18° = \sqrt{(5 + \sqrt{5})/8}$, $\sin 36° = \sqrt{(5 - \sqrt{5})/8}$

25 (i) $a = -\sqrt{3}/2, b = -1/2$ **26** $\pi/2 + 2k\pi + i\cosh^{-1} 2$

27 $\theta = 0, \pi, 2\pi, \ldots$

28 Z moves on the same locus, W lies on the real line between $-2a$ and $+2a$

29 $(\sqrt{3} + i)/2$ **31** i **32** $2.032 + i\,3.052$

33 $A = (\cosh^2 s - \sin^2 s)^{-1/2}$, $\cot\alpha = \tan s \tanh s$

35 (i) $1.241 - 0.219i$, $-0.431 + 1.184i$, $-0.810 - 0.965i$

(ii) $\pm(1.621 + 1.396i)$, $\pm(1.369 - 1.621i)$

36 (i) $1 + 3i$ (ii) $1 - i, -2, -2, a = 2, b = -2$

37 (i) 0, $\pm(1/\sqrt{2})(1 + i)$, $\pm(1/\sqrt{2})(1 - i)$

(ii) 1.587, 1.587 $\angle(\pm 2\pi/3)$, -1.442, 1.442 $\angle(\pm \pi/3)$

38 (i) $\sqrt[3]{2}\angle 10°$, $\sqrt[3]{2}\angle 130°$, $\sqrt[3]{2}\angle -110°$ (ii) $\sqrt[4]{5}\angle(13°17 + k \,.\, 90°)$

41 $1\angle(k \,.\, 72°)$; $1 + \delta + \delta^2 + \delta^3 + \delta^4 = 0$

Section 11.4

43 (i) $-(3 + 2\sqrt{3}) + i(2\sqrt{3} - 1)$, $(2\sqrt{3} - 3) - i(2\sqrt{3} + 1)$ **44** (4, –4)

45 (i) interior of circle (ii) area between two lines making angle ±30° with the x-axis
(iii) area to the left of $x = 3/2$
(iv) area outside circle on $x = 1$ and 2 as diameter, $y \geq 0$ and inside circle, $y \leq 0$

47 (b) $\sqrt{5}/2$; circle, centre $1 + \frac{1}{2}i$, radius $\sqrt{5}/2$

48 Area inside ellipse, foci at (1,0) and (0,1), major axis = 4

49 $z\bar{z} + z + \bar{z} - 4 = 0$, centre (–1,0), radius $\sqrt{5}$

50 $(\cos 80° + i \sin 80°)/2$ or $(\cos 40° + i \sin 40°)/2$ **51** $y = 1$

Section 11.5

52 Circuit impedance $= (R + j\omega L)/[(1 - \omega^2 LC) + j\omega RC]$, $i_1 = E/(R + j\omega L)$, $i_2 = jE\omega C$

Chapter 12

Section 12.1

1 (i) $3x^2 \sin x + x^3 \cos x$ (ii) $e^x(\cos x - \sin x)$ (iii) $x^3(4 \ln x + 1)$
(iv) $(2x + x^2)/(1 + x)^2$ (v) $\sec x(x \tan x - 1)/x^2$ (vi) $-10(3 - 2x)^4$
(vii) $-x/\sqrt{1 - x^2}$ (viii) $1 - x \sin^{-1} x/\sqrt{1 - x^2}$ (ix) $3 \cos 3x$
(x) $4 \sin 8x$ (xi) $-3 \cos 3x \sin(\sin 3x)$ (xii) $3e^{3x+2}$
(xiii) $2x \tan^2 2x + 4x^2 \tan 2x \sec^2 2x$ (xiv) $(2x - 1)/(1 - x + x^2)$
(xv) $2/(1 - x^2)$ (xvi) $-1/(\sin x \cos x)$ (xvii) $e^{2x}[2 \ln(1 + x^2) + 2x/(1 + x^2)]$

2 –34 cm/sec; –14 cm/sec^2 **3** 0.362 (3 d.p.)

4 $-e^{-0.2t}(2.32 \sin 4t + 1.72 \cos 4t)$ **6** 7500π cm^3/sec

7 $-24 \sin 6t/(a^2 - 16 \sin^2 3t)^{1/2}$ **8** $(40/9)\pi h^2$

9 $dy/dx = 3\sqrt[3]{x}(1 + 2x^3)/2\sqrt{1 + x^3}$

10 (i) –1 (ii) $1/x[1 + (\ln x)^2]$
(iii) $2(x + y)\cos[(x + y)^2]/\{1 - 2(x + y)\cos[(x + y)^2]\}$
(iv) $(25 - x - 2x^2)/\sqrt{25 - x^2}$
(v) $[x + (1/x)]^x [\ln[x + (1/x)] + (x^2 - 1)/(x^2 + 1)]$
(vi) $\cot(a/b + 1)t/2$ (vii) $-\cot\theta$ (viii) $(1 + y^2)/(1 + 2y^2)$
(ix) $-(2x + 3)/2(1 + y)$ (x) $(x^2 - 7)/(1 + x)^{3/2}(3 - x)^{1/2}$

(xi) $\sqrt{3}/(2 + \sin x)$ (xii) $2/\sqrt{1 - x^2}$

13 (i) $-6(2x - 3)^{-3/2} (2x + 3)^{-1/2}$ (ii) $e^x(1 + x)(9 - x - 2x^2)/2(1 - x)^{5/2}$

14 (i) $-(2x + 3y)/(3x + 2y)$ (ii) $3x^2/10y$
(iii) $[\cos(x + y) - 3y]/[3x - \cos(x + y)]$
(iv) $y \cos x/(1 + y \sin y)$ (v) $e^{(x-y)} (x - 1)/2x^2$ or $1 - 1/x$

15 (i) $\lambda \cos \lambda t/\cos t$ (ii) $-b \cot \theta/a$ (iii) $4/3t$ (iv) $\sec \theta/4a$

17 (i) $4/(3 \cos x + 5)$ (ii) $4x/(x^4 - 1)$ (iii) $4e^{-2x} (1 - x^2)$ (iv) $2/(1 + x^2)$

18 (i) $-\pi e$ (ii) 2/5 (iii) 26/5

19 (i) $2e^{-2}$ (ii) $\frac{1}{2}$ (iii) 0 (iv) 1

21 (ii) $(3 - x)^2 (12 - 11x + 4x^2)/(2 - x)^3$

23 (i) (a) $\sin^{-1} (\frac{1}{2})$ (b) $-4/\sqrt{3}$ (c) $1 - \ln 2$ (ii) $x/(1 + x)$

24 $(y + b \sin b\theta)(\theta + a \sin ax)/(b \cos b\theta - x)(a \cos ay - \theta)$

Section 12.2

26 (i) $10(x^2 + 3xy + y^2)/(3x + 2y)^3$ (ii) $3x(20y^2 - 3x^3)/100y^3$
(iii) $-[9(x - y)^2 \sin(x + y) + 6\{\cos(x + y) - 3y\}\{3x - \cos(x + y)\}]/\{3x - \cos(x + y)\}^3$
(iv) $\{(1/y + \sin y)^2 \sin x + (1/y^2 - \cos y)\cos^2 x\}/(1/y + \sin y)^3$
(v) $1/x^2$ (i) $(-\lambda^2 \sin \lambda t \cos t + \lambda \cos \lambda t \sin t)/\cos^3 t$
(ii) $-b \operatorname{cosec}^3 \theta/a^2$ (iii) $-4/9t^4$ (iv) $-\sec^3 \theta/16a^2$

27 $(4p^3k^3/27V^2) - F$ **28** $r^6 = 2A/B$ **32** (i) $-3t/2 - t^3/2,\ 3(1 + t^2)^3/4t$

36 Stable at $x = 0$, unstable at $|x| = a$ **37** (i) $-12/x^4$ (ii) 2e

41 (a) $\dfrac{dy}{dx} = \dfrac{3x^2 y - 4x^3}{4y^3 - x^2}$ (b) $\dfrac{1}{t}, -\dfrac{1}{2t^3}$

Section 12.3

42 (i) [−3,3], 20, −1/4; [−3,0], 20, 2; [1,2], 0, −1/4
(ii) [0,4], $2\frac{1}{4}$, −10; [−3,3], $2\frac{1}{4}$, −10; [−3,0], 2, −10; [1,2], 2, 0

43 (i) local max at $x = 6/5$, local min at $x = 4$ (ii) local min at $x = 0$
(iii) local min at $x = 3$ (iv) local max at $x = 3.281$, local min at $x = 1.219$
(v) turning points when $\theta \cot \theta = h$

44 $r = e^{-1}$ **45** $c\sqrt{2}$ **46** $(gT_0/3w)^{1/2}$ **47** $(CL)^{-1/2}$ **48** $R/2$

49 $\ln x = 1 + 1/x$; 3.591 **51** 2, 2, −2, −1

52 Symmetry about Oy, x-axis is asymptote

53 Symmetry about Ox, $x = -1$ is asymptote, loop between $x = 0$ and 1

54 (i) $f'(x) = x/(1 - x)^2$ (ii) max at $x = 1$, $2e^{-2}$, min at $x = -2$, $-e^4$

55 Local maximum at $x = -2$, point of inflexion at $x = 1$

56 (a) $y_{max} = 1$ at $t = 0.6931$, $y = 0$ at $t = 0.1438$, y_{inf} at $t = 1.242$
(b) Min, $x = 0$; Max, $x = 2$; Inflections, $x = 0.585$, 3.414

Section 12.4

58 $c \cosh^2 (x/c)$

59 (a) $\sqrt{12}$ (b) $5\sqrt{10}/3, \infty$ (c) $-\sqrt{2}$ (d) $-87\sqrt{29}/80$

60 $-\{1 + (2 - 8x_1)^2\}^{3/2}/8$ **62** $-ka/(a^2 + k^2)^{3/2}$, $2a + k^2/a$, $-(a^2 + k^2)/k$

63 Min = 0, max = $-16.6^3/5^5$, $1/\sqrt{2}$

Chapter 13

Section 13.1

1 (i) One root $-3 < x < -2$ (ii) one root $2 < x < 3$
(iii) two roots approx 0 and 100 (iv) two roots approx 2 and 6
(v) infinite number of roots, $0 < x < \pi/2$ and near $n\pi$, $n = 1, 2, 3, \ldots$
(vi) one root near $\pi/2$

2 (i) one positive root (ii) and (iii) one positive, two negative roots
(iv) two positive, two negative roots

Section 13.2

3 Roots near (i) 3.0032 (ii) 0.461 (iii) 4.9206 (iv) 2.9505
(v) 0.5175 (vi) 1.822

Section 13.3

5 1.2022, 0.8526, 1.7456 **6** Roots are ±2 **7** $x = 1, 2$

8 $x_0 = -2$ converges to -1, $x_0 = 0$ to -1, $x_0 = 1$ to 2, $x_0 = 3$ to 2

9 3.5782, 0, −3.5782 **10** $(3/2A) > x_0 > (1/2A)$

Section 13.4

12 $x_{n+1} = \frac{1}{2}(x_n + A/x_n)$; $x_{n+1} = (1 - 1/r)x_n + A/rx_n^{r-1}$ **13** 4.493

14 Repeated roots at $x = 2.1, 3.9$; both roots close to $x = 2$

15 (i) 0.6155 (ii) 0.9643 **16** 2.36 **18** 0.87

Section 13.5

21 (i) $-1.171 \pm 0.765i$, $1.171 \pm 0.399i$.
(ii) $0.55 \pm 0.59i$, $-0.55 \pm 1.12i$

22 $x^2 - 4.006x + 5.004$

23 First root = 0.4, quadratic factors $x^2 + 0.625x + 0.146$ and $x^2 - 0.225x + 0.155$

Chapter 14

Section 14.1

1 (i) Hyperboloid, all x, y (ii) Cone, vertex downward at (1,0,0), all x, y
(iii) Cone, vertex downward at (1,0,1), all x, y
(iv) Sphere, centre (2,0,1), radius 2, $0 \le x \le 4$, $-2 \le y \le 2$, $1 \le z \le 3$

2 (i) 13,10,1 (ii) 2/5, 2/5, not defined (iii) ln 6, ln 3, 0 (iv) $7e^2$, $e^{-4} - 6e^{-1}$, 0

3 (i) $z = (y + 1)^2 + 1$, parabola (ii) $z = (y + 1)^2 + 4$, parabola

(iii) $z = (y+1)^2 + a^2$, parabola (iv) $z = x^2 + 4$ (v) $z = x^2 + 9$

(vi) $z = x^2 + (b+1)^2$ (vii) $(x + \frac{1}{2})^2 + (y + 3/2)^2 = 5/2$, circle

4 $-5 \sin 4$

Section 14.2

5 (i) 1/7 (ii) (a) 0 (b) 0

6 (a) -1 (b) 1 (c) 0 (d) 3/5 (e) 1 (f) $\cos 2\theta$

7 (i) 0 (ii) ∞ (iii) ∞ (iv) $2/\sin 2\theta$

8 (i) continuous everywhere (ii) continuous everywhere except (0,0)
(iii) continuous except at $x = -y$ (iv) continuous except at $x = y$
(v) continuous except at (0,0), $f(0,0) = 0$

9 (i) $2x, 3y^2, 4z$ (ii) $2xy^3/z, 3x^2y^2/z, -x^2y^3/z^2$
(iii) $2xye^{3z}, x^2 e^{3z}, 3(1 + x^2y)e^{3z}$
(iv) $z \cos(xz + y), \cos(xz + y), x \cos(xz + y)$
(v) $ye^{(xy+2y^2)}, (x + 4y) e^{(xy+2y^2)}$ (vi) $\frac{1}{2}y \sin z, \frac{1}{2}x \sin z, \frac{1}{2}xy \cos z$
(vii) $y/(x^2 + y^2), -x/(x^2 + y^2)$ (viii) $(x^3 + 2y^3)/x^3y^2, -(2x^3 + y^3)/x^2y^3$

10

	f_x	f_y		f_x	f_y
(i)	$3x^2y^4$	$4x^3y^3$	(vi)	e^x	$-\sin y$
(ii)	$2x/y^3$	$-3x^2/y^4$	(vii)	$2xy\, e^{3y}$	$(3 + 3x^2y + x^2)e^{3y}$
(iii)	$2 \cos y$	$-2x \sin y$	(viii)	$-3y/(9x^2 + y^2)$	$3x/(9x^2 + y^2)$
(iv)	$\cos(x + y)$	$\cos(x + y)$	(ix)	$x(x^2 - y^2)^{-1/2}$	$-y(x^2 - y^2)^{-1/2}$
(v)	0	2			

	$f_x(0,0)$	$f_y(0,0)$	$f_x(1,2)$	$f_y(1,2)$	$f_x(-1,0)$	$f_y(-1,0)$
(i)	0	0	48	32	0	0
(ii)	–	–	1/4	–3/16	–	–
(iii)	2	0	2 cos 2	–2 sin 2	2	0
(iv)	1	1	cos 3	cos 3	cos 1	cos 1
(v)	0	2	0	2	0	2
(vi)	1	0	e	–sin 2	1/e	0
(vii)	0	3	$4e^6$	$10e^6$	0	4
(viii)	–	–	–6/13	3/13	0	–1/3
(ix)	–	–	–	–	–1	0

11 (iii) $n = 2, -3$ **14** (a) $0, a/V^2$

Section 14.3

16 $-6x + 11y + 14z = 2$

17 (i) (0,0) local minimum (ii) (–3,2) local minimum
(iii) (–5/3, 1) saddle point, (–1,1) local minimum
(iv) (0,0) saddle point, (1, –1) local minimum, (–1, 1) local minimum
(v) (2/5, 4/5) saddle point, (0,0) local minimum
(vi) (1,1), (–1, –1) saddle points (vii) (0,0) local maximum

(viii) (0,0), $(-\frac{1}{4}, 1)$, $(0, -\frac{1}{2})$ saddle points, $(-1/12, -1/6)$ local maximum

(ix) (0,0) local minimum

19 (i) (0,0) saddle point, $(6a, 18a)$ minimum (iii) maximum value = $1/e$

20 [16/9, −32/9, 32/9] **21** $\theta = 30°$, $x = l/6$

Section 14.4

23 (a) −1% (b) ±2.5% **24** 0.054π in^3 **25** −2%

26 (i) $[-2/(x^3y) + y]dx + [-1/(x^2y^2) + x]dy$

(ii) $e^{-(x+y)}[\cos(x+y) - \sin(x+y)](dx + dy)$

(iii) $[(-y/\sqrt{x^2+y^2}) + x\tan^{-1}(y/x)/\sqrt{x^2+y^2}]dx + [(x/\sqrt{x^2+y^2} + y\tan^{-1}(y/x)/\sqrt{x^2+y^2}]dy$ (iv) $(3x^2y + 3y^4)dx + (x^3 + 12xy^3)dy$

(v) $2e^{(x^2+y^2)}(xdx + ydy)$ (vi) $-\sin(x+2y)(dx + 2dy)$

28 −90 watts; exact answer − 85.3 watts (1 d.p.) **29** 1.8%

30 $\pi/2\,[8 + 77/\sqrt{41}]$ cm^2/sec **31** −14 cm/sec

32 $k/\sqrt{kTBR}\;[BR\,dT/dt + TR\,dB/dt + TB\,dR/dt]$

33 $3C\sqrt{60 + 2\mu}\,(0.3 + 0.1\mu)$ **34** $32(V + T/10)/V^2$

35 (i) $-x^2/y^2$ (ii) $-[\sin xy + y(x+y)\cos xy]/[\sin xy + x(x+y)\cos xy]$

(iii) $(y\sin x - \cos y)/(\cos x - x\sin y)$

(iv) $[6y(x+y)/x^2 - e^{(y/x)}]/[e^{(y/x)} + 6(x+y)/x]$

36 (i) $27y^2 + 4x^3 = 0$

(ii) $2x^2 - x^4 + 2y^2 - y^4 + 2x^2y^2 = 1$ or $[(x+y)^2 - 1][(x-y)^2 - 1] = 0$

(iii) $4y = (y + 1 - x)^2$

37 Circle is $x^2 + y^2 = 16$ **41** $(3y^2z - 4x)/3(z^2 - xy^2)$, $2xyz/(z^2 - xy^2)$

42 (i) 1 (ii) $2t(1 + 2t^2 + t^3)/(1 + t)^3$

43 (i) x/z, $2y/z$ (ii) $z/(1-z)$, $z/(1-z)$ (iii) $2x + 8y$, $8x$

(iv) $-(x+z)/(y+x)$, $-z/(y+x)$

45 (a) R (b) ±0.00025 cm (c) 1.95 to 2.04

46 (a) $2x + 3y + (3x - 4y)\left[\dfrac{-2(x + y\cos 2xy)}{4y + 2x + 2x\cos 2xy}\right]$ **48** (b) ±3.41%

Chapter 15

Section 15.1

1 $\Sigma e_i = 0.3$, $\Sigma|e_i| = 0.7$, $\Sigma e_i^2 = 0.19$; reject (2, 4.4); $\Sigma e_i = -0.1$, $\Sigma|e_i| = 0.3$, $\Sigma e_i^2 = 0.03$

2 $a_0 = 3$, $a_1 = 1$; $S = 0$; exact fit **3** $a_0 = 3.03$, $a_1 = 1.1$, $S \neq 0$

4 $y = a_0 + a_1 x$, $S = \sum_{i=1}^{n}(y_i - a_0 - a_1 x_i)^2$

Section 15.2

5 (i) $0 = \Sigma y - na - b\Sigma x^3 = \Sigma x^3 y - a\Sigma x^3 - b\Sigma x^6$

(ii) $0 = \Sigma y - na - b\Sigma x^2 - c\Sigma x^3 = \Sigma yx^2 - a\Sigma x^2 - b\Sigma x^4 - c\Sigma x^5 =$
$\Sigma yx^3 - a\Sigma x^3 - b\Sigma x^5 - c\Sigma x^6$

(iii) $0 = \Sigma xy - \Sigma x - a\Sigma x^2$

(iv) $0 = \Sigma xye^{bx} - a\Sigma xe^{2bx} = \Sigma ye^{bx} - a\Sigma e^{2bx}$

(v) $0 = \Sigma y/(a + bx) - \Sigma \ln(a + bx)/(a + bx) = \Sigma xy/(a + bx) - \Sigma x \ln(a + bx)/(a + bx)$

(vi) $0 = \Sigma y/x - a\Sigma 1/x^2$

6 $y = 461.44 - 48.39x$, $y = 487.82 - 75.12x + 3.06x^2$

7 $D = 15 + 0.329V + 0.057V^2$

8 $y = 13.26 - 18.90x$, $y = 13.07 - 18.43x$

9 (i) $y = 2.05x + 2.7/x$ (ii) $y = 2x^2 - 7\sqrt{x}$ (iii) $y = 0.0224/x^2$

10 $y = 1.524e^{3.014/x}$

11 $y = 3.66 - 2.08x + 0.42x^2$

12 $S = 3.743 + 1.7T - 0.071T^2$ (ii) $S = 3.6 + 1.7T$
(iii) $S = 3.743 + 1.417T - 0.071T^2 + 0.083T^3$

13 $s = 86.9\, t^{-1.407}$

Section 15.3

14 3.3244

15 0.72654

Section 15.4

16 12.706, 22.904

17 $f(5)$ should be 156, $f(2.5) = 25.375$

18 0.6088 (4 d.p.)

19 1.60379, 2.28962

20 Error in $y(0.6)$, should be -0.8432, $y(0.45) = -0.2919$

21 $f(1.5) = 7.125$, $f(7) = 399$

22 (a) 1.73 (b) 1.80 (c) 1.76

23 $f(1) = 4$, $f(2) = -2$

Section 15.5

25 $f'(0.5) = -0.4795$, estimated error = 0.0002

26 -0.6400

Chapter 16

Section 16.1

1 0.011

2 0.01, 0.0099

Section 16.2

3 Quadratic approximations

(a) $1 + x + x^2/2$, $e^{x_0}[1 + (x - x_0) + \frac{1}{2}(x - x_0)^2]$

(b) $x - x^2/2$, $\ln(1 + x_0) + (x - x_0)/(1 + x_0) - (x - x_0)^2/2(1 + x_0)^2$

(c) $1 - x^2/2$, $\cos x_0 - (\sin x_0)(x - x_0) - (\cos x_0)(x - x_0)^2/2$

(d) $-2x^2 + 7x - 5$, $(3x_0 - 2)x^2 + (7 - 3x_0^2)x + x_0^3 - 5$

4 Quadratic approximations

	$x = 0$			$x = x_0$		
	0.10	0.25	0.50	0.10	0.25	0.50
(a)	1.1050	1.2813	1.6250	1.1054	1.2841	1.6490
(b)	0.0950	0.2188	0.3750	0.0955	0.2232	0.4055
(c)	0.9950	0.9688	0.8750	0.9951	0.9689	0.8776
(d)	−4.3200	−3.3750	−2.0000	−4.1380	−3.3593	−1.8740

	Actual		
	0.10	0.25	0.50
(a)	1.1052	1.2840	1.6487
(b)	0.0953	0.2231	0.4053
(c)	0.9950	0.9691	0.8776
(d)	−4.1390	−3.3594	−1.8750

Section 16.3

5 (i) converges to $e - 1$ (ii) ratio $\to 1/3$, diverges (iii) ratio $\to \sqrt{3}$, converges
(iv) ratio $\to 2$, converges

6 (i) converges (ii) converges (iii) diverges (iv) diverges

7 250

8 (i) (1,3) (ii) (−2,0) (iii) (−1,1) (iv) (−1,1)

10 all x

Section 16.4

15 $-x^2/2 - x^4/12$

16 $-x^2/6 - x^4/180, -x^2/2 - x^4/12$

17 $|y| \leq 1$

18 $cx^3/2 - acx^4/4 + (3a^2c - 4bc)x^5/16$

19 1.00061

20 2.3026

22 $\ln 2 + a/8 - a^2/64$

24 (a) $\omega x - k\omega x^2 + \omega(3k^2 - \omega^2)\dfrac{x^3}{6} + \omega k(\omega^2 - k^2)\dfrac{x^4}{6}$

(b) (i) 0.7855 (ii) 0.277

27 $\dfrac{x}{2} + \dfrac{x^2}{6}$

Section 16.5

28 (a) $e[1 + (x - 1) + (x - 1)^2/2! + ...]$, all x;

(b) $2/3 - (2/9)(x - 2) + (4/27)(x - 2)^2/2! + ...$, $|x| < 1$;

(c) $\frac{1}{2} + (\sqrt{3}/2)(x - \pi/6) - (x - \pi/6)^2/4 - \sqrt{3}(x - \pi/6)^3/12 + ...$, all x

29 (ii) $\sin 51° = 0.7771$ (4 d.p.), estimate = 0.7773

30 $\sqrt{2}/10$, ${}^3\sqrt{60}/10$

31 Estimated error = 0.0008, actual error = 0.0002

32 0.9325

Section 16.6

35 (i) 0; (ii) 1; (iii) 1; (iv) $\frac{1}{2}$ **36** –3% **37** $(5/6\pi) \pm 2\%$ gm/cm^3

38 6% **39** (i) 2; (ii) $1/2a$; (iii) –2 **40** $\tan\theta = \theta + \theta^3/3 + 2\theta^5/15$

41 Et/L

43 (a) $d^2y/dx^2 = \left[(3y^2 + 2)(6\cos x - 3x\sin x) - (3\sin x + 3x\cos x)6y\frac{dy}{dx}\right]/(3y^2 + 2)^2$

(b) 8

Chapter 17

Section 17.3

8 Limit is 1 **9** –0.247; 0.1176

10 0.7854 analytically, $I_S = 0.7854$, error = 0.0000 (4 d.p.)

11 0.2927; error bound is 4.2×10^{-4} **12** 0.6565 and 0.6689 **13** 0.51444

14 4.109 (3 d.p.) **15** 0.403 (3 d.p.) either method **16** 1.418 a

19 0.825, 0.674, 0.0193, 0.00244, 0.000156, 0.0000381, 0.0000195

Section 17.4

21 (i) $1/\sqrt{2}$ (ii) $\pi/3 + \sqrt{3}/2$

22 $\frac{1}{2}e^x[x\cos x + (x-1)\sin x] + C$ and $\frac{1}{2}e^x[x\sin x + (1-x)\cos x] + C$

23 $\frac{1}{2}x\sqrt{x^2 - 4} - 2\cosh^{-1}(x/2) + C$ and $(1/5)(x^2-4)^{5/2} + (4/3)(x^2-4)^{3/2} + C$

24 (i) $\frac{1}{2}(x^2+1)\tan^{-1}x - \frac{1}{2}x + C$ (ii) 1/6

25 (i) (a) $\frac{1}{2}\ln\frac{3}{4} + \pi/3\sqrt{3}$ (b) $4/5\,[e^\pi + \frac{1}{2}]$ (c) $\ln\{(2+e)/3\}$

(ii) $-2a/VRT - \ln(V-b) + b/(V-b) + C$

26 $a^2\sigma/\varepsilon_0 x$ **27** $2/3x^{3/2}\ln x - (4/9)x^{3/2} + C$

28 (i) (a) $\pi/4$ (b) 8/105 (ii) $C = \frac{1}{2}e^{-x}[x(\sin x - \cos x) + \sin x] + C$

$S = \frac{1}{2}e^{-x}[-x(\sin x + \cos x) - \cos x] + C$

29 $(x^2 + x + 1)^{1/2} - 1/2\sinh^{-1}[(1/\sqrt{3})(2x+1)\} + C$ and

$\frac{1}{2}\ln(x^2 + x + 1) - (1/\sqrt{3})\tan^{-1}\{(1/\sqrt{3})(2x+1)\} + C$

30 (i) $(1/3)\ln(25/8)$; (ii) $\sinh^{-1}2$

31 (i) $e^{-x}(-x^2 - 2x - 2) + C$; (ii) $\sin^{-1}x + \sqrt{1-x^2} + C$

32 (i) $\frac{1}{2}\ln\frac{3}{4} + \pi/2\sqrt{3}$; (ii) $\frac{1}{2}\tan x\sec x + \frac{1}{2}\sinh^{-1}(\tan x) + C$

33 $3\pi/32 - \frac{1}{4}$ **34** $k[(2/\sqrt{5}) - (1/\sqrt{2})/a^2$

35 (i) $(e^{-\pi} + 1)/10$ (ii) $(1/\sqrt{2}) \cosh^{-1}\{(4x + 3)/\sqrt{17}\} + C$
(iii) $(1/3) \cosh^3 x - \cosh x + C$

37 (a) 0.275 (b) 0.00033

38 (i) $\dfrac{1}{a} e^{ax} \sin bx - \dfrac{b}{a} \int e^{ax} \cos bx \, dx$ (ii) $\dfrac{e^{ax}}{a^2 + b^2}(a \cos bx + b \sin bx)$

(iii) 0.23

39 (a) 1 (b) $0.5x\sqrt{1 + x^2} - \frac{1}{2} \sinh^{-1} + C$

40 0.65 **42** 0.406

43 (a) $x \sin^{-1} x + \sqrt{1 - x^2} + C$

(b) $x \tan^{-1} x + \ln \cos(\tan^{-1} x) + C$ or $x \tan^{-1} x - \ln\sqrt{1 + x^2} + C$

44 (i) $\frac{1}{4} \ln\{(2 + \tan x/2)/(2 - \tan x/2)\} + C$ (ii) $-\ln[1 + \cot x/2] + C$
(iii) $(1/\sqrt{2}) \tan^{-1}(\sqrt{2} \tan t) + C$ (iv) $(1/5) \tan^{-1}[(1/5) \tan t] + C$
(v) $(1/5)x + (14/15) \ln\{(2 \tan (x/2) - 1)/(\tan (x/2) - 2)] + C$

45 $-\frac{1}{4}$ **46** $1/a^2$ **47** (i) $\pi/2$ (ii) 2 (iii) 1/5

Chapter 18

Section 18.1

1 (i) 256/15 (ii) $(9\pi/2) \ln\left[\dfrac{9}{7}\right]$ **2** 0.35

3 16/3 **4** $3\pi a^2/2$ **6** $\pi^2/2, 2\pi^2$ **8** 2/3

9 $(2\pi/3)(t_2 - t_1)(r_2^2 + r_2 r_1 + r_1^2) + \pi(r_2 + r_1)(r_2 t_1 + r_1 t_2)$

Section 18.2

11 $8a^2/15\sqrt{3}, 4a/\sqrt{3}$ **12** $256a^2/15$

14 $\ln(2 + \sqrt{3})$ **15** $4/\sqrt{3}, 8/15\sqrt{3}$

Section 18.3

16 $\pi a^2(5^{3/2} - 1)/6$ **17** $2\pi rh$ **18** $8\pi(3\sqrt{3} - 1)/3$

19 (i) $\ln(2 + \sqrt{3})$

20 (i) $\left[\dfrac{15}{4} + \ln 2\right]\pi$ (ii) $8\pi(4 + 2 \cosh 2)$ (iii) $\dfrac{64}{3}\pi a^2$

Section 18.4

21 (i) $\bar{x} = 0, \bar{y} = \frac{1}{2} + 3\sqrt{3}/8\pi$ **22** $(e^{\pi} - 1)/8$

24 $\bar{x} = 5/6, \bar{y} = 109/30$ **25** 113

26 $11\pi a^2$; centroid at polar position $(20a/11, \pi)$

27 $4a^2/15$ (i) $\pi a^3/12$ (ii) $32\pi a^3/105$; centroid is $(4a/7, 5a/32)$
28 $(1/3)\,a^3$
29 $(3/2, 9/10)$

Section 18.5
30 1/20, 1/28, $\pi/6$
31 M is mass: (i) $2048M/35$, $16M/5$ (ii) $4Ma^2/5$, $3Ma^2/7$ (iii) $M/4$, M
32 M is mass: (i) $36M/5$ (ii) $8M/35$ (iii) $5M$
33 V is volume: $2V/5$
34 $\sqrt{2}$
35 $c[1 + \sinh 1 - \cosh 1]/\sinh 1$

Section 18.6

37 (i) 45000 (ii) $3\pi\sqrt{E_1^2 + E_3^2}\,/2\sqrt{2}(3E_1 + E_3)$

38 (i) πa^2 (ii) $\pi/4$ (iii) $6\pi a^2$

39 (i) $\left[\dfrac{16a}{9\pi},\ -\dfrac{5a}{6}\right]$ (ii) $\left[\dfrac{17}{18},\ \dfrac{80}{27\pi}\right]$

40 $8a$
41 10^{-4} m^3/s

42 $GM\alpha\left[\dfrac{1}{\sqrt{h^2 + a^2}} - \dfrac{1}{\sqrt{h^2 + b^2}}\right]$

43 (i) $\dfrac{2\pi rT\,\rho\, mL\,\delta r}{(L^2 + r^2)^{1/2}}$ (iii) $F = 2\pi\, T\rho m\left[1 - \dfrac{L}{\sqrt{L^2 + R^2}}\right]$

Chapter 19

Section 19.1
1 (i) Not a p.f. $\Sigma p_i \neq 1$; (ii) (a) 6/10; (b) 9/10; (c) 5/10.
(iii) Not defined at $x = 0$ (iv) $C = 5/9$; (a) 1/3; (b) 4/9; (c) 7/18.
(v) $p_0, p_1 < 0$; (vi) $C = 60/77$; (a) 32/77; (b) 47/77; (c) 35/77.
3 11, 136; 91/323, 455/969, 70/323, 10/323, 1/969.
4 Take $k = 20/29$
5 $(1-\theta)^3$, $3\theta(1-\theta)^4$, $3\theta^2(1-\theta)^3(3-2\theta)$, $\theta^3(16 - 33\theta + 24\theta^2 - 6\theta^3)$

Section 19.2
6 £5/18
7 $p(x) = 1/6$ for each x; 7/2, 35/12.
8 £2.30
10 1.04

Section 19.3
12 0.3
13 (i) 0.599 (ii) 0.787
14 $(3/5)^6.\,(169/25) \cong 0.315$
15 0.396
16 (i) 0.031 (ii) 0.018
17 0.651
19 18.2, 84.8, 148.3, 115.2, 33.5.
20 (i) 0.285 (ii) 0.608
21 (i) $(0.75)^{20}$ (ii) $5(0.75)^{19}$
22 (i) 1/125 (ii) 128/625

23 243/1024, 405/1024, 270/1024, 90/1024, 15/1024, 1/1024; 5/4; 15/16
25 $p(3) = p(4) = {}^{15}C_3(^1/_4)^3(^3/_4)^{12}$
26 (a) 0.09874 (b) 0.9798 **27** 0.02; 0.00376

Section 19.4

28 (i) 0.3235 (ii) (a) Poisson [0.94] (b) $s^2 = 1.15, \sigma^2 = 0.94$
29 $e^{-0.75} = 0.472$ **30** 0.185, stock 6 **31** 0.185
32 7 **33** 0.036; 1 or 2 **34** 0.337
36 0.189, 0.315, 0.262, 0.145, 0.061; Yes.
37 (i) 0.0902 (ii) 300 **38** $e^{-0.1} \cong 0.905$ **39** 443
40 0.8208; 0.07 **41** (i) 0.3327 (ii) 0.0779
42 (a) 0.1429 (b) (i) 0.0907 (ii) 0.2177 (iii) 0.4304

Chapter 20

Section 20.1

1 $\rho(x) \equiv 1/3$; (i) 0; (ii) 1/3; (iii) 1/3; (iv) 2/3; (v) 0; (vi) 0; (vii) 1/3
2 1/6; 1/3; 7/8, 5/24. **3** 1.582; 0.623; 0.165; 0.212.
4 $x - \frac{1}{4}x^2$; 3/4, 1/16, 27/64; 5 and 6
5 0.80, 0.64, 0.51, 0.41, 0.33, 0.26, 0.21, 0.17, 0.14, 0.11, ; $1/\lambda$

Section 20.2

6 0.966, 0.281, 0.979, 0.004, 0.027, 0.491, 0.021, 0.119, 0.069, 0.296, 0.675, 0.253, – 0.524, 1.414, –1.735, 0.988, 0.747, 2733, ≅ 264.
7 0.032, 0.1416 **8** 3.95 (4 years), 6.68% **9** ≅ 62
10 Yes, qualitatively **11** 0.39%
12 mean = 71 100, s.d. = 2040
13 (a) 683; (b) 0.091; (c) 87.9; (d) 67.4, 92.6; (e) 105; (f) 69.9, 80, 90.1
14 4.27, 1.24

Time	1	2	3	4	5	6	7
Frequency	1	5	16	26	22	11	3

15 (i) 0.0688 (ii) 62
16 (a) 0.4332, 0.0098, 0.0478, 800 hours (b) 0.3414, 0.3413
17 0.0268, 5.02%, 98.695 and 105.905, 103.26 mm
18 11% **19** 7.8p

Section 20.3

20 Binomial 0.121, Normal 0.119
21 (i) 0.01 (ii) 0.04 (iii) 0.97 (iv) 0.54 (v) 0.13×10^{-8} (vi) 0.21×10^{-4}

Section 20.4

22 (a) 4.5, 2.63 (b) 1.5, 2, 3.5, 4, 4.5, 2.5, 4, 4.5, 5, 4.5, 5, 5.5, 6.5, 7, 7.5; (c) 4.5 (d) 1.66.
23 (a) 0.534 (b) 0.997; **24** (a) 0.124 (b) 0.124. **25** 0.37 ± 0.0074
26 (a) 2% (b) 0.1. **27** (a) 0.998 (b) 0.0003 **28** 16%

29 0.16 **30** 10%, 6.59 kN

Section 20.5

31 0.179 **32** 5.15 – 6.25 kg
33 (i) 0.221; (ii) 0.990 **34** 0.824 ± 0.108 **35** 2280, 2592

Section 20.6

37 101.512 ± 0.720, 507.56 ± 1.61, 10.6%, 102.43
38 (a) 0.16 (b) yes (0.5% level)

Section 20.7

39 Yes **40** No **41** 0.145
42 Yes **43** 1.50 ± 0.882 **44** 0.02
45 Significant at 0.5% level.
46 Confidence interval [0.41, 0.55]; yes, significantly higher at 5% level.
47 No, there is a significant difference at the 1% level.

Chapter 21

Section 21.2

1 (a) 1, 1 (b) 2,1 (c) 2,1 (d) 2,1 (e) 3,1 (f) 1,3
2 (i) $d^2y/dx^2 = 0$ (ii) $2x dy/dx = y$ (iii) $d^2x/dy^2 = -4y$;
(iv) $d^2y/dx^2 + 4dy/dx + 4y = 0$ (v) $d^2x/dt^2 + 2(dx/dt)^2 = 0$
3 $y = x^2 + 1$, no **5** $y = 9/4 - (3/2)x$

Chapter 22

Section 22.1

1 (i) $y = \frac{1}{2}\ln(1 + x^2)$ (ii) $(1 + y^{3/2})^{2/3} = A\sqrt{1 + x^2}/x$
(iii) $\cos y = Ae^{-x \tan x}$ (iv) $y = Ae^{1/2(\ln x)^2}$
(v) $(y^2 - 1)/(y^2 + 1) = Ae^{2x^2}$ (vi) $y/(1 - y) = x/2$ or $y = x/(2 + x)$
(vii) $\tan y = x + x^2 + A$ (viii) $y^2 = x^2 + 2\ln x + A$
(ix) $1 - y^2 = A(1 - x^2)/x^2$

2 (i) $y = \ln \sec x$ (ii) $y = -\ln(9/4 - \sqrt{1 + x^2})$ (iii) $\theta = \sqrt{t^2 - 1}/\sqrt{t^2 + 1}$

3 $n = Ne^{-\lambda t}$; $(1/\lambda)\ln(4/3)$, $(1/\lambda)\ln 2$
4 $p = p_0 + r_1^2(p_1 - p_0)(r^2 - r_0^2)/[r^2(r_1^2 - r_0^2)]$
5 $t = (\pi/15k)(20a - 6h_0)h_0^{3/2}$
6 $(4/3)h^{3/2}R - (2/5)h^{5/2} = (14R^{5/2}/15) - a^2\sqrt{(2g)}\, t$

8 $$M = \frac{M_1 M_0 e^{kM_1 t}}{M_1 - M_0 + M_0 e^{kM_1 t}}$$

9 $q = q_0 + \frac{E}{R^2 + L^2\omega^2}\left[\frac{R}{\omega}\sin\omega t - L\cos\omega t + L\exp\left(\frac{-Rt}{L}\right)\right]$

10 $v = \sqrt{(g/k)}\tan[\sqrt{(gk)}\,(c-t)]$ (up); $v = (g/k)(e^{2\sqrt{(gk)}(t+c)} - 1)/(e^{2\sqrt{(gk)}(t+c)} + 1)$ (down); $1/(\sqrt{gk})\tan^{-1}(\sqrt{(k/g)}\,v_0)$.

11 $z = (4e^{2t} - 3)/(2e^{2t} - 1)$

12 3.39 pm

Section 22.2

13 (i) $E;\ x^3y - x^4/4 + y^2/2 = C$ (ii) $E;\ x^2/2 + 2xy + y^2 = C$ (iii) $E;\ r^2/2 + r(\sin\theta + \cos\theta) = C$ (iv) Not exact.

14 $x/y - x^2/y^3 = C$

15 $x^2y^2/2 + x^4/4 + x^3/3 = C$

16 $x^4/2 + 3xy + y^2/2 - y = C$

17 $(y + 1 + \ln x)/x = C$

18 $x^3y - xy^3 = C$

19 (i) S.V., I.F. = $e^{-\sin x}$ (ii) S.V. (iii) I.F. = $e^{-(1/2)e^{2t}}$ (iv) S.V. (v) Neither (vi) S.V., I.F. = $e^{(e^{-2t})}$ or $e^{(1/2e^{-2t})}$ (vii) I.F. = x

20 (i) $y = (1/3)x^2 + C/x$ (ii) $y = (1/4)(x + 1) + C/(x + 1)^3$ (iii) $y = (x + C)\operatorname{cosec} x$ (iv) $\theta = 0.5(1 + t)[1 - e^{(2-2t)}]$ (v) $\theta = t(\sin t - \cos t)$

21 (i) $y = (x/2 + 1 - \pi/4)\operatorname{cosec} x - (\frac{1}{2})\cos x$ (ii) $y = x + Cx(x^2 - 1)^{-1/2}$ (iii) $y = \frac{1}{3}e^x + Ce^{-2x}$ (iv) $y = \frac{1}{2} - \operatorname{cosec}^2 x$ (v) $y = (\frac{1}{4})(x^2 + 1/x^2)$ (vi) $z = e^{2t}/6 - Ce^{-t} + D$ (vii) $y = (3x - 4)/4 + C(x + 4)^{-3}$ (viii) $y = e^x + 1 - x^2/2 - x$

23 $y = (x + C)e^{-x^2/2}$

25 $[n\omega HA/(R^2/L + \omega^2 L)]\,[(R/L)\cos\omega t + \omega\sin\omega t - (R/L)e^{-Rt/L}]$

26 $1/(2k)\ln[1 + kv^2/\mu g]$

27 $\theta = 30 - 2t + 40e^{-0.1t}$, 68.804

28 $y = \sqrt{Cx^2 - \frac{2}{3x}}$

29 $y = \frac{2x^3 + 3.5x^2 + 2x + 2}{(2x + 1)^2}$; $y = x^2 + 2x\ln x - 1 + Ax$; $y = \frac{2t}{3} + \frac{13}{3t^2}$, 2.417

32 $i = E_0\,[1/R - \{1/(R - L\alpha)\}e^{-\alpha t}] + [i_0 + E_0 L\alpha/R(R - L\alpha)]e^{-Rt/L}$

33 380 days

34 $c = c_0 e^{-Qt/V} + (Qc_f + 10^4 Q_p)(1 - e^{-Qt/V})/Q$, $N = 14$

Section 22.4

35 (i) $2^1/8$ ($2^1/3$ analytically) (ii) $6^1/2$ (4.58 analytically) (iii) 0.6554 (0.7358 analytically) (iv) 1.5 (2.1353 analytically) (v) 1.9578 (1.744 analytically) (vi) 0.2654 (0.37 analytically)

36 1.285×10^4 ($g = 10, R = 6 \times 10^6$) **37** 0.8412
38 (i) $2^3/8$ (ii) $9^7/8$ (iii) 0.7414 (iv) 3/2 (v) 1.7503 (vi) 0.3688
39 $y(0.1) = 1.1005$ (1.1003 analytically), $y(0.2) = 1.203$ (1.203 analytically)
40 $y(0.1) = 2.400$, $y(0.2) = 2.803$

Section 22.5

41 0.9802, 0.9231, 0.7262 **42** Analytical values, 2.9144, 2.8561, 2.8224
43 First term neglected is $80x^7/7! = x^7/63$ **45** $x(0.10) = 1.078$, $x(0.15) = 1.104$

Section 22.6

47 0.021499, 0.042999
48 (i) $y = -x - 1 + (^1/2)(\cos x + \sin x) + (3/2)e^x$ (ii) $x^2 = y^2 - 3y$;
(iii) $v^2 = (^1/3)x^3 + (^3/4)x^2 - 14/3$
49 In one step $y(1) = 6.0417$ (Analytically 6.1549)
50 Difficulty in differentiation; 0.1628
51 $y(1) = 4$, $y(2) = 9$; With $h = 1$; 3.9444, 8.8784; With $h = 0.25$; 3.9994, 8.9984
53 $t = 2.521 \ln(100 - 4c) - 3.624 \ln(100 - 7c) + 4.934$

Chapter 23

Section 23.2

1 (i) $y = \alpha e^{-6t} + \beta e^t$ (ii) $y = (\alpha t + \beta)e^{8t}$ (iii) $y = e^{-3t}(\alpha \cos t + \beta \sin t)$
(iv) $y = (\alpha t + \beta)e^{-t/2}$ (v) $y = \alpha e^{2t} + e^{-(1/2)t}[\beta \cos(\sqrt{3}\, t/2) + \gamma \sin(\sqrt{3}\, t/2)]$
(vi) $y = e^{-t/2}[\alpha \cos(3\sqrt{3}\, t/2) + \beta \sin(3\sqrt{3}\, t/2)]$ (vii) $y = \alpha e^{-13t/10} + \beta e^{t/2}$
2 $y = H \cosh(wx/H)/w$
4 $y = \alpha e^{-t} + e^{-3t/2}[\beta e^{\sqrt{5}\, t/2} + \gamma e^{-\sqrt{5}\, t/2}]$
5 (i) moment of inertia, elastic constant, damping constant (ii) L, $1/C$, R
6 (i) $x = x_0 \cos kt$ (ii) $x = (v_0 \sin kt)/k$, k constant
7 $y = Ae^{2x} + Be^{-2x} + C \cos 2x + D \sin 2x$

Section 23.4

8 (i) $y = e^{-3x}$ (ii) $y = x^2 + 7x + 22$ (iii) $y = (3 \cos 2x - \sin 2x)/20$
(iv) $y = e^{-2x} + 3x^2/2 - 3x$ (v) $y = x^2e^{-2x}$ (vi) $y = -(x \cos 4x)/4$
9 (i) $y = e^{-2x} \cos x + \cos x + \sin x$ (ii) $y = (2e^{-x} - 2e^{2x} + 6xe^{2x})/9$
(iii) $x = (13 \sin 2t - 4 \cos 2t + 4e^t)/10$
(iv) $x = A \sin(t/3) + B \cos(t/3) - t \cos(t/3)/6$
(v) $x = e^{-2t}[Ae^{\sqrt{3}t} + B e^{-\sqrt{3}t}] + t^3 - 12t^2 + 90t - 338$
(vi) $y = [57e^{-4x} + e^{-x}(\sin x - 7 \cos x)]/50$
10 (a) $y = (Ax + B + x^2)e^{-x}$ (b) $y = (x/8 + B)e^{4x} + A$
11 $y = (w/k)[1 - e^{-\beta a}(\cos \beta a \cosh \beta x \cos \beta x + \sin \beta a \sinh \beta x \sin \beta x)]$, $(0 \le x \le a)$
$y = (w/k)\, e^{-\beta x}(\sinh \beta a \cos \beta a \cos \beta x + \cosh \beta a \sin \beta a \sin \beta x)$, $(x \ge a)$
12 $y = (5e^x - e^{-5x} - 4e^{-2x})/18$
13 (a) $s = Ae^{-t} + Be^{-4t} + (3 \cos t + 5 \sin t)/34$
(b) $s = (A + Bt)e^{-2t} + (3 \cos t + 4 \sin t)/25$

(c) $s = e^{-t}(A \cos \sqrt{3}t + B \sin \sqrt{3}t) + (3 \cos t + 2 \sin t)/13$

14 (i) $y = Ae^{-4x} + Be^{-x} - e^{-2x}/2 + (3 \sin x - 5 \cos x)/34$;

(ii) $y = e^{3x}(A \cos x + B \sin x) + x/10 + 4/25$

15 $I = 1/10 - e^{-2t}(25 \cos 50t + \sin 50t)/250, 0$

16 $s = \lambda(\sin 2t + 2 \sin t)/6$

18 $x = -\frac{2}{9}\sin 3t + \frac{11}{3}t + 2;\ t = \frac{\pi}{3},\ \dot{x}_{max} = \frac{13}{3}$

19 (i) $x = e^{-2t}(A \cos 2t + B \sin 2t) + \sin 2t - 2 \cos 2t$

(ii) $A = 2, B = 1$ (iv) $t = 1.339 + 1.57\,n, n = 0,1,2,\ldots$

20 (i) $q = Ae^{-t} + Be^{-4t}$ $q = e^{-t}(A \cos \sqrt{3}t + B \sin \sqrt{3}t)$

$q = (At + B)e^{-2t}$

(ii) $q = e^{-t}(A \cos \sqrt{3}t + B \sin \sqrt{3}t) + \frac{1}{13}(2 \sin t + 3 \cos t)$

$A = -\frac{3}{13}, B = -\frac{5}{13\sqrt{13}}$ (iii) $R = \frac{1}{\sqrt{13}}, \tan \alpha = \frac{3}{2}$

21 $q = e^{-1000t}(4 \times 10^{-3} \cos 1000t + 2 \times 10^{-3} \sin 1000t);\ \sqrt{20} \times 10^{-3}$

22 $x = e^{\lambda t}\left(b \cos nt + \frac{\lambda b}{n}\sin nt\right);\ R = \sqrt{2}\,b/4,\ \alpha = \pi/4$

23 $\omega^2 = (2IC - \alpha^2)/2I^2$

24 $y = Ae^{-2x} + Be^{-3x} + \frac{1}{4}(\sin 2x - 5 \cos 2x) + x - \frac{5}{6};\ A = 19/4,\ B = -8/3$

25 $EIy = 0.0833x^4 + 0.0205l^4 \sin\left[\frac{\pi x}{l}\right] - 0.167lx^3 + 0.148l^2x^2 - 0.0645l^3x$

Section 23.5

26 $\theta(0.5) = -0.722738,\ \dot{\theta}(0.5) = -1.964725,\ \theta(1.0) = 0.426372,\ \dot{\theta}(1.0) = 2.870328,$

$\theta(1.5) = -0.178261,\ \dot{\theta}(1.5) = -3.117539,\ \theta(2.0) = -0.017978,\ \dot{\theta}(2.0) = 2.936374,$

$\theta(2.5) = 0.163391,\ \dot{\theta}(2.5) = -2.486598$

27 Selected values, $\phi(0.5) = 4.9988, \phi(1.0) = 5.1663, \phi(2.0) = 5.1706, \phi(3.0) = 5.1706$

28 Selected values, $x(1) = 0.3203, x(2) = 0.0543, x(3) = -0.0765, x(4) = -0.0802$;

$\dot{x}(1) = -0.2850, \dot{x}(2) = -0.2106, \dot{x}(3) = -0.0557, \dot{x}(4) = 0.0338$

$\ddot{x}(1) = -0.0646, \ddot{x}(2) = 0.1556, \ddot{x}(3) = 0.1319, \ddot{x}(4) = 0.0466$

29 Selected values, $x(1) = 2.9986, x(2) = 2.9975, x(3) = 2.9964, x(4) = 2.9953,$
$x(5) = 2.9940$

Section 23.6

31 $y = w(x^4 - 2lx^3 + l^3x)/24EI$

32 $\theta = 30\,e^{-4\sqrt{15}x/3} + 70$

Chapter 24

Section 24.1

1 (i) $a/[(s+b)^2-a^2]$ (ii) $(s+b)/[(s+b)^2-a^2]$
(iii) $(a\cos b - s\sin b)/(s^2+a^2)$ (iv) $1/[s^2(s^2+a^2)]$

(v) $s/[(s^2+a^2)(s^2+b^2)$; (vi) $(s^2-a^2)/(s^2+a^2)^2$ (vii) $1/(s+2)^2$
(viii) $2/(s+2)^3$

Section 24.2

2 (i) $1/(s^2+a^2)^2$ (ii) $1/(s+a)^n$ (iii) $(s^2-k^2)/(s^2+k^2)^2$
(iv) $2ks/(s^2-k^2)^2$ (v) $(s+2)/[(s+2)^2+k^2]$ (vi) $2(s+1)/[(s+1)^2+1]^2$

Section 24.3

3 (i) $i = 40\cos t + 20\sin t - 40e^{-2t}$ (ii) $r = (9e^{-t} - 5e^{-3t})/2$
(iii) $x = 2e^{-3t} + 10t\,e^{-3t}$ (iv) $x = \cos 3t + (17\sin 3t + 3t)/27$
(v) $x = x_0\cos nt + (x_1\sin nt)/n + (t\sin nt)/2n$
(vi) $x = \cos 2t + (3\sin 2t)/2 + (5t\sin 2t)/4$

4 (i) $y = (5e^{3t} + 44e^{-t/2} + 7te^{3t})/49$ (ii) $y = [1 - e^{-2t}(\cos 2t + \sin 2t)]/8$
(iii) $\theta = [2(2+5t)e^{-t} - 3\sin 2t - 4\cos 2t]/25$
(iv) $x = (t\sin 3t + 12\cos 3t - 10\sin 3t)/6$
(v) $y = (t^3+30)e^{-t}/3$; (vi) $u = (e^{-t}\sin^2 t)/2$ (vii) $y \equiv 0$.

Section 24.4

5 Analytical solution $y = (\ln\cosh\sqrt{(gk)}\;t/k$
Runge-Kutta ($k = 1$, $g = 9.81$), $y(2.0) = 5.5710$, $y(4.0) = 11.8352$, $y(6.0) = 18.0994$, $y(8.0) = 24.3636$, $y(10.0) = 30.6278$

Section 24.5

6 $x = (e^{-3t} + e^{-t})/2$, $y = (e^{-3t} - e^{-t})/2$

7 $x = (1 + 3t - e^{-3t} - 6te^{-3t})/27$, $y = (4 - 6t - 4e^{-3t} - 6te^{-3t})/27$

8 $x = (3 + 10\cos t + 5\cos 2t)/6$, $y = (10\sin t - 5\sin 2t)/6$

10 $x_1 = [(x_1{}^0 - x_2{}^0)\cos\sqrt{k}\,t + (x_1{}^0 + x_2{}^0)\cos\sqrt{(3k)}t]/2$
$x_2 = [-(x_1{}^0 - x_2{}^0)\cos\sqrt{k}\,t + (x_1{}^0 + x_2{}^0)\cos\sqrt{(3k)}\,t]/2$

12 $x = 3e^t - e^{-t} - 2$, $y = e^t + e^{-t} - 1$

Section 24.6

13 (i) $2[u(t) - u(t-4)]$; $2(1 - e^{-4s})/s$
(ii) $2[u(t-2) - 2u(t-4) + u(t-6)]$; $2(e^{-2s} - 2e^{-4s} + e^{-6s})/s$
(iii) $u(t) - u(t-1) + u(t-2) - u(t-3)$; $(1 - e^{-s} + e^{-2s} - e^{-3s})/s$

14 (i) $2(1 - e^{-as})/s^2$ (ii) $2(-e^{-as} + 2e^{-2as} + e^{-3as})/s^2$

15 (i) $f(t) = 1$, $1 < t < 2$; $f(t) = 0$ otherwise
(ii) $f(t) = 2(t-1)$, $t > 1$; and 0, $t < 1$

(iii) $f(t) = \frac{1}{2}(t-1)^2$, $t > 1$; $= 0$, $t < 1$

(iv) $f(t) = t - 1,\ 1 < t < 2;\ = 1,\ 2 < t < 3;\ = 4 - t, 3 < t < 4;\ = 0$ otherwise

16 (i) e^{-as}/s^2 (ii) $(1/s^2 + 2/s)e^{-2s}$ (iii) $2e^{-as}/s^3$
(iv) $-e^{-\pi s}/(s^2 + 1)$ (v) $e^{-a(s-1)}/(s + a)$

17 (i) $4[1 - (1 + 16s)e^{-4s}]/s^2$ (ii) $2[2 - (2 + 10s + 25s^2)e^{-5s}]/s^3$
(iii) $A\alpha(1 + e^{-\pi s/\alpha})/(s^2 + \alpha^2)$ (iv) $Bs(1 + e^{-\pi s/\alpha})/(s^2 + \alpha^2)$
(v) $Bs(1 - e^{-2\pi s/\alpha})/(s^2 + \alpha^2)$

18 (i) Sine wave starting at $t = 2\pi$, zero before that
(ii) $e^{-(a-t)} \sin t,\ t > \pi;\ 0$ elsewhere

19 (i) $\dfrac{1}{t}e^{-\omega t}$ (ii) $\dfrac{1}{18}t \sinh 3t$ (iii) $(e^t - e^{2t})/t$

20 (i) $1/[s(1 - e^{-2\pi s})]$ (ii) $\dfrac{\pi s - 1 + (\pi s + 1)e^{-2\pi s}}{s^2(1 - e^{-2\pi s})}$ (iii) $\dfrac{e^{2(1-s)\pi} - 1}{(1 - s)(1 - e^{-2\pi s})}$

(iv) $\omega/[(s^2 + \omega^2)(1 - e^{-2\pi s})]$ (v) $e^{-\pi s}/(s(1 - e^{-2\pi s})]$ (vi) $\dfrac{1}{s}\tanh\dfrac{s}{2}$

(vii) $\left[\dfrac{\pi}{s}e^{-\pi s}(e^{-\pi s} - 1) + \dfrac{1}{s^2}(e^{-\pi s} - 1)^2\right] \Big/ (1 - e^{-2\pi s})$

INDEX